T0342457

ENGINEERING DYNAMICS

ENGINEERING DYNAMICS

A Comprehensive Introduction

N. Jeremy Kasdin and Derek A. Paley

PRINCETON UNIVERSITY PRESS

PRINCETON AND OXFORD

Published by Princeton University Press, 41 William Street, Princeton,
New Jersey 08540

In the United Kingdom: Princeton University Press, 6 Oxford Street, Woodstock,
Oxfordshire OX20 1TW

press.princeton.edu

Library of Congress Cataloging-in-Publication Data

Kasdin, N. Jeremy.
    Engineering dynamics : a comprehensive introduction / N. Jeremy Kasdin and
Derek A. Paley.
        p.   cm.
    Includes bibliographical references and index.
    ISBN 978-0-691-13537-3 (hardback : alk. paper)   1. Dynamics.   2. Mechanics.
I. Paley, Derek A., 1974–   II. Title.
    TA352.K375 2010
    620.1′04—dc22   2010023802

British Library Cataloging-in-Publication Data is available

This book has been composed in Times Roman and Avenir using ZzTEX
by Princeton Editorial Associates, Inc., Scottsdale, Arizona.

Printed on acid-free paper.   ∞

Printed in the United States of America

10 9 8 7 6 5 4 3

To our wives Kef and Robyn and our children Alex, Izzy, Ethan, and Adyn—your dynamics make it all worthwhile.

Philosophy is written in this grand book—I mean the universe—which stands continually open to our gaze, but the book cannot be understood unless one first learns to comprehend the language and read the letters in which it is composed. It is written in the language of mathematics.

Galileo Galilei, *The Controversy on the Comets of 1618*

CONTENTS

PREFACE

Dynamics is difficult. There is no getting around that. This is particularly true for undergraduates just starting their engineering and science education, when they are beginning to wrestle with the physics and mathematics needed to gain facility with dynamics. We find that simply acknowledging this fact goes a long way toward increasing confidence. Nevertheless, the pedagogical solution is not to simplify the material to make it more manageable. Rather, we feel quite strongly that students are best served by employing careful rigor and emphasizing deep understanding of the concepts as well as by using precise mathematics. In this way, they are provided with tools and concepts that will serve them throughout their educational and professional careers. The proper response to the admitted difficulty of the subject is to slow down the presentation, perhaps stretching it over multiple quarters or semesters, and gradually building complexity rather than simplifying in a way that lacks rigor and care. To that end, we have included extensive appendices covering the mathematical skills needed to understand all material in the book.

Most students who will use this book have had an introduction to mechanics in their freshman physics courses. It is our goal to reintroduce them to the material with the added sophistication of vector calculus and differential equations. Our approach to ensuring both understanding and confidence is to emphasize careful notation and rigor. Although some students complain about the pedantry and others want to jump to the end, it is our experience that the way to ensure competence is to enforce a rigorous and careful problem-solving process. Unfortunately, too close an adherence to this principle can lead to a course—and textbook—that is dry, uninviting, and presented in a way that is inconsistent with how students learn. The challenge we undertook in writing this book was to maintain rigor (and rigorous notation) while making the material sufficiently approachable and informal that students will spend time reading it and wrestling with it.

Certainly there are many good books available that treat the subject of dynamics with complete rigor. We confess that we like a good number of them and are attracted to the top-down approach of developing the material from first principles, starting with geometry, moving on to fully three-dimensional vector kinematics, and then continuing through particle and rigid-body dynamics. In fact, we use this approach in our graduate classes, where we also include Lagrangian and Hamiltonian methods. However, we have found that undergraduates (especially sophomores and juniors) have difficulty learning the material this way. Rather, a bottom-up approach that develops skills and techniques on simpler problems—without sacrificing rigor—and gradually increases sophistication—without losing sight of the basic physics—seems to best capitalize on the way these students learn. In that sense our approach can be likened to learning to play a musical instrument. We begin with the essential fundamentals and, through repeated problem solving (practice), develop "muscle memory" as new and more difficult pieces are tackled. Yet the notations we use from the beginning—the notes, chords, and time signatures—remain the same and return again and again.

We thus take a unique approach in this book. We introduce Newton's laws and start solving important problems even before beginning a discussion of vector kinematics. We seek to maintain student interest and present key notations and skills in the context of real problems. An overemphasis on the mathematics, without maintaining a connection to the physical objectives, can cause confusion and diminish enthusiasm among students. For this reason, in some chapters we defer more detailed or complex derivations to the end of the chapter, so as not to interrupt the physical picture. Kinematics is developed slowly, always in the context of dynamics problems. Yet we insist on a very careful notation, inspired by Thomas Kane's wonderful books. We always specify reference frames and are careful to maintain the distinction between vectors, components, and scalars. The emphasis on using and understanding reference frames (and specifying the inertial frame when solving problems) is something we are particularly wedded to and find lacking in many introductory dynamics texts. *In our experience, the best thing students can do to avoid errors and enhance learning is be compulsive about notation from the start.*

We also emphasize finding equations of motion. Before computers became commonplace, dynamics education (as reflected in older textbooks) tended to emphasize finding accelerations and treating dynamics problems as slightly more complicated statics problems. Dynamics, however, is about finding equations of motion and determining trajectories. We thus introduce students early on to the idea of using ordinary differential equations to describe the motion of systems and to the use of a computer to integrate these equations. Where possible, analytical solutions to the equations of motion are presented.

We have made every effort to include examples spanning a range of difficulty and covering the most important concepts and techniques. We have tried to connect the examples to real physical systems. Certain examples regularly repeat throughout the book, so that students can see how new concepts are used on familiar problems and how new insights can be gained from increasingly sophisticated analysis.

Our approach of distinguishing examples from tutorials allows us to employ simple problems to highlight specific ideas just after they are introduced (examples) while reserving problems that synthesize many concepts for the end of the chapter (tutorials). Some tutorials can be quite difficult, and instructors may want to judiciously select among them; however, we felt presenting a wide range of difficulty and depth resulted in a text that may prove useful for years after the course is taken.

We have also chosen to adopt an informal conversational style. Although purists may be put off by this tone in a technical work, our feedback from students—after trying a number of different textbooks—is that they appreciate the approachability of conversational writing and find the material more accessible. We directly address the reader and attempt to guide him or her through the difficult task of learning dynamics.

Acknowledgments

We owe a debt of gratitude to many people who aided in myriad ways, both direct and behind the scenes, to make this book a reality. First and foremost, we thank our students, who eagerly engaged in learning the material and refining the text. Without them, their curiosity, and their desire to learn, this text would not exist. We also thank our colleagues at Princeton and Maryland, who proved willing to engage in endless discussions about dynamics. In particular, we are grateful to Phil Holmes, Sean Humbert, Naomi Leonard, Michael Littman, Clancy Rowley, Ben Shapiro, Rob Stengel, and Bob Vanderbei for regularly teaching us new things.

Prof. Kasdin is also indebted to the many professors, colleagues, and friends at Stanford whose knowledge of dynamics, wisdom, and incredible insight are captured throughout the book, particularly Art Bryson, Bob Cannon, Dan DeBra, Tom Kane, Brad Parkinson, and Steve Rock.

We thank the many teaching assistants who worked above and beyond to make sure the book was correct, that examples worked, and that problems had solutions. They include Kevin Anderson, Tyler Groff, Ben Jorns, Jason Kay, Adele Lim, Ben Nabet, Laurent Pueyo, Harinder Singh, Andy Stewart, and Nitin Sydney. We can't thank Dmitry Savransky enough; beyond being an amazing teaching assistant, he has put in countless hours creating figures, examples, and problems and supporting almost every aspect of the book's production. Without him, the book would not be what it is.

We thank the manuscript reviewers, whose thorough reading and insightful suggestions improved the manuscript enormously. We thank John Lienhard for generously providing his images of the Watt flyball governor. We thank Peter Strupp, Cyd Westmoreland, and the staff at Princeton Editorial Associates, whose tireless editing made this a book worth reading, and Mark Bellis at Princeton University Press, who shepherded the book through its final stages. We are especially grateful to our editor, Ingrid Gnerlich, who showed more patience than anyone should expect in guiding us through the process of writing, editing, and production. She was always uplifting and always encouraging.

Most importantly, we thank our families: Kef, Alex, Izzy, Robyn, Ethan, and Adyn. Their love and support through the many days and nights we worked on the book made this text possible.

ENGINEERING DYNAMICS

CHAPTER ONE

Introduction

1.1 What Is Dynamics?

Dynamics is the science that describes the motion of bodies. Also called *mechanics* (we use the terms interchangeably throughout the book), its development was the first great success of modern physics. Much notation has changed, and physics has grown more sophisticated, but we still use the same fundamental ideas that Isaac Newton developed more than 300 years ago (using the formulation provided by Leonhard Euler and Joseph Louis Lagrange). The basic mathematical formulation and physical principles have stood the test of time and are indispensable tools of the practicing engineer.

Let's be more precise in our definition. Dynamics is the discipline that determines the position and velocity of an object under the action of forces. Specifically, it is about finding a set of differential equations that can be solved (either exactly or numerically on a computer) to determine the trajectory of a body.

In only the second paragraph of the book we have already introduced a great number of terms that require careful, mathematical definitions to proceed with the physics and eventually solve problems (and, perhaps, understand our admittedly very qualitative definition): *position*, *velocity*, *orientation*, *force*, *object*, *body*, *differential equation*, and *trajectory*. Although you may have an intuitive idea of what some of these terms represent, all have rigorous meanings in the context of dynamics. This rigor—and careful notation—is an essential part of the way we approach the subject of dynamics in this book. If you find some of the notation to be rather burdensome and superfluous early on, trust us! By the time you reach Part Two, you will find it indispensable.

We begin in this chapter and the next by providing qualitative definitions of the important concepts that introduce you to our notation, using only relatively simple ideas from geometry and calculus. In Chapter 3, we are much more careful and present the precise mathematical definitions as well as the full vector formulation of dynamics.

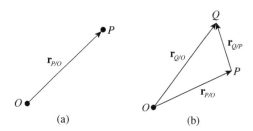

Figure 1.1 (a) Vector $\mathbf{r}_{P/O}$ from reference point O to point P represents the position of the point P relative to O. (b) The addition of two vectors, $\mathbf{r}_{P/O}$ and $\mathbf{r}_{Q/P}$, to get the resultant vector $\mathbf{r}_{Q/O}$.

1.1.1 Vectors

We live in a three-dimensional Euclidean[1] universe; we can completely locate the *position* of a point P relative to a reference point O in space by its relative distance in three perpendicular directions. (In Part One we talk about points rather than extended bodies and, consequently, don't have to keep track of the orientation of a body, as is necessary when discussing rigid bodies in Parts Three and Four.) We often call the reference point O the *origin*. An abstract quantity, the *vector*, is defined to represent the position of P relative to O, both in distance and direction.

Qualitative Definition 1.1 A **vector** is a geometric entity that has both magnitude and direction in space.[2]

A position vector is denoted by a boldface, roman-type letter with subscripts that indicate its head and tail. For example, the position $\mathbf{r}_{P/O}$ of point P relative to the origin O is a vector (Figure 1.1). An important geometric property of vectors is that they can be added to get a new vector, called the *resultant* vector. Figure 1.1b illustrates how two vectors are added to obtain a new vector of different magnitude and direction by placing the summed vectors "head to tail."

When the position of point P changes with time, the position at time t is denoted by $\mathbf{r}_{P/O}(t)$. In this case, the *velocity* of point P with respect to O is also a vector. However, to define the velocity correctly, we need to introduce the concept of a *reference frame*.

1.1.2 Reference Frames, Coordinates, and Velocity

We have all heard about reference frames since high school, and you may already have an idea of what one is. For example, on a moving train, objects that are stationary on the train—and thus with respect to a reference frame fixed to the train—move with respect to a reference frame fixed to the ground (as in Figure 1.2). To successfully use dynamics, such an intuitive understanding is essential. Later chapters discuss how reference frames fit into the physics and how to use them mathematically; for that

---

[1] Euclid of Alexandria (ca. 325–265 BCE) was a Greek mathematician considered to be the father of geometry. In his book *The Elements*, he laid out the basic foundations of geometry and the axiomatic method.

[2] In this book, a *qualitative* definition is typically followed by an *operational* or *mathematical* definition of the same term, although the latter definition may come in a later chapter.

Figure 1.2 Qualitative definition of a reference frame.

reason, we revisit the topic again in Chapter 3. For now, we summarize our intuition in the following qualitative definition of a reference frame.

Qualitative Definition 1.2 **A reference frame** is a point of view from which observations and measurements are made regarding the motion of a system.

It is impossible to overemphasize the importance of this concept. Solving a problem in dynamics always starts with defining the necessary reference frames.

From basic geometry, you may be used to seeing a reference frame written as three perpendicular axes meeting at an origin O, as illustrated in Figure 1.3. This representation is standard, as it highlights the three orthogonal Euclidean directions. However, this recollection should not be confused with a *coordinate system*. The reference frame and the coordinate system are not the same concept, but rather complement one another. It is necessary to introduce the reference frame to define a coordinate system, which we do next.

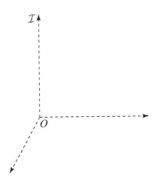

Figure 1.3 Reference frame $\mathcal{I}$ is represented by three mutually perpendicular axes meeting at origin O.

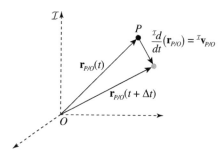

Figure 1.4 Velocity ${}^{\mathcal{I}}\mathbf{v}_{P/O}$ is the instantaneous rate of change of position $\mathbf{r}_{P/O}$ with respect to frame $\mathcal{I}$. That is, ${}^{\mathcal{I}}\mathbf{v}_{P/O} = (\mathbf{r}_{P/O}(t + \Delta t) - \mathbf{r}_{P/O}(t))/\Delta t$, in the limit $\Delta t \to 0$.

Definition 1.3 A **coordinate system** is the set of scalars that locate the position of a point relative to another point in a reference frame.

In our three-dimensional Euclidean universe, it takes three scalars to specify the position of a point P in a reference frame. The most natural set of scalars (the three numbers usually labeled x, y, and z) are *Cartesian coordinates*.[3] These coordinates represent the location of P in each of the three orthogonal directions of the reference frame. (Recall the discussion of vectors in the previous section stating that the position of P relative to O is specified in three perpendicular directions.) Cartesian coordinates, however, are only one possible set of the many different scalar coordinates, a number of which are discussed later in the book. Nevertheless, we begin the study of dynamics with Cartesian coordinates because they have a one-to-one correspondence with the directions of a reference frame. It is for this reason that the Cartesian-coordinate directions are often thought to define the reference frame (but don't let this lure you into forgetting the distinction between a coordinate system and a reference frame). We return to the concepts of reference frames and coordinate systems and discuss the relationship between a coordinate system and a vector in Chapter 3.

Throughout the book, reference frames are always labeled. Later we will be solving problems that employ many different frames, and these labels will become very important. Thus we often write the three Cartesian coordinates as $(x, y, z)_{\mathcal{I}}$, explicitly noting the reference frame—here labeled $\mathcal{I}$—in which the coordinates are specified. (The reason for the letter $\mathcal{I}$ will become apparent later.)

Likewise, the change in time of a point's position (i.e., the velocity) only has meaning when referred to some reference frame (recall the train example). For that reason, we always explicitly point out the appropriate reference frame when writing the velocity. A superscript calligraphic letter is used to indicate the frame. Figure 1.4 shows a schematic picture of the velocity ${}^{\mathcal{I}}\mathbf{v}_{P/O} \triangleq \frac{{}^{\mathcal{I}}d}{dt}(\mathbf{r}_{P/O})$ as the instantaneous rate of change in time of the position $\mathbf{r}_{P/O}$ with respect to the frame $\mathcal{I}$.[4]

We can also express the velocity of point P with respect to O as the rate of change $(\dot{x}, \dot{y}, \dot{z})_{\mathcal{I}} \triangleq \frac{d}{dt}(x, y, z)_{\mathcal{I}}$, where $\dot{x} \triangleq \frac{dx}{dt}$, $\dot{y} \triangleq \frac{dy}{dt}$, and $\dot{z} \triangleq \frac{dz}{dt}$. (Appendix A reviews

---

[3] Named after René Descartes (1596–1650), the celebrated French philosopher, who founded analytic geometry and invented the notation.

[4] In this book, the symbol $\triangleq$ denotes a definition as opposed to an equality.

some basic rules of calculus if you are rusty.) Because the variables x, y, and z are scalars, their time derivatives do not need a frame identification. We maintain the notation $(\dot{x}, \dot{y}, \dot{z})_{\mathcal{I}}$, however, to remind you that these three scalars are the rates of change of the three position coordinates in frame $\mathcal{I}$. The rate of change of a scalar Cartesian coordinate is called the *speed* to distinguish it from the velocity. We return to this topic and discuss it in depth and more formally in Chapter 3.

1.1.3 Equations of Motion

We now return to the definition of dynamics. *Trajectory* signifies the complete specification of the three positions and three speeds of a point in a reference frame as a function of time. It takes six quantities in our three-dimensional universe to completely specify the motion of a point. This is not necessarily obvious. Why six quantities and not three? Isn't the position enough (since we can always find the velocity by differentiating)? The answer is no, because dynamics is about more than just specifying the position and velocity. It is about finding equations, based on Newton's laws, that allow us to predict the complete trajectory of an object given only its *state* at a single moment in time. By state we mean the three positions and three speeds of the point. These six quantities, defined at a single moment in time, are called the *initial conditions*. The tools of dynamics allow us to find a set of differential equations that can be solved—using these initial conditions—for the position and velocity at any later time. These differential equations are called *equations of motion*.[5]

Definition 1.4 The **equations of motion of a point** are three second-order differential equations[6] whose solution is the position and velocity of the point as a function of time.

To see this a bit more clearly, imagine that we know the three position variables $x(t)$, $y(t)$, and $z(t)$ of a point in frame $\mathcal{I}$ at some time t and wish to know the position some short time later, $t + \Delta t$. Without the velocity at t we are lost; the point could move anywhere. However, with the three speeds $\dot{x}(t)$, $\dot{y}(t)$, and $\dot{z}(t)$, we know everything; the new position of the point in $\mathcal{I}$ is $(x(t) + \dot{x}(t)\Delta t, y(t) + \dot{y}(t)\Delta t, z(t) + \dot{z}(t)\Delta t)_{\mathcal{I}}$. The equations of motion allow us to find the speeds at time $t + \Delta t$. The six positions and speeds are sufficient to find the complete trajectory.

As an example, one of the simplest equations of motion is that for a mass on a spring. The position of the mass is given by the Cartesian coordinate x, and the force due to the spring is given by $-kx$ (see Figure 2.7c). The position thus satisfies the following second-order differential equation, obtained by equating the force with the mass times acceleration and solving for the acceleration:

$$\ddot{x} = -\frac{k}{m}x.$$

This differential equation is an equation of motion. Its solution gives $x(t)$ and $\dot{x}(t)$, the trajectory of the mass point. Don't worry if you didn't follow how the equation was obtained; that is covered in Chapter 2.

---

[5] Appendix C supplies a brief review of differential equations.
[6] Or, equivalently, six first-order differential equations.

Many equations of motion cannot be solved exactly; a computer is required to find numerical trajectories. You will have an opportunity to do this many times in this book. However, often we skip solving for the trajectory and find special solutions or conditions on the states by setting the time equal to a specific value, finding certain conditions on the forces, or setting the acceleration to a constant or zero (sometimes called a *steady state*). One particularly useful such solution is known as an *equilibrium point*. The mathematical details of equilibrium solutions are presented in Chapter 12, but it is useful to have a qualitative understanding now, as we will be finding equilibrium solutions of many systems here and there throughout the book.

Qualitative Definition 1.5 An **equilibrium point** of a dynamic system is a specific solution of the equations of motion in which the rates of change of the states are all zero.

In other words, an equilibrium point is a configuration in which the system is at rest. For the mass-spring system, for example, there is one equilibrium point, which corresponds to the mass situated at precisely the rest length of the spring. Mathematically, if $x(t)$ is an equilibrium point, then $\dot{x}(t) = 0$ and $\ddot{x}(t) = 0$. Thus $x(t) = x(0)$, where $x(0)$ is the initial condition at time $t = 0$. So an equilibrium point is a solution whose value over time remains equal to its initial value.

In summary, dynamics is about finding three second-order differential equations that can be solved for the complete trajectory of an object. The equations can be solved—using the six initial conditions—either analytically (by hand) or numerically (by a computer). It is true that other scalar quantities can be used to specify the position rather than Cartesian coordinates; we will begin to study alternate coordinate systems in detail in Chapter 3. However, we will always need six independent scalars. The remainder of this book describes methods for finding equations of motion—first for a point (particle) and later for extended (rigid) bodies—and presents various techniques for completely or partially solving them.

1.2 Organization of the Book

The next chapter reviews the physics of mechanics, covering Newton's laws in depth. We also start to solve simple problems. All the essential physical concepts that form the foundation for the rest of the book are presented in that chapter. Our approach is slightly unconventional in that we begin solving dynamics problems at the outset—in Chapter 2—to highlight the meaning of Newton's laws and how we incorporate the underlying postulates[7] into our methodology.

The remainder of the book is divided into five parts plus a set of four appendices. We divide the book into parts to highlight the logical separation of main topics and show how rigid-body motion builds on the key concepts of particle motion. The material could be covered in one semester by leaving out certain topics or stretched over multiple semesters or quarters. In Part One we restrict ourselves to studying only the planar motion of single particles. Thus motion in only two dimensions is studied; we thus need only four scalars to specify a particle's state

---

[7] A postulate is a basic assumption that is accepted without proof.

rather than six. We do this to simplify the mathematics and focus on the key physical concepts, allowing you to develop an understanding of the procedures used to solve dynamics problems. You will solve an amazing array of real and important problems in Part One.

Chapter 3 returns to first principles and lays out the mathematical framework for a full vector treatment of kinematics and dynamics in the plane. Our focus is on the use of various coordinate systems and approaches to treating velocity and acceleration. Throughout the chapter we return to the same example: the simple pendulum. While this example may seem a bit academic, our approach is to focus repeatedly on this relatively simple system to emphasize the various new techniques presented and explain how they interrelate and add value. At the end of the chapter these new concepts are used to solve a selection of more difficult problems.

Chapters 4 and 5 present the concepts of momentum and energy, respectively, for a particle. It is here that we begin to solve equations of motion for the *characteristics* of trajectories (also called *integrals of the motion*). These ideas will be useful throughout the remainder of your study of dynamics and form the foundation of modern physics.

Part Two presents an introduction to multiparticle systems (Chapters 6 and 7). The previous concepts are generalized to simultaneously study many, possibly interacting, particles. In Chapter 6 we introduce two important examples of multiparticle systems—collisions and variable-mass systems. Chapter 7 sets the stage for the rigid-body discussions in Parts Three and Four by analyzing angular momentum and energy for many particles.

Part Three introduces rigid-body dynamics in the plane. We show (Chapters 8 and 9) how to specialize our tools to study a rigid collection of particles (i.e., particles whose relative positions are fixed). In particular, the definition of equations of motion is expanded to include the differential equations that describe the orientation of a rigid body. We use these ideas to study a variety of important engineering systems. We still confine our study to motion in the plane, however, to focus on the physical concepts without being burdened by the complexity of three-dimensional kinematics. It is here that we introduce the moment of inertia and, most importantly, the separation of angular momentum.

Part Four develops the full three-dimensional equations that describe the motion of multiparticle systems and rigid bodies. Part Four (Chapter 10) begins with the study of the general orientation of reference frames, three-dimensional angular velocity, and the full vector kinematics of particles and rigid bodies. Chapter 11 completes the discussion by developing the equations of motion for three-dimensional rigid-body motion. It is here that we find the amazing and beautiful motion associated with rotation and spin, such as the gyroscope and the bicycle wheel.

Part Five—Advanced Topics—allows for greater exploration of important ideas and serves to whet the appetite for later courses in dynamics. Chapter 12 treats three important problems in dynamics more deeply, exploring how the concepts in the book are used to understand and synthesize engineered systems. This introduction is useful for future coursework in dynamics and dynamical systems. Chapter 13 includes a brief introduction to Lagrange's method and Kane's method. It serves as a bridge to your later, more advanced classes in dynamics and provides a first look at alternative techniques for finding equations of motion.

We have organized the book in a way that maximizes the use of problems and examples to enhance learning. Throughout the text we solve specific *examples*—sometimes repeated using different methods—to illustrate key concepts. Toward the end of each chapter we include a *tutorials* section. Tutorials are slightly longer than examples; they synthesize the material of the chapter and illustrate the important ideas on real systems. The tutorials are an essential learning tool to introduce useful techniques that may reappear later in the book. The tutorials vary widely in length, depth, and difficulty. You may want to skim the longer or more difficult tutorials on the first read and return later for reinforcement of key concepts or for practice on difficult problems. We have intentionally incorporated this range of tutorials to maximize the utility of the text for the widest possible audience and to make it a practical and helpful reference throughout your career.[8]

We also include computation in many of our examples, tutorials, and problems. Computation is central to modern engineering and an important skill to be learned. It is integral to the learning and practice of dynamics. To simplify our presentation and make it consistent throughout the book, we have exclusively used MATLAB for all numerical work. There are many excellent numerical packages available (and some students may want to code their own). We chose MATLAB because of its ubiquity, its ease of use, and the transparent nature of its programming language. Our goal, however, is not to teach the use of a particular programming tool but for you to become comfortable with the full problem-solving process, from model building through solution.

We end each chapter with a summary of *key ideas*, which contains a short list of the main topics of the chapter. We intentionally minimize the prose in these sections to make it as easy as possible to use for reference and review. Reading these sections does not replace reading the chapters; they are meant only to serve as helpful references.

We used many sources in preparing this book and are indebted to a large number of authors that preceded us. Our primary references are listed in the Bibliography. In some cases, however, we highlight a particularly important result and direct you to other references with more in-depth discussions or additional insights. Thus each chapter has a Notes and Further Reading section, where we point out these sources.

Finally, we end each chapter in Parts One to Four with a problems section that includes problems that address each of the topics of the chapter. We have tried to provide problems of varying levels of difficulty and those that require computation. We have not included problems sections in Part Five, as Chapters 12 and 13 are intended as only an introduction to more advanced topics.

1.3 Key Ideas

- A **vector** is a quantity with both magnitude and direction in space. The position of point P relative to point O is the vector $\mathbf{r}_{P/O}$.

- A **reference frame** provides the perspective for observations regarding the motion of a system. A reference frame contains three orthogonal directions.

---

[8] Because Chapters 12 and 13 are similar to extended tutorials and are meant as only an introduction to more advanced material, we do not include tutorials or problems in them.

- The **velocity** is the change in time of a position with respect to a particular reference frame. The velocity of point P relative to frame $\mathcal{I}$ is ${}^{\mathcal{I}}\mathbf{v}_{P/O} \triangleq \frac{{}^{\mathcal{I}}d}{dt}(\mathbf{r}_{P/O})$.

- A **coordinate system** is the set of scalars used to locate a point relative to another point in a reference frame. **Cartesian coordinates** x, y, and z constitute the most common coordinate system. We usually use $(x, y, z)_{\mathcal{I}}$ to represent the Cartesian coordinates with respect to frame $\mathcal{I}$. The rates of change $(\dot{x}, \dot{y}, \dot{z})_{\mathcal{I}}$ of the Cartesian coordinates are called *speeds*.

- The **state** of a particle consists of its position and velocity in a reference frame at a given time.

- The **equations of motion** are the three second-order differential equations for the particle state whose solution provides the trajectory of a point.

- An **equilibrium point** is a special solution of the equations of motion for which the rates of change of all states are zero.

1.4 Notes and Further Reading

The modern formulation of dynamics is the culmination of more than two centuries of development. For instance, while Newton presented the fundamental physics, the concept of equations of motion and the formulation of the second law we know today were given by Euler.[9] The modern concept of a vector was introduced by Hamilton in the mid-nineteenth century.[10] A good, concise discussion of the early history of dynamics can be found in Tenenbaum (2004). A more thorough treatment of the history of mechanics is in Dugas (1988). We also recommend the book of essays by Truesdell (1968) for insightful discussions of important historical developments.

Careful notation is essential for both learning dynamics and solving problems in your professional career. Unfortunately, no universally accepted notation is in use. In fact, there is much discussion among educators and practitioners over how to balance simplicity and clarity. Our notation—particularly the use of reference frames in derivatives—is closest to that of Kane (1978) and Kane and Levinson (1985). A similar notational approach is used by Tenenbaum (2004) and Rao (2006). Our notation for position is also used in Tongue and Sheppard (2005) with a variation in Beer et al. (2007). Our qualitative definition of reference frames is similar to that in Rao (2006). Other good discussions of the importance of reference frames in dynamics can be found in Greenwood (1988), Kane and Levinson (1985), and Tenenbaum (2004). Tenenbaum also has a similar and insightful discussion regarding the distinction between coordinate systems and reference frames.

---

[9] Leonhard Euler (1707–1783) was a Swiss mathematician and physicist. He is known for his seminal contributions in mathematics, dynamics, optics, and astronomy. Much of our current notation is attributable to Euler. He is probably best known for the identity $e^{i\pi} + 1 = 0$, often called the most beautiful equation in mathematics.

[10] Sir William Rowan Hamilton (1805–1865) was an Irish mathematician and physicist. He made fundamental contributions to dynamics and other related fields. His energy-based formulation is the foundation of modern quantum mechanics.

1.5 Problems

1.1   What are the Cartesian coordinates of point P in frame $\mathcal{I}$, as shown in Figure 1.5?

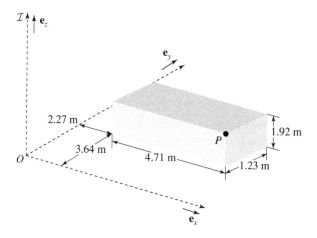

Figure 1.5 Problem 1.1.

1.2   Sketch and label the vectors $\mathbf{r}_{P/O}$, $\mathbf{r}_{P/Q}$, $\mathbf{r}_{Q/P}$ in Figure 1.6.

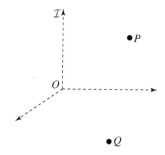

Figure 1.6 Problem 1.2.

1.3   Match each of the following definitions to the appropriate term below:

    a. A perspective for observations regarding the motion of a system
    b. A mathematical quantity with both magnitude and direction
    c. Second-order differential equations whose solution is the trajectory of a point
    d. A set of scalars used to locate a point relative to another point

    • Vector
    • Reference frame
    • Coordinate system
    • Equations of motion

CHAPTER TWO

-- -- -- -- -- -- -- -- -- -- -- -- -- -- -- -- --

Newtonian Mechanics

In this chapter we reintroduce the physical principles that underlie the study of dynamics. (We assume that you remember a bit from your physics classes.) In all that we do, Newtonian methods are used for solving dynamics problems. That is, we will solve problems using Newton's three laws of motion to relate forces and acceleration. This approach differs from the methods of Lagrange and Hamilton, which rely on energy techniques. (A brief introduction to Lagrangian methods is provided in Chapter 13.) To be sure, Lagrangian and Hamiltonian methods are important, are used frequently, and may become a cornerstone of your subsequent dynamics education. However, Newtonian methods have stood the test of time, are used regularly by practicing engineers, and provide an important foundation for the study and practice of dynamics. Using Newtonian methods, you can solve an amazing array of real engineering problems—some of which are actually more difficult to solve with other methods.

The purpose of this chapter is to instill a deep understanding of the physics of motion, which is interchangeably called *dynamics* or *mechanics.* You will finish this chapter with the basic skills needed for solving most dynamics problems. It is interesting to note that almost all of the physics in the book is contained in this chapter; no new laws of motion or other basic physical principles appear again (with the exception of one assumption about rigid bodies discussed in Chapter 9). In principle, the reader with great mathematical skills and insights could stop reading at the end of this chapter and solve any dynamics problem. This is, of course, an exaggeration, and we will provide many useful techniques and insights throughout the book, but the fact remains that there is no new physics beyond Newton's three laws of motion.

2.1 Newton's Laws

In 1687 Isaac Newton published the *Principia,* one of the greatest events—if not the greatest—in the history of science. With this single book he overturned centuries of misconceptions embedded in the Aristotelian idea that forces are necessary for bodies to remain in motion.[1] In the *Principia*, Newton provided not only a new philosophy but also specific tools for solving real problems. (In particular, Newton developed differential and integral calculus, though it is Leibniz's notation that we use today.) Here, then, are Newton's three laws of motion as translated in 1729 by Andrew Motte:[2]

Law I Every body perseveres in its state of rest, or of uniform motion in a right line, unless it is compelled to change that state by forces impressed thereon.

Law II The alteration of motion is ever proportional to the motive force impressed; and is made in the direction of the right line in which that force is impressed.

Law III To every action there is always opposed an equal reaction; or the mutual actions of two bodies upon each other are always equal, and directed to contrary parts.

Newton's first law is the explicit rejection of the Aristotelian idea that bodies must be acted on by a force to remain in motion (even at constant velocity). Newton elevated Galileo's experimental observations to his first law of motion. This idea is fundamental to modern dynamics and to the Newtonian method: forces cause the motion of bodies to change.

Newton's second law is the familiar statement that the rate of change of *linear momentum* of a body—represented by the symbol **p**—is equal to the *net force* acting on the body—represented by **F**.[3] Newton's second law is a vector relationship; hence we use boldface letters for the force and linear momentum vectors.

The net force **F** is the sum, or *resultant,* of all force vectors acting on the body (see Section 1.1).[4] Newton recognized the not necessarily obvious fact that the net force may produce an action in a direction different from any of the individual forces. (In fact, this observation is Corollary I of his laws.) When solving a dynamics problem, we begin by drawing each object and graphically illustrating every force vector acting on it; the net force is just their geometric sum. You should remember this diagram from statics; it is called a *free-body diagram.* Free-body diagrams are used throughout our study of dynamics as well. A simple example of a free-body diagram is shown in Figure 2.1. This vector nature of Newton's second law is extremely important; we explore it in depth in the next chapter.

Newton's third law, which is often called the *law of action and reaction,* is probably the most forgotten or misunderstood. Yet this law is of utmost importance to everyday experience and engineered systems. Simply put, if an object A exerts a force on object

---

[1] It was actually Galileo Galilei (1564–1642) who showed that, in the absence of forces, bodies stay at rest or in uniform motion, giving us the science of kinematics and the experimental method. However, Newton placed the new science on a firm mathematical foundation and developed the tools for solving problems.
[2] As with most scientific works of the day, the original was written in Latin.
[3] Note that Newton's second law actually only states proportionality, so to be absolutely correct we should include an arbitrary constant in the equation relating force and the rate of change of linear momentum. However, by convention, the units of force and linear momentum are selected to make this constant unity. Thus, in the International System (SI) of units the *newton* is a derived unit equal to 1 kg-m/s^2.
[4] For a brief review of vector algebra, consult Appendix B.

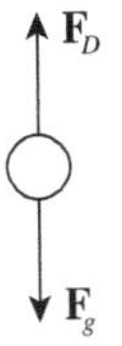

Figure 2.1 A simple example of a free-body diagram, consisting of a falling particle acted on by two forces: gravity in the downward direction and aerodynamic drag resisting its fall in the upward direction.

B, then object B exerts an equal and opposite force on object A. The recoil in a gun is a common example. Another example is the force on our wrist when we hit a tennis ball with a racket. Newton's third law is extremely important for explaining flight: an airfoil at a non-zero angle of attack pushes against the airstream as it moves forward; by Newton's third law, the air pushes back, generating lift.

Newton's three laws—and the accompanying mathematical tools we will develop—suffice to solve all problems posed in this book. We ignore dynamics at the atomic scale, where Newtonian physics breaks down and quantum mechanics takes over. We also ignore motion at speeds near the speed of light, where special relativity must be taken into account.

Although a complete vector treatment of Newton's second law is premature, we can still solve meaningful problems with only a scalar representation. Recall from Chapter 1 our discussion of position and velocity as described by three scalars and their derivatives. Newton's second law applies independently to each of these scalar coordinates. For example, in the x-direction, we write

$$f_x = \dot{p}_x = m\ddot{x},$$

where f_x is the sum of all forces in the x-direction, $p_x \triangleq m\dot{x}$ is the linear momentum in the x-direction, and $\ddot{x}$ is the second derivative of x with respect to time (acceleration). (Appendix A supplies a brief review of some essential results from calculus if you are rusty.)

Example 2.1 Straight-Line Motion with No Force

The simplest possible system in dynamics is one in which there are no net forces, that is, $f_x = 0$. In this case Newton's second law reduces to the simple equation of motion

$$\ddot{x} = 0. \tag{2.1}$$

To solve the equation of motion for $x(t)$ and $\dot{x}(t)$, we use the definition $\ddot{x} \triangleq \frac{d\dot{x}}{dt}$. Multiplying Eq. (2.1) by dt and integrating yields

$$\int d\dot{x} = \dot{x}(t) + C_1 = 0, \tag{2.2}$$

where C_1 is an integration constant. Multiplying both sides of Eq. (2.2) by dt and integrating a second time yields

$$\int dx + \int C_1 dt = x(t) + C_1 t + C_2 = 0, \qquad (2.3)$$

where C_2 is another integration constant. Setting $t = 0$ in Eqs. (2.2) and (2.3), we find that the constants are $C_1 = -\dot{x}(0)$ and $C_2 = -x(0)$, where $x(0)$ and $\dot{x}(0)$ are the initial conditions. The solution is

$$\dot{x}(t) = \dot{x}(0)$$

$$x(t) = x(0) + \dot{x}(0)t.$$

This example is a simple, vivid demonstration of our problem-solving methodology. We use Newton's second law to find an equation of motion (in this case, Eq. (2.1)) and then solve the equation of motion to find a trajectory in terms of initial conditions. In this case, the equation of motion was particularly easy to solve using direct integration (unfortunately, that is unusual!).

Example 2.2 Straight-Line Motion with Constant Force

In this example we consider f_x to be a constant force, in which case the equation of motion is[5]

$$\ddot{x} = \frac{f_x}{m}. \qquad (2.4)$$

Integrating Eq. (2.4) as in Example 2.1, we obtain

$$\int d\dot{x} = \dot{x}(t) + C_1 = \frac{f_x}{m}t \qquad (2.5)$$

and

$$\int dx + \int C_1 dt = x(t) + C_1 t + C_2 = \frac{f_x}{2m}t^2. \qquad (2.6)$$

Setting $t = 0$ in Eqs. (2.5) and (2.6), we find that the constants are $C_1 = -\dot{x}(0)$ and $C_2 = -x(0)$. The solution trajectory is

$$\dot{x}(t) = \dot{x}(0) + \frac{f_x}{m}t$$

$$x(t) = x(0) + \dot{x}(0)t + \frac{f_x}{2m}t^2.$$

You may recognize the equation for $x(t)$ from introductory physics; it describes the displacement of a particle undergoing constant acceleration f_x/m.

---

[5] Note that when writing equations of motion, we solve for the second-order variables (e.g., $\ddot{x}$), which are the unknowns in the sense of a system of algebraic equations.

Example 2.3 Straight-Line Motion with Position-Dependent Force

Consider a force $f_x(x)$ that is a function of position x. Thus Eq. (2.4) becomes

$$\ddot{x} = \frac{f_x(x)}{m} = a(x). \tag{2.7}$$

The quantity $f_x/m = a(x)$ has units of acceleration and is often referred to as the acceleration of the mass. You can also view it as the *specific* force acting on mass m.[6] Eq. (2.7) is a separable differential equation (see Appendix C). To integrate it, first multiply both sides by $dx = \dot{x}dt$ and then use the definition $\ddot{x} \triangleq \frac{d\dot{x}}{dt}$ on the left side. Integrating from time t_1 to time t_2 yields

$$\int_{t_1}^{t_2} \ddot{x}\dot{x}dt = \int_{t_1}^{t_2} \dot{x}d\dot{x} = \frac{1}{2}(\dot{x}(t_2))^2 - \frac{1}{2}(\dot{x}(t_1))^2 = \int_{x(t_1)}^{x(t_2)} a(x)dx. \tag{2.8}$$

Replacing $\dot{x}(t_i)$ by the velocity $v(t_i)$, for $i = 1, 2$, allows us to write Eq. (2.8) a bit more compactly:

$$v^2(t_2) = v^2(t_1) + 2\int_{x(t_1)}^{x(t_2)} a(x)dx. \tag{2.9}$$

Eq. (2.9) is an elegant and compact expression that relates the velocity directly to position. This expression can be helpful and convenient for some problems, particularly when the acceleration (specific force) is an integrable function of position. For instance, consider the case of constant acceleration. In this case, Eq. (2.9) yields an alternative velocity equation:

$$v^2(t_2) = v^2(t_1) + 2a(x(t_2) - x(t_1)).$$

In addition to forces that depend on position, we often encounter forces that depend on velocity (see Tutorial 2.2).

2.2 A Deeper Look at Newton's Second Law

It is helpful to pause for a moment to contemplate the significance and, more importantly, the assumptions underlying Newton's laws. This exercise is more than mere pedagogy. Newton postulated important universal facts that form the foundation for his and that will inform all we do. Understanding these postulates is essential for understanding dynamics and solving problems.

For instance, why are Newton's laws called *laws* and not *theories?* Why is it not Newton's *Theory of Motion?* A useful explanation appears in a publication by the National Academy of Sciences.

Definition 2.1 A **law** is a descriptive generalization about how some aspect of the natural world behaves under stated circumstances.

---

[6] In this context, the word *specific* means *divided by a quantity representing an amount of material.* The specific force is the force divided by mass.

Contrast this definition to the definition of a *theory*, the most important endpoint of a scientific endeavor.[7]

Definition 2.2 In science, a **theory** is a well-substantiated explanation of some aspect of the natural world that can incorporate facts, laws, inferences, and tested hypotheses.

Newton's three laws—most importantly, his second law—provide a predictive description of the behavior of objects subjected to forces. They allow us to analyze our observed universe (the motion of the planets being the most obvious—and first—successful application of his laws) and to synthesize engineered devices. They do not, however, explain why or how objects behave the way they do. In particular, Newton did not explain what a force is or how the concept of force arises from first principles. In fact, this omission is one of the greatest objections to Newton's *Principia*. Newton did not explain *mass* or *inertia*. Nor did he explain the meaning of acceleration or the absolute space relative to which acceleration is measured. All these ideas are important for understanding how to use Newton's second law—and its limitations— so we explore them in more detail in the following subsections.

2.2.1 The Concept of Force

The concept of force in Newton's second law is the most ill-defined and philosophically difficult concept in classical mechanics. What, in fact, did Newton mean by "force"? What is a force? How does it arise? As pointed out earlier, Newtonian mechanics consists of laws, not theories. In fact, there is no accepted explanation for the concept of a force. From almost the moment the *Principia* was published, scientists and philosophers criticized the concept of force as devoid of meaning. Modern physicists have completely eliminated the concept of force from the formulation of almost all statements of the laws of physics (such as quantum mechanics and general relativity). In fact, in Chapter 13 we will discuss how classical mechanics can be reformulated to avoid the use of forces entirely (at least most of the time).

Nevertheless, Newtonian mechanics was—and remains—profoundly successful and forms the foundation of what we will study in this book. We need only accept that a force is simply an abstract concept that causes objects to behave in a predictable way. A remarkable array of problems can be solved using Newtonian methods. We will introduce a variety of forces without worrying about a precise physical explanation. By using forces in Newton's second law, we can describe and predict the macroscopic motion of objects. This approach is a compact and elegant tool for engineering design and analysis.

2.2.2 The Concept of a Point Mass

Newton's use of the term "body" in his statement of the three laws is a bit misleading. In fact, his laws apply to one and only one sort of object: the point mass. What is a point mass? Unfortunately, there is no good explanation in Newton's laws. (They are

---

[7] For example, Darwin's Theory of Evolution or Einstein's Theory of Relativity.

laws, not theories!) In fact, there is no good explanation for the concept of mass—Newton defined it as the quantity of matter. For our purposes, it does not matter. A point mass is an infinitesimally small body, or *particle,* of mass m that behaves as predicted by Newton's second law. In fact, we measure mass by observing a particle's behavior under Newton's second law. We use the term "particle" interchangeably with the term "point mass."

An extended body is modeled as a collection of point masses. However, throughout Part One of the book we often treat an extended body as if it were a single point mass. This approximation turns out to be fine, though the rigorous proof is not given until Chapter 6. Part Three shows how to treat the motion of extended bodies by examining Newton's second law for each of the (possibly infinite) constituent point masses.

The main idea is that Newton's laws only apply to point masses. Remembering this point can help you avoid many pitfalls. To make it easier to remember, we embed our notation with little reminders. For example, we assign to every point mass a label, such as P or Q. The net force acting on particle P is written as $\mathbf{F}_P$. For clarity, we also add the subscript P to the particle mass m_P. For a collection of N particles, we use an index, such as $i = 1, \ldots, N$, to label each particle and m_i to denote its mass. In general, to find the equations of motion of the collection, Newton's laws are applied to each particle separately. Part Three shows how to simplify this procedure when the particles in the collection have fixed relative positions (that is, when they form a rigid body).

2.2.3 Acceleration and Absolute Space

Our statement of Newton's second law emphasized that the rate of change of momentum is relative to a reference frame $\mathcal{I}$. However, we failed to indicate what this reference frame is or why it matters. There is an inherent assumption in Newton's second law: the second law applies only when the acceleration is relative to a special frame of reference. Newton called this special frame *absolute space.* We today call it the *inertial frame.* We often use the letter $\mathcal{I}$ to label an inertial reference frame.

Unfortunately, there is no good definition of absolute space, which is another weakness of Newton's laws. It is common to refer to the "fixed stars" as absolute space. But, of course, the stars are not fixed. One of Einstein's great accomplishments in his development of the Theory of General Relativity was the Equivalence Principle. This principle removed the need for absolute space from Newton's laws. Einstein showed that a frame of reference falling freely in a gravitational field is an inertial frame, which implies Newton's laws hold in such a frame (e.g., the interior of a space station in orbit).

For our purposes, the inertial frame is an essential abstraction that need not be explained physically. It usually suffices to choose a reference frame whose acceleration is relatively small compared to the accelerations of the particles of interest. In this case, any errors introduced are negligible. (We will give mathematical substance to this approximation in Chapter 3.)

Nonetheless, the concept of an inertial frame is essential. The first thing we do in every dynamics problem is draw the inertial frame to remind ourselves that the laws of motion apply only in this frame. As in Chapter 1, we specify the (inertial) frame to which the velocity is referred. The same notation is used for the acceleration vector

$^{\mathcal{I}}\mathbf{a}_{P/O}$, which is the instantaneous rate of change in time with respect to the inertial frame of the velocity, that is, $^{\mathcal{I}}\mathbf{a}_{P/O} \triangleq \frac{^{\mathcal{I}}d}{dt}(^{\mathcal{I}}\mathbf{v}_{P/O})$. Using Cartesian coordinates with respect to frame $\mathcal{I}$, the acceleration is $(\ddot{x}, \ddot{y}, \ddot{z})_{\mathcal{I}}$, where $\ddot{x} \triangleq \frac{d\dot{x}}{dt}$, $\ddot{y} \triangleq \frac{d\dot{y}}{dt}$, and $\ddot{z} \triangleq \frac{d\dot{z}}{dt}$.

It is important to remember the concept of *Newtonian relativity*. This concept states that any reference frame moving at constant velocity but not rotating relative to an inertial frame is also an inertial frame. This can be proven rather succinctly using the ideas and notation of Chapter 3 (see Section 3.6).

2.2.4 Anatomy of Newton's Second Law

At this point we are in a position to write Newton's second law in the form to be used throughout the book. Let $^{\mathcal{I}}\mathbf{p}_{P/O} \triangleq m_P{}^{\mathcal{I}}\mathbf{v}_{P/O}$ denote the linear momentum vector of particle P with mass m_P. Newton's second law states the time derivative with respect to the inertial frame of the linear momentum of a point mass P is equal to the total force $\mathbf{F}_P$ acting on P, that is,

$$\boxed{\mathbf{F}_P = \frac{^{\mathcal{I}}d}{dt}(^{\mathcal{I}}\mathbf{p}_{P/O}) = m_P{}^{\mathcal{I}}\mathbf{a}_{P/O}.} \tag{2.10}$$

Figure 2.2 graphically identifies the various notational elements in this representation of Newton's second law.

Note that, to arrive at the form of Newton's second law in which force equals mass times acceleration, we factored the mass out of the derivative $\frac{^{\mathcal{I}}d}{dt}(^{\mathcal{I}}\mathbf{p}_{P/O})$. It is not correct to add a term $\dot{m}_P{}^{\mathcal{I}}\mathbf{v}_{P/O}$, which arises from using the product rule to evaluate the derivative. Recall again that Newton's second law applies only to point masses. If a point mass has no extent, it can't gain or lose mass. Thus in Newtonian mechanics, $\dot{m}_P = 0$. Including a mass derivative term is okay only at relativistic speeds where the

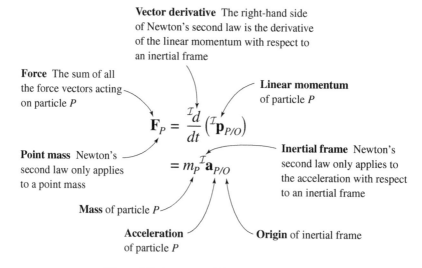

Figure 2.2 Anatomy of Newton's second law.

mass changes, and then all sorts of other problems arise. (If you are wondering how rockets fly by expelling mass, we cover that in detail in Chapter 6.)

2.2.5 Conservation Laws

You may have noticed that Newton's first and second laws are intimately related. Letting the force in Eq. (2.10) equal zero and integrating with respect to time results in a simple restatement of Newton's first law:

$$^{\mathcal{I}}\mathbf{p}_{P/O} = \text{constant}. \tag{2.11}$$

In words, Eq. (2.11) states that when the total force on a particle is zero, the linear momentum of the particle is a constant of the motion. This is just Newton's first law! We often say that the momentum is *conserved*. The concept of conserved quantities is an important one in dynamics and we return to it often in the book. In fact, it is useful to carefully define it here.

Definition 2.3 A scalar or vector function of the state of a particle or multiparticle system is **conserved** if it remains constant throughout the trajectory of the particle or system.

We introduce a *conservation law* when a quantity is conserved under some general set of circumstances. In fact, Eq. (2.11) already states our first conservation law: conservation of linear momentum of a particle. Although this law is just a restatement of Newton's first law, it is useful to start our practice of highlighting important conservation laws here.

Law 2.1 The law of **conservation of linear momentum of a particle** states that, when the total force acting on a particle is zero, the linear momentum of the particle is a constant of the motion.

Conservation laws can be incredibly useful—they may reduce the complexity of the equations of motion, provide important constraints on the trajectory, or provide tools for checking the accuracy of our analysis and simulations. Don't forget the constraints or conditions required to invoke the conservation law. When we find a conservation law by integrating the equations of motion once, as in this case, we call the resulting conserved quantity a *first integral of the motion*.

2.3 Building Models and the Free-Body Diagram

Since Newton's second law only applies to a point mass, it is sensible to ask how to use it to solve for the motion of more complex objects. In engineering practice we may want to understand how cars move, airplanes fly, or submarines maneuver. These objects seem rather different from point masses. The art of dynamics comes in building representative models for non-point masses out of elements we understand and can use to find equations of motion. In this book, we derive trajectories for point masses, collections of point masses, and rigid bodies of basic shapes (rods, disks, spheres, etc.). Where practicable we show that finding the equations of motion for these simple elements is equivalent to finding the equations of motion for the original system.

The first step of any dynamics problem is to formulate a model—using basic modeling elements and various connective abstractions—for the system being studied. The challenge is to find the simplest model that provides meaningful results. Simplicity is essential. To achieve it, we often render some model components as massless. However, oversimplification can lead to trouble if you are not careful. For example, a massless component does not satisfy Newton's third law!

Once you have modeled a system, the reference frames, coordinates, and forces need to be identified. Only then can Newton's second law be used to find the equations of motion and trajectories. We start by drawing for each mass a free-body diagram that explicitly identifies the relevant force vectors. You may be familiar with free-body diagrams from statics or physics. In statics, you vectorially add all the forces in the free-body diagram and set the sum equal to zero in each orthogonal direction—as required for a static equilibrium. In dynamics, the vector sum of the forces is proportional to the acceleration. In every problem you solve, draw a free-body diagram before writing down Newton's second law. This habit is essential for solving dynamics problems. It should soon become second nature.

Figure 2.3 depicts a physical system and a suitable model constructed from point masses. The Watt flyball governor, shown in Figure 2.3a, was one of the first examples of a feedback control system. Widely used in the control of mills during the seventeenth century, the first steam governor was designed by James Watt in 1788. The governor is a beautiful example of using simple dynamics in design. The actual governor pictured in Figure 2.3a is reasonably complicated, with many gears, masses, and linkages. However, it can be reduced to a very simple model using only point masses and massless rods, as shown in Figure 2.3b. The free-body diagram corresponding to this model is shown in Figure 2.3c. Even this bare-bones model produces a remarkable richness of motion and displays the fundamental principles underlying the flyball governor. We explore it in more depth in Chapter 10.

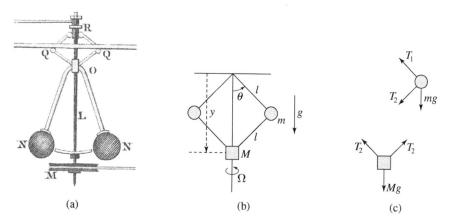

(a)   (b)   (c)

Figure 2.3 The Watt flyball governor, designed in 1788, was used to control the rotational speed of a steam engine drive shaft. Though a complex mechanical system, it is relatively simple to model using three point masses and four massless rods. (a) Flyball governor. (b) Governor model. (c) Free-body diagram.

2.4 Constraints and Degrees of Freedom

In Examples 2.1 and 2.2 we treated motion in one dimension only. A dynamicist would say that the particle had only a single *degree of freedom*. As discussed in Chapter 1, a particle trajectory is defined by three positions and three speeds. In the three-dimensional world, particles have three degrees of freedom: they are free to move in three orthogonal directions. However, not every problem we solve will consist of particles free to move in every direction. A particle sliding on top of a table, for example, or a hockey puck sliding on a flat ice rink, as in Figure 2.4, moves in only two directions. Such a particle has two degrees of freedom.

We use the idea of degrees of freedom often in dynamics to gain understanding of a problem and to avoid unnecessary work. In other words, if we know from the beginning that a particle has fewer than three degrees of freedom, then we should need fewer equations of motion. We associate a single scalar coordinate with each degree of freedom and produce a single equation of motion for each coordinate.[8] Thus, a two degree-of-freedom problem will have only two nontrivial equations of motion. We call out this important point in the following definition.

Definition 2.4 The number of **degrees of freedom** of a collection of particles is the number of independent coordinates needed to describe the position of every particle. For a collection of rigid bodies, the number of degrees of freedom is equal to the number of independent coordinates needed to describe the position and orientation of every rigid body.

If there are more equations of motion than degrees of freedom, we have to simultaneously solve the *constraint* equations. Each constraint equation represents a reduction of one degree of freedom. Constraints can be either explicit or implied. Either way, the presence of a constraint in a dynamics problem implies that there exists a coordinate that does not play a role in the solution. For instance, for the hockey puck in Figure 2.4, we can use Cartesian coordinates to track its position. If x and y describe the horizontal position, then a mathematical statement of the constraint is $z = 0$. Examples 2.1 and 2.2 are one-degree-of-freedom problems—in each example, there are implied constraints that prevent motion in the other two directions. Throughout Parts One, Two, and Three we treat only one- and two-dimensional systems. That is, we assume an implied constraint $z = 0$ and sometimes $y = 0$ as well.

For simple examples like the hockey puck, labeling the degrees of freedom and constraints may seem to be unnecessary. However, understanding degrees of freedom and constraints is an important part of understanding dynamics. Not all constraints are so simple, and the degrees of freedom do not always align nicely with simple Cartesian coordinates. For instance, if a particle is constrained to move on a curved surface, then there is still one constraint equation—albeit more complicated than $z = 0$—and two degrees of freedom. For instance, a hemispherical table has the constraint $\sqrt{x^2 + y^2 + z^2} = $ constant. When we treat multiple particles, the number of degrees of freedom can become quite large—each particle nominally has three degrees of freedom—or the number of constraints limiting the relative motion of the particles can become large. And, to further complicate things, we may have constraints on the

---

[8] Up until now we have discussed only Cartesian coordinates—other coordinate systems are introduced in Chapter 3.

Figure 2.4 A hockey puck sliding on the surface of an ice rink is an example of a two-degree-of-freedom system. Image courtesy of Shutterstock.

motion rather than on the configuration; for example, an ice skate can only move along the direction of the blade, not perpendicular to it.[9]

Counting degrees of freedom is an important habit that ensures you have the correct number of equations of motion. In certain problems—we will see some soon—the constraint equations need to be used explicitly to obtain the same number of equations of motion as there are degrees of freedom. In general, the number M of degrees of freedom in a system of N particles is

$$M = 3N - K, \qquad (2.12)$$

where K is the number of constraints. Note that the number of degrees of freedom in a system of N rigid bodies and K constraints is $M = 6N - K$, since the orientation of each body counts as three additional degrees of freedom.

In problems with reduced degrees of freedom, explicit constraints imply the presence of a force—called a *constraint force*. This observation results from a direct application of Newton's third law. Consider again a particle on a table. Assuming that the force of gravity keeps the particle on the table, then the table must apply an equal and opposite force on the particle. In this case, the force of constraint is a *normal* force—a force orthogonal to the direction of motion. Usually, we are not interested in the constraint forces and seek to eliminate them from the equations of motion. Sometimes we may solve for these forces in terms of the particle trajectory to be certain our system is engineered correctly. (We do not want the table to break under the weight of the particle.) These ideas are demonstrated in a classic example at the beginning of Chapter 3.

---

[9] Motion constraints are often called *nonholonomic constraints*. We discuss these in Chapter 13.

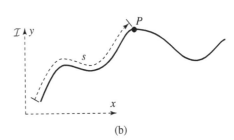

(a)   (b)

Figure 2.5 A roller coaster and the corresponding model. Image (a) courtesy of Shutterstock.

Example 2.4 Modeling a Roller Coaster

Suppose we want to analyze the dynamics of a roller coaster like the one shown in Figure 2.5a. The simplest model would confine the tracks to a horizontal or vertical plane and model the car as a single point mass, as shown in Figure 2.5b. How many degrees of freedom does this model possess?

The roller-coaster car's position can be specified using Cartesian coordinates $(x, y)_\mathcal{I}$. However, we know that the car must remain on the track. Suppose the track shape can be defined by a function $y = f(x)$. This is then a constraint on the geometry, reducing the number of degrees of freedom from two to one. We can solve for the motion of this single-degree-of-freedom model by finding equations for $(x, y)_\mathcal{I}$ subject to the constraint $y = f(x)$ or, alternatively, we can use a single coordinate, s, to represent the distance the car has traveled along the track. Remember, since the model has one degree of freedom, we only need one scalar coordinate. We call s a *path coordinate*. Path coordinates are discussed at some length in Chapter 3.

Example 2.5 Modeling a Three-Link Robot Arm

Figure 2.6a is a picture of a common three-link robot arm used to move and place objects. It consists of a shoulder joint, upper arm, elbow, lower arm, wrist joint, and hand. Suppose we wish to model the arm and develop equations describing its motion (presumably to develop a control system for it). The goal would be to determine the position of the hand. We discuss this problem in more detail in Chapter 8. Here, however, we are interested in modeling the system, determining how many degrees of freedom it has, and choosing appropriate coordinates to describe the system.

Figure 2.6b shows our model of the robot arm consisting of three linked rigid bodies (the upper and lower arms and hand). Each joint can rotate about only a single axis perpendicular to the plane of motion. We can thus define the angle of each joint by a single coordinate. By inspection, then, this system has three degrees of freedom. Thus three scalar coordinates suffice to describe the state of the system. In Figure 2.6b we use the three angular coordinates $(\theta_1, \theta_2, \theta_3)$ to describe the angles of each joint relative to the line through the previous arm. However, as mentioned earlier, the choice of scalar coordinates in a problem is not unique. For instance, in this problem we could equally well choose as coordinates the angles of each joint relative to the inertial x-axis. Although the motion of the hand would be the same (it would follow the

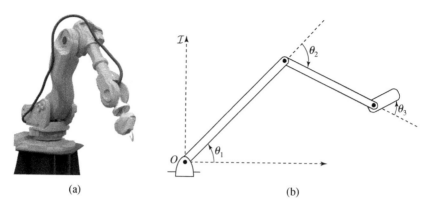

(a)   (b)

Figure 2.6 A three-link industrial robot arm and the corresponding model. Image (a) courtesy of Shutterstock.

same trajectory $x(t)$ and $y(t)$), how we describe it using scalar coordinates would be different. Part of the art of dynamics is finding the best set of scalar coordinates for a problem.

2.5 A Discussion of Units

We end this chapter with a brief discussion of units. For all problems in this book we use SI units.[10] This set of units is the internationally accepted system that uses metric measures of length and mass. All SI units and subunits in a category (length, say) are related by factors of 10, and the power of ten is identified explicitly by the prefix. Thus a kilometer is equal to 1,000 meters. We supply a list of the common prefixes in Table 2.1.

The SI system is an *absolute* system. It is based on three standards: length (the meter (m)), mass (the kilogram (kg)), and time (the second (s)). Each of these fundamental units can be related to a well-defined physical phenomenon and is independent of the location of the measurement (hence the term "absolute"). No matter where we solve problems or take measurements, the meter is the meter, the second is the second, and the kilogram is the kilogram. In case you were wondering, the second is defined to be the duration of 9,192,631,770 periods of the electromagnetic wave emanated during the transition between the two hyperfine energy levels of the ground state of the cesium 133 atom. The meter is the length of the path traveled by light in vacuum during a time interval of 1/299,792,458 of a second.[11]

The unit of force in SI—the newton—is a derived unit: 1 newton is equal to 1 kg-m/s^2. In words, under 1 N of force, a mass of 1 kg will accelerate at a rate of 1 m/s^2. The

---

[10] The common abbreviation for the system, SI, comes from the French: *Le Système International d'Unités.*
[11] The kilogram is less well defined; it is the only base unit still defined by a physical object. The kilogram is defined as the mass of a platinum-iridium bar being kept at the Bureau International des Poids et Mesures near Paris. Unfortunately, the kilogram was recently discovered to have lost 50 micrograms over the past 100 years for unknown reasons. Efforts are under way to develop an atomic standard for the kilogram.

TABLE 2.1
The common International System (SI) prefixes and conversions between
U.S. Customary (USC) and SI units

Multiple	Symbol	Prefix	SI	USC
10^9	G	giga	1 m	3.28 ft
10^6	M	mega	1 km	0.621 mi
10^3	k	kilo	2.54 cm	1 in
10	da	deca	1 mm^3	6.10×10^{-5} in^3
10^{-1}	d	deci	1 m^3	35.3 ft^3
10^{-2}	c	centi	1 m/s	3.28 ft/s
10^{-3}	m	milli	1 km/h	0.621 mi/h
10^{-6}	μ	micro	1 m/s^2	3.28 ft/s^2
10^{-9}	n	nano	1 kg	0.0685 slug
10^{-12}	p	pico	1 N	0.225 lb
			1 J	0.738 ft-lb
			1 W	1.34×10^{-3} hp
			1 N/m^2 (Pa)	1.45×10^{-4} lb/in^2 (psi)

weight of an object is a force equal to the mass of an object times the acceleration of gravity g at sea level ($g \approx 9.81$ m/s^2). Weight is thus equivalent to the force of gravity. A 1 kg mass weighs 9.81 N.

Despite governmental efforts to move the United States to the SI system, many practicing engineers still use the U.S. Customary (USC) system of units. It is thus helpful to be familiar with USC units, though we avoid using them in this book. In the USC system, the fundamental standards are length (the foot (ft)), time (the second (s)), and force (the pound (lb) or pound-force (lbf)). The second is the same as the SI second (the only common unit between the two systems). The foot is equal to 0.3048 m. The pound is equal to the sea level weight of a standard pound located at the National Institute of Standards and Technology in Boulder, Colorado. Because force, rather than mass, is the fundamental standard, USC units are not an absolute system. The definition of a pound depends on location—in this case, the mean sea level. The unit of mass (the slug) is thus a derived unit. To form a set of units consistent with Newton's second law, 1 slug is thus defined as the mass that accelerates 1 ft/s^2 when acted on by 1 lb of force (or 1 lbf). The slug is thus a derived unit equivalent to 1 lb-s^2/ft. Thus the standard pound has a mass of 1/32.2 slugs.

2.6 Tutorials

Even without the notational tools of the next chapter we can solve interesting and important problems. As in earlier examples, in this section we study problems in one dimension, where Newton's second law simplifies to the scalar equation $f_x = m\ddot{x}$

(or $f_y = m\ddot{y}$), where f_x (or f_y) refers to the total force on the point mass in the x-direction (or y-direction). In other words, we treat only problems with one degree of freedom here, except for the final problem, which examines a simple two-dimensional (two-degree-of-freedom) problem as an introduction to the next chapter. Many of the concepts and solutions presented here pop up again and again throughout the book. You should finish this section with an understanding of what our objectives are in all dynamics problems, the process we go through to find equations of motion, and even how we solve those equations (when possible) to develop insight into the motion of a body. The techniques introduced here are the same as those used for more complicated problems later in the book. For example, we always start by drawing reference frames and a free-body diagram.

The tutorials are sorted by the type of force acting on the point mass. It may surprise you, but there are only a few types of force encountered in the book (and in practice). We try to cover most of them here, albeit in only one dimension. Forces generally fall into one of two categories: *contact* forces and *field* forces. As the name implies, contact forces arise from the contact between two objects (or an object and a stationary surface). We often call a contact force perpendicular to the direction of motion a *normal* force. A contact force in the direction of motion is a *tangential* force. The main tangential contact forces we will be studying are *friction* forces (see Tutorials 2.1 and 2.2). Another example of a tangential contact force is *aerodynamic drag* (see Tutorial 2.5). The main normal force encountered is the *constraint* force. We discuss it at length in the next chapter.

Field forces act on a body at a distance. Three examples of a field force are gravity, electrostatics, and magnetism. In this book, the primary field force encountered is gravity. For objects that remain close to a fixed distance from the gravitating center (the center of the earth, say), we approximate the acceleration due to gravity as constant, represented by g. For objects farther away from the gravitational center, we use Newton's inverse-square law of gravity (see Tutorial 2.5).

Tutorial 2.1 Sliding Friction

Consider a (point) mass m sliding on a horizontal surface, where x denotes the position coordinate, as depicted in Figure 2.7a, and $\dot{x}$ is the particle speed.[12] Let N denote the magnitude of the normal force exerted by the table on the particle in the vertical direction. One simple model of the interaction force between the surface and the particle is called *kinetic* or *Coulomb friction*.[13] Assuming $\dot{x} > 0$, we have

$$f_x = -\mu_c N,$$

where μ_c is the (unitless) coefficient of Coulomb friction. The minus sign indicates that friction exerts a force in the opposite direction of the particle velocity, which is to the right. Since the particle is not moving in the vertical direction, we can write

$$m\ddot{y} = 0 = N - mg.$$

---

[12] Remember, even if it looks like a box, we treat mass m as a particle!

[13] Named for Charles-Augustin de Coulomb (1736–1806), the French physicist who developed the model.

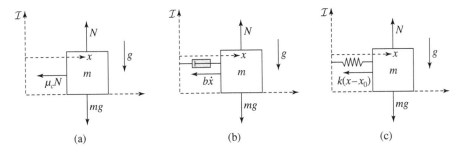

Figure 2.7 Free-body diagrams for (a) sliding friction, (b) viscous friction, and (c) a linear spring acting on mass m.

The normal force, N, is an example of a constraint force. The constraint equation is $y = 0$, which reduces the number of degrees of freedom from two to one. (There is also the implicit constraint that the box motion is confined to the plane of the page.)

In the horizontal direction, the equation of motion is

$$\ddot{x} = \frac{f_x}{m} = -\frac{\mu_c N}{m} = -\mu_c g.$$

As in previous examples with a constant force, the trajectory solution is easily obtained by integration:

$$\dot{x}(t) = \dot{x}(0) - \mu_c g t$$

$$x(t) = x(0) + \dot{x}(0)t - \frac{1}{2}\mu_c g t^2.$$

Note that this model is valid only as long as the particle is sliding to the right. In the case where the particle is not moving, that is, $\dot{x} = 0$, we can use the *static friction* model. The coefficient of static friction μ_s is typically greater than or equal to μ_c. The static friction model is valid only when the particle is not moving relative to the surface—in which case there is no equation of motion. In other words, an applied force must exceed the static friction for motion to occur. Once the particle is moving, the kinetic friction model applies.

Tutorial 2.2 Viscous Friction

Another type of friction, known as *viscous friction,* is proportional to the speed of the particle and is directed opposite the particle velocity. Consider a sliding (point) mass m attached to a *damper,* as shown in Figure 2.7b. A damper is a device that reduces oscillations or vibrations using viscous friction (e.g., an automobile shock absorber). In this system, the force from the damper on the mass depends on the velocity:

$$f_x = -b\dot{x},$$

and the equations of motion are

$$\ddot{x} = -\frac{b}{m}\dot{x}$$

$$m\ddot{y} = N - mg.$$

Note that we also have the constraint that $y = \dot{y} = \ddot{y} = 0$; that is, we assume the motion is one dimensional. This assumption yields the expression $N = mg$ for the normal force. (The normal force is an example of a constraint force.) The equation of motion in the x-direction can be written as a first order, separable differential equation (see Appendix C) using the change of coordinates $z = \dot{x}$. We have

$$\dot{z} = \frac{dz}{dt} = -\frac{b}{m}z. \tag{2.13}$$

Separating variables in Eq. (2.13) and then integrating the result yields

$$\frac{dz}{z} = -\frac{b}{m}dt \quad \text{which implies} \quad \int \frac{dz}{z} = -\frac{b}{m}t.$$

The solution is

$$\ln z + C_1 = -\frac{b}{m}t, \tag{2.14}$$

where $C_1 = -\ln z(0)$. Plugging C_1 into Eq. (2.14) and taking the exponential of both sides gives

$$z(t) = z(0)e^{-\frac{b}{m}t}$$

or, in the original coordinates,

$$\dot{x}(t) = \dot{x}(0)e^{-\frac{b}{m}t}. \tag{2.15}$$

Thus the speed of mass m subject to viscous friction decreases exponentially to zero.
Next we integrate Eq. (2.15) to find the position $x(t)$:

$$x(t) + C_2 = \int \dot{x}(0)e^{-\frac{b}{m}t}dt = -\frac{m\dot{x}(0)}{b}e^{-\frac{b}{m}t},$$

where $C_2 = -x(0) - \frac{m\dot{x}(0)}{b}$. The solution is

$$x(t) = x(0) + \frac{m\dot{x}(0)}{b}\left(1 - e^{-\frac{b}{m}t}\right).$$

Taking the limit $t \to \infty$, we see that the position of mass m asymptotically approaches the final position $x(0) + \frac{m\dot{x}(0)}{b}$.

Tutorial 2.3 Simple Harmonic Motion (The Ubiquitous Spring)

Consider a force that is a linear function of position:

$$f_x = -k(x - x_0), \tag{2.16}$$

where k is a proportionality constant. Eq. (2.16) is a model of the force exerted on a particle by a *linear spring* with spring constant k kg/s^2 and unstretched length x_0. The spring-mass system is illustrated in Figure 2.7c. In this system, the equation of motion

$$\ddot{x} = -\frac{k}{m}(x - x_0) \tag{2.17}$$

is a separable differential equation (see Appendix C). Multiplying Eq. (2.17) by $dx = \dot{x}dt$ and integrating yields

$$\int \dot{x}\,d\dot{x} = \frac{1}{2}(\dot{x}(t))^2 + C_1 = -\frac{k}{2m}(x(t) - x_0)^2. \tag{2.18}$$

Setting $t = 0$ in Eq. (2.18), we find that the constant C_1 is a function of the initial conditions and the spring parameters, that is,

$$C_1 = -\frac{k}{2m}(x(0) - x_0)^2 - \frac{1}{2}(\dot{x}(0))^2.$$

Solving for $\dot{x}(t) = \frac{dx}{dt}$ in Eq. (2.18), we obtain

$$\frac{dx}{dt} = \sqrt{-\frac{k}{m}(x - x_0)^2 - 2C_1}. \tag{2.19}$$

By separating the variables x and t in Eq. (2.19) and integrating, we obtain

$$\int \frac{dx}{\sqrt{-\frac{k}{m}(x - x_0)^2 - 2C_1}} = \int dt = t + C_2.$$

In principle, we can use an integral table (or trigonometric substitution) to integrate the left-hand side to obtain an implicit equation for the position x. Fortunately, there is an easier approach.

First, make the change of coordinates $z = x - x_0$ and let $\omega_0 = \sqrt{\frac{k}{m}}$. Note that $\dot{z} = \dot{x}$ and $\ddot{z} = \ddot{x}$, since x_0 is constant. Under this change of coordinates, the equation of motion Eq. (2.17) becomes

$$\boxed{\ddot{z} = -\omega_0^2 z.} \tag{2.20}$$

You may recognize Eq. (2.20) as the equation for *simple harmonic motion* with *natural frequency* ω_0. To solve Eq. (2.20), we assume a solution of the form

$$z(t) = A\cos(\omega_0 t) + B\sin(\omega_0 t), \tag{2.21}$$

so that

$$\dot{z}(t) = -A\omega_0 \sin(\omega_0 t) + B\omega_0 \cos(\omega_0 t)$$

and

$$\ddot{z}(t) = -A\omega_0^2 \cos(\omega_0 t) - B\omega_0^2 \sin(\omega_0 t) = -\omega_0^2 z(t).$$

Thus our assumed solution is indeed correct.

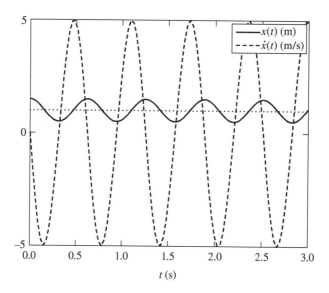

Figure 2.8 Position and velocity responses of a mass-spring system with natural frequency $\omega_0 = 10$ rad/s and rest length $x_0 = 1$ m. The initial conditions are $x(0) = 1.5$ m and $\dot{x}(0) = 0$.

We can solve for the value of the constants A and B by setting $t = 0$ in the expressions for $z(t)$ and $\dot{z}(t)$ to find that $A = z(0)$ and $B = \frac{\dot{z}(0)}{\omega_0}$. The solution in the original coordinate x becomes

$$x(t) = x_0 + (x(0) - x_0) \cos(\omega_0 t) + \frac{\dot{x}(0)}{\omega_0} \sin(\omega_0 t) \tag{2.22}$$

$$\dot{x}(t) = -(x(0) - x_0)\omega_0 \sin(\omega_0 t) + \dot{x}(0) \cos(\omega_0 t). \tag{2.23}$$

The solution is plotted in Figure 2.8.

This result is extremely important. The solution has the form of a sinusoidal oscillation about the unstretched length x_0 with amplitude and phase given by the initial conditions. Eq. (2.20) for simple harmonic motion occurs frequently and in a variety of contexts other than linear springs. Often we can reduce the equations of motion of much more complicated problems to Eq. (2.20), in which case we immediately know the solution has the form of Eqs. (2.22) and (2.23).

We conclude with the common situation of a simple harmonic oscillator with a constant driving force:

$$\ddot{x} + \omega_0^2 x = \frac{F_0}{m}.$$

This has the same form as Eq. (2.17) if we identify the unstretched length with the constant driving force:

$$x_0 = \frac{F_0}{k}.$$

Thus a simple harmonic oscillator with a constant driving force has the same solution as shown in Figure 2.8, a sinusoidal oscillation about a mean offset given by F_0/k.

Tutorial 2.4 Damped Simple Harmonic Motion

This tutorial combines Tutorials 2.2 and 2.3. That is, we find the equation of motion (and solution) for a mass connected to both a linear spring and a viscous damper. The total force on the mass is thus

$$f_x = -k(x - x_0) - b\dot{x}.$$

The resulting equation of motion from Newton's second law is

$$\ddot{x} = -\frac{k}{m}(x - x_0) - \frac{b}{m}\dot{x}.$$

This problem is most easily solved using the second approach from the previous tutorial: guessing a solution. First we make the change of variables $z = x - x_0$ and the substitution $\omega_0 = \sqrt{\frac{k}{m}}$. We also make another common substitution by replacing b with $2m\zeta\omega_0$. The dimensionless parameter ζ is called the *damping coefficient*. These substitutions yield the equation of motion

$$\ddot{z} + 2\zeta\omega_0\dot{z} + \omega_0^2 z = 0. \tag{2.24}$$

Our approach to solving this system is to use two facts from Appendix C: (1) if we have two solutions to a linear ordinary differential equation (ODE) then the sum is also a solution, and (2) the solution of any linear ODE is an exponential. (The sine and cosine solutions from the previous example can be written as a sum of complex exponentials.) Since Eq. (2.24) is a second-order ODE, its solution must have the form

$$z(t) = Az_1(t) + Bz_2(t),$$

where A and B are two arbitrary constants that are functions of the initial conditions. Since Eq. (2.24) is linear, $z_1(t)$ must have the form

$$z_1(t) = e^{\lambda t},$$

with

$$\dot{z}_1(t) = \lambda e^{\lambda t}$$

and

$$\ddot{z}_1(t) = \lambda^2 e^{\lambda t}.$$

The variable λ is a function—as yet undetermined—of the physical parameters of the problem. Substituting $Az_1(t)$ into Eq. (2.24) yields

$$Ae^{\lambda t}\left(\lambda^2 + 2\zeta\omega_0\lambda + \omega_0^2\right) = 0. \tag{2.25}$$

Properly choosing λ to satisfy Eq. (2.25) will validate our guess. Then we find the constants A and B in terms of the initial conditions by setting $t = 0$.

The only way for Eq. (2.25) to be true for all A and all time is if the quantity in parentheses is zero:

$$\lambda^2 + 2\zeta\omega_0\lambda + \omega_0^2 = 0.$$

The solution to this quadratic equation is easily found from the quadratic formula. We find two possible solutions:

$$\lambda_{1,2} = -\zeta\omega_0 \pm \omega_0\sqrt{\zeta^2 - 1}.$$

We have thus determined λ for both of the guessed solutions, $z_1(t)$ and $z_2(t)$. For most systems of interest, $\zeta < 1$. Thus it is convenient and common practice to write the two values of λ as complex numbers:

$$\lambda_{1,2} = -\zeta\omega_0 \pm i\omega_0\sqrt{1 - \zeta^2}.$$

The general solution becomes

$$z(t) = Ae^{\left(-\zeta\omega_0 + i\omega_0\sqrt{1-\zeta^2}\right)t} + Be^{\left(-\zeta\omega_0 - i\omega_0\sqrt{1-\zeta^2}\right)t}$$

$$= e^{-\zeta\omega_0 t}\left(Ae^{i\omega_0 t\sqrt{1-\zeta^2}} + Be^{-i\omega_0 t\sqrt{1-\zeta^2}}\right).$$

There are three types of damped oscillators of interest, depending on the value of ζ (or equivalently, b). The most common case of interest is $\zeta < 1$. In this case, we have an *underdamped* oscillation of the mass about the rest position. Since the physical motion of the spring must, of course, be real, the constants A and B are complex conjugates. Using Euler's equation on the complex exponentials, we can rewrite the above general solution as a real equation to make it easier to work with:

$$z(t) = e^{-\zeta\omega_0 t}\left((A + B)\cos\omega_d t + i(A - B)\sin\omega_d t\right),$$

where $A + B$ is purely real, $A - B$ is purely imaginary (implying that $i(A - B)$ is real), and we made the substitution

$$\omega_d \triangleq \omega_0\sqrt{1 - \zeta^2}.$$

The quantity ω_d is called the *damped frequency of oscillation*. We can now replace the complex constants A and B with the constants $c_1 = A + B$ and $c_2 = i(A - B)$ to find the underdamped general solution:

$$z(t) = e^{-\zeta\omega_0 t}\left(c_1\cos(\omega_d t) + c_2\sin(\omega_d t)\right).$$

The constants c_1 and c_2 can be found in terms of the initial conditions. First set $t = 0$ to obtain

$$c_1 = z(0).$$

Differentiating the general solution and setting $t = 0$ yields

$$c_2 = \frac{\dot{z}(0) + z(0)\zeta\omega_0}{\omega_0\sqrt{1 - \zeta^2}}.$$

This solution is plotted as a solid line in Figure 2.9.

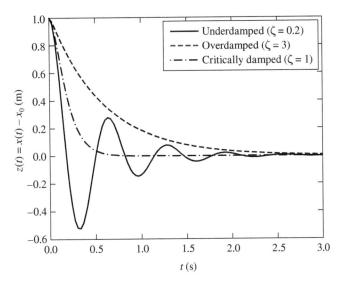

Figure 2.9 Position response of a damped mass-spring system with natural frequency $\omega_0 = 10$ rad/s. The solutions for three different values of the damping coefficient are shown: $\zeta = 0.2$ (underdamped), $\zeta = 1$ (critically damped), and $\zeta = 3$ (overdamped). The initial conditions are $z(0) = 1$ and $\dot{z}(0) = 0$.

The second case of interest is $\zeta > 1$. The solution to this case is called *overdamped*. In the overdamped situation, there are no oscillations, and the mass position gradually decays to the rest length of the spring. Mathematically, this occurs because λ is real. The general solution for an overdamped system is

$$z(t) = Ae^{-\omega_0\left(\zeta + \sqrt{\zeta^2 - 1}\right)t} + Be^{-\omega_0\left(\zeta - \sqrt{\zeta^2 - 1}\right)t}.$$

To find the constants A and B, set $t = 0$ to obtain

$$A + B = z(0).$$

Differentiating the general solution and setting $t = 0$ in the result yields

$$B - A = \frac{\dot{z}(0) + z(0)\omega_0\zeta}{\omega_0\sqrt{\zeta^2 - 1}}.$$

After a bit of algebra, we find

$$A = -\frac{\dot{z}(0) + z(0)\omega_0\left(\zeta - \sqrt{\zeta^2 - 1}\right)}{2\omega_0\sqrt{\zeta^2 - 1}}$$

$$B = \frac{\dot{z}(0) + z(0)\omega_0\left(\zeta + \sqrt{\zeta^2 - 1}\right)}{2\omega_0\sqrt{\zeta^2 - 1}}.$$

The overdamped solution is plotted as a dashed line in Figure 2.9.

For $\zeta = 1$ the system is called *critically damped*. In the critically damped case, $\lambda_1 = \lambda_2 = -\omega_0$. Like the overdamped response, the critically damped response has

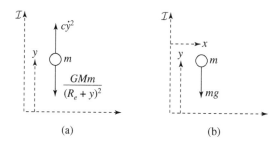

(a)   (b)

Figure 2.10 Free-body diagrams for (a) particle falling under gravity and (b) the tossed point mass.

no oscillation; it decays directly to the rest length. In fact, critical damping is the boundary between the damped oscillatory response and the overdamped response. We leave it as a problem to show that the general solution to the critically damped system is

$$z(t) = z(0)e^{-\omega_0 t} + (\dot{z}(0) + z(0)\omega_0)te^{-\omega_0 t}. \tag{2.26}$$

The critically damped solution is plotted as a dash-dot line in Figure 2.9.

Tutorial 2.5 Particle Falling under Gravity with Air Resistance

This tutorial treats a third type of friction, which in this case is proportional to speed squared, and introduces a more exact expression for gravity. In addition to formulating his laws of motion, Newton also postulated a universal law of gravity. In this law, the force of gravity between two masses is an inverse square law: the magnitude of the force is inversely proportional to the square of the distance between the two masses. The direction of the force is along the line connecting the two masses and acts to pull them together. Let y denote the height of a particle of mass m, as shown in Figure 2.10a; $\dot{y}$ denotes speed, which is negative because the particle is falling.

Mass m is falling toward a larger mass M. Assume that the second (point) mass is the earth, so $M = 5.9742 \times 10^{24}$ kg. The distance between m and M is $r = R_e + y$, where $R_e = 6,378,100$ m is the earth's radius. In SI units the gravitational proportionality constant is $G = 6.673 \times 10^{-11}$ m^{3}kg^{-1}s^{-2}. The total force on mass m is the sum of the force of gravity (which pulls the particle downward) and the force due to air resistance. We assume that the force of air resistance acts in the opposite direction to the particle velocity and is proportional to speed squared. The force acting on mass m is

$$f_y = -\frac{GMm}{(R_e + y)^2} + c\dot{y}^2,$$

where c is the *coefficient of air resistance* with units kg/m.[14] Using Newton's second law, we obtain the equation of motion:

---

[14] $c = (1/2)\rho C_D A$, where C_D is the dimensionless drag coefficient, A is the area of the "point" mass in contact with the air flow, and ρ is the air density.

$$\ddot{y} = -\frac{GM}{(R_e + y)^2} + \frac{c}{m}\dot{y}^2. \tag{2.27}$$

Eq. (2.27) looks a bit tricky to integrate by hand, so we proceed by integrating numerically using the MATLAB function ODE45. (See Appendix C for an introduction to numerical integration using MATLAB.) We start by putting the equations of motion in first-order form using the change of coordinates

$$Z = \begin{bmatrix} z_1 \\ z_2 \end{bmatrix} = \begin{bmatrix} y \\ \dot{y} \end{bmatrix},$$

which implies

$$\dot{z}_1 = z_2$$
$$\dot{z}_2 = -\frac{GM}{(R_e + z_1)^2} + \frac{c}{m}z_2^2. \tag{2.28}$$

In MATLAB, we create the function file falling_ode.m, which contains the first-order equations of motion from Eq. (2.28):

```
function zdot = falling_ode(t,z,G,M,m,Re,c)
zdot(1,1) = z(2);
zdot(2,1) = -G*M/(Re+z(1))^2+c/m*z(2)^2;
```

To integrate these equations, we use the MATLAB Command Window to define all relevant input variables and then call ODE45:

```
>> [t,z] = ode45('falling_ode',tspan,z0,[],G,M,m,Re,c);
```

Figure 2.11 shows a plot of $z_1(t) = y(t)$ and $z_2(t) = \dot{y}(t)$ for $c = 0.05$ kg/m, $y(0) = 1,000$ m, $\dot{y}(0) = 0$ m/s, $m = 10$ kg, and $t = [0, 20]$ s. Note that the velocity $\dot{y}(t)$ asymptotically approaches a lower limit, known as the *terminal velocity*.

We can obtain an analytical estimate of the terminal velocity by rewriting the equation of motion Eq. (2.27) using the simplifying approximation $R_e \gg y$, that is, by assuming that the earth's radius is much larger than the height of the mass. In this case, the inverse square gravity force simplifies to a constant gravitational acceleration, $g = \frac{GM}{R_e^2} = 9.81$ m/s^2. Under this assumption, the approximate equation of motion is

$$\ddot{y} \approx -g + \frac{c}{m}\dot{y}^2. \tag{2.29}$$

Note that the approximate equation of motion can also be obtained by assuming that the force of gravity is equal to the sea-level weight of the mass. Solving for the (constant) terminal velocity $\dot{y}^*$ using Eq. (2.29) by setting $\ddot{y}^* = 0$ gives

$$\dot{y}^* = -\sqrt{\frac{gm}{c}}.$$

(Note we take the negative root, since $\dot{y} < 0$.) This estimate is plotted as a dashed line in the lower panel of Figure 2.11 and shows good agreement with the terminal velocity obtained numerically.

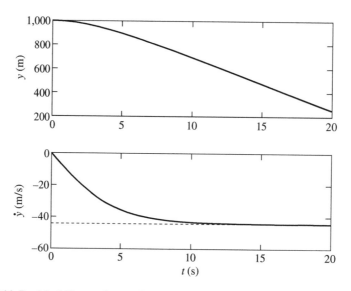

Figure 2.11 Particle falling under gravity with air resistance. The solid curves are the result of numerically integrating the equation of motion (Eq. (2.27)). The dashed line in the lower panel is the analytical estimate of the terminal velocity.

Tutorial 2.6 The Tossed Point Mass

This tutorial introduces two-dimensional motion. In Cartesian coordinates this simply means finding equations of motion for the x and y directions separately. Let m represent the mass of a tossed particle with position $\mathbf{r}_{P/O} = (x, y)_{\mathcal{I}}$. Rather than modeling the force that launches the mass, we solve for the in-flight trajectory of the mass as a function of its initial conditions immediately after the toss. The only force acting on the mass in mid-air (ignoring drag) is gravity, as shown in Figure 2.10b. Using Newton's second law, we obtain two equations of motion (one for each degree of freedom):

$$\ddot{x} = 0 \tag{2.30}$$

$$\ddot{y} = -g. \tag{2.31}$$

Integrating the equations of motion (assuming $x(0) = y(0) = 0$) gives the following trajectory for the position of the mass:

$$x(t) = \dot{x}(0)t$$

$$y(t) = \dot{y}(0)t - \frac{1}{2}gt^2.$$

We can solve for the trajectory $y(x)$ by eliminating time t to obtain

$$y(x) = \dot{y}(0)\frac{x}{\dot{x}(0)} - \frac{1}{2}g\left(\frac{x}{\dot{x}(0)}\right)^2. \tag{2.32}$$

Eq. (2.32) shows that the tossed mass takes a parabolic trajectory whose shape is determined by the initial velocities $\dot{x}(0)$ and $\dot{y}(0)$. One such trajectory is illustrated in Figure 2.12 for $\dot{x}(0) = 1$ m/s and $\dot{y}(0) = 10$ m/s.

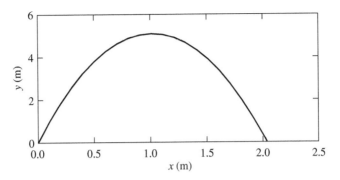

Figure 2.12 The tossed point mass. A particle tossed from the origin (0, 0) in the presence of gravity with initial horizontal speed 1 m/s and initial vertical speed 10 m/s travels in the parabolic trajectory shown.

In this tutorial, we took the approach of ignoring the details of the force that tosses the mass, since it acts for a short time only. Instead this force was treated as "setting" the initial conditions of the equations of motion. This idea is very common—it comes up frequently in this book.

2.7 Key Ideas

- **Newton's second law of motion** in vector form is

$$\mathbf{F}_P = \frac{{}^{\mathcal{I}}d}{dt}\left({}^{\mathcal{I}}\mathbf{p}_{P/O}\right) = m_P\,{}^{\mathcal{I}}\mathbf{a}_{P/O}.$$

Newton's second law describes the inertial acceleration ${}^{\mathcal{I}}\mathbf{a}_{P/O}$ of a point mass P with mass m_P under the influence of the total force $\mathbf{F}_P$. The quantity $\mathbf{F}_P$ is the vector sum of the forces acting on the particle P. When solving a dynamics problem, the forces acting on P should be hand-sketched in a **free-body diagram**.

- A **conserved quantity** in dynamics is a scalar or vector function of a particle's state that remains constant throughout the trajectory. When such a quantity exists, it is described by a **conservation law**.

- The number of **degrees of freedom** of a collection of particles is the total number of independent coordinates needed to describe the position of every particle. Each particle has three degrees of freedom minus the number of constraints. That is, if there are N particles and K constraints, then the total number of degrees of freedom is $M = 3N - K$. The total number of degrees of freedom of a system of particles represents the number of equations of motion needed to describe the motion of the system.

- For every constraint there is a **constraint force**. These forces are called **normal forces** if they are perpendicular to the direction of motion.

- All units in this text are in **SI**; that is, we use the following units and abbreviations: meters (m), seconds (s), and kilograms (kg). The units of force are newtons (N).

- In this chapter we introduced forces used throughout the book—**Coulomb friction**, **viscous friction**, **air drag**, **spring (restoring) forces**, and **gravity**.

- The trajectory resulting from the one-dimensional equation of motion

$$\ddot{z} = -\omega_0^2 z$$

is known as **simple harmonic motion**. This equation arises frequently in dynamics.

2.8 Notes and Further Reading

The historical references in this and later chapters come from a variety of sources. A particularly comprehensive discussion of the history of mechanics can be found in Dugas (1988). For a popular treatment, more information on Galileo and his contribution to mechanics can be found in Dava Sobel's wonderful book *Galileo's Daughter* (Sobel 1999). There have been many editions of Newton's *Principia* over the centuries. We used the 1934 University of California edition (Newton 1934). A particularly good popular book on Newton's science and the history of the *Principia* is Ivar Peterson's *Newton's Clock, Chaos in the Solar System* (Peterson 1993). James Gleick also has an engaging biography of Newton (Gleick 2003). Our discussion of the terms "theory" and "law" comes from a publication on evolution and creationism (National Academy of Sciences 1999).

Many dynamics and physics texts discuss in varying degrees of detail Newton's laws and the important concepts of force, mass, acceleration, and absolute space. For example, introductory descriptions can be found in Meriam and Kraige (2001), Bedford and Fowler (2002), Tenenbaum (2004), and Thornton and Marion (2004). A more detailed and particularly clear discussion is found in Greenwood (1988). An insightful discussion of the concept of force in physics can be found in the 2004 *Physics Today* column by Frank Wilczek (Wilczek 2004).

Most introductory dynamics texts also discuss units and conversions. For our discussion, we referred to Meriam and Kraige (2001), Bedford and Fowler (2002), Hibbeler (2003), Tongue and Sheppard (2005), and Beer et al. (2007). An excellent reference on standards and units is the website http://www.nist.gov of the National Institute of Standards and Technology.

2.9 Problems

2.1   Newton's second law applies to a point mass with the acceleration relative to what type of frame?

2.2   In Newton's second law, what does m_p refer to?

2.3   Consider the straight-line motion of a particle of mass m_p acted on only by air resistance, $F_p = -b\dot{x}^2$. Find analytically an expression for the velocity of the particle as a function of time if it starts with initial velocity v_0 at time t_0.

2.4 In Example 2.4 we showed that the number of degrees of freedom for a car on a planar roller coaster reduces to one. What is the corresponding constraint force?

2.5 Consider the luge shown in Figure 2.13. Treating the rider and sled as a point mass, how many degrees of freedom does it have? What coordinates might you use?

Figure 2.13 Problem 2.5. Image courtesy of Shutterstock.

2.6 Sketch a planar model of a weightlifter's barbell (shown in Figure 2.14) using two point masses and a rigid massless rod. How many degrees of freedom are there in your model? (HINT: Find the constraint equation(s).)

Figure 2.14 Problem 2.6. Image courtesy of Shutterstock.

2.7 In Example 2.5 we found that the three-link robot arm in Figure 2.6a has three degrees of freedom. We described them by three angles in Figure 2.6b. Suppose, instead, you desired to describe the system by the six Cartesian coordinates for the end of each link, $(x_1, y_1)_\mathcal{I}$, $(x_2, y_2)_\mathcal{I}$, and $(x_3, y_3)_\mathcal{I}$. How many constraint equations would be necessary and what are they?

2.8 Draw the free-body diagram for the barbell model developed in Problem 2.6.

2.9 Draw the free-body diagram for the luge model in Problem 2.5.

2.10 A rocket lifting off from the ground is acted on by a variety of forces, including thrust, gravity, drag, and wind. Treating the rocket as a point mass, draw the free-body diagram of the rocket during ascent.

2.11 Despite Galileo's claim, large, heavy skiers have an unfair advantage: they get down a mountain faster than lighter skiers. In fact, in the four-person bobsled, for instance, ballast weights are added to the sled to ensure that every sled, including the drivers, has the same weight (again giving an advantage to larger racers, as the lighter competitors have to push a heavier sled at the start).

   Where did Galileo go wrong? That is, using Newton's second law and your knowledge of the forces at work, show why it is that heavier skiers get down the mountain faster.

   (You may assume that the skier is acted on by gravity; friction, $F = \mu N$; and aerodynamic drag, $F = \frac{1}{2} C_D A \rho v^2$.)

2.12 Find the equations of motion for a mass m suspended vertically from a spring as shown in Figure 2.15, assuming that the mass is constrained to move only vertically and that it is subject to the force of gravity. Draw a free-body diagram, choose a coordinate system, and use Newton's second law to find the equation of motion.

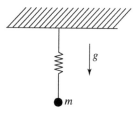

Figure 2.15 Problem 2.12.

2.13 Many springs are what are called *nonlinear springs*. That is, the force depends nonlinearly on position. A common nonlinear-spring force is $F = -k(x - x_0) - c(x - x_0)^3$, where x_0 is the unstretched length of the spring, and k and c are constants. One way to think about F is that the spring gets stiffer as it is stretched.

   a. Find the equation of motion for a mass on a frictionless surface attached to a linear spring ($c = 0$) and then for a nonlinear spring, and put each equation in first-order form.

   b. Numerically integrate the equations of motion over the time interval $[0, 20]$ s. Assume the unstretched length of the spring x_0 is 0.25 m.

Assume an initial stretched position of 0.4 m. Use $k = 1$ N/m, $c = 5$ N/m^3, and $m = 1$ kg. Plot the resulting trajectories on the same axes and compare the responses.

2.14 Show that the general solution for a critically damped harmonic oscillator is given by Eq. (2.26). Find expressions for the constants A and B in terms of the initial conditions $z(0)$ and $\dot{z}(0)$.

2.15 Find expressions for the position $x(t)$ as a function of time for a critically damped spring and an overdamped one in terms of the initial conditions $x(0)$ and $\dot{x}(0)$.

2.16 A rock dropped from rest from the top of a cliff will experience an acceleration due to gravity along with a deceleration due to drag. The downward acceleration is $g - \frac{c}{m}\dot{y}^2$, and the downward speed is $\dot{y}$. If the rock is dropped from a height of y m, calculate the rock's impact velocity. How high will the cliff have to be for the rock to hit the ground at 99% of its terminal velocity? (This problem should be solved analytically, that is, without the use of numerical integration.)

2.17 Consider the equation of motion of a driven simple harmonic oscillator

$$\ddot{x} = a - \frac{k}{m}x,$$

where a is a constant.

    a. Integrate the equation to find the solution analytically.
    b. Solve the equation of motion using MATLAB ODE45, and plot $x(t)$ and $\dot{x}(t)$. Let $a = 1$, $m = 0.5$, and $k = 3$. Be sure to label your axes (with units!) and add a legend to the plot.

2.18 In Tutorial 2.5 we solved for the equation of motion describing a particle falling under Newton's law of gravity and acted on by air resistance. We found that the particle reached terminal velocity. One approximation made, however, was that the density of the earth's atmosphere is constant over the trajectory. A better model of atmospheric density includes exponential decay,

$$\rho = \rho_0 e^{-y/h},$$

where ρ_0 is the atmospheric density at sea level (roughly 1.2 kg/m^3), and h is called the *scale height* of the atmosphere. Near the surface, h is approximately 7,000 m (the actual value can vary with temperature and season). Using the same parameter values as in Tutorial 2.5 and assuming the given value of c at sea level, find the new equation of motion incorporating the exponential atmosphere model and simulate the resulting height and speed trajectories. Compare the terminal velocity you find to that in Figure 2.11.

2.19 Two masses are released from a height h above the ground as shown in Figure 2.16. Mass P is released from rest and mass Q is launched horizontally with speed v_0. Which mass lands first? Draw a free-body diagram for each

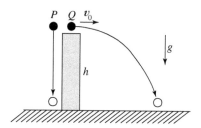

Figure 2.16 Problem 2.19.

mass, choose a coordinate system, use Newton's second law to find the equations of motion for each mass, and solve the equations of motion for the time until impact with the ground.

PART ONE

Particle Dynamics in the Plane

CHAPTER THREE

-- -- -- -- -- -- -- -- -- -- -- -- -- -- -- -- -- --

Planar Kinematics and Kinetics of a Particle

Chapter 2 introduced the fundamental physical concepts and tools for solving all dynamics problems. In it, we also solved complete and important problems in one dimension. To whet your appetite, toward the end of the chapter we had a first look at generalizing Newton's second law to two dimensions, though using only a scalar notation (as opposed to the vector notation developed here), by solving for the motion of a point mass under gravity. Although it is true that all particle dynamics problems could be solved using only Cartesian coordinates and simple extensions of the scalar treatment used in the last chapter, that would be unnecessarily limiting. This chapter introduces the vector formalism used for modern treatments of dynamics and lays out the basic mathematical foundation we use throughout the book. By the end you will have almost all the tools you need to solve any dynamics problem. We limit the discussion to only two dimensions, however, to focus on the important concepts without overly complicating the mathematics. Part Four shows that only a few new ideas are necessary to extend this analysis to three dimensions.

3.1 The Simple Pendulum

The limitations of what we have learned up to now are easily demonstrated by an only slightly more complicated example—the *simple pendulum,* illustrated in Figure 3.1.[1] In this example, we provide a first look at the problem-solving notation and techniques that are used throughout the rest of the book. Don't worry if these tools seem a bit mysterious at first; things should clarify by the end of the chapter.

---

[1] The pendulum is simply a swinging mass in a uniform gravity field. Its periodic motion was first discovered by Galileo Galilei in 1602. Pendulums have been incorporated into many devices since, most notably clocks and accelerometers. The *simple pendulum* refers to a pendulum consisting of a point mass on a massless rod or string, whereas the *compound pendulum* refers to a pendulum consisting of an extended rigid body. Figure 3.1 is a simple pendulum in a configuration that is sometimes referred to as an *inverted pendulum* because it is directed up rather than down.

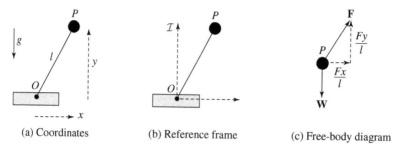

(a) Coordinates   (b) Reference frame   (c) Free-body diagram

Figure 3.1 Solving the pendulum using Cartesian coordinates in an inertial frame.

Example 3.1 Solving the Pendulum Using Cartesian Coordinates in the Inertial Frame

Consider a mass m_P fixed to a stiff (massless) rod of length l that pivots freely (in the plane of the page) about the point O, as shown in Figure 3.1. We approach this problem in the same manner as the tossed point mass in Tutorial 2.6. The Cartesian coordinates of the pendulum mass m_P with respect to the inertial frame $\mathcal{I}$ are $(x, y)_{\mathcal{I}}$. Just as in Tutorial 2.6, we write two equations of motion, one in the x-direction and one in the y-direction, in terms of the forces acting in those directions and the speeds and accelerations, $(\dot{x}, \dot{y})_{\mathcal{I}}$ and $(\ddot{x}, \ddot{y})_{\mathcal{I}}$, respectively. Using the free-body diagram in Figure 3.1c, Newton's second law applied to each of the two planar directions is then

$$\frac{Fx}{l} = m_P \ddot{x} \tag{3.1}$$

$$\frac{Fy}{l} - m_P g = m_P \ddot{y}. \tag{3.2}$$

The force **W** in Figure 3.1c is the weight of the mass, which acts downward with magnitude $m_P g$. The force **F** is the internal tension or compression in the rod connecting the mass to the pivot. (Recall that by Newton's third law, the rod experiences an equal and opposite force.) Because the rod is massless and can pivot freely about point O, there is no rod-mass interaction force that is perpendicular to the rod—the only force from the rod on the mass is along the line between point O and the mass.[2] (In general, F can be positive or negative.) To write Eqs. (3.1) and (3.2), we have used our knowledge of vector sums and free-body diagrams to write the tension as the sum of two other forces of magnitudes Fx/l and Fy/l, acting in the x- and y-directions, respectively. Nevertheless, there is only one unknown magnitude, F.

To derive the equations of motion, we eliminate F from Eqs. (3.1) and (3.2) to obtain the single equation

$$\ddot{x}y - \ddot{y}x = gx. \tag{3.3}$$

---

[2] This observation does not hold if we were to model the pendulum rod as having finite mass (i.e., as a rigid body).

The pendulum is an example of a single-degree-of-freedom system (see Section 2.4) and thus has a constraint on the position,

$$x^2 + y^2 = l^2. \tag{3.4}$$

Differentiating the constraint in Eq. (3.4) with respect to time and using the fact that l is constant yields

$$x\dot{x} + y\dot{y} = 0. \tag{3.5}$$

Differentiating Eq. (3.5) a second time yields

$$\dot{x}^2 + \dot{y}^2 + x\ddot{x} + y\ddot{y} = 0. \tag{3.6}$$

Solving Eqs. (3.3) and (3.6) for the two unknowns $\ddot{x}$ and $\ddot{y}$, we obtain

$$\ddot{x} = (-x\dot{x}^2 - x\dot{y}^2 + xyg)/l^2$$
$$\ddot{y} = (-y\dot{x}^2 - y\dot{y}^2 - x^2 g)/l^2. \tag{3.7}$$

Note that we explicitly used the constraint Eq. (3.6) to arrive at Eq. (3.7), so the solution automatically satisfies Eq. (3.4).

Solving Eq. (3.7) will certainly give the correct motion for the pendulum in Cartesian coordinates, but not only are these equations complicated (there is no exact solution), they may also be very poorly behaved numerically. You have a chance to explore their numerical behavior in Problem 3.22 at the end of the chapter. The equations of motion also provide no obvious qualitative insight into the behavior of the pendulum. For instance, your experience should tell you that the way we drew the pendulum in Figure 3.1 results in *unstable* motion; that is, the pendulum will fall until hanging downward. We call this configuration an *inverted pendulum.*

All is not lost, however. With a few new concepts and some careful notation, we can dramatically simplify the approach to solving this problem and many others. These new concepts and notation are the subject of the remainder of this chapter.

We start by introducing some important concepts from vector algebra and vector calculus. Pay particular attention to the notation—it will be used throughout the book, and being comfortable with it is essential. Although some of these concepts may be familiar to you, we still carefully define each one to develop a consistent approach.

3.2 More on Vectors and Reference Frames

3.2.1 The Vector

In Chapter 1 we gave a qualitative definition of the term "vector." Understanding this concept and how we use it in mechanics is essential. The vector is the fundamental geometric quantity used to tie the physical world to the mathematics of dynamics. The vector is the foundation for all that we do in solving mechanics problems—it thus helps to be familiar with the algebraic properties of vectors. For example, the concepts of total force and the free-body diagram discussed in Chapter 2 use the additive properties of vectors. In case you are rusty on the properties and rules for vectors, Appendix B summarizes some useful identities.

We make frequent use in the book of the *unit* vector. The unit vector is a vector that is constrained to have unit length. That is, a unit vector has length equal to one. Introducing the unit vector lets us write an arbitrary vector as a scalar magnitude times a unit vector pointing in the same direction, which can be convenient for many problems. We label all unit vectors with lowercase, boldface, roman type, and we often use the letter **e**. Unit vectors often have a subscript arabic numeral or letter (e.g., $\mathbf{e}_1$ or $\mathbf{e}_x$). A carat above a vector identifies a unit vector pointing in the same direction. For example, given a vector **r**, the unit vector in the same direction is

$$\hat{\mathbf{r}} \triangleq \frac{\mathbf{r}}{\|\mathbf{r}\|},$$

where the notation $\|\mathbf{r}\|$ denotes the *magnitude* of the vector **r**.

We also often label unit vectors by the lowercase letter corresponding to the uppercase label of the reference frame in which they are fixed, such as the unit vectors $\mathbf{b}_1$, $\mathbf{b}_2$, $\mathbf{b}_3$, which are fixed in frame $\mathcal{B}$. In fact, the most important use of a unit vector is to define the directions in a reference frame, which we discuss next.

3.2.2 The Reference Frame

Chapter 1 presented a qualitative definition of the reference frame. That qualitative definition allowed us to introduce what we mean by dynamics and to discuss the important concept of absolute space in the context of Newton's laws. The qualitative definition also made it possible to give abstract meaning to the rate of change of a vector. Here we present two equivalent definitions of a reference frame—a *physical* definition and an *operational* one. These new definitions of a reference frame permit solving a broader class of problems by tying physical concepts to the mathematical tools used.

Physical Definition 3.1 A **reference frame** is a collection of at least three non-collinear points that are rigidly connected or rigidly attached; that is, the distance between any two points in the collection does not change with time.

The physical definition of a reference frame points out that frames are attached to entities and can thus move and rotate with respect to one another. For example, an inertial frame is fixed with respect to absolute space. A *rigid body*, which is a collection of particles with fixed relative spacing from one another, can be used to define a reference frame. When we study rigid bodies in Parts Three and Four, the interchangeability of a rigid body and the reference frame associated with it will be indispensable. Like rigid bodies, reference frames cannot change shape. They don't distort, expand, or contract. The orthogonal unit vectors defining the frame stay fixed with respect to one another, which leads us to the third equivalent definition of a reference frame.

Operational Definition 3.2 A **reference frame** is defined by three orthogonal unit vectors and one point (the origin).

In this text, a reference frame is labeled by a calligraphic capital letter and defined according to Definition 3.2. For example, we write $\mathcal{I} = (O, \mathbf{e}_1, \mathbf{e}_2, \mathbf{e}_3)$ to define

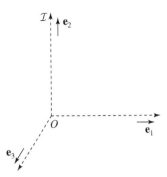

Figure 3.2 Reference frame $\mathcal{I} = (O, \mathbf{e}_1, \mathbf{e}_2, \mathbf{e}_3)$ is defined by its origin O and three orthogonal unit vectors $\mathbf{e}_1$, $\mathbf{e}_2$, and $\mathbf{e}_3$.

reference frame $\mathcal{I}$ using the origin O and the unit vectors $\mathbf{e}_1$, $\mathbf{e}_2$, and $\mathbf{e}_3$. Frame $\mathcal{I}$ is illustrated in Figure 3.2, where $\mathbf{e}_3 = \mathbf{e}_1 \times \mathbf{e}_2$ (see Appendix B).

The operational definition of a reference frame provides the elements needed to perform operations on vectors. As discussed in Chapter 1, the concept of a reference frame is essential to dynamics. How we describe a vector—and how its motion is defined—differs depending on the frame in which we are interested. Most importantly, Newton's laws of dynamics imply a specific reference frame. Recall from Chapter 2 that Newtonian dynamics applies only when the acceleration is referred to a special reference frame called the *inertial frame*. When using Newton's second law to solve for the motion of mass P, we use the acceleration $^{\mathcal{I}}\mathbf{a}_{P/O}$—the second time derivative of the position $\mathbf{r}_{P/O}$ with respect to the inertial frame. The vector quantity $^{\mathcal{I}}\mathbf{a}_{P/O}$ indicates how the velocity $^{\mathcal{I}}\mathbf{v}_{P/O}$ evolves relative to the inertial frame. *Newton's second law can be written and solved in coordinates defined in any frame, but the vector acceleration must refer to the time rate of change with respect to an inertial frame.* We return to this important point later.

Parts One through Three consider dynamics problems in which the motion is confined to a plane spanned by two of the three unit vectors. Nonetheless, it is important to keep track of the third unit vector, as it plays a key role in solving problems using Newton's method.[3] We treat three-dimensional motion in Part Four.

Once you deeply understand the idea of the reference frame, you can move on to the concept of *vector components*.

3.2.3 Components of a Vector

We often write a vector as the weighted sum of the unit vectors of a reference frame. For instance, if reference frame $\mathcal{A}$ is described by the three unit vectors $\mathbf{a}_1$, $\mathbf{a}_2$, and $\mathbf{a}_3$, then the position of point P relative to O can be written

$$\mathbf{r}_{P/O} = a_1\mathbf{a}_1 + a_2\mathbf{a}_2 + a_3\mathbf{a}_3.$$

---

[3] If the third unit vector is not provided in the frame definition, we assume that it equals the cross product of the first unit vector with the second unit vector.

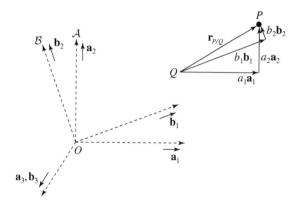

Figure 3.3 Vector components of $\mathbf{r}_{P/Q}$ in reference frames $\mathcal{A} = (O, \mathbf{a}_1, \mathbf{a}_2, \mathbf{a}_3)$ and $\mathcal{B} = (O, \mathbf{b}_1, \mathbf{b}_2, \mathbf{b}_3)$.

These weighted unit vectors—$a_1\mathbf{a}_1$, $a_2\mathbf{a}_2$, and $a_3\mathbf{a}_3$—are the *components* of the vector $\mathbf{r}_{P/O}$ in frame $\mathcal{A}$. The scalar a_i, $i = 1, 2, 3$, is the *magnitude* of the component of $\mathbf{r}_{P/O}$ in the direction $\mathbf{a}_i$. We can also write $\mathbf{r}_{P/O}$ in terms of components in a second reference frame $\mathcal{B}$:

$$\mathbf{r}_{P/O} = b_1\mathbf{b}_1 + b_2\mathbf{b}_2 + b_3\mathbf{b}_3.$$

Figure 3.3 shows this idea graphically for the case $\mathbf{a}_3 = \mathbf{b}_3$.

We can also write the magnitudes of each component of a vector in *matrix notation*. For example, we can describe the vector $\mathbf{r}_{P/O}$ in frame $\mathcal{A}$ as

$$[\mathbf{r}_{P/O}]_\mathcal{A} = \begin{bmatrix} a_1 \\ a_2 \\ a_3 \end{bmatrix}_\mathcal{A}.$$

Note that we indicate the reference frame to which the scalars refer. Matrix notation eliminates the need to explicitly write the unit vectors—these vectors are implied by the frame definition. Matrix notation also allows replacement of some vector operations with matrix operations. Matrix operations become particularly useful when discussing three-dimensional motion of rigid bodies in Part Four.

The ability to write a vector as a weighted sum of the unit vectors of a specific reference frame is extremely useful. You should become very comfortable with this notation.

Example 3.2 The Pendulum Position Expressed in Vector Components

We are now in a position to treat the pendulum a bit more carefully using vectors. In this example the position of the pendulum is represented as a vector, rather than just using the scalar Cartesian coordinates as in Example 3.1. Figure 3.4 shows the simple pendulum and two frames of reference with origin O: the inertial frame $\mathcal{I} = (O, \mathbf{e}_x, \mathbf{e}_y, \mathbf{e}_z)$ and a second frame $\mathcal{B} = (O, \mathbf{b}_1, \mathbf{b}_2, \mathbf{b}_3)$, oriented so the $\mathbf{b}_1$ axis

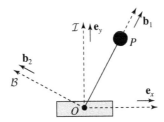

Figure 3.4 The pendulum represented by components in two different frames.

is always pointed at the pendulum bob. The position of the pendulum bob P can be written as components in the inertial frame $\mathcal{I}$:

$$\mathbf{r}_{P/O} = x\mathbf{e}_x + y\mathbf{e}_y,$$

where as before, $(x, y)_\mathcal{I}$ must satisfy the constraint that $x^2 + y^2 = l^2$.

However, when written as components in frame $\mathcal{B}$, the position is

$$\mathbf{r}_{P/O} = l\mathbf{b}_1.$$

As shown later, writing the pendulum as components in this frame can significantly simplify the problem of finding the equations of motion.

Example 3.3 Satellite Tracking

Figure 3.5 shows an inertial frame $\mathcal{I}$ located at the center of the earth. A second frame, $\mathcal{B}$, is located at a position on the surface of the earth at 45° latitude, such that $\mathbf{b}_1$ and $\mathbf{e}_1$ are aligned, $\mathbf{b}_3$ is aligned with the vertical at O', and $\mathbf{b}_2$ is along the horizon to complete an orthogonal right-handed set ($\mathbf{b}_2 = \mathbf{b}_3 \times \mathbf{b}_1$). Suppose we are tracking a satellite from a station located at O', and we wish to know the position of the satellite, S, relative to the center of the earth, O, expressed as components in the inertial frame $\mathcal{I}$.

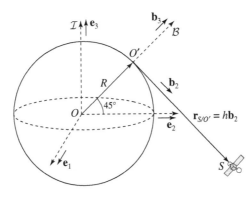

Figure 3.5 Satellite tracking from the earth's surface.

As shown in Figure 3.5, at the time of measurement the satellite is just on the horizon, so its position in terms of components in frame $\mathcal{B}$ is

$$\mathbf{r}_{S/O'} = h\mathbf{b}_2,$$

where h is the range to the satellite.

The position of the satellite relative to O is given by the addition rule for vectors,

$$
\begin{aligned}
\mathbf{r}_{S/O} &= \mathbf{r}_{O'/O} + \mathbf{r}_{S/O'} \\
&= R\mathbf{b}_3 + h\mathbf{b}_2,
\end{aligned}
$$

where R is the radius of the earth.

To find the satellite position entirely as components in frame $\mathcal{I}$, we need to write the vectors $\mathbf{b}_2$ and $\mathbf{b}_3$ as components in $\mathcal{I}$. This is done by inspection:

$$\mathbf{b}_2 = \frac{1}{\sqrt{2}}\mathbf{e}_2 - \frac{1}{\sqrt{2}}\mathbf{e}_3 \quad \text{and} \quad \mathbf{b}_3 = \frac{1}{\sqrt{2}}\mathbf{e}_2 + \frac{1}{\sqrt{2}}\mathbf{e}_3. \tag{3.8}$$

The final position of the satellite as components in $\mathcal{I}$ is thus

$$\mathbf{r}_{S/O} = \frac{1}{\sqrt{2}}(R+h)\,\mathbf{e}_2 + \frac{1}{\sqrt{2}}(R-h)\,\mathbf{e}_3.$$

3.2.4 Coordinate Systems

The next concept of great importance is the *coordinate system*. Recall that we defined a coordinate system in Chapter 1 as a set of scalars that locates the position of a point. We strongly emphasized the distinction between a coordinate system and a reference frame—they are not the same thing! A coordinate system locates a point in a specific reference frame. This section introduces three essential coordinate systems that are used to describe the configuration of points in the plane. We introduce two more coordinate systems when discussing three-dimensional motion in Part Four.

Cartesian Coordinates

The most common coordinate system is the one we have already been using—Cartesian coordinates. The Cartesian coordinates of an arbitrary frame $\mathcal{I}$ are equal to the magnitudes of the components of a vector expressed in frame $\mathcal{I}$. The letters x, y, and z are often used to label Cartesian coordinates. For example, the position of point P relative to O expressed using Cartesian coordinates in reference frame $\mathcal{I} = (O, \mathbf{e}_x, \mathbf{e}_y, \mathbf{e}_z)$ is

$$\mathbf{r}_{P/O} = x\mathbf{e}_x + y\mathbf{e}_y + z\mathbf{e}_z.$$

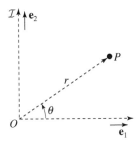

Figure 3.6 Polar coordinates $(r, \theta)_{\mathcal{I}}$ locate the position of point P relative to point O.

In matrix notation, we have

$$[\mathbf{r}_{P/O}]_{\mathcal{I}} = \begin{bmatrix} x \\ y \\ z \end{bmatrix}_{\mathcal{I}}.$$

Cartesian coordinates are just one possible choice of coordinates. For some problems, particularly problems with constraints, another coordinate system may simplify the equations of motion. A judicious choice of coordinate system (and corresponding reference frame) may allow the use of fewer scalars to represent the motion. *Part of the art of solving dynamics problems is selecting the coordinate systems and reference frames that maximize insight and minimize algebra.*

Polar Coordinates

One alternative to the Cartesian coordinate system is the *polar* coordinate system. In polar coordinates a point P is located in a plane by the polar radius r and the polar angle θ, as shown in Figure 3.6. We can still write the vector $\mathbf{r}_{P/O}$ in terms of components in the $\mathbf{e}_1$ and $\mathbf{e}_2$ directions in reference frame $\mathcal{I}$, only we use the scalars associated with polar coordinates:

$$\mathbf{r}_{P/O} = r \cos \theta \mathbf{e}_1 + r \sin \theta \mathbf{e}_2. \qquad (3.9)$$

In matrix notation, we have

$$[\mathbf{r}_{P/O}]_{\mathcal{I}} = \begin{bmatrix} r \cos \theta \\ r \sin \theta \end{bmatrix}_{\mathcal{I}}.$$

Path Coordinates

Another type of coordinate system is the *path* coordinate system. Path coordinates are also called *curvilinear* coordinates or *normal/tangential* coordinates. Unlike the Cartesian and polar coordinate systems, which use two scalars to locate a point P in a plane, the path coordinate system uses only a single scalar s to locate P along a curve. Thus path coordinates are typically used for problems that have only one degree of freedom. Clearly, the value of s is not enough information to completely express the position of P relative to O. Some mathematical description of the path is

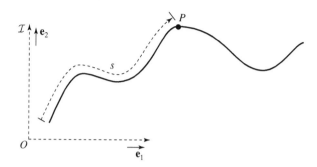

Figure 3.7 An arbitrary path in the fixed reference frame $\mathcal{I}$ and path coordinate s measured along the path.

also necessary (e.g., the oval curve that describes a racetrack). Figure 3.7 sketches a situation in which path coordinates may be the best choice.

Example 3.4 The Simple Pendulum Using Different Coordinates

Example 3.2 described the position of a pendulum bob in Cartesian coordinates in an inertial reference frame $\mathcal{I} = (O, \mathbf{e}_x, \mathbf{e}_y, \mathbf{e}_z)$ (and found the corresponding equations of motion). Here we look again at the pendulum and describe its position using polar and path coordinates.

Figure 3.8 shows a pendulum and an inertial reference frame $\mathcal{I} = (O, \mathbf{e}_1, \mathbf{e}_2, \mathbf{e}_3)$. Note that we have changed the definition of the inertial frame from that in Example 3.2: the directions of the axes are arbitrary as long as the frame remains inertial. (We often move the inertial frame around to find the most convenient coordinate system for a problem.) This configuration is more typical for the simple pendulum in contrast to the inverted pendulum shown in Figure 3.1. The pendulum-mass position in Cartesian coordinates is still given by $(x, y)_{\mathcal{I}}$, where x is the coordinate in the $\mathbf{e}_1$ direction and y is the coordinate in the $\mathbf{e}_2$ direction, and the length constraint is still $x^2 + y^2 = l^2$.

A Cartesian coordinate system is not the most convenient for the simple pendulum, however, because it uses two scalar coordinates for the single-degree-of-freedom system; we should be able to find a single coordinate to describe the pendulum's configuration. Figure 3.8 uses polar coordinates to describe the configuration. With polar coordinates, we can completely describe the position of the pendulum bob by

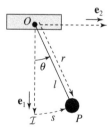

Figure 3.8 The simple pendulum described by polar and path coordinates in reference frame $\mathcal{I}$. The position of pendulum bob P is entirely determined by polar angle θ or path length s.

the polar angle θ. The polar coordinate r is equal to the (fixed) length of the pendulum rod. (We solve the pendulum problem using polar coordinates in Example 3.9.) The polar coordinate θ is related to the Cartesian coordinates by

$$\theta = \arctan\left(\frac{y}{x}\right).$$

We can also describe the position of the pendulum as the distance along the arc of a circle of radius l, as indicated by the single path coordinate s in Figure 3.8. The path length s is related to the polar angle θ by

$$s = l\theta.$$

In each case, we are still describing the position of the pendulum mass in reference frame $\mathcal{I}$. These are three equivalent ways of doing so using the three coordinate systems introduced so far.

Example 3.5 An Inclined Plane

Figure 3.9 shows a box of mass m_P sliding down an inclined plane of height h and base b. An inertial frame $\mathcal{I}$ is aligned with the base and has its origin at the lower left corner. What is the position of the box in the three coordinate systems studied so far?

The box position in Cartesian coordinates in $\mathcal{I}$ is $(x, y)_{\mathcal{I}}$. However, this problem has one degree of freedom, since the box is able to slide only along the incline. This implies the following constraint:

$$\frac{y}{b - x} = \tan 30° = \frac{1}{\sqrt{3}}.$$

We could also specify the position of the box in polar coordinates, also shown in Figure 3.9. The polar coordinates are related to the Cartesian coordinates by

$$r = \sqrt{x^2 + y^2}$$

$$\theta = \arctan\left(\frac{y}{x}\right),$$

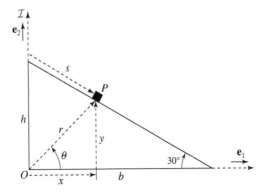

Figure 3.9 Block on an inclined plane in three coordinate systems.

The constraint in polar coordinates becomes

$$r = \frac{b}{\sqrt{3} \sin \theta + \cos \theta}.$$

The single degree of freedom also implies that we can locate the box using only one scalar coordinate. In this case, the most convenient coordinate is a path coordinate s, specifying the distance the block has slid down the incline, as shown in Figure 3.9. The relationship of the Cartesian coordinates to s is

$$x = s \cos 30° = s \frac{\sqrt{3}}{2}$$

$$y = h - s \sin 30° = h - \frac{s}{2},$$

where l is the length of the incline, $l = \sqrt{h^2 + b^2}$. We could likewise find similar relationships between the polar coordinates and the path coordinate (see Problem 3.4).

It is worth noting that the reason the relationships between s and the other coordinates (and the constraints) are complicated is due to our choice of reference frame. We could have instead chosen a frame located at the top of the incline and rotated so that the $\mathbf{e}_1$ direction is along the incline. In that case the path coordinate would be the same as the Cartesian x coordinate and the polar coordinate r. The constraints would be trivial: $y = 0$ for Cartesian coordinates and $\theta = 0$ for polar coordinates. This example highlights the importance of a careful choice of reference frames.

3.3 Velocity and Acceleration in the Inertial Frame

Now that you are comfortable with vectors, reference frames, and coordinate systems, we are ready to discuss the *vector derivative*. The vector derivative is fundamental to the topic of *kinematics,* which describes the motion of particles (and rigid bodies) without reference to the forces acting on them. In this section we carefully define the vector derivative and derive certain properties. Then we use the vector derivative to find the velocity and acceleration of a point in the inertial frame in terms of the three coordinate systems studied so far.

3.3.1 The Vector Derivative

Chapter 1 qualitatively described the vector derivative as the change with respect to time of the magnitude and direction of a vector in a reference frame. Figure 3.10 again illustrates this qualitative description, now using our notations for the unit vectors and origin of a reference frame. Remember that the time derivative of a vector is itself a vector. In the sections that follow, we provide a mathematical definition of the vector derivative, state its properties (we prove them in Section 3.8), and show its equivalence to our geometric ideas. If you are rusty, you can review the definition of the scalar derivative of a function in Appendix A.

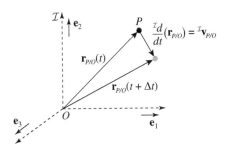

Figure 3.10 Schematic representation of the vector derivative.

Definition 3.3 Let $\mathbf{r}(t) = r_1(t)\mathbf{a}_1 + r_2(t)\mathbf{a}_2 + r_3(t)\mathbf{a}_3$ be a vector function of time expressed as components in reference frame $\mathcal{A} = (O, \mathbf{a}_1, \mathbf{a}_2, \mathbf{a}_3)$. If the scalar functions $r_1(t)$, $r_2(t)$, and $r_3(t)$ are all differentiable at $t = a$, then the vector function $\mathbf{r}(t)$ is **differentiable** in reference frame $\mathcal{A}$ at $t = a$ with the **vector derivative** in $\mathcal{A}$ given by

$$\frac{{}^{\mathcal{A}}d}{dt}\mathbf{r}\bigg|_{t=a} \triangleq \frac{d}{dt}r_1\bigg|_{t=a}\mathbf{a}_1 + \frac{d}{dt}r_2\bigg|_{t=a}\mathbf{a}_2 + \frac{d}{dt}r_3\bigg|_{t=a}\mathbf{a}_3 \qquad (3.10)$$

or, more compactly,

$$\boxed{\frac{{}^{\mathcal{A}}d}{dt}\mathbf{r}\bigg|_{t=a} = \dot{r}_1(a)\mathbf{a}_1 + \dot{r}_2(a)\mathbf{a}_2 + \dot{r}_3(a)\mathbf{a}_3.}$$

As emphasized in Chapter 1, it is essential to remember that the derivative of a vector may be different in different frames. The change in a vector—magnitude or direction—has no meaning unless a reference frame is specified, since the frame may also be moving. (The change in a scalar, however, does not need a frame specification.) This stipulation should hopefully be clear from our definition. Since the unit vectors $\mathbf{a}_1$, $\mathbf{a}_2$, $\mathbf{a}_3$ define frame $\mathcal{A}$, they are fixed in it and thus the change with time of $\mathbf{r}(t)$ in $\mathcal{A}$ is given by Eq. (3.10). However, if we asked how $\mathbf{r}$ changed in a different frame, say $\mathcal{B}$, then the $\mathbf{a}_i$ unit vector might be changing because $\mathcal{B}$ might be rotating with respect to $\mathcal{A}$.

As an example, consider a point P traveling around a circle centered at O, as in Figure 3.11. The rate of change of $\mathbf{r}_{P/O}$ with respect to a fixed reference frame $\mathcal{I}$ with origin O is simply the change in direction of $\mathbf{r}_{P/O}$ as P travels around O. The corresponding velocity is written ${}^{\mathcal{I}}\mathbf{v}_{P/O}$ and is always tangent to the circle. However, the rate of change of $\mathbf{r}_{P/O}$ with respect to a rotating frame $\mathcal{B} = (O, \mathbf{b}_1, \mathbf{b}_2, \mathbf{b}_3)$, where $\mathbf{b}_1$ is always aligned with $\mathbf{r}_{P/O}$, is zero. Point P is not moving in $\mathcal{B}$!

This example highlights the importance of labeling a vector derivative with the appropriate reference frame. We always specify the frame to which a velocity is related by using a superscript capital letter for that frame (e.g., ${}^{\mathcal{I}}\mathbf{v}_{P/O} = \frac{{}^{\mathcal{I}}d}{dt}\mathbf{r}_{P/O}$). Likewise

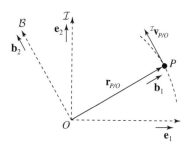

Figure 3.11 Point P traveling around a fixed circle with velocity $^{\mathcal{I}}\mathbf{v}_{P/O}$ tangent to the circle. In a frame $\mathcal{B}$ rotating so that $\mathbf{b}_1$ always points at P, the velocity is zero: $^{\mathcal{B}}\mathbf{v}_{P/O} = 0$.

TABLE 3.1
Properties of the vector derivative

Addition rule	$\frac{^{A}d}{dt}(\mathbf{r}(t) + \mathbf{s}(t)) = \frac{^{A}d}{dt}\mathbf{r}(t) + \frac{^{A}d}{dt}\mathbf{s}(t)$
Product rule	$\frac{^{A}d}{dt}(f(t)\mathbf{r}(t)) = \frac{df}{dt}(t)\mathbf{r}(t) + f(t)\frac{^{A}d}{dt}\mathbf{r}(t)$
	$\frac{d}{dt}(\mathbf{r}(t) \cdot \mathbf{s}(t)) = \frac{^{A}d}{dt}\mathbf{r}(t) \cdot \mathbf{s}(t) + \mathbf{r}(t) \cdot \frac{^{A}d}{dt}\mathbf{s}(t)$
	$\frac{^{A}d}{dt}(\mathbf{r}(t) \times \mathbf{s}(t)) = \frac{^{A}d}{dt}\mathbf{r}(t) \times \mathbf{s}(t) + \mathbf{r}(t) \times \frac{^{A}d}{dt}\mathbf{s}(t)$
Chain rule	$\frac{^{A}d}{dt}\mathbf{r}(f(t)) = \frac{^{A}d}{df}\mathbf{r}(f(t))\frac{df}{dt}$
	$\frac{^{A}d}{dt}f(\mathbf{r}(t)) = \nabla f \cdot \frac{^{A}d}{dt}\mathbf{r}$

for an acceleration, we write $^{\mathcal{I}}\mathbf{a}_{P/O} = \frac{^{\mathcal{I}}d}{dt}(^{\mathcal{I}}\mathbf{v}_{P/O})$. Although this notation may seem burdensome, writing vector derivatives this way is an important habit to begin now. Remember that Newton's second law only applies when the acceleration (and the velocity it is differentiating) is defined with respect to an inertial frame.

It turns out that all important properties of the scalar derivative also apply to vector derivatives; they are summarized in Table 3.1. (If you are rusty on the scalar versions, you can review them in Appendix A.) These properties are often used in our later work, particularly when finding expressions for the inertial velocity and inertial acceleration of a particle in terms of the three coordinate systems studied so far. These rules are derived in Section 3.8.

Example 3.6 The Vector Derivative on an Inclined Plane

This example returns to the block sliding on the inclined plane of Example 3.5, as shown again in Figure 3.12. Here we have labeled the block by the letter P and included three additional frames, an inertial frame $\mathcal{A}$ aligned along the plane and fixed to the upper corner, a frame $\mathcal{B}$ fixed at O whose $\mathbf{b}_1$ direction always points at P, and a frame $\mathcal{C}$ fixed to the block and moving with it with direction $\mathbf{c}_1$ along the incline.

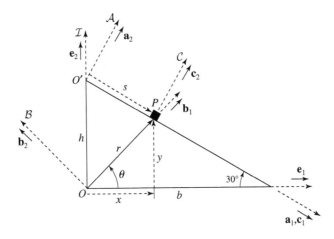

Figure 3.12 Block on an inclined plane and four different reference frames.

The origin of $\mathcal{C}$ is $O'' = P$. From Example 3.5, the position of the block expressed as components in each frame and in terms of the different coordinates is

$$\mathbf{r}_{P/O} = x\mathbf{e}_1 + y\mathbf{e}_2 = r\cos\theta\mathbf{e}_1 + r\sin\theta\mathbf{e}_2$$

$$= r\mathbf{b}_1$$

and

$$\mathbf{r}_{P/O'} = s\mathbf{a}_1.$$

We now use the definition of the vector derivative to find the velocity of P in each frame. From the geometry it is obvious that the velocity of P in $\mathcal{C}$ is zero, as it always remains at the origin of $\mathcal{C}$:

$$\frac{^{\mathcal{C}}d}{dt}\left(\mathbf{r}_{P/O''}\right) = 0.$$

The velocity of P in $\mathcal{A}$ is given by the vector derivative

$$\frac{^{\mathcal{A}}d}{dt}\left(\mathbf{r}_{P/O'}\right) = \dot{s}\mathbf{a}_1.$$

The velocity of P in $\mathcal{I}$ is given in terms of the Cartesian coordinates as

$$\frac{^{\mathcal{I}}d}{dt}\left(\mathbf{r}_{P/O}\right) = \dot{x}\mathbf{e}_1 + \dot{y}\mathbf{e}_2$$

and in terms of polar coordinates as

$$\frac{^{\mathcal{I}}d}{dt}\left(\mathbf{r}_{P/O}\right) = (\dot{r}\cos\theta - r\dot{\theta}\sin\theta)\mathbf{e}_1 + (\dot{r}\sin\theta + r\dot{\theta}\cos\theta)\mathbf{e}_2,$$

where we have used the chain rule for vector derivatives. Note that we can use the constraints described in Example 3.5 to rewrite the vector derivative in $\mathcal{I}$ in terms of the single coordinate s describing the single degree of freedom:

$$\frac{{}^{\mathcal{I}}d}{dt}(\mathbf{r}_{P/O}) = \dot{s}\frac{\sqrt{3}}{2}\mathbf{e}_1 - \dot{s}\frac{1}{2}\mathbf{e}_2.$$

Finally, the velocity of P in frame $\mathcal{B}$ is again given by the definition of the vector derivative:

$$\frac{{}^{\mathcal{B}}d}{dt}(\mathbf{r}_{P/O}) = \dot{r}\mathbf{b}_1.$$

3.3.2 Velocity Is the Vector Derivative of Position

Now that we have a formal mathematical definition of the vector derivative, it is useful to show how the qualitative description follows from our formal definition. We use the definition of the scalar derivative in Definition A.2 to rewrite the vector derivative in Eq. (3.10) as

$$\frac{{}^{\mathcal{A}}d}{dt}\mathbf{r}\Big|_{t=a} \triangleq \left(\lim_{h\to0}\frac{r_1(a+h)-r_1(a)}{h}\right)\mathbf{a}_1 + \left(\lim_{h\to0}\frac{r_2(a+h)-r_2(a)}{h}\right)\mathbf{a}_2$$
$$+ \left(\lim_{h\to0}\frac{r_3(a+h)-r_3(a)}{h}\right)\mathbf{a}_3. \tag{3.11}$$

We can now use the distributive property of the limit operator and of vector addition to obtain

$$\frac{{}^{\mathcal{A}}d}{dt}\mathbf{r}\Big|_{t=a} = \lim_{h\to0}\frac{1}{h}\big[\left(r_1(a+h)\mathbf{a}_1 + r_2(a+h)\mathbf{a}_2 + r_3(a+h)\mathbf{a}_3\right)$$
$$- \left(r_1(a)\mathbf{a}_1 + r_2(a)\mathbf{a}_2 + r_3(a)\mathbf{a}_3\right)\big].$$

Thus the vector derivative is simply the limit of the difference between the vector evaluated at time $t = a$ in frame $\mathcal{A}$ and the vector a short time $t = a + h$ later—also expressed in frame $\mathcal{A}$—just as illustrated in Figure 3.10. Again, it is important to remember that we evaluate the change in the vector by expressing it in terms of components in frame $\mathcal{A}$ at both times. *The vector derivative—or any change in a vector—is frame dependent.*

The vector derivative of the position of point P with respect to an inertial frame $\mathcal{I}$ is the inertial velocity of P, illustrated in Figure 3.10. Eq. (3.11) shows that the velocity always points along the tangent to the trajectory of P, as illustrated in Figure 3.11. The trajectory of P—traced by the tip of the vector $\mathbf{r}_{P/O}$—is defined by the three scalar functions $r_1(t)$, $r_2(t)$, and $r_3(t)$. From Appendix A and Eq. (3.11), the magnitudes of the components of the velocity ${}^{\mathcal{I}}\mathbf{v}_{P/O}$ are the slopes of those three functions. The velocity is thus directed along the line defined by these three slopes, that is, the tangent to the trajectory.

We could go through the same development by replacing $\mathbf{r}_{P/O}(t)$ with ${}^{\mathcal{I}}\mathbf{v}_{P/O}$ in Definition 3.3 to derive the acceleration ${}^{\mathcal{I}}\mathbf{a}_{P/O}$ and show that it is tangent to the curve traced by the tip of the velocity. It is not necessary, however, because our development of the vector derivative was completely general—it applies to any vector, including the velocity. This is important, since we use the acceleration in Newton's second law to obtain the equations of motion needed to solve for the trajectory of P.

3.3.3 Velocity and Acceleration in Various Coordinates

The previous two subsections supplied all the tools you need for vector calculus—in particular, the vector derivative and its properties. We also showed geometrically how it is used to define the velocity and acceleration. As discussed in Chapter 1, once the position of a particle (as a function of time) and its velocity have been determined, we know everything about its state at all times and can draw its trajectory. The acceleration allows us to solve for that trajectory using the physics of Newton's second law.

In reality, although the geometric description is exceptionally important, it is not as useful as it might at first seem. This is because to actually write down or draw the position and velocity of a particle we need coordinates. Thus deriving Newton's second law for a particle is really about finding differential equations for the coordinates that describe the position of a particle in a reference frame. In this subsection we complete the picture by using the definition of the vector derivative to find expressions for the inertial velocity and acceleration of a particle in the three coordinate systems discussed so far—Cartesian, polar, and path coordinates.

Kinematics in Cartesian Coordinates

As discussed in Section 3.2.4, Cartesian coordinates are the most common and fundamental of our coordinate systems. Cartesian coordinates of a point in a reference frame directly correspond to the magnitudes of the components of the position in that frame. Thus it is trivial to write the velocity and acceleration of the particle in Cartesian coordinates, since they follow directly from the definition of the vector derivative:

$$^{\mathcal{I}}\mathbf{v}_{P/O} \triangleq \frac{^{\mathcal{I}}d}{dt}\left(\mathbf{r}_{P/O}\right) = \dot{x}\mathbf{e}_x + \dot{y}\mathbf{e}_y + \dot{z}\mathbf{e}_z$$

$$^{\mathcal{I}}\mathbf{a}_{P/O} \triangleq \frac{^{\mathcal{I}}d}{dt}\left(^{\mathcal{I}}\mathbf{v}_{P/O}\right) = \ddot{x}\mathbf{e}_x + \ddot{y}\mathbf{e}_y + \ddot{z}\mathbf{e}_z.$$

Many problems can be solved using Cartesian coordinates. The examples in Chapter 2 all used such coordinates in an inertial frame. Here we have introduced a useful formalism for thinking about these problems using vectors. While it may seem that this is superfluous, hang on. To solve more complicated and difficult problems, we rely on the careful definitions and notations of this chapter.

Example 3.7 The Tossed Point Mass, Revisited

Tutorial 2.6 introduced our first two-dimensional dynamics problem, the tossed point mass. However, we cheated a bit and avoided the use of vectors, as the important definitions and notations of this chapter had not yet been introduced. We simply solved it by means of two separate, scalar one-dimensional systems. Here we revisit it

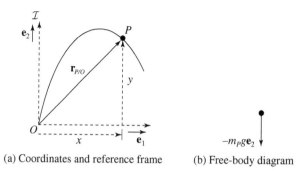

(a) Coordinates and reference frame   (b) Free-body diagram

Figure 3.13 Solving for the vector motion of a tossed point mass.

and solve it using vector notation. By tackling such an admittedly straightforward problem, we hope to get you comfortable with the general vector approach to Newton's second law.

Figure 3.13a shows a picture of the geometry for the example. The position of the point mass relative to origin O of inertial frame $\mathcal{I}$ is given by Cartesian coordinates, so that $\mathbf{r}_{P/O} = x\mathbf{e}_1 + y\mathbf{e}_2$. The point mass P is thrown with an initial velocity ${}^{\mathcal{I}}\mathbf{v}_{P/O}(0) = \dot{x}(0)\mathbf{e}_1 + \dot{y}(0)\mathbf{e}_2$. The objective is to solve for the trajectory of P as a function of time in Cartesian coordinates, $\mathbf{r}_{P/O}(t) = x(t)\mathbf{e}_1 + y(t)\mathbf{e}_2$.

The acceleration of the mass in Cartesian coordinates is

$$
{}^{\mathcal{I}}\mathbf{a}_{P/O} = \frac{{}^{\mathcal{I}}d}{dt}(\dot{x}\mathbf{e}_1 + \dot{y}\mathbf{e}_2) = \ddot{x}\mathbf{e}_1 + \ddot{y}\mathbf{e}_2. \tag{3.12}
$$

The only force acting on the point mass, shown in the free-body diagram in Figure 3.13b, is the downward action of gravity,

$$
\mathbf{F}_P = -m_P g\mathbf{e}_2.
$$

Using Newton's second law, then, we have the vector equation of motion:

$$
{}^{\mathcal{I}}\mathbf{a}_{P/O} = -g\mathbf{e}_2. \tag{3.13}
$$

Combining Eq. (3.12) with Eq. (3.13) gives the two scalar differential equations for the Cartesian coordinates of the point mass:

$$
\mathbf{e}_1: \quad \ddot{x} = 0
$$
$$
\mathbf{e}_2: \quad \ddot{y} = -g.
$$

These are the same two scalar equations as Eqs. (2.30) and (2.31). The rest of the example is solved as in Tutorial 2.6.

Example 3.8 Solving the Pendulum Using Cartesian Coordinates in the Inertial Frame, Revisited

We now return to Example 3.1 and use our new notational tools on the simple pendulum. As shown in Example 3.2, the position of the pendulum bob m_P with respect to O can be written in terms of components in frame $\mathcal{I}$ as

$$\mathbf{r}_{P/O} = x\mathbf{e}_1 + y\mathbf{e}_2. \tag{3.14}$$

Taking the time derivative with respect to the inertial frame of the position in Eq. (3.14) by using Definition 3.3 gives the velocity of P with respect to the inertial frame:

$$^{\mathcal{I}}\mathbf{v}_{P/O} = \frac{^{\mathcal{I}}d}{dt}(x\mathbf{e}_1 + y\mathbf{e}_2) = \dot{x}\mathbf{e}_1 + \dot{y}\mathbf{e}_2. \tag{3.15}$$

Similarly, taking the time derivative with respect to the inertial frame of the velocity in Eq. (3.15) gives the acceleration of P with respect to the inertial frame:

$$^{\mathcal{I}}\mathbf{a}_{P/O} = \frac{^{\mathcal{I}}d}{dt}(\dot{x}\mathbf{e}_1 + \dot{y}\mathbf{e}_2) = \ddot{x}\mathbf{e}_1 + \ddot{y}\mathbf{e}_2. \tag{3.16}$$

Next, using the inertial acceleration in Eq. (3.16) and the free-body diagram in Figure 3.1c, we can write down Newton's second law for the pendulum bob:

$$\frac{Fx}{l}\mathbf{e}_1 + \frac{Fy}{l}\mathbf{e}_2 - m_P g \mathbf{e}_2 = m_P(\ddot{x}\mathbf{e}_1 + \ddot{y}\mathbf{e}_2). \tag{3.17}$$

Vector equation Eq. (3.17) is equivalent to the two scalar Eqs. (3.1) and (3.2) in Example 3.1. The rest of this example can be solved as before.

Examples 3.1 and 3.8 showed that we can use Cartesian coordinates (differentiated in the inertial frame) to find the equations of motion of the simple pendulum, but the results can be messy. In fact, for the simple pendulum, the result was not simple at all. This motivates us to look for alternate coordinates to solve this and other problems.

Kinematics in Polar Coordinates

Of course Cartesian coordinates are not the only ones available for describing the position of a point in the plane. The complexity of the pendulum example is a strong motivator, for instance, for experimenting with different coordinates. Eq. (3.9) showed how to represent a vector in polar coordinates. We can thus easily find the inertial velocity and acceleration in terms of polar coordinates using Definition 3.3 (as was done in Example 3.6). By using the scalar product rule and chain rule on each component, we have:

$$^{\mathcal{I}}\mathbf{v}_{P/O} = (\dot{r}\cos\theta - r\dot{\theta}\sin\theta)\mathbf{e}_1 + (\dot{r}\sin\theta + r\dot{\theta}\cos\theta)\mathbf{e}_2 \tag{3.18}$$

$$^{\mathcal{I}}\mathbf{a}_{P/O} = (\ddot{r}\cos\theta - 2\dot{r}\dot{\theta}\sin\theta - r\ddot{\theta}\sin\theta - r\dot{\theta}^2\cos\theta)\mathbf{e}_1$$
$$+ (\ddot{r}\sin\theta + 2\dot{r}\dot{\theta}\cos\theta + r\ddot{\theta}\cos\theta - r\dot{\theta}^2\sin\theta)\mathbf{e}_2. \tag{3.19}$$

These equations look much more complicated than their counterparts in Cartesian coordinates. Have we really gained anything by moving to polar coordinates? For certain problems, yes! Let's take another look at the simple pendulum, now using polar coordinates.

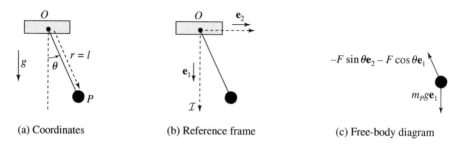

(a) Coordinates   (b) Reference frame   (c) Free-body diagram

Figure 3.14 Solving the pendulum using polar coordinates in the inertial frame.

Example 3.9 Solving the Pendulum Using Polar Coordinates in the Inertial Frame

As shown in Example 3.4, the simple pendulum is a single-degree-of-freedom system. We should thus be able to describe its motion with only one coordinate rather than the two Cartesian coordinates and a constraint. Here we explore using polar coordinates to solve it. Figure 3.14 shows that the pendulum mass position can be defined using the polar coordinates $(r, \theta)_{\mathcal{I}}$, where $\mathcal{I} = (O, \mathbf{e}_1, \mathbf{e}_2, \mathbf{e}_3)$.

In polar coordinates, the position of the pendulum expressed as components in the inertial frame is

$$\mathbf{r}_{P/O} = r \cos \theta \mathbf{e}_1 + r \sin \theta \mathbf{e}_2.$$

However, the constraint imposed by the fixed length of the pendulum rod is simply $r = l$, implying $\dot{r} = \dot{l} = 0$. This eliminates the r coordinate—leaving only θ to describe the one degree of freedom—part of our motivation for moving to polar coordinates in this example. Taking the time derivative with respect to the inertial frame of the position gives

$$^{\mathcal{I}}\mathbf{v}_{P/O} = \frac{^{\mathcal{I}}d}{dt}(l \cos \theta \mathbf{e}_1 + l \sin \theta \mathbf{e}_2) = -l\dot{\theta} \sin \theta \mathbf{e}_1 + l\dot{\theta} \cos \theta \mathbf{e}_2.$$

Taking the time derivative again yields the inertial acceleration:

$$^{\mathcal{I}}\mathbf{a}_{P/O} = (-l\ddot{\theta} \sin \theta - l\dot{\theta}^2 \cos \theta)\mathbf{e}_1 + (l\ddot{\theta} \cos \theta - l\dot{\theta}^2 \sin \theta)\mathbf{e}_2. \tag{3.20}$$

Using Eq. (3.20) and the free-body diagram in Figure 3.14c, we can write down Newton's second law for the pendulum bob:

$$-F \sin \theta \mathbf{e}_2 - F \cos \theta \mathbf{e}_1 + m_P g \mathbf{e}_1 = m_P(-l\ddot{\theta} \sin \theta - l\dot{\theta}^2 \cos \theta)\mathbf{e}_1$$
$$+ m_P(l\ddot{\theta} \cos \theta - l\dot{\theta}^2 \sin \theta)\mathbf{e}_2.$$

This vector equation is equivalent to the two scalar equations,

$$-F \cos \theta + m_P g = m_P(-l\ddot{\theta} \sin \theta - l\dot{\theta}^2 \cos \theta) \tag{3.21}$$

$$-F \sin \theta = m_P(l\ddot{\theta} \cos \theta - l\dot{\theta}^2 \sin \theta). \tag{3.22}$$

As mentioned, part of our motivation for moving to polar coordinates is its compact description of this single-degree-of-freedom system using the polar angle θ. We find the equation of motion for θ by multiplying Eq. (3.21) by $-\sin\theta$ and Eq. (3.22) by $\cos\theta$ and then adding the result to obtain

$$\ddot{\theta} = -\frac{g}{l}\sin\theta. \qquad (3.23)$$

This is the equation of motion for coordinate θ describing the single degree of freedom. It is a far simpler equation of motion than the two expressions in Eq. (3.7) plus the constraint equation.

We can also easily solve for the constraint force in the rod by instead multiplying Eq. (3.21) by $\cos\theta$ and Eq. (3.22) by $\sin\theta$ and adding to obtain

$$F = m_P(g\cos\theta + l\dot{\theta}^2).$$

Having this expression for the constraint force in the rod is useful if we want to engineer a pendulum that won't break. As mentioned in Chapter 1, we will often solve for the constraint forces as well as the equations of motion. Remember, any time there is a constraint that reduces the number of degrees of freedom, there is an accompanying constraint force.

We can also solve for the equilibrium points of the simple pendulum by setting $\ddot{\theta} = 0$ in Eq. (3.23):

$$0 = -\frac{g}{l}\sin\theta.$$

There are two equilibrium angles for the pendulum: $\theta_0 = 0$ (the pendulum hanging down) and $\theta_0 = \pm\pi$ (the pendulum standing up). As shown later in the book, the inverted pendulum is an example of a system with an *unstable equilibrium point*: a slight perturbation will cause the pendulum to fall.

Eq. (3.23) is a classic equation that appears often in dynamics. Of special significance is an approximation of the equation when the angle θ is small (i.e., for small motion about the equilibrium $\theta_0 = 0$). In this case, we replace $\sin\theta$ by its Taylor series about $\theta_0 = 0$ and keep only the first-order term:[4]

$$\sin\theta = \theta - \frac{\theta^3}{3!} + \frac{\theta^5}{5!} + \cdots \approx \theta.$$

(This approximation is discussed in Appendix A.)

After making this approximation, the pendulum equation becomes

$$\ddot{\theta} = -\frac{g}{l}\theta. \qquad (3.24)$$

Eq. (3.24) is a simple harmonic motion equation with natural frequency $\omega_0 = \sqrt{g/l}$ (see Tutorial 2.3). Therefore, as long as θ is small, the pendulum mass oscillates sinusoidally about $\theta_0 = 0$ and the frequency of oscillation is inversely proportional to the square root of the length of the pendulum rod.

---

[4] Also called the *small-angle approximation*.

Not only were polar coordinates algebraically less cumbersome than Cartesian coordinates for the pendulum, but the final solution was also much more elegant and better behaved numerically. We have a single equation of motion for the degree of freedom represented by θ and we eliminated the constraint force in the rod. However, the algebra was still involved—we had to solve for two scalar equations and then manipulate these equations to find the equation of motion. Soon we will introduce another concept—the *polar frame*—that will make the pendulum problem even easier to solve.

Kinematics in Path Coordinates

Path coordinates don't lend themselves easily to differentiation in the inertial frame because there is no representation for the position of P with respect to the origin O of the inertial frame. The scalar coordinate s is not sufficient to specify the (planar) position of P without a functional form for the path. In fact, path coordinates are rarely used in cases where the position of a particle is of interest. Typically, it is only the particle motion (and the accompanying forces) we care about. Thus finding the acceleration and applying Newton's second law using path coordinates requires more sophisticated approaches that we delve into next.

3.4 Inertial Velocity and Acceleration in a Rotating Frame

If we were only interested in representing the position and velocity of a particle in terms of components in an inertial frame and in solving for its trajectory there, we would be done. We could end this chapter here and move on to other topics. But then why do we place so much emphasis on the reference-frame concept? The answer is that it is unnecessarily limiting and complicated to solve every problem using vectors expressed as components in an inertial frame. Many, many problems are more easily solved—or produce expressions that are easier to work with—when we introduce additional, noninertial reference frames. This is the subject of the remainder of this chapter.

Introducing one or more additional reference frames to a problem offers several advantages and will become a fundamental piece of our problem-solving machinery throughout the book. First, it is often important for engineers to build models that relate to actual measurements. Since these measurements are often obtained in moving or rotating frames, it is very convenient to have expressions for the dynamical trajectories in terms of coordinates in those frames. Even measurements from the ground, tracking an airplane, say, are in a moving and rotating frame—fixed to the earth—as opposed to an inertial frame fixed to the distant stars; see Figure 3.15. Second, many problems have constraints that reduce the number of degrees of freedom. For these problems, operating in absolute space often means carrying around more scalar coordinates than necessary, and thus we have to explicitly utilize the constraint equations (see Example 3.1). We can often simplify such problems by introducing a new reference frame. Finally, moving and rotating frames are an essential construct for keeping track of the various coordinates associated with multiple particles and/or

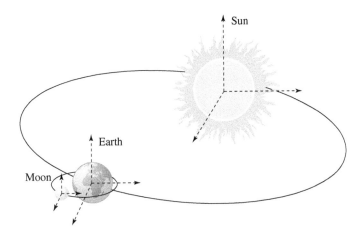

Figure 3.15 Illustration of multiple reference frames.

multiple rigid bodies. *The solution to almost every dynamics problem begins with the careful and judicious choice of appropriate and useful reference frames.*

In this section we discuss one of the key uses of rotating frames: to simplify finding the velocity and acceleration in different coordinates. For instance, while procedurally deriving Eqs. (3.18) and (3.19) was reasonably straightforward (though algebraically messy), the resulting expressions for the velocity and acceleration are rather involved. Introducing a new frame and expressing the inertial velocity and acceleration in it greatly reduces our work.

3.4.1 Relative Orientation

Before discussing the use of alternate reference frames in forming the velocity and acceleration, we must introduce the geometric tool used to describe two reference frames that differ in orientation. What, in fact, do we mean by the orientation of one frame relative to another? For now, we only consider planar orientation. That is, we consider only the relative orientation of two reference frames that share a common axis (typically, the third axis). Chapter 10 presents the more general description of three-dimensional relative orientation.

Figure 3.16 shows two reference frames, $\mathcal{A} = (O, \mathbf{a}_1, \mathbf{a}_2, \mathbf{a}_3)$ and $\mathcal{B} = (O, \mathbf{b}_1, \mathbf{b}_2, \mathbf{b}_3)$, where $\mathbf{a}_3 = \mathbf{b}_3$. Frame $\mathcal{B}$ is rotated with respect to frame $\mathcal{A}$ by an angle θ about their common axis. Since $\mathbf{b}_1$ is a vector, it can be expressed as components in $\mathcal{A}$:

$$\mathbf{b}_1 = \cos\theta\,\mathbf{a}_1 + \sin\theta\,\mathbf{a}_2.$$

Likewise,

$$\mathbf{b}_2 = -\sin\theta\,\mathbf{a}_1 + \cos\theta\,\mathbf{a}_2.$$

We call such a change in orientation about an axis fixed in both frames a *simple rotation*.

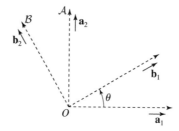

Figure 3.16 Reference frame $\mathcal{B} = (O, \mathbf{b}_1, \mathbf{b}_2, \mathbf{b}_3)$ is rotated with respect to frame $\mathcal{A} = (O, \mathbf{a}_1, \mathbf{a}_2, \mathbf{a}_3)$ by angle θ.

These relationships can be compactly represented by writing them as rows in a vector transformation table:

	$\mathbf{a}_1$	$\mathbf{a}_2$
$\mathbf{b}_1$	$\cos\theta$	$\sin\theta$
$\mathbf{b}_2$	$-\sin\theta$	$\cos\theta$,

where the row i and column j element is given by $\mathbf{b}_i \cdot \mathbf{a}_j$.

The transformation-table notation also makes it straightforward to write the $\mathcal{A}$ unit vectors in terms of the $\mathcal{B}$ unit vectors by summing each column of the table:

$$\mathbf{a}_1 = \cos\theta\,\mathbf{b}_1 - \sin\theta\,\mathbf{b}_2$$
$$\mathbf{a}_2 = \sin\theta\,\mathbf{b}_1 + \cos\theta\,\mathbf{b}_2.$$

So, for example, given a vector $\mathbf{r}_{P/O}$ expressed as components in $\mathcal{A}$ (i.e., $\mathbf{r}_{P/O} = a_1\mathbf{a}_1 + a_2\mathbf{a}_2$), we can use the transformation table to express the vector in terms of components in $\mathcal{B}$,

$$\mathbf{r}_{P/O} = a_1(\cos\theta\,\mathbf{b}_1 - \sin\theta\,\mathbf{b}_2) + a_2(\sin\theta\,\mathbf{b}_1 + \cos\theta\,\mathbf{b}_2)$$
$$= \underbrace{(a_1\cos\theta + a_2\sin\theta)}_{=b_1}\mathbf{b}_1 + \underbrace{(-a_1\sin\theta + a_2\cos\theta)}_{=b_2}\mathbf{b}_2,$$

where $(b_1, b_2)_{\mathcal{B}}$ are the scalar magnitudes of the components of $\mathbf{r}_{P/O}$ expressed in frame $\mathcal{B}$.

We can also use matrix notation to write the transformation of the component magnitudes in matrix form:

$$\begin{bmatrix} b_1 \\ b_2 \end{bmatrix}_{\mathcal{B}} = \underbrace{\begin{bmatrix} \cos\theta & \sin\theta \\ -\sin\theta & \cos\theta \end{bmatrix}}_{\triangleq\,^{\mathcal{B}}C^{\mathcal{A}}} \begin{bmatrix} a_1 \\ a_2 \end{bmatrix}_{\mathcal{A}},$$

where $^{\mathcal{B}}C^{\mathcal{A}}$ is the *transformation matrix* from frame $\mathcal{A}$ to frame $\mathcal{B}$. Its elements are likewise given by $C_{ij} = \mathbf{b}_i \cdot \mathbf{a}_j$. The inverse transformation is given by $^{\mathcal{A}}C^{\mathcal{B}} = (^{\mathcal{B}}C^{\mathcal{A}})^{-1} = (^{\mathcal{B}}C^{\mathcal{A}})^T$.

We use the transformation table throughout the book to describe the relative orientation of one frame with respect to another. The important thing to remember is that when presented with a vector, we have a variety of frame choices for writing the components. Expressing the components of a vector (particularly the acceleration) in one frame versus another can make problems significantly easier to solve. The transformation table introduced in this section is used to transform the components of a vector from the unit-vector directions of one frame to the unit-vector directions of another.

Note that Figure 3.16 may give the misleading impression that the transformation table only applies when the two frames share a common origin. This is not the case. Remember, the unit vectors are just vectors of unit magnitude. It does not matter where they are located. We can translate frame $\mathcal{B}$ anywhere and, as long as its relative orientation to $\mathcal{I}$ stays the same, the transformation table between the two sets of unit vectors is the same.

Example 3.10 Satellite Tracking, Revisited

Example 3.3 showed how to express a vector as components in two different frames. We repeat that example here using our new formalism for relative orientation. Recall that Example 3.3 defined two frames of reference: an inertial frame $\mathcal{I}$ located at the center of the earth O and a second frame $\mathcal{B}$ located on the surface of the earth at origin O' and sharing the $\mathbf{e}_1 = \mathbf{b}_1$ axis with frame $\mathcal{I}$. Even though the two frames do not share the same origin, the unit vectors defining them are still related by a transformation table using the fact that $\mathcal{B}$ is rotated by $-45°$ about $\mathbf{e}_1$ from $\mathcal{I}$:

	$\mathbf{b}_2$	$\mathbf{b}_3$
$\mathbf{e}_2$	$\cos 45°$	$\sin 45°$
$\mathbf{e}_3$	$-\sin 45°$	$\cos 45°$.

The transformation table shows that we can write the $\mathbf{b}_2$ unit vector in terms of the $\mathbf{e}_2$ and $\mathbf{e}_3$ unit vectors as

$$\mathbf{b}_2 = \cos 45° \mathbf{e}_2 - \sin 45° \mathbf{e}_3 = \frac{1}{\sqrt{2}}\mathbf{e}_2 - \frac{1}{\sqrt{2}}\mathbf{e}_3,$$

which is the same as Eq. (3.8) in Example 3.3. The rest of the example follows as before.

3.4.2 Two Rotating Frames

By now you should be convinced that finding equations of motion for a particle using Newton's second law requires expressing the acceleration relative to an inertial frame. This is fundamental. Earlier in the chapter we showed how to find the velocity and acceleration in terms of both Cartesian and polar coordinates (we postponed path coordinates) and how to use the resulting vector expressions to find equations of motion. This main point cannot be made often enough: to compute the inertial kinematics for use in Newton's second law, the vector derivative must be taken with respect to an inertial frame.

However, we are free to express the resulting velocity and acceleration as components in any frame we like. As noted for Eqs. (3.18) and (3.19), expressing them as components in the inertial frame can result in rather complicated and lengthy expressions. In the next two subsections we introduce two important reference frames.

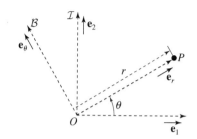

Figure 3.17 Polar coordinates $(r, \theta)_{\mathcal{I}}$ and polar frame $\mathcal{B} = (O, \mathbf{e}_r, \mathbf{e}_\theta, \mathbf{e}_3)$, where $\mathbf{e}_3 = \mathbf{e}_r \times \mathbf{e}_\theta$ points out of the page (not shown).

Either can be used for expressing the components of $^{\mathcal{I}}\mathbf{v}_{P/O}$ and $^{\mathcal{I}}\mathbf{a}_{P/O}$ and, for certain problems, these representations greatly simplify finding the motion. These frames can also be used to find a new and elegant approach for calculating the inertial kinematics.

The Polar Frame

The *polar frame* is a frame of reference that shares an axis with the inertial frame (usually the $\mathbf{e}_3$ or $\mathbf{e}_z$ axis) and has a unit vector defining a second axis (usually the $\mathbf{e}_r$ axis) directed from the frame's origin to a point P even as P moves. In the case where the polar frame $\mathcal{B}$ has the same origin O as the inertial frame $\mathcal{I}$, it is particularly simple to express the position of P with respect to O as components in the polar frame using the polar radius r:

$$\mathbf{r}_{P/O} = r\mathbf{e}_r. \tag{3.25}$$

The unit vector $\mathbf{e}_\theta$ points in the direction of increasing θ. Since $\mathbf{e}_r$ always points at P, $\mathbf{e}_\theta$ is not needed to describe the position of P relative to O as components in $\mathcal{B}$.

Although both the polar radius r and polar angle θ are needed to locate a point in the inertial frame, we can locate P in the polar frame using only r. Of course, we can't ignore θ altogether, as it defines the orientation of the polar frame with respect to the inertial frame (see Section 3.4.1). The polar angle is thus used to express the position $\mathbf{r}_{P/O}$ as components in the inertial frame. The polar frame is particularly convenient for constrained problems that have only a single degree of freedom, such as the simple pendulum that we study extensively in this chapter.

The polar frame is completed by $\mathbf{e}_\theta = \mathbf{e}_3 \times \mathbf{e}_r$, which is perpendicular to $\mathbf{e}_r$ in a right-hand sense. (See Appendix B for properties of the vector cross product.) Figure 3.17 shows the polar frame $\mathcal{B} = (O, \mathbf{e}_r, \mathbf{e}_\theta, \mathbf{e}_3)$. As P moves on some arbitrary path, frame $\mathcal{B}$ rotates relative to frame $\mathcal{I}$ and is thus not necessarily inertial, unless θ is constant.

The real value of introducing the polar frame is that the inertial acceleration of P in Newton's second law can be expressed in terms of the unit vectors $\mathbf{e}_r$ and $\mathbf{e}_\theta$: this representation can simplify—and enhance our understanding of—certain problems. We start by writing the kinematics—that is, the inertial velocity and acceleration— as components in the polar frame. In Eqs. (3.18) and (3.19), these kinematics were written in polar coordinates but as vector components in the inertial frame. Using the transformation table in Section 3.4.1, we can express the inertial unit vectors as components in the polar frame:

$$\mathbf{e}_1 = \cos\theta\mathbf{e}_r - \sin\theta\mathbf{e}_\theta$$

$$\mathbf{e}_2 = \sin\theta\mathbf{e}_r + \cos\theta\mathbf{e}_\theta.$$

Substituting these expressions into Eqs. (3.18) and (3.19), we obtain (after some algebra)

$$^{\mathcal{I}}\mathbf{v}_{P/O} = \dot{r}\mathbf{e}_r + r\dot{\theta}\mathbf{e}_\theta \tag{3.26}$$

$$^{\mathcal{I}}\mathbf{a}_{P/O} = (\ddot{r} - r\dot{\theta}^2)\mathbf{e}_r + (2\dot{r}\dot{\theta} + r\ddot{\theta})\mathbf{e}_\theta. \tag{3.27}$$

Eqs. (3.26) and (3.27) are important equations and are much simpler than Eqs. (3.18) and (3.19). These equations also have important physical consequences that are discussed in Section 3.5. It is important to remember that the velocity and acceleration vectors given in Eqs. (3.18) and (3.19) and the vectors in Eqs. (3.26) and (3.27) are the same. We have simply expressed them as vector components in two different reference frames.

The process by which we obtained the kinematics in the polar frame, however, was a bit involved. And this process doesn't always work out so cleanly, as discussed in Section 3.3 with regard to path coordinates. It would be much more useful to directly find the inertial derivative of a vector expressed as vector components in the rotating frame. This alternative is conceptually challenging, however, because the unit vector associated with each component may be changing with time (recall that the polar frame is rotating with respect to the inertial frame). The advantage, though, is having a procedure that provides a completely general formalism for finding the time derivative of a vector expressed in a rotating frame. We do this next for the polar frame. Chapter 8 generalizes this approach to any (planar) rotating frame.

Starting with Eq. (3.25) for $\mathbf{r}_{P/O}$ in frame $\mathcal{B}$, the inertial velocity is written as

$$^{\mathcal{I}}\mathbf{v}_{P/O} = \frac{^{\mathcal{I}}d}{dt}\left(\mathbf{r}_{P/O}\right) = \frac{^{\mathcal{I}}d}{dt}(r\mathbf{e}_r) = \dot{r}\mathbf{e}_r + r\frac{^{\mathcal{I}}d}{dt}(\mathbf{e}_r),$$

where we have used the product rule of vector derivatives.

What is $\frac{^{\mathcal{I}}d}{dt}(\mathbf{e}_r)$? If we write the unit vector $\mathbf{e}_r$ as

$$\mathbf{e}_r = \cos\theta\mathbf{e}_1 + \sin\theta\mathbf{e}_2,$$

then we can use Definition 3.3 to find its derivative in $\mathcal{I}$:

$$\frac{^{\mathcal{I}}d}{dt}(\mathbf{e}_r) = -\dot{\theta}\sin\theta\mathbf{e}_1 + \dot{\theta}\cos\theta\mathbf{e}_2$$

$$= \dot{\theta}(-\sin\theta\mathbf{e}_1 + \cos\theta\mathbf{e}_2).$$

Using the vector transformation table in Section 3.4.1 to convert back to the polar frame, we find

$$\frac{^{\mathcal{I}}d}{dt}(\mathbf{e}_r) = \dot{\theta}\mathbf{e}_\theta \triangleq {}^{\mathcal{I}}\omega^{\mathcal{B}}\mathbf{e}_\theta. \tag{3.28}$$

Here we introduced an important new symbol, $^{\mathcal{I}}\omega^{\mathcal{B}} \triangleq \dot{\theta}$, to represent the *time rate of change of the orientation of frame $\mathcal{B}$ in frame $\mathcal{I}$*. The inertial velocity of P expressed as components in $\mathcal{B}$ is

$$^{\mathcal{I}}\mathbf{v}_{P/O} = \dot{r}\mathbf{e}_r + r\dot{\theta}\mathbf{e}_\theta, \tag{3.29}$$

which is the same as the velocity given in Eq. (3.26).

We can follow the same procedure to find the derivative of the $\mathbf{e}_\theta$ unit vector:

$$\frac{^{\mathcal{I}}d}{dt}(\mathbf{e}_\theta) = -\dot{\theta}\mathbf{e}_r \triangleq -^{\mathcal{I}}\omega^{\mathcal{B}}\mathbf{e}_r. \tag{3.30}$$

Eqs. (3.28) and (3.30) can be written more compactly by introducing a vector quantity—the *angular velocity*—defined as

$$\boxed{^{\mathcal{I}}\boldsymbol{\omega}^{\mathcal{B}} \triangleq {}^{\mathcal{I}}\omega^{\mathcal{B}}\mathbf{e}_3,} \tag{3.31}$$

where $\mathbf{e}_3$ is a unit vector mutually perpendicular to $\mathbf{e}_r$ and $\mathbf{e}_\theta$, forming a right-hand set. We call an angular velocity with a fixed inertial direction a *simple angular velocity*.

Definition 3.4 A **simple angular velocity** is a vector aligned with the common, fixed axis of rotation when a reference frame rotates relative to another reference frame about a direction fixed in both frames. Typically that axis is one of the unit vector directions of both frames. The magnitude of the simple angular velocity is the rate of rotation about the axis.

The angular velocity points along the axis of rotation of frame $\mathcal{B}$. Its magnitude is the instantaneous rate of rotation about that axis. Because we are only treating planar motion in this chapter, the angular velocity takes a particularly simple form— it is always directed along the axis perpendicular to the plane containing the particle motion. We postpone until Chapter 10 the more general three-dimensional treatment.

Inspection of Eqs. (3.28) and (3.30) reveals that the angular velocity in Eq. (3.31) can be used to write the following compact expressions for the inertial derivative of the unit vectors of the polar frame:

$$\boxed{\begin{array}{rcl} \dfrac{^{\mathcal{I}}d}{dt}(\mathbf{e}_r) & = & {}^{\mathcal{I}}\boldsymbol{\omega}^{\mathcal{B}} \times \mathbf{e}_r \\[2mm] \dfrac{^{\mathcal{I}}d}{dt}(\mathbf{e}_\theta) & = & {}^{\mathcal{I}}\boldsymbol{\omega}^{\mathcal{B}} \times \mathbf{e}_\theta. \end{array}} \tag{3.32}$$

You can check these formulas by substituting Eq. (3.31) into Eq. (3.32) and comparing the result to Eqs. (3.28) and (3.30). The formulas in Eq. (3.32) are extremely important and are used throughout the book. MEMORIZE THEM!

Eq. (3.32) is true for any unit vector fixed in a frame $\mathcal{B}$ that is rotating in another frame $\mathcal{I}$,

$$\frac{^{\mathcal{I}}d}{dt}\mathbf{b} = {}^{\mathcal{I}}\boldsymbol{\omega}^{\mathcal{B}} \times \mathbf{b},$$

where $\mathbf{b}$ is a unit vector fixed in $\mathcal{B}$. One helpful way to remember this formula is that the superscript to the left of $\boldsymbol{\omega}$ is the same as the superscript on the derivative to the

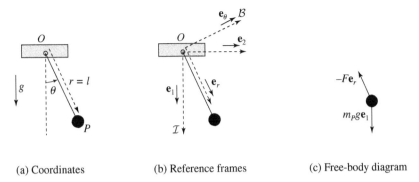

(a) Coordinates   (b) Reference frames   (c) Free-body diagram

Figure 3.18 Solving the pendulum using polar coordinates in the polar frame.

left; it is the frame in which $\mathcal{B}$ is rotating. The superscript to the right of $\boldsymbol{\omega}$ is the frame in which the unit vector is fixed (not rotating).

Using Eqs. (3.29) and (3.32), the acceleration of point P can be computed with respect to reference frame $\mathcal{I}$ and expressed as components in the polar frame much more easily than before (see if you can do this):

$$^{\mathcal{I}}\mathbf{a}_{P/O} = (\ddot{r} - r\dot{\theta}^2)\mathbf{e}_r + (2\dot{r}\dot{\theta} + r\ddot{\theta})\mathbf{e}_\theta. \tag{3.33}$$

Again, this result is the same as in Eq. (3.27), but we obtained it much more directly.

Remember that we are computing the time derivative of a vector with respect to a reference frame. In this case, it is the rate of change of the vector with respect to the inertial frame. The only difference between Eqs. (3.18) and (3.19) and Eqs. (3.29) and (3.33) is the frame in which we have expressed the components of the respective vectors. Although this distinction may seem minor, it makes solving many problems much more straightforward. To demonstrate, we solve for the equations of motion of the simple pendulum once again.

Example 3.11 Solving the Pendulum Using Polar Coordinates and the Polar Frame

The polar frame $\mathcal{B}$ illustrated in Figure 3.18b has angular velocity

$$^{\mathcal{I}}\boldsymbol{\omega}^{\mathcal{B}} = \dot{\theta}\mathbf{e}_3.$$

We relate frame $\mathcal{B}$ to the inertial frame $\mathcal{I}$ by using the polar angle θ to generate the transformation table

	$\mathbf{e}_r$	$\mathbf{e}_\theta$
$\mathbf{e}_1$	$\cos\theta$	$-\sin\theta$
$\mathbf{e}_2$	$\sin\theta$	$\cos\theta$.

We use this table to convert each unit vector from one reference frame to the other:

$$\mathbf{e}_1 = \cos\theta\mathbf{e}_r - \sin\theta\mathbf{e}_\theta \tag{3.34}$$

$$\mathbf{e}_2 = \sin\theta\mathbf{e}_r + \cos\theta\mathbf{e}_\theta \tag{3.35}$$

and

$$\mathbf{e}_r = \cos\theta\mathbf{e}_1 + \sin\theta\mathbf{e}_2 \tag{3.36}$$

$$\mathbf{e}_\theta = -\sin\theta\mathbf{e}_1 + \cos\theta\mathbf{e}_2. \tag{3.37}$$

The position $\mathbf{r}_{P/O}$ of the pendulum bob P with respect to O is

$$\mathbf{r}_{P/O} = l\mathbf{e}_r. \tag{3.38}$$

Taking the time derivative with respect to the inertial frame of the position in Eq. (3.38) and using Eq. (3.32) gives the inertial velocity of P expressed as vector components in the polar frame:

$$^{\mathcal{I}}\mathbf{v}_{P/O} = \frac{^{\mathcal{I}}d}{dt}(l\mathbf{e}_r) = l^{\mathcal{I}}\boldsymbol{\omega}^{\mathcal{B}} \times \mathbf{e}_r = l\dot\theta\mathbf{e}_\theta. \tag{3.39}$$

Similarly, taking the time derivative with respect to the inertial frame of the velocity in Eq. (3.39), using Eq. (3.32) and the product rule, gives the inertial acceleration of P,

$$^{\mathcal{I}}\mathbf{a}_{P/O} = \frac{^{\mathcal{I}}d}{dt}(l\dot\theta\mathbf{e}_\theta) = \frac{d}{dt}(l\dot\theta)\mathbf{e}_\theta + l\dot\theta^{\mathcal{I}}\boldsymbol{\omega}^{\mathcal{B}} \times \mathbf{e}_\theta = l\ddot\theta\mathbf{e}_\theta - l\dot\theta^2\mathbf{e}_r. \tag{3.40}$$

Next, using Eq. (3.40) and the free-body diagram in Figure 3.18c, we write down Newton's second law for the pendulum bob:

$$-F\mathbf{e}_r + m_P g\mathbf{e}_1 = m_P(l\ddot\theta\mathbf{e}_\theta - l\dot\theta^2\mathbf{e}_r). \tag{3.41}$$

To derive the equations of motion, it is easiest—in this case—to equate vector components in the polar frame. We thus need to write Eq. (3.41) entirely as vector components in the polar frame. Using Eqs. (3.34) and (3.41), we obtain

$$-F\mathbf{e}_r + m_P g(\cos\theta\mathbf{e}_r - \sin\theta\mathbf{e}_\theta) = m_P(l\ddot\theta\mathbf{e}_\theta - l\dot\theta^2\mathbf{e}_r). \tag{3.42}$$

Eq. (3.42) is equivalent to the two scalar equations

$$-F + m_P g\cos\theta = -m_P l\dot\theta^2 \tag{3.43}$$

$$-m_P g\sin\theta = m_P l\ddot\theta. \tag{3.44}$$

Simplifying Eq. (3.44), we once again obtain the equation of motion for the pendulum bob in polar coordinates (see Eq. (3.23)),

$$\ddot\theta = -\frac{g}{l}\sin\theta.$$

As in Example 3.9, we can also solve for the force in the rod using Eq. (3.43) to verify that the pendulum can support this constraint force.

Example 3.12 A Particle on a Spring Free to Move in the Plane

Consider a simple two-degree-of-freedom problem in which a particle of mass m_P connected to the origin via a linear spring is free to move anywhere in the plane

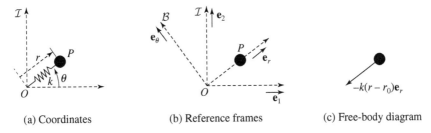

(a) Coordinates   (b) Reference frames   (c) Free-body diagram

Figure 3.19 A two-degree-of-freedom problem consisting of a mass on a spring free to move in the plane.

(Figure 3.19). From the free-body diagram (Figure 3.19c), we see that the spring applies a force only when it is stretched radially. Thus polar coordinates are a sensible choice to describe the particle's position in the plane. By using a polar frame, the force on the particle is easily written as

$$\mathbf{F}_P = -k(r - r_0)\mathbf{e}_r,$$

where k is the spring constant and r_0 is the unstretched length of the spring. The acceleration of the particle in polar coordinates using the polar frame is given by Eq. (3.33). Newton's second law then becomes

$$^{\mathcal{I}}\mathbf{a}_{P/O} = (\ddot{r} - r\dot{\theta}^2)\mathbf{e}_r + (2\dot{r}\dot{\theta} + r\ddot{\theta})\mathbf{e}_\theta = -\frac{k}{m_P}(r - r_0)\mathbf{e}_r. \qquad (3.45)$$

Eq. (3.45) is equivalent to the two scalar equations

$$\ddot{r} + \frac{k}{m_P}(r - r_0) - r\dot{\theta}^2 = 0 \qquad (3.46)$$

$$\ddot{\theta} + 2\frac{\dot{r}}{r}\dot{\theta} = 0. \qquad (3.47)$$

These two scalar equations can be integrated numerically for various initial conditions to find the trajectory of the pendulum described by $(r, \theta)_{\mathcal{I}}$.

Two special analytical solutions stand out. If the particle is given no initial speed in the $\mathbf{e}_\theta$ direction, so that $\dot{\theta}(0) = 0$, then $\ddot{\theta} = 0$ for all time and the particle undergoes simple harmonic motion in the radial direction with angle $\theta(0)$; it is the same solution as the important one-degree-of-freedom problem studied in Tutorial 2.3.

There is a second interesting solution with $\dot{\theta} = $ constant for all time: the particle travels in a circle at constant angular velocity. Setting $\ddot{r} = \dot{r} = \ddot{\theta} = 0$ in Eqs. (3.46) and (3.47), the condition for circular motion is

$$\frac{k}{m_P}(r - r_0) = r\dot{\theta}(0)^2.$$

This equation can be solved for the steady-state radial offset of the particle,

$$r_{ss} = \frac{\frac{k}{m_P} r_0}{\left(\frac{k}{m_P} - \dot{\theta}(0)^2 \right)}.$$

Finally, the equations of motion (Eqs. (3.46) and (3.47)) can be simplified by recognizing that Eq. (3.47) is a perfect differential,

$$\frac{d}{dt} \left(r^2 \dot{\theta} \right) = 0,$$

or

$$r^2 \dot{\theta} = h_O, \tag{3.48}$$

where h_O is some constant depending on initial conditions. Eq. (3.48) lets us rewrite Eq. (3.46) as

$$\ddot{r} + \frac{k}{m_P}(r - r_0) - \frac{h_O^2}{r^3} = 0.$$

Making this observation reduces the two-degree-of-freedom problem to a single differential equation in r, thus simplifying the analysis. Once the radial trajectory is found, the angular trajectory $\theta(t)$ is found from Eq. (3.48). Such situations are further discussed in Chapter 4.

The Path Frame

A rotating frame is especially helpful when finding the kinematics of a point using path coordinates. Path coordinates are normally used when we are not interested in the location of a point relative to an origin but instead want to know how its velocity and acceleration vary along a predefined trajectory (e.g., to determine whether the acceleration of a car traveling on a curved track leads to skidding). This motivates us to define the *path frame,* whose origin is the point itself.

We define the path frame as $\mathcal{B} = (P, \mathbf{e}_t, \mathbf{e}_n, \mathbf{e}_3)$. The *tangent* unit vector $\mathbf{e}_t$ is directed along the (instantaneous) velocity $^{\mathcal{I}}\mathbf{v}_{P/O}$ and is thus always tangent to the path. The *normal* unit vector $\mathbf{e}_n$ is perpendicular (normal) to the path (Figure 3.20). We use the convention that $\mathbf{e}_n$ always points inward from the path curvature. That is, the unit vector $\mathbf{e}_3 = \mathbf{e}_t \times \mathbf{e}_n$ points out of the plane of the page when the path turns counterclockwise, and it points into that plane when the path turns clockwise. In a moment we will see how this convention is established by the instantaneous angular velocity $^{\mathcal{I}}\boldsymbol{\omega}^{\mathcal{B}}$.

The inertial velocity of point P written as components in $\mathcal{B}$ is

$$\boxed{^{\mathcal{I}}\mathbf{v}_{P/O} = \dot{s}\mathbf{e}_t \triangleq v\mathbf{e}_t.}$$

Note that we have kept the O in the subscript here, to represent the origin of the frame in which the path is defined.

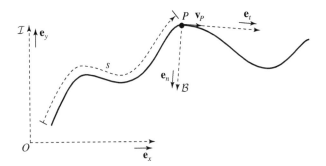

Figure 3.20 Path coordinate s measures the position of point P along an arbitrary path in inertial frame $\mathcal{I}$. The path frame is $\mathcal{B} = (P, \mathbf{e}_t, \mathbf{e}_n, \mathbf{e}_3)$, where $\mathbf{e}_3$ is into the page at the instant shown.

The acceleration is then found the usual way by differentiating the velocity with respect to the inertial frame. We obtain

$$^{\mathcal{I}}\mathbf{a}_{P/O} = \dot{v}\mathbf{e}_t + v\frac{^{\mathcal{I}}d}{dt}(\mathbf{e}_t). \tag{3.49}$$

We are again faced with having to find the derivative of a unit vector. To find $\frac{^{\mathcal{I}}d}{dt}(\mathbf{e}_t)$, observe that the path frame $\mathcal{B}$ behaves just like the polar frame—or any frame that rotates with respect to $\mathcal{I}$. That is, as the point P moves a small amount ds in a small time dt, the frame $\mathcal{B}$ rotates a small amount $d\theta$ due to the curvature of the path. At that instant, the frame is rotating with respect to $\mathcal{I}$ with angular velocity $^{\mathcal{I}}\boldsymbol{\omega}^{\mathcal{B}} \triangleq \dot{\theta}\mathbf{e}_3 = \frac{d\theta}{dt}\mathbf{e}_3$. The time rate of change of the unit vector $\mathbf{e}_t$ is thus given by Eq. (3.32) with $\mathbf{e}_r$ replaced by $\mathbf{e}_t$:

$$\frac{^{\mathcal{I}}d}{dt}(\mathbf{e}_t) = {}^{\mathcal{I}}\boldsymbol{\omega}^{\mathcal{B}} \times \mathbf{e}_t = \dot{\theta}\mathbf{e}_n.$$

As mentioned above, the direction of $\mathbf{e}_n$ depends on the path curvature. But what is the relationship between the path curvature and $\dot{\theta}$? We find this relationship by approximating the path as a series of circular segments. An infinitesimal displacement ds along the path can be modeled by an infinitesimal arc of a circle, as illustrated in Figure 3.21. The angle subtended by the circular arc ds in time dt is

$$d\theta = \frac{ds}{R_c},$$

where R_c is the *radius of curvature* of the path at that point.

Dividing by dt, we find

$$\frac{d\theta}{dt} = \dot{\theta} = \frac{1}{R_c}\frac{ds}{dt} = \frac{v}{R_c},$$

where $v \triangleq \dot{s}$ is the (instantaneous) speed along the path. Remember that R_c is not a constant but varies along the path as the curvature changes—it is a property of the path geometry. Substituting this result into Eq. (3.49) allows us to write the inertial

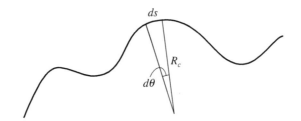

Figure 3.21 The radius of curvature of a small segment ds of the path.

acceleration of P in terms of its instantaneous speed along the path and the local radius of curvature:

$$^{\mathcal{I}}\mathbf{a}_{P/O} = \dot{v}\mathbf{e}_t + \frac{v^2}{R_c}\mathbf{e}_n.$$

The final result needed is a relationship between the radius of curvature and the path geometry. If the path can be defined by a function $y(x)$ relating the Cartesian coordinates x and y, then the radius of curvature of the path is

$$R_c = \frac{\left[1 + \left(\frac{dy}{dx}\right)^2\right]^{\frac{3}{2}}}{\left|\frac{d^2y}{dx^2}\right|}. \tag{3.50}$$

You have an opportunity to derive this relationship in Problem 3.13.

Example 3.13 Charged Particle in a Magnetic Field

This example uses the path frame $\mathcal{B} = (P, \mathbf{e}_t, \mathbf{e}_n, \mathbf{e}_3)$ to find the equations of motion of a charged particle in a uniform and constant magnetic field (Figure 3.22). Let $q \neq 0$ denote the electrical charge of particle P. Orient the reference frame such that the magnetic field is given by $B\mathbf{e}_z$,[5] where $B > 0$ and $\mathcal{I} = (O, \mathbf{e}_x, \mathbf{e}_y, \mathbf{e}_z)$ is an inertial frame. We define $\mathcal{B}$ so that $\mathbf{e}_3 = \mathbf{e}_z$ if the particle is turning counterclockwise, and $\mathbf{e}_3 = -\mathbf{e}_z$ if the particle is turning clockwise. According to the laws of electromagnetism, the force on the particle is given by the *Lorentz force*

$$\mathbf{F}_P = q\,^{\mathcal{I}}\mathbf{v}_{P/O} \times B\mathbf{e}_z.$$

The kinematics of P in terms of vector components in $\mathcal{B}$ are

$$^{\mathcal{I}}\mathbf{v}_{P/O} = \dot{s}\mathbf{e}_t$$

$$^{\mathcal{I}}\mathbf{a}_{P/O} = \ddot{s}\mathbf{e}_t + \frac{\dot{s}^2}{R_c}\mathbf{e}_n.$$

---

[5] A magnetic field is a *vector field*, that is, at each point in space, the local magnetism has both magnitude and direction.

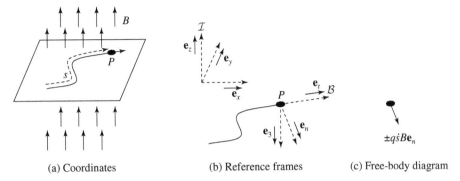

(a) Coordinates   (b) Reference frames   (c) Free-body diagram

Figure 3.22 Charged particle in a magnetic field.

Using Newton's second law, we have

$$m_P \left(\ddot{s}\mathbf{e}_t + \frac{\dot{s}^2}{R_c}\mathbf{e}_n \right) = q\dot{s}\mathbf{e}_t \times (\pm B\mathbf{e}_3) = \pm q\dot{s}B\mathbf{e}_n.$$

We obtain the two scalar equations

$$\ddot{s} = 0$$

$$\dot{s} = \pm \frac{qBR_c}{m_P}.$$

The first equation—the equation of motion—reveals that the speed of the particle is constant: $\dot{s}(t) = \dot{s}(0) > 0$. Using the second equation, we compute the radius of curvature

$$R_c = \frac{m_P}{|q|B}\dot{s}(0) > 0,$$

where, by convention, we adopt the positive solution. Thus a charged particle in a magnetic field travels in a circle in a plane perpendicular to the direction of the field. The radius of the circle is proportional to the mass and initial speed of the particle and is inversely proportional to the magnitude of the electrical charge and the magnitude of the magnetic field. The direction of rotation depends on the charge of the particle. For example, a negatively charged particle rotates in the direction of the fingers of your right hand when you curl them so that your thumb points along the magnetic field vector $B\mathbf{e}_z$.

3.5 The Polar Frame and Fictional Forces

It is worthwhile to pause at this point to discuss in more depth the formula for acceleration in Eq. (3.33) (repeated here for convenience):

$$^\mathcal{I}\mathbf{a}_{P/O} = (\ddot{r} - r\dot{\theta}^2)\mathbf{e}_r + (2\dot{r}\dot{\theta} + r\ddot{\theta})\mathbf{e}_\theta.$$

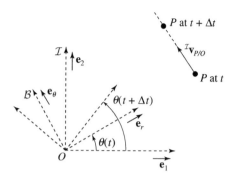

Figure 3.23 Point P traveling in a straight line in absolute space with velocity $^{\mathcal{I}}\mathbf{v}_{P/O}$ and the rotating polar frame $\mathcal{B}$.

This expression often causes great confusion. When expressed in polar coordinates and the polar frame, the acceleration is no longer a simple second-derivative of a scalar coordinate—it includes two additional terms, one in each of the polar-frame directions. It is helpful to discuss the physical meaning of Eq. (3.33) and how it results from the geometry of motion.

First, remember that the acceleration $^{\mathcal{I}}\mathbf{a}_{P/O}$ in Eq. (3.33) is still the acceleration of point P with respect to absolute space. We have just used polar coordinates rather than Cartesian coordinates and expressed it as vector components in the polar frame. Using polar coordinates and the polar frame does not change the fundamental nature of the acceleration. Consider, for example, the simple situation of a point P traveling at a constant velocity in the inertial frame $\mathcal{I}$ with origin O, as shown in Figure 3.23. The inertial acceleration of P is zero, of course, and its velocity could be written quite simply in Cartesian coordinates as $\dot{x}\mathbf{e}_1 + \dot{y}\mathbf{e}_2$, where $(\dot{x},\ \dot{y})_{\mathcal{I}}$ are both constant. The second derivatives of the Cartesian coordinates are zero.

Now use polar coordinates $(r,\ \theta)_{\mathcal{I}}$ to locate P and add a polar frame $\mathcal{B} = (O,\ \mathbf{e}_r,\ \mathbf{e}_\theta,\ \mathbf{e}_3)$, where the $\mathbf{e}_r$ unit vector points at P (Figure 3.23). As P travels along a constant-velocity straight-line path in absolute space, its radial position r increases, and $\mathcal{B}$ must rotate for $\mathbf{e}_r$ to continue pointing at P. Since the inertial acceleration is zero, Newton's second law becomes

$$\mathbf{F}_P = 0 = m_P(\ddot{r} - r\dot{\theta}^2)\mathbf{e}_r + m_P(2\dot{r}\dot{\theta} + r\ddot{\theta})\mathbf{e}_\theta. \tag{3.51}$$

Since the magnitude of each vector component in Eq. (3.51) equals zero, the second derivatives $\ddot{r}$ and $\ddot{\theta}$ of the polar coordinates are not zero!

We call the acceleration-like term $-r\dot{\theta}^2\mathbf{e}_r$ the *centripetal acceleration* and $2\dot{r}\dot{\theta}\mathbf{e}_\theta$ the *Coriolis acceleration*. For instance, Eq. (3.51) can be written as the two scalar equations

$$\ddot{r} = r\dot{\theta}^2$$

$$\ddot{\theta} = -2\frac{\dot{r}\dot{\theta}}{r}.$$

These equations look a lot like Newton's second law for polar coordinates. As a result, the terms on the right are frequently called *fictional forces*. We want to emphasize that these are kinematic terms and not forces. No forces are actually acting; they just look like forces from the way they appear in these equations.

What is going on here? Cartesian coordinates really are special when it comes to dynamics. Only in Cartesian coordinates do we expect to see simple expressions of Newton's second law in terms of the second derivative of each coordinate. In other coordinate systems, the scalar derivatives are coupled and there may be extra terms. The appearance of these terms is a purely kinematic effect.

In Problem 3.9 you have the opportunity to explore this simple example a bit further, solving for $\ddot{r}$ and $\ddot{\theta}$. You will find that the polar angle θ approaches an asymptotic value as the radial unit vector $\mathbf{e}_r$ gradually aligns with the particle's path.

In the remainder of this section we further explore the kinematic source of these two terms in the accelerations.

3.5.1 Centripetal Acceleration

Consider next a slightly different scenario. Here point P is traveling on a circle of radius r at angular rate $\dot{\theta}$, as shown in Figure 3.24. In other words, the position of P is fixed in a rotating polar frame $\mathcal{B} = (O, \mathbf{e}_r, \mathbf{e}_\theta, \mathbf{e}_3)$ whose origin O is at the center of the circle. The fact that the radius r is constant implies $\ddot{r} = \dot{r} = 0$. The velocity $^{\mathcal{I}}\mathbf{v}_{P/O}$ at time t is tangent to the circle and has magnitude $r\dot{\theta}(t)$. At some short time Δt later, the velocity, while still of the same magnitude, has changed by a small amount $^{\mathcal{I}}\Delta\mathbf{v}_{P/O} \triangleq \,^{\mathcal{I}}\mathbf{v}_{P/O}(t + \Delta t) - \,^{\mathcal{I}}\mathbf{v}_{P/O}(t)$. Any time the velocity changes magnitude or direction relative to a particular frame, there must be, by definition, an acceleration relative to that frame. The inertial acceleration $^{\mathcal{I}}\Delta\mathbf{v}_{P/O}/\Delta t$ accounts for the change in direction of the velocity. This acceleration is exactly equal to the centripetal acceleration $-r\dot{\theta}^2\mathbf{e}_r$, which points inward toward O.

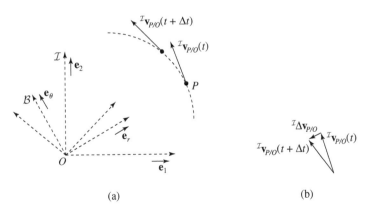

(a)   (b)

Figure 3.24 (a) Point P is traveling in a circle in the inertial frame $\mathcal{I}$. Polar frame $\mathcal{B}$ stays aligned with the vector toward P. (b) Over a short time interval the velocity changes by an amount $^{\mathcal{I}}\Delta\mathbf{v}_{P/O} \triangleq \,^{\mathcal{I}}\mathbf{v}_{P/O}(t + \Delta t) - \,^{\mathcal{I}}\mathbf{v}_{P/O}(t)$ in the radial direction, resulting in centripetal acceleration.

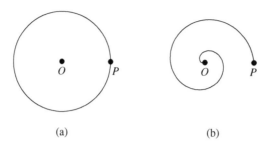

(a)   (b)

Figure 3.25 Particle trajectories in the inertial frame. (a) Centripetal force with $\dot{r}(0) = 0$ yields circular motion in the inertial frame. (b) Centripetal and Coriolis forces with $\dot{r}(0) < 0$ yield spiral motion in the inertial frame.

Since the velocity of P is changing as it travels in a circle, there must be a force accelerating P with respect to absolute space. From Eq. (3.33), that force satisfies

$$\mathbf{F}_P \cdot \mathbf{e}_r = m_P(\ddot{r} - r\dot{\theta}^2) = -m_P r\dot{\theta}^2. \tag{3.52}$$

The radial force component in Eq. (3.52) is called a *centripetal force*. The circular trajectory of a particle acted on by a constant centripetal force is shown in Figure 3.25a.

3.5.2 Coriolis Acceleration

Now consider particle P traveling around O at constant angular rate $\dot{\theta}$ and moving radially inward at constant rate $\dot{r} < 0$. In the polar frame, the particle will move in a straight line at constant speed toward the origin. In absolute space, however, the particle trajectory is a spiral, as shown in Figure 3.25b. For the particle to follow such a trajectory, clearly a force must be acting. Newton's second law tells us that an additional force must be acting in the $\mathbf{e}_\theta$ direction:

$$\mathbf{F}_P \cdot \mathbf{e}_\theta = m_P(2\dot{r}\dot{\theta} + r\ddot{\theta}) = 2m_P \dot{r}\dot{\theta}. \tag{3.53}$$

The tangential force component in Eq. (3.53) is called the *Coriolis force*.

The implications of the Coriolis force are more subtle than those of the centripetal force and thus tend to cause more confusion. Physically, the Coriolis acceleration arises from two sources (hence the factor of 2 in Eq. (3.33)). The first source comes from the fact that, as the point P moves inward (outward), it must move slower (faster) in the $\mathbf{e}_\theta$ direction relative to absolute space to continue rotating at a constant angular rate $\dot{\theta}$. Thus over a small time period, its speed decreases (increases), hence the acceleration. The second source arises because, over this small time interval, the direction of the inertial velocity changes as the frame $\mathcal{B}$ rotates with respect to $\mathcal{I}$. Each source contributes a factor of $\dot{r}\dot{\theta}$ to the acceleration.

We revisit centripetal and Coriolis accelerations in Chapter 8 when discussing motion relative to a rotating frame.

3.6 An Introduction to Relative Motion

Another situation where we need to introduce additional frames of reference is when describing the motion of a particle relative to a moving platform. Rather than use scalar coordinates to specify the position of P in a fixed inertial frame, we locate the particle in a moving frame $\mathcal{B}$. This frame can be either translating or rotating. For instance, many problems on the surface of the earth must account for the fact that a reference frame fixed to the earth is both rotating and translating (because of its orbit about the sun) in inertial space.

For most problems involving relative motion (e.g., an accelerating or rotating frame), we cannot employ Newton's second law in a moving frame, even though we may want to solve for trajectories in that frame; remember, Newton's second law works only when the acceleration is relative to absolute space (i.e., an inertial frame). We need to be more careful. This section introduces the problem of relative motion by solving for the equations of motion of a particle relative to a reference frame translating (but not rotating) in inertial space. One outcome is the observation that all reference frames traveling at constant velocity are inertial. The more complicated situation of a particle moving relative to a frame both translating and rotating is deferred to Chapter 8.

Consider the situation depicted in Figure 3.26. Frame $\mathcal{I} = (O, \mathbf{e}_1, \mathbf{e}_2, \mathbf{e}_3)$ is a fixed inertial frame while frame $\mathcal{B} = (O', \mathbf{e}_x, \mathbf{e}_y, \mathbf{e}_z)$ is translating relative to $\mathcal{I}$ with acceleration $^{\mathcal{I}}\mathbf{a}_{O'/O}$. That is, we have

$$^{\mathcal{I}}\mathbf{v}_{O'/O} = \frac{^{\mathcal{I}}d}{dt}\left(\mathbf{r}_{O'/O}\right)$$

$$^{\mathcal{I}}\mathbf{a}_{O'/O} = \frac{^{\mathcal{I}}d}{dt}\left(^{\mathcal{I}}\mathbf{v}_{O'/O}\right).$$

Frame $\mathcal{B}$ is not rotating with respect to $\mathcal{I}$.

Consider the motion of point P in Figure 3.26 with position $\mathbf{r}_{P/O'}$ in $\mathcal{B}$. Using Cartesian coordinates $(x, y)_\mathcal{B}$ to describe the position of P in $\mathcal{B}$, we have

$$\mathbf{r}_{P/O'} = x\mathbf{e}_x + y\mathbf{e}_y$$

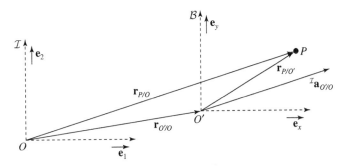

Figure 3.26 Translating reference frame $\mathcal{B} = (O', \mathbf{e}_x, \mathbf{e}_y, \mathbf{e}_z)$ accelerates in $\mathcal{I} = (O, \mathbf{e}_1, \mathbf{e}_2, \mathbf{e}_3)$ by $^{\mathcal{I}}\mathbf{a}_{O'/O}$.

or in matrix notation,

$$[\mathbf{r}_{P/O'}]_\mathcal{B} = \begin{bmatrix} x \\ y \end{bmatrix}_\mathcal{B}.$$

Our goal is to find the equations of motion for scalar coordinates $(x, y)_\mathcal{B}$ describing the trajectory of P in $\mathcal{B}$ (of course, any other scalar coordinates would work as well).

Since $\mathcal{B}$ may not be inertial, we cannot simply use Newton's second law on ${}^\mathcal{B}\mathbf{a}_{P/O'}$. Instead we must first find the motion of P relative to $\mathcal{I}$. Using vector geometry, the position of P relative to O is

$$\mathbf{r}_{P/O} = \mathbf{r}_{O'/O} + \mathbf{r}_{P/O'}. \tag{3.54}$$

From the vector triad in Eq. (3.54) we have

$${}^\mathcal{I}\mathbf{v}_{P/O} = \frac{{}^\mathcal{I}d}{dt}\left(\mathbf{r}_{O'/O}\right) + \frac{{}^\mathcal{I}d}{dt}\left(\mathbf{r}_{P/O'}\right) = {}^\mathcal{I}\mathbf{v}_{O'/O} + {}^\mathcal{I}\mathbf{v}_{P/O'}. \tag{3.55}$$

Assume that ${}^\mathcal{I}\mathbf{v}_{O'/O}$ is known. What is ${}^\mathcal{I}\mathbf{v}_{P/O'}$? Using the expression for $\mathbf{r}_{P/O'}$ in Cartesian coordinates in $\mathcal{B}$ yields

$${}^\mathcal{I}\mathbf{v}_{P/O'} = \frac{{}^\mathcal{I}d}{dt}(x\mathbf{e}_x + y\mathbf{e}_y).$$

We can now use the product rule on each component as before. However, if $\mathcal{B}$ is only translating, the two unit vectors $\mathbf{e}_x$ and $\mathbf{e}_y$ do not change in $\mathcal{I}$—there is no angular velocity of $\mathcal{B}$ in $\mathcal{I}$ (remember Eq. (3.32)?). Since these are unit vectors, if they don't change their magnitude or direction in $\mathcal{I}$, then they have no derivative in $\mathcal{I}$. Consequently, we find

$${}^\mathcal{I}\mathbf{v}_{P/O'} = \dot{x}\mathbf{e}_x + \dot{y}\mathbf{e}_y,$$

which implies, by the definition of the vector derivative,

$${}^\mathcal{I}\mathbf{v}_{P/O'} = {}^\mathcal{B}\mathbf{v}_{P/O'}.$$

This result may seem counterintuitive. How can the derivative of the position (the velocity) be the same relative to either frame? Hopefully, a bit of thought makes it clear. Translation of the vector $\mathbf{r}_{P/O'}$ does not change its magnitude or direction in space. Since these are the only ways a vector can change, its derivative is the same whether referred to $\mathcal{B}$ or $\mathcal{I}$. Of course, this is only true for a translating frame. Chapter 8 shows that things get more complicated in a rotating frame.

We now return to the original problem of finding the inertial acceleration of P in $\mathcal{I}$ so that Newton's second law can be applied. Take the derivative with respect to time of Eq. (3.55) to obtain (using ${}^\mathcal{I}\mathbf{a}_{P/O'} = {}^\mathcal{B}\mathbf{a}_{P/O'}$)

$${}^\mathcal{I}\mathbf{a}_{P/O} = {}^\mathcal{I}\mathbf{a}_{O'/O} + {}^\mathcal{B}\mathbf{a}_{P/O'} = {}^\mathcal{I}\mathbf{a}_{O'/O} + \ddot{x}\mathbf{e}_x + \ddot{y}\mathbf{e}_y. \tag{3.56}$$

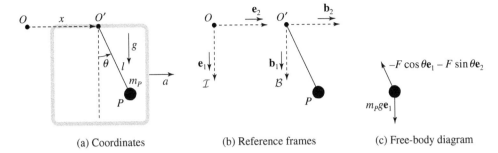

(a) Coordinates   (b) Reference frames   (c) Free-body diagram

Figure 3.27 Pendulum in an accelerating box.

The acceleration of P relative to $\mathcal{I}$ is the sum of the acceleration of the origin of frame $\mathcal{B}$ relative to $\mathcal{I}$ and that of P relative to $\mathcal{B}$. We can use Eq. (3.56) to find the variant of Newton's second law that determines the motion of P in $\mathcal{B}$. Multiplying Eq. (3.56) by the mass m_P and rearranging yields

$$m_P{}^{\mathcal{B}}\mathbf{a}_{P/O'} = \mathbf{F}_P - m_P{}^{\mathcal{I}}\mathbf{a}_{O'/O}, \tag{3.57}$$

where we have substituted $\mathbf{F}_P$ for $m_P{}^{\mathcal{I}}\mathbf{a}_{P/O}$ using Newton's second law.

Thus to find the equations of motion for a particle relative to a translating frame, we have to subtract the acceleration of the translating frame from the right side of Newton's second law. If frame $\mathcal{B}$ is moving at a constant velocity, then ${}^{\mathcal{I}}\mathbf{a}_{P/O} = {}^{\mathcal{B}}\mathbf{a}_{P/O'}$ and it makes no difference whether we write Newton's second law in frame $\mathcal{I}$ or frame $\mathcal{B}$. This scenario is a demonstration of *Newtonian relativity*: *Every nonrotating frame moving at constant velocity in absolute space is an inertial frame.*

Example 3.14 Pendulum in an Accelerating Box

Consider again the simple pendulum, but now let the attachment point be fixed to a box that is accelerating in the inertial $\mathbf{e}_2$ direction (Figure 3.27). The goal is to find the equation of motion for the pendulum relative to a frame fixed to the box. We will also see that the steady-state equilibrium angle of the pendulum is offset from the vertical when the box is accelerating.

Figure 3.27a shows the coordinates describing the motion. As in the stationary pendulum, θ is the angle of the pendulum from the vertical. Let x denote the position of the attachment point O' with respect to origin O of the inertial frame. The velocity and acceleration of the box are thus $\dot{x}\mathbf{e}_2$ and $\ddot{x}\mathbf{e}_2 \triangleq a\mathbf{e}_2$, respectively. We wish to find the equation of motion for θ, the angle of the pendulum inside the box. To do so, we will use Eq. (3.57), the relative-motion form of Newton's second law.

The position of pendulum bob P in moving frame $\mathcal{B} = (O', \mathbf{b}_1, \mathbf{b}_2, \mathbf{b}_3)$ is

$$\mathbf{r}_{P/O'} = l\cos\theta\mathbf{b}_1 + l\sin\theta\mathbf{b}_2.$$

Following the approach of Example 3.9, we find the velocity and acceleration of the pendulum bob in the frame fixed to the box:

$$^B\mathbf{v}_{P/O'} = \frac{^Bd}{dt}\left(\mathbf{r}_{P/O'}\right) = -l\dot{\theta}\sin\theta\mathbf{e}_1 + l\dot{\theta}\cos\theta\mathbf{e}_2$$

$$^B\mathbf{a}_{P/O'} = \frac{^Bd}{dt}\left(^\mathcal{I}\mathbf{v}_{P/O'}\right) = (-l\ddot{\theta}\sin\theta - l\dot{\theta}^2\cos\theta)\mathbf{e}_1 + (l\ddot{\theta}\cos\theta - l\dot{\theta}^2\sin\theta)\mathbf{e}_2.$$

Now use Eq. (3.57) and the free-body diagram in Figure 3.27c to obtain a modification of the two scalar equations in Example 3.9,

$$-F\sin\theta - m_P a = m_P(l\ddot{\theta}\cos\theta - l\dot{\theta}^2\sin\theta)$$

$$-F\cos\theta + m_P g = m_P(-l\ddot{\theta}\sin\theta - l\dot{\theta}^2\cos\theta).$$

By carefully considering the acceleration of the box in inertial space, we have an additional term in the equations of motion. If the velocity of the box were constant ($a = 0$), then these equations would be identical to Eqs. (3.21) and (3.22) in Example 3.9 because a frame moving at constant velocity is an inertial frame.

These equations can be manipulated as before to find the equation of motion for θ:

$$\ddot{\theta} = -\frac{g}{l}\sin\theta - \frac{a}{l}\cos\theta. \tag{3.58}$$

The equation of motion is more complicated because of the acceleration of the box. There are two situations of interest. First, like the simple pendulum in Example 3.9, we seek the equilibrium points for constant θ. To find the equilibrium angle θ_0, set $\ddot{\theta} = 0$ in Eq. (3.58) and solve for θ_0:

$$\theta_0 = -\arctan\left(\frac{a}{g}\right).$$

Thus for a box accelerating at a, the pendulum hangs to the rear by an angle θ_0, as shown in Figure 3.28. This example is a model of a *pendulous accelerometer*—a device used to measure acceleration. (It also describes well what happens to something hanging from your car's rear-view mirror when you accelerate.)

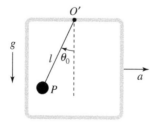

Figure 3.28 Equilibrium angle of a hanging pendulum in an accelerating box.

We can also examine the motion for small θ, as in Example 3.9. Using the small-angle approximation, the linear equation for the accelerating pendulum is

$$\ddot{\theta} + \frac{g}{l}\theta = -\frac{a}{l}. \tag{3.59}$$

Eq. (3.59) is the equation for a simple harmonic oscillator with constant driving force (discussed at the end of Tutorial 2.3). The solution is a sinusoidal motion about the offset angle θ_0.

3.7 How to Solve a Dynamics Problem

There are often many ways to solve a given dynamics problem. We recommend starting with a general, rigorous approach to solve most problems, no matter how hard (or easy) they appear. This approach, which is outlined below, utilizes tools from our growing toolkit—vectors, reference frames, vector derivatives, and Newton's laws. Although we may not specifically refer to the following steps in examples and tutorials, we seek to adhere to this process throughout the book.

1. **Model the system.** All dynamics problems start with a physical system to be described. The first step is to produce a model of the system that incorporates elements that can be handled mathematically, that is, point masses and/or rigid bodies. Include any springs, strings, (massless) rods, hinges, and surfaces that are relevant. There is no such thing as an "exact" model—as engineers we are always making approximations. The key is to find the right set of approximations that render the problem tractable but not trivial. Often this step is the most important step. *You should produce a drawing of the model.*

2. **Choose the coordinate(s).** There should be at least one coordinate for each degree of freedom of the system. Add the coordinates to your sketch of the model or make a new sketch of the coordinates. For all coordinates, be sure to indicate (e.g., with an arrow) the direction in which the coordinate is increasing.

3. **Define the reference frame(s).** Define the inertial frame and any additional frames that may be useful for solving the problem. Note that *sometimes you have to repeat this step!* (The frames you initially choose may not turn out to be the most useful.) For each frame, identify its origin and unit vectors. For each rotating frame, write down its angular velocity with respect to the inertial frame. If applicable, write down transformation tables among the unit vectors in the various reference frames.

4. **Draw the free-body diagram(s).** For each particle and/or rigid body, draw the free-body diagram and identify all forces and their lines of action. *Label the forces as vector components in your reference frames.*

5. **Calculate the kinematics.** Write down the position of each particle (and the position of the center of mass of each rigid body). Take the vector derivative with respect to the inertial frame of the position to obtain the inertial velocity.

Take the vector derivative with respect to the inertial frame of the velocity to obtain the inertial acceleration.

6. **Write down Newton's second law.** For each particle, write down Newton's second law in vector form using the free-body diagram and inertial acceleration. Using the transformation tables (Section 3.4.1) to convert the vector components into a common frame, *rewrite Newton's second law as a system of scalar algebraic equations.*

7. **Find the equations of motion.** Compare the number of unknown variables— forces and second-order time derivatives—to the number of equations. If necessary, write down the constraint equation(s) and differentiate with respect to time so that the number of scalar equations equals the number of unknowns. *Solve the system of algebraic equations for the unknown variables* and write down the equations of motion of the system.

8. **Integrate the equations of motion.** Integrate—analytically if you can, otherwise numerically—the equations of motion to *find the trajectories of the system.*

This list may seem to have a lot of steps, but don't despair. Once you start solving problems, these steps will become automatic. The list is meant as a general guideline and may not be applicable to every problem. However, it is important to always use rigorous (vector) notation!

3.8 Derivations—Properties of the Vector Derivative

In this section we derive the various rules of vector derivatives shown in Table 3.1 using Definition 3.3. We also use them to derive the vector Taylor series.

3.8.1 The Addition Rule

The addition rule of scalar differentiation states that the derivative of the sum of two functions is equal to the sum of their derivatives. It is easy to show that this rule also holds for vector derivatives. Consider the vector derivative in a reference frame $\mathcal{A}$ of the vector sum $\mathbf{r}(t) + \mathbf{s}(t)$,

$$\frac{^{\mathcal{A}}d}{dt}\left(\mathbf{r}(t) + \mathbf{s}(t)\right) = \frac{^{\mathcal{A}}d}{dt}\left(r_1\mathbf{a}_1 + r_2\mathbf{a}_2 + r_3\mathbf{a}_3 + s_1\mathbf{a}_1 + s_2\mathbf{a}_2 + s_3\mathbf{a}_3\right)$$

$$= \frac{^{\mathcal{A}}d}{dt}\left((r_1 + s_1)\mathbf{a}_1 + (r_2 + s_2)\mathbf{a}_2 + (r_3 + s_3)\mathbf{a}_3\right),$$

where we have used the properties of vector addition in Appendix B.
Applying Definition 3.3 yields

$$\frac{^{\mathcal{A}}d}{dt}\left(\mathbf{r}(t) + \mathbf{s}(t)\right) = \frac{d}{dt}(r_1 + s_1)\mathbf{a}_1 + \frac{d}{dt}(r_2 + s_2)\mathbf{a}_2 + \frac{d}{dt}(r_3 + s_3)\mathbf{a}_3$$

$$= (\dot{r}_1 + \dot{s}_1)\mathbf{a}_1 + (\dot{r}_2 + \dot{s}_2)\mathbf{a}_2 + (\dot{r}_3 + \dot{s}_3)\mathbf{a}_3,$$

where we have used the distributive property of the scalar derivative. Finally, we can again use the distributive property of vector addition from Appendix B and Definition 3.3 to find

$$\frac{^{A}d}{dt}(\mathbf{r}(t) + \mathbf{s}(t)) = \frac{^{A}d}{dt}\mathbf{r}(t) + \frac{^{A}d}{dt}\mathbf{s}(t), \qquad (3.60)$$

which is the addition rule.

3.8.2 The Product Rule

Recall from calculus the scalar product rule:

$$\frac{d}{dt}(f(t)g(t)) = \dot{f}(t)g(t) + f(t)\dot{g}(t).$$

For vectors, there are three possible products: the product of a vector and a scalar function of time, the scalar *dot product* of two vectors, and the vector *cross product* of two vectors.

Using the vector properties in Appendix B and the definition of the vector derivative, the derivative of a scalar times a vector is

$$\begin{aligned}
\frac{^{A}d}{dt}(f(t)\mathbf{r}(t)) &= \frac{^{A}d}{dt}\left(f(t)r_1(t)\mathbf{a}_1 + f(t)r_2(t)\mathbf{a}_2 + f(t)r_3(t)\mathbf{a}_3\right) \\
&= \frac{d}{dt}(f(t)r_1(t))\mathbf{a}_1 + \frac{d}{dt}(f(t)r_2(t))\mathbf{a}_2 + \frac{d}{dt}(f(t)r_3(t))\mathbf{a}_3.
\end{aligned}$$

From the scalar product rule we thus have

$$\begin{aligned}
\frac{^{A}d}{dt}(f(t)\mathbf{r}(t)) &= (\dot{f}(t)r_1(t) + f(t)\dot{r}_1(t))\mathbf{a}_1 + (\dot{f}(t)r_2(t) \\
&\quad + f(t)\dot{r}_2(t))\mathbf{a}_2 + (\dot{f}(t)r_3(t) + f(t)\dot{r}_3(t))\mathbf{a}_3 \\
&= \dot{f}(t)(r_1(t)\mathbf{a}_1 + r_2(t)\mathbf{a}_2 + r_3(t)\mathbf{a}_3) + f(t)(\dot{r}_1(t)\mathbf{a}_1 \\
&\quad + \dot{r}_2(t)\mathbf{a}_2 + \dot{r}_3(t)\mathbf{a}_3),
\end{aligned}$$

where the distributive property of vector addition has again been applied.

Finally, applying Definition 3.3 gives

$$\frac{^{A}d}{dt}(f(t)\mathbf{r}(t)) = \frac{df}{dt}\mathbf{r}(t) + f(t)\frac{^{A}d}{dt}\mathbf{r}(t), \qquad (3.61)$$

which is the first product rule.

The derivative of the dot product of two vectors is just a scalar derivative,

$$\frac{d}{dt}(\mathbf{r}(t)\cdot\mathbf{s}(t)) = \frac{d}{dt}\left(r_1(t)s_1(t) + r_2(t)s_2(t) + r_3(t)s_3(t)\right),$$

where we have used the definition of the dot product and expressed each vector as components in the same frame. Using the distributive property of scalar differentiation and the scalar product rule gives

$$\frac{d}{dt}(\mathbf{r}(t) \cdot \mathbf{s}(t)) = \dot{r}_1(t)s_1(t) + r_1(t)\dot{s}_1(t) + \dot{r}_2(t)s_2(t) + r_2(t)\dot{s}_2(t)$$
$$+ \dot{r}_3(t)s_3(t) + r_3(t)\dot{s}_3(t).$$

Rearranging leaves

$$\frac{d}{dt}(\mathbf{r}(t) \cdot \mathbf{s}(t)) = \dot{r}_1(t)s_1(t) + \dot{r}_2(t)s_2(t) + \dot{r}_3(t)s_3(t) + r_1(t)\dot{s}_1(t)$$
$$+ r_2(t)\dot{s}_2(t) + r_3(t)\dot{s}_3(t),$$

which, from the definition of the vector derivative and the dot product, gives the second product rule:

$$\boxed{\frac{d}{dt}(\mathbf{r}(t) \cdot \mathbf{s}(t)) = \frac{^A d}{dt}\mathbf{r}(t) \cdot \mathbf{s}(t) + \mathbf{r}(t) \cdot \frac{^A d}{dt}\mathbf{s}(t).}$$ (3.62)

Of course, because the derivative of a dot product is a scalar derivative, the frame in which we choose to take the vector derivatives is arbitrary—it just needs to be the same for the two vectors.

The derivation of the product rule for the vector cross product follows a similar approach; we leave it to the problems (Problem 3.11). For completeness we state it here:

$$\boxed{\frac{^A d}{dt}(\mathbf{r}(t) \times \mathbf{s}(t)) = \frac{^A d}{dt}\mathbf{r}(t) \times \mathbf{s}(t) + \mathbf{r}(t) \times \frac{^A d}{dt}\mathbf{s}(t).}$$ (3.63)

3.8.3 The Chain Rule

Now we consider the vector chain rule. Recall that, for a scalar function $g(f(t))$ of another scalar function $f(t)$, the derivative is

$$\frac{d}{dt}g(f(t)) = \frac{dg}{df}\frac{df}{dt}.$$

We consider three cases: a vector function of a scalar function of time, a scalar function of a vector function of time, and a scalar function of a vector function of time that has an explicit time dependence. The chain rule for the first case is a simple generalization of the scalar chain rule,

$$\boxed{\frac{^A d}{dt}\mathbf{r}(f(t)) = \frac{^A d}{df}\mathbf{r}(f(t))\frac{df}{dt}.}$$ (3.64)

The derivation again relies on Definition 3.3. We have

$$\frac{^A d}{dt}\mathbf{r}(f(t)) \triangleq \frac{d}{dt}\left[r_1(f(t))\right]\mathbf{a}_1 + \frac{d}{dt}\left[r_2(f(t))\right]\mathbf{a}_2 + \frac{d}{dt}\left[r_3(f(t))\right]\mathbf{a}_3.$$

Using the scalar chain rule on each component yields

$$\frac{^{A}d}{dt}\mathbf{r}(f(t)) = \frac{dr_1}{df}\frac{df}{dt}\mathbf{a}_1 + \frac{dr_2}{df}\frac{df}{dt}\mathbf{a}_2 + \frac{dr_3}{df}\frac{df}{dt}\mathbf{a}_3$$

$$= \dot{f}(t)\left(\frac{dr_1}{df}\mathbf{a}_1 + \frac{dr_2}{df}\mathbf{a}_2 + \frac{dr_3}{df}\mathbf{a}_3\right),$$

where we have again used the vector properties in Appendix B. The quantity in large parentheses is the definition of the vector derivative with respect to f, which results in Eq. (3.64).

The chain rule for a scalar function of a vector function of time is more subtle. To begin, we have to be clear about what is meant by the notation $f(\mathbf{r}(t))$. Since f is a scalar function, to perform computations, its argument must be a set of scalars. Thus f is operating on the scalar components of $\mathbf{r}(t)$ in some frame. Let $\mathbf{r}(t)$ in frame $\mathcal{A}$ be

$$\mathbf{r}(t) = r_1(t)\mathbf{a}_1 + r_2(t)\mathbf{a}_2 + r_3(t)\mathbf{a}_3.$$

Then the function f is written as

$$f(r_1(t)\mathbf{a}_1 + r_2(t)\mathbf{a}_2 + r_3(t)\mathbf{a}_3) = f(r_1(t), r_2(t), r_3(t)). \qquad (3.65)$$

Now, even though f is a scalar function, its derivative is frame dependent, because how it is written as a function of the scalar component magnitudes may differ depending on the frame we choose. In other words, taking the derivative of f with respect to t is equivalent to asking how f depends on changes to $\mathbf{r}(t)$. But how $\mathbf{r}(t)$ varies depends on which frame you choose, as seen throughout this chapter. Thus the definition of the scalar derivative of $f(\mathbf{r}(t))$ with respect to time is similar to that of the vector derivative. Taking the derivative with respect to frame $\mathcal{A}$ means we write $\mathbf{r}$ in terms of components in $\mathcal{A}$ and f as a function of those components (as in Eq. (3.65)) and then use the scalar chain rule on each of the component magnitudes:

$$\frac{^{A}d}{dt}f(\mathbf{r}(t)) = \frac{^{A}d}{dt}f(r_1(t)\mathbf{a}_1 + r_2(t)\mathbf{a}_2 + r_3(t)\mathbf{a}_3) = \frac{\partial f}{\partial r_1}\dot{r}_1 + \frac{\partial f}{\partial r_2}\dot{r}_2 + \frac{\partial f}{\partial r_3}\dot{r}_3$$

$$= \underbrace{\left(\frac{\partial f}{\partial r_1}\mathbf{a}_1 + \frac{\partial f}{\partial r_2}\mathbf{a}_2 + \frac{\partial f}{\partial r_3}\mathbf{a}_3\right)}_{=\nabla f} \cdot \underbrace{\left(\dot{r}_1\mathbf{a}_1 + \dot{r}_2\mathbf{a}_2 + \dot{r}_3\mathbf{a}_3\right)}_{=\frac{^{A}d}{dt}\mathbf{r}(t)}.$$

This calculation yields the following compact expression in terms of the gradient ∇f:

$$\boxed{\frac{^{A}d}{dt}f(\mathbf{r}(t)) = \nabla f \cdot \frac{^{A}d}{dt}\mathbf{r}(t).} \qquad (3.66)$$

Eq. (3.66) is the derivative of f in the direction of the vector $\frac{^{A}d}{dt}\mathbf{r}$. (See the discussion of the directional derivative in Appendix A.) Note that, when taking the dot product, the frame in which one expresses the components of each vector doesn't matter.

For a scalar function with an explicit time dependence (i.e., $f(\mathbf{r}(t), t)$), the chain rule becomes

$$\frac{^A d}{dt} f(\mathbf{r}(t), t) = \nabla f \cdot \frac{^A d}{dt}(\mathbf{r}(t)) + \frac{\partial f}{\partial t}.$$

As a consequence, for any scalar function $f(t)$ that depends only on time, the time derivative is frame independent:

$$\frac{^A d}{dt} f(t) = \frac{^B d}{dt} f(t) = \frac{\partial f}{\partial t}.$$

It is the latter expression that we are implicitly using when differentiating a coordinate such as $x(t)$ or $\theta(t)$.

Example 3.15 Calculating a Derivative Using the Vector Chain Rule

The chain rule derivation is subtle enough that it is helpful to see a simple example of its use. Consider a point P in the plane with Cartesian coordinates $(x, y)_{\mathcal{I}}$ and polar coordinates $(r, \theta)_{\mathcal{I}}$. Suppose the scalar function f is equal to the x-coordinate of P,

$$f(\mathbf{r}_{P/O}) = x = r \cos \theta, \tag{3.67}$$

where we have written f in terms of both Cartesian and polar coordinates. We now compute the time derivative of f. It is, in this case, easy to just take the derivative of the function in Eq. (3.67) to obtain $\dot{f} = \dot{x} = \dot{r} \cos \theta - r\dot{\theta} \sin \theta$. However, let us instead find it using the chain rule, Eq. (3.66). First find the gradient of $f(\mathbf{r}_{P/O})$ (see Appendix A):

$$\nabla f = \frac{\partial f}{\partial x} \mathbf{e}_x + \frac{\partial f}{\partial y} \mathbf{e}_y = \mathbf{e}_x$$

$$= \frac{\partial f}{\partial r} \mathbf{e}_r + \frac{1}{r} \frac{\partial f}{\partial \theta} \mathbf{e}_\theta = \cos \theta \mathbf{e}_r - \sin \theta \mathbf{e}_\theta.$$

The time derivative of f in $\mathcal{I}$, from Eq. (3.66), is

$$\frac{^{\mathcal{I}} d}{dt} f(\mathbf{r}_{P/O}(t)) = \nabla f \cdot \frac{^{\mathcal{I}} d}{dt}(\mathbf{r}_{P/O})$$

$$= \mathbf{e}_x \cdot (\dot{x} \mathbf{e}_x + \dot{y} \mathbf{e}_y) = \dot{x}$$

$$= (\cos \theta \mathbf{e}_r - \sin \theta \mathbf{e}_\theta) \cdot (\dot{r} \mathbf{e}_r + r\dot{\theta} \mathbf{e}_\theta) = \dot{r} \cos \theta - r\dot{\theta} \sin \theta = \dot{x}$$

$$= \mathbf{e}_x \cdot (\dot{r} \mathbf{e}_r + r\dot{\theta} \mathbf{e}_\theta) = \dot{r} \cos \theta - r\dot{\theta} \sin \theta = \dot{x}.$$

Note that, as pointed out, the frame chosen for the components of the gradient or the vector derivative doesn't matter, as long as we are careful with the dot product.

Although the results agree with intuition, we are not quite done. What is the time derivative of $f(\mathbf{r}_{P/O}(t))$ in the polar frame $\mathcal{B}$? Using Eq. (3.66), we have

$$\frac{^{\mathcal{B}} d}{dt} f(\mathbf{r}_{P/O}(t)) = \nabla f \cdot \frac{^{\mathcal{B}} d}{dt}(\mathbf{r}_{P/O})$$

$$= (\cos \theta \mathbf{e}_r - \sin \theta \mathbf{e}_\theta) \cdot (\dot{r} \mathbf{e}_r)$$

$$= \dot{r} \cos \theta.$$

This is a different result! The polar frame is defined such that P only changes its position in one coordinate direction. Thus, if we are asking how f changes as $\mathbf{r}_{P/O}$ changes *in* $\mathcal{B}$, the answer is that the change in f depends only on how r—but not θ—changes.

3.8.4 The Vector Taylor Series

We wish to find a Taylor series expansion (Appendix A.4) of the vector function $\mathbf{r}(t)$ about the point $t = a$. To do so, use the scalar Taylor series on each component. First, write the vector $\mathbf{r}(t)$ as components in a frame $\mathcal{A} = (O, \mathbf{a}_1, \mathbf{a}_2, \mathbf{a}_3)$:

$$\mathbf{r}(t) = r_1(t)\mathbf{a}_1 + r_2(t)\mathbf{a}_2 + r_3(t)\mathbf{a}_3. \tag{3.68}$$

Then take the Taylor series of the scalar magnitude of each component. For $i = 1, 2, 3$, we have

$$r_i(t) = r_i(a) + \left.\frac{dr_i}{dt}\right|_{t=a}(t-a) + \frac{1}{2!}\left.\frac{d^2r_i}{dt^2}\right|_{t=a}(t-a)^2 + \cdots.$$

Expanding and rearranging each coefficient in Eq. (3.68) yields

$$\mathbf{r}(t) = r_1(a)\mathbf{a}_1 + r_2(a)\mathbf{a}_2 + r_3(a)\mathbf{a}_3$$
$$+ \left.\frac{dr_1}{dt}\right|_{t=a}(t-a)\mathbf{a}_1 + \left.\frac{dr_2}{dt}\right|_{t=a}(t-a)\mathbf{a}_2 + \left.\frac{dr_3}{dt}\right|_{t=a}(t-a)\mathbf{a}_3$$
$$+ \frac{1}{2!}\left.\frac{d^2r_1}{dt^2}\right|_{t=a}(t-a)^2\mathbf{a}_1 + \frac{1}{2!}\left.\frac{d^2r_2}{dt^2}\right|_{t=a}(t-a)^2\mathbf{a}_2 + \frac{1}{2!}\left.\frac{d^2r_3}{dt^2}\right|_{t=a}(t-a)^2\mathbf{a}_3$$
$$+ \cdots.$$

Using Definition 3.3, we obtain the vector Taylor series:

$$\mathbf{r}(t) = \mathbf{r}(a) + \left.\frac{{}^{\mathcal{A}}d}{dt}\mathbf{r}\right|_{t=a}(t-a) + \frac{1}{2!}\left.\frac{{}^{\mathcal{A}}d^2}{dt^2}\mathbf{r}\right|_{t=a}(t-a)^2 + \cdots$$

$$+ \frac{1}{n!}\left.\frac{{}^{\mathcal{A}}d^n}{dt^n}\mathbf{r}\right|_{t=a}(t-a)^n + \cdots.$$

3.9 Tutorials

Tutorial 3.1 A Torsion Spring

This tutorial examines the *torsion spring*—a mechanism that provides a restoring force in a rotational direction. The torsion spring is an important modeling construct analogous to a linear spring. Consider the configuration illustrated in Figure 3.29a. Particle P is attached to a massless rod of fixed length l. The mass is acted on (by means of the rod) by a torsion spring with restoring force $-k(\theta - \theta_0)$, where θ_0 is

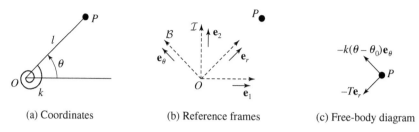

(a) Coordinates   (b) Reference frames   (c) Free-body diagram

Figure 3.29 A torsion spring.

the rest angle of the spring.[6] Using Figure 3.29c and the polar frame $\mathcal{B}$ illustrated in Figure 3.29b, the total force on P is

$$\mathbf{F}_P = -T\mathbf{e}_r - k(\theta - \theta_0)\mathbf{e}_\theta,$$

where T is the tension in the rod. The kinematics of particle P are

$$\mathbf{r}_{P/O} = l\mathbf{e}_r$$
$$^I\mathbf{v}_{P/O} = l\dot{\theta}\mathbf{e}_\theta$$
$$^I\mathbf{a}_{P/O} = -l\dot{\theta}^2\mathbf{e}_r + l\ddot{\theta}\mathbf{e}_\theta.$$

Using Newton's second law, we obtain

$$-T\mathbf{e}_r - k(\theta - \theta_0)\mathbf{e}_\theta = m_P(-l\dot{\theta}^2\mathbf{e}_r + l\ddot{\theta}\mathbf{e}_\theta).$$

The tension in the rod,

$$T = m_P l\dot{\theta}^2,$$

acts as a centripetal force (see Section 3.5). The equation of motion,

$$\ddot{\theta} = -\frac{k}{m_P l}(\theta - \theta_0),$$

exhibits simple harmonic motion with natural frequency $\omega_0 = \sqrt{\frac{k}{m_P l}}$ (see Tutorial 2.3 for the solution). As you might expect, this rotational mass-spring system oscillates about the rest angle of the spring.

Tutorial 3.2 A Particle Sliding on a Parabola

This tutorial studies the dynamics of a particle sliding on a parabola, as shown in Figure 3.30. This could be a model for a snowboarder in a half-pipe, as shown

[6] Note that k here has different units than the spring constant of the linear spring studied in Tutorial 2.3.

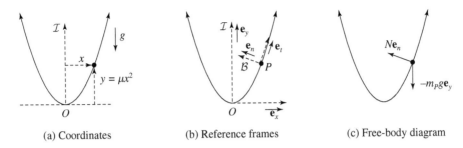

(a) Coordinates   (b) Reference frames   (c) Free-body diagram

Figure 3.30 A particle sliding on a parabola.

Figure 3.31 A simple model for a snowboarder in a half-pipe is a particle sliding on a parabola, the dynamics of which are explored in Tutorial 3.2. Image courtesy of Shutterstock.

in Figure 3.31. Define the (fixed) inertial frame $\mathcal{I} = (O, \mathbf{e}_x, \mathbf{e}_y, \mathbf{e}_z)$ and the path frame $\mathcal{B} = (P, \mathbf{e}_t, \mathbf{e}_n, \mathbf{e}_z)$. Using the equation $y(x) = \mu x^2$ to describe the parabola and the Cartesian coordinates shown in Figure 3.30a, the kinematics of P are

$$\mathbf{r}_{P/O} = x\mathbf{e}_x + \mu x^2 \mathbf{e}_y$$

$$^{\mathcal{I}}\mathbf{v}_{P/O} = \dot{x}\mathbf{e}_x + 2\mu x\dot{x}\mathbf{e}_y$$

$$^{\mathcal{I}}\mathbf{a}_{P/O} = \ddot{x}\mathbf{e}_x + 2\mu(\dot{x}^2 + x\ddot{x})\mathbf{e}_y.$$

Note that this system has one degree of freedom, x, since the y-position of the particle is a function of x ($y = \mu x^2$ is a constraint equation).

Using the free-body diagram in Figure 3.30c, we write Newton's second law for the mass m_P as

$$N\mathbf{e}_n - m_P g\mathbf{e}_y = m_P \ddot{x}\mathbf{e}_x + 2m_P \mu(\dot{x}^2 + x\ddot{x})\mathbf{e}_y. \tag{3.69}$$

To separate this vector equation into two scalar equations and solve for the equation of motion in $\ddot{x}$, we must write $\mathbf{e}_n$ in terms of $\mathbf{e}_x$ and $\mathbf{e}_y$ (or vice versa). By definition, we have

$$\mathbf{e}_t = \frac{{}^{\mathcal{I}}\mathbf{v}_{P/O}}{\|{}^{\mathcal{I}}\mathbf{v}_{P/O}\|} = \frac{\dot{x}\mathbf{e}_x + 2\mu x\dot{x}\mathbf{e}_y}{\sqrt{\dot{x}^2 + 4\mu^2 x^2 \dot{x}^2}} = \frac{\mathbf{e}_x + 2\mu x\mathbf{e}_y}{\sqrt{1 + 4\mu^2 x^2}}$$

$$\mathbf{e}_n = \mathbf{e}_z \times \mathbf{e}_t = \frac{-2\mu x\mathbf{e}_x + \mathbf{e}_y}{\sqrt{1 + 4\mu^2 x^2}}. \tag{3.70}$$

Substituting Eq. (3.70) into Eq. (3.69), we obtain the two scalar equations

$$\frac{-2\mu x N}{\sqrt{1 + 4\mu^2 x^2}} = m_P \ddot{x}$$

$$\frac{N}{\sqrt{1 + 4\mu^2 x^2}} - m_P g = 2m_P \mu(\dot{x}^2 + x\ddot{x}),$$

which yield the normal force

$$N = -\frac{m_P \ddot{x}}{2\mu x}\sqrt{1 + 4\mu^2 x^2} = \frac{m_P(g + 2\mu\dot{x}^2)}{\sqrt{1 + 4\mu^2 x^2}}$$

and the equation of motion

$$\ddot{x} = -\frac{2\mu x(g + 2\mu\dot{x}^2)}{1 + 4\mu^2 x^2}. \tag{3.71}$$

We integrate the equation of motion using the MATLAB command ODE45. MATLAB can also be used to calculate the normal force as a function of the position of P. Since ODE45 integrates differential equations in first-order form—and the equation of motion in Eq. (3.71) is a second-order differential equation—we rewrite the equation of motion using the change of variables $Z = [z_1, z_2]^T \triangleq [x, \dot{x}]^T$:

$$\dot{z}_1 = z_2$$

$$\dot{z}_2 = -\frac{2\mu z_1(g + 2\mu z_2^2)}{1 + 4\mu^2 z_1^2}. \tag{3.72}$$

In MATLAB, we create the function MYODEFUN to define the equations of motion (Eq. (3.72)):

```
function zdot = myodefun(t,z,g,mu)
zdot(1,1) = z(2);
zdot(2,1) = -2*mu*z(1)*(g+2*mu*z(2)^2)/(1+4*mu^2*z(1)^2);
```

Next we define the constants g and mu in the Command Window and call ODE45 to integrate the equations of motion from $x(0) = 1$ m and $\dot{x}(0) = 0$ m/s for 10 s:

```
>> [t,z] = ode45(@myodefun,[0 10],[1 0],[],g,mu);
```

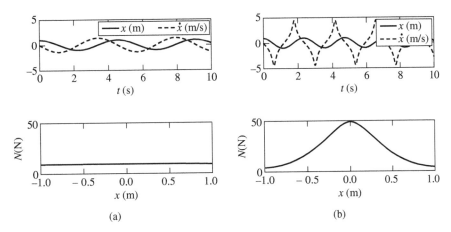

Figure 3.32 Particle sliding on a parabola. The top panels show position $x(t)$ and speed $\dot{x}(t)$; the bottom panels show normal force N as a function of position x. (a) $\mu = 0.1$ m^{-1}. (b) $\mu = 1.0$ m^{-1}.

The first column of the output matrix z contains the position $x(t)$ and the second column of z contains the speed $\dot{x}(t)$. To compute the normal force N, enter

```
>> N = m*(g+2*mu*z(:,2).^2)./sqrt(1+4*mu^2*z(:,1).^2);
```

Note the use of the . for pointwise matrix operations.

The results are plotted in Figure 3.32 for $x(0) = 1$ m, $\dot{x}(0) = 0$ m/s, $m_P = 1$ kg, and two values of the parabola parameter—$\mu = 0.1$ m^{-2} and $\mu = 1.0$ m^{-2}. For the initial position $x(0) = 1$ m, the initial height satisfies $y(0) = \mu x(0)^2 = \mu$ m. For $\mu = 0.1$ m^{-2}, both the position and the speed oscillate sinusoidally, and the normal force only slightly exceeds the weight $m_P g$ of the particle. For $\mu = 1$ m^{-2}, the position still oscillates, but the speed curve has sharper peaks than a sinusoid; the normal force exceeds five times the weight of the mass.

Tutorial 3.3 A Particle Sliding on a Hemisphere

This tutorial examines the dynamics of a particle sliding on a frictionless hemisphere of radius R, as shown in Figure 3.33. We solve analytically for the angular displacement of the particle at the instant it falls off the hemisphere and use MATLAB to compute the time at which the particle falls off.

We use the (fixed) inertial frame $\mathcal{I} = (O, \mathbf{e}_x, \mathbf{e}_y, \mathbf{e}_z)$ and the rotating polar frame $\mathcal{B} = (O, \mathbf{e}_r, \mathbf{e}_\theta, \mathbf{e}_3)$, where $\mathbf{e}_3 = -\mathbf{e}_z$. The unit vectors in frames $\mathcal{I}$ and $\mathcal{B}$ are related by the coordinate transformation table

	$\mathbf{e}_x$	$\mathbf{e}_y$
$\mathbf{e}_r$	$\sin\theta$	$\cos\theta$
$\mathbf{e}_\theta$	$\cos\theta$	$-\sin\theta$.

The particle starts at $\theta = 0$ and is given a very small push in the direction of increasing θ. At some time $t = t^*$ later, the particle will lose contact with the surface at angular

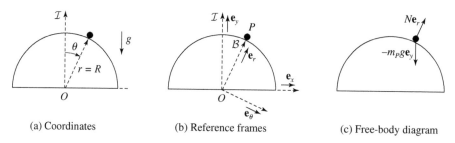

(a) Coordinates   (b) Reference frames   (c) Free-body diagram

Figure 3.33 A particle sliding on a hemisphere.

displacement $\theta^* = \theta(t^*)$. Before it falls off the surface, the kinematics of particle P are

$$\mathbf{r}_{P/O} = R\mathbf{e}_r$$

$$^{\mathcal{I}}\mathbf{v}_{P/O} = R\dot{\theta}\mathbf{e}_\theta$$

$$^{\mathcal{I}}\mathbf{a}_{P/O} = R\ddot{\theta}\mathbf{e}_\theta - R\dot{\theta}^2\mathbf{e}_r,$$

where we used $\dot{R} = \ddot{R} = 0$.

As usual, let m_P denote the mass of particle P. Using the free-body diagram in Figure 3.33c and Newton's second law yields

$$N\mathbf{e}_r - m_P g\mathbf{e}_y = m_P(R\ddot{\theta}\mathbf{e}_\theta - R\dot{\theta}^2\mathbf{e}_r).$$

Using the coordinate transformation table to write $\mathbf{e}_y$ in terms of $\mathbf{e}_r$ and $\mathbf{e}_\theta$, we can solve for the normal force N and the equation of motion of the system:

$$N = m_P g \cos\theta - m_P R\dot{\theta}^2 \qquad (3.73)$$

$$\ddot{\theta} = \frac{g}{R}\sin\theta. \qquad (3.74)$$

Note that the equation of motion Eq. (3.74) looks like the equation for simple harmonic motion with natural frequency $\sqrt{g/R}$, except that the sign of the right-hand side is reversed—if it was physically realizable, the particle would oscillate about $\theta = \pi$ instead of $\theta = 0$.

Perhaps counterintuitively, the particle falls off the surface at the instant the normal force satisfies $N = 0$ (or, equivalently, $N/m_P = 0$). At this instant, the angular rate $\dot{\theta} = \dot{\theta}^*$ satisfies the equation

$$0 = m_P g \cos\theta^* - m_P R(\dot{\theta}^*)^2,$$

that is,

$$\dot{\theta}^* = \sqrt{\frac{g}{R}\cos\theta^*}. \qquad (3.75)$$

We solve analytically for the angle θ^* by integrating Eq. (3.74) from 0 to t. That is, using separation of variables (see Section C.2.1) and the initial conditions $\theta(0) = 0$

and $\dot{\theta}(0) = \dot{\theta}_0 > 0$ gives

$$\frac{1}{2}\dot{\theta}^2 - \frac{1}{2}\dot{\theta}_0^2 = -\frac{g}{R}(\cos\theta - 1),$$

which implies

$$\dot{\theta} = \sqrt{\dot{\theta}_0^2 + \frac{2g}{R}(1 - \cos\theta)}. \tag{3.76}$$

Evaluating Eq. (3.76) at t^* and equating it to Eq. (3.75), we obtain the relation

$$\cos\theta^* = \frac{2}{3} + \frac{R}{3g}\dot{\theta}_0^2,$$

which implies

$$\theta^* = \arccos\left(\frac{2}{3} + \frac{R}{3g}\dot{\theta}_0^2\right). \tag{3.77}$$

In the limit $\dot{\theta}_0 \to 0.01$ rad/s (i.e., if the initial push is very small), $\theta^* \approx \arccos(2/3)$.

Next solve for the time t^* by numerically integrating Eq. (3.76) assuming $R = 1$ m and $\dot{\theta}_0 = 0.01$ m/s (note that we cannot integrate Eq. (3.75), which holds only at the instant $t = t^*$). Using separation of variables, we obtain

$$t^* = \int_0^{\theta^*} \frac{d\theta}{\sqrt{\dot{\theta}_0^2 + \frac{2g}{R}(1 - \cos\theta)}}, \tag{3.78}$$

where θ^* is given by Eq. (3.77). To integrate the right-hand side, use the MATLAB command QUAD, which is short for *quadrature*.[7] Just like ODE45, which integrates equations of motion contained in a separate function, the QUAD function uses an integrand contained in a separate function QUADFUN. The QUAD function is invoked by passing the function handle @QUADFUN using the syntax Q = QUAD(@(X)QUADFUN(X,P1, P2, . . .),A,B), where X is the argument of the integrand function FUN, P1 and P2 are parameters to FUN, and A and B are the integral limits. In our case, FUN contains the integrand in Eq. (3.78):

```
function dt = myquadfun(theta,g,R,dottheta0)
dt = 1./sqrt(dottheta0^2+2*g/R*(1-cos(theta)));
```

Note that we need to use the array divide ./ in this function because theta will be an array of values from 0 to θ^*. We define the following variables in the MATLAB workspace: g=9.81 m/s^2, R=1 m, dottheta0=0.01 rad/s, and thetastar=acos(2/3+R/3/g*dottheta0^2) rad. Finally, call QUAD from the command line

```
>> quad(@(theta)myquadfun(theta,g,R,dottheta0),0,thetastar),
```

which returns $t^* = 2.0$ s.

---

[7] "Quadrature" is a term that can mean solving an integral analytically or numerically.

3.10 Key Ideas

- A **vector** indicates a magnitude and a direction in space. A **unit vector** is a vector with magnitude one. Three orthogonal unit vectors—along with an origin—define a **reference frame**, for example, $\mathcal{A} = (O, \mathbf{a}_1, \mathbf{a}_2, \mathbf{a}_3)$.

- A vector can be written as the sum of weighted unit vectors—called **vector components**. For example, the position of point P with respect to O is

$$\mathbf{r}_{P/O} = a_1\mathbf{a}_1 + a_2\mathbf{a}_2 + a_3\mathbf{a}_3.$$

A vector can be written in **matrix notation** using the magnitude of the vector components:

$$[\mathbf{r}_{P/O}]_{\mathcal{A}} = \begin{bmatrix} a_1 \\ a_2 \\ a_3 \end{bmatrix}_{\mathcal{A}}.$$

- If frame $\mathcal{B} = (O, \mathbf{b}_1, \mathbf{b}_2, \mathbf{b}_3)$ is rotated about the $\mathbf{b}_3$ axis by an angle θ with respect to frame $\mathcal{A}$, then the unit vectors of $\mathcal{A}$ and $\mathcal{B}$ are related by the **transformation table**

	$\mathbf{a}_1$	$\mathbf{a}_2$
$\mathbf{b}_1$	$\cos\theta$	$\sin\theta$
$\mathbf{b}_2$	$-\sin\theta$	$\cos\theta$.

- A **coordinate system** is a set of scalars that specifies the position of a point in a reference frame. Three examples of a coordinate system are **Cartesian coordinates**, **polar coordinates**, and **path coordinates**.

- The rate of change with time of a vector is reference-frame dependent. The **velocity** $^{\mathcal{A}}\mathbf{v}_{P/O} \triangleq \frac{^{\mathcal{A}}d}{dt}(\mathbf{r}_{P/O})$ is the rate of change of the position $\mathbf{r}_{P/O}$ in reference frame $\mathcal{A}$. Likewise, the **acceleration** $^{\mathcal{A}}\mathbf{a}_{P/O} \triangleq \frac{^{\mathcal{A}}d}{dt}(^{\mathcal{A}}\mathbf{v}_{P/O})$ is the rate of change of the velocity in reference frame $\mathcal{A}$. Since the unit vectors $\mathbf{a}_1, \mathbf{a}_2, \mathbf{a}_3$ are fixed in frame $\mathcal{A}$, the velocity (or acceleration) of P with respect to frame $\mathcal{A}$ is given by differentiating the scalar weights of each component:

$$^{\mathcal{A}}\mathbf{v}_{P/O} = \dot{a}_1\mathbf{a}_1 + \dot{a}_2\mathbf{a}_2 + \dot{a}_3\mathbf{a}_3.$$

- The rate of rotation of the polar frame (or path frame) $\mathcal{B}$ with respect to frame $\mathcal{A}$ is given by the simple **angular velocity** $^{\mathcal{A}}\omega^{\mathcal{B}}$. Since $\mathcal{B}$ rotates with respect to $\mathcal{A}$ about a common axis $\mathbf{b}_3 = \mathbf{a}_3$,

$$^{\mathcal{A}}\boldsymbol{\omega}^{\mathcal{B}} = {}^{\mathcal{A}}\omega^{\mathcal{B}}\mathbf{a}_3 \triangleq \dot{\theta}\mathbf{a}_3,$$

where $^{\mathcal{A}}\omega^{\mathcal{B}} \triangleq \dot{\theta}$ is the **angular speed**.

- The rate of a change in frame $\mathcal{A}$ of unit vector $\mathbf{b}_i$, $i = 1, 2$, that is fixed in frame $\mathcal{B}$ is

$$\frac{^{\mathcal{A}}d}{dt}(\mathbf{b}_i) = {}^{\mathcal{A}}\boldsymbol{\omega}^{\mathcal{B}} \times \mathbf{b}_i,$$

where the angular velocity $^A\omega^B$ corresponds to rotation of B about a common axis $\mathbf{b}_3 = \mathbf{a}_3$.

- Two important frames are the **polar frame** $B = (O, \mathbf{e}_r, \mathbf{e}_\theta, \mathbf{e}_3)$ and the **path frame** $B = (P, \mathbf{e}_t, \mathbf{e}_n, \mathbf{e}_3)$. The inertial velocity and inertial acceleration expressed in the polar frame using polar coordinates $(r, \theta)_\mathcal{I}$ are

$$^\mathcal{I}\mathbf{v}_{P/O} = \dot{r}\mathbf{e}_r + r\dot{\theta}\mathbf{e}_\theta$$

$$^\mathcal{I}\mathbf{a}_{P/O} = (\ddot{r} - r\dot{\theta}^2)\mathbf{e}_r + (2\dot{r}\dot{\theta} + r\ddot{\theta})\mathbf{e}_\theta.$$

- The inertial velocity and inertial acceleration expressed in the **path frame** using path coordinate s are

$$^\mathcal{I}\mathbf{v}_{P/O} = \dot{s}\mathbf{e}_t$$

$$^\mathcal{I}\mathbf{a}_{P/O} = \ddot{s}\mathbf{e}_t + \frac{\dot{s}^2}{R_c}\mathbf{e}_n,$$

where R_c is the instantaneous radius of curvature of the path.

- The trajectory of a particle, P, can be found relative to frame B with origin O' even if O' is accelerating in the inertial frame by using the modified form of Newton's second law

$$m_P{}^B\mathbf{a}_{P/O'} = \mathbf{F}_P - m_P{}^\mathcal{I}\mathbf{a}_{O'/O}.$$

- The **vector derivative** obeys many of the same properties as the scalar derivative, including the **addition rule**, **product rule**, and **chain rule**.

3.11 Notes and Further Reading

Nearly all introductory (and advanced) dynamics texts treat kinematics of particles, but there is great variation in the notation and level of detail. Very few emphasize reference frames as we do here; those that do (e.g., Kane and Levinson 1985; Greenwood 1988; Tenenbaum 2004) are typically used as graduate texts. Rao (2006) has a particularly nice discussion of the importance and definition of reference frames and was a reference for our mathematical and operational definitions. As noted in Chapter 1, our notation is similar to that developed by Kane (1978) and Kane and Levinson (1985) and used by Rao (2006) and Tenenbaum (2004).

Our use of the unit-vector transformation array is modeled after the approach taken in Kane (1978), Kane and Levinson (1985), and Tongue and Sheppard (2005). Tutorial 3.2 is based on a similar example in Rao (2006). Many texts take a geometric approach to the definition of the vector derivative. We follow the analytical approach in Kane (1978).

3.12 Problems

3.1 Express $\mathbf{r}_{P/Q}$ shown in Figure 3.34 using vector components in frame $\mathcal{A}$. Now express $\mathbf{r}_{P/Q}$ using vector components in frame $\mathcal{B}$.

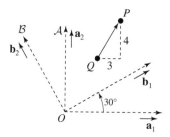

Figure 3.34 Problem 3.1.

3.2 Three forces act on particle P: $\mathbf{F}_1 = 2\mathbf{e}_x - 5\mathbf{e}_y$ N, $\mathbf{F}_2 = 10(\cos\theta\mathbf{e}_x + \sin\theta\mathbf{e}_y)$ N, and $\mathbf{F}_3 = (5 - 4\sqrt{6})\mathbf{e}_y$ N. If the resultant force is zero, find θ and the force vector $\mathbf{F}_2$. Sketch the three forces acting on P.

3.3 Consider frames $\mathcal{I} = (O, \mathbf{e}_1, \mathbf{e}_2, \mathbf{e}_3)$ and $\mathcal{A} = (O, \mathbf{a}_1, \mathbf{a}_2, \mathbf{a}_3)$, where $\mathbf{e}_3 = \mathbf{a}_3$, as shown in Figure 3.35. Find the position of P with respect to O in the following coordinates:

   a. Cartesian coordinates in $\mathcal{I}$, $(x, y)_{\mathcal{I}}$.
   b. Cartesian coordinates in $\mathcal{A}$, $(a_1, a_2)_{\mathcal{A}}$.
   c. Polar coordinates in $\mathcal{I}$, $(r, \theta)_{\mathcal{I}}$.
   d. Polar coordinates in $\mathcal{A}$, $(\rho, \beta)_{\mathcal{A}}$.

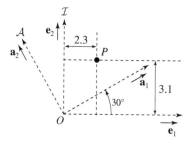

Figure 3.35 Problem 3.3.

3.4 In Example 3.5, find expressions for the polar coordinates $(r, \theta)_{\mathcal{I}}$ in terms of the path coordinate s.

3.5 The position of P shown in Figure 3.36 is described by the path coordinate s on a circular path with radius ρ and whose center Q is located at $\mathbf{r}_{Q/O} = a\mathbf{e}_1$. Find the position of P with respect to O using Cartesian coordinates in frame $\mathcal{I}$, $(x, y)_{\mathcal{I}}$.

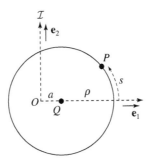

Figure 3.36 Problem 3.5.

3.6 Consider a two-link arm shown in Figure 3.37 with frames $\mathcal{I} = (O, \mathbf{e}_1, \mathbf{e}_2, \mathbf{e}_3)$, $\mathcal{A} = (O, \mathbf{a}_1, \mathbf{a}_2, \mathbf{a}_3)$, and $\mathcal{B} = (O, \mathbf{b}_1, \mathbf{b}_2, \mathbf{b}_3)$, where $\mathbf{e}_3 = \mathbf{a}_3 = \mathbf{b}_3$. Construct three transformation tables, relating the unit vectors of frames $\mathcal{I}$ and $\mathcal{A}$, $\mathcal{A}$ and $\mathcal{B}$, and $\mathcal{I}$ and $\mathcal{B}$.

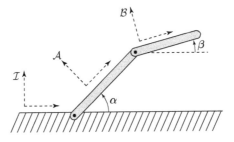

Figure 3.37 Problem 3.6.

3.7 Sketch an idealized planar model of the crane shown in Figure 3.38 using a point mass, a rigid, massless rod, and a massless cable. How many degrees of

Figure 3.38 Problem 3.7. Image courtesy of Shutterstock.

freedom are there in your model? Derive the inertial kinematics of a point on the hook of the crane.

3.8   Consider inertial frame $\mathcal{A} = (O, \mathbf{a}_1, \mathbf{a}_2, \mathbf{a}_3)$ and polar frame $\mathcal{B} = (O, \mathbf{b}_1, \mathbf{b}_2, \mathbf{b}_3)$, as shown in Figure 3.39, where $\mathbf{a}_3 = \mathbf{b}_3$ and $\mathcal{B}$ is rotating with respect to $\mathcal{A}$ with angular velocity ${}^{\mathcal{A}}\boldsymbol{\omega}^{\mathcal{B}} = \dot{\theta}\mathbf{b}_3$. The position of P with respect to O expressed as vector components in frame $\mathcal{B}$ is $\mathbf{r}_{P/O}(t) = b_1(t)\mathbf{b}_1$. Find ${}^{\mathcal{B}}\mathbf{v}_{P/O}$ and ${}^{\mathcal{A}}\mathbf{v}_{P/O}$.

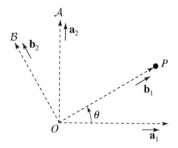

Figure 3.39 Problem 3.8.

3.9   Consider the example discussed in the beginning of Section 3.5 and shown in Figure 3.23. Suppose the constant inertial velocity is given in Cartesian coordinates as:

$$^{\mathcal{I}}\mathbf{v}_{P/O} = v_1\mathbf{e}_1 + v_2\mathbf{e}_2$$

a. Calculate the rates of change $\dot{r}$ and $\dot{\theta}$ of the polar coordinates in terms of v_1, v_2, r, and θ.

b. Using Eq. (3.51), find the second derivatives $\ddot{r}$ and $\ddot{\theta}$ of the polar coordinates in terms of v_1, v_2, r, and θ.

c. Show that asymptotically (as time goes to infinity) $\dot{\theta}$ and $\ddot{\theta}$ go to zero and the polar frame reaches a fixed orientation in the inertial frame with the radial unit vector $\mathbf{e}_r$ parallel to the velocity.

3.10   Use the angular velocity to derive the expression for the acceleration of a particle in polar coordinates in Eq. (3.33).

3.11   Show that the product rule for vector differentiation (see Table 3.1) holds for the vector cross product.

3.12   Derive the vector derivative of $\mathbf{e}_\theta$ in Eq. (3.30).

3.13   Show that the radius of curvature of a path defined by the function $y(x)$ in Cartesian coordinates is given by Eq. (3.50).

3.14   Consider the inertial frame $\mathcal{I} = (O, \mathbf{e}_1, \mathbf{e}_2, \mathbf{e}_3)$ and a point P that is constrained to move in the $(\mathbf{e}_2, \mathbf{e}_3)$ plane, as shown in Figure 3.40.

a. Sketch a polar frame $\mathcal{B}$ and derive the transformation table between $\mathcal{B}$ and $\mathcal{I}$.

b. Write the position $\mathbf{r}_{P/O}$ of P with respect to O in polar coordinates using vector components in frame $\mathcal{B}$.

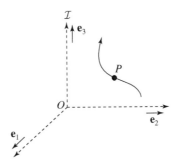

Figure 3.40 Problem 3.14.

    c. What is the direction of angular velocity $^{\mathcal{I}}\boldsymbol{\omega}^{\mathcal{B}}$ of $\mathcal{B}$ with respect to $\mathcal{I}$? What is its magnitude?

    d. Derive the inertial kinematics of P with respect to O in polar coordinates using vector components in frame $\mathcal{B}$.

3.15 Consider a bead P that is constrained to move along a path $x = y^2$, as shown in Figure 3.41, where $(x, y)_{\mathcal{I}}$ are Cartesian coordinates in frame $\mathcal{I}$. Assuming the bead is forced to move at constant rate v_0 in the $\mathbf{e}_2$ direction, write $^{\mathcal{I}}\mathbf{v}_{P/O}$ and $^{\mathcal{I}}\mathbf{a}_{P/O}$ in Cartesian coordinates as vector components in frame $\mathcal{I}$.

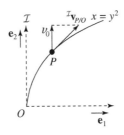

Figure 3.41 Problem 3.15.

3.16 Consider a particle P accelerating from rest on a curved path with $\ddot{s} = a$, where a is a constant.

    a. Find the radius of curvature $R_c(t)$ such that the magnitude of the particle's acceleration $\|^{\mathcal{I}}\mathbf{a}_{P/O}\|$ is equal to $2g$ for all time.

    b. Sketch the path of the particle.

3.17 Let $\mathcal{A} = (O, \mathbf{a}_1, \mathbf{a}_2, \mathbf{a}_3)$ be an inertial reference frame and $\mathcal{B} = (O, \mathbf{b}_1, \mathbf{b}_2, \mathbf{b}_3)$ be a polar reference frame with $^{\mathcal{A}}\boldsymbol{\omega}^{\mathcal{B}} = \dot{\theta}\mathbf{b}_3$, as shown in Figure 3.42. Derive the following in polar coordinates $(r, \theta)_{\mathcal{A}}$:

    a. The transformation table between $\mathcal{A}$ and $\mathcal{B}$.

    b. The position $\mathbf{r}_{P/O}$ of P with respect to O expressed as components in $\mathcal{B}$.

    c. The position $\mathbf{r}_{P/O}$ of P with respect to O expressed as components in $\mathcal{A}$.

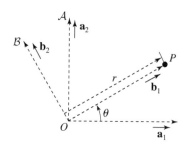

Figure 3.42 Problem 3.17.

    d. The velocity ${}^{B}\mathbf{v}_{P/O}$ and acceleration ${}^{B}\mathbf{a}_{P/O}$ of P with respect to O in $\mathcal{B}$ expressed as components in $\mathcal{B}$.

    e. The velocity ${}^{A}\mathbf{v}_{P/O}$ and acceleration ${}^{A}\mathbf{a}_{P/O}$ of P with respect to O in $\mathcal{A}$ expressed as components in $\mathcal{B}$.

3.18 Let $\mathcal{I} = (O, \mathbf{e}_1, \mathbf{e}_2, \mathbf{e}_3)$ be an inertial reference frame and frame $\mathcal{A} = (O', \mathbf{a}_1, \mathbf{a}_2, \mathbf{a}_3)$ be aligned with $\mathcal{I}$ and translating with velocity ${}^{\mathcal{I}}\mathbf{v}_{O'/O} = v_0\mathbf{e}_1$. Frame $\mathcal{B} = (O', \mathbf{b}_1, \mathbf{b}_2, \mathbf{b}_3)$ is a polar frame with angular velocity ${}^{\mathcal{I}}\boldsymbol{\omega}^{\mathcal{B}} = \dot{\theta}\mathbf{b}_3$. ($P$ is free to move in $\mathcal{A}$.) Derive the following using the coordinates shown in Figure 3.43:

    a. The transformation table between $\mathcal{A}$ and $\mathcal{B}$.

    b. The position $\mathbf{r}_{P/O'}$ of P with respect to O' expressed as components in $\mathcal{B}$.

    c. The position $\mathbf{r}_{P/O}$ of P with respect to O expressed as components in $\mathcal{B}$.

    d. The velocity ${}^{B}\mathbf{v}_{P/O'}$ and acceleration ${}^{B}\mathbf{a}_{P/O'}$ of P with respect to O' in $\mathcal{B}$ expressed as components in $\mathcal{B}$.

    e. The velocity ${}^{A}\mathbf{v}_{P/O}$ and acceleration ${}^{A}\mathbf{a}_{P/O}$ of P with respect to O in $\mathcal{A}$ expressed as components in $\mathcal{B}$.

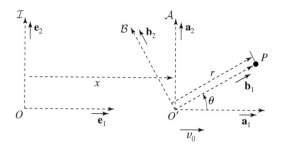

Figure 3.43 Problem 3.18.

3.19 To simulate the zero-gravity conditions of space, the National Aeronautics and Space Administration (NASA) regularly flies an airplane on a parabolic trajectory. At the apex of the trajectory, gravity seems to disappear and everything in the cabin floats for about 30 s. Consider such a parabolic trajectory given by $y = -cx^2$. While the plane is flying up the parabola, you are stationary, that is, you feel a normal force N that cancels your weight. At the top of the parabola, when you begin to float, the normal force vanishes and you (and the plane) are in free fall as if you were in orbit.

    a. Draw a free-body diagram of you in the plane.

    b. Find your acceleration as a function of the speed v of the airplane, which you may assume is constant, and the shape c of the parabola.

    c. Find an expression for c in terms of v, so that zero-gravity conditions are achieved at the apex of the trajectory.

3.20 Consider a collar of mass m sliding on a frictionless shaft, as depicted in Figure 3.44. The collar can slide along the shaft and is connected to a spring of spring constant k, the other end of which is connected to the pivot point O a distance l from the shaft. The unstretched length of the spring is l.

    a. How many degrees of freedom are there in this problem?

    b. Find the equation of motion for the collar.

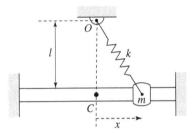

Figure 3.44 Problem 3.20: Sliding collar on a spring.

3.21 A snowboarder of mass m_B and cross-sectional area A is sliding down the terrain park, which has lots of powder. Draw a free-body diagram of the boarder identifying all forces acting on her. If you were asked to figure out whether she was going to leave the ground on a particular hill, what coordinate system would you choose to solve the problem?

3.22 Using the equations of motion for the simple pendulum in Cartesian coordinates (Eq. (3.7)), numerically integrate the trajectory for the initial conditions $(\theta_0, \dot{\theta}_0) = (0.1 \text{ rad}, 0 \text{ rad/s})$. Plot the pendulum angle $\theta(t) = \arctan\left(\frac{y(t)}{x(t)}\right)$ and the pendulum length $l(t) = \sqrt{x(t)^2 + y(t)^2}$ as a function of time. What do you notice about the length?

3.23 Consider the simple planar pendulum discussed at length in this chapter, only now, rather than a solid massless rod, suppose the pendulum bob is attached to a spring with spring constant k (Figure 3.45). Note that the spring does not bend or twist.

    a. How many degrees of freedom are there in this problem?

    b. What coordinates would you use to describe the configuration of the pendulum bob?

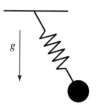

Figure 3.45 Problem 3.23.

     c. Solve for the equations of motion in the coordinates you chose in (b).
       Be sure to draw a free-body diagram.

3.24 Solve the equations of motion for the spring-pendulum system in the previous
problem using MATLAB ODE45 and plot the pendulum angle $\theta(t)$ and pendulum
length $r(t)$ versus time. Assume that the pendulum is released with initial
conditions $\theta(0) = \pi/4$, $\dot{\theta}(0) = 0$, $r(0) = l_0$, and $\dot{r}(0) = 0$, where l_0 is the rest
length of the spring. Let $m = 1\,\text{kg}$, $l_0 = 0.1\,\text{m}$, and $k = 2\,\text{N/m}$. Be sure to label
your plots with units and include a legend.

3.25 Derive the inertial kinematics of the tip of a two-link arm, shown in Figure 3.37.

3.26 Consider the planar spring in Example 3.12. Let $k/m_P = 10\ \text{s}^{-2}$ and $r_0 =$
1 m. Computationally simulate a 5-s trajectory of the system for the initial
conditions $r(0) = 5\,\text{m}$, $\dot{r}(0) = 0$, $\theta(0) = \pi/4$ rad, and $\dot{\theta}(0) = \pi/6$ rad/s.

3.27 Rederive the equations of motion for the planar spring in Example 3.12
including a Coulomb friction force acting opposite the velocity.

3.28 Suppose a rocket with mass 10^4 kg propels itself at an angle of 30° above the
horizontal using a thrust force of 2×10^7 N, as shown in Figure 3.46. Use a
path frame to compute the magnitude of the tangential and normal components
of the rocket's acceleration at the instant shown in the figure, assuming the
rocket's velocity is aligned with the thrust vector.

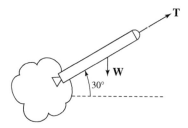

Figure 3.46 Problem 3.28.

3.29 Sketch a planar model of a two-cord bungie jump (Figure 3.47) in which you
represent the person of mass m_P using a point mass P and the bungie cords as
simple springs with identical spring constants k and rest lengths l_0. Assume
the distance between the attachment points is $2l_0$.

Figure 3.47 Problem 3.29. Image courtesy of Shutterstock.

    a. How many degrees of freedom does your model have?
    b. Find the equations of motion of P.
    c. Integrate the equations of motion in MATLAB using $m_P = 50$ kg, $k = 10$ N/m, and $l_0 = 5$ m, assuming that the person falls from rest.
    d. Plot the tension in the bungie cords versus time. How strong do they have to be so they won't snap?

3.30 Consider a lunar rover of mass $m = 200$ kg traveling at constant speed over a semicircular hill of radius $\rho = 100$ m. The acceleration due to gravity on the moon is $g = 1.6$ m/s^2. How fast can the rover travel without leaving the moon's surface anywhere on the hill?

3.31 Consider a mass m attached to the end of a rope of length l. One end of the rope is fixed. If the mass travels at a constant speed v_0 in a circle centered on the fixed end of the rope (with the rope taut), what is the magnitude of the centripetal acceleration on the mass?

3.32 Consider a race car being transported in a moving trailer, as shown in Figure 3.48. Suppose the trailer is traveling at 30 mph when it hits a bump that jostles the race car off of its blocks and opens the trailer ramp. The race car starts to accelerate down the trailer ramp, which is 1 m tall and 5 m long. What

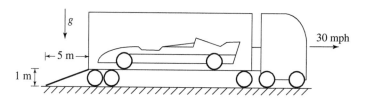

Figure 3.48 Problem 3.32.

is the inertial velocity of the car when it hits the ground? In what direction does it roll (left or right)? [HINT: Model the car as a point mass.]

3.33 Consider a mass-spring system in a falling box (Figure 3.49). Let $m_P = 1$ kg, the spring constant $k = 2$ N/m, and the rest length of the spring $l_0 = 0.2$ m. Find the equations of motion of the mass.

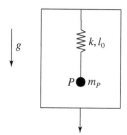

Figure 3.49 Problem 3.33.

3.34 Consider a helium balloon attached by a taut string to an inertially fixed point (Figure 3.50). Assume that the buoyancy force is equal to twice the weight of

Figure 3.50 Problem 3.34.

the balloon. Find the equations of motion of the balloon relative to the fixed point using the following steps:

    a. Sketch a point-mass model of the balloon and string.
    b. Choose coordinates and define reference frame(s).
    c. Draw a free-body diagram of the balloon.
    d. Derive the inertial kinematics of the balloon.
    e. Write down Newton's second law for the balloon and solve for its equation(s) of motion.

3.35 Suppose you forget to tie down a load of mass m on a flatbed truck of mass M, as in Figure 3.51. Suppose the truck starts to move forward with acceleration a. What does the coefficient of static friction have to be for the load not to slide off the back of the truck?

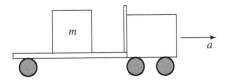

Figure 3.51 Problem 3.35.

3.36 Suppose a batter hits a fly ball that leaves his bat with speed v_0 at angle θ_0 with the horizontal (Figure 3.52). Assuming that the ball is initially at a height of h_0 above the ground and there is no air resistance, how high, h, does it go? How far, d, does it go?

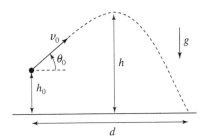

Figure 3.52 Problem 3.36.

3.37 A mass m is on an inclined plane at angle θ, as shown in Figure 3.53. Attached to the mass is a spring of spring constant k that is fixed at the top of the incline. There is gravity in this problem.

    a. Assuming a coefficient of friction μ, find the equation of motion for this system using Cartesian coordinates in the $\mathcal{A}$ frame. [HINT: Use the sgn($\cdot$) function to extract the sign of the speed.]
    b. Solve the equation of motion assuming no friction with initial conditions $x(0) = x_0$ and $\dot{x}(0) = \dot{x}_0$.

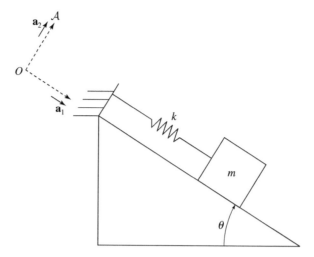

Figure 3.53 Problem 3.37.

3.38 Consider the classic problem of boring a tunnel through the earth (say, from Tokyo to New York). Figure 3.54 shows such a tunnel bored at colatitude λ (the colatitude is the angle from the north). Assume a mass P can move freely in the tunnel, there is no friction, and the earth is not rotating. The only force acting on P is gravity. Show that the equation of motion for P is simple harmonic motion. What is the frequency of the oscillation?

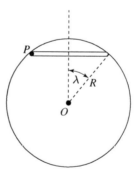

Figure 3.54 Problem 3.38.

[HINT: R is the radius of the earth (6,378.145 km). It is a well-known result (that we haven't discussed) that the gravitational attraction on a particle inside the earth is due only to the mass inside the radial position of the particle. In other words, there is no gravitational force inside a spherical shell. Assume the earth has uniform density ρ, and ignore the effect of the missing material from the tunnel (it is small).]

CHAPTER FOUR

Linear and Angular Momentum of a Particle

Chapter 3 provided all of the basic tools you need to find the equations of motion for any (planar) particle dynamics problem starting from Newton's second law. Nevertheless, there are many additional concepts that can make finding equations of motion easier. Using these concepts, we can sometimes solve for the trajectory of a particle— or properties of the trajectory—without having to find and integrate the equations of motion. This chapter explores two of these additional concepts that lend new insights into the motion and provide new tools for solving specific problems. Nevertheless, the only physics is still Newton's three laws.

4.1 Linear Momentum and Linear Impulse

Recall from Chapter 2 the linear momentum form of Newton's second law

$$\mathbf{F}_P = \frac{^{\mathcal{I}}d}{dt}\left(^{\mathcal{I}}\mathbf{p}_{P/O}\right),$$

where $^{\mathcal{I}}\mathbf{p}_{P/O} \triangleq m_P{}^{\mathcal{I}}\mathbf{v}_{P/O}$ is the linear momentum. Integrating this equation with respect to time for an arbitrary force $\mathbf{F}_P$ yields

$$\int_{t_1}^{t_2} \mathbf{F}_P\,dt = \int_{t_1}^{t_2}{}^{\mathcal{I}}d\left(^{\mathcal{I}}\mathbf{p}_{P/O}\right) = {}^{\mathcal{I}}\mathbf{p}_{P/O}(t_2) - {}^{\mathcal{I}}\mathbf{p}_{P/O}(t_1) \tag{4.1}$$

or, after rearranging and using the definition of linear momentum,

$$\boxed{m_P{}^{\mathcal{I}}\mathbf{v}_{P/O}(t_2) = m_P{}^{\mathcal{I}}\mathbf{v}_{P/O}(t_1) + \int_{t_1}^{t_2} \mathbf{F}_P\,dt.} \tag{4.2}$$

Eqs. (4.1) and (4.2) state that the change in linear momentum of a particle P is equal to the integrated force acting on P. Unfortunately, the integral on the right side of

Eq. (4.2) usually is too hard to solve or depends on the position or velocity (the state of the particle), making Eq. (4.2) a differential equation. However, there are three situations where this equation is useful:

a. $\mathbf{F}_P$ *is identically zero.* That is, the total force acting on particle P is zero. In this case, the magnitude and direction of $^T\mathbf{p}_{P/O}$ are constant over the trajectory of P. This case is simply the law of conservation of linear momentum (or Newton's first law) discussed in Section 2.2.5.

b. $\mathbf{F}_P$ *is constant or is a known function of time.* That is, $\mathbf{F}_P \neq 0$ is independent of the position and velocity of the particle. In this case, the integral in Eq. (4.2) can be solved, either explicitly or numerically, to relate the initial and final velocities at t_1 and t_2, respectively. Constant, uniform gravity is an example of such a force (e.g., Tutorial 2.6).

c. $\Delta t \triangleq t_2 - t_1$ *is extremely short.* That is, $\mathbf{F}_P$ acts over a very small time interval. We call such a force an *impulsive force.* In this case—which is the subject of the rest of this section—we can safely ignore the details of the dependence of $\mathbf{F}_P$ on the particle state to solve Eq. (4.2).

We begin our discussion of the third case with the definition of *linear impulse.*

Definition 4.1 The **linear impulse** $\overline{\mathbf{F}}_P(t_1, t_2)$ acting on particle P is the integrated force on P over a (short) time interval $\Delta t \triangleq t_2 - t_1$,

$$\boxed{\overline{\mathbf{F}}_P(t_1, t_2) \triangleq \int_{t_1}^{t_2} \mathbf{F}_P \, dt.}$$

Figure 4.1a shows a schematic of the magnitude $\|\mathbf{F}_P\|$ of an impulsive force and the area underneath its plot with respect to time—which is the magnitude $\|\overline{\mathbf{F}}_P(t_1, t_2)\|$ of the impulse. Remember that the impulse is a vector and thus also has a direction.

Linear impulse is an extremely useful concept. Rather than try and understand the details of what might be a very complex force, we simply ignore the motion during application of the force and treat the problem as an instantaneous change in the momentum, via Eq. (4.2), due to the linear impulse. In this model, the impulse is so short that we assume the change in linear momentum of particle P occurs without significant motion of P. This assumption is illustrated in Figure 4.1b and c. In fact, this assumption was implicit in the solution to the tossed point-mass problem in Tutorial 2.6. Recall that we neglected the detailed modeling of the force during the toss and simply computed the trajectory as a function of the initial conditions. The tossing force can be treated as an impulse whose effect is to change the initial conditions. The goal of the thrower is to apply just the right impulse to achieve the desired initial velocity. After the release of the particle, we can solve the equations of motion for the trajectory under gravity alone.

Why is it important to assume the force acts only over a short interval? Isn't Eq. (4.2) exact? Yes, it is. However, most forces depend on the state of the particle. If the time interval Δt is small, then we can treat the force as integrable and avoid solving

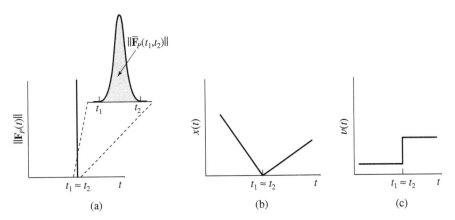

Figure 4.1 Linear impulse $\overline{\mathbf{F}}_P(t_1, t_2)$ is the integral from $t = t_1$ to $t = t_2$ of the force $\mathbf{F}_P$ that produces an instantaneous change of direction of motion and a discontinuous change in velocity. (a) An impulsive force. (b) Instantaneous change of direction. (c) Discontinuous change in velocity.

a differential equation. It is rare that a force is independent of position and velocity over long periods of time. However, when the force on a particle is impulsive and acts only over a very short interval, modeling it as state independent—and assuming the particle position doesn't change—is an excellent approximation.

Example 4.1 Impulsive Spring

One of the simplest examples of a linear impulse is the sudden striking of a mass on a spring (see Tutorial 2.3 for the solution of the motion of a mass on a spring). This example is an excellent model of, for example, the motion of a car tire hitting a bump in the road. The car is massive enough that the tires can be considered connected to a fixed wall by means of a spring (and, possibly, a shock absorber).

Consider the model illustrated in Figure 4.2. Suppose the mass is initially at rest. It is struck by a force that produces a linear impulse of magnitude $\overline{F}_P(t_1, t_2)$ in the negative x-direction. This impulse has units of N-s. In our idealized model of the impulse, the mass does not move until after time t_2. Using Eq. (4.2), the mass has horizontal speed $\dot{x}(t_2) = -\overline{F}_P(t_1, t_2)/m_P$ after the impulse. Assuming that the

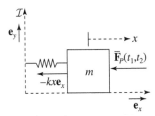

Figure 4.2 A mass on a spring hit by impulse $\overline{\mathbf{F}}_P(t_1, t_2)$ in the $-\mathbf{e}_x$ direction.

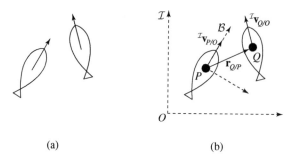

(a)   (b)

Figure 4.3 Synchronized swimming. (a) Top view. (b) Reference frames.

position of the mass at $t = t_2$ is $x(t_2) = 0$, then we can simply substitute the speed $\dot{x}(t_2)$ into the trajectory Eqs. (2.22) and (2.23) to find the final motion of the mass:

$$x(t) = \frac{-\overline{F}_P(t_1, t_2)}{m_P \omega_0} \sin(\omega_0(t - t_2)) \tag{4.3}$$

$$\dot{x}(t) = \frac{-\overline{F}_P(t_1, t_2)}{m_P} \cos(\omega_0(t - t_2)). \tag{4.4}$$

Thus the linear impulse $\overline{F}_P(t_1, t_2)$ sends the mass-spring system into oscillatory motion with amplitude proportional to the magnitude of the impulse.

Example 4.2 Synchronized Swimming

Consider the two fish shown in Figure 4.3a. Each fish glides through the water, beating its tail intermittently. Suppose that, between tail beats, each fish glides at constant velocity. The interaction between a flapping tail and the water is quite complicated; nonetheless we can study the fish motion using the concept of linear impulse.

Model the fish using two point masses P and Q. Assume Q is gliding at a constant velocity $^{\mathcal{I}}\mathbf{v}_{Q/O}$, and suppose fish P seeks to align with fish Q, knowing only the velocity of Q relative to its velocity. That is, fish P cannot estimate the inertial velocity $^{\mathcal{I}}\mathbf{v}_{Q/O}$, only its relative velocity $^{\mathcal{B}}\mathbf{v}_{Q/P}$ in a path coordinate frame fixed to P. Let's find the linear impulse that fish P needs to exert to match inertial velocities with fish Q. Remember, by Newton's third law, if a fish exerts an impulse on the water with its tail, the water then exerts an equal and opposite linear impulse that accelerates the fish.

From Figure 4.3b we have

$$\mathbf{r}_{Q/O} = \mathbf{r}_{Q/P} + \mathbf{r}_{P/O},$$

which implies

$$^{\mathcal{I}}\mathbf{v}_{Q/O} = {}^{\mathcal{I}}\mathbf{v}_{Q/P} + {}^{\mathcal{I}}\mathbf{v}_{P/O}. \tag{4.5}$$

Between tail beats, the path frame $\mathcal{B}$ is inertial, which implies $^{\mathcal{I}}\mathbf{v}_{Q/P} = {}^{\mathcal{B}}\mathbf{v}_{Q/P}$. If fish P beats its tail from time t_1 to t_2, it seeks to satisfy

$$^{\mathcal{I}}\mathbf{v}_{P/O}(t_2) = {}^{\mathcal{I}}\mathbf{v}_{Q/O}(t_1)$$

$$= {}^{\mathcal{B}}\mathbf{v}_{Q/P}(t_1) + {}^{\mathcal{I}}\mathbf{v}_{P/O}(t_1).$$

Using Eq. (4.2) and Definition 4.1, the linear impulse required to instantaneously change the velocity of P to align with Q is

$$\overline{\mathbf{F}}_P(t_1, t_2) = m_P\left({}^{\mathcal{I}}\mathbf{v}_{P/O}(t_2) - {}^{\mathcal{I}}\mathbf{v}_{P/O}(t_1)\right)$$

$$= m_P {}^{\mathcal{B}}\mathbf{v}_{Q/P}(t_1). \tag{4.6}$$

Note the smaller the speed of Q relative to P, the smaller the magnitude of linear impulse required to align them. (Of course a real fish is not a point mass and thus would also want to rotate its body while changing its velocity direction.)

Eq. (4.6) also looks a bit like a linear momentum. To examine this, we observe that

$$^{\mathcal{I}}\mathbf{v}_{P/O} = {}^{\mathcal{I}}\mathbf{v}_{P/Q} + {}^{\mathcal{I}}\mathbf{v}_{Q/O},$$

which, using Eq. (4.5), implies $^{\mathcal{I}}\mathbf{v}_{Q/P} = -{}^{\mathcal{I}}\mathbf{v}_{P/Q}$. Therefore an equivalent expression for the linear impulse in Eq. (4.6) is

$$\overline{\mathbf{F}}_P(t_1, t_2) = -m_P {}^{\mathcal{B}}\mathbf{v}_{P/Q}(t_1) = -{}^{\mathcal{B}}\mathbf{p}_{P/Q}(t_1).$$

As we might have expected, the applied linear impulse eliminates the linear momentum of fish P relative to fish Q.

The concept of the impulse as setting initial conditions is a powerful and commonly used one. It is rare that the detailed description of an impulsive force is available; normally we work only with its overall effect on the motion. The examples above are vivid demonstrations of this principle. We return to linear impulses again in Chapter 6 when addressing collisions and impacts.

4.2 Angular Momentum and Angular Impulse

This section introduces *angular momentum*. Angular momentum is not a new physical concept; it is a mathematical manipulation that provides both new insights and new ways of solving problems. Although its value may not be obvious now, angular momentum and its integrated form—the angular impulse—are enormously useful for solving many classes of problems, particularly those involving extended rigid bodies. In this chapter, we introduce the definition of the angular momentum of a particle and use it to solve some important examples. We return to it repeatedly in later chapters, as it is a key tool for understanding and modeling complex multiparticle systems. The key thing to remember about angular momentum is that it depends on a reference point. We often—but not always—select the reference point as the origin of an inertial reference frame.

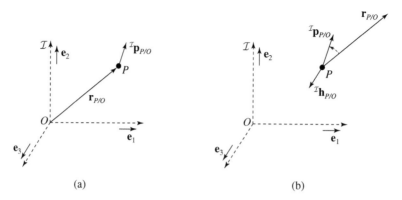

Figure 4.4 (a) Assume position $\mathbf{r}_{P/O}$ and linear momentum $^{\mathcal{I}}\mathbf{p}_{P/O}$ are confined to the plane spanned by $\mathbf{e}_1$ and $\mathbf{e}_2$. (b) Angular momentum $^{\mathcal{I}}\mathbf{h}_{P/O}$ is a vector in the $\mathbf{e}_3$ direction formed by applying the right-hand rule to the cross product $\mathbf{r}_{P/O} \times {}^{\mathcal{I}}\mathbf{p}_{P/O}$.

4.2.1 Angular Momentum Relative to an Inertially Fixed Point

The angular momentum $^{\mathcal{I}}\mathbf{h}_{P/O}$ represents the motion of particle P relative to reference point O. We always explicitly include the reference point in our notation for angular momentum.

Definition 4.2 The **angular momentum** $^{\mathcal{I}}\mathbf{h}_{P/O}$ of a point P relative to point O fixed in frame $\mathcal{I}$ is

$$
{}^{\mathcal{I}}\mathbf{h}_{P/O} \triangleq \mathbf{r}_{P/O} \times {}^{\mathcal{I}}\mathbf{p}_{P/O} = \mathbf{r}_{P/O} \times m_P {}^{\mathcal{I}}\mathbf{v}_{P/O}.
$$

Definition 4.2 is illustrated in Figure 4.4b. In this section we assume that the reference point O is the origin of an inertial frame.

What have we gained by defining angular momentum? Let's look at its inertial rate of change. Differentiating Definition 4.2 using the product rule results in

$$
\frac{{}^{\mathcal{I}}d}{dt}\left({}^{\mathcal{I}}\mathbf{h}_{P/O}\right) = \frac{{}^{\mathcal{I}}d}{dt}\left(\mathbf{r}_{P/O}\right) \times m_P {}^{\mathcal{I}}\mathbf{v}_{P/O} + \mathbf{r}_{P/O} \times m_P \frac{{}^{\mathcal{I}}d}{dt}\left({}^{\mathcal{I}}\mathbf{v}_{P/O}\right)
$$

$$
= m_P\big(\underbrace{{}^{\mathcal{I}}\mathbf{v}_{P/O} \times {}^{\mathcal{I}}\mathbf{v}_{P/O}}_{=0} + \mathbf{r}_{P/O} \times {}^{\mathcal{I}}\mathbf{a}_{P/O}\big).
$$

Note that the first term in parentheses is zero, since the cross product of any vector with itself is zero (see Appendix B). Using Newton's second law to further simplify the result, we obtain

$$
\frac{{}^{\mathcal{I}}d}{dt}\left({}^{\mathcal{I}}\mathbf{h}_{P/O}\right) = m_P\big(\mathbf{r}_{P/O} \times {}^{\mathcal{I}}\mathbf{a}_{P/O}\big) = \mathbf{r}_{P/O} \times m_P {}^{\mathcal{I}}\mathbf{a}_{P/O} = \mathbf{r}_{P/O} \times \mathbf{F}_P.
$$

This expression yields a new definition.

Definition 4.3 The **moment** acting on point P relative to point O due to the net force $\mathbf{F}_P$ is

$$\mathbf{M}_{P/O} \triangleq \mathbf{r}_{P/O} \times \mathbf{F}_P.$$

The final result is the angular momentum form of Newton's second law:

$$\frac{^{\mathcal{I}}d}{dt}\left(^{\mathcal{I}}\mathbf{h}_{P/O}\right) = \mathbf{M}_{P/O}. \tag{4.7}$$

We have chosen our words carefully here. There is no new physics in Eq. (4.7). The result was obtained simply by using vector algebra and Newton's second law. We are still studying the motion of particles and how they behave according to Newton's laws. Using some new definitions, we derived a new formula that is enormously useful—especially when considering multiple particles and rigid bodies. In fact, this new statement of the second law can already make solving some problems easier, as shown in the following examples.

Example 4.3 Calculating the Angular Momentum of a Moving Point Mass

In this example we compute the angular momentum of a particle P with mass $m_P = 1\,\text{kg}$ relative to the origin O, as shown in Figure 4.5. Using Cartesian coordinates, the instantaneous position of the particle as components in frame $\mathcal{I}$ is

$$\mathbf{r}_{P/O} = 4\mathbf{e}_1 + 3\mathbf{e}_2 \text{ m}.$$

The instantaneous velocity relative to $\mathcal{I}$ is also given as components in $\mathcal{I}$:

$$^{\mathcal{I}}\mathbf{v}_{P/O} = -6\mathbf{e}_1 + 8\mathbf{e}_2 \text{ m/s}. \tag{4.8}$$

Using Definition 4.2, the instantaneous angular momentum of P relative to O is

$$\begin{aligned}
^{\mathcal{I}}\mathbf{h}_{P/O} &= (1)(4\mathbf{e}_1 + 3\mathbf{e}_2) \times (-6\mathbf{e}_1 + 8\mathbf{e}_2) \\
&= 32(\mathbf{e}_1 \times \mathbf{e}_2) - 18(\mathbf{e}_2 \times \mathbf{e}_1) \\
&= 32\mathbf{e}_3 + 18\mathbf{e}_3 \\
&= 50\mathbf{e}_3 \text{ kg-m}^2/\text{s}^2.
\end{aligned}$$

It is also interesting to compute the angular momentum using the polar frame, also shown in Figure 4.5. In terms of components in $\mathcal{B}$ the instantaneous position of P is

$$\mathbf{r}_{P/O} = 5\mathbf{e}_r.$$

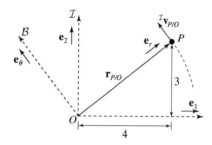

Figure 4.5 Point P located at Cartesian coordinates $(4, 3)_\mathcal{I}$ in reference frame $\mathcal{I}$ traveling at velocity $^\mathcal{I}\mathbf{v}_{P/O}$.

Referring to Section 3.4.1, we can write the instantaneous transformation array between the two frames as

	$\mathbf{e}_1$	$\mathbf{e}_2$
$\mathbf{e}_r$	$\frac{4}{5}$	$\frac{3}{5}$
$\mathbf{e}_\theta$	$-\frac{3}{5}$	$\frac{4}{5}$

The instantaneous inertial velocity is then found from Eq. (4.8) and the transformation array,

$$^\mathcal{I}\mathbf{v}_{P/O} = -6\left(\frac{4}{5}\mathbf{e}_r - \frac{3}{5}\mathbf{e}_\theta\right) + 8\left(\frac{3}{5}\mathbf{e}_r + \frac{4}{5}\mathbf{e}_\theta\right)$$

$$= 10\mathbf{e}_\theta.$$

The instantaneous angular momentum is then again found from the definition:

$$^\mathcal{I}\mathbf{h}_{P/O} = (1)(5\mathbf{e}_r) \times (10\mathbf{e}_\theta)$$

$$= 50(\mathbf{e}_r \times \mathbf{e}_\theta)$$

$$= 50\mathbf{e}_3.$$

Example 4.4 Solving the Pendulum Using Angular Momentum

This example shows how using angular momentum makes the simple pendulum even easier to solve. We again refer to Figure 3.18b, which illustrates the pendulum and the polar frame $\mathcal{B} = (O, \mathbf{e}_r, \mathbf{e}_\theta, \mathbf{e}_3)$. As in Eq. (3.38), the position of the pendulum bob is

$$\mathbf{r}_{P/O} = l\mathbf{e}_r$$

and the velocity of the pendulum bob given in Eq. (3.39) is

$$^\mathcal{I}\mathbf{v}_{P/O} = \frac{^\mathcal{I}d}{dt}(l\mathbf{e}_r) = l^\mathcal{I}\boldsymbol{\omega}^\mathcal{B} \times \mathbf{e}_r = l\dot{\theta}\mathbf{e}_\theta.$$

Using Definition 4.2, the angular momentum of the pendulum bob about the origin O is

$$^\mathcal{I}\mathbf{h}_{P/O} = \mathbf{r}_{P/O} \times m_P{}^\mathcal{I}\mathbf{v}_{P/O} = l\mathbf{e}_r \times m_P l\dot{\theta}\mathbf{e}_\theta = m_P l^2 \dot{\theta}\mathbf{e}_3.$$

Therefore, since $\mathbf{e}_3$ is fixed in $\mathcal{I}$, we have

$$\frac{^{\mathcal{I}}d}{dt}\left(^{\mathcal{I}}\mathbf{h}_{P/O}\right) = m_P l^2 \ddot{\theta} \mathbf{e}_3. \tag{4.9}$$

As before, the total force on the pendulum bob is $\mathbf{F}_P = -F\mathbf{e}_r + m_P g \mathbf{e}_1$. Using $\mathbf{e}_r \times \mathbf{e}_r = 0$, the moment of the pendulum bob about O is

$$\mathbf{M}_{P/O} = \mathbf{r}_{P/O} \times \mathbf{F}_P = l\mathbf{e}_r \times m_P g(\cos\theta \mathbf{e}_r - \sin\theta \mathbf{e}_\theta)$$

$$= -lm_P g \sin\theta \mathbf{e}_3. \tag{4.10}$$

Note that the tension force F does not contribute to the moment $\mathbf{M}_{P/O}$. Substituting Eqs. (4.9) and (4.10) into Eq. (4.7) produces

$$m_P l^2 \ddot{\theta} \mathbf{e}_3 = -lm_P g \sin\theta \mathbf{e}_3,$$

which can be solved to find the equation of motion

$$\ddot{\theta} + \frac{g}{l}\sin\theta = 0.$$

This approach is even more compact than the approach taken in Example 3.11. By using angular momentum, we eliminated the force of constraint from the problem and worked immediately with only a single equation for the single degree of freedom.

We cannot emphasize enough that the angular momentum of a particle depends on a reference point. The following example reinforces this fact.

Example 4.5 Angular Momentum of a Particle Relative to Two Different Points

This example compares the angular momentum $^{\mathcal{I}}\mathbf{h}_{P/O'}$ of particle P with respect to point O', the origin of a polar frame $\mathcal{B} = (O', \mathbf{e}_r, \mathbf{e}_\theta, \mathbf{e}_z)$, to the angular momentum $^{\mathcal{I}}\mathbf{h}_{P/O}$ of P with respect to O, the origin of an inertial frame $\mathcal{I} = (O, \mathbf{e}_x, \mathbf{e}_y, \mathbf{e}_z)$. Particle P moves along a circular trajectory centered at O' with radius l, as shown in Figure 4.6a. The vector $\mathbf{r}_{O'/O}$ is oriented along the $\mathbf{e}_x$ axis and has (fixed) magnitude d. The reference frames $\mathcal{I}$ and $\mathcal{B}$, illustrated in Figure 4.6b, are related by the transformation table

$$\begin{array}{c|cc} & \mathbf{e}_x & \mathbf{e}_y \\ \hline \mathbf{e}_r & \cos\theta & \sin\theta \\ \mathbf{e}_\theta & -\sin\theta & \cos\theta \end{array} \tag{4.11}$$

The kinematics of P with respect to O' are

$$\mathbf{r}_{P/O'} = l\mathbf{e}_r$$

$$^{\mathcal{I}}\mathbf{v}_{P/O'} = l\dot{\theta}\mathbf{e}_\theta.$$

Using Definition 4.2, the angular momentum of P with respect to O' is

$$^{\mathcal{I}}\mathbf{h}_{P/O'} = l\mathbf{e}_r \times m_P l\dot{\theta}\mathbf{e}_\theta = m_P l^2 \dot{\theta} \mathbf{e}_z. \tag{4.12}$$

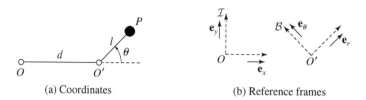

(a) Coordinates   (b) Reference frames

Figure 4.6 Angular momentum of a particle relative to two different points.

The kinematics of P with respect to O are

$$\mathbf{r}_{P/O} = d\mathbf{e}_x + l\mathbf{e}_r$$

$$^{\mathcal{I}}\mathbf{v}_{P/O} = l\dot{\theta}\mathbf{e}_\theta,$$

where we used the fact that $\dot{d} = 0$. Using Definition 4.2 and the transformation table Eq. (4.11), the angular momentum of P with respect to O is

$$^{\mathcal{I}}\mathbf{h}_{P/O} = (d\mathbf{e}_x + l\mathbf{e}_r) \times m_P l\dot{\theta}\mathbf{e}_\theta = m_P(dl\dot{\theta}\mathbf{e}_x \times (-\sin\theta\mathbf{e}_x + \cos\theta\mathbf{e}_y) + l^2\dot{\theta}\mathbf{e}_z)$$

$$= m_P l\dot{\theta}(d\cos\theta + l)\mathbf{e}_z. \tag{4.13}$$

Comparing Eqs. (4.12) and (4.13), we observe that $^{\mathcal{I}}\mathbf{h}_{P/O} \neq {}^{\mathcal{I}}\mathbf{h}_{P/O'}$, as $^{\mathcal{I}}\mathbf{h}_{P/O}$ depends on the polar angle θ, whereas $^{\mathcal{I}}\mathbf{h}_{P/O'}$ does not.

We conclude this section with a discussion of angular momentum conservation. Recall from Section 2.2.5 that Newton's first law was equivalent to the law of conservation of linear momentum. In the absence of forces on a particle, the linear momentum is a constant of the motion. We can state a similar law for angular momentum by noting from Eq. (4.7) that, in the absence of moments acting on a particle, its angular momentum is a constant of the motion:

$$^{\mathcal{I}}\mathbf{h}_{P/O} = \text{constant.}$$

This observation gives us a new law.

Law 4.1 The law of **conservation of angular momentum of a particle** states that, when the total moment acting on a particle relative to point O is zero, the inertial angular momentum of the particle relative to O is a constant of the motion.

It is important to note that angular momentum can be conserved even when there is a nonzero force acting on a particle, as long as the total moment about O is zero. This is in contrast to linear momentum, where the total force must be zero for the momentum to be a constant. We discuss an important example of this in Section 4.2.4.

4.2.2 Angular Momentum Relative to an Arbitrary Point

Our discussion in the previous section is a special case of the more general scenario in which the angular momentum is defined relative to an arbitrary point that may be accelerating in absolute space.

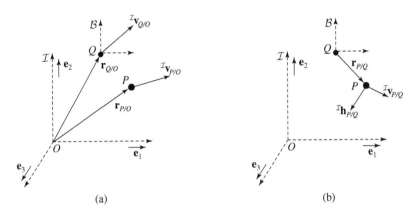

Figure 4.7 Angular momentum relative to an arbitrary point. (a) Points P and Q move in the plane spanned by $\mathbf{e}_1$ and $\mathbf{e}_2$. There is a nonrotating frame $\mathcal{B}$ attached to Q. (b) Angular momentum $^{\mathcal{I}}\mathbf{h}_{P/Q}$ is the vector in the $\mathbf{e}_3$ direction formed by applying the right-hand rule to the cross product $\mathbf{r}_{P/Q} \times m_P{}^{\mathcal{I}}\mathbf{v}_{P/Q}$, where $^{\mathcal{I}}\mathbf{v}_{P/Q}$ is equal to $^{\mathcal{I}}\mathbf{v}_{P/O} - {}^{\mathcal{I}}\mathbf{v}_{Q/O}$.

Figure 4.7 illustrates the angular momentum of point P relative to an arbitrary point Q moving in the inertial reference frame $\mathcal{I}$. We also introduce a nonrotating but translating frame B fixed to Q. Using Definition 4.2, the angular momentum relative to Q is

$$^{\mathcal{I}}\mathbf{h}_{P/Q} = \mathbf{r}_{P/Q} \times m_P{}^{\mathcal{I}}\mathbf{v}_{P/Q}, \qquad (4.14)$$

where $^{\mathcal{I}}\mathbf{v}_{P/Q} = \frac{^{\mathcal{I}}d}{dt}\left(\mathbf{r}_{P/Q}\right)$.[1] We substitute in Eq. (4.14) for both the position and velocity of P relative to Q using the appropriate vector triads from Figure 4.7 to obtain

$$^{\mathcal{I}}\mathbf{h}_{P/Q} = (\mathbf{r}_{P/O} - \mathbf{r}_{Q/O}) \times m_P({}^{\mathcal{I}}\mathbf{v}_{P/O} - {}^{\mathcal{I}}\mathbf{v}_{Q/O}).$$

The inertial rate of change of the angular momentum is

$$\frac{^{\mathcal{I}}d}{dt}\left(^{\mathcal{I}}\mathbf{h}_{P/Q}\right) = \underbrace{\frac{^{\mathcal{I}}d}{dt}\left(\mathbf{r}_{P/O} - \mathbf{r}_{Q/O}\right) \times m_P\left(^{\mathcal{I}}\mathbf{v}_{P/O} - {}^{\mathcal{I}}\mathbf{v}_{Q/O}\right)}_{=0}$$

$$+ (\mathbf{r}_{P/O} - \mathbf{r}_{Q/O}) \times m_P\frac{^{\mathcal{I}}d}{dt}\left(^{\mathcal{I}}\mathbf{v}_{P/O} - {}^{\mathcal{I}}\mathbf{v}_{Q/O}\right)$$

$$= \underbrace{(\mathbf{r}_{P/O} - \mathbf{r}_{Q/O})}_{=\mathbf{r}_{P/Q}} \times m_P\left(^{\mathcal{I}}\mathbf{a}_{P/O} - {}^{\mathcal{I}}\mathbf{a}_{Q/O}\right),$$

---

[1] As in the discussion of relative motion in Chapter 3, we could replace the velocity $^{\mathcal{I}}\mathbf{v}_{P/Q}$ in Eq. (4.14) with $^{\mathcal{B}}\mathbf{v}_{P/Q}$, since the reference frame B is only translating and not rotating. However, subsequent chapters consider both rotating and translating frames, and this notation would introduce much confusion. Therefore we continue to use $^{\mathcal{I}}\mathbf{v}_{P/Q}$.

where the first term is zero because a vector crossed into itself is zero. Finally, from Newton's second law, we substitute $m_P{}^{\mathcal{I}}\mathbf{a}_{P/O} = \mathbf{F}_P$ to find

$$\frac{{}^{\mathcal{I}}d}{dt}\left({}^{\mathcal{I}}\mathbf{h}_{P/Q}\right) = \underbrace{\mathbf{r}_{P/Q} \times \mathbf{F}_P}_{\triangleq \mathbf{M}_{P/Q}} -\mathbf{r}_{P/Q} \times m_P{}^{\mathcal{I}}\mathbf{a}_{Q/O}.$$

The term $\mathbf{M}_{P/Q}$ is the moment acting on P relative to Q. This calculation establishes the final expression for the inertial rate of change of the angular momentum of P relative to Q:

$$\boxed{\frac{{}^{\mathcal{I}}d}{dt}\left({}^{\mathcal{I}}\mathbf{h}_{P/Q}\right) = \mathbf{M}_{P/Q} - \mathbf{r}_{P/Q} \times m_P{}^{\mathcal{I}}\mathbf{a}_{Q/O},} \tag{4.15}$$

where

$$\mathbf{M}_{P/Q} \triangleq \mathbf{r}_{P/Q} \times \mathbf{F}_P.$$

Eq. (4.15) is the most general equation for the inertial rate of change of angular momentum relative to an arbitrary point Q. By comparison to Eq. (4.7), observe that writing the angular momentum form of Newton's second law relative to an accelerating point Q adds a correction term $-\mathbf{r}_{P/Q} \times m_P{}^{\mathcal{I}}\mathbf{a}_{Q/O}$, just as writing the linear form in an accelerating frame added a similar correction term in Chapter 3, Eq. (3.57). The correction term is sometimes called the *inertial moment*. Note also that, if point Q is not accelerating, that is, if ${}^{\mathcal{I}}\mathbf{a}_{Q/O} = 0$, then frame $\mathcal{B}$ is an inertial frame and Eq. (4.15) equals Eq. (4.7).

Example 4.6 Solving the Pendulum Using Angular Momentum in an Accelerating Frame

Here we consider again the pendulum in an accelerating box of Example 3.14. This time, however, we use angular momentum and Eq. (4.15) to solve for the equation of motion. Figure 4.8 is the same as Figure 3.27 except that we have added an additional polar frame $\mathcal{C}$. The origin of $\mathcal{C}$ is O', the base of the pendulum.

Following Example 4.4, we write the position of the pendulum bob relative to O' using polar coordinates as a vector component in the polar frame $\mathcal{C}$,

$$\mathbf{r}_{P/O'} = l\mathbf{e}_r.$$

The corresponding velocity is

$${}^{\mathcal{I}}\mathbf{v}_{P/O'} = l\dot{\theta}\mathbf{e}_\theta.$$

The angular momentum of P relative to the (accelerating) attachment point O' is

$${}^{\mathcal{I}}\mathbf{h}_{P/O'} = \mathbf{r}_{P/O'} \times m_P{}^{\mathcal{I}}\mathbf{v}_{P/O'} = m_P l^2 \dot{\theta}\mathbf{e}_3.$$

Using Eq. (4.10), the moment on the pendulum bob about O' is

$$\mathbf{M}_{P/O'} = -l m_P g \sin\theta\mathbf{e}_3.$$

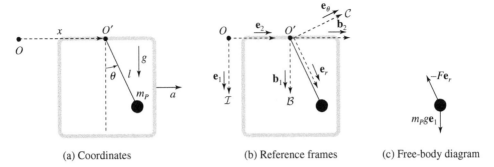

(a) Coordinates   (b) Reference frames   (c) Free-body diagram

Figure 4.8 Solving the pendulum problem using angular momentum in an accelerating frame.

Because the point O' is accelerating at ${}^{\mathcal{I}}\mathbf{a}_{O'/O} = a\mathbf{e}_2$, we use Eq. (4.15) for the equation of motion. The inertial moment term in Eq. (4.15) is

$$\mathbf{r}_{P/O'} \times m_P{}^{\mathcal{I}}\mathbf{a}_{O'/O} = m_P l \mathbf{e}_r \times a\mathbf{e}_2 = m_P l a \cos\theta \mathbf{e}_3,$$

where we used $\mathbf{r}_{P/O} - \mathbf{r}_{O'/O} = \mathbf{r}_{P/O'}$ and $\mathbf{e}_r \times \mathbf{e}_2 = \cos\theta \mathbf{e}_3$ (see the transformation table in Example 3.11).

Using the fact that $\mathbf{e}_3$ is fixed in $\mathcal{I}$, Eq. (4.15) becomes

$$\frac{{}^{\mathcal{I}}d}{dt}\left({}^{\mathcal{I}}\mathbf{h}_{P/O'}\right) = m_P l^2 \ddot{\theta} \mathbf{e}_3 = -l m_P g \sin\theta \mathbf{e}_3 - l m_P a \cos\theta \mathbf{e}_3.$$

This vector equation is equivalent to the single scalar equation of motion

$$\ddot{\theta} + \frac{g}{l}\sin\theta = -\frac{a}{l}\cos\theta. \tag{4.16}$$

Eq. (4.16) is the same equation of motion as in Eq. (3.58).

4.2.3 Angular Impulse

Just as we defined a linear impulse for short impulsive forces, we can define a corresponding angular impulse when a moment acts only over a short interval.

Definition 4.4 The **angular impulse** $\overline{\mathbf{M}}_{P/O}(t_1, t_2)$ acting on particle P is the integrated moment on P relative to point O over a (short) time interval $\Delta t \triangleq t_2 - t_1$:

$$\overline{\mathbf{M}}_{P/O}(t_1, t_2) \triangleq \int_{t_1}^{t_2} \mathbf{M}_{P/O}\, dt.$$

As for the linear form of Newton's second law, we compute the integral with respect to time of the angular momentum form of Newton's second law in Eq. (4.7),

$${}^{\mathcal{I}}\mathbf{h}_{P/O}(t_2) = {}^{\mathcal{I}}\mathbf{h}_{P/O}(t_1) + \overline{\mathbf{M}}_{P/O}(t_1, t_2). \tag{4.17}$$

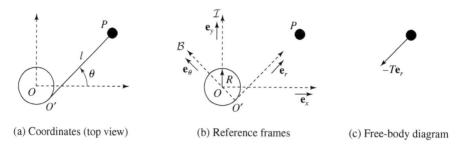

(a) Coordinates (top view)   (b) Reference frames   (c) Free-body diagram

Figure 4.9 Tetherball.

Our discussion in the previous section about linear impulses directly applies to angular impulses and impulsive moments.

Example 4.7 Tetherball

Consider the (planar) tetherball apparatus shown in Figure 4.9. In this game, two players hit the ball P in opposite directions and seek to be the first to wrap the tether around the column centered at O. In this example we find the angular impulse that reverses the velocity of the tetherball using a polar frame $\mathcal{B} = (O', \mathbf{e}_r, \mathbf{e}_\theta, \mathbf{e}_z)$. Assume the polar angle θ satisfies $\theta = 0$ when the tether is completely unwrapped, which implies that $l = L - R\theta$, where L is the total length of the tether and R is the radius of the column. We ignore the thickness of the tether. The kinematics of the ball are

$$\mathbf{r}_{P/O} = -R\mathbf{e}_\theta + l\mathbf{e}_r$$
$$= -R\mathbf{e}_\theta + (L - R\theta)\mathbf{e}_r$$
$$^\mathcal{I}\mathbf{v}_{P/O} = R\dot{\theta}\mathbf{e}_r - R\dot{\theta}\mathbf{e}_r + (L - R\theta)\dot{\theta}\mathbf{e}_\theta$$
$$= (L - R\theta)\dot{\theta}\mathbf{e}_\theta. \tag{4.18}$$

Using Definition 4.2, the angular momentum of P about O is

$$^\mathcal{I}\mathbf{h}_{P/O} = \left(-R\mathbf{e}_\theta + (L - R\theta)\mathbf{e}_r\right) \times m_P(L - R\theta)\dot{\theta}\mathbf{e}_\theta$$
$$= m_P(L - R\theta)^2\dot{\theta}\mathbf{e}_z. \tag{4.19}$$

To reverse the ball's velocity by hitting it at time t_1, we seek

$$^\mathcal{I}\mathbf{h}_{P/O}(t_2) = \mathbf{r}_{P/O}(t_1) \times m_P(-^\mathcal{I}\mathbf{v}_{P/O}(t_1)) = -^\mathcal{I}\mathbf{h}_{P/O}(t_1),$$

where t_2 is the time immediately after the impact. Using Eq. (4.17), we obtain

$$\overline{\mathbf{M}}_{P/O}(t_1, t_2) = {^\mathcal{I}\mathbf{h}}_{P/O}(t_2) - {^\mathcal{I}\mathbf{h}}_{P/O}(t_1)$$
$$= -2{^\mathcal{I}\mathbf{h}}_{P/O}(t_1) = -2m_P(L - R\theta)^2\dot{\theta}\mathbf{e}_z.$$

Next we find the equation of motion for θ after the impact. Using Definition 4.3 and Figure 4.9c, the moment on P about O after the impact is

$$\mathbf{M}_{P/O} = (-R\mathbf{e}_\theta + l\mathbf{e}_r) \times (-T\mathbf{e}_r) = -RT\mathbf{e}_z, \tag{4.20}$$

where T is the tension in the tether. Eq. (4.20) implies that the angular momentum of P about O is not conserved. Using Eq. (4.7), Eq. (4.20), and the time derivative of Eq. (4.19), the equation of motion is

$$\ddot{\theta} = \frac{2R\dot{\theta}^2}{L - R\theta} - \frac{RT}{m_P(L - R\theta)^2}. \tag{4.21}$$

But what is the tension T? To find T, we use Newton's second law again. Using Figure 4.9c and the inertial derivative of Eq. (4.18), we have

$$-T\mathbf{e}_r = -m_P(L - R\theta)\dot{\theta}^2\mathbf{e}_r + m_P(-R\dot{\theta}^2 + (L - R\theta)\ddot{\theta})\mathbf{e}_\theta. \tag{4.22}$$

Equating the $\mathbf{e}_r$ terms yields $T = m_P(L - R\theta)\dot{\theta}^2$. Thus the equation of motion in Eq. (4.21) becomes

$$\ddot{\theta} = \frac{2R\dot{\theta}^2}{L - R\theta} - \frac{m_P R(L - R\theta)\dot{\theta}^2}{m_P(L - R\theta)^2} = \frac{R\dot{\theta}^2}{L - R\theta}. \tag{4.23}$$

(Note that the same equation of motion follows from the $\mathbf{e}_\theta$ term in Eq. (4.22).) We illustrate the tetherball trajectory in Figure 4.10 by integrating Eq. (4.23) in MATLAB using ODE45, with $L = 10$ m, $R = 1$ m, $\theta(0) = 0$ rad, and $\dot{\theta}(0) = 0.1$ rad/s.

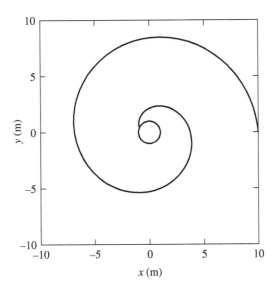

Figure 4.10 Trajectory of the tetherball as it winds around the post after an impact. (The rope is not shown.)

4.2.4 Central-Force Problems

The angular momentum form of Newton's second law in Eq. (4.7) and the law of conservation of angular momentum reveal an important insight into a broad class of problems with so-called *central forces*. The concept of a central force, and the use of angular momentum in solving a central-force problem, is so important that we devote a section here to introducing it.

Definition 4.5 A **central force** is a force $\mathbf{F}_P$ acting on particle P that satisfies the relationship

$$\boxed{\mathbf{M}_{P/O} = \mathbf{r}_{P/O} \times \mathbf{F}_P = 0,}$$

where $\mathbf{r}_{P/O} \neq 0$.

Definition 4.5 implies that the force $\mathbf{F}_P$ acts parallel to the line between the origin O and the particle P. An important consequence of Definition 4.5 is that the angular momentum of P about O is conserved since it satisfies the zero-moment condition in the law of conservation of angular momentum of a particle (as long as O is not accelerating in absolute space).

The time invariance of the angular momentum implies that the motion of point P is confined to a plane perpendicular to the angular momentum—an implication of Definition 4.2 of angular momentum. Since $^{\mathcal{I}}\mathbf{h}_{P/O} = \mathbf{r}_{P/O} \times m_P{}^{\mathcal{I}}\mathbf{v}_{P/O}$, and $^{\mathcal{I}}\mathbf{h}_{P/O}$ is fixed in the inertial frame, then $\mathbf{r}_{P/O}$ and $^{\mathcal{I}}\mathbf{v}_{P/O}$ must be confined to a plane so that they complete a right-hand set with the fixed $^{\mathcal{I}}\mathbf{h}_{P/O}$. We call the motion of P in the plane perpendicular to $^{\mathcal{I}}\mathbf{h}_{P/O}$ an *orbit*. We can thus reduce any central-force problem to a two-dimensional particle dynamics problem similar to the ones solved in Chapter 3. This reduction is a profound result. Also, because of the radial form of the central force, we often solve such problems using polar coordinates in a polar frame.

Example 4.8 A Particle on a Planar Spring, Revisited

Example 3.12 considered the motion of particle P of mass m_P connected to a linear spring and free to move anywhere in the plane. We described this two-degree-of-freedom problem with polar coordinates and found the two scalar equations of motion to be

$$\ddot{r} + \frac{k}{m_P}(r - r_0) - r\dot{\theta}^2 = 0$$

$$\ddot{\theta} + 2\frac{\dot{r}}{r}\dot{\theta} = 0.$$

This is an example of a central-force problem, as the only force acting on P is due to the spring and always acts in the $\mathbf{e}_r$ direction. Thus the trajectory of P is an orbit in the plane determined by these equations of motion. We also know that the angular momentum of the particle relative to the origin O must be conserved. This angular momentum is

$$
\begin{aligned}
{}^{\mathcal{I}}\mathbf{h}_{P/O} &= \mathbf{r}_{P/O} \times m_P{}^{\mathcal{I}}\mathbf{v}_{P/O} \\
&= r\mathbf{e}_r \times m_P(\dot{r}\mathbf{e}_r + r\dot{\theta}\mathbf{e}_\theta) \\
&= m_P r^2 \dot{\theta}\mathbf{e}_3.
\end{aligned}
$$

Since the magnitude and direction of the angular momentum are fixed, we use the constant h_O to represent the specific magnitude of the angular momentum:

$$
h_O \triangleq \frac{\|{}^{\mathcal{I}}\mathbf{h}_{P/O}\|}{m_P} = r^2\dot{\theta}.
$$

This constant is the same found in Example 3.12 (Eq. (3.48)) simply by reducing the equations of motion! Thus we can make a substitution in the equations of motion, leaving us with what we found in Example 3.12,

$$
\ddot{r} + \frac{k}{m_P}(r - r_0) - \frac{h_O^2}{r^3} = 0. \tag{4.24}
$$

This result is remarkable and important. Because of the conservation of angular momentum, we were able to decouple the radial and angular equations of motion. Given a set of initial conditions that determine h_O, we need only integrate the single second-order differential equation for r in Eq. (4.24). The angular position θ can then be found simply from the constancy of angular momentum,

$$
\theta(t) = \int_0^t \frac{h_O}{r(t)^2}dt + \theta(0).
$$

This is true for all central force problems.

Example 4.9 Simple Satellite Equations of Motion

The most important central force we will study is gravity—described by Newton's universal law of gravitation (see Tutorial 2.5), which in vector form is

$$
\mathbf{F}_P = -\frac{Gm_O m_P}{\|\mathbf{r}_{P/O}\|^3}\mathbf{r}_{P/O}. \tag{4.25}
$$

Here we have assumed a gravitating body of mass m_O located at the fixed origin O and a corresponding satellite P of mass m_P at $\mathbf{r}_{P/O}$, as shown in Figure 4.11.[2] Eq. (4.25) allows us to write the vector form of the satellite equation of motion, using the free-body diagram in Figure 4.11c and Newton's second law:

$$
\frac{{}^{\mathcal{I}}d^2}{dt^2}\mathbf{r}_{P/O} + \frac{Gm_O}{\|\mathbf{r}_{P/O}\|^3}\mathbf{r}_{P/O} = 0. \tag{4.26}
$$

---

[2] A model with a fixed origin does not exactly represent two bodies in space, since the sun—or the earth—also moves. It is, however, an exact description of a central-force problem with a fixed attracting origin and is a remarkably good approximation of a real satellite orbit problem. Chapter 7 treats the exact two-body problem more carefully.

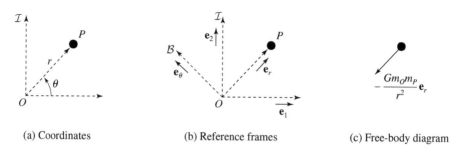

(a) Coordinates   (b) Reference frames   (c) Free-body diagram

Figure 4.11 A simple satellite.

Although Eq. (4.26) looks like a formidable equation of motion, it turns out to be solvable. One nice way to approach it is by using polar coordinates. Let $\mathcal{B} = (O, \mathbf{e}_r, \mathbf{e}_\theta, \mathbf{e}_z)$ be a polar frame, where $\mathbf{e}_r$ points from O to P, and $\mathbf{e}_r$ and $\mathbf{e}_\theta$ span the orbit plane. (Because of the constancy of angular momentum, we know that the orbit is confined to a plane and that we can treat the motion in two dimensions.) The kinematics of P expressed as vector components in frame $\mathcal{B}$ are

$$\mathbf{r}_{P/O} = r\mathbf{e}_r$$

$$^{\mathcal{I}}\mathbf{v}_{P/O} = \dot{r}\mathbf{e}_r + r\dot{\theta}\mathbf{e}_\theta$$

$$^{\mathcal{I}}\mathbf{a}_{P/O} = (\ddot{r} - r\dot{\theta}^2)\mathbf{e}_r + (2\dot{r}\dot{\theta} + r\ddot{\theta})\mathbf{e}_\theta.$$

Again using the free-body diagram in Figure 4.11c and the gravitational force in Eq. (4.25), we write Newton's second law in Eq. (4.26) in terms of components in $\mathcal{B}$,

$$-\frac{Gm_Om_P}{r^2}\mathbf{e}_r = m_P(\ddot{r} - r\dot{\theta}^2)\mathbf{e}_r + m_P(2\dot{r}\dot{\theta} + r\ddot{\theta})\mathbf{e}_\theta.$$

This results in two scalar equations of motion

$$\ddot{\theta} = -\frac{2\dot{r}\dot{\theta}}{r} \tag{4.27}$$

$$\ddot{r} = -\frac{Gm_O}{r^2} + r\dot{\theta}^2. \tag{4.28}$$

We can now use the conservation of angular momentum to reduce the two equations of motion for the polar coordinates $(r, \theta)_{\mathcal{I}}$ to a single differential equation for r. Since the angular momentum is constant, it is determined by the initial radial position and initial velocity—these initial conditions thus determine the characteristics of the orbit. Using Definition 4.2, the angular momentum of satellite P about origin O is

$$^{\mathcal{I}}\mathbf{h}_{P/O} = r\mathbf{e}_r \times m_P(\dot{r}\mathbf{e}_r + r\dot{\theta}\mathbf{e}_\theta) = m_Pr^2\dot{\theta}\mathbf{e}_z.$$

Again, since the magnitude and direction of the angular momentum are fixed, we use the constant h_O to represent the specific magnitude of the angular momentum:

$$h_O \triangleq \frac{\|^{\mathcal{I}}\mathbf{h}_{P/O}\|}{m_P} = r^2\dot{\theta}.$$

Solving for $\dot\theta = h_O/r^2$ and substituting this expression into the satellite equations of motion in Eqs. (4.27) and (4.28) yields

$$\ddot\theta = -\frac{2\dot r h_O}{r^3} \tag{4.29}$$

$$\ddot r = -\frac{Gm_O}{r^2} + \frac{h_O^2}{r^3}. \tag{4.30}$$

Eqs. (4.29) and (4.30) are significant because the equation of motion in r is uncoupled from the equation of motion in θ. In fact, Eq. (4.30) can be analytically integrated (using a surprisingly simple change of coordinates) to find the satellite trajectory $r(\theta)$ (see Tutorial 4.2). Even without integrating, Eqs. (4.29) and (4.30) demonstrate the existence of closed, circular orbits about O. If we consider the steady-state condition $\ddot r = \ddot\theta = 0$ for circular orbit, the equations of motion reduce to

$$\dot\theta = \sqrt{\frac{Gm_O}{r_0^3}}$$

$$\dot r = 0,$$

which is a circular orbit with radius r_0. Using the kinematics of P, the satellite velocity in a circular orbit of radius r_0 is

$$^{\mathcal{I}}\mathbf{v}_{P/O} = \sqrt{\frac{Gm_O}{r_0}}\,\mathbf{e}_\theta.$$

It is also interesting to point out another special trajectory. If the initial velocity is zero, then $h_O = 0$ and there is no angular motion. In this case, the particle simply accelerates on a linear trajectory toward the origin, as in Tutorial 2.5.

4.3 Tutorials

Tutorial 4.1 A Passive Walker

A recent discovery in biomechanics is that robotic bipeds can stably walk down an incline—that is, walk without falling—without any control of the hips or legs from the brain. Walking without active control of the hips or legs is generally referred to as *passive dynamic walking* and has important implications in the study of neuro-mechanics. With only a few simplifying assumptions, the tools of this chapter can be used to model and study this remarkable result.

Figure 4.12 shows a simple model of a passive walking robot. We model the robot using a point mass P to represent the body and two massless rods to represent the legs. The robot "walks" down a ramp by tipping forward, like an inverted pendulum, and transferring its weight from one leg to the other. The point of contact of each leg with the ramp is called a *foot*. Each stride starts and ends with both feet on the ramp. We assume that the dynamics essential to walking are confined to a plane.

Let l denote the length of the legs and assume that every stride has a fixed length. The angle $0 < \alpha < \pi$ denotes the angular separation of the legs at the instant when both

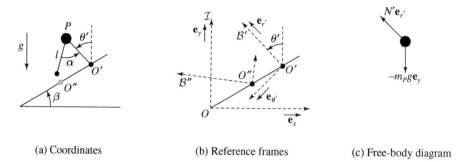

| (a) Coordinates | (b) Reference frames | (c) Free-body diagram |

Figure 4.12 A passive walker. The coordinates, reference frames, and free-body diagram are shown for a single stride. Stable walking consists of an infinite series of strides.

feet are on the ramp.[3] Let $0 < \beta < \frac{\pi}{2}$ denote the ramp inclination angle. Intuition tells us that, if the ramp is not steep enough, then the robot's momentum may be insufficient to sustain continuous motion down the ramp, in which case the robot falls backward. Alternatively, if the ramp is too steep, then the robot may gain too much momentum and lose contact with (or topple off) the ramp. In this tutorial we model the walking dynamics using linear impulse and, for a given α, find the sufficient conditions on β that support walking.

Equations of Motion

Let $\mathcal{I} = (O, \mathbf{e}_x, \mathbf{e}_y, \mathbf{e}_z)$ denote an inertial frame, as shown in Figure 4.12b. Let $\mathcal{B}' = (O', \mathbf{e}_{r'}, \mathbf{e}_{\theta'}, \mathbf{e}_z)$ denote the polar frame fixed to one leg with origin O' located at the foot of the leg. The corresponding angle θ' is the angle of the leg relative to vertical. Let $\mathcal{B}'' = (O'', \mathbf{e}_{r''}, \mathbf{e}_{\theta''}, \mathbf{e}_z)$ and θ'' denote the polar frame and angle associated with the other leg. The three reference frames are related by

	$\mathbf{e}_x$	$\mathbf{e}_y$	$\mathbf{e}_{r''}$	$\mathbf{e}_{\theta''}$
$\mathbf{e}_{r'}$	$-\sin\theta'$	$\cos\theta'$	$\cos\alpha$	$\sin\alpha$
$\mathbf{e}_{\theta'}$	$-\cos\theta'$	$-\sin\theta'$	$-\sin\alpha$	$\cos\alpha$,

where the transformation between $\mathcal{B}'$ and $\mathcal{B}''$ is valid only when both feet are touching the ramp (when the angle between the legs is α).

Consider, for the moment, the geometry of the robot with both feet on flat ground, that is, with $\beta = 0$. In this configuration, the angle between the "back" leg and $\mathbf{e}_y$ is $\alpha/2$, whereas the angle between the "front" leg and $\mathbf{e}_y$ is $-\alpha/2$. If the robot is walking

---

[3] We do not model the mechanism at the hip joint that controls the leg placement. Our model is equivalent to assuming the robot has $2\pi/\alpha$ massless legs uniformly splayed in all directions, so that the robot "rolls" bumpily down the ramp or, equivalently, that the rear leg instantaneously rotates to the front after leaving the ramp. In a more detailed model we might include the mass of the feet and treat the legs as pendulums.

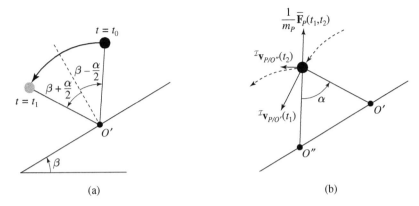

Figure 4.13 (a) Front leg during the duration of a single stride. The stride starts at $t = t_0$ when $\theta' = \beta - \alpha/2$ and the rear leg has just lost contact with the ramp. The stride ends at $t = t_1$ when $\theta' = \beta + \alpha/2$ and the new front leg has just made contact with the ramp. (b) As the new front foot is planted, a linear impulse redirects the velocity to be tangent to a circular trajectory centered at O''.

stably down an inclined ramp, then a stride supported by the first leg starts at time $t = t_0$ at an angle of

$$\theta'(t_0) = \beta - \frac{\alpha}{2} \tag{4.31}$$

and ends at $t = t_1$ at the angle

$$\theta'(t_1) = \beta + \frac{\alpha}{2}. \tag{4.32}$$

Such a stride is shown in Figure 4.13a.

During a stride, the kinematics of P are

$$\mathbf{r}_{P/O'} = l\mathbf{e}_{r'}$$

$$^{\mathcal{I}}\mathbf{v}_{P/O'} = l\dot{\theta}'\mathbf{e}_{\theta'}$$

$$^{\mathcal{I}}\mathbf{a}_{P/O'} = l\ddot{\theta}'\mathbf{e}_{\theta'} - l(\dot{\theta}')^2\mathbf{e}_{r'}.$$

Let N' denote the magnitude of the normal force from the ramp during the stride. Using the free-body diagram in Figure 4.12c, the dynamics of P are

$$N'\mathbf{e}_{r'} - m_P g\mathbf{e}_y = m_P l\ddot{\theta}'\mathbf{e}_{\theta'} - m_P l(\dot{\theta}')^2\mathbf{e}_{r'},$$

which yields the force equation

$$N = m_P g \cos\theta' - m_P l(\dot{\theta}')^2 \tag{4.33}$$

and the equation of motion

$$\ddot{\theta}' = \frac{g}{l} \sin\theta'. \tag{4.34}$$

Eq. (4.34) is the equation of motion for an inverted pendulum (see Problem 4.3) and is the same equation of motion as in Tutorial 3.3. (Note the sign change from the equation of motion for the simple pendulum.)

For the robot to remain in contact with the ramp throughout the stride and to not topple off, the normal force N' must satisfy $N' > 0$ or, equivalently, the specific normal force N'/m_P must satisfy $(N'/m_P) > 0$. We call this inequality the *toppling condition*.

Impulsive Stride Transfer

During a stride, the velocity $^T\mathbf{v}_{P/O'}$ is in the $\mathbf{e}_{\theta'}$ direction, which implies that the trajectory of P is tangent to a circle with radius l centered at O'. When the stride ends and the next stride begins, the new velocity $^T\mathbf{v}_{P/O''}$ is in the $\mathbf{e}_{\theta''}$ direction. Therefore the velocity of P changes direction from $\mathbf{e}_{\theta'}$ to $\mathbf{e}_{\theta''}$ during the short time interval $\Delta t = t_2 - t_1$ when both feet are in contact with the ramp. This situation is illustrated in Figure 4.13b.

We study the change in the velocity using the impulse equation

$$m_P{}^T\mathbf{v}_{P/O''}(t_2) = m_P{}^T\mathbf{v}_{P/O'}(t_1) + \underbrace{m_P{}^T\mathbf{v}_{O'/O''}(t_1)}_{=0} + \overline{\mathbf{F}}_P(t_1, t_2),$$

which, after substituting the kinematics, becomes

$$m_P l\dot{\theta}''(t_2)\mathbf{e}_{\theta''} = m_P l\dot{\theta}'(t_1)\mathbf{e}_{\theta'} + \overline{\mathbf{F}}_P(t_1, t_2). \tag{4.35}$$

The relationship

$$\mathbf{e}_{\theta'} = -\sin\alpha\,\mathbf{e}_{r''} + \cos\alpha\,\mathbf{e}_{\theta''} \tag{4.36}$$

is satisfied during the time interval t_1 to t_2. Using Eq. (4.36), we compute the linear impulse that eliminates the $\mathbf{e}_{r''}$ component of the velocity:

$$\overline{\mathbf{F}}_P(t_1, t_2) = m_P l\dot{\theta}'(t_1)\sin\alpha\,\mathbf{e}_{r''}. \tag{4.37}$$

We find the impulse magnitude $\|\overline{\mathbf{F}}_P(t_1, t_2)\|$ by solving the equation of motion during the stride for the final angular velocity. Substituting Eq. (4.37) into Eq. (4.35) and using Eq. (4.36) yields

$$\dot{\theta}''(t_2) = \dot{\theta}'(t_1)\cos\alpha. \tag{4.38}$$

Eq. (4.38) provides the initial condition for the second (nth) stride in terms of the final condition of the first (($n - 1$)th) stride.

Sufficient Conditions for Walking

A sufficient condition for walking expressed in terms of the parameters α, β, and l can be derived from

$$\dot{\theta}''(t_2) = \dot{\theta}'(t_0). \tag{4.39}$$

This condition states that the angular momentum of P about the contact point at the start of each stride is the same.[4] Let h_0 denote the magnitude of the specific angular momentum at the start of each stride:

$$h_0 \triangleq \frac{1}{m_P} \|{}^{\mathcal{I}}\mathbf{h}_{P/O'}(t_0)\| = \|l\mathbf{e}_{r'} \times l\dot{\theta}'(t_0)\mathbf{e}_{\theta'}\| = l^2\dot{\theta}'(t_0). \tag{4.40}$$

Multiplying Eq. (4.34) by $\dot{\theta}'dt$ and integrating from t_0 to $t_0 < t \le t_1$ yields

$$\int_{\dot{\theta}'(t_0)}^{\dot{\theta}'(t)} \dot{\theta}d\dot{\theta} = \frac{g}{l} \int_{\theta'(t_0)}^{\theta'(t)} \sin\theta d\theta,$$

which, using Eqs. (4.31) and (4.40), implies

$$\dot{\theta}'(t) = \sqrt{(\dot{\theta}'(t_0))^2 + \frac{2g}{l}(\cos(\theta'(t_0)) - \cos(\theta'(t)))}$$

$$= \sqrt{\frac{h_0^2}{l^4} + \frac{2g}{l}\left(\cos\left(\beta - \frac{\alpha}{2}\right) - \cos(\theta'(t))\right)}. \tag{4.41}$$

When $t = t_1$ (the end time of the stride), we find, using Eqs. (4.38) and (4.41), that

$$\dot{\theta}''(t_2) = \cos\alpha\sqrt{\frac{h_0^2}{l^4} + \frac{2g}{l}\left(\cos\left(\beta - \frac{\alpha}{2}\right) - \cos\left(\beta + \frac{\alpha}{2}\right)\right)}$$

$$= \cos\alpha\sqrt{\frac{h_0^2}{l^4} + \frac{4g}{l}\sin\beta\sin\frac{\alpha}{2}}, \tag{4.42}$$

where we also used Eqs. (4.31) and (4.32) and the trigonometric identity $\cos(\beta \pm \frac{\alpha}{2}) = \cos\beta\cos\frac{\alpha}{2} \mp \sin\beta\sin\frac{\alpha}{2}$. We now substitute for $\dot{\theta}''(t_2)$ from the walking condition in Eq. (4.39) and replace $\dot{\theta}'(t_0)$ by the angular momentum in Eq. (4.40) to obtain

$$\frac{h_0}{l^2} = \cos\alpha\sqrt{\frac{h_0^2}{l^4} + \frac{4g}{l}\sin\beta\sin\frac{\alpha}{2}}.$$

After some algebra, this expression simplifies to

$$h_0 = 2l\cot\alpha\sqrt{gl\sin\beta\sin\frac{\alpha}{2}}. \tag{4.43}$$

Suppose t^* satisfies $\theta'(t^*) = 0$, which is the time that P is vertical. For the robot to tip forward and not backward, we require that $\dot{\theta}'(t^*) > 0$. Using Eq. (4.41), a *falling condition* is then

$$\frac{h_0^2}{l^4} + \frac{2g}{l}\left(\cos\left(\beta - \frac{\alpha}{2}\right) - 1\right) > 0,$$

---

[4] Another mode of walking is, for example, when the angular momentum of P about the contact point at the start of alternate strides is the same.

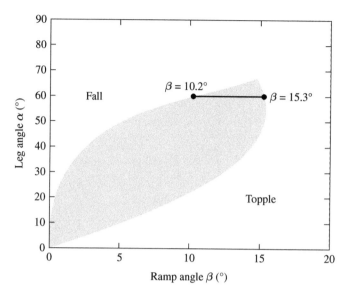

Figure 4.14 A passive walker. Values of α and β shown in the shaded region are sufficient but not necessary for stable walking with $l = 1$ m. For example, for $\alpha = 60°$, the robot falls when $\beta < 10.2°$ and topples when $\beta > 15.3°$.

where we have used $\cos(\theta'(t^*)) = 1$. Using Eq. (4.43), this inequality is equivalent to

$$2 \cot^2 \alpha \sin \beta \sin \frac{\alpha}{2} + \cos(\beta - \frac{\alpha}{2}) - 1 > 0. \tag{4.44}$$

If the robot doesn't fall, then the angular momentum of P about the contact point is maximum just before the footfall at time t_1. Thus from Eq. (4.33) the normal force N is a minimum at t_1. Evaluating Eq. (4.33) at t_1 using Eqs. (4.32) and (4.38), the toppling condition becomes

$$g \cos \theta'(t_1) - l(\dot\theta'(t_1))^2 = g \cos(\beta + \frac{\alpha}{2}) - l \frac{(\dot\theta''(t_2))^2}{\cos^2 \alpha} > 0,$$

which, using Eqs. (4.39) and (4.43), is equivalent to

$$\cos(\beta + \frac{\alpha}{2}) - \frac{h_0^2}{gl^3 \cos^2 \alpha} = \cos(\beta + \frac{\alpha}{2}) - \frac{4 \sin \beta \sin \frac{\alpha}{2}}{\sin^2 \alpha} > 0. \tag{4.45}$$

Using Eqs. (4.45) and (4.44), we plot in Figure 4.14 the values of α and β that, for $l = 1$ m, are sufficient for walking. Interestingly, these values are independent of g, which means that these results are independent of the acceleration of gravity.

Tutorial 4.2 The Orbit Equation

This tutorial picks up where we left off in Example 4.9 and derives the general equation for a satellite orbit. Recall the observation made in Example 4.9. The angular momentum of satellite P about origin O is constant, since the only force acting on P is gravity—which is a central force. The magnitude and direction of the angular momentum are constants determined by the initial position and initial velocity of P. We denoted by h_O the specific magnitude of the angular momentum:

$$h_O = \frac{\|{}^{\mathcal{I}}\mathbf{h}_{P/O}\|}{m_P} = r^2 \dot{\theta}.$$

The satellite equations of motion (Eqs. (4.29) and (4.30)) are

$$\ddot{\theta} = -\frac{2\dot{r}h_O}{r^3} \tag{4.46}$$

$$\ddot{r} = -\frac{\mu}{r^2} + \frac{h_O^2}{r^3}, \tag{4.47}$$

where $\mu \triangleq Gm_O$.

Unfortunately, there is no known solution of Eqs. (4.46) and (4.47) for the two remaining degrees of freedom, $r(t)$ and $\theta(t)$, in terms of elementary functions. This is a well-known problem in celestial mechanics referred to as *Kepler's problem*. It is beyond our scope to discuss Kepler's problem. However, with a simple change of coordinates, we can integrate Eq. (4.47) to find an implicit expression for the satellite trajectory $r(\theta)$.

Let $y = 1/r$. Then we have $h_O = \dot{\theta}/y^2$ and

$$\dot{r} = \frac{d}{dt}\left(\frac{1}{y}\right) = -\frac{1}{y^2}\dot{y} = -\frac{1}{y^2}\frac{dy}{d\theta}\dot{\theta} = -h_O\frac{dy}{d\theta},$$

which implies

$$\ddot{r} = -h_O\frac{d^2y}{d\theta^2}\dot{\theta} = -h_O^2 y^2\frac{d^2y}{d\theta^2}. \tag{4.48}$$

Substituting Eq. (4.48) in Eq. (4.47) and switching to the y coordinate yields

$$\frac{d^2y}{d\theta^2} + y = \frac{\mu}{h_O^2}. \tag{4.49}$$

Eq. (4.49) is the equation for a simple harmonic oscillator with unit oscillation frequency and constant driving force! Tutorial 2.3 showed that the solution to this equation is sinusoidal about a constant offset, that is,

$$y(\theta) = \frac{\mu}{h_O^2} + A\cos\theta + B\sin\theta.$$

For this problem it is more convenient to write the solution in terms of an amplitude and phase rather than the integration constants A and B (in both cases the constants are given by initial conditions):

$$y(\theta) = \frac{\mu}{h_O^2} + \frac{\mu}{h_O^2}e\cos(\theta - \theta_0),$$

where

$$e = \frac{h_O^2}{\mu}\sqrt{A^2 + B^2}$$

TABLE 4.1
The orbit shape corresponding to
eccentricity e

Eccentricity e	Orbit shape
$e = 0$	circle
$0 < e < 1$	ellipse
$e = 1$	parabola
$e > 1$	hyperbola

and

$$\theta_0 = \arctan\left(\frac{B}{A}\right).$$

Therefore, in terms of the r coordinate, we have

$$r(\theta) = \frac{h_O^2/\mu}{1 + e\cos(\theta - \theta_0)}. \tag{4.50}$$

Eq. (4.50) is the polar equation of a *conic section*[5] with eccentricity e. It is not difficult to show that the eccentricity is nonnegative. The dependence of orbit shape on eccentricity is described in Table 4.1.

Many orbit problems can be solved using Eq. (4.50). The satellite's radial position r for any given angular position θ follows a conic section determined by e with the gravitating body at one focus, as shown in Figure 4.15. The specific size and shape of the orbit is given by the initial conditions. That is, if we consider the radial position and angular velocity at the reference angle θ_0, polar radius $r(\theta_0)$, and polar angle rate $\dot{\theta}(\theta_0)$, we have

$$h_O = r^2(\theta_0)\dot{\theta}(\theta_0) \tag{4.51}$$

$$e = \frac{r^3(\theta_0)\dot{\theta}^2(\theta_0)}{\mu} - 1. \tag{4.52}$$

Eq. (4.50) shows that the polar radius is smallest when $\theta = \theta_0$. Thus the angle reference θ_0 corresponds to the point of closest approach—known as *periapsis*. The radius at this point is

$$r_p \triangleq r(\theta_0) = \frac{h_O^2}{\mu(1 + e)}. \tag{4.53}$$

---

[5] A conic section is the intersection of a plane with a cone. Depending on the intersection angle, the section is a circle, an ellipse, a parabola, or a hyperbola.

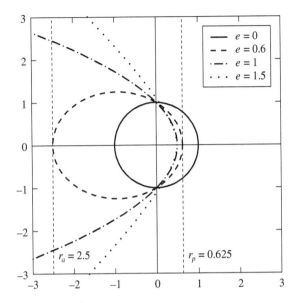

Figure 4.15 Eq. (4.50) with $h_O^2 = \mu$ plotted for three values of eccentricity e. For $e = 0.6$, the trajectory is an ellipse; the periapsis r_p and apoapsis r_a denote the closest and farthest, respectively, points of approach to the origin at $(0, 0)$.

The farthest point of approach—known as *apoapsis*—corresponds to $\theta = \theta_0 + \pi$, which implies

$$r_a \triangleq r(\theta_0 + \pi) = \frac{h_O^2}{\mu(1 - e)}. \tag{4.54}$$

Finally, the *semimajor axis* a of an ellipse (or circle) is defined as half the length of the long axis:

$$a = \frac{1}{2}(r_a + r_p) = \frac{h_O^2}{\mu(1 - e^2)}. \tag{4.55}$$

Tutorial 4.3 Impulsive Orbital Transfer

The concept of angular impulse is used regularly in satellite orbital transfer. Consider the situation sketched in Figure 4.16 in which a satellite is in a circular orbit of radius r_A and needs to be transferred to a larger circular orbit of radius r_B (e.g., the satellite may have been left in a low-earth orbit by a launch vehicle). To do so, the satellite changes the eccentricity of its orbit to achieve a *transfer orbit*—an elliptical orbit with periapsis equal to r_A and apoapsis equal to r_B.

For the satellite to move from the initial circular orbit to the eccentric transfer orbit, it must change its angular momentum and eccentricity, given by Eqs. (4.51) and (4.52), respectively. Satellites are normally equipped with propulsion devices that can produce large forces over short intervals. The firing time is so brief compared to the orbit period that it is very accurately modeled as an impulsive force. (We call

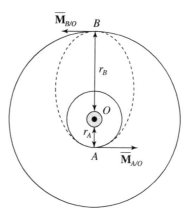

Figure 4.16 Sketch of a Hohmann transfer orbit between two circular orbits of radii r_A and r_B. Two angular impulses are required, one of $\overline{M}_{A/O}$ at periapsis of the elliptical transfer orbit and one of $\overline{M}_{B/O}$ at apoapsis.

the resulting change in velocity a $\Delta\mathbf{v}$, pronounced "delta-vee.") The impulsive force produces an impulsive moment about O. According to Eq. (4.17), a thrust applied at time t_1 and ending at time t_2 produces a change in angular momentum of

$$^{\mathcal{I}}\mathbf{h}_{P/O}(t_2) = {}^{\mathcal{I}}\mathbf{h}_{P/O}(t_1) + \overline{\mathbf{M}}_{P/O}(t_1, t_2).$$

In the initial circular orbit, the satellite's angular velocity is

$$\dot{\theta}_A = \sqrt{\frac{\mu}{r_A^3}}.$$

Thus the specific angular momentum before the impulse has magnitude $h_O(t_1) = \sqrt{\mu r_A}$. We wish the satellite to enter into the elliptic transfer orbit with periapsis r_A and apoapsis r_B. From Eqs. (4.53) and (4.54) we find that the magnitude of the specific angular momentum of the transfer orbit (i.e., after the impulse) is

$$h_O(t_2) = \sqrt{\frac{2\mu r_A r_B}{r_A + r_B}}. \tag{4.56}$$

The eccentricity of the transfer orbit is also found from Eqs. (4.53) and (4.54),

$$e = \frac{r_B - r_A}{r_B + r_A}.$$

Because the transfer orbit is tangent to the circular orbit at periapsis, the velocity after the impulse is in the same direction as the velocity before the impulse, implying that the impulsive force is tangent to both orbits. The magnitude of the specific angular impulse required at point $P = A$ is

$$\frac{\|\overline{\mathbf{M}}_{A/O}(t_1, t_2)\|}{m_P} = h_O(t_2) - h_O(t_1) = \sqrt{\mu r_A}\left[\sqrt{\frac{2r_B}{r_A + r_B}} - 1\right].$$

Once the satellite reaches the apoapsis of the transfer orbit at radius r_B, another impulsive firing is required to "circularize" the orbit—that is, to change the angular momentum to that of the final circular orbit. Switching from an elliptical orbit to a circular orbit requires the satellite to speed up. The angular momentum before this second impulse has magnitude h_O in Eq. (4.56). The angular momentum after the second impulse is equal to the angular momentum of the final circular orbit, which has magnitude $h_O(t_2) = \sqrt{\mu r_B}$. The required impulsive moment at point $P = B$ (again from a force tangent to both orbits) is the difference between these angular momenta,

$$\frac{\|\overline{\mathbf{M}}_{B/O}(t_1, t_2)\|}{m_P} = \sqrt{\mu r_B}\left[1 - \sqrt{\frac{2r_A}{r_A + r_B}}\right].$$

An orbital transfer using two impulses is by far the most commonly used and is called a *Hohmann transfer*.[6] A Hohmann transfer requires the least fuel of all possible two-impulse transfers.

4.4 Key Ideas

- Integrating Newton's second law with respect to time yields

$$m_P{}^{\mathcal{I}}\mathbf{v}_{P/O}(t_2) = m_P{}^{\mathcal{I}}\mathbf{v}_{P/O}(t_1) + \overline{\mathbf{F}}_P(t_1, t_2),$$

where $\overline{\mathbf{F}}_P(t_1, t_2)$ is the **linear impulse** acting on the particle:

$$\overline{\mathbf{F}}_P(t_1, t_2) \triangleq \int_{t_1}^{t_2} \mathbf{F}_P dt.$$

For sufficiently small $\Delta t \triangleq t_2 - t_1$, a linear impulse changes the linear momentum of a particle without changing its position.

- The **angular momentum** of a particle relative to point O is

$$^{\mathcal{I}}\mathbf{h}_{P/O} \triangleq \mathbf{r}_{P/O} \times m_P{}^{\mathcal{I}}\mathbf{v}_{P/O}.$$

Angular momentum always depends on a reference point. The rate of change of angular momentum when the reference point is inertially fixed is

$$\frac{^{\mathcal{I}}d}{dt}\left(^{\mathcal{I}}\mathbf{h}_{P/O}\right) = \mathbf{M}_{P/O},$$

where the **moment** acting on P due to the force $\mathbf{F}_P$ is

$$\mathbf{M}_{P/O} \triangleq \mathbf{r}_{P/O} \times \mathbf{F}_P.$$

---

[6] Named after Walter Hohmann (1880–1945), a German scientist, who first described this transfer in 1925.

- More generally, the rate of change of angular momentum when taken relative to an arbitrary point Q is

$$\frac{{}^{\mathcal{I}}d}{dt}\left({}^{\mathcal{I}}\mathbf{h}_{P/Q}\right) = \mathbf{M}_{P/Q} - \mathbf{r}_{P/Q} \times m_P\,{}^{\mathcal{I}}\mathbf{a}_{Q/O},$$

where ${}^{\mathcal{I}}\mathbf{a}_{Q/O}$ is the inertial acceleration of the reference point.

- Integrating with respect to time the angular momentum form of Newton's second law (referenced to an inertially fixed point) yields

$$^{\mathcal{I}}\mathbf{h}_{P/O}(t_2) = {}^{\mathcal{I}}\mathbf{h}_{P/O}(t_1) + \overline{\mathbf{M}}_{P/O}(t_1, t_2),$$

where $\overline{\mathbf{M}}_{P/O}(t_1, t_2)$ is the **angular impulse**,

$$\overline{\mathbf{M}}_{P/O}(t_1, t_2) = \int_{t_1}^{t_2} \mathbf{M}_{P/O}\, dt.$$

- When the total moment acting on P relative to point Q (or O) is zero, then the angular momentum of P relative to that point is **conserved**, even if there are forces acting on P. The angular momentum is constant in both magnitude and direction.

- A **central force** on P results in zero moment about O:

$$\mathbf{M}_{P/O} = \mathbf{r}_{P/O} \times \mathbf{F}_P = 0.$$

Under the action of a central force, the angular momentum of a particle is conserved and its motion is described by a two-dimensional orbit in a plane perpendicular to the angular momentum.

4.5 Notes and Further Reading

In this chapter we begin to see the extension of mechanics beyond the simple statements of Newton's laws. In fact, it was d'Alembert in 1742 who introduced the concept of the impulse and Euler who introduced angular momentum. The concepts of inertial forces and inertial moments were also introduced by Euler in his laws of mechanics and used by d'Alembert in his treatise. Details can be found in Dugas (1988) and Truesdell (1968).

Newton's formulation of the law of gravity—and his combining of it with his laws of motion to successfully solve for the motion of celestial bodies—was the greatest triumph of his laws of mechanics. (In 1705 Edmund Halley used them to correctly predict the return of the comet named after him, thus sealing the case for Newton's laws.) A rather nice, concise history is in Vallado (2001). A more thorough and popular treatment is in Peterson (1993). Our solution to the motion of a body in orbit under gravity is just a brief introduction. The subject of orbital motion is rich and complex, with many excellent texts devoted solely to it. We like many of them, including

Bate et al. (1971), Chobotov (1991), Kaplan (1976), and Vallado (2001). If interested in learning more about orbital maneuvers, such as the Hohmann transfer, you might consult Kaplan (1976) or Vallado (2001).

4.6 Problems

4.1 Let P represent mass m shown in Figure 4.17, which is traveling at ${}^{\mathcal{I}}\mathbf{v}_{P/O} = \dot{x}\mathbf{e}_x + \dot{y}\mathbf{e}_y$, where $\dot{x} > \dot{y} > 0$. Point Q is fixed with respect to O, and $\mathbf{r}_{Q/O} = a\mathbf{e}_x + b\mathbf{e}_y$.

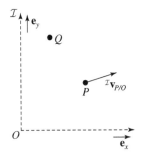

Figure 4.17 Problem 4.1.

a. Using Cartesian coordinates $(x, y, z)_{\mathcal{I}}$, compute the following vector quantities: ${}^{\mathcal{I}}\mathbf{p}_{P/O}, {}^{\mathcal{I}}\mathbf{h}_{P/O}, {}^{\mathcal{I}}\mathbf{p}_{P/Q}$, and ${}^{\mathcal{I}}\mathbf{h}_{P/Q}$.

b. True or false: At the instant shown in Figure 4.17, the vectors ${}^{\mathcal{I}}\mathbf{h}_{P/O}$ and ${}^{\mathcal{I}}\mathbf{h}_{P/Q}$ point in the same direction.

4.2 Consider a 10^4 kg spaceship in deep space. Suppose the spaceship's thruster produces 100 N of constant thrust. If the spaceship is moving at 50 m/s and the thruster is anti-aligned with the spaceship's velocity, how long should the thruster fire to bring the spaceship to rest with respect to absolute space?

4.3 Find the equation of motion for the inverted simple pendulum shown in Figure 4.18. How does it differ from the equation for the simple pendulum?

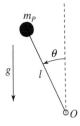

Figure 4.18 Problem 4.3.

4.4   Solve the equation of motion for the inverted simple pendulum from the previous problem for small deviations from vertical. How does the solution differ from that for the simple pendulum (i.e., simple harmonic motion)?

4.5   A hockey player hits a puck of mass m_P into a semicircular groove (when viewed from above) in the ice of radius R. The ice has coefficient of Coulomb friction μ_c. What is the impulse that must be imparted to the puck at the entrance to the groove such that it makes it around the semicircle exactly once before stopping (assuming the only source of friction arises from the vertical component of the normal force)?

4.6   Suppose frame $\mathcal{B} = (O', \mathbf{b}_1, \mathbf{b}_2, \mathbf{b}_3)$ is traveling in the $\mathbf{e}_1$ direction at a constant speed of v_0 with respect to stationary frame $\mathcal{I} = (O, \mathbf{e}_1, \mathbf{e}_2, \mathbf{e}_3)$, as shown in Figure 4.19. Mass m_P is connected to point O' by a spring with spring constant k and rest length r_0. The spring can freely pivot about O'. Assume that the positions of O and O' are the same at time $t = 0$.

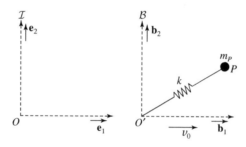

Figure 4.19 Problem 4.6.

a.   Using the coordinates of your choice, find the position $\mathbf{r}_{P/O'}$ and velocity $^{\mathcal{I}}\mathbf{v}_{P/O'}$ of the mass with respect to O' in $\mathcal{I}$. [HINT: Introduce a polar frame at O'.]

b.   Find the position $\mathbf{r}_{P/O}$ and velocity $^{\mathcal{I}}\mathbf{v}_{P/O}$ of the mass with respect to O in $\mathcal{I}$.

c.   Draw a free-body diagram for mass m_P. (There is no gravity in this problem.)

d.   Find the angular momentum $^{\mathcal{I}}\mathbf{h}_{P/O'}$ of the mass with respect to O' in $\mathcal{I}$.

e.   Show that the angular momentum of the mass with respect to O' in $\mathcal{I}$ is conserved, but the angular momentum with respect to O in $\mathcal{I}$ is not.

4.7   Figure 4.20 shows a simple model for the gravitational attraction of particle P to an object with nonuniform density. Here, rather than a single mass at the origin O, we have a dipole close to the origin separated by $2d$. Assuming that the motion of P starts in the plane of the dipole, then to a good approximation (for $r \gg d$) the force acting on the particle due to the dipole is

$$\mathbf{F}_P \approx -\left(\frac{2m_P\mu}{r^2} + \frac{3m_P\mu d^2}{r^4}(3\cos^2\theta - 1) \right) \mathbf{e}_r - \frac{3m_P\mu d^2 \sin(2\theta)}{r^4} \mathbf{e}_\theta,$$

where $\mu = Gm$.

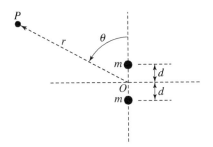

Figure 4.20 Problem 4.7.

Determine whether the angular momentum of P relative to O is conserved for this problem. Is the motion still a planar orbit, as it was for a single attracting particle? Write down the resulting equations of motion for P. Numerically compute the trajectory of P for the following parameters: $r_0 = 1.5625$ distance units, $\dot{r}_0 = 0$, $\theta_0 = 0.6435$ radians, $\dot{\theta}_0 = 0.6144$ radians/time unit, $\mu = 0.5$ distance units^3/time units^2, and $d = 0.3125$ distance units.[7] You should integrate for 100 time units. Also compute the trajectory for the central force due to a mass of $2m$ located at the origin and compare. Use the same initial conditions for both integrations. Note that distance is measured in earth radii and time is measured in fractions of an orbital period. [HINT: Numerical integration errors may cause the central-force trajectory to appear like a spiral rather than a closed orbit. Try reducing the error tolerances used by ODE45 using ODESET.]

4.8 Consider again Problem 3.20 from Chapter 3. What is the magnitude and direction of the angular momentum of the collar relative to the point O? Is it conserved? Explain.

4.9 Show that the eccentricity of an orbit in terms of the polar coordinates of the initial conditions is given by Eq. (4.52).

4.10 The International Space Station is in a circular orbit with an altitude of 250 mi (400 km). It has a mass of 1.8×10^5 kg. Here are some constants you may need to know:

$$\mu_{\text{earth}} = 3.986 \times 10^5 \text{ km}^3/\text{s}^2$$
$$R_{\text{earth}} = 6{,}378 \text{ km}.$$

  a. What is the velocity of the space station?
  b. What is the orbital period of the space station?
  c. If a crew-rescue vehicle leaves the station, how much must its velocity be changed so that its new orbit has its closest approach to the earth at R_{earth}? Ignore aerodynamic drag.

---

[7] The reason that d is not much, much less than r in this problem is that if more realistic numbers were used, it would take far more orbits than we could reasonably expect you to simulate to see any noticeable difference.

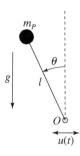

Figure 4.21 Problem 4.11

4.11 Consider again the inverted pendulum from Problem 4.3. This time the base can move back and forth at a known displacement $u(t)$, as shown in Figure 4.21.

    a. Using angular momentum, find the equations of motion including the displacement $u(t)$ of the base.

    b. Suppose you are holding the bottom of the stick and can move it back and forth with a motion given by $u(t)$ to try to keep it balanced. If the pendulum starts at an offset angle θ_0, show that it can be returned to vertical if you accelerate your hand with the following function:

$$\ddot{u} = -\frac{g \sin \theta + k\theta + b\dot{\theta}}{\cos \theta}$$

for some constants $k > 0$ and $b > 0$. This is called a *feedback law*, as it depends on the instantaneous angle of the pendulum. How might you choose k and b?

4.12 A particle slides freely on a flat surface, attached to a massless string that passes through a hole in the surface. At a given moment, the particle is moving in a circle of radius r_0 with angular velocity ω_0. Assume that the string is pulled through the hole at a constant rate. Find an expression for the tension in the string as a function of the length r of string left on the surface.

4.13 A particle of mass m is attached to a spring, as in Figure 4.22. Initially the spring is unstretched with length l_0. A linear impulse is applied to give the

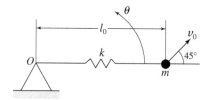

Figure 4.22 Problem 4.13

particle an initial velocity of magnitude v_0 in the direction shown in the figure. During the particle's trajectory, the spring stretches to a maximum length of $4l_0/3$. Assume there is no gravity acting on the system.

    a. What is the linear impulse about O applied to the mass to give it the initial conditions specified?

    b. Find the equations of motion of the mass as a function of l and θ.

    c. Solve for the spring stiffness k as a function of m, l_0, and v_0.

4.14   Reproduce Figure 4.14 in Tutorial 4.1 for five different values of the leg length l. Comment on the dependence on leg length of the admissible range for stable walking.

4.15   Numerically simulate the passive walker described in Tutorial 4.1. Use ODE45 in MATLAB to integrate the equation of motion in Eq. (4.34) with $l = 1$ m, $\alpha = 60°$, $\theta'(0) = 0°$, and $\dot{\theta}'(0) = 0.1°/s$. For each stride, you will need one call to ODE45; the initial conditions of each stride (other than the first one) are given by Eqs. (4.31) and (4.38). Use Eq. (4.33) to check for toppling, which occurs if the specific normal force goes to zero. Simulate the walker for 10 steps, or until it falls over, using ramp angles of $5°$, $12°$, and $18°$. See if you can find bounds on the ramp angle β for which the walker can move without toppling. Note that to do this problem, you will need to use the OPTIONS argument of ODE45 to set two terminal events to stop execution of ODE45. The first event, which is satisfied at the end of a normal stride, is if the leg angle equals the value given in Eq. (4.32). The second event, which corresponds to toppling, is if the specific normal force goes to zero. Look at the documentation for ODESET for information on creating an event function.

CHAPTER FIVE

- -

Energy of a Particle

In Chapter 4 we integrated Newton's second law with respect to time to find the change in linear (or angular) momentum of a particle. We call this result a *first integral of the motion* if the momentum is conserved. For impulsive forces, we found the approach very useful for solving certain problems. Although it does not allow us to compute entire trajectories (as done in Chapter 3 for some simple systems), it does provide a convenient formula for finding velocities at different points.

This chapter introduces another integral of the motion—energy—found by integrating Newton's second law with respect to distance. For certain classes of forces we can solve for the instantaneous position and/or velocity of a particle without finding the equations of motion or solving for trajectories. Nevertheless, we still do not introduce any new physics. We simply manipulate Newton's second law to gain insight and provide new tools for problem solving.

5.1 Work and Power

Our treatment of the energy of a particle begins with the concept of *work*, the SI units of which are joules ($kg\text{-}m^2/s^2$). The following definition is illustrated in Figure 5.1.

Definition 5.1 Let γ_P denote the trajectory of a particle P traced by the position $\mathbf{r}_{P/O}$ from $\mathbf{r}_{P/O}(t_1)$ to $\mathbf{r}_{P/O}(t_2)$. The **work** performed on P by force $\mathbf{F}_P$ along trajectory γ_P is

$$W_P^{(\mathbf{F}_P)}(\mathbf{r}_{P/O}; \gamma_P) \triangleq \int_{\gamma_P} \mathbf{F}_P \cdot {}^{\mathcal{I}}d\mathbf{r}_{P/O},$$

where ${}^{\mathcal{I}}d\mathbf{r}_{P/O} \triangleq {}^{\mathcal{I}}\mathbf{v}_{P/O}dt$, the differential displacement of P relative to O, is everywhere tangent to γ_P.

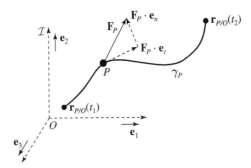

Figure 5.1 Particle P travels along trajectory γ_P from $\mathbf{r}_{P/O}(t_1)$ to $\mathbf{r}_{P/O}(t_2)$ under the influence of force $\mathbf{F}_P$. The work done on P by $\mathbf{F}_P$ along γ_P is the integral of the tangential component $\mathbf{F}_P \cdot \mathbf{e}_t$ from t_1 to t_2, where $\mathbf{e}_t$ and $\mathbf{e}_n$ are unit vectors of path frame $\mathcal{B} = (P, \mathbf{e}_t, \mathbf{e}_n, \mathbf{e}_3)$.

Qualitatively, work is the amount of energy it takes to move a particle along a certain path. The units, joules, are thus units of energy. Anyone who has gone to the beach knows that it is much easier to drag a wagon full of beach chairs over a boardwalk than over sand. Dragging the wagon over sand requires more work. This scenario also depends upon the path. A straight-line path to your desired spot on the beach involves less work than a circuitous route around various obstacles. Work is an important tool for finding certain properties of dynamical systems—even though we integrate out the details of the force.

The notation $W_P^{(\mathbf{F}_P)}(\mathbf{r}_{P/O}; \gamma_P)$ is clearly a bit cumbersome; nevertheless, it serves an important purpose. Work is described as a function of the argument $\mathbf{r}_{P/O}$, parametrized by the path γ_P, since it depends on the path of point P. Even though work is a scalar quantity, the integral is frame dependent. Because the integration takes place over changes in the position of P, we specify the frame $\mathcal{I}$ as a superscript on the differential vector displacement $^\mathcal{I}d\mathbf{r}_{P/O}$ (see Appendix A.6). We do not use a superscript to denote the frame on the work term itself because we assume that work is calculated in an inertial frame. We label the work with a superscript specifying a particular force, associating the generation of work with that force.

We can also write the work integral in an alternate form. As long as the particle doesn't stop moving along its trajectory, that is, if $\|^\mathcal{I}\mathbf{v}_{P/O}(t)\| > 0$ for all $t_1 \leq t \leq t_2$, then the work integral can be written in terms of velocity and time. Simply multiplying the integrand in Definition 5.1 by dt/dt yields

$$W_P^{(\mathbf{F}_P)}(\mathbf{r}_{P/O}; \gamma_P) = \int_{t_1}^{t_2} \mathbf{F}_P \cdot {}^\mathcal{I}\mathbf{v}_{P/O}\,dt. \tag{5.1}$$

Since the independent variable is now time, we replaced the path integral with a definite integral between t_1 and t_2. The specific path is implied by the velocity $^\mathcal{I}\mathbf{v}_{P/O}$, which is the velocity of P along the trajectory γ_P. This alternate form of the work integral will prove to be very useful.

It is important to remember that not all forces acting on a particle necessarily do work. In particular, since the definition of work depends on the dot product of the force $\mathbf{F}_P$ with the differential displacement $^\mathcal{I}d\mathbf{r}_{P/O}$, the component of $\mathbf{F}_P$ perpendicular to the trajectory does no work. An important consequence is that a constraint force in a system with reduced degrees of freedom does no work, as it is always, by definition, perpendicular to the path. Looking only at the work, then, tells us nothing about constraint forces.

The alternate form Eq. (5.1) of the work integral naturally leads us to write the work integral in the path coordinate s, defined by $ds/dt \triangleq \|^{\mathcal{I}}\mathbf{v}_{P/O}\|$, and the path frame $\mathcal{B} = (P, \mathbf{e}_t, \mathbf{e}_n, \mathbf{e}_3)$, where $\mathbf{e}_t \triangleq {}^{\mathcal{I}}\mathbf{v}_{P/O}/\|^{\mathcal{I}}\mathbf{v}_{P/O}\|$. The work in Eq. (5.1) is then

$$W_P^{(\mathbf{F}_P)}(\mathbf{r}_{P/O}; \gamma_P) = \int_{\gamma_P} \mathbf{F}_P \cdot \mathbf{e}_t \, ds.$$

We conclude this introduction to the energy of a particle with the definition of *power*. While work is a measure of energy, power is a measure of how work changes over time. Intuitively, power reflects the fact that, over the same period of time, it takes more work to produce a large force on a particle than it takes to produce a small force.

Definition 5.2 The **power** at time $t = a$ of the force $\mathbf{F}_P$ acting on particle P is the rate of change of the work $W_P^{(\mathbf{F}_P)}(\mathbf{r}_{P/O}; \gamma_P)$ evaluated at $t = a$:

$$P_P^{(\mathbf{F}_P)}(a) \triangleq \frac{d}{dt}\bigg|_{t=a} W_P^{(\mathbf{F}_P)}(\mathbf{r}_{P/O}; \gamma_P) = \mathbf{F}_P(a) \cdot {}^{\mathcal{I}}\mathbf{v}_{P/O}(a).$$

The SI units of power are watts (J/s). Note that we used Eq. (5.1) and the second fundamental theorem of calculus (see Appendix A) to arrive at the second equality in this definition. To save writing, we often drop the time notation and write

$$\boxed{P_P^{(\mathbf{F}_P)} = \mathbf{F}_P \cdot {}^{\mathcal{I}}\mathbf{v}_{P/O}.} \tag{5.2}$$

Notationally, power P is distinguished from particle P by the use of Roman type.

Example 5.1 Hoisting a Heavy Mass

Consider the situation illustrated in Figure 5.2a. The objective is to use a rope and pulley to hoist mass m_P to height h. From the free-body diagram in Figure 5.2b we see that, to lift the mass, the force in the rope T must be greater than or equal to the weight $m_P g$. Clearly, one criteria of a successful hoist is that the puller be strong enough to lift the mass. In this example we compute the work done and the power needed to lift the mass under two different force profiles.

The motion of P is confined to the $\mathbf{e}_y$ direction. The path γ_P of the mass is thus a straight vertical line. From the free-body diagram, Newton's second law for this problem is simply

$$(T - m_P g)\mathbf{e}_y = m_P \ddot{y}\mathbf{e}_y.$$

The first approach to hoisting the mass is initially to use the maximum possible force and then taper off the force until the mass is at the top. This approach is similar to the clean-and-jerk move used by professional weightlifters. In this scenario, we

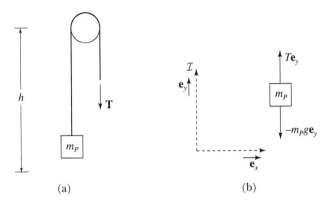

Figure 5.2 Hoisting mass m_P by means of a pulley to height h. (a) Hoist. (b) Reference frame and free-body diagram.

assume a linear model for the tension force,

$$T = m_P g + m_P b \left(1 - \frac{y}{h}\right),$$

where $b > 0$ is a parameter of the model. This force model results in acceleration

$$\ddot{y} = b \left(1 - \frac{y}{h}\right). \tag{5.3}$$

The force T reaches its maximum at the beginning of the pull when $y = 0$ and tapers off to the minimum $T = m_P g$ needed to hold the mass at the top of the hoist, when $y = h$. The work performed on P by T during the hoist is found from Definition 5.1:

$$W_P^{(T)}(y; \gamma_P) = \int_0^h T \, dy = \int_0^h m_P \left(g + b - \frac{b}{h} y\right) dy$$

$$= m_P h \left(g + \frac{b}{2}\right). \tag{5.4}$$

It is also interesting to find the power exerted by T during the hoist. To do this requires the speed of the mass as it rises, which we can then insert into Eq. (5.2). We find the speed by recognizing that Eq. (5.3) is the equation of motion for the mass,

$$\ddot{y} + \omega_0^2 y = b,$$

where $\omega_0 \triangleq \sqrt{b/h}$. This equation yields simple harmonic motion. Assuming P starts from rest, the solution is

$$y(t) = h \left(1 - \cos(\omega_0 t)\right)$$

$$\dot{y}(t) = \sqrt{bh} \sin(\omega_0 t).$$

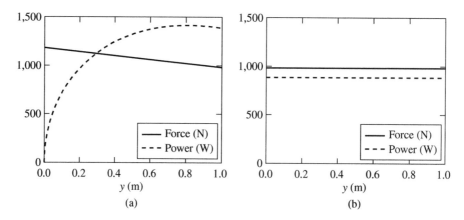

Figure 5.3 The force and power exerted as a 100 kg mass is hoisted to a height of 1 m. (a) Linear force model with an initial acceleration of 2 m/s². (b) Constant-velocity model using the average velocity from (a).

Observe that P reaches $y = h$ at time $t = (\pi/2)\sqrt{h/b}$, which implies the average speed is $v_0 = (2/\pi)\sqrt{bh}$. Substituting the tension force $T\mathbf{e}_y$, the position $y\mathbf{e}_y$, and the speed $\dot{y}\mathbf{e}_y$ into Eq. (5.2) yields

$$\mathrm{P}_P^{(T)}(t) = m_P \sqrt{bh}\, \sin(\omega_0 t)\left(g + b\cos(\omega_0 t)\right).$$

Figure 5.3a shows a plot of the power exerted during the hoist from $y = 0$ to $y = h = 1$ m for $m_P = 100$ kg and $b = 2$ m/s².

The second approach to lifting the mass is to assume a constant speed. In this case, $\ddot{y} = 0$ and, from Newton's second law, $T = m_P g$ during the hoist. In other words, the puller exerts the minimum force needed to keep the mass moving at constant velocity $v_0 \mathbf{e}_y$. Note that we are ignoring the initial impulse required to set the mass moving; we discuss that a bit later.

In this case, the work done by the tension in the rope,

$$W_P^{(T)}(y; \gamma_P) = \int_0^h m_P g\, dy = m_P g h,$$

is independent of the speed v_0. Furthermore, less work is done in this approach than in the jerk hoist. The power is also straightforward to calculate. Suppose the constant speed of the mass is given by the average speed during the jerk hoist, $v_0 = (2/\pi)\sqrt{bh}$. The power is then

$$\mathrm{P}_P^{(T)} = \frac{2m_P g}{\pi}\sqrt{bh},$$

which is constant during the hoist. Figure 5.3b shows a plot of the power exerted during the hoist from $y = 0$ to $y = h = 1$ m for $m_P = 100$ kg and $b = 2$ m/s².

Which strategy is better? It depends. Certainly, the constant-velocity hoist involves less total work, which has some appeal. However, it requires a constant exertion of power, which can be tiring. The jerk hoist has a varying power profile, with maximum power exerted for only a short period of time—interestingly, the maximum power is not exerted at the same time as the maximum force!

5.2 Total Work and Kinetic Energy

The work integral in Definition 5.1 looks qualitatively similar to the impulse integral in Definition 4.1 in Chapter 4—in both definitions, force is integrated over an interval. It is not surprising, then, that we can also find an equation that relates work to the change of a quantity evaluated at the endpoints of the particle trajectory.

First we need to make explicit the *total work* $W_P^{(tot)}(\mathbf{r}_{P/O}; \gamma_P)$ acting on particle P. The total work is the work done on P by the vector sum of all of the forces acting on P over the trajectory γ_P. This distinction allows us to use Newton's second law to replace the force $\mathbf{F}_P$ in the work integral in Definition 5.1 with the mass, m_p, times the acceleration $^\mathcal{I}\mathbf{a}_{P/O}$:

$$W_P^{(tot)}(\mathbf{r}_{P/O}; \gamma_P) = \int_{\gamma_P} m_P \frac{^\mathcal{I}d}{dt}\left(^\mathcal{I}\mathbf{v}_{P/O}\right) \cdot {^\mathcal{I}d}\mathbf{r}_{P/O} = m_P \int_{\gamma_P} {^\mathcal{I}d}{^\mathcal{I}}\mathbf{v}_{P/O} \cdot \frac{^\mathcal{I}d}{dt}\left(\mathbf{r}_{P/O}\right)$$

$$= m_P \int_{\gamma_P} {^\mathcal{I}d}{^\mathcal{I}}\mathbf{v}_{P/O} \cdot {^\mathcal{I}}\mathbf{v}_{P/O} = \frac{m_P}{2} \int_{^\mathcal{I}\mathbf{v}_{P/O}(t_1)}^{^\mathcal{I}\mathbf{v}_{P/O}(t_2)} {^\mathcal{I}d}\left(^\mathcal{I}\mathbf{v}_{P/O} \cdot {^\mathcal{I}}\mathbf{v}_{P/O}\right).$$

The last integral above contains a perfect differential $^\mathcal{I}d\|^\mathcal{I}\mathbf{v}_{P/O}\|^2$, which implies that the integral depends only on the endpoints of the path γ_P. Integrating results in

$$W_P^{(tot)}(\mathbf{r}_{P/O}; \gamma_P) = \frac{m_P}{2} \int_{^\mathcal{I}\mathbf{v}_{P/O}(t_1)}^{^\mathcal{I}\mathbf{v}_{P/O}(t_2)} {^\mathcal{I}d}\|^\mathcal{I}\mathbf{v}_{P/O}\|^2$$

$$= \underbrace{\frac{m_P}{2}\|^\mathcal{I}\mathbf{v}_{P/O}(t_2)\|^2}_{\triangleq T_{P/O}(t_2)} - \underbrace{\frac{m_P}{2}\|^\mathcal{I}\mathbf{v}_{P/O}(t_1)\|^2}_{\triangleq T_{P/O}(t_1)}. \tag{5.5}$$

We call the quantity $T_{P/O}$ the *kinetic energy* of P relative to O. We have shown that the total work on P along γ_P equals the change in the kinetic energy of P from t_1 to t_2. We cannot emphasize enough the importance of this result. Even though work, in general, depends on path, *the total work equals the change in kinetic energy.*

Definition 5.3 The **kinetic energy** of particle P relative to O at time t is

$$\boxed{T_{P/O}(t) \triangleq \frac{1}{2}m_P\left(^\mathcal{I}\mathbf{v}_{P/O}(t) \cdot {^\mathcal{I}}\mathbf{v}_{P/O}(t)\right) = \frac{1}{2}m_P\|^\mathcal{I}\mathbf{v}_{P/O}(t)\|^2,}$$

where $\mathcal{I}$ is an inertial frame.

Remember that the velocity used in Definition 5.3 must be an inertial velocity. *One of the most common mistakes in dynamics is to use a noninertial velocity in computing the kinetic energy.* For example, this mistake often occurs when computing the kinetic energy of a particle moving with respect to a noninertial frame (see Chapter 8). Also note that the kinetic energy depends on the reference point O. The following example illustrates both of these points.

Example 5.2 Kinetic Energy of a Pendulum

Consider again a pendulum suspended in a box translating relative to inertial frame $\mathcal{I} = (O, \mathbf{e}_1, \mathbf{e}_2, \mathbf{e}_3)$ (see Example 4.6 and Figure 4.8). Here we assume that the box is traveling at constant velocity, which implies that frame $\mathcal{B} = (O', \mathbf{b}_1, \mathbf{b}_2, \mathbf{b}_3)$ is also inertial. Polar frame $\mathcal{C} = (O', \mathbf{e}_r, \mathbf{e}_\theta, \mathbf{e}_3)$, where $\mathbf{e}_r$ points from O' to pendulum bob P, is not inertial. The kinematics of P relative to O' are

$$\mathbf{r}_{P/O'} = l\mathbf{e}_r$$

$$^{\mathcal{I}}\mathbf{v}_{P/O'} = {}^{\mathcal{B}}\mathbf{v}_{P/O'} = l\dot{\theta}\mathbf{e}_\theta,$$

whereas the kinematics of P relative to O are

$$\mathbf{r}_{P/O} = \mathbf{r}_{O'/O} + \mathbf{r}_{P/O'} = x\mathbf{b}_2 + l\mathbf{e}_r$$

$$^{\mathcal{I}}\mathbf{v}_{P/O} = \dot{x}\mathbf{b}_2 + l\dot{\theta}\mathbf{e}_\theta.$$

Using Definition 5.3, we find the kinetic energy of P relative to O' is

$$T_{P/O'} = \frac{m_P}{2}\|^{\mathcal{I}}\mathbf{v}_{P/O'}\|^2 = \frac{m_P}{2}l^2\dot{\theta}^2,$$

which is the typical expression for the kinetic energy of a simple pendulum. The kinetic energy of P relative to O is

$$T_{P/O} = \frac{m_P}{2}\|^{\mathcal{I}}\mathbf{v}_{P/O}\|^2 = \frac{m_P}{2}\|(\dot{x} + l\dot{\theta}\cos\theta)\mathbf{b}_2 + l\dot{\theta}\sin\theta\mathbf{b}_1\|^2$$

$$= \frac{m_P}{2}(l^2\dot{\theta}^2 + \dot{x}^2 + 2l\dot{\theta}\dot{x}\cos\theta).$$

That is, $T_{P/O}$ and $T_{P/O'}$ are different but equally valid measures of the kinetic energy of P.[1] Also, note that using the noninertial velocity $^{\mathcal{C}}\mathbf{v}_{P/O'} = 0$ to compute $T_{P/O'}$ yields the incorrect result.

Eq. (5.5) leads us to the following *work–kinetic-energy formula*, which equates the change in kinetic energy of a particle to the total work done on the particle:

$$T_{P/O}(t_2) = T_{P/O}(t_1) + W_P^{\text{(tot)}}(\mathbf{r}_{P/O}; \gamma_P). \tag{5.6}$$

----

[1] Because the kinetic energy is a quadratic function of the velocity, it does not obey the vector summation rule analogous to $^{\mathcal{I}}\mathbf{v}_{P/O} = {}^{\mathcal{I}}\mathbf{v}_{P/O'} + {}^{\mathcal{I}}\mathbf{v}_{O'/O}$. That is, even assuming point O' has mass m_P, the kinetic energy $T_{P/O}$ does *not* equal $T_{P/O'} + T_{O'/O}$.

What about the relationship between kinetic energy and power? This is easily found by taking the time derivative of the kinetic energy in Definition 5.3:

$$\frac{d}{dt}(T_{P/O}) = \frac{1}{2}m_P\left(\frac{^{\mathcal{I}}d}{dt}\left(^{\mathcal{I}}\mathbf{v}_{P/O}\right)\cdot {^{\mathcal{I}}\mathbf{v}}_{P/O} + {^{\mathcal{I}}\mathbf{v}}_{P/O}\cdot\frac{^{\mathcal{I}}d}{dt}\left(^{\mathcal{I}}\mathbf{v}_{P/O}\right)\right)$$

$$= \underbrace{m_P\frac{^{\mathcal{I}}d}{dt}\left(^{\mathcal{I}}\mathbf{v}_{P/O}\right)}_{=\mathbf{F}_P}\cdot {^{\mathcal{I}}\mathbf{v}}_{P/O}.$$

Using Newton's second law to replace $m_P\frac{^{\mathcal{I}}d}{dt}(^{\mathcal{I}}\mathbf{v}_{P/O})$ with the total force $\mathbf{F}_P$ acting on P, we find that

$$\frac{d}{dt}(T_{P/O}) = \mathbf{F}_P\cdot {^{\mathcal{I}}\mathbf{v}}_{P/O} = \mathrm{P}_P^{(\mathrm{tot})}.$$

By Definition 5.2, $\mathbf{F}_P\cdot {^{\mathcal{I}}\mathbf{v}}_{P/O}$ is the power of $\mathbf{F}_P$. To highlight the fact that this expression is the *total* power, we once again use the superscript (tot). In words, the total power acting on P is equal to the rate of change of the kinetic energy of P.

Example 5.3 Carrier Landing

Figure 5.4 shows a fighter jet landing on an aircraft carrier. In this example we use work to study the forces and acceleration needed to stop the jet before it reaches the end of the deck. Suppose the jet approaches the aircraft carrier at $v_0 = 125$ knots (64.3 m/s). Assume that, once the fighter jet touches down on the flight deck, the arresting cable decelerates it constantly over $d = 300$ ft (91.44 m) before coming to a stop. What is the acceleration of the fighter jet and what is the force required to stop its motion?

Figure 5.4 A fighter jet landing on an aircraft carrier as the tailhook captures the arresting cable. Image courtesy of Shutterstock.

First we solve this problem using Newton's second law, assuming that the aircraft-carrier flight deck is an inertial frame. Let $\mathbf{e}_1$ be the unit vector in the direction of the jet's initial velocity and $\mathbf{F}_P = -F\mathbf{e}_1$ be the (constant) force exerted on the jet by the carrier's arresting mechanism. Using the coordinate x for the position $\mathbf{r}_{P/O} = x\mathbf{e}_1$ of the jet relative to its position at touchdown, we have

$$-F\mathbf{e}_1 = m_P \ddot{x} \mathbf{e}_1.$$

Integrating $\ddot{x} = \frac{d\dot{x}}{dx}\dot{x} = a$ (constant) over the deceleration,

$$\int_{v_0}^{0} \dot{x}\, d\dot{x} = \int_{0}^{d} a\, dx,$$

gives

$$-\frac{1}{2}v_0^2 = ad,$$

or

$$^{\mathcal{I}}\mathbf{a}_{P/O} = -\frac{1}{2d}v_0^2 \mathbf{e}_1 = -22.6 \text{ m/s}^2 \mathbf{e}_1$$

$$\mathbf{F}_P = m_P\,{}^{\mathcal{I}}\mathbf{a}_{P/O} = -\frac{m_P}{2d}v_0^2 \mathbf{e}_1.$$

Alternatively, using the work–kinetic-energy formula (Eq. (5.6)) and Definition 5.1, we obtain

$$W_P^{(F)}(t_1, t_2) = T_{P/O}(t_2) - T_{P/O}(t_1)$$

$$\int_{0}^{d} (-F)\, dx = 0 - \frac{1}{2}m_P v_0^2,$$

which can be solved to get

$$\mathbf{F}_P = -F\mathbf{e}_1 = -\frac{m_P}{2d}v_0^2 \mathbf{e}_1.$$

Using Newton's second law results in

$$^{\mathcal{I}}\mathbf{a}_{P/O} = \frac{1}{m_P}\mathbf{F}_P = -\frac{1}{2d}v_0^2 \mathbf{e}_1,$$

as before.

Example 5.4 Take-off from an Aircraft Carrier

This example examines the opposite problem from Example 5.3—an airplane taking off from the deck of an aircraft carrier. To do so we utilize the expression for power in Eq. (5.2).

There are two types of planes we want to consider: a propeller-driven aircraft and a fighter jet. These two types are distinguished by the relationship between thrust and power in the engine. *Thrust* is the force applied to the airplane by the engine, causing it to accelerate. A jet engine is characterized by constant thrust at a given throttle setting. A propeller aircraft is characterized by constant power (determined by burning fuel in its internal combustion engine). The question is whether enough power is available in each plane for it to take off before it reaches the end of the carrier deck.

As in Example 5.3, we take $\mathbf{e}_1$ to be the unit vector in the direction of the plane's velocity along the deck. The thrust force on the plane is $\mathbf{F}_P = T\mathbf{e}_1$ in the direction of motion. We first consider the jet. Since the jet applies a constant thrust, the work done to travel a distance x along the deck (neglecting friction and aerodynamic drag) is

$$W_P^{(T)} = \int_0^x \mathbf{F}_P \cdot \mathbf{e}_1 dx = Tx.$$

From the work–kinetic-energy formula, the work to accelerate the jet a distance d to the end of the flight deck is equal to the change in kinetic energy

$$W_P^{(T)}(t_1, t_2) = Td = \frac{1}{2}m_P v_f^2 - 0,$$

where v_f is the final speed of the jet when it leaves the deck. We can thus solve for the final speed generated by the constant thrust T:

$$v_f = \sqrt{\frac{2Td}{m_P}}. \tag{5.7}$$

Alternatively, the thrust needed to achieve v_f over a distance d is

$$T = \frac{v_f^2 m_P}{2d}.$$

Using Eqs. (5.2) and (5.7), the power exerted by the jet engine as a function of distance x is

$$P = Tv = \sqrt{\frac{2T^3 x}{m_P}}.$$

A typical fighter jet must reach roughly 180 miles/hr in 200 ft to take off from the deck. The F-14 Tomcat, for example, weighs roughly 61,000 lb fully loaded (27,700 kg) and has a maximum thrust with afterburner of 27,800 lbf (124.7 kN). Plugging these values into Eq. (5.7) yields the final speed of the jet at the end of 200 ft of only 53 miles/hr, far below the requirement. That is why all aircraft carriers are equipped with catapults to increase the effective thrust and launch the fighters with the required speed.

We can perform a similar calculation for propeller-driven aircraft. These are characterized by a constant power P, so the thrust as a function of speed is

$$T = \frac{P}{v}.$$

The equation of motion for the velocity is given by Newton's second law,

$$\dot{v}\mathbf{e}_1 = \frac{T}{m_P}\mathbf{e}_1 = \frac{P}{m_P v}\mathbf{e}_1.$$

Using the same change of variables as in the previous example, $\dot{v} = \frac{dv}{dx}\frac{dx}{dt} = v\frac{dv}{dx}$, we can integrate the equation of motion to obtain

$$\int_0^{v_f} v^2 dv = \frac{P}{m_P} \int_0^d dx,$$

which yields

$$\frac{1}{3}v_f^3 = \frac{P}{m_P}d.$$

Thus the constant power required for the plane to reach the final speed v_f in a distance d is

$$P = \frac{m_P v_f^3}{3d}.$$

If we consider a plane of comparable weight to the F-14 and ask how much power is required in the engine to reach 100 miles/hr after 200 ft, rather than 180 miles/hr, we find that the power needed exceeds 20,000 hp, two orders of magnitude larger than a typical engine!

5.3 Work Due to an Impulse

What is the work done by an impulsive force? Recall that our model of an impulse is that the force acts over so short an interval of time that the particle doesn't move. From our definition of work, you might reach the conclusion that an impulsive force does no work. However, the impulse does instantaneously change the inertial velocity of the particle. By its definition, the kinetic energy is thus instantly changed. Applying the work–kinetic-energy formula (Eq. (5.6)) contradicts the zero-work conclusion.

The problem here is in our assumption of zero motion during an impulse. This is only an approximation; in real systems the particle will move, albeit by only a small amount. When combined with the very large force of an impulse, the result is that work is in fact done over the short time interval. However, computing the work based on the force and displacement is not possible, as we don't have access to the details of the force or the trajectory. Rather, we use the work–kinetic-energy formula to define the work done by an impulse.

Definition 5.4 The work $W_P^{(\overline{\mathbf{F}}_P)}(t_1, t_2)$ due to an impulse $\overline{\mathbf{F}}_P(t_1, t_2)$ is equal to the change in kinetic energy before and after the impulse:

$$W_P^{(\overline{\mathbf{F}}_P)}(t_1, t_2) \triangleq T_{P/O}(t_2) - T_{P/O}(t_1) = \frac{1}{2}m_P \|{}^{\mathcal{I}}\mathbf{v}_{P/O}(t_2)\|^2 - \frac{1}{2}m_P \|{}^{\mathcal{I}}\mathbf{v}_{P/O}(t_1)\|^2.$$

Again, the work done by an impulse is exactly equal to the change in kinetic energy between t_1 and t_2. The change in kinetic energy is found by using the velocities before and after the impulse from the linear-impulse form in Eq. (4.2) or the angular-impulse form in Eq. (4.17).

Example 5.5 Hoisting a Heavy Mass, Revisited

Recall in Example 5.1 that we brushed aside the impulse required to get the mass moving for the constant-velocity hoist. In truth, that makes a comparison with the jerk hoist unfair. Now we treat it more carefully and add in the work to get the mass moving.

Example 5.1 showed that the work done to lift the mass at constant speed v_0 was independent of the speed and equal to the work against gravity,

$$W_P^{(T)}(y; \gamma_P) = \int_0^h m_P g\, dy = m_P g h.$$

The work to lift the mass, however, should include the work to start the mass from rest. We model the force to initialize the mass at v_0 by an impulse; the resulting work is thus given by the difference in kinetic energy, which in this case is the kinetic energy of the mass as it rises. The work is

$$W_P^{(T)}(y; \gamma_P) = m_P g h + \frac{1}{2} m_P v_0^2.$$

It is now clear from comparing the work for the constant-velocity hoist to the work for the jerk hoist in Eq. (5.4) that the constant-velocity hoist involves less work than the jerk hoist as long as

$$v_0 < \sqrt{bh}.$$

5.4 Conservative Forces and Potential Energy

It is clear from Eq. (5.6) that, if the total work on a particle is zero, then the kinetic energy of the particle is conserved—this is why kinetic energy is sometimes called an integral of the motion. Using Definition 5.1, compare Eq. (5.6) to the momentum integral in Eq. (4.2). Although energy is a scalar and (linear) momentum is a vector, these two equations are quite similar. Both represent a general solution of Newton's second law that, when given a specific force, can be used to find the speed of a particle at two different points in time.

Recall that, in Chapter 4, we discussed how the momentum integral becomes extremely useful for certain special classes of forces. This is true for the work–kinetic-energy formula as well. There are again three classes of forces for which we make particular use of the work–kinetic-energy formula:

    a. $\mathbf{F}_P$ *is identically zero.* That is, the total force acting on particle P is zero. In this case the total work on P is zero and the kinetic energy is conserved.

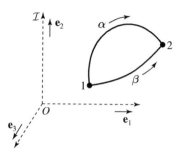

Figure 5.5 Two trajectories of a particle from point 1 to point 2. If the particle is acted on by a conservative force, then the work done along path α is the same as the work done along path β. If the force is nonconservative, then the work done along α may differ from that along β.

    b. $\mathbf{F}_P$ *is constant or is a known function of time.* That is, $\mathbf{F}_P$ is independent of the position and velocity of the particle. In this (rare) case, the integral in Definition 5.1 can be solved, either explicitly or numerically, to relate the kinetic energy at t_1 to the kinetic energy at t_2.

    c. *The work done by* $\mathbf{F}_P$ *is independent of the path of the particle.* We call such a force a *conservative force.* In this case—which is the subject of the rest of this section—Eq. (5.6) becomes particularly useful.

Conservative forces reduce the work integral in Definition 5.1 to a convenient analytic form. When combined with the kinetic energy in the work–kinetic-energy formula, we have an extremely useful tool for problem solving and a new conservation law.

5.4.1 Conservative Forces

We begin with a qualitative definition of a conservative force.

Qualitative Definition 5.5 The work performed by a **conservative force** acting on a particle depends only on the endpoints of the particle's path.

    The distinction between a conservative force and a nonconservative one is shown in Figure 5.5. A nonconservative force may perform more (or less) work along path α than along path β. For a conservative force, the work is the same, since paths α and β have the same endpoints.

    What is the significance of a conservative force? Consider the work done by a force $\mathbf{F}_P$ acting on particle P as it first moves on path α from 1 to 2 and then back to 1 along path $-\beta$. The work done while traveling on $-\beta$ is

$$W_P^{(\mathbf{F}_P)}(\mathbf{r}_{P/O}; -\beta) = \int_{-\beta} \mathbf{F}_P \cdot {}^{\mathcal{I}}d\mathbf{r}_{P/O} = \int_{t_2}^{t_1} \mathbf{F}_P \cdot {}^{\mathcal{I}}\mathbf{v}_{P/O}\,dt$$

$$= -\int_{t_1}^{t_2} \mathbf{F}_P \cdot {}^{\mathcal{I}}\mathbf{v}_{P/O}\,dt = -W_P^{(\mathbf{F}_P)}(\mathbf{r}_{P/O}; \beta), \qquad (5.8)$$

where we have used the alternate form of the work integral from Eq. (5.1). Thus *the work going backward on a path is simply the negative of the work going forward on the path.*

Now suppose the force $\mathbf{F}_P$ is conservative—that is, the work done by $\mathbf{F}_P$ is independent of the path. For such a force, $W_P^{(\mathbf{F}_P)}(\mathbf{r}_{P/O}; \beta) = W_P^{(\mathbf{F}_P)}(\mathbf{r}_{P/O}; \alpha)$ and, in fact, the work is the same along any path between points 1 and 2. The work done around the complete loop from 1 to 2 along α and returning from 2 to 1 along $-\beta$ is

$$W_P^{(\mathbf{F}_P)}(\mathbf{r}_{P/O}; \alpha) + W_P^{(\mathbf{F}_P)}(\mathbf{r}_{P/O}; -\beta) = W_P^{(\mathbf{F}_P)}(\mathbf{r}_{P/O}; \alpha) - W_P^{(\mathbf{F}_P)}(\mathbf{r}_{P/O}; \beta)$$

$$= 0 = \oint_1 \mathbf{F}_P \cdot {}^{\mathcal{I}} d\mathbf{r}_{P/O}, \tag{5.9}$$

where $\oint_1$ denotes the integral around any closed curve through point 1. To obtain Eq. (5.9), we have used Eq. (5.8) and the fact that the work on the two paths is the same for a conservative force. Physically, this result means that whatever work is done by the force in moving away from 1 is returned when coming back. We use the fact that this result holds for any closed curve to introduce the following, more mathematical, definition of a conservative force.

Definition 5.6 A **conservative force** is a force $\mathbf{F}_P$ that satisfies

$$\oint \mathbf{F}_P \cdot {}^{\mathcal{I}} d\mathbf{r}_{P/O} = 0,$$

which implies that $\mathbf{F}_P$ does no work on particle P on any closed trajectory.

The notation $W_P^{(c)}$ denotes the work performed on P by the total conservative force, and $W_P^{(nc)}$ denotes the work performed on P by the total nonconservative force. Since for a conservative force we do not need to indicate the path γ_P—the work depends only on the endpoints $\mathbf{r}_{P/O}(t_1)$ and $\mathbf{r}_{P/O}(t_2)$—we write $W_P^{(c)}(\mathbf{r}_{P/O}(t_1), \mathbf{r}_{P/O}(t_2))$ or simply $W_P^{(c)}(t_1, t_2)$. The total work $W_P^{(tot)}(\mathbf{r}_{P/O}; \gamma_P)$ is equal to the sum of the work produced by both conservative and nonconservative forces:

$$W_P^{(tot)}(\mathbf{r}_{P/O}; \gamma_P) = W_P^{(c)}(t_1, t_2) + W_P^{(nc)}(\mathbf{r}_{P/O}; \gamma_P). \tag{5.10}$$

Example 5.6 A Simple Conservative Force

The simplest conservative force we consider is a constant force field. That is, the force acting on the particle is independent of its position. A common example is the weight of a particle near the surface of the earth. The force of gravity is simply $\mathbf{F}_P = -m_P g \mathbf{e}_y$ everywhere (where the unit vector $\mathbf{e}_y$ is in the vertical direction). We now show that this force is conservative using Definition 5.6.

Consider a closed path consisting of the particle rising to height h, moving horizontally a distance l, dropping back down to $y = 0$, and then moving horizontally

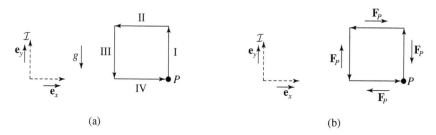

Figure 5.6 Conservative and nonconservative forces. A particle traveling around a closed loop acted on by (a) a constant gravity force and (b) a constant drag force.

again back to its original position. We label each segment of this square path I–IV, as shown in Figure 5.6a. The work along segment I is

$$W_P^{(\mathbf{F}_P)}(\mathbf{r}_{P/O}; \mathrm{I}) = \int_{\mathrm{I}} -m_P g \mathbf{e}_y \cdot dy \mathbf{e}_y = -m_P gh.$$

The work along II is

$$W_P^{(\mathbf{F}_P)}(\mathbf{r}_{P/O}; \mathrm{II}) = \int_{\mathrm{II}} -m_P g \mathbf{e}_y \cdot dx \mathbf{e}_x = 0.$$

The work along III is

$$W_P^{(\mathbf{F}_P)}(\mathbf{r}_{P/O}; \mathrm{III}) = \int_{\mathrm{III}} -m_P g \mathbf{e}_y \cdot dy \mathbf{e}_y = m_P gh.$$

Finally, the work along IV is also zero, since the force is perpendicular to the trajectory. Summing these together, we find that the work around the closed path is zero. The constant force thus satisfies Definition 5.6.[2]

Example 5.7 A Nonconservative Force

This example considers a different force. Let particle P travel the same trajectory as in Example 5.6 at a constant speed. We find the work done by a drag force $\mathbf{F}_P = -D^{\mathcal{I}}\hat{\mathbf{v}}_{P/O}$ acting against the velocity with magnitude $D = \frac{1}{2}\beta \|^{\mathcal{I}}\mathbf{v}_{P/O}\|^2$, as in Figure 5.6b. Since the drag force is constant, the work around the closed path is

$$W_P^{(\mathbf{F}_P)}(\mathbf{r}_{P/O}; \gamma_P) = -D \int_0^h dy + D \int_l^0 dx + D \int_h^0 dy - D \int_0^l dx$$

$$= -Dh - Dl - Dh - Dl = -2D(h + l).$$

---

[2] You may have noticed that, strictly speaking, we actually did not show that the constant gravity force is conservative, as we did not show that *any and every* loop integral was zero, only this specific one. Showing that every loop integral is zero is quite a bit more complicated; as shown later, we usually determine that a force is conservative by a different method.

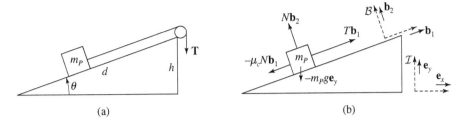

Figure 5.7 Mechanical advantage. Hoisting mass m_P up an inclined plane to height h using a pulley. (a) Hoist. (b) Reference frames and free-body diagram.

Thus the drag force is not conservative. Because the force is always directed against the velocity, thus impeding motion, the work done around the closed path is not zero. This is true for all retarding forces, including viscous damping and Coulomb friction. When the work done by a force around a closed path is negative, as here, we call the force *dissipative*.

Example 5.8 Mechanical Advantage

Consider again the problem of lifting a mass by a pulley (see Example 5.1). Suppose now, however, that the puller is unable to exert the force necessary to raise the mass (i.e., $T_{max} < m_P g$). In this case, the puller can use *mechanical advantage*. Rather than lift the mass directly, he or she places it on an inclined plane and pulls it up the slope, as in Figure 5.7a. Here we assume a constant-speed lift.

Figure 5.7b shows the reference frames and a free-body diagram for the mass on the inclined plane. In this problem there are two frames, $\mathcal{I}$ and $\mathcal{B}$, both inertial. The unit vectors of the two frames are related by

	$\mathbf{e}_x$	$\mathbf{e}_y$
$\mathbf{b}_1$	$\cos\theta$	$\sin\theta$
$\mathbf{b}_2$	$-\sin\theta$	$\cos\theta$.

For a constant-speed lift, the acceleration of the mass is zero and Newton's second law becomes

$$T\mathbf{b}_1 - \mu_c N\mathbf{b}_1 + N\mathbf{b}_2 - m_P g \sin\theta\mathbf{b}_1 - m_P g \cos\theta\mathbf{b}_2 = 0,$$

where N is the normal force and μ_c is the coefficient of friction.

As a simple case, we begin by ignoring friction ($\mu_c = 0$), which results in

$$T = m_P g \sin\theta.$$

Hence the advantage of using the ramp. The block can be raised to the same height with a force that decreases with the ramp angle. However, the block must be pulled a greater distance $d = h/\sin\theta$, and thus it will take longer to reach the desired height.

What is the work done by the tension in this case? Following Definition 5.1, we integrate over the linear path γ_P using the path coordinate s. We have

$$W_P^{(T)}(s; \gamma_P) = \int_0^d T \, ds = T d = m_P g h,$$

which is independent of θ. By pulling the block up an inclined plane, we gain the mechanical advantage at seemingly no cost. Because the constant gravity force is conservative, the work is always $m_P g h$, no matter the path angle.

However, a frictionless surface is rather unrealistic. If instead we include the friction between the block and the ramp, the tension force becomes

$$T = m_P g \sin \theta + \mu_c m_P g \cos \theta,$$

and the resulting work is

$$W_P^{(T)}(s; \gamma_P) = \int_0^d T \, ds = m_P g h (1 + \mu_c \cot \theta).$$

Thus there is a work penalty if we make the plane angle too small. Since friction is a nonconservative force, it depends on the path length. For smaller angles, the path is longer and more work is done to counteract friction. Thus to lift the block with minimal work, we want to pick the largest possible angle that satisfies the maximum force limit. Until the angle becomes quite small, the friction cost is very low for reasonable coefficients of friction.

5.4.2 Potential Energy

Since conservative forces are independent of the path, the line integral in the work definition can be replaced by a *definite* integral, which depends only on the endpoints of the path. We call this integral the *potential energy* and often write it in indefinite form.

Definition 5.7 Let $\mathbf{F}_P$ be a conservative force. The **potential energy** of a particle P associated with $\mathbf{F}_P$ is

$$U_{P/O}^{(\mathbf{F}_P)}(\mathbf{r}_{P/O}) \triangleq - \int \mathbf{F}_P \cdot {}^{\mathcal{I}} d\mathbf{r}_{P/O}.$$

The minus sign in Definition 5.7 is there to be consistent with convention (it makes later calculations a bit easier). We can always find a potential energy for a conservative force; Section 5.6 validates this claim.

You may have noticed that, since the potential energy is an indefinite integral, its value is arbitrary to within an additive constant. Typically, we determine the constant of integration by choosing a convenient reference point where the potential energy is zero. Since, as we'll see next, work is proportional to the difference in potential energy, any constant chosen simply cancels out.

How does the potential energy relate to the work done along a trajectory? Because the work done by a conservative force is a function only of the endpoints, we can write it more simply as

$$
W_P^{(\mathbf{F}_P)}(t_1, t_2) = \int_{\mathbf{r}_{P/O}(t_1)}^{\mathbf{r}_{P/O}(t_2)} \mathbf{F}_P \cdot {}^{\mathcal{I}} d\mathbf{r}_{P/O}.
$$

From Definition 5.7 and the rules of definite integration, this work can be written as

$$
W_P^{(\mathbf{F}_P)}(t_1, t_2) = U_{P/O}^{(\mathbf{F}_P)}(\mathbf{r}_{P/O}(t_1)) - U_{P/O}^{(\mathbf{F}_P)}(\mathbf{r}_{P/O}(t_2)). \tag{5.11}
$$

Thus, *the work performed by a conservative force is equal to the negative change in the potential energy.* We rearrange Eq. (5.11) to write the *work–potential-energy formula:*

$$
\boxed{U_{P/O}^{(\mathbf{F}_P)}(t_2) = U_{P/O}^{(\mathbf{F}_P)}(t_1) - W_P^{(\mathbf{F}_P)}(t_1, t_2).} \tag{5.12}
$$

Because the potential energy is a (time-varying) function of position, Eq. (5.12) includes the origin to which the position is referenced (O in Definition 5.7). This notation enables us to write, without loss of clarity, $U_{P/O}^{(\mathbf{F}_P)}(t)$ or $U_{P/O}^{(\mathbf{F}_P)}$ to refer to $U_{P/O}^{(\mathbf{F}_P)}(\mathbf{r}_{P/O}(t))$. When it is clear from context, we may also drop the superscript $\mathbf{F}_P$. As with work, later we use the total potential energy $U_{P/O}$ associated with the vector sum of all conservative forces acting on P. Note that we dropped the superscript (tot) on the potential energy for brevity; it is understood that $U_{P/O}$ represents the total potential energy associated with the total force on P.

Example 5.9 The Potential Energy of a Uniform Gravitational Field

When measured close to the surface of the earth, we typically model the gravitational force as a constant with magnitude $m_P g$ (where g is the acceleration due to gravity), rather than use the exact $1/r^2$ form of Newton's universal law of gravity. Many examples of this appeared earlier in the book. We call this a *uniform gravitational field* because the acceleration due to gravity is the same everywhere. Here we compute the potential energy associated with this force model.

Introducing an inertial frame with unit vector $\mathbf{e}_y$ in the vertical direction, the force due to the constant gravity field on a particle P of mass m_P is simply $\mathbf{F}_P = -m_P g \mathbf{e}_y$. Using Definition 5.7, the potential energy is

$$
U_{P/O}^{(\mathbf{F}_P)} = -\int (-m_P g \mathbf{e}_y) \cdot dy \mathbf{e}_y = m_P g \int dy = m_P g y,
$$

where we have chosen the constant of integration to be zero so that the potential is zero at the surface of the earth ($y = 0$). Note that there is no need to consider the horizontal displacement of the mass (in the $\mathbf{e}_x$ direction) as it drops out of the integral when dotted with the gravity force.

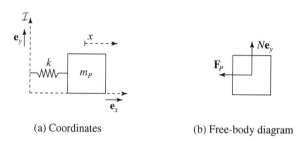

(a) Coordinates   (b) Free-body diagram

Figure 5.8 Potential energy of a spring-mass system.

Example 5.10 The Potential Energy of a Spring-Mass System

Consider the spring-mass system shown in Figure 5.8a. When the spring stretches or contracts, it exerts a force $\mathbf{F}_P = -kx\mathbf{e}_x$ on the mass, where k is the spring constant and x is the distance the spring is stretched (x is negative for contraction). The spring force $\mathbf{F}_P$ is an example of a conservative force. In this example, we compute the potential energy associated with $\mathbf{F}_P$.

The position of the mass is $\mathbf{r}_{P/O} = x\mathbf{e}_x$ and its velocity is $^{\mathcal{I}}\mathbf{v}_{P/O} = \dot{x}\mathbf{e}_x$. Let $x = 0$ denote the position of the mass when the spring is not stretched. We use $x = 0$ as the reference point for computing the potential energy associated with the spring force. Using Definition 5.7, we find

$$U_{P/O}^{(\mathbf{F}_P)} = -\int (-kx\mathbf{e}_x) \cdot dx\mathbf{e}_x = k\int x\,dx = \frac{1}{2}kx^2.$$

We have chosen the constant of integration to be zero so that the potential is zero when the spring is unstretched.

Table 5.1 summarizes some common and important conservative forces and the corresponding potential energies that arise throughout the book. These forces and their potentials come up again and again as we solve problems. In each case, we derive the potential energy $U_{P/O}^{(\mathbf{F}_P)}$ using Definition 5.7 and specify the reference point at which $U_{P/O}^{(\mathbf{F}_P)} = 0$.

An important consequence of Definition 5.7 is that *the gradient of the potential energy is a conservative force.* (See Appendix A for a review of the gradient operator ∇.) This result follows from the definition of the directional derivative in Appendix A.6,

TABLE 5.1
Common conservative forces, their corresponding potential energies, and the reference points where the potential energies are zero

	Force $\mathbf{F}_P$	Potential energy $U_{P/O}^{(\mathbf{F}_P)}$	Reference point
Spring	$-kx\mathbf{e}_x$	$-\int(-kx\mathbf{e}_x)dx \cdot \mathbf{e}_x = \frac{1}{2}kx^2$	$x = 0$
Weight	$-m_P g\mathbf{e}_y$	$-\int(-m_P g\mathbf{e}_y) \cdot dy\mathbf{e}_y = m_P g y$	$y = 0$
Gravity	$-\frac{Gm_O m_P}{r^2}\mathbf{e}_r$	$-\int\left(-\frac{Gm_O m_P}{r^2}\mathbf{e}_r\right) \cdot dr\mathbf{e}_r = -\frac{Gm_O m_P}{r}$	$r = \infty$

$$\boxed{\mathbf{F}_P = -\nabla U_{P/O}^{(\mathbf{F}_P)}.}$$

(5.13)

It is straightforward to verify Eq. (5.13) in Example 5.10. It is an important fact that every conservative force is the gradient of a scalar potential-energy function. We show this in the derivations Section 5.6.

It is instructive to verify that replacing the conservative force with the gradient of a potential leads to the same work–potential-energy formula. Inserting the expression for $\mathbf{F}_P$ in Eq. (5.13) into Definition 5.1 and using the fact that $\mathbf{F}_P$ is conservative yields

$$W_P^{(\mathbf{F}_P)}(t_1, t_2) = \int_{t_1}^{t_2} (-\nabla U_{P/O}^{(\mathbf{F}_P)}) \cdot {}^{\mathcal{I}}d\mathbf{r}_{P/O} = \int_{t_2}^{t_1} \underbrace{\nabla U_{P/O}^{(\mathbf{F}_P)} \cdot {}^{\mathcal{I}}d\mathbf{r}_{P/O}}_{={}^{\mathcal{I}}dU_{P/O}^{(\mathbf{F}_P)}}$$

$$= U_{P/O}^{(\mathbf{F}_P)}(\mathbf{r}_{P/O}(t_1)) - U_{P/O}^{(\mathbf{F}_P)}(\mathbf{r}_{P/O}(t_2)),$$

where we used the definition of the *total derivative* from Appendix A.6.

It is natural to ask how we know when a given force is conservative. One way, as shown in the previous section, is to confirm that the line integral around any closed path is zero. (It is easy to use this to show that a force is not conservative, but actually rather tricky to show that every loop integral is zero.) An easier way is to find a scalar potential-energy function whose gradient is equal to the force. If that can be done, then the force is guaranteed to be conservative (though this is not necessarily trivial to do either). Each force in Table 5.1 is conservative because we were able to find a potential energy for it.

The rate of change of the work associated with a conservative force $\mathbf{F}_P$ is the power $\mathrm{P}_P^{(\mathbf{F}_P)}$. Just as for kinetic energy, we can show that $\mathrm{P}_P^{(\mathbf{F}_P)}$ is equal to the rate of change of the potential energy:

$$\mathrm{P}_P^{(\mathbf{F}_P)} = -\frac{d}{dt} U_{P/O}^{(\mathbf{F}_P)}.$$

Example 5.11 Computing Work Using Potential Energy

Consider a collar sliding on a shaft connected to a spring, as depicted in Figure 5.9a. (This system is the subject of Problem 3.20.) The unstretched length of the spring is l, and the angle between the spring and the shaft is denoted by θ. Since this system has only one degree of freedom, we need only one coordinate. Let x represent the coordinate used to describe the position $\mathbf{r}_{P/O}$ of collar P on the shaft, where $x = 0$ at the center of the shaft. Assume that P starts at $\mathbf{r}_{P/O} = -d\mathbf{e}_x$. When the collar is released, the spring contracts and pulls the collar along the shaft in the $\mathbf{e}_x$ direction. We compute the work done by the spring to move the collar from its starting position, $x = -d$, at $t = t_1$, to the center, $x = 0$, at $t = t_2$.

First we compute the work directly from Definition 5.1. Using Figure 5.9b, the spring force is

$$\mathbf{F}_P = k\Delta l(\cos\theta\,\mathbf{e}_x + \sin\theta\,\mathbf{e}_y),$$

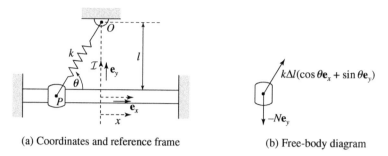

(a) Coordinates and reference frame   (b) Free-body diagram

Figure 5.9 Sliding collar attached to a spring.

where Δl is the amount of stretch in the spring and k is the spring constant. We have

$$W_P^{(\mathbf{F}_P)} = \int_{-d}^{0} k\Delta l(\cos\theta\mathbf{e}_x + \sin\theta\mathbf{e}_y) \cdot {}^{\mathcal{I}}dx\mathbf{e}_x = k\int_{-d}^{0} \Delta l\cos\theta dx. \quad (5.14)$$

Using the geometric relations

$$\Delta l = \sqrt{x^2 + l^2} - l$$

and

$$\cos\theta = -\frac{x}{\sqrt{x^2 + l^2}},$$

we can integrate Eq. (5.14) to obtain

$$W_P^{(\mathbf{F}_P)} = -k\int_{-d}^{0}\left(\sqrt{x^2 + l^2} - l\right)\frac{x}{\sqrt{x^2 + l^2}}dx$$

$$= k\int_{0}^{-d}\left(x - \frac{lx}{\sqrt{x^2 + l^2}}\right)dx$$

$$= \frac{1}{2}kx^2\Big|_{0}^{-d} - kl\sqrt{x^2 + l^2}\Big|_{0}^{-d} = \frac{1}{2}kd^2 - kl\sqrt{d^2 + l^2} + kl^2.$$

We can also compute the work using Eq. (5.12), which, in this problem, involves less algebra. As in Example 5.10, the potential energy associated with the spring force is

$$U_{P/O}^{(\mathbf{F}_P)} = \frac{1}{2}k\Delta l^2.$$

According to Eq. (5.12), we can compute the work by subtracting the value of the potential energy at t_2 from the value of the potential energy at t_1. The initial stretch in the spring is

$$\Delta l(t_1) = \sqrt{d^2 + l^2} - l.$$

When the mass is at the center, the stretch is zero and thus the potential energy is also zero. Thus the work done by the spring is

$$W_P^{(\mathbf{F}_P)}(t_1, t_2) = U_{P/O}^{(\mathbf{F}_P)}(t_1) - U_{P/O}^{(\mathbf{F}_P)}(t_2)$$

$$= \frac{1}{2}k\left(\sqrt{d^2 + l^2} - l\right)^2 - 0 = \frac{1}{2}kd^2 + kl^2 - kl\sqrt{d^2 + l^2},$$

as expected.

You might have guessed that the work is not the same as the change in potential energy of the spring because the shaft constrains the mass to move in a different direction than the spring. However, the normal force associated with the constraint does no work, as with all constraint forces, so the work is due only to the conservative force from the spring.

5.5 Total Energy

The two work-energy formulas, Eqs. (5.6) and (5.12), relate the work to the change in kinetic energy of a particle and, for a conservative force, to the change in potential energy. Eq. (5.10) shows that the total work is also equal to the sum of the conservative work (work done by conservative forces) plus the nonconservative work (work done by nonconservative forces). We now show how to combine the work-energy formulas and eliminate the conservative-work term.

We start by rewriting the work–kinetic-energy formula by separating out the conservative and nonconservative work:

$$T_{P/O}(t_2) = T_{P/O}(t_1) + W_P^{(c)}(t_1, t_2) + W_P^{(nc)}(\mathbf{r}_{P/O}; \gamma_P).$$

Next we replace the conservative work term using Eq. (5.12) and rearrange:

$$T_{P/O}(t_2) + U_{P/O}(t_2) = T_{P/O}(t_1) + U_{P/O}(t_1) + W_P^{(nc)}(\mathbf{r}_{P/O}; \gamma_P), \qquad (5.15)$$

where $U_{P/O}(t)$ is the total potential energy associated with the total conservative force on P at t. Eq. (5.15) can be written more compactly by defining the *total energy*.

Definition 5.8 The **total energy** of particle P at time t is the sum of the kinetic energy and the potential energy,

$$\boxed{E_{P/O}(t) \triangleq T_{P/O}(t) + U_{P/O}(t),}$$

where $U_{P/O}$ is the potential energy associated with the total conservative force acting on P.

Using Eq. (5.15) and Definition 5.8, we obtain our third and most important work-energy formula:

$$\boxed{E_{P/O}(t_2) = E_{P/O}(t_1) + W_P^{(nc)}(\mathbf{r}_{P/O}; \gamma_P).} \qquad (5.16)$$

It is worth noting that this formula is why the potential energy is defined as the negative of the work integral. It allows the total energy to be written as the sum of the kinetic and potential energies rather than their difference.

By taking the limit of Eq. (5.16) as $\Delta t = t_2 - t_1$ goes to zero and using Eq. (5.1) and Definition 5.2, we can write the work-energy formula in differential form as

$$\frac{d}{dt} E_{P/O}(t) = \mathbf{F}_P^{(\mathrm{nc})} \cdot {}^{\mathcal{I}}\mathbf{v}_{P/O}(t) \triangleq \mathrm{P}_P^{(\mathrm{nc})}.$$

In other words, the rate of change of the total energy of P is equal to the power $\mathrm{P}_P^{(\mathrm{nc})}$ of the total nonconservative force acting on P.

Eq. (5.16) provides a tool for computing the change in total energy due to nonconservative forces, just as Eq. (4.2) allowed us to compute the change in linear momentum due to an impulsive force. In the latter case, the conservation of linear momentum of a particle in the absence of forces is simply a restatement of Newton's first law. In the energy case, we have arrived at a new conservation law.

Law 5.1   The law of **conservation of total energy of a particle** states that, if the total nonconservative force is zero, then the total energy of a particle is a constant of the motion.

Law 5.1 follows directly from Eq. (5.16). If every force acting on particle P is conservative, then $W_P^{(\mathrm{nc})}(\mathbf{r}_{P/O}; \gamma_P) = 0$ and $E_{P/O}(t_2) = E_{P/O}(t_1)$.

Conservation of energy is an essential tool for solving many dynamics problems. In fact, it forms the foundation of much of modern physics (as well as thermodynamics). Without realizing it, we have already used energy conservation to solve problems, as in Tutorial 3.3. To see how it can be used, we return to our familiar problem, the simple pendulum.

Example 5.12 Solving the Simple Pendulum Using Energy

This example examines again the simple pendulum, but now using energy concepts. We use the same kinematics, only now we do not need the acceleration! Let $\mathcal{I} = (O, \mathbf{e}_1, \mathbf{e}_2, \mathbf{e}_3)$, where $\mathbf{e}_1$ is oriented downward (in the direction of gravity). In a rotating frame $\mathcal{B} = (O, \mathbf{e}_r, \mathbf{e}_\theta, \mathbf{e}_3)$ fixed to the pendulum (see, e.g., Figure 3.18), the kinematics are

$$\mathbf{r}_{P/O} = l\mathbf{e}_r$$

$${}^{\mathcal{I}}\mathbf{v}_{P/O} = l\dot{\theta}\mathbf{e}_\theta.$$

The pendulum is subject to two forces: the force of gravity $\mathbf{F}_P = m_P g \mathbf{e}_1$, which is a conservative force, and the tension in the rod, $-F\mathbf{e}_r$, which is a normal (constraint) force. The force of gravity is associated with a gravitational potential energy (see Table 5.1). Let $\theta = \pm\frac{\pi}{2}$ (i.e., when $\mathbf{e}_r = \pm\mathbf{e}_2$) represent the reference points where the potential energy is zero. The height of the pendulum relative to the reference points is $-\mathbf{r}_{P/O} \cdot \mathbf{e}_1$. Using Definition 5.8 and Table 5.1, the total energy of P is

$$E_{P/O} = T_{P/O} + U_{P/O} = \frac{1}{2} m_P \|{}^{\mathcal{I}}\mathbf{v}_{P/O}\|^2 - m_P g \mathbf{r}_{P/O} \cdot \mathbf{e}_1$$

$$= \frac{1}{2} m_P l^2 \dot{\theta}^2 - m_P g l \cos \theta.$$

Since the tension in the pendulum rod is a normal force (i.e., it acts in a direction orthogonal to the velocity of P), the work done by tension on particle P is zero. We verify this fact by evaluating the work $W_P^{(T\mathbf{e}_r)}$ over an arbitrary time interval t_1 to t_2:

$$\int_{t_1}^{t_2} (-T\mathbf{e}_r) \cdot {}^{\mathcal{I}}d\mathbf{r}_{P/O} = \int_{t_1}^{t_2} (-T\mathbf{e}_r) \cdot {}^{\mathcal{I}}\mathbf{v}_{P/O} dt = \int_{t_1}^{t_2} (-T\mathbf{e}_r) \cdot l d\theta \mathbf{e}_\theta = 0.$$

Since there are no nonconservative forces, $W_P^{(\mathrm{tot})}(\mathbf{r}_{P/O}; \gamma_P) = W_P^{(c)}(t_1, t_2)$, and the total energy remains fixed at the initial value

$$E_{P/O}(0) = \frac{1}{2} m_P l^2 \dot{\theta}(0)^2 - m_P g l \cos \theta(0).$$

Using conservation of energy, we can solve for

$$\dot{\theta}^2 = \frac{2}{m_P l^2} E_{P/O}(0) + 2\frac{g}{l} \cos \theta \qquad (5.17)$$

in terms of the initial energy $E_{P/O}(0)$. Note that the result is an equation for $\dot{\theta}$ rather than $\ddot{\theta}$, which is why the energy formula is an integral of the motion (we have effectively solved the equation of motion). We came to a similar result in Tutorial 3.3 by directly integrating the equation of motion.

Observe that we obtain the familiar equation of motion for the pendulum simply by differentiating Eq. (5.17). For many single-degree-of-freedom problems, this approach to finding the equation of motion is less cumbersome than using Newton's second law because you don't need to compute the acceleration. However, for multiple-degree-of-freedom problems, this approach will not work, as there is not enough information in the (scalar) energy equation to find the equations of motion for more than one coordinate.[3]

Of perhaps more interest is the use of the constant total energy to solve for the trajectory. Solving Eq. (5.17) for $\dot{\theta}$ and integrating with respect to time (using separation of variables) gives θ as an implicit function of time in the form of the elliptic integral

$$t = \int_{\theta(0)}^{\theta(t)} \frac{d\eta}{\sqrt{\frac{2}{m_P l^2} E_{P/O}(0) + 2\frac{g}{l} \cos \eta}}. \qquad (5.18)$$

---

[3] Chapter 13 discusses an alternate approach to finding equations of motion for any number of degrees of freedom by starting with the kinetic and potential energies.

It would be great if we could complete this integral to find some function of θ that could be inverted to find the trajectory $\theta(t)$ in terms of the initial angle and initial energy. Unfortunately, there is no simple closed-form solution for this integral (as already shown, the pendulum equation of motion has no exact solution in terms of elementary functions, except for the small-angle case). However, values of this elliptic integral have been tabulated, and most software packages have built-in elliptic functions. For instance, this equation is often used to find the period of the simple pendulum for large initial angles.

5.6 Derivations—Conservative Forces and Potential Energy

In this section, we validate the assertions that (a) for every conservative force, there is a corresponding potential energy; and (b) if we can find such a scalar potential, then the force must be conservative.

Theorem 5.1 A force $\mathbf{F}_P$ associated with a potential energy defined in Definition 5.7 is conservative if and only if it is equal to the negative gradient of the potential energy:

$$\mathbf{F}_P = -\nabla U_{P/O}^{(\mathbf{F}_P)}.$$

We justify this theorem using *Stokes' theorem*. Let Σ be a surface enclosed by a curve S and $\mathbf{n}$ be a unit vector normal to Σ. We have

$$\oint_S \mathbf{F} \cdot d\mathbf{r} = \int_\Sigma (\nabla \times \mathbf{F}) \cdot \mathbf{n}dA,$$

where dA is a differential area element. Compare the integral on the left-hand side to the integral in Definition 5.6; if this integral is zero, then $\mathbf{F}$ is a conservative force.

Assume that $U_{P/O}^{(\mathbf{F}_P)}$ is a potential energy and $\mathbf{F}_P = -\nabla U_{P/O}^{(\mathbf{F}_P)}$. It is a property of the gradient operator that the *curl* (i.e., $\nabla \times$) of the gradient of any scalar function U is identically zero. (You have the opportunity to show that in the problems.) We have

$$\nabla \times \nabla U = 0,$$

which implies

$$\nabla \times \mathbf{F}_P = -\nabla \times \nabla U_{P/O}^{(\mathbf{F}_P)} = 0.$$

Applying Stokes' theorem, the loop integral of $\mathbf{F}_P$ is zero, because

$$\oint_S \mathbf{F}_P \cdot d\mathbf{r} = \int_\Sigma \underbrace{(\nabla \times \mathbf{F}_P)}_{=0} \cdot \mathbf{n}dA = 0.$$

Therefore, $\mathbf{F}_P$ is conservative.

Now we validate the "only if" claim. Assume that $\mathbf{F}_P$ is conservative and define a potential energy $U_{P/O}^{(\mathbf{F}_P)}$ as in Definition 5.7:

$$U_{P/O}^{(\mathbf{F}_P)}(\mathbf{r}_{P/O}) = -\int \mathbf{F}_P \cdot {}^{\mathcal{I}}d\mathbf{r}_{P/O}.$$

Then, from the definition of the directional derivative in Section A.6, we have

$$\nabla U_{P/O}^{(\mathbf{F}_P)}(\mathbf{r}_{P/O}) = -\nabla \int \mathbf{F}_P \cdot {}^{\mathcal{I}}d\mathbf{r}_{P/O} = -\mathbf{F}_P,$$

which completes the proof.

Theorem 5.1 provides another method to determine whether a force is conservative —the curl of a conservative force is always zero.

5.7 Tutorials

Tutorial 5.1 Computing the Potential Energy of a Conservative Force

Computing the potential energy of a conservative force is not always as easy as in the examples in Table 5.1, especially if the force depends on multiple coordinates. This tutorial solves for the potential energy $U_{P/O}^{(\mathbf{F}_P)}$ associated with the conservative force

$$\mathbf{F}_P = (-x + y)\mathbf{e}_x + (x - y + y^2)\mathbf{e}_y. \tag{5.19}$$

Using Cartesian coordinates in reference frame $\mathcal{I} = (O, \mathbf{e}_x, \mathbf{e}_y, \mathbf{e}_z)$, the position of particle P is $\mathbf{r}_{P/O} = x\mathbf{e}_x + y\mathbf{e}_y + z\mathbf{e}_z$. Using Definition 5.7, we compute the potential energy (relative to the origin O) as

$$
\begin{aligned}
U_{P/O}^{(\mathbf{F}_P)} &= -\int ((-x+y)\mathbf{e}_x + (x - y + y^2)\mathbf{e}_y) \cdot (dx\mathbf{e}_x + dy\mathbf{e}_y + dz\mathbf{e}_z) \\
&= -\int (-x + y)dx + (x - y + y^2)dy \\
&= \int x\,dx + (y - y^2)dy - \underbrace{(y\,dx + x\,dy)}_{d(xy)} \\
&= \frac{1}{2}x^2 + \frac{1}{2}y^2 - \frac{1}{3}y^3 - xy.
\end{aligned}
$$

Again, how do we know $\mathbf{F}_P$ is conservative? Since a potential-energy function exists, the force must be conservative. We can verify this simply by taking the gradient of $U_{P/O}^{(\mathbf{F}_P)}$. Alternatively, we can check the curl of the force and verify that it is zero, since the curl of any conservative force is zero:

$$\nabla \times \mathbf{F}_P = 0.$$

In Cartesian coordinates using matrix notation, the curl of the position-dependent force given in Eq. (5.19) is

$$
\begin{bmatrix} \frac{\partial}{\partial x} \\ \frac{\partial}{\partial y} \\ \frac{\partial}{\partial z} \end{bmatrix}_{\mathcal{I}} \times \begin{bmatrix} -x + y \\ x - y + y^2 \\ 0 \end{bmatrix}_{\mathcal{I}} = \begin{bmatrix} 0 - \frac{\partial}{\partial z}(x - y + y^2) \\ \frac{\partial}{\partial z}(-x + y) - 0 \\ \frac{\partial}{\partial x}(x - y + y^2) - \frac{\partial}{\partial y}(-x + y) \end{bmatrix}_{\mathcal{I}} = \begin{bmatrix} 0 \\ 0 \\ 0 \end{bmatrix}_{\mathcal{I}},
$$

which implies that $\mathbf{F}_P$ is indeed conservative.

Tutorial 5.2 Energy in Orbital Motion

Now we look again at the orbit of a particle under a central force $\mathbf{F}_P$. Chapter 4 showed that the equations of motion for the orbit of a particle under gravitational attraction is a conic section whose eccentricity (a constant of integration) determines the shape of the orbit. The approach taken in Chapter 4 used conservation of the angular momentum of the particle.

Because gravity is a conservative force, the total energy of an orbit should also be a constant of the motion (as there are no nonconservative forces acting). As usual, the energy is

$$
E_{P/O} = \frac{1}{2} m_P \|{}^{\mathcal{I}}\mathbf{v}_{P/O}\|^2 + U_{P/O}.
$$

Using ${}^{\mathcal{I}}\mathbf{v}_{P/O} = \dot{r}\mathbf{e}_r + r\dot{\theta}\mathbf{e}_\theta$ and $U_{P/O} = -Gm_O m_P/r$ (see Table 5.1), the specific energy (i.e., energy per unit mass) is

$$
\varepsilon_P \triangleq \frac{1}{m_P} E_{P/O} = \frac{1}{2}(\dot{r}^2 + r^2 \dot{\theta}^2) - \frac{\mu}{r},
$$

where $\mu \triangleq Gm_O$.

Recall from Tutorial 4.2 that the specific angular momentum $h_O \triangleq r^2 \dot{\theta}$ of the orbit is constant. This allows us to write the energy entirely in terms of r:

$$
\varepsilon_P = \frac{1}{2}\dot{r}^2 + \frac{h_O^2}{2r^2} - \frac{\mu}{r}. \tag{5.20}
$$

This equation looks like the energy of a single-degree-of-freedom system with co-ordinate r, kinetic energy $\frac{1}{2}\dot{r}^2$, and an *effective potential*

$$
U_{\text{eff}}(r) = \frac{h_O^2}{2r^2} - \frac{\mu}{r}.
$$

The concept of the effective potential arises frequently in dynamics. It often allows us to reduce a complicated system to what looks like a single degree of freedom. For instance, for a given set of initial conditions, we could compute the energy and angular momentum and then find $\dot{r}(t)$ at any $r(t)$ from Eq. (5.20).

The effective potential also determines the *turning points* of the orbit, that is, the points where $\dot{r}$ goes to zero and thus the trajectory turns around. The trajectory must turn around at such points, since the kinetic-energy–like term in Eq. (5.20) is always positive; the effective potential is a maximum when $\dot{r} = 0$. (This is another way of showing that $\dot{r} = 0$ at periapsis and apoapsis.) We can write the energy in terms of r_p, the periapsis radius:

$$\varepsilon_P = U_{\text{eff}}(r_p) = \frac{h_O^2}{2r_p^2} - \frac{\mu}{r_p}.$$

Using the polar equation for the periapsis distance in Eq. (4.53), we can also write the energy in terms of the angular momentum and eccentricity,

$$\varepsilon_P = -\frac{\mu^2}{2h_O^2}(1 - e^2). \tag{5.21}$$

Note this implies $\varepsilon_P < 0$ when $e < 1$ (circular and elliptical orbits).

This result can be used to find a new expression for the eccentricity of the orbit in terms of the specific angular momentum and energy:

$$e = \sqrt{1 + \frac{2\varepsilon_P h_O^2}{\mu^2}}. \tag{5.22}$$

Thus the angular momentum and energy can be used as constants of the trajectory in lieu of the initial conditions. Eq. (5.22) shows how the eccentricity of the orbit is determined by the specific angular momentum and the specific energy.

Finally, we can substitute from Eq. (4.55) for the semimajor axis into Eq. (5.21) to find

$$a = -\frac{\mu}{2\varepsilon_P},$$

which implies that the semimajor axis a of an elliptical orbit is entirely determined by the specific energy, ε_P! The new expressions for a and e show that the two constants of the motion, ε_P and h_O, determine the size and shape of the orbit.

Tutorial 5.3 Energy of a Two-Degree-of-Freedom System

Up to now, every example we have examined using energy ideas was a single-degree-of-freedom system. Energy is particularly useful for such systems because the work-energy formula can be used to determine the state of the system at some time given its state at an earlier time (such as the speed of a falling ball when it hits the ground). We could even obtain the equation of motion by differentiating the total energy.

Yet, many (actually, most) systems have more than one degree of freedom and are thus described by multiple coordinates. But we have only one formula for conservation of energy (a scalar equation). That does not provide enough information to completely determine the state of the system. We can also no longer find the equations of motion by simply differentiating the work-energy formula. Is energy only a useful concept for single-degree-of-freedom problems? How might it help for this broader class of problems?

Conservation of energy is still an extremely helpful property and something you should always examine. As problems become more complicated, keeping track of energy and verifying that it remains conserved provides a helpful check that you are staying on course. We often use it in numerical integration to check the performance of the code. It also can provide useful information on the global behavior of a system,

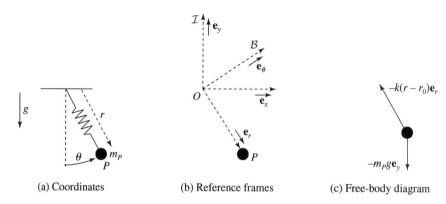

(a) Coordinates   (b) Reference frames   (c) Free-body diagram

Figure 5.10 Pendulum attached to a spring.

even when specific trajectories might be very complex. We show such a case in this tutorial, where we introduce the concept of a *zero-velocity curve*.

Problem 3.23 asked you to solve for the equations of motion for the simple spring pendulum. Figure 5.10 shows this system again. This is a two-degree-of-freedom problem best solved using polar coordinates. Unlike the simple pendulum, where the radial position is a fixed length, here r varies along with θ. The position and velocity of the pendulum are thus given by the usual polar coordinate equations,

$$\mathbf{r}_{P/O} = r\mathbf{e}_r$$

$$^{\mathcal{I}}\mathbf{v}_{P/O} = \dot{r}\mathbf{e}_r + r\dot{\theta}\mathbf{e}_\theta.$$

We can use Newton's second law in polar coordinates to find the equations of motion,

$$\ddot{r} = r\dot{\theta}^2 + g\cos\theta - \frac{k}{m_P}(r - r_0)$$

$$\ddot{\theta} = -\frac{2\dot{r}\dot{\theta}}{r} - \frac{g\sin\theta}{r},$$

where r_0 is the unstretched length of the spring. These equations are quite complicated and have no exact solution. We show a particular simulated trajectory in Figure 5.11.

Looking at the total energy gives us information about all possible trajectories of P. The kinetic energy of P is found from the velocity:

$$T_{P/O} = \frac{1}{2}m_P{}^{\mathcal{I}}\mathbf{v}_{P/O} \cdot {}^{\mathcal{I}}\mathbf{v}_{P/O} = \frac{1}{2}m_P(\dot{r}^2 + r^2\dot{\theta}^2).$$

The potential energy of the particle arises from the spring and from gravity,

$$U_{P/O} = -m_P gr\cos\theta + \frac{1}{2}k(r - r_0)^2.$$

The total energy is thus

$$E_{P/O} = \frac{1}{2}m_P(\dot{r}^2 + r^2\dot{\theta}^2) - m_P gr\cos\theta + \frac{1}{2}k(r - r_0)^2. \tag{5.23}$$

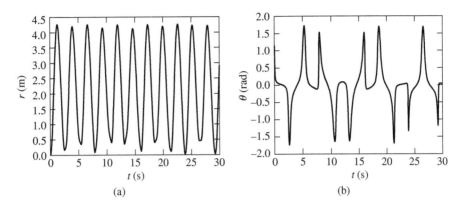

Figure 5.11 The r and θ trajectories for a simple spring pendulum of unstretched length $r_0 = 0.5$ m, spring constant $k = 6$ N/m, and initial conditions $r(0) = 0.01$ m and $\theta(0) = \pi/3$ rad. (a) Radial position. (b) Angle of pendulum.

The trajectories of P must be such that, given the initial conditions, $E_{P/O}$ is a constant. We can say more, however. Observe from Eq. (5.23) that the kinetic energy term is always positive or zero; it is quadratic in the rates of both coordinates, $\dot{r}$ and $\dot{\theta}$. Thus for any point on the trajectory where $\dot{r} = 0$ and $\dot{\theta} = 0$, the energy is a minimum. We plot potential energy contours as a function of $(r, \theta)_{\mathcal{I}}$ and know that, whatever the trajectory, it must always be on one side of the potential contour equal to the initial total energy. These lines are called zero-velocity curves. They represent the configuration of the system at zero velocity.

For example, Figure 5.12 shows the zero-velocity curves for the simple spring pendulum using the same parameters as in Figure 5.11. We also overlay a plot of $(r, \theta)_{\mathcal{I}}$ for the trajectory from Figure 5.11. When the two coordinate rates go to zero, the trajectory touches the zero-velocity curve. The touch points are called *turning points* because the trajectory must turn around at these points, since $\dot{r}$ and $\dot{\theta}$ both go to zero. Plots such as these yield a great deal of information about where the trajectories must reside and what form they take without our actually solving the equations of motion.

Contour plots such as Figure 5.12 are particularly useful for nonlinear systems that exhibit *chaotic* or almost chaotic behavior. These systems are characterized by extreme sensitivity to initial conditions. In other words, changing the initial conditions by a small amount results in a dramatically different trajectory. For instance, for the simple spring pendulum simulated here, a small change in the initial angle produces a trajectory indistinguishable from the one in Figure 5.11. However, if we modify the system slightly by making the spring nonlinear such that the spring force is

$$\mathbf{F}_p = -k_1(r - r_0)\mathbf{e}_r - k_2(r - r_0)^3\mathbf{e}_r,$$

very small changes in the initial angle result in significantly different trajectories. For example, Figure 5.13 shows two such trajectories for $r(0) = 0.46$ m and two different initial angles, $\theta(0) = 1.05$ rad and $\theta(0) = 1.06$ rad. Even though the resulting motion

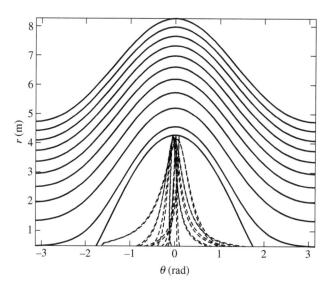

Figure 5.12 Zero-velocity curves (solid lines) associated with a simple spring pendulum with parameters $r_0 = 0.5$ m and $k = 6$ N/m. Dashed lines are the trajectories of the system in Figure 5.11.

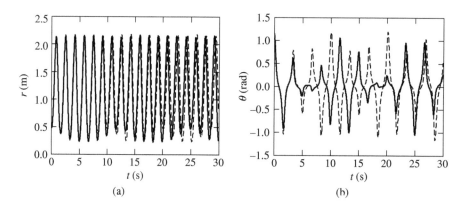

Figure 5.13 The r and θ trajectories for a nonlinear spring pendulum of unstretched length $r_0 = 0.305$ m, linear spring constant $k_1 = 6$ N/m, nonlinear spring constant $k_2 = 3$ N/m^3, and initial conditions $r(0) = 0.46$ m, $\theta(0) = 1.05$ rad (solid curves), and $\theta(0) = 1.06$ rad (dashed curves). (a) Radial position. (b) Angle of pendulum.

is quite different, the total energy for the two sets of trajectories is very close. The zero-velocity curves thus provide useful information on the behavior of families of trajectories without having to integrate specific ones. Figure 5.14 shows the constant-energy contours for the nonlinear-spring system overlaid with the two trajectories shown in Figure 5.13.

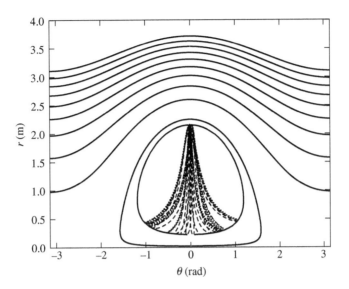

Figure 5.14 Zero-velocity curves (solid lines) associated with a nonlinear spring pendulum with parameters $r_0 = 0.305$ m, $k_1 = 6$ N/m, and $k_2 = 3$ N/m^3. Dashed lines are the trajectories of the system in Figure 5.13.

5.8 Key Ideas

- The **work** performed on particle P by force $\mathbf{F}_P$ along trajectory γ_P is

$$W_P^{(\mathbf{F}_P)}(\mathbf{r}_{P/O}; \gamma_P) \triangleq \int_{\gamma_P} \mathbf{F}_P \cdot {}^{\mathcal{I}}d\mathbf{r}_{P/O}.$$

  The work performed on P by the total force acting on P is denoted $W_P^{(\text{tot})}(\mathbf{r}_{P/O}; \gamma_P)$.

- The **power** exerted by force $\mathbf{F}_P$ acting on P at time t is the instantaneous rate of change of the work,

$$\mathrm{P}_P^{(\mathbf{F}_P)}(t) \triangleq \mathbf{F}_P(t) \cdot {}^{\mathcal{I}}\mathbf{v}_{P/O}(t).$$

- The **kinetic energy** of particle P at time t is

$$T_{P/O}(t) \triangleq \frac{1}{2}m_P \left({}^{\mathcal{I}}\mathbf{v}_{P/O}(t) \cdot {}^{\mathcal{I}}\mathbf{v}_{P/O}(t) \right) = \frac{1}{2}m_P \| {}^{\mathcal{I}}\mathbf{v}_{P/O}(t) \|^2.$$

  The change in kinetic energy from t_1 to t_2 equals the total work on P along trajectory γ_P from $\mathbf{r}_{P/O}(t_1)$ to $\mathbf{r}_{P/O}(t_2)$:

$$T_{P/O}(t_2) = T_{P/O}(t_1) + W_P^{(\text{tot})}(\mathbf{r}_{P/O}; \gamma_P).$$

- The work performed by impulse $\overline{\mathbf{F}}_P(t_1, t_2)$ is equal to the change in kinetic energy across the impulse,

$$W_P^{(\overline{\mathbf{F}}_P)}(t_1, t_2) = T_{P/O}(t_2) - T_{P/O}(t_1).$$

- A **conservative force** does no work on any closed trajectory:

$$\oint \mathbf{F}_P \cdot {}^{\mathcal{I}}d\mathbf{r}_{P/O} = 0.$$

- The work generated by a **conservative force** on particle P depends only on the endpoints of γ_P. We decompose the total work into the sum of the conservative work and the nonconservative work:

$$W_P^{(\text{tot})}(\mathbf{r}_{P/O}; \gamma_P) = W_P^{(\text{c})}(t_1, t_2) + W_P^{(\text{nc})}(\mathbf{r}_{P/O}; \gamma_P).$$

- The **potential energy** of particle P associated with conservative force $\mathbf{F}_P$ is

$$U_{P/O}^{(\mathbf{F}_P)}(\mathbf{r}_{P/O}) \triangleq -\int \mathbf{F}_P \cdot {}^{\mathcal{I}}d\mathbf{r}_{P/O}.$$

A conservative force can be represented as the gradient of a potential energy,

$$\mathbf{F}_P^{(\text{c})} = -\nabla U_{P/O}^{(\mathbf{F}_P)},$$

and the negative change in the potential energy from t_1 to t_2 equals the conservative work,

$$U_{P/O}(\mathbf{r}_{P/O}(t_2)) = U_{P/O}(\mathbf{r}_{P/O}(t_1)) - W_P^{(\text{c})}(t_1, t_2).$$

- The **total energy** of particle P at time t is the sum of the kinetic energy and the potential energy

$$E_{P/O}(t) = T_{P/O}(t) + U_{P/O}(t).$$

- The **work-energy formula,**

$$E_{P/O}(t_2) = E_{P/O}(t_1) + W_P^{(\text{nc})}(\mathbf{r}_{P/O}; \gamma_P),$$

implies that total energy is conserved if the total nonconservative force acting on P is zero.

5.9 Notes and Further Reading

The material in this chapter is exceptionally important and forms the foundation of much of modern physics. In particular, an understanding of conservative forces and potential energy is critical. In addition to providing important tools for solving problems, they are the foundation of your later courses in dynamics. In fact, there is a fascinating historical debate over what is more fundamental, forces or energy.

As noted in Chapter 2, the entire concept of force is problematic. Most modern approaches to mechanics (and quantum mechanics) begin with kinetic and potential energy as the fundamental physical quantities. We introduce you to this approach in Chapter 13. The historical implications are discussed in the article by Wilczek (2004).

By now it is clear that we have a particular fondness for the simple pendulum, both for the elegance of the problem and its pedagogical value. In fact, you probably have noticed that the simple pendulum is not so simple (there is a lovely article with that name that discusses the complexity of the pendulum by Antman [1998]). This chapter introduced a solution approach to the pendulum using energy. For readers interested in delving further into the large-angle solution of the pendulum, Greenwood (1988) has an excellent and concise discussion involving elliptic functions. The pendulum is also often used as the foundation for studying and teaching chaos theory. An introductory, though mathematically advanced, book on chaos is the one by Guckenheimer and Holmes (2002).

5.10 Problems

5.1   You are driving a car down a road at constant speed v_0 when you see a deer in your headlights. After a reaction time t_0 you slam on your brakes and your wheels lock. There is a constant friction force $\mathbf{F}_f$ between your wheels and the road. Ignoring air drag, find how far the car will travel after you see the deer before coming to a stop. Assume that $t_0 = 0.75$ s, $v_0 = 45$ mph, and $\mu_k = 0.7$. State any additional assumptions you make while solving the problem.

5.2   Consider a child on a sled going down a hill of height 5 m, as depicted in Figure 5.15. You can assume the hill is frictionless. The child and sled have a mass m of 20 kg. Once on the ground, the friction brings the sled to a stop within 10 m. What is the coefficient of friction?

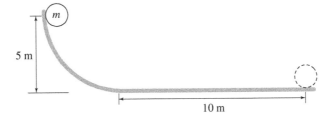

Figure 5.15 Problem 5.2.

5.3   In the log flume you are dropped from height h, and at the very bottom you are turned horizontal and brought to a stop by the water. Assume that you and the log have mass m. Is the system conservative? What is the work done by the water to bring you to a stop?

5.4   Consider again Problem 3.20. If the collar starts at a distance x_0 from the center C of the shaft, what is its speed when it passes through C?

5.5   Show that the potential energy of the gravity force from Newton's universal law of gravity, $-Gm_0 m_P/r^2$, is correct as given in Table 5.1.

5.6   Recall Problem 2.13. Is such a nonlinear-spring force conservative? If so, then what is the potential-energy function? If not, then explain why not.

5.7   Consider the attracting dipole of Problem 4.7. Using the expression for the gravitational potential of a particle, show that the potential due to the dipole at a position $(r, \theta)_{\mathcal{I}}$ is

$$U_{P/O}^{(\mathbf{F}_P)}(r, \theta) \approx -\frac{m_P \mu}{r}\left[2 + \left(\frac{d}{r}\right)^2 (3\cos^2\theta - 1)\right],$$

where $\mu = Gm$ and P is far from the dipole ($r \gg d$). Retain terms only to the first and second powers in (d/r). Show also that this potential corresponds to the force given in Problem 4.7. Finally, is the energy of the resulting motion in Problem 4.7 conserved?

5.8   Consider a mass-spring system with mass m of 1.5 kg and spring constant k of 0.2 N/m, as shown in Figure 5.16. The spring rest length is 0.5 m. Suppose the mass is initially located at $x = 0.2$ m. What is the potential energy of the mass due to the spring force? If the mass is released from rest, what will its velocity be when it reaches $x = 0.6$ m?

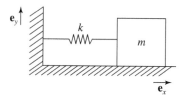

Figure 5.16 Problem 5.8.

5.9   Mass m is projected up a fixed ramp with initial speed v_0, as shown in Figure 5.17. The ramp angle is θ and its coefficient of friction is μ. What is the maximum height that the mass will reach?

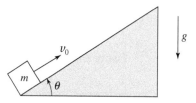

Figure 5.17 Problem 5.9.

5.10   Mass m on a string of length l is released from rest at $\theta = 30°$, as shown in Figure 5.18. The string encounters a fixed obstacle that is half as long as the string, but the mass keeps swinging. What is the angle that the mass will reach at the leftmost point of its swing?

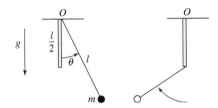

Figure 5.18 Problem 5.10: Initial (black) and final (white) mass positions.

5.11   Suppose satellite P is in a circular orbit of radius ρ around a large fixed mass O, as shown in Figure 5.19. How much energy is required to change to an elliptical orbit of eccentricity e and semimajor axis a? [HINT: See Tutorial 5.2.]

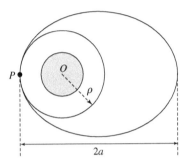

Figure 5.19 Problem 5.11.

5.12   Mass m slides in a frictionless parabolic bowl defined by $y = x^2$, where $(x, y)_{\mathcal{I}}$ are Cartesian coordinates in an inertial reference frame $\mathcal{I} = (O, \mathbf{e}_x, \mathbf{e}_y, \mathbf{e}_z)$, as shown in Figure 5.20.

   a. Draw a free-body diagram for mass m. (There *is* gravity in this problem.)
   b. Using only the Cartesian coordinate x, find the position $\mathbf{r}_{P/O}$ and inertial velocity ${}^{\mathcal{I}}\mathbf{v}_{P/O}$ of the mass with respect to O. [HINT: Remember $y = x^2$.]
   c. Find the potential energy $U_{P/O}$ of the mass with respect to O.
   d. Find the kinetic energy $T_{P/O}$ of the mass with respect to O.
   e. Show that the total energy of the mass with respect to O is conserved.

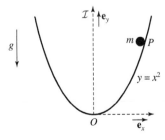

Figure 5.20 Problem 5.12.

5.13 Figure 5.21 shows two tracks connecting points A and B. A car can be started from point A with zero initial velocity and will follow either track to point B, propelled only by gravity (friction should be taken as negligible). The first track connects the points in a straight line (making it the shortest-distance route) and has length l and a vertical drop of h. The second track is a very special shape called a *cycloid* and is the shortest-time route between points A and B. The shortest-time trajectory is the solution to the *brachistochrone* problem (from Greek—*brachistos*, meaning "the shortest," and *chronos*, meaning "time"), which was one of the earliest problems solved using the calculus of variations.

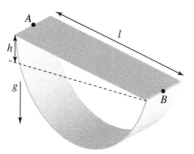

Figure 5.21 Problem 5.13.

    a. For the straight track, find the time it would take the car to travel from point A to point B and its velocity at B. Do so using Newton's laws and then by using conservation of energy. Ensure that the results from your two methods match. *All calculations for this part should be done analytically, without the use of* MATLAB *or any other computer tools.*

    b. Write out the total energy of a car traveling on a cycloid track in terms of its height y and speed $\left\| {}^{\mathcal{I}}\mathbf{v}_{P/O} \right\| \triangleq v$. If the car starts at rest with zero height, we can define the total energy of the car to be zero. Show, then, that the speed of the car as a function of height can be written as

$$v = \sqrt{-2gy}.$$

    c. Find the velocity at point B of a car traveling on the cycloid track.

5.14 The cycloid discussed in Problem 5.13 can be represented parametrically as

$$x(\phi) = \frac{C}{2}(\phi - \sin \phi)$$

$$y(\phi) = -\frac{C}{2}(1 - \cos \phi),$$

where ϕ is the parametric parameter (varying between 0 and 2π), and C is a constant of the cycloid, determined by the endpoints.

    a. We can express the velocity of a car traveling on the cycloid track as $v = \frac{ds}{dt}$ because the velocity is constrained to be along the path. We can also write $ds^2 = dx^2 + dy^2$. Equate the two expressions for ds,

$$ds = vdt = \sqrt{dx^2 + dy^2},$$

and divide both sides by $d\phi$ to get

$$v\frac{dt}{d\phi} = \frac{\sqrt{dx^2 + dy^2}}{d\phi}.$$

Because $x(\phi)$ and $y(\phi)$ are known, we are able to find $\frac{dt}{d\phi}$ and integrate. Using $v = \sqrt{-2gy}$, show that the relationship between ϕ and t is

$$t = \sqrt{\frac{C}{2g}}\phi.$$

b. If the length of the straight track l is 1 m and its vertical drop h is 12 cm, the constant of the cycloid, C, will equal 0.33 and ϕ will equal 1.59π at point B (assuming that it is 0 at point A). Using these values, find how much time it would take a car to travel from A to B when following either the straight track or the cycloid track in Figure 5.21.

c. Using MATLAB, numerically integrate the equations of motion for a car following the cycloid track and verify the time you calculated in part (b) for the car to reach point B.

d. Using MATLAB, see if you can find the values of C and ϕ at point B given in part (a). You may find the function fsolve useful here, but there are many different ways of doing this.

5.15 Show that the curl of the gradient of any scalar function is identically zero.

5.16 Consider a simple circular loop-the-loop. The car enters the circular loop of radius R_c with horizontal speed v_0.

a. Find an expression for the speed of the car as a function of the angle around the loop.

b. Find the inertial acceleration of the car as a function of v_0 and θ and express it in a frame fixed to the car.

c. What is the normal force applied to the car by the track?

d. What is the minimum speed v_0 needed to complete the loop?

5.17 Consider the pile driver shown in Figure 5.22. Mass M of 50 kg is dropped on top of the pile, where it stays. The pile (mass $m = 10$ kg) is stationary until

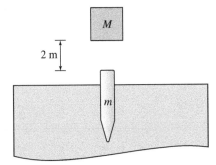

Figure 5.22 Problem 5.17.

M hits it. The large mass is dropped from a height of 2 m. After the impact the pile is driven into the ground by an additional 5 cm. What is the force of resistance of the ground to the combined pile/mass motion?

5.18   A block of mass m_P is initially sitting at rest at the bottom of a frictionless, circular loop of radius R, as shown in Figure 5.23. Gravity acts downward. An impulse to the right is then imparted to the block.

   a. What minimum impulse $\overline{\mathbf{F}}_P^*$ must be imparted to the block to ensure that it can make it all the way around the hoop without losing contact?
   b. If an impulse of $\overline{\mathbf{F}}_P = \frac{2}{\sqrt{5}}\overline{\mathbf{F}}_P^*$ is applied to the block (still to the right), at what angle θ (measured counterclockwise from the downward direction) does the block lose contact with the hoop?

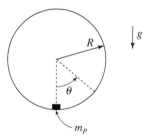

Figure 5.23 Problem 5.18.

5.19   In Problem 5.18, at what angle θ does the block strike the hoop again if given the impulse in part (b)?

5.20   Consider particle P of mass m_P starting at height h and attached to a slack string of length $2h/3$ a distance $h/3$ away. The mass falls vertically downward under gravity. When the string goes taut, the mass will start to swing (assume that the string does not stretch).

   a. What is the linear impulse imparted to the mass at the moment the string pulls taut?
   b. How high will P rise on the other side as it swings as a pendulum?

5.21   Suppose you are driving your new SUV that gets 25 mpg highway at 55 mph (the speed for which the mpg was calculated). You drive from Baltimore to Philadelphia and back, roughly 140 miles. First calculate how many gallons of gas you use. Then estimate how many gallons you burn if you drive to Philadelphia and back at 70 mph and at 80 mph. Be sure to state your assumptions and modeling steps. Assume the road is flat for the entire drive and ignore any frictional losses.

PART TWO

Planar Motion of a Multiparticle System

CHAPTER SIX

Linear Momentum of a Multiparticle System

Up to now we have focused on the dynamics of one particle only. In fact, we made a particular effort in Chapter 2 to point out that Newton's second law applies only to a single point mass. What to do, then, when faced with many particles? This chapter answers that question. We examine dynamics problems involving pairs of and sometimes many interacting particles. Does that mean there is a need for new physical concepts? In short, no. Treating multiple particles simply entails using Newton's second law on each particle individually. The result, of course, is that the number of degrees of freedom increases with the number of particles, as does the corresponding number of equations of motion. Nevertheless, studying the dynamics of many particles is no more complicated than solving Newton's second law (or its integrated form) for each particle. However, as the number of particles increases, this task can quickly become unwieldy. We therefore introduce some important concepts concerning the linear momentum of multiparticle systems that enhance understanding and simplify certain problems. These ideas are essential to the developments in the remainder of the book.

6.1 Linear Momentum of a System of Particles

We begin by generalizing to many particles the study of linear momentum and Newton's laws from Chapters 3 and 4. As before, we use Newton's second law, keeping track of the degrees of freedom and coordinates for each particle. The main addition is that we carefully consider the forces among the particles (called *internal* forces). This concept is examined in the first subsection below.

For some multiparticle systems we gain understanding and/or reduce complexity by studying the linear momentum of the entire collection of particles as a whole. The resulting tools sometimes dramatically simplify problems and prove useful when attacking more complicated systems. Sections 6.1.2 and 6.1.3 examine these ideas.

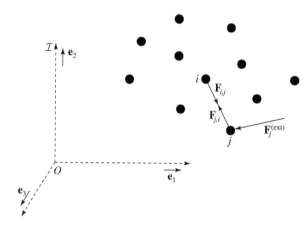

Figure 6.1 Internal and external forces acting on a collection of many particles.

6.1.1 Newton's Second Law for Multiple Particles

There are several notational conveniences adopted here that will simplify the treatment. In most problems with more than one particle, we separate the forces acting on the particles into external and internal forces (see Figure 6.1). External forces are field or contact forces coming from outside the system that act on one or more of the particles independently of the other particles. Internal forces are those that act among the various particles (e.g., gravitational or electrostatic attraction), usually along the lines between particles. The notation $\mathbf{F}_{j,i}$ indicates the force acting on particle j due to particle i.[1] $\mathbf{F}_j^{(\text{ext})}$ is the total external force acting on particle j.

Using multiparticle notation, Newton's second law for particle j is

$$m_j\,{}^{\mathcal{I}}\mathbf{a}_{j/O} = \mathbf{F}_j^{(\text{ext})} + \sum_{i=1}^{N} \mathbf{F}_{j,i}. \qquad (6.1)$$

We make explicit the assumption that a particle cannot exert a force on itself by setting $\mathbf{F}_{j,j} \equiv 0$.

To solve many multiparticle problems, Eq. (6.1) is all that is needed. We simply solve a system of N vector equations.

Example 6.1 The N-Body Problem

One of the most famous (and earliest) multiparticle dynamics problems is the collective motion of the sun and planets under their mutual gravitational attraction. The general problem of finding the equations of motion for N gravitationally bound particles is known as the N-body problem. This problem is illustrated in Figure 6.1 with $\mathbf{F}_j^{(\text{ext})} = 0$ and the internal forces given by Newton's universal law of gravitation,

---

[1] We often label particles using lowercase letters, such as i and j, which represent a number from one to N, where N denotes the total number of particles.

$$\mathbf{F}_{j,i} = -\frac{Gm_j m_i}{\|\mathbf{r}_{i/O} - \mathbf{r}_{j/O}\|^3}\mathbf{r}_{j/i},$$

where $\mathbf{r}_{j/i} = \mathbf{r}_{j/O} - \mathbf{r}_{i/O}$. The equations of motion for each particle (planet or star) are thus given by Eq. (6.1):

$$\frac{{}^{\mathcal{I}}d^2}{dt^2}(\mathbf{r}_{j/O}) + \sum_{i=1, i \neq j}^{N} \frac{Gm_i}{\|\mathbf{r}_{i/O} - \mathbf{r}_{j/O}\|^3}\mathbf{r}_{j/i} = 0, \quad j = 1, \ldots, N.$$

This is a system of N coupled, vector differential equations that must be solved simultaneously. Unfortunately, unlike the case studied earlier of a single body attracted toward a fixed origin, there is no solution of this problem in terms of elementary functions. The best we can do for this multiparticle problem is to write down the separate equations of motion and study them numerically (or analytically under certain approximating assumptions).

In some cases we can reduce the number of scalar equations of motion by recognizing explicit constraints that reduce the number of degrees of freedom. For example, in this chapter, as before, we assume that each particle is constrained to move in a plane only. In some problems, the particles are rigidly connected (forming a rigid body, discussed in detail later) or semirigidly connected (e.g., by means of a connection that allows motion in only one direction, like a linear spring). A common approach for these problems is to "break" the connection, introduce internal constraint forces (as often done with single-particle problems), write the equations of motion for each particle, and then eliminate the internal constraint forces as before. The next example illustrates this procedure.

Example 6.2 The Crane

Consider a simple pendulum P hanging from block Q, which is free to slide horizontally (Figure 6.2). There are two degrees of freedom in this problem: the horizontal motion of the block and the swing of the pendulum. We thus use the scalar coordinates x and θ to describe, respectively, the position of the crane and the angle of the pendulum relative to the vertical (Figure 6.2a). In this problem we find the equations of motion for x and θ by breaking apart the block and pendulum and accounting for the internal tension in the rod.

Let $\mathcal{I} = (O, \mathbf{e}_x, \mathbf{e}_y, \mathbf{e}_z)$ denote the inertial frame and $\mathcal{B} = (Q, \mathbf{e}_r, \mathbf{e}_\theta, \mathbf{e}_z)$ denote a polar frame fixed to the pendulum. The frames $\mathcal{B}$ and $\mathcal{I}$ are related by the transformation

	$\mathbf{e}_x$	$\mathbf{e}_y$
$\mathbf{e}_r$	$\sin\theta$	$-\cos\theta$
$\mathbf{e}_\theta$	$\cos\theta$	$\sin\theta$.

(6.2)

We start by writing the kinematics of the two masses P and Q:

$$\mathbf{r}_{Q/O} = x\mathbf{e}_x$$

$${}^{\mathcal{I}}\mathbf{v}_{Q/O} = \dot{x}\mathbf{e}_x$$

$${}^{\mathcal{I}}\mathbf{a}_{Q/O} = \ddot{x}\mathbf{e}_x,$$

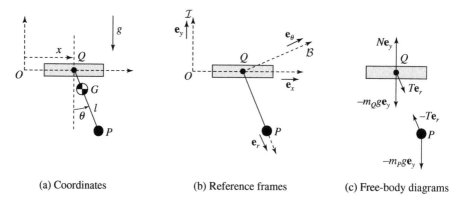

(a) Coordinates   (b) Reference frames   (c) Free-body diagrams

Figure 6.2 A crane. G is defined in Example 6.6.

and, using $\mathbf{r}_{P/O} = \mathbf{r}_{Q/O} + \mathbf{r}_{P/Q}$,

$$\mathbf{r}_{P/O} = x\mathbf{e}_x + l\mathbf{e}_r$$

$${}^{\mathcal{I}}\mathbf{v}_{P/O} = \dot{x}\mathbf{e}_x + l\dot{\theta}\mathbf{e}_\theta$$

$${}^{\mathcal{I}}\mathbf{a}_{P/O} = \ddot{x}\mathbf{e}_x + l\ddot{\theta}\mathbf{e}_\theta - l\dot{\theta}^2\mathbf{e}_r.$$

Using the free-body diagram Figure 6.2c and Newton's second law on each mass gives

$$(N - m_Q g)\mathbf{e}_y + T\mathbf{e}_r = m_Q \ddot{x}\mathbf{e}_x \tag{6.3}$$

$$-m_P g\mathbf{e}_y - T\mathbf{e}_r = m_P(\ddot{x}\mathbf{e}_x + l\ddot{\theta}\mathbf{e}_\theta - l\dot{\theta}^2\mathbf{e}_r). \tag{6.4}$$

Using the transformation in Eq. (6.2), the two vector expressions in Eqs. (6.3) and (6.4) yield four scalar equations relating the unknown quantities $\ddot{x}, \ddot{\theta}, N$, and T. These four equations can be solved algebraically for the equations of motion in x and θ:

$$\left(1 - \frac{m_P}{m_G}\cos^2\theta\right)\ddot{x} - \frac{m_P}{2m_G}g\sin 2\theta - \frac{m_P}{m_G}l\dot{\theta}^2\sin\theta = 0 \tag{6.5}$$

$$\left(1 - \frac{m_P}{m_G}\cos^2\theta\right)\ddot{\theta} + \frac{m_P}{2m_G}\dot{\theta}^2\sin 2\theta + \frac{g}{l}\sin\theta = 0, \tag{6.6}$$

where $m_G = m_P + m_Q$.

These equations can be integrated to find the motion of the pendulum mass and the block. They can also be used to design a controller for an overhead crane.

6.1.2 Total Linear Momentum and Momentum Conservation

This subsection develops an idea that can be used to study the motion of a collection of particles without necessarily resorting to simultaneously solving the equations of

motion for each particle. We begin by integrating Eq. (6.1) to obtain the impulse form
of Newton's second law, Eq. (4.2), for each particle:

$$m_j {}^{\mathcal{I}}\mathbf{v}_{j/O}(t_2) = m_j {}^{\mathcal{I}}\mathbf{v}_{j/O}(t_1) + \overline{\mathbf{F}}_j^{(\text{ext})}(t_1, t_2) + \sum_{i=1}^{N} \overline{\mathbf{F}}_{j,i}(t_1, t_2), \qquad (6.7)$$

where $\overline{\mathbf{F}}_{j,i}(t_1, t_2) = \int_{t_1}^{t_2} \mathbf{F}_{j,i} dt$ is the linear impulse acting on particle j from t_1 to t_2 due
to the internal force of particle i. Again, there is nothing new here, other than being
careful to indicate which particle we are considering. There are N different (vector)
equations for the change in momentum of each particle. However, what happens if
we sum over all particles? This leads us to define a new quantity, the *total linear
momentum*.

Definition 6.1 The **total linear momentum** ${}^{\mathcal{I}}\mathbf{p}_O$ of a collection of particles is the
vector sum of the individual linear momenta:

$$\boxed{{}^{\mathcal{I}}\mathbf{p}_O \triangleq \sum_{j=1}^{N} m_j {}^{\mathcal{I}}\mathbf{v}_{j/O}.} \qquad (6.8)$$

Next we use Definition 6.1 to find the integrated form of Newton's second law for
the total linear momentum by summing over all particles in Eq. (6.7):

$$ {}^{\mathcal{I}}\mathbf{p}_O(t_2) = {}^{\mathcal{I}}\mathbf{p}_O(t_1) + \sum_{j=1}^{N} \overline{\mathbf{F}}_j^{(\text{ext})}(t_1, t_2) + \sum_{j=1}^{N} \sum_{i=1}^{N} \overline{\mathbf{F}}_{j,i}(t_1, t_2). \qquad (6.9)$$

Now we use Newton's third law of motion, which states that $\mathbf{F}_{i,j} = -\mathbf{F}_{j,i}$. The
implication of this (and the assumption $\mathbf{F}_{j,j} \equiv 0$) is that $\sum_{j=1}^{N} \sum_{i=1}^{N} \overline{\mathbf{F}}_{j,i}(t_1, t_2) = 0$.
(To see this, imagine creating an $N \times N$ matrix with the (i, j)th entry equal to $\mathbf{F}_{i,j}$
for all pairs i and j. Summing over the rows and columns of the matrix yields zero
because the diagonal entries are zero and all of the off-diagonal entries cancel out.)
Thus, Eq. (6.9) becomes

$$\boxed{{}^{\mathcal{I}}\mathbf{p}_O(t_2) = {}^{\mathcal{I}}\mathbf{p}_O(t_1) + \overline{\mathbf{F}}^{(\text{ext})}(t_1, t_2),} \qquad (6.10)$$

where $\overline{\mathbf{F}}^{(\text{ext})}(t_1, t_2)$ is the total external force acting on the system from t_1 to t_2. This
result is extremely important; the total linear momentum of the collection of particles
obeys Newton's second law when acted on by an external force. The internal forces
completely cancel out.

Eq. (6.10) also leads us to a new conservation law. Consider the case when there
are no external forces ($\mathbf{F}_j^{(\text{ext})} = 0$). Eq. (6.10) becomes

$$\boxed{{}^{\mathcal{I}}\mathbf{p}_O(t_2) = \sum_{j=1}^{N} m_j {}^{\mathcal{I}}\mathbf{v}_{j/O}(t_2) = \sum_{j=1}^{N} m_j {}^{\mathcal{I}}\mathbf{v}_{j/O}(t_1) = {}^{\mathcal{I}}\mathbf{p}_O(t_1).} \qquad (6.11)$$

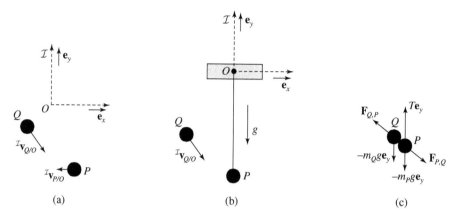

Figure 6.3 Sticky impacts.

In other words, in the absence of external forces, the total linear momentum of all the particles stays the same throughout their trajectories, regardless of the forces between each pair. This is a profound result and a broader statement of conservation of linear momentum than in Chapter 4; it proves very useful in the next section and throughout the book. It is so important that we elevate it to a new law.

Law 6.1 The law of **conservation of total linear momentum** states that, if the total external force is zero, the total inertial linear momentum of a system of particles is a constant of the motion.

Note that this law is a direct consequence of Newton's third law and is true for any internal forces acting among the particles. Thus total linear momentum can be conserved even if there are forces acting on the individual particles as long as the total external force is zero.

Example 6.3 Sticky Impact

Consider particles P and Q, as shown in Figure 6.3a. Suppose the particles collide and stick together to form a new particle labeled PQ. This is a good model of a collision between two cars, for instance. Since there are no external forces, we can use the conservation of total linear momentum in Eq. (6.11) to solve for the final velocity of PQ. From conservation of total linear momentum, we have

$$m_P{}^{\mathcal{I}}\mathbf{v}_{P/O}(t_2) + m_Q{}^{\mathcal{I}}\mathbf{v}_{Q/O}(t_2) = m_P{}^{\mathcal{I}}\mathbf{v}_{P/O}(t_1) + m_Q{}^{\mathcal{I}}\mathbf{v}_{Q/O}(t_1).$$

Using ${}^{\mathcal{I}}\mathbf{v}_{PQ/O}(t_2) = {}^{\mathcal{I}}\mathbf{v}_{P/O}(t_2) = {}^{\mathcal{I}}\mathbf{v}_{Q/O}(t_2)$,

$$^{\mathcal{I}}\mathbf{v}_{PQ/O}(t_2) = \frac{m_P{}^{\mathcal{I}}\mathbf{v}_{P/O}(t_1) + m_Q{}^{\mathcal{I}}\mathbf{v}_{Q/O}(t_1)}{m_P + m_Q},$$

which completes the example. We found the final velocity of the joint particle PQ in terms of the initial velocities of particles P and Q regardless of what the force between P and Q might be, as long as they collide and stick (we treat bouncy collisions in Section 6.2).

Example 6.4 Sticky Impact with a Pendulum

Consider particles P and Q, as shown in Figure 6.3b, where P is now attached to a pendulum and gravity is present. Suppose Q collides and sticks to P, which is initially at rest. Let us determine the velocity of the joint particle PQ after the collision using the velocity of Q before the collision. This is an example of a problem that requires some care, since the assumptions used to find Eq. (6.11) no longer apply. We cannot use the conservation of total linear momentum in Eq. (6.11) because the pendulum rod exerts an external force on P and gravity exerts an external force on both particles. Nonetheless, using the same approach that led to Eq. (6.11) will show that the horizontal component of momentum is conserved during the collision.

Let t_1 be the time of impact and t_2 be the short time later when the particles are firmly stuck together. The integral form of Newton's second law, Eq. (4.2), for each particle is

$$m_P{}^{\mathcal{I}}\mathbf{v}_{P/O}(t_2) = m_P{}^{\mathcal{I}}\mathbf{v}_{P/O}(t_1) + \int_{t_1}^{t_2} \mathbf{F}_P^{(\text{ext})}dt + \overline{\mathbf{F}}_{P,Q}(t_1, t_2)$$

$$m_Q{}^{\mathcal{I}}\mathbf{v}_{Q/O}(t_2) = m_Q{}^{\mathcal{I}}\mathbf{v}_{Q/O}(t_1) + \int_{t_1}^{t_2} \mathbf{F}_Q^{(\text{ext})}dt + \overline{\mathbf{F}}_{Q,P}(t_1, t_2).$$

We also have that ${}^{\mathcal{I}}\mathbf{v}_{P/O}(t_1) = 0$ and ${}^{\mathcal{I}}\mathbf{v}_{PQ/O}(t_2) = {}^{\mathcal{I}}\mathbf{v}_{P/O}(t_2) = {}^{\mathcal{I}}\mathbf{v}_{Q/O}(t_2)$. By Newton's third law, the linear impulses between the particles are equal and opposite. Therefore, adding these two equations together yields

$$(m_P + m_Q){}^{\mathcal{I}}\mathbf{v}_{PQ/O}(t_2) = m_Q{}^{\mathcal{I}}\mathbf{v}_Q(t_1) + \int_{t_1}^{t_2} (\mathbf{F}_P^{(\text{ext})} + \mathbf{F}_Q^{(\text{ext})})dt.$$

From the free-body diagram Figure 6.3c, the external forces on particles P and Q during the collision are $\mathbf{F}_P^{(\text{ext})} = (T - m_P g)\mathbf{e}_y$ and $\mathbf{F}_Q^{(\text{ext})} = -m_Q g\mathbf{e}_y$.[2] Let $\dot{x}$ and $\dot{y}$ denote the Cartesian coordinates for speed in the inertial frame $\mathcal{I}$. That is, let ${}^{\mathcal{I}}\mathbf{v}_{PQ/O} = \dot{x}_{PQ}\mathbf{e}_x + \dot{y}_{PQ}\mathbf{e}_y$. In terms of frame $\mathcal{I}$ components we have

$$(m_P + m_Q)\dot{x}_{PQ}(t_2) = m_Q\dot{x}_Q(t_1)$$

$$(m_P + m_Q)\dot{y}_{PQ}(t_2) = m_Q\dot{y}_Q(t_1) + \int_{t_1}^{t_2} (T - (m_P + m_Q)g)dt.$$

We solve for $\dot{x}_{PQ}(t_2)$ from the first equation. In fact, this equation is the horizontal component of the law of conservation of total linear momentum. (The second equation could be used to find the magnitude of the tension force, which would be useful for sizing the rod if we were building such a system.) We also have the constraint imposed by the pendulum rod on the final velocity of particle PQ,

$$\dot{y}_{PQ}(t_2) = 0.$$

---

[2] The assumption that the collision is very short implies that the pendulum force T on particle P stays approximately vertical during the collision. In reality, it will have a very small horizontal component as well.

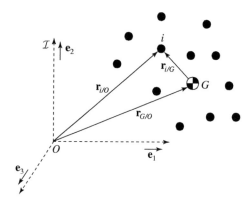

Figure 6.4 Center of mass G of a collection of many particles.

In words, this constraint states that, immediately after the collision, the joint particle can only move horizontally. Now we know the velocity of particle PQ immediately after the collision:

$$^{\mathcal{I}}\mathbf{v}_{PQ/O}(t_2) = \dot{x}_{PQ}(t_2)\mathbf{e}_x = \frac{m_Q}{m_P + m_Q}\dot{x}_Q(t_1)\mathbf{e}_x, \qquad (6.12)$$

which depends only on the particle masses and the initial horizontal speed of particle Q.

6.1.3 The Center of Mass

This subsection introduces the concept of the *center of mass* of a collection of particles. The center of mass leads to important conclusions about the motion of a collection of particles. The center of mass also becomes extremely important later when treating the angular momentum of a multiparticle system and when treating rigid bodies in Part Three.

Definition 6.2 The position of the **center of mass** $\mathbf{r}_{G/O}$ of a collection of N particles, as shown in Figure 6.4, is the vector sum of the mass-weighted particle positions:

$$\boxed{\mathbf{r}_{G/O} \triangleq \frac{1}{m_G}\sum_{i=1}^{N} m_i \mathbf{r}_{i/O},} \qquad (6.13)$$

where $m_G \triangleq \sum_{i=1}^{N} m_i$ is the total mass of all particles.

A consequence of this definition is that the sum of the mass-weighted positions of all of the particles in the collection relative to the center of mass is zero. To see this, start with the vector triad

$$\mathbf{r}_{i/O} = \mathbf{r}_{G/O} + \mathbf{r}_{i/G}. \qquad (6.14)$$

Using the definition of the center of mass, we find

$$m_G \mathbf{r}_{G/O} = \sum_{i=1}^{N} m_i (\mathbf{r}_{G/O} + \mathbf{r}_{i/G}) = m_G \mathbf{r}_{G/O} + \sum_{i=1}^{N} m_i \mathbf{r}_{i/G},$$

which implies

$$\sum_{i=1}^{N} m_i \mathbf{r}_{i/G} = 0. \tag{6.15}$$

We will use this result a lot! We call it the *center-of-mass corollary.*

Motion of the Center of Mass

Why is the concept of the center of mass helpful? One reason is that, in the absence of external forces, the center of mass is fixed (or moves at a constant velocity) in an inertial frame. To see this, return again to Eq. (6.1). Instead of using the integrated form, we can sum Newton's second law for each particle over all of the particles to get

$$\sum_{j=1}^{N} m_j \frac{{}^{\mathcal{I}} d^2}{dt^2} (\mathbf{r}_{j/O}) = \sum_{j=1}^{N} \mathbf{F}_j^{(\text{ext})},$$

where, as before, the double sum over the internal forces is zero. Again we substitute for $\mathbf{r}_{j/O}$ from the vector triad in Eq. (6.14):

$$m_G \frac{{}^{\mathcal{I}} d^2}{dt^2} (\mathbf{r}_{G/O}) + \sum_{j=1}^{N} m_j \frac{{}^{\mathcal{I}} d^2}{dt^2} (\mathbf{r}_{j/G}) = \sum_{j=1}^{N} \mathbf{F}_j^{(\text{ext})},$$

where we have used the fact that $m_G \triangleq \sum_{j=1}^{N} m_j$. Using the center-of-mass corollary in Eq. (6.15), the second term evaluates to

$$\sum_{j=1}^{N} m_j \frac{{}^{\mathcal{I}} d^2}{dt^2} (\mathbf{r}_{j/G}) = \frac{{}^{\mathcal{I}} d^2}{dt^2} \underbrace{\left(\sum_{j=1}^{N} m_j \mathbf{r}_{j/G} \right)}_{=0} = 0.$$

This calculation yields the important result,

$$\frac{{}^{\mathcal{I}} d}{dt} \left({}^{\mathcal{I}} \mathbf{p}_{G/O} \right) = \sum_{j=1}^{N} \mathbf{F}_j^{(\text{ext})} \triangleq \mathbf{F}_G^{(\text{ext})}, \tag{6.16}$$

where we have introduced a new quantity,

$$ {}^{\mathcal{I}} \mathbf{p}_{G/O} \triangleq m_G \frac{{}^{\mathcal{I}} d}{dt} (\mathbf{r}_{G/O}). \tag{6.17}$$

Eq. (6.17) looks like the linear momentum of a single particle of mass m_G located at the center of mass G. Comparing to Eq. (6.8), *we observe that $^I\mathbf{p}_{G/O}$ is equal to the total linear momentum $^I\mathbf{p}_O$ of the collection of particles.*

What have we found? Summing Newton's second law over all particles yields a single equation of motion for the center of mass of the collection. The vector sum of all external forces acts as if it were a single force applied to a single particle located at the center of mass. In other words, *to study the translational dynamics of a collection of particles, we can treat the collection as a single equivalent particle of mass m_G located at the center of mass G and ignore the internal forces!* The center of mass follows a trajectory determined by solving Newton's second law independently of the relative motion of the particles.

This result justifies many of the examples done so far in which we treated the translational motion of an extended body as if it were a point mass. It also gives us a new way of thinking about a multiparticle system. For many problems, the motion of the particles relative to the center of mass is of more interest than that of the center of mass itself. (Consider, e.g., the motion of astronauts relative to the center of mass of the space station as the station orbits the earth.) This result shows that we can separate the motion of the center of mass from the motion of the system about the center of mass.

Eq. (6.16) also implies that, if $\mathbf{F}_G^{(\text{ext})} = 0$, then $^I\mathbf{p}_{G/O}$ is constant, which is another statement of the law of conservation of total linear momentum. In the absence of external forces—or if the external forces sum to zero—the center of mass of the collection travels at a constant velocity. Keeping track of the center of mass can thus lead to important insights or act as an important check to avoid mistakes.

Finally, we could easily write the impulse form of Eq. (6.16), as done in Chapter 4 for the momentum of a single particle. All results in Chapter 4 describing an impulse on a single particle apply here to the center of mass of a particle collection treated as a single particle.

Example 6.5 A Falling Collection of Particles

Consider the collection of particles falling in a uniform gravitational field (with acceleration $g\mathbf{e}_y$), as shown in Figure 6.5a. Each particle has a force $-m_i g\mathbf{e}_y$ acting on it, as shown in the free-body diagram Figure 6.5b. Thus each particle obeys Newton's second law,

$$m_i \frac{^I d^2}{dt^2}\left(\mathbf{r}_{i/O}\right) = \mathbf{F}_i = -m_i g\mathbf{e}_y.$$

What is the equation of motion of the center of mass? From Eq. (6.16) we consider the sum

$$\frac{^I d}{dt}\left(^I\mathbf{p}_{G/O}\right) = \sum_{i=1}^{N} -m_i g\mathbf{e}_y,$$

which simplifies to

$$m_G \frac{^I d^2}{dt^2}\left(\mathbf{r}_{G/O}\right) = -m_G g\mathbf{e}_y.$$

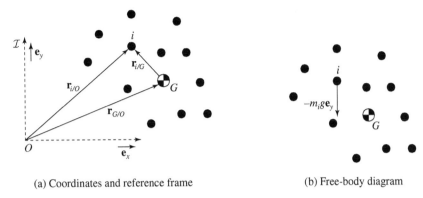

(a) Coordinates and reference frame   (b) Free-body diagram

Figure 6.5 A collection of particles falling freely in a uniform gravitational field.

The center-of-mass equation of motion is

$$\frac{^{\mathcal{I}}d^2}{dt^2}(\mathbf{r}_{G/O}) = -g\mathbf{e}_y.$$

The center of mass constantly accelerates at g in the $-\mathbf{e}_y$ direction. We treated the collection of particles as a single one located at the center of mass and found the equation of motion for it. This is possible whether the particles are free, as in this example, or connected in a flexible or rigid body.

This example shows that the acceleration of the center of mass is independent of the arrangement of particles. It is a function of the center-of-mass position only (here, the acceleration is a constant). Unfortunately, that is not always the case. Sometimes the force on the center of mass depends on the relative position of the particles, coupling the motion of the center of mass to the motion relative to the center of mass. Tutorial 6.1 discusses such a situation.

Example 6.6 The Center of Mass of an Overhead Crane

Consider again the crane of Example 6.2. Suppose the system is released from rest at $x(0) = 0$ and $\theta(0) \neq 0$. This example examines the motion of the center of mass G of the crane.

Using the kinematics of the pendulum and the block derived in Example 6.2 and Definition 6.2, we can compute the position of the center of mass:

$$\mathbf{r}_{G/O} = \frac{m_Q x\mathbf{e}_x + m_P(x\mathbf{e}_x + l\mathbf{e}_r)}{m_G}$$

$$= x\mathbf{e}_x + \frac{m_P}{m_G}l\mathbf{e}_r.$$

Let $\tilde{l} \triangleq \frac{m_P}{m_G} l$. The kinematics of G are

$$\mathbf{r}_{G/O} = x\mathbf{e}_x + \tilde{l}\mathbf{e}_r$$

$$^{\mathcal{I}}\mathbf{v}_{G/O} = \dot{x}\mathbf{e}_x + \tilde{l}\dot{\theta}\mathbf{e}_\theta$$

$$^{\mathcal{I}}\mathbf{a}_{G/O} = \ddot{x}\mathbf{e}_x + \tilde{l}\ddot{\theta}\mathbf{e}_\theta - \tilde{l}\dot{\theta}^2\mathbf{e}_r.$$

Note the similarities between the kinematics of G and P. In fact, in the limit $m_Q \to 0$, the kinematics of G and P are identical.

Using Eq. (6.16) and the free-body diagram Figure 6.2c, we find that

$$\frac{^{\mathcal{I}}d}{dt}\left(^{\mathcal{I}}\mathbf{p}_{G/O}\right) = m_G(\ddot{x}\mathbf{e}_x + \tilde{l}\ddot{\theta}\mathbf{e}_\theta - \tilde{l}\dot{\theta}^2\mathbf{e}_r) = (N - m_G g)\mathbf{e}_y \qquad (6.18)$$

Eq. (6.18) yields two scalar equations in terms of three unknowns: $\ddot{x}$, $\ddot{\theta}$, and N. Therefore, we cannot directly solve this algebraic system of equations without introducing another equation, such as Eq. (6.15). Nevertheless, we can gain insight into the motion of the system from the observation that the total external force acts only in the vertical direction.

In the absence of horizontal external forces, the horizontal component of $^{\mathcal{I}}\mathbf{p}_{G/O}$ is conserved. And, since the particles are initially at rest, the horizontal component of $^{\mathcal{I}}\mathbf{p}_{G/O}$ is zero for all time and the center of mass does not move horizontally. We express this observation mathematically by writing $^{\mathcal{I}}\mathbf{p}_{G/O}$ in terms of components in the $\mathcal{I}$ frame, using Eq. (6.2):

$$^{\mathcal{I}}\mathbf{p}_{G/O} = m_G(\dot{x} + \tilde{l}\dot{\theta}\cos\theta)\mathbf{e}_x + m_G\tilde{l}\dot{\theta}\sin\theta\mathbf{e}_y.$$

We have the conservation equation

$$\dot{x} + \tilde{l}\dot{\theta}\cos\theta = 0,$$

which can be integrated by inspection to find

$$x = -\tilde{l}\sin\theta + x(0). \qquad (6.19)$$

Even without solving for the equations of motion of x and θ, we found that their trajectories must satisfy Eq. (6.19). If $|\theta| < \pi/2$, this equation shows that, as the pendulum P swings one way, the block Q slides the other way. In fact, as you might expect, P oscillates like a pendulum; Eq. (6.19) implies that Q must also oscillate in such a way as to keep the horizontal position of G fixed.

Motion Relative to the Center of Mass

To complete the picture of a multiparticle system, we write the equations of motion for the position of each particle relative to the center of mass, which was examined in Section 3.6. We attach a nonrotating frame $\mathcal{B}$ to the center of mass, as in Figure 6.6, and consider the trajectory $\mathbf{r}_{i/G}$ of each particle in this frame. The equations of motion for each particle relative to $\mathcal{B}$ are given by Eq. (3.57), with P replaced by i, O'

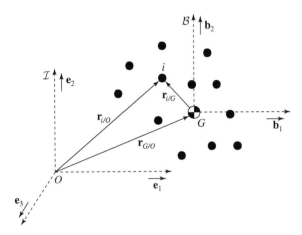

Figure 6.6 Center of mass G of a collection of many particles with a nonrotating frame $\mathcal{B}$ attached to G.

replaced by G, and the force $\mathbf{F}_i$ separated into external and internal components:

$$m_i\,{}^{\mathcal{B}}\mathbf{a}_{i/G} = \mathbf{F}_i^{(\text{ext})} + \sum_{j=1}^{N} \mathbf{F}_{i,j} - m_i\,{}^{\mathcal{I}}\mathbf{a}_{G/O}.$$

We can also use the angular-momentum form of Newton's second law discussed in Section 4.2.2 to find equations of motion for each particle relative to the center of mass. The result is given by Eq. (4.15), with P replaced by i and Q replaced by G:

$$\frac{{}^{\mathcal{I}}d}{dt}\left({}^{\mathcal{I}}\mathbf{h}_{i/G}\right) = \mathbf{M}_{i/G} - \mathbf{r}_{i/G} \times m_i\,{}^{\mathcal{I}}\mathbf{a}_{G/O}, \tag{6.20}$$

where

$$\mathbf{M}_{i/G} = \mathbf{r}_{i/G} \times \mathbf{F}_i^{(\text{ext})} + \mathbf{r}_{i/G} \times \sum_{j=1}^{N} \mathbf{F}_{i,j}.$$

Chapter 7 examines Eq. (6.20) in more detail.

We now have equations of motion for the trajectory of the center of mass and for the relative position of each particle about the center of mass. It may seem like we have more equations than degrees of freedom, but the inclusion of Eq. (6.15) as a constraint reconciles this imbalance. Chapter 7 returns to the concept of separating the motion of the center mass from that of the particles relative to the center of mass.

It is also worth noting that, in the absence of external forces, when the linear momentum of the center of mass is a constant (i.e., it is stationary or moves at constant velocity), Newton's second law for each particle relative to the center of mass reduces to

$$ {}^{\mathcal{B}}\mathbf{a}_{i/G} = \frac{1}{m_i} \sum_{j=1}^{N} \mathbf{F}_{i,j}. \tag{6.21}$$

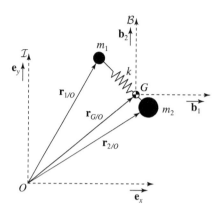

Figure 6.7 Two particles sliding in the plane connected by a linear spring with spring constant k.

Example 6.7 Two Sliding Masses Connected by a Spring

Consider two particles of equal mass connected by a linear spring with spring constant k and rest length l_0. The masses are constrained to slide on a frictionless table, as shown in Figure 6.7. This is a four-degree-of-freedom problem: each particle can move in two directions on the table. The only external force is gravity (directed into the page), which is countered by the normal force from the table (directed out of the page).

In this example we find the equations of motion for the center of mass of the system and for the motion of the particles relative to the center of mass after one particle is struck by an impulse $\overline{\mathbf{F}}_1(t_1, t_2)$, assuming the system begins at rest.

Because there are only two masses, the center of mass is always on the line joining them. From Eq. (6.16) the equation of motion of the center of mass is

$$m_G \frac{{}^\mathcal{I}d^2}{dt^2}(\mathbf{r}_{G/O}) = \mathbf{F}_G^{(\text{ext})}.$$

Using Eqs. (4.2) and (6.16) with $t \geq t_2$, we find

$$^\mathcal{T}\mathbf{v}_{G/O}(t) = \frac{1}{m_G}\overline{\mathbf{F}}_1(t_1, t_2). \tag{6.22}$$

Thus, our results from Chapter 4 show that the center of mass travels at constant velocity in the direction of the impulse.

For motion relative to the center of mass after the impulse we use Eq. (6.21) and the vector triads $\mathbf{r}_{1/O} = \mathbf{r}_{G/O} + \mathbf{r}_{1/G}$ and $\mathbf{r}_{2/O} = \mathbf{r}_{G/O} + \mathbf{r}_{2/G}$ to obtain

$$^\mathcal{B}\mathbf{a}_{1/G} = \frac{\mathbf{F}_{1,2}}{m_1} = \frac{-k}{m_1}\left(\|\mathbf{r}_{1/G} - \mathbf{r}_{2/G}\| - l_0\right)\hat{\mathbf{r}}_{1/2}$$

$$^\mathcal{B}\mathbf{a}_{2/G} = \frac{\mathbf{F}_{2,1}}{m_2} = \frac{-k}{m_2}\left(\|\mathbf{r}_{1/G} - \mathbf{r}_{2/G}\| - l_0\right)\hat{\mathbf{r}}_{2/1},$$

where l_0 is the rest length of the spring and, as usual, the spring force acts only along the line connecting the particles and is proportional to the compression (or extension) of the spring. The unit vector directed from particle 2 to particle 1 is given by

$$\hat{\mathbf{r}}_{1/2} = \frac{\mathbf{r}_{1/O} - \mathbf{r}_{2/O}}{\|\mathbf{r}_{1/O} - \mathbf{r}_{2/O}\|} = \frac{\mathbf{r}_{1/G} - \mathbf{r}_{2/G}}{\|\mathbf{r}_{1/G} - \mathbf{r}_{2/G}\|} = -\hat{\mathbf{r}}_{2/1}.$$

Because there are no external forces, the frame $\mathcal{B}$ attached to the center of mass is also an inertial frame.

These equations, combined with the center-of-mass equation of motion, provide six scalar equations of motion for the system. However, the system has only four degrees of freedom. Thus there must be two scalar constraint equations. The constraints are given in vector form by the center-of-mass corollary (Eq. (6.15)):

$$m_1 \mathbf{r}_{1/G} + m_2 \mathbf{r}_{2/G} = 0.$$

For this simple problem it is perhaps more interesting to find an equation of motion for the relative position between the particles, $\mathbf{r}_{2/1} = \mathbf{r}_{2/G} - \mathbf{r}_{1/G}$, thus eliminating the need to carry the center of mass constraint. This is done by subtracting the two equations of motion relative to the center of mass:

$$\frac{{}^{\mathcal{B}}d^2}{dt^2}(\mathbf{r}_{2/1}) = {}^{\mathcal{B}}\mathbf{a}_{2/G} - {}^{\mathcal{B}}\mathbf{a}_{1/G}$$

$$= \frac{-k(m_1 + m_2)}{m_1 m_2}(\|\mathbf{r}_{2/1}\| - l_0)\hat{\mathbf{r}}_{2/1}. \tag{6.23}$$

The quantity $m_1 m_2 / (m_1 + m_2)$ is often called the *reduced mass*.

Eq. (6.23) has the form of an equation for simple harmonic motion; the particle separation will oscillate at a frequency $\omega = \sqrt{k(m_1 + m_2)/m_1 m_2}$, while the center of mass travels at the constant velocity given by Eq. (6.22). For multiparticle systems such as this one we often associate the degrees of freedom with what are called the *modes* of motion, rather than the trajectories of the individual particles. In this problem, the four modes are the two translational degrees of freedom of the center of mass, the rotation of the line connecting the particles about the center of mass, and the oscillation of the particles along the line connecting them.

The initial conditions for the oscillation are also found from the impulse equation. Since particle 1 is hit by the impulse, we have

$$^{\mathcal{I}}\mathbf{v}_{1/O}(t_2) = \frac{1}{m_1}\bar{\mathbf{F}}_1(t_1, t_2)$$

$$^{\mathcal{I}}\mathbf{v}_{2/O}(t_2) = 0$$

or, in terms of the relative velocity,

$$^{\mathcal{B}}\mathbf{v}_{2/1}(t_2) = -\frac{1}{m_1}\bar{\mathbf{F}}_1(t_1, t_2).$$

This is another example where the motion relative to the center of mass is completely independent of the motion of the center of mass. In this case, it is because there are no external forces, so the linear momentum of the center of mass is conserved.

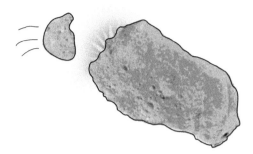

Figure 6.8 Deflecting an asteroid using an explosive to split it apart.

Example 6.8 Deflecting an Asteroid

Consider the inverse of Example 6.3: rather than two objects colliding and sticking, we have two objects traveling together at a constant velocity before bursting apart. This is a simple model for myriad phenomena, from explosions, to propulsion, to the recoil in a gun. One particularly interesting application is asteroid deflection, shown in Figure 6.8. There is rising concern that some time in the next few centuries a large asteroid may collide with the earth, causing devastating damage. To avoid this scenario, some scientists have proposed setting off a large explosive (probably nuclear) on the surface of the asteroid to redirect it away from the earth.

Consider again two particles P and Q that are initially joined and traveling at the velocity $^{\mathcal{I}}\mathbf{v}_{PQ/O}(t_0)$. The total mass of the joint particle PQ is $m_{PQ} \triangleq m_P + m_Q$. Suppose particle Q is ejected with relative velocity $^{\mathcal{B}}\mathbf{v}_{Q/P}$,[3] where $\mathcal{B}$ is a nonrotating frame fixed to the center of mass G. What is the final velocity of particle P?

Since no external forces are acting, total linear momentum is conserved and G travels at a constant velocity $^{\mathcal{I}}\mathbf{v}_{PQ/O}(t_0)$. From conservation of total linear momentum, we have

$$m_Q \, {}^{\mathcal{I}}\mathbf{v}_{Q/O}(t_f) + m_P \, {}^{\mathcal{I}}\mathbf{v}_{P/O}(t_f) = m_{PQ} \, {}^{\mathcal{I}}\mathbf{v}_{PQ/O}(t_0), \qquad (6.24)$$

where t_f is some time after P is ejected. Differentiating the vector triad

$$\mathbf{r}_{Q/O} = \mathbf{r}_{Q/P} + \mathbf{r}_{P/O}$$

yields the relative motion:

$$^{\mathcal{I}}\mathbf{v}_{Q/O}(t_f) = {}^{\mathcal{I}}\mathbf{v}_{Q/P}(t_f) + {}^{\mathcal{I}}\mathbf{v}_{P/O}(t_f)$$

$$= {}^{\mathcal{B}}\mathbf{v}_{Q/P}(t_f) + {}^{\mathcal{I}}\mathbf{v}_{P/O}(t_f).$$

The latter equality holds because frame $\mathcal{B}$ is not rotating (see Section 3.6). Substituting $^{\mathcal{I}}\mathbf{v}_{Q/O}(t_f)$ into Eq. (6.24), we find

$$m_Q \, {}^{\mathcal{B}}\mathbf{v}_{Q/P}(t_f) + m_{PQ} \, {}^{\mathcal{I}}\mathbf{v}_{P/O}(t_f) = m_{PQ} \, {}^{\mathcal{I}}\mathbf{v}_{PQ/O}(t_0),$$

which can be solved for the final velocity of particle P,

---

[3] The ejected particle Q represents the portion of the asteroid broken off by the explosive.

$$^{\mathcal{I}}\mathbf{v}_{P/O}(t_f) = {}^{\mathcal{I}}\mathbf{v}_{PQ/O}(t_0) - \frac{m_Q}{m_P + m_Q}{}^{\mathcal{B}}\mathbf{v}_{Q/P}(t_f).$$

Normally $m_Q \ll m_P$, so the ejected particle makes only a slight change to the velocity of the remaining mass. Fortunately, if caught early enough, only a very small change in the asteroid velocity is needed to cause it to miss the earth.

There is one caveat that should be kept in mind here. Although we have rigorously shown that we can combine the particles and write a single equation of motion for the center of mass in terms of the net force on it, it is not always the case that we can solve this equation without also considering the relative motion of the particles. This is because in many common situations the net force $\mathbf{F}_G^{(\text{ext})}$ cannot be determined as a function of the center-of-mass position only but may (and often does) depend on the positions (or velocities) of the individual particles. This situation produces a coupling that can't always be reconciled. Tutorial 6.1 illustrates such a situation.[4]

6.2 Impacts and Collisions

This section examines one of the most commonly encountered dynamics problems: collisions. Be it cars crashing, balls bouncing, baseball bats hitting home runs, or many other phenomena, we experience two objects colliding with each other on a regular basis. Our goal is to understand the dynamics of the objects during and after the collisions. Remarkably, despite its ubiquity, the collision is one of the most difficult dynamic problems to model. Accurately representing the complete physics would require detailed modeling of the shapes of the objects, their material properties, the deformations each undergoes during contact, and any energy losses during deformation. This task is clearly formidable and certainly beyond our scope.

However, all is not lost. By making a few simplifying assumptions, we can use the tools acquired so far—namely, linear impulse and the conservation of total linear momentum—to form remarkably effective predictive models of many impacts.

6.2.1 Planar Collision between Two Particles

The overall objective of this subsection can be summed up quite succinctly: *given the velocities of two particles before a collision, find the velocities of the two particles after the collision.*

This objective is shown schematically in Figure 6.9. Although easy to state, the task is difficult to execute. In fact, it is hopelessly complex without first making a few simplifying assumptions. Even though the following list of assumptions may seem rather restrictive, they are, in fact, met by a wide variety of situations. For some cases

---

[4] It is an interesting historical footnote that this problem prevented Newton from publishing his laws of dynamics and the differential calculus for more than 30 years (thus ceding the notation to Liebnitz). He found that he could predict the motion of the moon about the earth as long as he treated each as a point mass. Proving it was okay to ignore the distribution of mass in each was a formidable problem and required the invention of the integral calculus. As it turns out, it is only correct to treat the gravitational force as coming from a point mass if the earth is a sphere. For any other shape, the relative orientation of the earth and the moon affects the trajectory of the moon's center of mass.

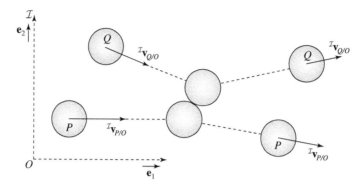

Figure 6.9 A collision between two particles changes both the magnitude and direction of the velocity of each particle.

when one or more of the following assumptions are not met, we can often apply the general approach developed below, but the specific results are not guaranteed to hold.

Assumption 6.1 *The collision occurs in a very short time interval,* during which the objects do not move significantly. We thus do not need to model the details of the force acting between the two objects during the impact; instead, the details are lumped into the linear impulse imparted to each object. An implication of this assumption is that the collision instantaneously changes the particle velocities.

Assumption 6.2 *There are no external forces acting on either particle during the collision,* which implies that the total linear momentum is conserved. If there are no external forces acting on either particle at any time, then their trajectories before and after the collision are straight lines.

Assumption 6.3 *The colliding objects are infinitesimally small particles.* We model each object as an infinitesimally small sphere of uniform density located at its center of mass. An implication of this assumption is that the pre-collision trajectories are such that the collision would occur no matter how small the objects are.

Assumption 6.4 *There are no frictional forces between the colliding objects.* An implication of this assumption and the previous one is that the (internal) forces generated by the collision act along the line through the center of mass of each body.[5]

  Chapter 12 examines a more general model of impacts in which Assumption 6.3 is relaxed. Note that sticky collisions in which particles collide and stick (as in Example 6.3) violate Assumption 6.4.

  For brevity, we continue our two-dimensional treatment when studying collisions and only treat impacts that occur in a plane. However, the analysis is completely general and extends to three dimensions. The remainder of this section develops a

---

[5] This line is called the *line of impact* and our model of a collision is sometimes called a *central impact* collision.

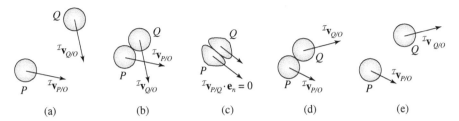

Figure 6.10 The five phases of the collision of two particles. (a) Initial phase before the collision; (b) start of deformation phase; (c) compressed phase; (d) end of restitution phase; (e) final phase after the collision.

model of the impact between two bodies following the process outlined in Chapter 3; that is, we define the appropriate reference frames and coordinate systems, draw free-body diagrams, and work out the particles' motion using Newton's laws.

Modeling the Impact

First we describe the physical model of the collision between two spherical particles, labeled P and Q, using Assumptions 6.1–6.4. If the particles collide and stick together, then Assumption 6.4 is violated; we find the final velocities using the law of conservation of total linear momentum in Eq. (6.11). If the particles collide and bounce apart, then we need a more detailed model. In fact, the final equations developed for the bouncy collision are not valid for the perfectly sticky collision (because of Assumption 6.4), as we shall see.

Using Assumption 6.1 (that the collision occurs over a very short, but finite, time interval) we label the start and end times of the impact, respectively, by t_1 and t_2. The particle velocities just before the collision are ${}^{\mathcal{I}}\mathbf{v}_{P/O}(t_1)$ and ${}^{\mathcal{I}}\mathbf{v}_{Q/O}(t_1)$ and the final velocities are ${}^{\mathcal{I}}\mathbf{v}_{P/O}(t_2)$ and ${}^{\mathcal{I}}\mathbf{v}_{Q/O}(t_2)$. For the short time during the collision, the particles squish together and then spring apart.[6] Let t_c denote the time at which the particles stop compressing and start expanding, where $t_1 < t_c < t_2$. (At $t = t_c$, the *normal* components of the particle velocities are the same—we explain what we mean by "normal" below.) The collision is thus divided into five sequential phases, as shown in Figure 6.10: (a) the *initial phase, $t < t_1$,* before the collision; (b) the *deformation phase,* from t_1 to t_c, when each particle compresses; (c) the *compressed phase,* which is the time t_c of the maximum compression when there is no relative motion between the particles along the line of impact; (d) the *restitution phase,* from t_c to t_2, when each particle returns to its original shape; and (e) the *final phase, $t > t_2$,* after the collision.

Next we define two frames of reference and draw the free-body diagram for each particle so that we can apply Newton's second law to the collision. Figure 6.11a shows the inertial frame $\mathcal{I} = (O, \mathbf{e}_1, \mathbf{e}_2, \mathbf{e}_3)$ and the *collision frame* $\mathcal{C} = (O', \mathbf{e}_n, \mathbf{e}_t, \mathbf{e}_3)$. The

---

[6] We recognize that there appears to be an inconsistency here. If all particles are point masses (Assumption 6.3), with no extent, as we insisted earlier in the application of Newton's second law, then how can they "squish"? In a detailed treatment, this would be a problem and each object would have to be treated as a collection of particles (i.e., a nonrigid body). However, as we are ignoring the details of the motion during the impact, we can safely ignore this problem. Before and after the impact we still treat the bodies as non-squishy particles.

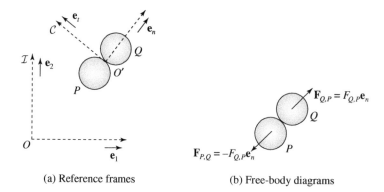

(a) Reference frames   (b) Free-body diagrams

Figure 6.11 Reference frames and free-body diagrams for the collision of two particles.

collision frame is defined by the point of impact O' and the two unit vectors, $\mathbf{e}_t$ and $\mathbf{e}_n$, which lie in the same plane as the particle velocities. Because we have assumed the particles are small spherical particles (Assumption 6.3), O' always lies on the line of impact connecting the centers. For large, and unequal sized, particles, finding O' and the normal unit vector $\mathbf{e}_n$ involves consideration of the trajectory of both particles and their relative positions. Chapter 12 treats this more complex case.

The situation simplifies greatly for small particles, however. If the particles have no extent and collide, then the line of impact is parallel to the velocity of P relative to the velocity of Q; that is, $\mathbf{e}_n$ is parallel to ${}^{\mathcal{I}}\mathbf{v}_{P/Q}(t_1) = {}^{\mathcal{I}}\mathbf{v}_{P/O}(t_1) - {}^{\mathcal{I}}\mathbf{v}_{Q/O}(t_1)$.[7] Since $\mathbf{e}_n$ has unit length, we have

$$
\mathbf{e}_n \triangleq \frac{{}^{\mathcal{I}}\mathbf{v}_{P/Q}(t_1)}{\|{}^{\mathcal{I}}\mathbf{v}_{P/Q}(t_1)\|} = {}^{\mathcal{I}}\hat{\mathbf{v}}_{P/Q}(t_1), \tag{6.25}
$$

where the caret denotes a vector of unit length. The tangent unit vector completes the frame: $\mathbf{e}_t = \mathbf{e}_3 \times \mathbf{e}_n$.

Figure 6.11b shows the free-body diagrams for the two particles during the collision. According to Assumptions 6.2–6.4, the only force on each particle during the collision is from the other particle; this force acts along the line of impact and is not due to friction. *The internal force between P and Q pushes the particles apart during the collision.* The direction of the collision force is fixed along the line of impact, but its magnitude varies during the deformation and restitution phases. We write the force on particle P due to particle Q as $\mathbf{F}_{P,Q}$. By Newton's third law, $\mathbf{F}_{P,Q} = -\mathbf{F}_{Q,P}$. According to the definition of frame $\mathcal{C}$ shown in Figure 6.11a, $\mathbf{F}_{P,Q}$ acts in the $-\mathbf{e}_n$ direction.

Figure 6.12 shows a schematic plot of the strength of the force during the collision. Let $F_{P,Q} \triangleq \|\mathbf{F}_{P,Q}\| \geq 0$ denote the magnitude of $\mathbf{F}_{P,Q}$ and likewise for $F_{Q,P}$. By Newton's third law, $F_{P,Q} = F_{Q,P}$. Both forces satisfy the criteria for a linear impulse,

---

[7] Why? Since the colliding particles are infinitesimally small spheres, then the line of impact is parallel to $\mathbf{r}_{Q/P}(t_1)$, and $\mathbf{r}_{Q/P}(t_1)$ must be parallel to ${}^{\mathcal{I}}\mathbf{v}_{P/Q}(t_1)$ or else the particles would not collide. (See Chapter 12.)

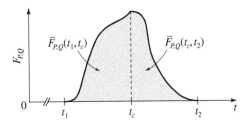

Figure 6.12 Magnitude $F_{P,Q}$ of the interaction force versus time. The area under the curve between t_1 and t_c is the magnitude of the deformation linear impulse. The area under the curve between t_c and t_2 is the magnitude of the restitution linear impulse. The deformation and restitution linear impulses are not necessarily the same. (The restitution impulse cannot be the larger of the two because of energy considerations.)

since we assume that $\Delta t \triangleq t_2 - t_1$ is very short. The linear impulse imparted to particle P from the collision with particle Q is $\overline{\mathbf{F}}_{P,Q}(t_1, t_2)$, which has magnitude $\overline{F}_{P,Q}(t_1, t_2)$ and direction $-\mathbf{e}_n$. We distinguish between the linear impulse $\overline{\mathbf{F}}_{P,Q}(t_1, t_c)$ during the deformation phase and the linear impulse $\overline{\mathbf{F}}_{P,Q}(t_c, t_2)$ during the restitution phase. By definition, the magnitude of the deformation linear impulse on particle P is the area under the force magnitude curve during the deformation phase. (The analogous statement holds for the restitution phase.) Just like the interaction forces, the linear impulses on P and Q are equal in magnitude and opposite in direction.

Conservation of Linear Momentum

To solve for the velocities after the collision, we begin with the impulse form of Newton's second law in Eq. (4.2), which yields

$$m_P\,^{\mathcal{I}}\mathbf{v}_{P/O}(t_2) = m_P\,^{\mathcal{I}}\mathbf{v}_{P/O}(t_1) + \overline{\mathbf{F}}_{P,Q}(t_1, t_2) \tag{6.26}$$

$$m_Q\,^{\mathcal{I}}\mathbf{v}_{Q/O}(t_2) = m_Q\,^{\mathcal{I}}\mathbf{v}_{Q/O}(t_1) + \overline{\mathbf{F}}_{Q,P}(t_1, t_2). \tag{6.27}$$

Letting u and v denote the components of the particle speed in the collision frame $\mathcal{C}$, we can write the velocities as

$$^{\mathcal{I}}\mathbf{v}_{P/O}(t) = u_P(t)\mathbf{e}_t + v_P(t)\mathbf{e}_n \tag{6.28}$$

$$^{\mathcal{I}}\mathbf{v}_{Q/O}(t) = u_Q(t)\mathbf{e}_t + v_Q(t)\mathbf{e}_n. \tag{6.29}$$

Thus, written as components in frame $\mathcal{C}$, Eqs. (6.26) and (6.27) become

$$m_P u_P(t_2)\mathbf{e}_t + m_P v_P(t_2)\mathbf{e}_n = m_P u_P(t_1)\mathbf{e}_t + m_P v_P(t_1)\mathbf{e}_n - \overline{F}_{P,Q}(t_1, t_2)\mathbf{e}_n$$

$$m_Q u_Q(t_2)\mathbf{e}_t + m_Q v_Q(t_2)\mathbf{e}_n = m_Q u_Q(t_1)\mathbf{e}_t + m_Q v_Q(t_1)\mathbf{e}_n + \overline{F}_{Q,P}(t_1, t_2)\mathbf{e}_n.$$

(Recall that $F_{P,Q} = F_{Q,P}$ and act in the opposite directions; hence the choice of signs above.)

Since there is no linear impulse in the tangential direction, the $\mathbf{e}_t$ component of linear momentum is conserved for each particle:

$$u_P(t_2) = u_P(t_1) \tag{6.30}$$

$$u_Q(t_2) = u_Q(t_1). \tag{6.31}$$

This observation, of course, implies that the tangential component of the total linear momentum is also conserved during the collision, which we can see by adding Eqs. (6.30) and (6.31).

For the normal direction, we have

$$m_P v_P(t_2) = m_P v_P(t_1) - \overline{F}_{P,Q}(t_1, t_2) \tag{6.32}$$

$$m_Q v_Q(t_2) = m_Q v_Q(t_1) + \overline{F}_{Q,P}(t_1, t_2). \tag{6.33}$$

Since the linear impulses acting on P and Q are equal and opposite, the normal component of total linear momentum is conserved. Adding Eqs. (6.32) and (6.33) and using $\overline{F}_{P,Q}(t_1, t_2) = \overline{F}_{Q,P}(t_1, t_2)$, we obtain the conservation equation

$$\boxed{m_P v_P(t_2) + m_Q v_Q(t_2) = m_P v_P(t_1) + m_Q v_Q(t_1).} \tag{6.34}$$

Eq. (6.34) and the corresponding tangential-direction equation (obtained by adding Eqs. (6.30) and (6.31)) are simply a rederivation of the conservation law for the total linear momentum (see Eq. (6.11)) since, by Assumption 6.2, the only forces acting during the collision are internal.

While Eqs. (6.30) and (6.31) allow us to trivially solve for the tangential components of the final velocities, Eq. (6.34) is a single equation with two unknowns, $v_P(t_2)$ and $v_Q(t_2)$. To find another equation in these unknowns, we need to define a new quantity, called the *coefficient of restitution*.

Coefficient of Restitution

The second equation needed to find $v_P(t_2)$ and $v_Q(t_2)$ from Eq. (6.34) comes from using the integrated form of Newton's second law in Eq. (1.2) for each particle again, but this time we distinguish between the deformation and restitution phases:

$$m_P{}^{\mathcal{I}}\mathbf{v}_{P/O}(t_c) = m_P{}^{\mathcal{I}}\mathbf{v}_{P/O}(t_1) + \overline{\mathbf{F}}_{P,Q}(t_1, t_c) \tag{6.35}$$

$$m_Q{}^{\mathcal{I}}\mathbf{v}_{Q/O}(t_c) = m_Q{}^{\mathcal{I}}\mathbf{v}_{Q/O}(t_1) + \overline{\mathbf{F}}_{Q,P}(t_1, t_c) \tag{6.36}$$

$$m_P{}^{\mathcal{I}}\mathbf{v}_{P/O}(t_2) = m_P{}^{\mathcal{I}}\mathbf{v}_{P/O}(t_c) + \overline{\mathbf{F}}_{P,Q}(t_c, t_2) \tag{6.37}$$

$$m_Q{}^{\mathcal{I}}\mathbf{v}_{Q/O}(t_2) = m_Q{}^{\mathcal{I}}\mathbf{v}_{Q/O}(t_c) + \overline{\mathbf{F}}_{Q,P}(t_c, t_2). \tag{6.38}$$

By assumption, the two particles have the same normal speed at time t_c: ${}^{\mathcal{I}}\mathbf{v}_{P/Q}(t_c) \cdot \mathbf{e}_n = 0$. In addition, the impulses on P and Q are equal and opposite. (Note that, if we add all four of these equations, we recover the conservation equation for total linear momentum.) Nonetheless, we still do not have enough equations to solve for

$v_P(t_2)$ and $v_Q(t_2)$ because there is no explicit form for the deformation and restitution impulses. To avoid having to do the detailed physics, we define a new quantity, the *coefficient of restitution.*

Definition 6.3 The **coefficient of restitution** e of the collision between particles P and Q is the ratio of the magnitudes of the restitution and deformation linear impulses:

$$e \triangleq \frac{\overline{F}_{P,Q}(t_c, t_2)}{\overline{F}_{P,Q}(t_1, t_c)} = \frac{\overline{F}_{Q,P}(t_c, t_2)}{\overline{F}_{Q,P}(t_1, t_c)}.$$

The coefficient of restitution satisfies $0 \leq e \leq 1$.

The coefficient of restitution is a joint property of the materials of the colliding objects. It must be a non-negative number less than or equal to one to satisfy energy conservation (the restitution force cannot do more work than the deformation force). We use it to predict the behavior of objects after a collision. There are simple experiments for measuring e for different collisions without having to directly measure the deformation and restitution forces (see, e.g., Example 6.12). Note that the coefficient of restitution depends on the properties of the materials of both particles involved and thus differs for every collision model.

It may seem like we haven't gained much by defining e; however, we have actually gained a great deal. This quantity gives a measure of the elasticity of the collision. If e is one, then the collision is perfectly *elastic;* that is, all of the linear momentum that deformed the bodies is "returned" during restitution. Conversely, if e is zero, then the collision is perfectly *inelastic* and the two bodies remain together after the collision (also called a *plastic* deformation, which violates Assumption 6.4). For values between zero and one, the particles rebound with reduced normal velocities (recall that we showed the tangential speeds were conserved). You may rightfully be wondering where the energy goes when $e < 1$. If the particles leave with lower normal speeds than before the collision, then energy must have been lost during the collision. Indeed, the coefficient of restitution is, in one sense, a measure of the energy loss; the excess energy goes into heating up or deforming the material. The deformation and restitution forces are not conservative.

Using Eqs. (6.28), (6.29), (6.35)–(6.38), and Definition 6.3, we can solve for the coefficient of restitution in terms of the various speeds:

$$e = \frac{v_P(t_2) - v_P(t_c)}{v_P(t_c) - v_P(t_1)} = \frac{v_Q(t_2) - v_Q(t_c)}{v_Q(t_c) - v_Q(t_1)}. \tag{6.39}$$

Note that the coefficient of restitution is independent of the masses and tangential speeds of the two particles; it depends only on their speeds in the normal direction. Eq. (6.39) is really a pair of equations. These two equations can be used to eliminate the unknown intermediate speed, $v_P(t_c) = v_Q(t_c)$. After a little algebra, we reach our final expression:

$$\boxed{e = \frac{v_P(t_2) - v_Q(t_2)}{v_Q(t_1) - v_P(t_1)}.} \tag{6.40}$$

In words, the coefficient of restitution is the ratio of the difference in magnitude of the normal speeds before and after the collision.

Final Velocities

We now have two equations, Eq. (6.34) and Eq. (6.40), for the two unknowns, $v_P(t_2)$ and $v_Q(t_2)$. Given the initial speed in the normal direction of each of the two particles (i.e., $v_P(t_1)$ and $v_Q(t_1)$), we can solve for the final speeds in terms of the coefficient of restitution. Although it is usually easiest to use these two equations directly when solving a problem, it is possible to find a general expression for the final velocities by simultaneously solving the following pair of equations:

$$m_P v_P(t_2) + m_Q v_Q(t_2) = m_P v_P(t_1) + m_Q v_Q(t_1) \tag{6.41}$$

$$v_P(t_2) - v_Q(t_2) = e(v_Q(t_1) - v_P(t_1)). \tag{6.42}$$

By multiplying Eq. (6.42) by m_Q and adding Eqs. (6.41) and (6.42) and, likewise, multiplying Eq. (6.42) by m_P and subtracting Eqs. (6.41) and (6.42), we find the final normal speeds:

$$v_P(t_2) = \frac{m_P - e m_Q}{m_P + m_Q} v_P(t_1) + \frac{m_Q(1+e)}{m_P + m_Q} v_Q(t_1) \tag{6.43}$$

$$v_Q(t_2) = \frac{m_Q - e m_P}{m_P + m_Q} v_Q(t_1) + \frac{m_P(1+e)}{m_P + m_Q} v_P(t_1). \tag{6.44}$$

Combining Eqs. (6.43) and (6.44) for the normal direction with Eqs. (6.30) and (6.31) for the tangential direction, we obtain the final velocities:

$$^{\mathcal{I}}\mathbf{v}_{P/O}(t_2) = u_P(t_1)\mathbf{e}_t + \left(\frac{m_P - e m_Q}{m_P + m_Q} v_P(t_1) + \frac{m_Q(1+e)}{m_P + m_Q} v_Q(t_1) \right) \mathbf{e}_n \tag{6.45}$$

$$^{\mathcal{I}}\mathbf{v}_{Q/O}(t_2) = u_Q(t_1)\mathbf{e}_t + \left(\frac{m_Q - e m_P}{m_P + m_Q} v_Q(t_1) + \frac{m_P(1+e)}{m_P + m_Q} v_P(t_1) \right) \mathbf{e}_n. \tag{6.46}$$

As noted earlier, this model does not hold for a perfectly plastic (i.e., sticky) collision, for which $e = 0$. In that case, Eqs. (6.45) and (6.46) show that the normal speeds of P and Q are equal but their tangential speeds are not! The no-friction assumption in the tangential direction is violated for a perfectly sticky collision. For such collisions, we use Newton's second law and the law of conservation of total linear momentum, as in Example 6.3. Eqs. (6.45) and (6.46) are only valid when the coefficient of restitution e is greater than zero (that is, a bouncy collision).

Example 6.9 A Simple Bouncy Collision

A simple impact is where two objects of equal mass whose velocities are aligned collide and rebound in a straight line, as shown in Figure 6.13. This is an excellent

Figure 6.13 Two equal-mass particles traveling in a straight line collide and bounce with different velocities. (a) Before collision ($t_0 < t_1$). (b) After collision ($t_f > t_2$).

model of two pool balls colliding head-on. The tangential speeds are zero ($u_Q = u_P = 0$). Using Eqs. (6.43) and (6.44) and the initial speeds $v_Q(t_1)$ and $v_P(t_1)$, the final speeds of the two balls are

$$v_P(t_2) = \frac{1}{2}(1 - e)v_P(t_1) + \frac{1}{2}(1 + e)v_Q(t_1)$$

$$v_Q(t_2) = \frac{1}{2}(1 - e)v_Q(t_1) + \frac{1}{2}(1 + e)v_P(t_1).$$

Supposing the initial speed of Q is zero—corresponding, for instance, to the cue ball, P, hitting a stationary ball, Q—the final speeds simplify to

$$v_P(t_2) = \frac{1}{2}(1 - e)v_P(t_1)$$

$$v_Q(t_2) = \frac{1}{2}(1 + e)v_P(t_1).$$

In this case, if e is one (a perfectly elastic collision), the linear momentum is completely transferred from P to Q: the cue ball stops, and the target ball moves off at the initial speed of the cue ball. For $e < 1$, the coefficient of restitution is a measure of the amount of speed that is transferred from one ball to the other.

Example 6.10 An Oblique Bouncy Collision

This example demonstrates how to use Eqs. (6.45) and (6.46) to find the final velocities of two arbitrarily small particles P and Q after the oblique collision shown in Figure 6.14a. We treat the collision as bouncy, that is, the coefficient of restitution e is greater than zero. Since there are no external forces, the particle velocities are constant before and after the collision. The main challenge is to write the initial velocities in the collision frame.

We begin by assuming that the initial velocities are known:

$$^\mathcal{I}\mathbf{v}_{P/O}(0) = \dot{x}_P(0)\mathbf{e}_x + \dot{y}_P(0)\mathbf{e}_y$$
$$^\mathcal{I}\mathbf{v}_{Q/O}(0) = \dot{x}_Q(0)\mathbf{e}_x + \dot{y}_Q(0)\mathbf{e}_y. \tag{6.47}$$

We next use the particle velocities at impact to determine the tangential and normal directions of the collision frame $\mathcal{C} = (O', \mathbf{e}_n, \mathbf{e}_t, \mathbf{e}_z)$:

$$\mathbf{e}_n = \frac{^\mathcal{I}\mathbf{v}_{P/Q}(t_1)}{\|^\mathcal{I}\mathbf{v}_{P/Q}(t_1)\|} = \frac{^\mathcal{I}\mathbf{v}_{P/O}(0) - ^\mathcal{I}\mathbf{v}_{Q/O}(0)}{\|^\mathcal{I}\mathbf{v}_{P/O}(0) - ^\mathcal{I}\mathbf{v}_{Q/O}(0)\|}. \tag{6.48}$$

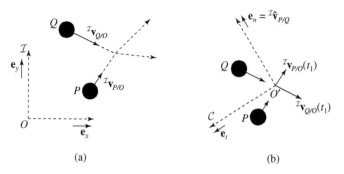

Figure 6.14 Oblique bouncy collision.

Since the normal unit vector makes a right-hand triad with $\mathbf{e}_t$ and $\mathbf{e}_3$ (the vector pointing out of the collision plane), we have

$$\mathbf{e}_t = \mathbf{e}_z \times \mathbf{e}_n. \tag{6.49}$$

Using Eqs. (6.48) and (6.49), we solve for the tangential and normal unit vectors from the initial particle velocities in terms of the unit vectors $\mathbf{e}_x$ and $\mathbf{e}_y$:

$$\dot{s}(0)\mathbf{e}_n = \dot{x}_{P/Q}(0)\mathbf{e}_x + \dot{y}_{P/Q}(0)\mathbf{e}_y$$
$$\dot{s}(0)\mathbf{e}_t = -\dot{y}_{P/Q}(0)\mathbf{e}_x + \dot{x}_{P/Q}(0)\mathbf{e}_y, \tag{6.50}$$

where $x_{P/Q} = x_P - x_Q$, $y_{P/Q} = y_P - y_Q$, and $\dot{s} = \sqrt{\dot{x}_{P/Q}^2 + \dot{y}_{P/Q}^2}$. To write the initial particle velocities in the collision frame, we invert Eq. (6.50) to find

$$\mathbf{e}_x = \frac{1}{\dot{s}(0)} \left(\dot{x}_{P/Q}(0)\mathbf{e}_n - \dot{y}_{P/Q}(0)\mathbf{e}_t \right)$$
$$\mathbf{e}_y = \frac{1}{\dot{s}(0)} \left(\dot{y}_{P/Q}(0)\mathbf{e}_n + \dot{x}_{P/Q}(0)\mathbf{e}_t \right) . \tag{6.51}$$

Eq. (6.51) allows us to write the initial velocities (Eq. (6.47)) in the collision frame $\mathcal{C}$. We can then use Eqs. (6.45) and (6.46) to solve for the final velocities in frame $\mathcal{C}$. Finally, using Eq. (6.50), we can write the final velocities as components in frame $\mathcal{I}$.

6.2.2 Collision between a Particle and a Surface

This section examines the (bouncy) collision between a particle and a surface. The goal is similar to that of the previous section: *given the velocity of the particle before a collision with the surface, find the velocity of the particle after the collision.*

The presence of an external force holding up the wall means that the total linear momentum of the particle-surface system is not conserved. As a result, we cannot directly use the results of Section 6.2.1 to study a surface impact. Instead, we must go back to first principles. In doing so, we follow steps similar to those in Section 6.2.1 using the following simplifying assumptions:

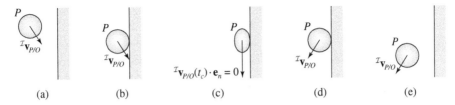

Figure 6.15 The five phases of the collision of a particle and a surface. (a) Initial phase before the collision; (b) start of deformation phase; (c) compressed phase; (d) end of restitution phase; (e) final phase after the collision.

Assumption 6.1′ *The collision occurs over a very short time interval, during which the particle does not move.* Thus the collision instantaneously changes the particle velocity.

Assumption 6.2′ *There are no external forces acting on the particle during the collision, other than from the surface.* For some cases where this assumption is not met, we can still apply the same approach developed here, but the general result is not guaranteed to hold (see Tutorial 6.2).

Assumption 6.3′ *The moving object is a spherical point mass;* if it is an extended body, we treat it as if it is a point mass located at the center of mass.

Assumption 6.4′ *There are no frictional forces between the particle and the surface.* Thus the collision impulse acts along the line from the point of impact through the center of the body.

Assumption 6.5′ *In the vicinity of the impact point, the surface is smooth, rigid, and fixed;* that is, the normal direction is well defined, the surface does not compress, and there is some external force holding the surface in place.

To model the impact of particle P with a surface, we use the same terminology and notation as in Section 6.2.1. We assume that there are five sequential phases during the collision (see Figure 6.15): (a) the initial phase, $t < t_1$, before the collision; (b) the deformation phase, from t_1 to t_c, when P compresses; (c) the compressed phase, which is the time t_c of maximum compression of the particle; (d) the restitution phase, from t_c to t_2, when P returns to its original shape; and (e) the final phase, $t > t_2$, after P leaves the surface. According to Assumption 6.1′, $\Delta t \triangleq t_2 - t_1$ is very small.

Next we describe the inertial and collision frames illustrated in Figure 6.16a. We assume that the inertial frame is determined by the orientation of the surface near the impact point and the (planar) trajectory of the particle. That is, let $\mathcal{I} = (O, \mathbf{e}_x, \mathbf{e}_y, \mathbf{e}_z)$, where the surface lies in the plane spanned by the unit vectors $\mathbf{e}_y$ and $\mathbf{e}_z$, and the trajectory of the particle lies in the plane spanned by the unit vectors $\mathbf{e}_x$ and $\mathbf{e}_y$. Because the position and orientation of the surface are fixed, there is a natural choice for the collision frame. Namely, let $\mathcal{C} = (O', \mathbf{e}_n, \mathbf{e}_t, \mathbf{e}_3)$, where O' is the point of impact, $\mathbf{e}_t = \mathbf{e}_y$, $\mathbf{e}_n = -\mathbf{e}_x$, and $\mathbf{e}_3 = -\mathbf{e}_z$.

By Assumption 6.4′, the only force acting on the particle during the collision is the interaction force $\mathbf{F}_P$, which acts normal to the surface (in the $\mathbf{e}_n$ direction). We denote the magnitude of the force by F_P, so that $\mathbf{F}_P = F_P\mathbf{e}_n$. As in Section 6.2.1, we split the collision impulse $\overline{\mathbf{F}}_P(t_1, t_2) = \overline{F}_P(t_1, t_2)\mathbf{e}_n$ from the surface to the particle

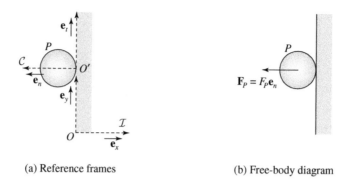

(a) Reference frames   (b) Free-body diagram

Figure 6.16 Collision of particle P and a fixed surface.

into two components. The first, $\overline{F}_P(t_1, t_c)$, corresponds to the impulse during the deformation phase and the second, $\overline{F}_P(t_c, t_2)$, corresponds to the impulse during the restitution phase. Figure 6.16b illustrates the interaction force $\mathbf{F}_P$. (The magnitude of the interaction force and the deformation and compression impulses are still adequately described by Figure 6.12 with $\overline{F}_{P,Q}$ replaced by $\overline{F}_P$.)

To solve for the final velocity of particle P after the collision, we begin again with the impulse form of Newton's second law:

$$m_P{}^\mathcal{I}\mathbf{v}_{P/O}(t_2) = m_P{}^\mathcal{I}\mathbf{v}_{P/O}(t_1) + \overline{\mathbf{F}}_P(t_1, t_2)$$
$$= m_P{}^\mathcal{I}\mathbf{v}_{P/O}(t_1) + \overline{F}_P(t_1, t_2)\mathbf{e}_n. \qquad (6.52)$$

Let u_P and v_P denote the Cartesian coordinates for the speed of P in collision frame $\mathcal{C}$, that is, ${}^\mathcal{I}\mathbf{v}_{P/O} = u_P\mathbf{e}_t + v_P\mathbf{e}_n$. Eq. (6.52) becomes

$$m_P u_P(t_2) = m_P u_P(t_1) \qquad (6.53)$$

$$m_P v_P(t_2) = m_P v_P(t_1) + \overline{F}_P(t_1, t_2). \qquad (6.54)$$

We observe that the tangential component of the particle's linear momentum is conserved during the collision, but the normal component is not. Thus the final tangential speed $u_P(t_2) = u_P(t_1)$ is known, but the final normal speed $v_P(t_2)$ is not.

To find $v_P(t_2)$, we once again use the impulse form of Newton's second law, this time considering separately the deformation and restitution phases. Since the deformation and restitution impulses act only in the normal direction, we only write the normal component of Newton's second law:

$$m_P v_P(t_c) = m_P v_P(t_1) + \overline{F}_P(t_1, t_c) \qquad (6.55)$$

$$m_P v_P(t_2) = m_P v_P(t_c) + \overline{F}_P(t_c, t_2). \qquad (6.56)$$

Eqs. (6.55) and (6.56) represent two equations in terms of four unknowns: $v_P(t_2)$, $v_P(t_c)$, $\overline{F}_P(t_1, t_c)$, and $\overline{F}_P(t_c, t_2)$. Hence we need two more equations to algebraically solve the system of equations. The first equation comes from the following intuitive observation: the normal speed of the particle reverses direction at time t_c. That is, time t_c denotes the instant in time before which the particle is still moving toward the surface and after which the particle moves away from the surface. This means

$$v_P(t_c) = 0, \tag{6.57}$$

which is our third equation.[8] The final equation emerges by once again introducing a coefficient of restitution.

Definition 6.4 The **coefficient of restitution** e of the collision between particle P and a rigid, fixed surface is the ratio of the magnitude of the restitution linear impulse to the magnitude of the deformation linear impulse:

$$e \triangleq \frac{\overline{F}_P(t_c, t_2)}{\overline{F}_P(t_1, t_c)}.$$

The coefficient of restitution satisfies $0 \leq e \leq 1$.

Definition 6.4 essentially replaces the two unknowns $\overline{F}_P(t_1, t_c)$ and $\overline{F}_P(t_c, t_2)$ with a single unknown e. Recall that we assume the collision is bouncy, which means that $e > 0$ ($e = 0$ corresponds to a sticky or plastic collision and $e = 1$ corresponds to a perfectly bouncy or perfectly elastic collision). Using Eqs. (6.55)–(6.57) and Definition 6.4, we can solve for the final normal speed:

$$v_P(t_2) = -ev_P(t_1). \tag{6.58}$$

Using Eqs. (6.53) and (6.58), we find the final velocity in terms of the initial velocity. The final velocity in terms of components in the collision frame is

$$^{\mathcal{I}}\mathbf{v}_{P/O}(t_2) = u_P(t_1)\mathbf{e}_t - ev_P(t_1)\mathbf{e}_n. \tag{6.59}$$

Perhaps as important as Eq. (6.59) is Eq. (6.58), which yields a practical definition of the coefficient of restitution in terms of the initial and final normal speeds:

$$\boxed{e = -\frac{v_P(t_2)}{v_P(t_1)}.} \tag{6.60}$$

Note that Eq. (6.60) is equivalent to Eq. (6.40) with $v_Q(t_2) = v_Q(t_1) = 0$. Thus the coefficient of restitution of a collision between a particle and a fixed surface is the same as that of a collision between a moving particle and a fixed particle. Example 6.12 describes a procedure for measuring the coefficient of restitution for a surface impact.

Example 6.11 A Carom in Billiards

Rather than two pool balls colliding, consider a carom, or bank shot off the rail, as shown in Figure 6.17. The goal is to find the initial conditions—as a function of the coefficient of restitution—that generate a rebound of the desired angle between the ball and the rail. (We postpone considering rotation of the ball until Example 9.11.)

---

[8] Note the resemblance of this formula to its counterpart in the collision between two particles: $v_{P/Q}(t_c) = 0$.

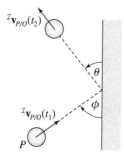

Figure 6.17 A billiard ball impacting the rail at angle ϕ and banking off the rail at angle θ.

The incoming velocity is $^{\mathcal{I}}\mathbf{v}_{P/O}(t_1) = u_P(t_1)\mathbf{e}_t + v_P(t_1)\mathbf{e}_n$. The incoming angle with the rail is thus

$$\tan \phi = \frac{-v_P(t_1)}{u_P(t_1)}.$$

After the bank off the rail, the final velocity is given by Eq. (6.59). The bank angle satisfies

$$\tan \theta = \frac{-ev_P(t_1)}{u_P(t_1)} = e \tan \phi.$$

For a perfectly elastic collision, the ball leaves the rail at exactly the same angle as it arrived (making bank shots much easier to align). However, the felt-covered rail results in $e < 1$, reducing the rebound angle and making the carom more challenging.

Example 6.12 Measuring e for a Surface Impact

This example describes an experimental procedure for determining the coefficient of restitution of the collision between a ball (particle P) and the floor (the fixed surface). The procedure is quite simple: we drop the ball from a known height and measure how high it bounces. However, the presence of gravity violates Assumption 6.2′, which was used to derive Eq. (6.60). Thus we have to revisit the derivation of Eq. (6.60) to determine the effect of gravity. We show that, even in the presence of gravity, Eq. (6.60) is a good approximation of e, as long as the bounce duration $\Delta t \triangleq t_2 - t_1$ is very short, where t_1 and t_2 denote the start and end times of the impact, respectively.

Let $\mathcal{I} = (O, \mathbf{e}_x, \mathbf{e}_y, \mathbf{e}_z)$ and $\mathcal{C} = (O', \mathbf{e}_n, \mathbf{e}_t, \mathbf{e}_3)$, where $\mathcal{I} = \mathcal{C}$, as shown in Figure 6.18b. Consider the free-body diagrams of particle P shown in Figure 6.18c. The only force acting on P before and after the collision is the force of gravity $-m_P g \mathbf{e}_y = -m_P g \mathbf{e}_n$. The total force acting on P during the collision is the sum of the force from the wall $\mathbf{F}_P = F_P \mathbf{e}_n$ and the force of gravity. Since we drop the ball from rest at time $t = 0$, the initial horizontal speed is zero and, since there is never any force in the horizontal direction, the final horizontal speed is also zero.

We find the vertical speed of the particle at time t_1 using conservation of energy, since, prior to the impact, the only force acting on the particle is gravity (which is conservative). When the particle is at height y_0, it has potential energy $m_P g y_0$. Just

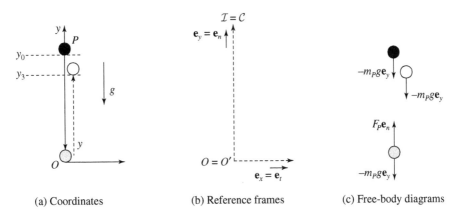

(a) Coordinates   (b) Reference frames   (c) Free-body diagrams

Figure 6.18 Measuring e for a surface impact. Particle P is colored black before impact, gray during impact, and white after impact.

before it hits the floor the height is zero, so all of the potential energy is converted to kinetic energy. Conservation of energy yields

$$\frac{1}{2} m_P (v_P(t_1))^2 = m_P g y_0,$$

which results in a speed before impact of

$$v_P(t_1) = -\sqrt{2 y_0 g}. \tag{6.61}$$

We have used the negative root because the velocity is in the $-\mathbf{e}_n$ direction.

Let $t_3 > t_2$ denote the time at which the ball reaches its maximum bounce height y_3. We find the vertical speed of the particle at t_2 the same way as we found the vertical speed at t_1, that is, by using energy conservation. Thus, if the particle reaches height y_3 with potential energy $m_P g y_3$, then its initial speed after impact is

$$v_P(t_2) = \sqrt{2 y_3 g}. \tag{6.62}$$

If we were to ignore the effect of gravity during the impact, we could find e by substituting Eqs. (6.61) and (6.62) into Eq. (6.60). However, let us see how big an effect gravity has on the previous analysis. As before, we use the impulse form of Newton's second law for both the deformation and restitution phases, which has the normal components

$$m_P v_P(t_c) = m_P v_P(t_1) + \overline{F}_P(t_1, t_c) - \int_{t_1}^{t_c} m_P g \, dt \tag{6.63}$$

$$m_P v_P(t_2) = m_P v_P(t_c) + \overline{F}_P(t_c, t_2) - \int_{t_c}^{t_2} m_P g \, dt. \tag{6.64}$$

Since m_P and g are constant during the interval t_1 to t_2, the integrals in Eqs. (6.63) and (6.64) evaluate trivially to the small quantities $m_P g (t_c - t_1)$ and $m_P g (t_2 - t_c)$, respectively. Let $O(\Delta t)$ denote a small quantity that is less than or equal to a constant times Δt. Since $t_1 < t_c < t_2$, the integrals in Eqs. (6.63) and (6.64) are both $O(\Delta t)$. Using this observation, Eq. (6.57), and Definition 6.4, we find that

$$e = \frac{v_P(t_2) + O(\Delta t)}{-v_P(t_1) + O(\Delta t)} = \frac{v_P(t_2)}{-v_P(t_1) + O(\Delta t)} + O(\Delta t). \qquad (6.65)$$

We now use the *binomial expansion* described in Example A.1 of Appendix A. That is, the Taylor series expansion of the first term on the right-hand side in Eq. (6.65) can be expanded using the fact that $O(\Delta t)$ is a small quantity:

$$\frac{1}{-v_P(t_1) + O(\Delta t)} \approx \frac{-1}{v_P(t_1)}(1 + O(\Delta t)).$$

This estimate lets us rewrite Eq. (6.65) as

$$e = -\frac{v_P(t_2)}{v_P(t_1)} + O(\Delta t).$$

Thus ignoring gravity and using Eq. (6.60) produces an error proportional to the impact duration Δt.

Using Eqs. (6.61) and (6.62) to substitute for the speeds, we find the coefficient of restitution is

$$e = \sqrt{\frac{y_3}{y_0}} + O(\Delta t).$$

In the limit $\Delta t \to 0$, $e \approx \sqrt{y_3/y_0}$, which is what we would have obtained using Eqs. (6.61) and (6.62) in Eq. (6.60) directly. This approximation is only as good as the assumption of a short impact duration.

6.3 Mass Flow

This section shifts from considering only two particles to the other extreme and examines the behavior of a system with infinitely many particles. Parts Three and Four examine rigid collections of many particles, or rigid bodies, in great detail. Here we study the specialized problem of a *variable-mass system,* that is, a collection of particles that is gaining or losing mass (or both) by means of particles flowing in and/or out of the system. It is far beyond our scope to study fluid dynamics in any detail. However, as with collisions, we need only make a few simplifying assumptions to be able to solve a number of useful problems using only a particle-dynamics model.

All analyses of variable-mass systems begin with the concept of the *control volume.* The control volume is simply a geometric shape that encloses the particles for which we desire to find trajectories (see Figure 6.19). Recall how we approached the study of multiparticle systems in Section 6.1. As long as we keep track of every particle, we can use conservation of total linear momentum or the equations for linear impulse and momentum on the total linear momentum of the entire collection. Here we isolate a subset of the particles, draw a control volume around them, and examine the total linear momentum of the particles in the control volume. Our goal is to find an equation

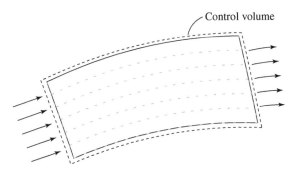

Figure 6.19 An example of a control volume isolating a vessel and a collection of particles as part of a stream.

of motion for a point fixed to the control volume. The only requirement is that we keep track of the total linear momentum of *the same collection of particles* before and after a particle is ejected or added.

In the two subsections that follow we study two situations: (a) a steady stream of particles entering and leaving a system such that the total mass stays the same and (b) a system that is ejecting or collecting particles such that its mass is increasing or decreasing while it is moving.

6.3.1 Steady Streams

Consider a system consisting of a stream of fluid passing through an open vessel, as shown in Figure 6.20. We take as our control volume the boundary of the vessel; it thus contains the particles making up the vessel and the particles making up the fluid passing through it. The control volume has center of mass G and attached body frame $\mathcal{B}$. The control volume is acted on by an external force $\mathbf{F}_{cv}^{(ext)}$. Our goal is to find an equation of motion for G. The system of particles we examine consists of the mass in the control volume plus the small differential mass Δm in the stream about to enter the control volume.

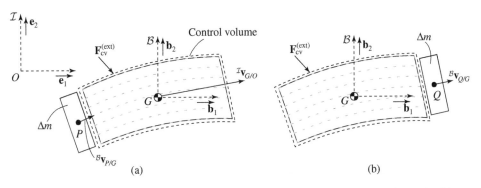

Figure 6.20 A system of mass m_G consisting of a steady stream of particles entering and exiting a control volume. (a) $t = t_1$. (b) $t = t_2 = t_1 + \Delta t$.

A general solution to this problem involves sophisticated fluid mechanics. We therefore make a few restrictive assumptions, as we did for collisions, that allow us to use a particle model to find the equations of motion. Although many problems can be solved with these assumptions, it is important to remember them to avoid using the resulting equation on problems for which it doesn't apply.

Assumption 6.6 *The particles making up the fluid are evenly distributed in the control volume and maintain the same distribution throughout the motion of G.* Thus the center of mass G has a fixed position in the control volume.

Assumption 6.7 *The fluid maintains a constant velocity relative to the frame $\mathcal{B}$ fixed to the control volume throughout the motion of G.*

Assumption 6.8 *Fluid mass enters the control volume at the same rate as it leaves,* so that the total mass of particles in the control volume is constant.

Assumption 6.9 *The control volume does not rotate relative to the inertial frame* $\mathcal{I}$. That is, we only study translational motion of the control volume.

Let $^{\mathcal{I}}\mathbf{v}_{G/O}$ represent the inertial velocity of the center of mass of the control volume (and thus also of the vessel or any other stationary mass in the control volume). The velocity of each fluid particle in the control volume relative to frame $\mathcal{B}$ is given by $^{\mathcal{B}}\mathbf{v}_{j/G}$. The differential mass about to enter the control volume has a velocity relative to the control volume of $^{\mathcal{B}}\mathbf{v}_{P/G}$. The total linear momentum of the system of particles consisting of the control volume and the mass P about to enter at time t_1 is thus given by (see Definition 6.1):

$$^{\mathcal{I}}\mathbf{p}_O(t_1) = m_v\,^{\mathcal{I}}\mathbf{v}_{G/O}(t_1) + \sum_{j=1}^{N} m_j(^{\mathcal{B}}\mathbf{v}_{j/G}(t_1) + {}^{\mathcal{I}}\mathbf{v}_{G/O}(t_1))$$

$$+ \Delta m(^{\mathcal{B}}\mathbf{v}_{P/G}(t_1) + {}^{\mathcal{I}}\mathbf{v}_{G/O}(t_1)),$$

where m_v is the mass of the vessel in the control volume and m_j is the mass of the jth particle in the control volume. In writing the inertial velocity of Δm as the sum of the inertial velocity of the control volume plus the relative velocity, $^{\mathcal{B}}\mathbf{v}_{P/G}$, we have used Assumption 6.9 (the control volume is not rotating) and the results of Section 3.6.

Note that G here is not the center of mass of the entire system of particles and thus does not obey Newton's second law as in Eq. (6.16). Nevertheless, we use the fact that we can write the integrated form of Newton's second law for the total system of particles (including the differential mass about to enter the control volume) as in Eq. (6.10) to find an equation of motion for the control volume. We thus write the impulse form of Newton's second law for all particles between times t_1 and $t_2 \triangleq t_1 + \Delta t$:

$$m_v\,^{\mathcal{I}}\mathbf{v}_{G/O}(t_2) + \sum_{j=1}^{N} m_j(^{\mathcal{B}}\mathbf{v}_{j/G}(t_2) + {}^{\mathcal{I}}\mathbf{v}_{G/O}(t_2)) + \Delta m(^{\mathcal{B}}\mathbf{v}_{Q/G}(t_2) + {}^{\mathcal{I}}\mathbf{v}_{G/O}(t_2))$$

$$= m_v\,^{\mathcal{I}}\mathbf{v}_{G/O}(t_1) + \sum_{j=1}^{N} m_j(^{\mathcal{B}}\mathbf{v}_{j/G}(t_1) + {}^{\mathcal{I}}\mathbf{v}_{G/O}(t_1))$$

$$+ \Delta m(^{\mathcal{B}}\mathbf{v}_{P/G}(t_1) + {}^{\mathcal{I}}\mathbf{v}_{G/O}(t_1)) + \overline{\mathbf{F}}_{\text{cv}}^{(\text{ext})}(t_1, t_2),$$

where ${}^{\mathcal{B}}\mathbf{v}_{Q/G}$ is the relative velocity of mass Δm which left the control volume at t_2 (by Assumption 6.8) and $\overline{\mathbf{F}}_{cv}^{(ext)}(t_1, t_2)$ is the impulse corresponding to the external force acting on the control volume over Δt.

From Assumption 6.7 we have

$$ {}^{\mathcal{B}}\mathbf{v}_{j/G}(t_2) = {}^{\mathcal{B}}\mathbf{v}_{j/G}(t_1). $$

This allows us to simplify and combine terms to find

$$ m_G \left({}^{\mathcal{I}}\mathbf{v}_{G/O}(t_2) - {}^{\mathcal{I}}\mathbf{v}_{G/O}(t_1) \right) + \Delta m \left({}^{\mathcal{I}}\mathbf{v}_{G/O}(t_2) - {}^{\mathcal{I}}\mathbf{v}_{G/O}(t_1) \right) $$
$$ = \Delta m \left({}^{\mathcal{B}}\mathbf{v}_{P/G}(t_1) - {}^{\mathcal{B}}\mathbf{v}_{Q/G}(t_2) \right) + \overline{\mathbf{F}}_{cv}^{(ext)}(t_1, t_2), \tag{6.66} $$

where $m_G = m_v + \sum_{j=1}^{N} m_j$ is the total mass of material inside the control volume.

Our final step is to divide Eq. (6.66) by Δt and take the limit $\Delta t \to 0$. This allows us to use the definitions of differentiation and integration (see Appendix A) to find

$$ \lim_{\Delta t \to 0} \frac{1}{\Delta t} \left({}^{\mathcal{I}}\mathbf{v}_{G/O}(t_2) - {}^{\mathcal{I}}\mathbf{v}_{G/O}(t_1) \right) = \frac{{}^{\mathcal{I}}d}{dt} \left({}^{\mathcal{I}}\mathbf{v}_{G/O} \right)(t_1) = {}^{\mathcal{I}}\mathbf{a}_{G/O}(t_1) $$

and

$$ \lim_{\Delta t \to 0} \frac{1}{\Delta t} \overline{\mathbf{F}}_{cv}^{(ext)}(t_1, t_2) = \lim_{\Delta t \to 0} \frac{1}{\Delta t} \int_{t_1}^{t_2} \mathbf{F}_{cv}^{(ext)} dt = \mathbf{F}_{cv}^{(ext)}(t_1). $$

We also define the *mass flow rate* $\dot{m}$:

$$ \dot{m} \triangleq \lim_{\Delta t \to 0} \frac{\Delta m}{\Delta t}. $$

The mass flow rate is the rate at which the mass of the fluid enters and leaves the control volume. It is taken to be a positive number.

Finally, the second term in Eq. (6.66) after dividing by Δt is zero in the limit $\Delta t \to 0$ because it equals $\lim_{\Delta t \to 0} \dot{m}{}^{\mathcal{I}}\mathbf{a}_{G/O}\Delta t$. This leaves the final equation of motion for the center of mass of the control volume:

$$ \boxed{ m_G {}^{\mathcal{I}}\mathbf{a}_{G/O} = \mathbf{F}_{cv}^{(ext)} + \dot{m} \left({}^{\mathcal{B}}\mathbf{v}_{in/G} - {}^{\mathcal{B}}\mathbf{v}_{out/G} \right), } \tag{6.67} $$

where we have introduced the new notation

$$ {}^{\mathcal{B}}\mathbf{v}_{in/G} \triangleq {}^{\mathcal{B}}\mathbf{v}_{P/G} $$
$$ {}^{\mathcal{B}}\mathbf{v}_{out/G} \triangleq {}^{\mathcal{B}}\mathbf{v}_{Q/G} $$

for the *inflow* and *outflow* velocities, respectively. Note that it is common practice to put the effect of the mass flow on the right of Newton's second law, as in Eq. (6.67), to make it look like a force (such as the thrust in a jet engine or rocket), even though it arises from momentum balance. As we'll see in the next example, the mass flow has the effect of a thrust on the mass inside the control volume.

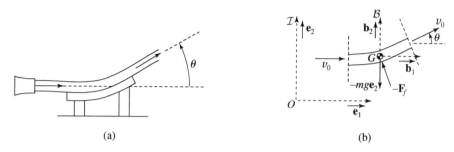

Figure 6.21 Fluid flowing onto a diverter at angle θ with mass flow rate $\dot{m}$. (a) Flow into diverter. (b) Free-body diagram.

Example 6.13 Force Due to Mass Flow

This example examines the force on a barrier due to a flowing fluid, as shown in Figure 6.21a. The diverter changes the direction of the constant flow, which has mass flow rate $\dot{m}$, by angle θ. The flow in turn exerts a force $\mathbf{F}_f$ on the diverter; or, conversely, by Newton's third law, the diverter exerts a force $-\mathbf{F}_f$ on the fluid.

Figure 6.21b shows a free-body diagram of the fluid (collection of particles) inside the control volume in contact with the diverter. Assuming a steady stream of constant cross-sectional area, the speed of the flow into and out of the volume must be the same for the mass to stay constant (this is true by the condition of *continuity*) and thus satisfies Assumptions 6.6 and 6.7. We label this speed v_0. Since we are assuming a steady state, for which the center of mass of the fluid on the diverter is fixed in absolute space, frame $\mathcal{B}$ and the inertial frame $\mathcal{I}$ are the same. This also means that the acceleration of the control volume is zero: $\frac{^\mathcal{I}d}{dt}\left(^\mathcal{I}\mathbf{v}_{G/O}\right) = 0$. Written in terms of components in the inertial frame, the velocities of the inflow and outflow are

$$^\mathcal{B}\mathbf{v}_{\text{in}/G} = v_0\mathbf{e}_1$$

$$^\mathcal{B}\mathbf{v}_{\text{out}/G} = v_0 \cos\theta\mathbf{e}_1 + v_0 \sin\theta\mathbf{e}_2.$$

These velocities are substituted into Eq. (6.67) to find the total external force on the mass of fluid in the steady state:

$$\mathbf{F}_f = \dot{m}(v_0 - v_0 \cos\theta)\mathbf{e}_1 - (mg + \dot{m}v_0 \sin\theta)\mathbf{e}_2.$$

6.3.2 Systems Gaining and Losing Mass

The other common situation we now examine is a control volume that is continually ejecting or collecting particles, as in Figure 6.22. The classic example is the rocket, which produces thrust by ejecting gas at high velocity. We study the rocket problem in detail in Tutorial 6.3.

As for the study of a steady stream of fluid, we need to make certain simplifying assumptions to make our particle-dynamics approach tractable:

Assumption 6.6′ *The particles inside the control volume have zero velocity relative to frame $\mathcal{B}$ fixed to the control volume at point C.*

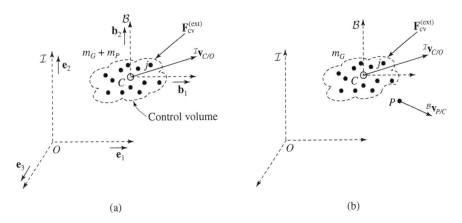

Figure 6.22 A variable-mass system of mass m_G before and after ejecting a single particle P of mass m_P. (a) $t = t_1$. (b) $t = t_2 = t_1 + \Delta t$.

Assumption 6.7' *The control volume does not rotate relative to inertial frame* $\mathcal{I}$. That is, we only study translational motion of the control volume.

Note that we relaxed the assumptions from the steady-stream example. We no longer require that the particles be evenly distributed and thus that the center of mass be fixed in the control volume. This allows for mass to be added or subtracted arbitrarily. The consequence is that we will be deriving equations of motion for a geometric point fixed to the control volume rather than for the center of mass of the material inside the control volume; the center of mass can potentially move in the control volume if mass is added or subtracted unevenly. We also now assume that the particles are not moving in the control volume prior to ejection (or that they are brought instantaneously to zero relative velocity once captured). Although this is a strong assumption, it works quite well and simplifies problem formulation. Finally, we have also removed the restriction that the control volume has constant mass. In this case, we allow the total mass of particles in the control volume to increase or decrease over time as particles are added or removed.

The procedure for finding the equation of motion of the control volume is almost exactly the same as for the steady stream, except here we consider the motion of a point C fixed to the control volume that is not necessarily the center of mass. We first consider only the case of mass being ejected from the control volume. Since all of the material inside the control volume (the vessel and the fluid particles) are traveling together at the same velocity by Assumption 6.6', the total linear momentum prior to ejection is

$$^{\mathcal{I}}\mathbf{p}_O(t_1) = m_v\,^{\mathcal{I}}\mathbf{v}_{C/O}(t_1) + \sum_{j=1}^{N} m_j\,^{\mathcal{I}}\mathbf{v}_{C/O}(t_1) + m_P\,^{\mathcal{I}}\mathbf{v}_{C/O}(t_1),$$

where m_P is the mass of the particle being ejected. The total linear momentum after ejection is

$$^{\mathcal{I}}\mathbf{p}_O(t_2) = m_v\,^{\mathcal{I}}\mathbf{v}_{C/O}(t_2) + \sum_{j=1}^{N} m_j\,^{\mathcal{I}}\mathbf{v}_{C/O}(t_2) + m_P(^{\mathcal{B}}\mathbf{v}_{P/C}(t_2) + {}^{\mathcal{I}}\mathbf{v}_{C/O}(t_2)).$$

Substituting the total linear momentum at t_1 and t_2 into the impulse form of Newton's second law for a collection of particles (Eq. (6.10)) gives

$$m_G{}^{\mathcal{I}}\mathbf{v}_{C/O}(t_2) + m_P{}^{\mathcal{B}}\mathbf{v}_{P/C}(t_2) + m_P({}^{\mathcal{I}}\mathbf{v}_{C/O}(t_2) - {}^{\mathcal{I}}\mathbf{v}_{C/O}(t_1))$$

$$= m_G{}^{\mathcal{I}}\mathbf{v}_{C/O}(t_1) + \overline{\mathbf{F}}_{\text{cv}}^{(\text{ext})}(t_1, t_2),$$

where, as before, $m_G = m_v + \sum_{j=1}^{N} m_j$. Once again, we divide by Δt and take the limit $\Delta t \to 0$. The second-order term again goes to zero, and we are left with the equation of motion:

$$\boxed{m_G{}^{\mathcal{I}}\mathbf{a}_{C/O} = \mathbf{F}_{\text{cv}}^{(\text{ext})} - \dot{m}_{\text{out}}{}^{\mathcal{B}}\mathbf{v}_{\text{out}/C},}$$ (6.68)

where, as before, we used the notation ${}^{\mathcal{B}}\mathbf{v}_{\text{out}/C} \triangleq {}^{\mathcal{B}}\mathbf{v}_{P/C}$ and we introduced $\dot{m}_{\text{out}} = \lim_{\Delta t \to 0} m_P/\Delta t$.

Recall in Chapter 2 when we emphasized that Newton's law does not contain an $\dot{m}_P$ term? That is still true. The $\dot{m}_{\text{out}}$ here refers to the rate that mass is ejected from the collection. It only arises in the study of systems of particles. The rate of change of mass term appeared because we were careful to keep track of the total linear momentum of all particles during the time interval—a requirement of the impulse form of Newton's second law and the law of conservation of total linear momentum.

What about systems collecting mass? Following the same procedure, except now considering a particle Q entering the control volume, leads to the same equation of motion for the control volume but with the opposite sign on the mass flow. (Deriving this result is a useful exercise.) In fact, in some cases there is mass both entering and leaving the system (such as in a jet engine) but with different velocities. In that case, we just include two thrust terms in Eq. (6.68). Using the convention that the scalar inflow and outflow mass rates are both positive ($\dot{m}_{\text{out}}, \dot{m}_{\text{in}} > 0$), we can rewrite Eq. (6.68) for both inflow and outflow as

$$\boxed{m_G{}^{\mathcal{I}}\mathbf{a}_{C/O} = \mathbf{F}_{\text{cv}}^{(\text{ext})} - \dot{m}_{\text{out}}{}^{\mathcal{B}}\mathbf{v}_{\text{out}/C} + \dot{m}_{\text{in}}{}^{\mathcal{B}}\mathbf{v}_{\text{in}/C},}$$ (6.69)

where $\dot{m}_{\text{in}}$ is the *mass inflow rate* and ${}^{\mathcal{B}}\mathbf{v}_{\text{in}/C}$ is the *inflow velocity*. Note that both flow velocities are relative to frame $\mathcal{B}$ fixed to point C of the control volume. Also note that the mass m_G of the collection varies with time according to $\dot{m}_G = -\dot{m}_{\text{out}} + \dot{m}_{\text{in}}$.

Example 6.14 Loading a Dump Truck

Consider a dump truck of unloaded mass $M = 2,000$ kg being filled from a chute that is at angle $\theta = 30°$ from the horizontal, as shown in Figure 6.23a. The material leaves the chute with a speed of $u = 2$ m/s and a mass flow rate of 75 kg/s. Suppose that the driver forgot to put on the parking brake and left the truck in neutral; the truck will thus accelerate as material continues to be added. We will calculate the speed of the truck as a function of time.

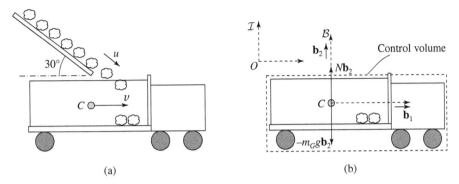

(a) (b)

Figure 6.23 (a) Material falling into a dump truck from a chute at angle θ. (b) An attached control volume and reference frames.

Figure 6.23b shows the frames of reference and control volume surrounding the truck. The truck is constrained to move only in the $\mathbf{b}_1$ direction, so its velocity is $^{\mathcal{I}}\mathbf{v}_{C/O} = v\mathbf{b}_1$. The inflow velocity of material is given by

$$^{\mathcal{B}}\mathbf{v}_{\text{in}/C} = {}^{\mathcal{I}}\mathbf{v}_{\text{in}/C} = {}^{\mathcal{I}}\mathbf{v}_{\text{in}/O} - {}^{\mathcal{I}}\mathbf{v}_{C/O} = (u\cos\theta - v)\mathbf{b}_1 - u\sin\theta\mathbf{b}_2.$$

Using $^{\mathcal{I}}\mathbf{a}_{C/O} = \dot{v}\mathbf{b}_1$, we can substitute into Eq. (6.69):

$$(M + \dot{m}_{\text{in}}t)\dot{v}\mathbf{b}_1 = (N - (M + \dot{m}_{\text{in}}t)g)\mathbf{b}_2 + \dot{m}_{\text{in}}((u\cos\theta - v)\mathbf{b}_1 - u\sin\theta\mathbf{b}_2),$$

where we have used the fact that $m_G = M + \dot{m}_{\text{in}}t$, the total mass in the control volume, and $\mathbf{F}_{\text{cv}}^{(\text{ext})} = (N - (M + \dot{m}_{\text{in}}t)g)\mathbf{b}_2$ (we are ignoring any frictional forces on the tires and assuming only a vertical normal force). Considering the $\mathbf{b}_1$ components gives the differential equation of motion for the truck:

$$\dot{v} = \frac{\dot{m}_{\text{in}}(u\cos\theta - v)}{M + \dot{m}_{\text{in}}t}.$$

This equation is integrated using $\dot{v} = dv/dt$ and rearranging terms:

$$\int_0^{v(t)} \frac{1}{u\cos\theta - v}\,dv = \int_0^t \frac{\dot{m}_{\text{in}}}{M + \dot{m}_{\text{in}}s}\,ds.$$

Completing both integrals yields

$$\ln\frac{u\cos\theta}{u\cos\theta - v(t)} = \ln\frac{M + \dot{m}_{\text{in}}t}{M}.$$

This equation can be solved for $v(t)$:

$$v(t) = u\cos\theta - \left(\frac{M}{M + \dot{m}_{\text{in}}t}\right)u\cos\theta = \frac{\dot{m}_{\text{in}}tu\cos\theta}{M + \dot{m}_{\text{in}}t}.$$

For instance, the speed of the truck after 2 s is

$$v(2) = 0.12 \text{ m/s.}$$

In the problems you have the opportunity to find the final speed of the truck when it passes the chute.

6.4 Tutorials

Tutorial 6.1 Two Falling Masses Connected by a Spring

This tutorial considers the simple model of two particles falling vertically under the influence of gravity, the lower one of mass M and the upper one of mass $m < M$. This model illustrates a situation when the motion of the center of mass cannot be treated independently of the motion relative to the center of mass. The particles are connected by a spring with spring constant k and unstretched length l, as shown in Figure 6.24. Since the particles are falling in the $\mathbf{e}_y$ direction only, each has one degree of freedom and the problem thus has two degrees of freedom, represented by the coordinates y_M (the height of the lower mass) and y_m (the height of the upper mass). We consider two cases: (a) the gravitational force acting on the two masses is a constant (Mg for the lower mass and mg for the upper one) and (b) the gravitational force is the more realistic $1/r^2$ force (see Example 4.9). These cases are shown in the two free-body diagrams, Figure 6.24b and c.

From Eq. (6.13) we find the center-of-mass position $\mathbf{r}_{G/O}$ of the two particles,

$$\mathbf{r}_{G/O} = \frac{1}{M+m} \left(My_M + my_m \right) \mathbf{e}_y \triangleq y_G \mathbf{e}_y.$$

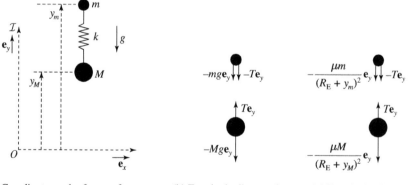

(a) Coordinates and reference frame   (b) Free-body diagram 1   (c) Free-body diagram 2

Figure 6.24 (a) Two masses connected by a spring falling vertically under two different gravity models. (b) Constant-gravity model. (c) Model with $1/r^2$ gravity.

Consider for a moment the case where the masses are rigidly connected by a rod rather than by a spring. Then the position of the center of mass is fixed relative to the two particles (a function of y_G and l only). For the constant-gravity case, the equation of motion for the center of mass is given by Eq. (6.16):

$$\ddot{y}_G = -g.$$

For the more realistic gravity case, we use Figure 6.24c and $y_m - y_M = l$ to find

$$\ddot{y}_G = \frac{-1}{M + m}\left[\frac{\mu M}{\left(R_E + y_G - \frac{m}{m+M}l\right)^2} + \frac{\mu m}{\left(R_E + y_G + \frac{M}{m+M}l\right)^2}\right], \quad (6.70)$$

where $\mu = Gm_E$, and m_E and R_E are the mass and radius of the earth, respectively.

In both cases, the equations of motion are in terms of the center-of-mass position only; it doesn't matter that the forces act on each particle away from the center of mass. As expected, the two rigidly connected particles behave as a single particle of mass $m + M$.

Nevertheless, in the more realistic gravity case, the force is more complicated than just that on a single particle. It is possible to find another point between the two masses where the force of gravity of a virtual particle of mass $M + m$ would be the same as in Eq. (6.70). If we call that point Q, its location can be found for this simple geometry by solving for y_Q from

$$\frac{\mu(M + m)}{(R_E + y_Q)^2} = \frac{\mu M}{\left(R_E + y_G - \frac{m}{m+M}l\right)^2} + \frac{\mu m}{\left(R_E + y_G + \frac{M}{m+M}l\right)^2}.$$

The point y_Q is where the force of gravity on an effective particle of the same total mass is equal to the net force of gravity on all the particles; we call it the *center of gravity*. We could formulate the translational equation of motion in terms of the center of gravity but, as shown in the next chapter, we would no longer have separation of translation and rotation. For a general collection of particles (or a rigid body), the center of gravity is not the same as the center of mass. It is a common error to refer to the center of gravity instead of the center of mass when discussing the translational motion of a collection of particles.

The situation is different when the masses are connected by a spring. For the case of a constant gravity field, we can use the free-body diagram in Figure 6.24b to write the equations of motion for each mass:

$$\ddot{y}_m + \frac{k}{m}y_m - \frac{k}{m}y_M = -g + \frac{kl}{m}$$

$$\ddot{y}_M + \frac{k}{M}y_M - \frac{k}{M}y_m = -g - \frac{kl}{M},$$

where the spring force in Figure 6.24b is given by $T = -k\left[l - (y_m - y_M)\right]$.

We can use the definition of the center of mass to take the weighted sum of these equations and find, as expected, that the internal spring force cancels out and the equation of motion for the center of mass is the same as for the rigid connection:

$$\ddot{y}_G = -g.$$

Again, the center of mass behaves like a single particle acted on by the total external gravity force. In this case, the center-of-mass motion separates completely from the oscillatory motion of the two particles about the center of mass (which, for the constant-gravity case, is zero if the masses start a distance l apart).

For the more realistic gravity force, the two equations of motion are

$$\ddot{y}_m + \frac{k}{m} y_m - \frac{k}{m} y_M + \frac{\mu}{(R_\mathrm{E} + y_m)^2} = \frac{kl}{m}$$

$$\ddot{y}_M + \frac{k}{M} y_M - \frac{k}{M} y_m + \frac{\mu}{(R_\mathrm{E} + y_M)^2} = -\frac{kl}{M}.$$

These equations of motion are slightly more complicated. Taking the weighted sum yields the center-of-mass equation of motion,

$$\ddot{y}_G + \frac{\mu}{m + M} \left(\frac{m}{(R_\mathrm{E} + y_m)^2} + \frac{M}{(R_\mathrm{E} + y_M)^2} \right) = 0. \tag{6.71}$$

The internal spring force still cancels out (as expected from Eq. (6.16)). However, for the more complicated gravity force, the center-of-mass equation of motion, while still a function of the total external force, *depends on the position of the individual particles*. Eq. (6.70) is the same as Eq. (6.71) only when $y_m - y_M = l$. There is no way to make the center-of-mass equation depend only on y_G. This is an example of when the external force introduces a dependence on the relative position of the particles. Eq. (6.16) does not imply that the center-of-mass equation always separates completely from the individual particle motions.

It is interesting to note that formulating the equation of motion for the center of mass looks like we added an additional degree of freedom (that of a "virtual" particle located at the center of mass), leaving us with three equations rather than two. However, at the outset of this example we noted that it was a two-degree-of-freedom problem. It still is; the three equations represent two degrees of freedom with the added constraint from the center-of-mass corollary in Eq. (6.15).

Tutorial 6.2 Bouncing Ball

This tutorial studies the dynamics of a falling ball that bounces off a rigid, fixed inclined plane. The flight of the ball has three phases: before, during, and after the bounce. We solve for the trajectory of the ball in each phase and find the bounce distance as a function of the inclined plane angle. We also numerically compute the plane inclination angle that maximizes the bounce distance.

Let particle P represent the ball. We use the Cartesian coordinates x_P and y_P to describe the position of P, as shown in Figure 6.25a. The inertial frame $\mathcal{I} = (O, \mathbf{e}_x, \mathbf{e}_y, \mathbf{e}_z)$ and the collision frame $\mathcal{C} = (O', \mathbf{e}_n, \mathbf{e}_t, \mathbf{e}_3)$ are shown in Figure 6.25b, where $O = O'$. The collision frame and the inertial frame are related by the transformation

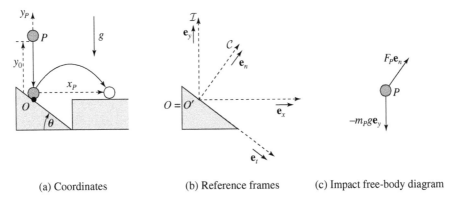

(a) Coordinates (b) Reference frames (c) Impact free-body diagram

Figure 6.25 Bouncing ball. During freefall, the only force acting on P is gravity.

$$\begin{array}{c|cc} & \mathbf{e}_x & \mathbf{e}_y \\ \hline \mathbf{e}_t & \cos\theta & -\sin\theta \\ \mathbf{e}_n & \sin\theta & \cos\theta \; , \end{array} \qquad (6.72)$$

where we assume $0 < \theta < \pi/4$.[9]

We write the velocity of P in frame $\mathcal{C}$ as ${}^{\mathcal{I}}\mathbf{v}_{P/O} = u_P\mathbf{e}_t + v_P\mathbf{e}_n$. Assume P starts falling (from rest) at $t = 0$. Let t_1 and t_2 denote the times immediately before and after the collision, where $\Delta t \triangleq t_2 - t_1$ is very short. Also assume that we have previously conducted an experiment to estimate the coefficient of restitution e for the impact of the ball and the plane (see Example 6.12).

As in Example 6.12, we use energy to find that the initial velocity before impact is

$$ {}^{\mathcal{I}}\mathbf{v}_{P/O}(t_1) = -\sqrt{2gy_0}\mathbf{e}_y. \qquad (6.73)$$

Eq. (6.73) expressed as components in the collision frame is

$$ {}^{\mathcal{I}}\mathbf{v}_{P/O}(t_1) = \sqrt{2gy_0}(\sin\theta\mathbf{e}_t - \cos\theta\mathbf{e}_n). \qquad (6.74)$$

From Example 6.12 we know that Eq. (6.60) is a valid approximation of the coefficient of restitution e for the collision between a particle and a rigid fixed surface (with the error being proportional to $\Delta t \ll 1$). We thus drop the $O(\Delta t)$ terms to find

$$ v_P(t_2) = -ev_P(t_1) = e\sqrt{2gy_0}\cos\theta. \qquad (6.75)$$

But what about the tangential speed $u_P(t_2)$ after impact? Since the plane is held fixed by external forces, we cannot use conservation of total linear momentum.

---

[9] The upper bound on θ is a consequence of the following observation: if the coefficient of restitution satisfies $e = 1$ and $\theta = \pi/4$, then the ball bounces in the $\mathbf{e}_x$ direction. Hence if $e < 1$ and $\theta = \pi/4$, the ball bounces in the $\mathbf{e}_x$ and $-\mathbf{e}_y$ direction (to the right and down). Since we would like to measure the horizontal bounce distance at the height of the impact, we need the upper bound $\theta < \pi/4$. The lower bound ensures the ball bounces to the right.

Nonetheless, we know that the impulse $\overline{\mathbf{F}}_P(t_1, t_2)$ acts only in the normal direction, which implies that the only tangential force on P during the collision is gravity. Using the impulse form of Newton's second law, we observe

$$^{\mathcal{I}}\mathbf{v}_{P/O}(t_2) = {}^{\mathcal{I}}\mathbf{v}_{P/O}(t_1) + \overline{\mathbf{F}}_P(t_1, t_2) + \int_{t_1}^{t_2} (-m_P g \mathbf{e}_y)$$

$$= {}^{\mathcal{I}}\mathbf{v}_{P/O}(t_1) + \left(\overline{F}_P(t_1, t_2) - m_P g \cos\theta \Delta t\right) \mathbf{e}_n + m_P g \sin\theta \Delta t \mathbf{e}_t,$$

which implies

$$u_P(t_2) = u_P(t_1) + \underbrace{m_P g \sin\theta \Delta t}_{=O(\Delta t)}. \tag{6.76}$$

At impact the tangential speed of P increases by $O(\Delta t)$. In the limit $\Delta t \to 0$, the tangential speed of P is conserved. Thus we take $u_P(t_2) = u_P(t_1)$ with a small error on the same order as in the normal direction. Using Eqs. (6.74)–(6.76), we obtain the velocity of P after the impact:

$$^{\mathcal{I}}\mathbf{v}_{P/O}(t_2) = \left(\sqrt{2gy_0} \sin\theta\right) \mathbf{e}_t + \left(e\sqrt{2gy_0} \cos\theta\right) \mathbf{e}_n$$

$$= (1+e)\sqrt{2gy_0} \sin\theta \cos\theta \mathbf{e}_x + \tag{6.77}$$

$$\sqrt{2gy_0}(e\cos^2\theta - \sin^2\theta)\mathbf{e}_y.$$

We now compute the trajectory of P during the final phase of flight by integrating the equation of motion

$$-m_P g \mathbf{e}_y = m_P {}^{\mathcal{I}}\mathbf{a}_{P/O}$$

from t_2 to $t > t_2$ using the initial conditions $^{\mathcal{I}}\mathbf{v}_{P/O}(t_2)$ given by Eq. (6.77):

$$x_P(t) = \dot{x}_P(t_2)(t - t_2) \tag{6.78}$$

$$y_P(t) = -\frac{1}{2}g(t - t_2)^2 + \dot{y}_P(t_2)(t - t_2). \tag{6.79}$$

At $t = t_f$, which is the time when P lands after the bounce, $y_P(t_f) = 0$ and Eqs. (6.78) and (6.79) yield

$$x_P(t_f) = \frac{2\dot{x}_P(t_2)\dot{y}_P(t_2)}{g}$$

$$= 2(1+e)y_0 \sin 2\theta (e\cos^2\theta - \sin^2\theta).$$

For $y_0 = 1$ m, Figure 6.26 shows $x_P(t_f)$ as a function of the plane incline angle θ and several values of e. Note that the bounce distance decreases to zero as the coefficient of restitution decreases to zero. We also plot the optimal angle that maximizes x_P as a function of e. For example, if $e = 1$, then the optimal angle is $\theta \approx 22.5°$, which corresponds to an initial flight angle of $45°$ relative to $\mathbf{e}_x$. Notice that the optimal angle also decreases to zero as the coefficient of restitution decreases to zero.

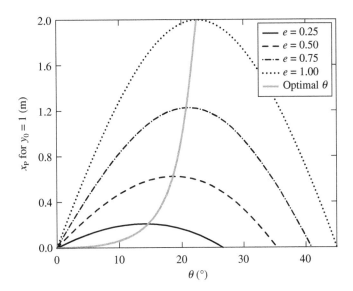

Figure 6.26 Bouncing ball. The optimal incline angle maximizes the bounce distance x_P for a given value of the coefficient of restitution e.

Tutorial 6.3 The Rocket Problem

A common example of a variable-mass system is a rocket. A rocket expels mass to provide thrust, which accelerates the rocket. Figure 6.27 shows the appropriate frames and a free-body diagram of a rocket. The inertial frame $\mathcal{I} = (O, \mathbf{e}_1, \mathbf{e}_2, \mathbf{e}_3)$ is fixed to the ground with $\mathbf{e}_1$ pointing upward along the path of the rocket. The frame $\mathcal{B} = (G, \mathbf{e}_x, \mathbf{e}_y, \mathbf{e}_z)$ is fixed to the rocket at its center of mass G and aligned with the inertial frame. The inertial velocity of the rocket is

$$^{\mathcal{I}}\mathbf{v}_{G/O} = v_G \mathbf{e}_x.$$

The exhaust velocity of the gas is $^{\mathcal{B}}\mathbf{v}_{\text{out}/G} = -u_e \mathbf{e}_x$. Using Figure 6.27 and Eq. (6.69) in the $\mathbf{e}_x$ direction, with C replaced by G, we find

$$m_G \dot{v}_G = A_e \Delta P - m_G g + \dot{m}_{\text{out}} u_e, \tag{6.80}$$

where $A_e \Delta P$ is the external force on the rocket due to the difference between atmospheric pressure and the pressure of the exhaust gas right at the nozzle opening (A_e is the exit area of the nozzle).

Eq. (6.80) is known as the *rocket equation*. It applies to all propulsion systems, from the small jets used to control a satellite to the enormous chemical engines used to loft the Saturn V to the moon.[10] The primary design parameters available to the engineer are the mass (out)flow rate $\dot{m}_{\text{out}}$ and the exhaust velocity u_e. For large chemical rockets, the mass flow rate is usually constant and determined by the pressure in

---

[10] A chemical rocket is an engine that employs combustion among volatile reactants and oxygen to heat up the resulting gas and accelerate it out through the nozzle.

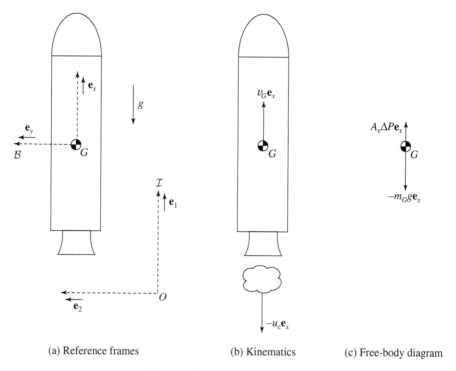

(a) Reference frames   (b) Kinematics   (c) Free-body diagram

Figure 6.27 The rocket problem.

the fuel tank, so the exhaust velocity is the main driver. This can be seen by solving Eq. (6.80) using $\dot{m}_{out} = -\dot{m}_G$ (the mass loss of the rocket due to the outflow of burning fuel). For large engines with substantial thrust, the pressure term is usually very small and can be ignored. Thus, to find the final velocity of the rocket, we need to solve

$$\frac{dv_G}{dt} = -u_e \frac{1}{m_G} \frac{dm_G}{dt} - g, \tag{6.81}$$

which is a separable differential equation (see Appendix C). Integrating Eq. (6.81) from t_0 to t_f yields

$$\Delta v = u_e \ln \frac{m_0}{m_f} - g(t_f - t_0),$$

where $\Delta v \triangleq v_G(t_f) - v_G(t_0)$ is the change in rocket velocity, $m_0 \triangleq m_G(t_0)$ is the initial mass of the rocket, and $m_f \triangleq m_G(t_f)$ is the final mass of the rocket.

For a constant mass flow rate, the solution can be written in terms of the change in mass of the rocket only by substituting $m_0 - m_f = \dot{m}_{out}(t_f - t_0)$:

$$\Delta v = u_e \ln \frac{m_0}{m_f} - \frac{g}{\dot{m}_{out}}(m_0 - m_f). \tag{6.82}$$

Note that this solution is an approximation for rockets launched from the surface of the earth, as we have assumed that the acceleration g due to gravity is a constant. The second term in Eq. (6.82), however, is normally small compared to the first (and essentially zero for deep-space propulsion), so we can ignore it and still get reasonable estimates for rocket performance. Ignoring this term and inverting Eq. (6.82) yields the rocket equation of Tsiolkovsky,[11]

$$\frac{m_f}{m_0} = e^{-\Delta v/u_e}.$$ (6.83)

Eq. (6.83) is an extremely useful relationship between the fraction of mass brought to altitude as a function of the needed velocity change and the exhaust velocity of the rocket. Recall from Tutorial 4.3 that we typically treat orbital maneuvers by means of impulsive changes in velocity. Eq. (6.83) indicates how much fuel is needed for a given Δv, assuming a specific type of rocket characterized by u_e.

Now we can plug in some numbers for launch from the surface of the earth to orbit. The best large chemical rockets have an exhaust velocity in the range of 2–3 km/sec. Getting from the surface of the earth to low-earth orbit requires a velocity increment on the order of 6–7 km/sec. Thus a single rocket engine going all the way to orbit would have a final payload mass fraction of (best case) $e^{-2} \approx 0.15$. That means that at least 85% of the rocket mass is fuel! Add in all structural mass and fuel tanks, and very little is left for payload. It is extremely difficult to launch satellites into space on a single rocket, usually referred to as a *single-stage-to-orbit vehicle*. That is why most rockets today have multiple stages.

6.5 Key Ideas

- The **total linear momentum** of a collection of particles is the sum of the individual inertial momenta:

$$^{\mathcal{I}}\mathbf{p}_O \triangleq \sum_{j=1}^{N} m_j \,{}^{\mathcal{I}}\mathbf{v}_{j/O}.$$

- The total linear momentum of a collection of particles satisfies Newton's second law, which in integrated form is

$$^{\mathcal{I}}\mathbf{p}_O(t_2) = {}^{\mathcal{I}}\mathbf{p}_O(t_1) + \overline{\mathbf{F}}^{(\text{ext})}(t_1, t_2),$$

where $\overline{\mathbf{F}}^{(\text{ext})}(t_1, t_2)$ is the total external impulse from t_1 to t_2.

- The **law of conservation of total linear momentum** states that, if the total external force is zero, the total inertial linear momentum of a system of particles is a constant of the motion:

$$\sum_{j=1}^{N} m_j \,{}^{\mathcal{I}}\mathbf{v}_{j/O}(t_2) = \sum_{j=1}^{N} m_j \,{}^{\mathcal{I}}\mathbf{v}_{j/O}(t_1).$$

[11] Konstantin Tsiolkovsky (1857–1935) was a pioneering Russian rocket scientist. He was among the first to develop the theory of rocketry, publishing the famous "rocket equation" in 1903. Often considered the father of modern rocketry, he was visionary and prescient in both his fiction and nonfiction writings regarding satellites and manned space travel.

- The **center of mass** G of a collection of particles is the vector sum of the particle's mass-weighted positions, normalized by the total mass $m_G = \sum_{j=1}^{N} m_j$:

$$\mathbf{r}_{G/O} = \frac{1}{m_G} \sum_{i=1}^{N} m_i \mathbf{r}_{i/O}.$$

- The **center-of-mass corollary** states that the vector sum of the mass-weighted positions of each particle relative to the center of mass is zero:

$$\sum_{i=1}^{N} m_i \mathbf{r}_{i/G} = 0.$$

- The center of mass obeys an equation of motion corresponding to an equivalent particle of mass m_G located at the center of mass,

$$\frac{^{\mathcal{I}}d}{dt} \left(^{\mathcal{I}}\mathbf{p}_{G/O} \right) = \mathbf{F}_G^{(\text{ext})}.$$

In the absence of external forces, the center of mass is either stationary or moving at a constant inertial velocity.

- The motion of a particle i in a multiparticle system relative to a frame $\mathcal{B}$ fixed to the center of mass is

$$m_i {}^{\mathcal{B}}\mathbf{a}_{i/G} = \mathbf{F}_i^{(\text{ext})} + \sum_{j=1}^{N} \mathbf{F}_{i,j} - m_i {}^{\mathcal{I}}\mathbf{a}_{G/O},$$

where $\mathbf{F}_{i,j}$ is the internal force on particle i due to particle j.

- In a collision between two particles, P and Q, we find the velocities after the collision ($t \geq t_2$) using the velocities before the collision ($t \leq t_1$) expressed in a **collision frame** $C = (O', \mathbf{e}_n, \mathbf{e}_t, \mathbf{e}_3)$. The origin of the collision frame is the point of impact and the normal direction is

$$\mathbf{e}_n = {}^{\mathcal{I}}\hat{\mathbf{v}}_{P/Q}(t_1).$$

In general the tangential components of the velocities are constant and the normal components obey the law of conservation of total linear momentum,

$$m_P v_P(t_2) + m_Q v_Q(t_2) = m_P v_P(t_1) + m_Q v_Q(t_1).$$

The normal components also obey the coefficient-of-restitution equation,

$$e = \frac{v_P(t_2) - v_Q(t_2)}{v_Q(t_1) - v_P(t_1)}.$$

- In a collision between a particle and a fixed surface, the tangential components are constant, but the total linear momentum is no longer conserved for the normal components. The coefficient-of-restitution equation is

$$e = -\frac{v_P(t_2)}{v_P(t_1)}.$$

- For a **variable-mass system** with mass flowing in a steady stream at a rate $\dot{m}$, the equation of motion for the center of mass is

$$m_G{}^{\mathcal{I}}\mathbf{a}_{G/O} = \overline{\mathbf{F}}_{\text{cv}}^{(\text{ext})} + \dot{m}\left({}^{\mathcal{B}}\mathbf{v}_{\text{in}/G} - {}^{\mathcal{B}}\mathbf{v}_{\text{out}/G}\right),$$

where ${}^{\mathcal{B}}\mathbf{v}_{\text{in}/G}$ and ${}^{\mathcal{B}}\mathbf{v}_{\text{out}/G}$ are the inflow and outflow velocities.

- For a **variable-mass system** with mass flowing in at a rate $\dot{m}_{\text{in}}$ and out at a rate $\dot{m}_{\text{out}}$, the equation of motion for the center of mass is

$$m_G(t){}^{\mathcal{I}}\mathbf{a}_{G/O} = \mathbf{F}_G - \dot{m}_{\text{out}}{}^{\mathcal{B}}\mathbf{v}_{\text{out}/G} + \dot{m}_{\text{in}}{}^{\mathcal{B}}\mathbf{v}_{\text{in}/G},$$

where ${}^{\mathcal{B}}\mathbf{v}_{\text{out}/G}$ is the relative velocity of the outflowing mass and ${}^{\mathcal{B}}\mathbf{v}_{\text{in}/G}$ is the relative velocity of the inflowing mass.

6.6 Notes and Further Reading

Although Newton formulated the basic laws of mechanics, it was Euler who showed they could be applied to collections of particles independently and first developed the techniques of multiparticle systems we presented here. (He also defined the center of mass and formulated the N-body problem.) In modern texts it forms the basis of the development of rigid-body motion, which we discuss in Chapter 9. Almost all introductory texts have some discussion of multiparticle systems and impacts.

Most introductory texts (e.g., Greenwood 1988; Pytel and Kiusalaas 1999; Meriam and Kraige 2001; Bedford and Fowler 2002; Hibbeler 2003; Tongue and Sheppard 2005; Rao 2006; Beer et al. 2007) also introduce simple models of mass flow or variable mass systems, though the level of detail varies widely. In our treatment we have tried to be explicit about the assumptions being made and the problems to which they apply. A more careful treatment requires the development of fluid mechanics, of which this is only the briefest introduction. An excellent starting text is Smits (1999). Versions of Example 6.13 appear in almost every text; ours is most similar to that in Bedford and Fowler (2002). Likewise, Example 6.14 is similar to problems in Pytel and Kiusalaas (1999), Bedford and Fowler (2002), Tongue and Sheppard (2005), and Beer et al. (2007).

6.7 Problems

6.1   Consider four masses whose Cartesian coordinates $(x, y)_{\mathcal{I}}$ are given in Figure 6.28. Find the position of the center of mass.

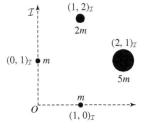

Figure 6.28 Problem 6.1.

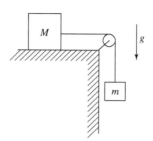

Figure 6.29 Problem 6.2.

6.2 Two masses are connected by a string, as shown in Figure 6.29. The larger mass M slides on a surface with coefficient of friction μ. Find the equation of motion for each mass.

6.3 Consider a double mass-spring system with two masses of M and m on a frictionless surface, as shown in Figure 6.30. Mass m is connected to M by a spring of constant k and rest length l_0. Mass M is connected to a fixed wall by a spring of constant k and rest length l_0 and a damper with constant b. Find the equations of motion of each mass. [HINT: See Tutorial 2.1.]

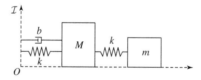

Figure 6.30 Problem 6.3.

6.4 Use MATLAB to find and plot the solution to the equations of motion in Problem 6.3, assuming $M = 10$ kg, $m = 2$ kg, $k = 0.5$ N/m, $b = 0.2$ N-s/m, $l_0 = 0.5$ m, and each mass starts from rest. Set the initial distance between mass M and the wall to be $l_0/2$ and the initial distance between mass m and M to be l_0. Be sure to label your axes with units and add a legend.

6.5 Masses P and Q collide and stick together to form mass PQ, as shown in Figure 6.31. Assuming mass P weighs twice as much as mass Q, find the

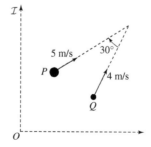

Figure 6.31 Problem 6.5.

speed and orientation of the velocity of PQ, measured relative to the initial heading of mass Q.

6.6 Three seconds remain in the fourth quarter of the Super Bowl, and the offense is down by four points (Figure 6.32). On fourth and goal, the wide-receiver lines up on the 15 yard line. Sensing a defensive blitz, the quarterback calls an audible that directs the receiver to sprint directly down the sideline toward the endzone. If the receiver runs at 8 yd/s and the quarterback immediately throws the ball at 12 yd/s, at what angle θ should the quarterback aim to connect with his receiver? If the ball is caught, will he catch the ball in the endzone?

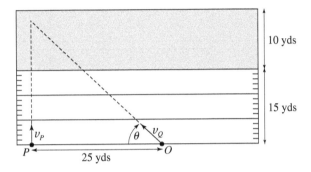

Figure 6.32 Problem 6.6.

6.7 Show that the linear momentum of the center of mass $^{\mathcal{I}}\mathbf{p}_{G/O}$ of a collection of N particles is equal to the total linear momentum of the collection, $^{\mathcal{I}}\mathbf{p}_O = \sum_{i=1}^{N} m_i {}^{\mathcal{I}}\mathbf{v}_{i/O}$. Please state any assumptions that you make.

6.8 Consider a model of two cars colliding, as in Figure 6.33. Car B was traveling south with speed v_B and car A was traveling northeast at angle θ_1 and at speed v_A when they collided. After the crash, the two cars stuck together and skidded off at angle θ_2. Each driver claimed to be traveling at the speed limit of 30 mph (48 km/hr) and each claimed to have slowed down but collided anyway because the other car was speeding. Assume the mass of car A is m_A and the mass of car B is m_B.

a. Assume that $m_B/m_A = 0.9$. If $\theta_1 = 40°$ and $\theta_2 = 15°$, which car was going faster?

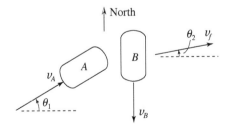

Figure 6.33 Problem 6.8.

b. What was the speed of the faster of the two cars if the slower car was indeed going at the speed limit?

6.9   Masses P and Q are attached by a massless rigid rod to form a dumbbell, as shown in Figure 6.34. The dumbbell is tossed in the presence of gravity so that the masses may spin about the center of mass G. Show that the trajectory of the center of mass is a parabola.

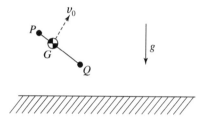

Figure 6.34 Problem 6.9.

6.10 An Olympic biathlete must fire a rifle at a target while standing on cross-country skis, as shown in Figure 6.35. Suppose a bullet of mass m is fired with speed v_0, and the biathlete and rifle have a combined mass of M.

a. If the ski/snow interface is frictionless, how fast does the biathlete slide backward after firing at the target?

b. Now suppose that there is friction between the skis and the snow. If the bullet leaves the rifle barrel in a time interval of Δt after the trigger is pulled, what is the minimum coefficient of friction that will prevent the biathlete from sliding?

Figure 6.35 Problem 6.10.

6.11 A planar, *quarter-car model* of an automobile suspension system is shown in Figure 6.36. The model consists of two masses: mass M represents one-quarter of the car and mass m represents the tire. The masses M and m are connected by a spring with constant k_1 and rest length l_1 and a damper with constant b. Mass m is connected to the ground by a spring with constant k_2 and rest length l_2. Find the equations of motion for x_1 and x_2. [HINT: The damper force on M is $-b\dot{x}_1\mathbf{e}_2$.]

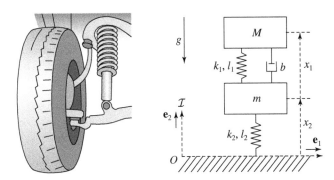

Figure 6.36 Problem 6.11.

6.12 An inverted simple pendulum of mass m and length l is attached to a cart of mass M that slides without friction on a horizontal track subject to an unspecified force F (Figure 6.37). Find the equations of motion of x and θ.

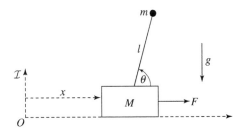

Figure 6.37 Problem 6.12.

6.13 Figure 6.38 shows two balls falling, a larger bottom ball P and a smaller top ball Q, just resting on top of the bigger ball (e.g., a tennis ball and a basketball). The lower ball has mass m_P and the upper one has mass m_Q. Assume that e_1 is the coefficient of restitution between the larger ball and the ground and e_2 is the coefficient of restitution between the smaller and larger balls. Suppose that they fall from height h, such that their joint speed just before the lower ball hits the ground is v_0.

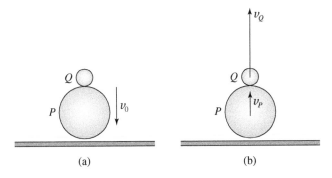

Figure 6.38 Problem 6.13.

a. Find an expression for the upward speed of ball Q after the collision with the ground in terms of e_1, e_2, m_P, m_Q, and v_0.

b. Show that for $m_P \gg m_Q$ and $e_1 \approx e_2 \approx 1$ (a perfectly elastic collision), the top ball triples its speed, $v_Q \approx 3v_0$. (This is a fun experiment to do, but do it outside!)

6.14 Consider a model for two stationary rail cars, B and C, of masses m_B and m_C, connected together by a spring coupling of spring constant k. A third car A, of mass m_A, is traveling along the frictionless track at a constant speed v_A until it collides with the two coupled cars. Model the three rail cars as point masses, arranged as in Figure 6.39. Assume that the coupling is broken on car A and that the collision of car A with car B is perfectly elastic ($e = 1$). What is the speed of the center of mass of cars B and C after the collision?

Figure 6.39 Problem 6.14.

6.15 Figure 6.40 shows two masses on springs arranged to collide with each other when the springs are at their unstretched lengths l_0. At time $t = 0$, both springs are compressed an identical amount x_0. Assuming each mass has identical mass m, each spring has the same spring constant k, and the coefficient of restitution between the masses is e, find an expression for the maximum spring compression after the nth collision. Using MATLAB, plot maximum spring compression versus collision number for the first 10 collisions. Use the following parameters: $x_0 = 5$ m, $l_0 = 10$ m, $e = 0.9, 0.75, 0.5$, and 0.2. Plot the results on a single graph for all values of e (so there should be four plots on the graph).

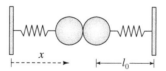

Figure 6.40 Problem 6.15.

6.16 Two small balls, each of mass $m = 0.2$ kg, hang from strings of length $L = 1.5$ m. The left ball is released from rest with $\theta = -45°$. As a result of the initial collision the right ball swings through a maximum angle of $30°$. Determine the coefficient of restitution between the two balls.

6.17 A jet of water of cross-section A and density ρ moves horizontally at inertial speed v_0 and hits a block of mass M, which is connected to a wall with a spring of constant k, as in Figure 6.41. Assume that the collision is inelastic (i.e., the water leaves with a zero horizontal component of velocity relative to the block) and that the spring is initially at its rest length l_0. Find the final length of the spring.

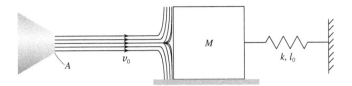

Figure 6.41 Problem 6.17.

6.18 A jet of water of cross-section A and density ρ moves horizontally at an inertial velocity v_0 and hits a block of mass M, as in Figure 6.42. Assume that the collision is inelastic (i.e., the water leaves with a zero horizontal component of velocity relative to the block) and that there is a coefficient of friction μ between the block and the surface on which it slides. Find the final speed v_f of the block.

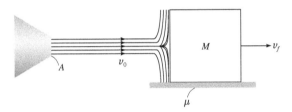

Figure 6.42 Problem 6.18.

6.19 A nozzle with exit area A emits a column of air of density ρ, which lifts a ball of mass M, as shown in Figure 6.43. If the inertial speed of the air column is v_0, find the height h to which the ball is raised.

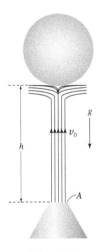

Figure 6.43 Problem 6.19.

6.20 A turbojet (shown in Figure 6.44) uses a steady stream of air to accelerate an airplane. It operates by taking in air via a compressor and accelerating it by means of a combustion and compression process. The air is assumed stationary in the inertial frame and the mass flow rate $\dot{m}$ is largely independent

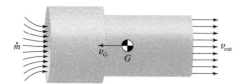

Figure 6.44 Problem 6.20.

of the speed of the plane. Assuming that the jet accelerates the air to a speed v_{rel} relative to the jet's center of mass G, draw a control volume and find an expression for the thrust produced by the engine.

6.21 Suppose a jet engine takes in air at a mass flow rate of 100 kg/s. Also assume that the exhaust velocity of the air is 500 m/s relative to the plane. Determine the magnitude of the thrust on the plane for two different cruising speeds: (a) 460 km/h and (b) 980 km/h.

6.22 Suppose the container on the dump truck in Example 6.14 has a length of 5 m. What is the final speed of the dump truck when it clears the chute?

6.23 A "kinetic kill" missile destroys its target through conversion of kinetic energy to mechanical work. That is, there is no explosion—it destroys the target only by means of the collision. Consider a missile with a rocket engine flying a straight-line path to a target. Suppose the engine has a mass flow rate of 0.5 kg/sec and an exhaust velocity of 2.75 km/s relative to the rocket. Assuming that the missile body has a mass of 1,000 kg, find the total mass of fuel necessary to achieve a final kinetic energy of 5×10^6 J.

6.24 A thrust test stand is used to directly measure the thrust of a rocket. One simple arrangement is to attach the rocket engine to a spring/damper system and measure the spring compression over time, as shown in Figure 6.45. Consider a rocket engine of dry mass of 30 kg loaded with 10 kg of fuel. It has a mass flow rate of 0.5 kg/s and an exhaust velocity $u_e = 20$ m/s. It is attached to a spring of spring constant $k = 50$ N/m and rest length $l_0 = 1$ m, and to a damper with constant $b = 25$ N·s/m. Find the equations of motion and, using MATLAB, plot the rocket's position $x(t)$ as a function of time. Integrate for the amount of time that there is any fuel remaining in the rocket, starting with the spring at its rest length. How might you use this measurement to find the thrust $T = \dot{m}_{out} u_e$?

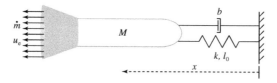

Figure 6.45 Problem 6.24.

6.25 How high will your model rocket fly? Assume the model rocket weighs 2 kg and the fuel weighs 3 kg. Assume a mass flow rate of 0.1 kg/s and an exhaust velocity of 500 m/s. What is the final altitude of the rocket when all the fuel is consumed? How good was the constant-gravity assumption?

CHAPTER SEVEN

Angular Momentum and Energy of a Multiparticle System

This chapter continues our study of multiparticle systems by examining the total angular momentum and energy of the collection. Recall the following important result from Chapter 6: we can treat a collection of particles as a single equivalent particle located at the center of mass, whose translational motion obeys Newton's second law. If the external forces on the collection are independent of the position of the particles relative to the center of mass, then the motion of the center of mass is independent of the relative motion of the particles. This result is used in combination with equations describing the motion of the particles relative to the center of mass, in terms of either linear or angular momentum, to find the entire trajectory.

In this chapter we take a slightly different approach to the motion of the particles about the center of mass. By studying the total angular momentum and total energy of the collection, we obtain results that not only make some problems easier to analyze, but also open the door to the study of rigid bodies in Parts Three and Four. The most important discovery in this chapter is that the angular momentum of the collection dynamically separates; that is, we can treat the angular momentum of the center of mass of the system (relative to the origin of the inertial frame) independently from the motion of the particles about the center of mass. This result is profound and is critical to our later treatment of rigid bodies.

7.1 Angular Momentum of a System of Particles

Consider again a system of N particles, as shown in Figure 7.1, in which each particle $i = 1, \ldots, N$, is subject to external force $\mathbf{F}_i^{(\text{ext})}$ and internal force $\mathbf{F}_i^{(\text{int})} \triangleq \sum_{j=1}^{N} \mathbf{F}_{i,j}$, where $\mathbf{F}_{ii} = 0$. Recall that the internal forces $\mathbf{F}_{i,j}$ are equal and opposite, that is,

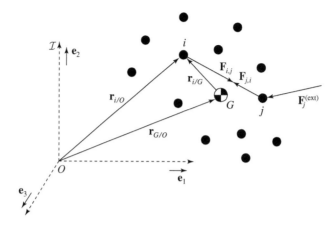

Figure 7.1 Collection of particles with center of mass G.

$\mathbf{F}_{i,j} = -\mathbf{F}_{j,i}$, by Newton's third law. The angular momentum of particle i relative to the origin O is then, by Definition 4.2,

$$^{\mathcal{I}}\mathbf{h}_{i/O} = m_i \mathbf{r}_{i/O} \times {}^{\mathcal{I}}\mathbf{v}_{i/O}.$$

As done in Chapter 6 for linear momentum, we can write Newton's second law in angular momentum form separately for each particle,

$$\frac{{}^{\mathcal{I}}d}{dt}\left({}^{\mathcal{I}}\mathbf{h}_{i/O}\right) = \mathbf{M}_{i/O}^{(\text{ext})} + \mathbf{M}_{i/O}^{(\text{int})},$$

where $\mathbf{M}_{i/O}^{(\text{ext})} = \mathbf{r}_{i/O} \times \mathbf{F}_i^{(\text{ext})}$ is the moment on particle i due to the total external force on i, and $\mathbf{M}_{i/O}^{(\text{int})} = \mathbf{r}_{i/O} \times \mathbf{F}_i^{(\text{int})}$ is the moment on particle i due to the total internal force on i from the other $N-1$ particles in the collection. We could thus compute the trajectories of the N particles by solving these N vector equations of motion instead of using Eq. (6.1). Although there are certainly problems that can be solved this way, this approach does not really provide any new tools for modeling the behavior of a multiparticle system. Instead, we introduce the *total angular momentum*, just as we did for the total linear momentum in Section 6.1.2.

7.1.1 The Total Angular Momentum

The definition of total angular momentum is similar to that of total linear momentum.

Definition 7.1 The **total angular momentum** of a system of particles about O is the sum of the individual angular momenta:

$$^{\mathcal{I}}\mathbf{h}_O \triangleq \sum_{i=1}^{N} {}^{\mathcal{I}}\mathbf{h}_{i/O} = \sum_{i=1}^{N} m_i \mathbf{r}_{i/O} \times {}^{\mathcal{I}}\mathbf{v}_{i/O}.$$

The utility of the total angular momentum becomes apparent when we examine its inertial derivative. Taking the derivative of $^{\mathcal{I}}\mathbf{h}_O$ and using the product rule gives

$$\frac{^{\mathcal{I}}d}{dt}\left(^{\mathcal{I}}\mathbf{h}_O\right) = \sum_{i=1}^{N} m_i \underbrace{^{\mathcal{I}}\mathbf{v}_{i/O} \times {}^{\mathcal{I}}\mathbf{v}_{i/O}}_{=0} + \sum_{i=1}^{N} m_i \mathbf{r}_{i/O} \times \frac{^{\mathcal{I}}d}{dt}\left(^{\mathcal{I}}\mathbf{v}_{i/O}\right). \qquad (7.1)$$

As with linear momentum, the second term can be rewritten using Newton's second law, since

$$m_i \frac{^{\mathcal{I}}d}{dt}\left(^{\mathcal{I}}\mathbf{v}_{i/O}\right) = \mathbf{F}_i^{(\text{ext})} + \sum_{j=1}^{N} \mathbf{F}_{i,j}.$$

This results in the following expression for the derivative of the total angular momentum,

$$\frac{^{\mathcal{I}}d}{dt}\left(^{\mathcal{I}}\mathbf{h}_O\right) = \sum_{i=1}^{N} \mathbf{r}_{i/O} \times (\mathbf{F}_i^{(\text{ext})} + \sum_{j=1}^{N} \mathbf{F}_{i,j}). \qquad (7.2)$$

We now examine the second term in the sum. Since the order in which the forces are summed is arbitrary, it must be true that $\sum_i \sum_j \mathbf{r}_{i/O} \times \mathbf{F}_{i,j} = \sum_j \sum_i \mathbf{r}_{j/O} \times \mathbf{F}_{j,i}$ (we just swapped the labels i and j). Thus

$$\sum_{i=1}^{N}\sum_{j=1}^{N} \mathbf{r}_{i/O} \times \mathbf{F}_{i,j} = \frac{1}{2}\sum_{i=1}^{N}\sum_{j=1}^{N}(\mathbf{r}_{i/O} \times \mathbf{F}_{i,j} + \mathbf{r}_{j/O} \times \mathbf{F}_{j,i}).$$

Again invoking Newton's third law, as we did with linear momentum, and making the substitution $\mathbf{F}_{i,j} = -\mathbf{F}_{j,i}$ yields

$$\sum_{i=1}^{N}\sum_{j=1}^{N} \mathbf{r}_{i/O} \times \mathbf{F}_{i,j} = \frac{1}{2}\sum_{i=1}^{N}\sum_{j=1}^{N} \underbrace{(\mathbf{r}_{i/O} - \mathbf{r}_{j/O})}_{=\,\mathbf{r}_{i/j}} \times \mathbf{F}_{i,j}. \qquad (7.3)$$

In Chapter 6 we were able to eliminate the double sum over internal forces by invoking Newton's third law. Here the situation is more complicated. To get an expression that depends only on external forces, we need the double sum over internal moments in Eq. (7.3) to be zero. We call this the *internal-moment assumption*.

Assumption 7.1 *Under the* **internal-moment assumption**, *the sum of all internal moments among the constituent particles of a multiparticle system is zero:*

$$\frac{1}{2}\sum_{i=1}^{N}\sum_{j=1}^{N} \mathbf{r}_{i/j} \times \mathbf{F}_{i,j} = 0. \qquad (7.4)$$

It is important to remember that not all multiparticle systems satisfy the internal-moment assumption; it is not implied by Newton's third law. When it is not satisfied, there is not much more that can be done to model the system other than applying Newton's second law separately to each particle; examining the total angular momentum of the collection in that situation may not be very helpful. Fortunately, it is difficult to find a system of particles that does not satisfy this assumption. For almost all the problems and examples in the book (except Problem 7.13, where you can study a simple counterexample), we assume that the internal-moment assumption is satisfied. In fact, many systems satisfy the more restrictive condition that the internal force between every pair of particles in the system is along the line connecting them, so that each term in the sum in Eq. (7.4) is zero (since the cross product of two parallel vectors is zero). We call this the *internal-force assumption*. As always in engineering, it is good practice to examine your assumptions when attacking a new problem.

When the internal-moment assumption is satisfied, the derivative of the total angular momentum in Eq. (7.2) simplifies to

$$\frac{^{\mathcal{I}}d}{dt}\left(^{\mathcal{I}}\mathbf{h}_O\right) = \sum_{i=1}^{N} \mathbf{r}_{i/O} \times \mathbf{F}_i^{(\text{ext})} \triangleq \mathbf{M}_O^{(\text{ext})}, \tag{7.5}$$

where we have used Definition 4.3 to define $\mathbf{M}_O^{(\text{ext})}$ as the *total external moment* about O. Eq. (7.5) can be very useful for certain multiparticle problems, particularly problems in which part of the collection is rotating very fast. It will arise again in the study of two- and three-dimensional motion of rigid bodies.

Chapter 6 showed that we can treat a collection of particles as a single equivalent particle of mass m_G located at the center of mass. This virtual particle obeys Newton's second law, where the total force is the vector sum of all external forces acting on the collection. The result in Eq. (7.5) is similar, in that it treats the collection as a single entity and shows that the behavior of its total angular momentum is determined solely by the total external moment.

BE CAREFUL!

For the translational dynamics of a multiparticle system, the total linear momentum $^{\mathcal{I}}\mathbf{p}_O$ is equal to the linear momentum $^{\mathcal{I}}\mathbf{p}_{G/O}$ of the center of mass relative to O. However, for the rotational dynamics of such a system, the total angular momentum $^{\mathcal{I}}\mathbf{h}_O$ in Eq. (7.5) is not the same as the angular momentum $^{\mathcal{I}}\mathbf{h}_{G/O}$ of the center of mass relative to O. Nor is $\mathbf{M}_O^{(\text{ext})}$ computed using the total force on the center of mass. That is why Eq. (7.5) is of less help than it might initially seem, since computing the total angular momentum and total external moment can be tricky. Also, it only provides one equation for what may be a multiple-degree-of-freedom problem. Nevertheless, there are many problems that can be solved with this equation. As shown in the next section, it is particularly useful when the total external moment is zero. In addition, the procedure used to derive Eq. (7.5) is quite general and will inform much of our future analysis.

Example 7.1 A Two-Particle Pendulum

In this example we turn to a variant of the simple pendulum, this time with two particles connected in an equilateral triangle, with the pivot point as shown in Figure 7.2. We use the midpoint of the line connecting the two particles (C) as a convenient reference point. As the particles are rigidly connected by massless rods, this system still has only one degree of freedom. We represent that degree of freedom

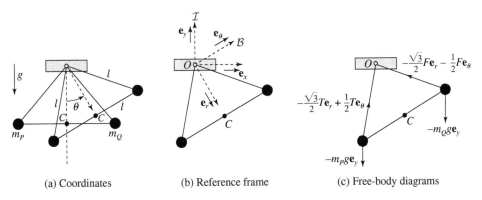

(a) Coordinates   (b) Reference frame   (c) Free-body diagrams

Figure 7.2 A two-particle pendulum configured in an equilateral triangle.

by the angle θ as defined in Figure 7.2a. We wish to find the equation of motion for θ.

One approach to this problem is to follow the procedure outlined in Chapter 6: separate the particles, write the equation of motion for each, including all constraint and internal forces, and then combine to eliminate the unknown forces and find a single equation of motion. This approach is challenging and involved. An easier approach is to consider the total angular momentum about O.

As usual, we introduce an inertial frame $\mathcal{I}$ located at O and a polar frame $\mathcal{B}$ with unit vector $\mathbf{e}_r$ directed toward C, as shown in Figure 7.2b. The transformation table between these two frames is

	$\mathbf{e}_x$	$\mathbf{e}_y$
$\mathbf{e}_r$	$\sin\theta$	$-\cos\theta$
$\mathbf{e}_\theta$	$\cos\theta$	$\sin\theta$.

We next write the kinematics of the two particles:

$$\mathbf{r}_{P/O} = \frac{\sqrt{3}}{2}l\mathbf{e}_r - \frac{1}{2}l\mathbf{e}_\theta$$

$$^{\mathcal{I}}\mathbf{v}_{P/O} = \frac{\sqrt{3}}{2}l\dot{\theta}\mathbf{e}_\theta + \frac{1}{2}l\dot{\theta}\mathbf{e}_r$$

$$\mathbf{r}_{Q/O} = \frac{\sqrt{3}}{2}l\mathbf{e}_r + \frac{1}{2}l\mathbf{e}_\theta$$

$$^{\mathcal{I}}\mathbf{v}_{Q/O} = \frac{\sqrt{3}}{2}l\dot{\theta}\mathbf{e}_\theta - \frac{1}{2}l\dot{\theta}\mathbf{e}_r.$$

The total angular momentum is found from Definition 7.1,

$$^{\mathcal{I}}\mathbf{h}_O = m_P\mathbf{r}_{P/O} \times {}^{\mathcal{I}}\mathbf{v}_{P/O} + m_Q\mathbf{r}_{Q/O} \times {}^{\mathcal{I}}\mathbf{v}_{Q/O}$$

$$= (m_Q + m_P)l^2\dot{\theta}\mathbf{e}_3. \tag{7.6}$$

We now use Eq. (7.5) to find the equation of motion. First we need to compute the total external moment acting on the pendulum, which can be done using the transformation array and the external forces in the free-body diagram in Figure 7.2c. Note that we have shown the tension in each rod (T and F), but by using a moment approach, these are irrelevant, as they do not produce a moment about O. We omit the internal force between the two particles, as this force cancels out in the angular momentum equation because of the internal-moment assumption. The relevant moments about O come only from the external gravity forces:

$$\mathbf{M}_{P/O} = \mathbf{r}_{P/O} \times \mathbf{F}_P = \frac{1}{2}l \cos\theta m_P g \mathbf{e}_3 - \frac{\sqrt{3}}{2}l \sin\theta m_P g \mathbf{e}_3$$

$$\mathbf{M}_{Q/O} = \mathbf{r}_{Q/O} \times \mathbf{F}_Q = -\frac{1}{2}l \cos\theta m_Q g \mathbf{e}_3 - \frac{\sqrt{3}}{2}l \sin\theta m_Q g \mathbf{e}_3.$$

The final equation of motion is now easily found from Eq. (7.5) by differentiating the total angular momentum in Eq. (7.6) and setting it equal to the sum of the above two moments:

$$\boxed{\ddot{\theta} = \frac{g}{l(m_P + m_Q)}\left(m_P \cos\left(\theta + \frac{\pi}{3}\right) - m_Q \sin\left(\theta + \frac{\pi}{6}\right)\right)} \qquad (7.7)$$

If $m_P \equiv m_Q$, Eq. (7.7) becomes $\ddot{\theta} = -\frac{\sqrt{3}}{2}\frac{g}{l}\sin\theta$, the equation of motion for a simple pendulum with length equal to $2l/\sqrt{3}$ (which is not equal to $\|\mathbf{r}_{C/O}\|$).

It is worth noting that Example 7.1 worked out so well because the pendulum was actually a rigid body and thus our single angular-momentum equation was adequate for the single degree of freedom. If we had allowed the masses to move relative to each other, it would have been more complicated, as more equations would be needed to fully describe the system. These could have been obtained using the results of Chapter 6 (by separating the particles). Soon we will explore an alternative approach to finding the angular momentum of multiparticle systems and discover one of our most important results.

7.1.2 Conservation of Total Angular Momentum

Section 6.1.2 showed that, in the absence of external forces on a collection of particles, the total linear momentum is conserved (Law 6.1). This exceptionally important result is used for solving many problems. For instance, it was fundamental to the analysis of collisions in Section 6.2. Here we find a parallel result for the total angular momentum of a collection of particles. Eq. (7.5) shows that, in the absence of external moments on the collection, the total angular momentum is a constant of the motion.

Law 7.1 Under the internal-moment assumption, the law of **conservation of total angular momentum** states that, if the total external moment acting on a system of particles about the origin O is zero, then the total inertial angular momentum of the collection about O is a constant of the motion.

It is worth noting that, unlike for conservation of total linear momentum, here we can still have a non-zero total external force, as long as it produces no moment about O. A central force is an example of such a situation (see Definition 4.5). It is also

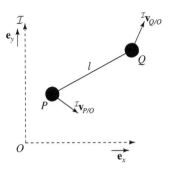

Figure 7.3 Two hockey pucks P and Q connected by a massless string of length l.

important to remember that Law 7.1 only applies when the internal-moment assumption holds. Otherwise, a system without any external moments may not conserve total angular momentum. You have the opportunity to study such a system in Problem 7.13. Again, always check your assumptions before embarking on a new problem!

Example 7.2 Two Hockey Pucks

As an example of using conservation of total angular momentum, consider the case of two sliding hockey pucks connected by a massless string, as shown in Figure 7.3. Using Cartesian coordinates in frame $\mathcal{I} = (O, \mathbf{e}_x, \mathbf{e}_y, \mathbf{e}_z)$, the position and velocity of each mass are

$$\mathbf{r}_{P/O} = x_P \mathbf{e}_x + y_P \mathbf{e}_y$$

$$^{\mathcal{I}}\mathbf{v}_{P/O} = \dot{x}_P \mathbf{e}_x + \dot{y}_P \mathbf{e}_y$$

$$\mathbf{r}_{Q/O} = x_Q \mathbf{e}_x + y_Q \mathbf{e}_y$$

$$^{\mathcal{I}}\mathbf{v}_{Q/O} = \dot{x}_Q \mathbf{e}_x + \dot{y}_Q \mathbf{e}_y.$$

The total angular momentum about O is the sum of the angular momenta of P and Q about O:

$$^{\mathcal{I}}\mathbf{h}_O = \mathbf{r}_{P/O} \times m_P\,^{\mathcal{I}}\mathbf{v}_{P/O} + \mathbf{r}_{Q/O} \times m_Q\,^{\mathcal{I}}\mathbf{v}_{Q/O}$$

$$= \left[m_P(x_P \dot{y}_P - y_P \dot{x}_P) + m_Q(x_Q \dot{y}_Q - y_Q \dot{x}_Q) \right] \mathbf{e}_z.$$

Despite the internal force connecting mass P and Q, the absence of any external forces leads to conservation of total angular momentum. (Since angular momentum is a vector, both its magnitude and direction are conserved.) Differentiating $^{\mathcal{I}}\mathbf{h}_O \cdot \mathbf{e}_z$ with respect to time and setting the result equal to zero yields

$$m_P(x_P \ddot{y}_P - y_P \ddot{x}_P) + m_Q(x_Q \ddot{y}_Q - y_Q \ddot{x}_Q) = 0.$$

In this case, conservation of total angular momentum provided only one equation—we would need three more scalar equations to solve for all four unknowns.

Example 7.3 Sticky Impact with a Pendulum, Revisited

Example 6.4 considered a sticky collision between a moving particle and a stationary pendulum. We asked for the velocity of the composite particle after the collision when the particles are stuck together (which is equivalent to asking for the initial swing rate of the pendulum). That example proved to be a bit involved because the system was acted on by external forces (gravity and the tension in the pendulum rod), which prevented a straightforward application of momentum conservation. We revisit it here using conservation of angular momentum.

We can fruitfully use conservation of angular momentum here because, right up until the end of the collision, the only forces in the problem act through attachment point O and thus produce no moment about O. (Of course, after the collision, when the particles are stuck together, we then have a simple-pendulum problem, which can be solved using the angular-momentum equation from Section 4.2.1 and the initial conditions determined here.) Before the collision, the only contribution to the angular momentum about O is from particle Q:

$$
\begin{aligned}
{}^{\mathcal{I}}\mathbf{h}_O(t_1) &= \mathbf{r}_{Q/O}(t_1) \times {}^{\mathcal{I}}\mathbf{v}_{Q/O}(t_1) \\
&= -l\mathbf{e}_y \times m_Q(\dot{x}_Q(t_1)\mathbf{e}_x + \dot{y}_Q(t_1)\mathbf{e}_y) \\
&= l m_Q \dot{x}_Q(t_1)\mathbf{e}_z,
\end{aligned}
$$

where l is the length of the pendulum rod. At the instant the particles collide and stick, the velocity of Q in the $\mathbf{e}_y$ direction makes no contribution to the angular momentum. Of course, at the time of impact the pendulum rod applies a significant impulse to make the $\mathbf{e}_y$ component of the velocity of Q zero, but it has no effect on the total angular momentum.

Right after the collision the total angular momentum is given by the same position at the bottom of the pendulum crossed with the new linear momentum:

$$
\begin{aligned}
{}^{\mathcal{I}}\mathbf{h}_O(t_2) &= \mathbf{r}_{PQ/O}(t_2) \times (m_P + m_Q){}^{\mathcal{I}}\mathbf{v}_{PQ/O}(t_2) \\
&= -l\mathbf{e}_y \times (m_P + m_Q)\dot{x}_{PQ}(t_2)\mathbf{e}_x \\
&= l(m_P + m_Q)\dot{x}_{PQ}(t_2)\mathbf{e}_z.
\end{aligned}
$$

Since this must be the same as the angular momentum before the collision, setting ${}^{\mathcal{I}}\mathbf{h}_O(t_2) = {}^{\mathcal{I}}\mathbf{h}_O(t_1)$ yields the velocity after the sticky impact:

$$
\dot{x}_{PQ}(t_2) = \frac{m_Q}{m_P + m_Q}\dot{x}_Q(t_1),
$$

which is the same result as Eq. (6.12).

7.2 Angular Momentum Separation

This section revisits the total angular momentum in Definition 7.1 to get an alternative, and immensely useful, form. Figure 7.1 shows that we can replace the position of each particle by the vector triad,

$$\mathbf{r}_{i/O} = \mathbf{r}_{G/O} + \mathbf{r}_{i/G},$$

just as in our linear-momentum discussion. We can also replace the velocity of the ith particle by taking the derivative of this triad to obtain

$$\frac{^{\mathcal{I}}d}{dt}(\mathbf{r}_{i/O}) = {}^{\mathcal{I}}\mathbf{v}_{i/O} = {}^{\mathcal{I}}\mathbf{v}_{G/O} + {}^{\mathcal{I}}\mathbf{v}_{i/G}.$$

The result is a new equation for the total angular momentum. Using Definition 7.1, we have

$$^{\mathcal{I}}\mathbf{h}_O = \sum_{i=1}^{N} m_i(\mathbf{r}_{G/O} + \mathbf{r}_{i/G}) \times ({}^{\mathcal{I}}\mathbf{v}_{G/O} + {}^{\mathcal{I}}\mathbf{v}_{i/G})$$

$$= \sum_{i=1}^{N} m_i\mathbf{r}_{G/O} \times {}^{\mathcal{I}}\mathbf{v}_{G/O} + \mathbf{r}_{G/O} \times \underbrace{\sum_{i=1}^{N} m_i{}^{\mathcal{I}}\mathbf{v}_{i/G}}_{=0} + \sum_{i=1}^{N} m_i\mathbf{r}_{i/G} \times {}^{\mathcal{I}}\mathbf{v}_{i/O}$$

$$= \underbrace{m_G\mathbf{r}_{G/O} \times {}^{\mathcal{I}}\mathbf{v}_{G/O}}_{\triangleq\,{}^{\mathcal{I}}\mathbf{h}_{G/O}} + \underbrace{\left(\sum_{i=1}^{N} m_i\mathbf{r}_{i/G}\right)}_{=0} \times {}^{\mathcal{I}}\mathbf{v}_{G/O} + \underbrace{\sum_{i=1}^{N} m_i\mathbf{r}_{i/G} \times {}^{\mathcal{I}}\mathbf{v}_{i/G}}_{\triangleq\,{}^{\mathcal{I}}\mathbf{h}_G},$$

where we twice set terms to zero using the center-of-mass corollary in Eq. (6.15).

Now we do have the angular momentum ${}^{\mathcal{I}}\mathbf{h}_{G/O}$ relative to the origin of an equivalent point mass of mass m_G traveling at the velocity of the center of mass. Although the total angular momentum in Eq. (7.5) does not equal this simple expression, we have nonetheless been able to pull it out.

The third term in the equation, ${}^{\mathcal{I}}\mathbf{h}_G$, is the total angular momentum of all particles about the center of mass G. We conclude that *the total angular momentum of the system of particles about O separates into the angular momentum of the center of mass relative to the origin O of the inertial frame plus the total angular momentum of all particles about their center of mass G.* We call this the *separation principle*. This result is so important, we rewrite it:

$$\boxed{{}^{\mathcal{I}}\mathbf{h}_O = {}^{\mathcal{I}}\mathbf{h}_{G/O} + {}^{\mathcal{I}}\mathbf{h}_G,} \tag{7.8}$$

where

$$\boxed{{}^{\mathcal{I}}\mathbf{h}_{G/O} \triangleq m_G\mathbf{r}_{G/O} \times {}^{\mathcal{I}}\mathbf{v}_{G/O}} \tag{7.9}$$

and

$$\boxed{{}^{\mathcal{I}}\mathbf{h}_G \triangleq \sum_{i=1}^{N} m_i\mathbf{r}_{i/G} \times {}^{\mathcal{I}}\mathbf{v}_{i/G}.} \tag{7.10}$$

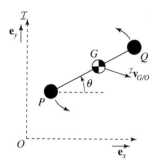

Figure 7.4 Two hockey pucks P and Q connected by a string of length l.

Thus, if we can find the two angular momenta in Eqs. (7.9) and (7.10), it is easy to find the total angular momentum relative to the origin of an inertial frame. These momenta are often much easier to calculate or measure than the total angular momentum.

Example 7.4 Two Hockey Pucks, Revisited

As an example of the separation of total angular momentum, consider again the case of two sliding hockey pucks connected by a massless string, as shown in Figure 7.4. Using Cartesian coordinates in frame $\mathcal{I} = (O, \mathbf{e}_x, \mathbf{e}_y, \mathbf{e}_z)$, the position and velocity of the center of mass G are

$$\mathbf{r}_{G/O} = \underbrace{\frac{m_P x_P + m_Q x_Q}{m_P + m_Q}}_{\triangleq x_G} \mathbf{e}_x + \underbrace{\frac{m_P y_P + m_Q y_Q}{m_P + m_Q}}_{\triangleq y_G} \mathbf{e}_y$$

and

$$^{\mathcal{I}}\mathbf{v}_{G/O} = \dot{x}_G \mathbf{e}_x + \dot{y}_G \mathbf{e}_y.$$

The angular momentum of the center of mass about the origin is

$$^{\mathcal{I}}\mathbf{h}_{G/O} = \mathbf{r}_{G/O} \times m_G{}^{\mathcal{I}}\mathbf{v}_{G/O}$$

$$= m_G(x_G \dot{y}_G - y_G \dot{x}_G)\mathbf{e}_z,$$

where $m_G = m_P + m_Q$.

To compute the angular momentum about the center of mass, we use polar frame $\mathcal{B} = (G, \mathbf{b}_1, \mathbf{b}_2, \mathbf{b}_3)$ attached to the center of mass with $\mathbf{b}_1$ directed from G to Q at angle θ relative to $\mathbf{e}_x$. Let $\mathbf{r}_{Q/G} = l_Q \mathbf{b}_1$, where

$$l_Q = \|\mathbf{r}_{Q/O} - \mathbf{r}_{G/O}\| = \sqrt{(x_Q - x_G)^2 + (y_Q - y_G)^2}$$

is constant. Similarly, let $\mathbf{r}_{P/G} = -l_P\mathbf{b}_1$. We have

$$
\begin{aligned}
{}^{\mathcal{I}}\mathbf{h}_G &= \mathbf{r}_{P/G} \times m_P {}^{\mathcal{I}}\mathbf{v}_{P/G} + \mathbf{r}_{Q/G} \times m_Q {}^{\mathcal{I}}\mathbf{v}_{Q/G} \\
&= (-l_P\mathbf{b}_1) \times m_P(-l_P\dot{\theta}\mathbf{b}_2) + (l_Q\mathbf{b}_1) \times m_Q(l_Q\dot{\theta}\mathbf{b}_2) \\
&= (m_P l_P^2 + m_Q l_Q^2)\dot{\theta}\mathbf{e}_z,
\end{aligned}
$$

where we used $\mathbf{b}_3 = \mathbf{e}_z$.

We are not yet done, however. Next we take the derivative of each term in Eq. (7.8), as we did to arrive at Eq. (7.5), to find equations of motion for the angular momenta.

7.2.1 Angular Momentum of the Center of Mass

We start by taking the derivative of the angular momentum of the center of mass about O:

$$
\begin{aligned}
\frac{{}^{\mathcal{I}}d}{dt}\left({}^{\mathcal{I}}\mathbf{h}_{G/O}\right) &= \frac{{}^{\mathcal{I}}d}{dt}\left(m_G\mathbf{r}_{G/O} \times {}^{\mathcal{I}}\mathbf{v}_{G/O}\right) \\
&= \underbrace{m_G {}^{\mathcal{I}}\mathbf{v}_{G/O} \times {}^{\mathcal{I}}\mathbf{v}_{G/O}}_{=0} + m_G\mathbf{r}_{G/O} \times \frac{{}^{\mathcal{I}}d}{dt}\left({}^{\mathcal{I}}\mathbf{v}_{G/O}\right).
\end{aligned}
$$

The first term is zero because the cross product between a vector and itself is zero. For the second term, we use the result for the center of mass of a collection of particles in Eq. (6.16) to replace the velocity derivative with the total external force:

$$
\boxed{\frac{{}^{\mathcal{I}}d}{dt}\left({}^{\mathcal{I}}\mathbf{h}_{G/O}\right) = \mathbf{r}_{G/O} \times \mathbf{F}_G^{(\text{ext})} \triangleq \mathbf{M}_{G/O}^{(\text{ext})}.} \tag{7.11}
$$

Eq. (7.11) is a reassuring result and again confirms that we can treat the collection of particles as an equivalent single particle of mass m_G located at the center of mass. Eq. (7.11) is the angular-momentum form of Newton's second law applied to the center of mass of a multiparticle system, just as we found in Chapter 4 for a single particle. Thus we can solve for the translational motion of the center of mass using either the force equation in Eq. (6.16) or the moment equation in Eq. (7.11), whichever is more convenient for the problem at hand.

The motion of the center of mass described by Eq. (7.11) has no dependence on the motion of the individual particles relative to the center of mass—unless the force or moment depends upon them, as in Tutorial 6.1. Amazing. What about the other term in Eq. (7.8)? Now we look at the total angular momentum about the center of mass.

7.2.2 Angular Momentum about the Center of Mass

As done for $^{\mathcal{I}}\mathbf{h}_{G/O}$, we now take the derivative of the total angular momentum of the system of particles about the center of mass:

$$
\frac{^{\mathcal{I}}d}{dt}\left(^{\mathcal{I}}\mathbf{h}_G\right) = \frac{^{\mathcal{I}}d}{dt}\left(\sum_{i=1}^{N} m_i \mathbf{r}_{i/G} \times {^{\mathcal{I}}\mathbf{v}_{i/G}}\right)
$$

$$
= \underbrace{\sum_{i=1}^{N} m_i {^{\mathcal{I}}\mathbf{v}_{i/G}} \times {^{\mathcal{I}}\mathbf{v}_{i/G}}}_{=0} + \sum_{i=1}^{N} m_i \mathbf{r}_{i/G} \times \frac{^{\mathcal{I}}d}{dt}\left(^{\mathcal{I}}\mathbf{v}_{i/G}\right).
$$

Now we have to be clever. Recall that $^{\mathcal{I}}\mathbf{v}_{i/O} = {^{\mathcal{I}}\mathbf{v}_{G/O}} + {^{\mathcal{I}}\mathbf{v}_{i/G}}$. Thus

$$
\sum_{i=1}^{N} m_i \mathbf{r}_{i/G} \times \frac{^{\mathcal{I}}d}{dt}\left(^{\mathcal{I}}\mathbf{v}_{i/G}\right) = \sum_{i=1}^{N} m_i \mathbf{r}_{i/G} \times \frac{^{\mathcal{I}}d}{dt}\left(^{\mathcal{I}}\mathbf{v}_{i/O}\right) - \underbrace{\left(\sum_{i=1}^{N} m_i \mathbf{r}_{i/G}\right)}_{=0} \times \frac{^{\mathcal{I}}d}{dt}\left(^{\mathcal{I}}\mathbf{v}_{G/O}\right),
$$

where we again used the center-of-mass corollary in Eq. (6.15). The result is

$$
\frac{^{\mathcal{I}}d}{dt}\left(^{\mathcal{I}}\mathbf{h}_G\right) = \sum_{i=1}^{N} \mathbf{r}_{i/G} \times m_i \frac{^{\mathcal{I}}d}{dt}\left(^{\mathcal{I}}\mathbf{v}_{i/O}\right).
$$

Now we can use Newton's second law on each particle as in Chapter 6. That is, we replace the inertial acceleration with the forces acting on particle i,

$$
m_i \frac{^{\mathcal{I}}d}{dt}\left(^{\mathcal{I}}\mathbf{v}_{i/O}\right) = \mathbf{F}_i^{(\text{ext})} + \sum_{j=1}^{N} \mathbf{F}_{i,j},
$$

which leaves

$$
\frac{^{\mathcal{I}}d}{dt}\left(^{\mathcal{I}}\mathbf{h}_G\right) = \sum_{i=1}^{N} \mathbf{r}_{i/G} \times \mathbf{F}_i^{(\text{ext})} + \sum_{i=1}^{N}\sum_{j=1}^{N} \mathbf{r}_{i/G} \times \mathbf{F}_{i,j}
$$

$$
= \sum_{i=1}^{N} \mathbf{r}_{i/G} \times \mathbf{F}_i^{(\text{ext})} + \underbrace{\frac{1}{2}\sum_{i=1}^{N}\sum_{j=1}^{N}(\mathbf{r}_{i/G} - \mathbf{r}_{j/G}) \times \mathbf{F}_{i,j}}_{=0}.
$$

We have once again used the internal-moment assumption. The final result is

$$
\boxed{\frac{^{\mathcal{I}}d}{dt}\left(^{\mathcal{I}}\mathbf{h}_G\right) = \sum_{i=1}^{N} \mathbf{r}_{i/G} \times \mathbf{F}_i^{(\text{ext})} = \sum_{i=1}^{N} \mathbf{M}_{i/G}^{(\text{ext})} \triangleq \mathbf{M}_G^{(\text{ext})}.}
\tag{7.12}
$$

As with the total angular momentum, as long as the internal-moment assumption holds, the effect of the internal forces cancel and the dynamics of the angular momentum of the collection about the center of mass is determined entirely by the total external moment about G.

Eq. (7.12) describes the motion of the system of particles about their center of mass. Compare Eq. (7.12) to Eq. (7.11). First we found that we could write the equation of motion for the center of mass of the system relative to the origin of the inertial frame, independently of the motion of the particles about their center of mass. We have now shown, in Eq. (7.12), that we can write the equation of motion for the total angular momentum of the system of particles relative to their center of mass solely in terms of the total external moment about the center of mass, independently of the translational motion of the center of mass. *There is complete separation of motion of the center of mass and motion about the center of mass!*

Eqs. (7.11) and (7.12) also imply a corollary to the law of conservation of angular momentum. If there are no external moments about the center of mass, either angular momentum of the center of mass about O is conserved or, more importantly, the angular momentum of the system about the center of mass is, even if the total angular momentum about O is not conserved. *That is, even if there are external forces acting, resulting in acceleration of the center of mass, and even if they result in a moment about O, $^{\mathcal{I}}\mathbf{h}_G$ is a constant of the motion as long as* $\mathbf{M}_G^{(\text{ext})} = 0$. This is so important that we highlight it as a separate law.

Law 7.2 Under the internal-moment assumption, the law of **conservation of total angular momentum about the center of mass** states that, if the total external moment acting on a system of particles about the center of mass G is zero, then the total inertial angular momentum about G is a constant of the motion.

The separation result is probably one of the most important results in the entire book and informs all of Parts Three and Four. It allows us to study the rotational motion of a rigid body (which we model as a collection of many particles with fixed position relative to the center of mass) without necessarily having to simultaneously solve for the translational motion of the body's center of mass. It also helps explain all sorts of motion, from tops to gyroscopes to satellites.[1]

Example 7.5 Tethered Satellites

Consider the motion of a pair of identical satellites connected by a massless tether, as shown in Figure 7.5. Suppose the satellites have mass m and the tether has length $2l$. Each satellite has a thruster that is capable of exerting force F perpendicular to the tether. If the thrusters fire simultaneously, and for identical durations, the total force on the center of mass of the satellite pair is zero and the total linear momentum is conserved. To determine the thruster's effect on the angular momentum about center of mass G, we refer to Eq. (7.12). In particular, suppose we would like to determine the maximum duration that both thrusters can fire without exceeding $2g$ centripetal acceleration onboard either satellite (which might be hazardous to the science cargo).

---

[1] One caveat. This separation is not always entirely complete because a dependence on translation can be hidden in the external moment in Eq. (7.12). We showed this in Tutorial 6.1 for translation. Another good example is orbital and attitude motion of a satellite, where a portion of the torque on the satellite can depend on where it is in earth orbit (particularly for an elliptical orbit). Nevertheless, there is no coupling in the angular momenta.

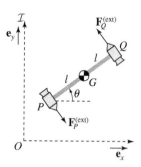

Figure 7.5 Identical tethered satellites P and Q.

Let $\mathcal{B} = (G, \mathbf{b}_1, \mathbf{b}_2, \mathbf{b}_3)$ be a polar frame with origin G and unit vector $\mathbf{b}_1$ directed from G to Q at angle θ relative to $\mathbf{e}_x$. The angular momentum about the center of mass is

$${}^{\mathcal{I}}\mathbf{h}_G = \mathbf{r}_{P/G} \times m\,{}^{\mathcal{I}}\mathbf{v}_{P/G} + \mathbf{r}_{Q/G} \times m\,{}^{\mathcal{I}}\mathbf{v}_{Q/G}$$

$$= (-l\mathbf{b}_1) \times m(-l\dot{\theta}\mathbf{b}_2) + (l\mathbf{b}_1) \times m(l\dot{\theta}\mathbf{b}_2) = 2ml^2\dot{\theta}\mathbf{b}_3.$$

The total external moment about the center of mass is

$$\mathbf{M}_G^{(\text{ext})} = \mathbf{r}_{P/G} \times \mathbf{F}_P^{(\text{ext})} + \mathbf{r}_{Q/G} \times \mathbf{F}_Q^{(\text{ext})}$$

$$= (-l\mathbf{b}_1) \times (-F\mathbf{b}_2) + (l\mathbf{b}_1) \times (F\mathbf{b}_2) = 2lF\mathbf{b}_3.$$

Therefore, the dynamics of the angular momentum about the center of mass are

$$2ml^2\ddot{\theta} = 2lF. \tag{7.13}$$

The centripetal acceleration of P is ${}^{\mathcal{I}}\mathbf{a}_{P/G} = l\dot{\theta}^2\mathbf{b}_1$. Rearranging Eq. (7.13) and integrating with $\dot{\theta}(0) = 0$ yields

$$\dot{\theta} = \frac{F}{ml}t.$$

To ensure $\|{}^{\mathcal{I}}\mathbf{a}_{P/G}\| < 2g$, the angular rate must satisfy $|\dot{\theta}| < \sqrt{2g/l}$, which implies

$$t < \frac{m}{|F|}\sqrt{2gl}.$$

One interesting application of a system like this is to produce artificial gravity during long space voyages by connecting a counter mass to a long tether and rotating the system such that the centripetal acceleration is equal to g.

Example 7.6 Two Hockey Pucks, Take 3

Here we return to the two rotating hockey pucks from Examples 7.2 and 7.4 and Figure 7.4. Since there are no external forces on the system, the total linear momentum and total angular momentum are conserved. In addition, the angular momentum of the center of mass about O is a constant of the motion:

$$^\mathcal{I}\mathbf{h}_{G/O} = m_G(x_G\dot{y}_G - y_G\dot{x}_G)\mathbf{e}_z = \text{constant},$$

where $m_G = m_P + m_Q$.

Likewise, because the only internal force is along the line connecting the pucks and there is no external moment, the angular momentum about the center of mass is conserved by Law 7.2, which implies

$$^\mathcal{I}\mathbf{h}_G = (m_P l_P^2 + m_Q l_Q^2)\dot{\theta}\mathbf{e}_z = \text{constant} = h_0. \tag{7.14}$$

Suppose the length of the string, $l = l_P + l_Q$, can change. For instance, we could imagine motors in the hockey pucks or a reel at the center of the string that draws it in or lets it out. Since the resulting tension in the string is still along the line connecting the hockey pucks, the internal-moment assumption still holds and the total angular momentum about G does not change. As a result, the rotation rate $\dot{\theta}$ must change to maintain constant magnitude of the angular momentum. When the string shortens the rotation rate speeds up and when the string lengthens it slows down. Every time you watch ice skaters spin faster as they bring in their arms you are seeing this effect.

In fact, it is rather simple to find an expression for the change in the rate of rotation for a constant rate of change in the separation $\dot{l}$. Since the center of mass must be fixed and the string doesn't stretch, we know that $\dot{l}_P = \dot{l}_Q = \dot{l}$. Thus we can write the two distances from the center of mass as

$$l_P(t) = l_P(t_0) - \dot{l}(t - t_0)$$

$$l_Q(t) = l_Q(t_0) - \dot{l}(t - t_0).$$

The rotation rate as a function of time is thus

$$\dot{\theta}(t) = \frac{h_0}{m_P\left(l_P(t_0) + \dot{l}(t - t_0)\right)^2 + m_Q\left(l_Q(t_0) + \dot{l}(t - t_0)\right)^2}.$$

The rotation rate of the hockey pucks increases (or decreases) as the inverse of time squared.

7.3 Total Angular Momentum Relative to an Arbitrary Point

Following the development in Section 4.2.2, this section examines the more general case of computing the total angular momentum of a system of particles relative to an arbitrary—and perhaps inertially accelerating—point $Q \neq G$. Figure 7.6 depicts the situation. As in Section 4.2.2, we can write the angular momentum of particle i relative to Q as

$$^\mathcal{I}\mathbf{h}_{i/Q} = \mathbf{r}_{i/Q} \times m_i\,{}^\mathcal{I}\mathbf{v}_{i/Q}.$$

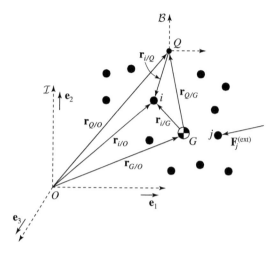

Figure 7.6 A collection of particles with center of mass G and an arbitrary reference point Q.

Next we sum the individual angular momenta relative to Q to find the total angular momentum of the collection relative to the point Q:

$$\boxed{{}^{\mathcal{I}}\mathbf{h}_Q \triangleq \sum_{i=1}^{N} \mathbf{r}_{i/Q} \times m_i \, {}^{\mathcal{I}}\mathbf{v}_{i/Q}.} \tag{7.15}$$

Eq. (7.15) is similar to the definition of the angular momentum relative to the center of mass in Eq. (7.10). The question is whether we can find a similarly compact expression for the dynamics of ${}^{\mathcal{I}}\mathbf{h}_Q$ as we did for ${}^{\mathcal{I}}\mathbf{h}_G$ in Eq. (7.12).

The first step is to rewrite the angular momentum about Q in terms of the angular momentum about G. Replacing $\mathbf{r}_{i/Q}$ and ${}^{\mathcal{I}}\mathbf{v}_{i/Q}$ using $\mathbf{r}_{i/G} = \mathbf{r}_{i/Q} + \mathbf{r}_{Q/G}$ (and its derivative) yields

$$\,^{\mathcal{I}}\mathbf{h}_Q = \sum_{i=1}^{N} m_i \left(\mathbf{r}_{i/G} - \mathbf{r}_{Q/G} \right) \times \left({}^{\mathcal{I}}\mathbf{v}_{i/G} - {}^{\mathcal{I}}\mathbf{v}_{Q/G} \right).$$

Distributing the cross product results in

$$\,^{\mathcal{I}}\mathbf{h}_Q = \underbrace{\sum_{i=1}^{N} m_i \mathbf{r}_{i/G} \times {}^{\mathcal{I}}\mathbf{v}_{i/G}}_{={}^{\mathcal{I}}\mathbf{h}_G} - \underbrace{\left(\sum_{i=1}^{N} m_i \mathbf{r}_{i/G} \right) \times {}^{\mathcal{I}}\mathbf{v}_{Q/G}}_{=0}$$

$$- \mathbf{r}_{Q/G} \times \underbrace{\sum_{i=1}^{N} m_i \,{}^{\mathcal{I}}\mathbf{v}_{i/G}}_{=0} + m_G \mathbf{r}_{Q/G} \times {}^{\mathcal{I}}\mathbf{v}_{Q/G},$$

where we have twice used the center-of-mass corollary.

Thus our final expression for the angular momentum relative to Q in terms of the angular momentum relative to the center of mass G is

$$\boxed{{}^{\mathcal{I}}\mathbf{h}_Q = {}^{\mathcal{I}}\mathbf{h}_G + m_G \mathbf{r}_{Q/G} \times {}^{\mathcal{I}}\mathbf{v}_{Q/G}.} \tag{7.16}$$

The usefulness of Eq. (7.16) comes from applying Newton's second law to it. Recall that taking the derivative of ${}^{\mathcal{I}}\mathbf{h}_G$ and applying Newton's second law resulted in Eq. (7.12); the dynamics of the collection can be considered relative to the center of mass without regard to the motion of the center of mass. What if, instead, we are interested in the motion relative to the arbitrary point Q? What are the dynamics of ${}^{\mathcal{I}}\mathbf{h}_Q$? To find out, we take the derivative of Eq. (7.16) to obtain

$$\frac{{}^{\mathcal{I}}d}{dt}\left({}^{\mathcal{I}}\mathbf{h}_Q\right) = \frac{{}^{\mathcal{I}}d}{dt}\left({}^{\mathcal{I}}\mathbf{h}_G\right) + m_G \underbrace{\frac{{}^{\mathcal{I}}d}{dt}\left(\mathbf{r}_{Q/G}\right) \times {}^{\mathcal{I}}\mathbf{v}_{Q/G}}_{=0}$$

$$+ m_G \mathbf{r}_{Q/G} \times \frac{{}^{\mathcal{I}}d}{dt}\left({}^{\mathcal{I}}\mathbf{v}_{Q/O} - {}^{\mathcal{I}}\mathbf{v}_{G/O}\right),$$

where the second term is zero because the cross product of a vector with itself is zero. Replacing the inertial derivative of the angular momentum about G with the moment definition from Eq. (7.12), we find

$$\frac{{}^{\mathcal{I}}d}{dt}\left({}^{\mathcal{I}}\mathbf{h}_Q\right) = \sum_{i=1}^{N} \mathbf{r}_{i/G} \times \mathbf{F}_i^{(\text{ext})} - \mathbf{r}_{Q/G} \times m_G{}^{\mathcal{I}}\mathbf{a}_{G/O} + \mathbf{r}_{Q/G} \times m_G{}^{\mathcal{I}}\mathbf{a}_{Q/O}.$$

We now use the finding from Chapter 6 that the total mass times the acceleration of the center of mass is equal to the sum of the external forces on the collection:

$$m_G{}^{\mathcal{I}}\mathbf{a}_{G/O} = \sum_{i=1}^{N} \mathbf{F}_i^{(\text{ext})}.$$

This equation allows us to write the angular-momentum derivative as

$$\frac{{}^{\mathcal{I}}d}{dt}\left({}^{\mathcal{I}}\mathbf{h}_Q\right) = \sum_{i=1}^{N} \left(\mathbf{r}_{i/G} - \mathbf{r}_{Q/G}\right) \times \mathbf{F}_i^{(\text{ext})} + \mathbf{r}_{Q/G} \times m_G{}^{\mathcal{I}}\mathbf{a}_{Q/O}.$$

Finally, we use the vector triad $\mathbf{r}_{Q/G} + \mathbf{r}_{i/Q} = \mathbf{r}_{i/G}$ again to obtain

$$\sum_{i=1}^{N} \left(\mathbf{r}_{i/G} - \mathbf{r}_{Q/G}\right) \times \mathbf{F}_i^{(\text{ext})} = \sum_{i=1}^{N} \mathbf{r}_{i/Q} \times \mathbf{F}_i^{(\text{ext})} \triangleq \mathbf{M}_Q^{(\text{ext})},$$

which results in our final expression for the angular-momentum derivative:

$$\boxed{\frac{{}^{\mathcal{I}}d}{dt}\left({}^{\mathcal{I}}\mathbf{h}_Q\right) = \mathbf{M}_Q^{(\text{ext})} + \mathbf{r}_{Q/G} \times m_G{}^{\mathcal{I}}\mathbf{a}_{Q/O}.} \tag{7.17}$$

Compare Eq. (7.17) to Eq. (7.12). The beauty of considering the angular momentum about the center of mass is that its inertial rate of change is equal solely to the external moment about G. This allows us to consider the orientation of the particle collection without worrying about the overall translation of the center of mass. If instead we consider the angular momentum relative to some arbitrary point Q, then Eq. (7.17) shows that the separation of orientation and translation no longer holds. The rate of change of $^{\mathcal{I}}\mathbf{h}_Q$ depends on the external moment about Q and the inertial acceleration of Q. Nevertheless, there are problems where this added complication is offset by the advantage of using a more geometrically convenient point.

Example 7.7 Crane, Take 2

This example revisits the simple model of an overhead crane solved in Example 6.2 and shown in Figure 6.2, but with the addition of a horizontal external force on the block, $\mathbf{F}_Q = u\mathbf{e}_x$. This force can be considered a control force used to move the block and center the pendulum (this is a common problem in automatic control). We wish to find the equations of motion for the block and the pendulum. In Chapter 6 we solved this problem by considering Newton's second law separately on the block and the pendulum mass and then eliminating the constraint forces. Here we use the angular momentum about Q and Eq. (7.17).

Since the block has no angular momentum about Q, the total angular momentum about Q is equal to the angular momentum of the pendulum mass P about Q,

$$^{\mathcal{I}}\mathbf{h}_Q = m_P \mathbf{r}_{P/Q} \times {}^{\mathcal{I}}\mathbf{v}_{P/Q}.$$

Using $\mathbf{r}_{P/Q} = l\mathbf{e}_r$ and its inertial derivative $^{\mathcal{I}}\mathbf{v}_{P/Q} = l\dot{\theta}\mathbf{e}_\theta$ gives

$$^{\mathcal{I}}\mathbf{h}_Q = m_P l^2 \dot{\theta}\mathbf{e}_3. \tag{7.18}$$

We now insert Eq. (7.18) into Eq. (7.17). The inertial derivative of the angular momentum about Q is

$$\frac{^{\mathcal{I}}d}{dt}\left(^{\mathcal{I}}\mathbf{h}_Q\right) = m_P l^2 \ddot{\theta}\mathbf{e}_3.$$

The moment about Q is

$$\mathbf{M}_Q = \mathbf{r}_{P/Q} \times \mathbf{F}_P = l\mathbf{e}_r \times (-m_P g)\mathbf{e}_y = -m_P g l \sin\theta\mathbf{e}_3.$$

To calculate the correction term in Eq. (7.17), $\mathbf{r}_{Q/G} \times m_G{}^{\mathcal{I}}\mathbf{a}_{Q/O}$, we use

$$\mathbf{r}_{Q/G} = \mathbf{r}_{Q/O} - \mathbf{r}_{G/O} = -\tilde{l}\mathbf{e}_r,$$

where $\tilde{l} \triangleq \frac{m_P}{m_G}l$, as defined in Example 6.6. The acceleration of Q in $\mathcal{I}$ is $^{\mathcal{I}}\mathbf{a}_{Q/O} = \ddot{x}\mathbf{e}_x$. Thus the angular momentum derivative in Eq. (7.17) becomes

$$m_P l^2 \ddot{\theta}\mathbf{e}_3 = -m_P g l \sin\theta\mathbf{e}_3 - m_G \tilde{l}\ddot{x}\mathbf{e}_r \times \mathbf{e}_x$$

$$= -m_P g l \sin\theta\mathbf{e}_3 - m_P l\ddot{x}\cos\theta\mathbf{e}_3.$$

This equation simplifies to the single scalar equation

$$l\ddot{\theta} + \ddot{x}\cos\theta + g\sin\theta = 0. \tag{7.19}$$

Eq. (7.19) is only a single equation in the two unknowns $\ddot{x}$ and $\ddot{\theta}$. Another equation comes from examining the center of mass. The results of Chapter 6 indicate that the acceleration of the center of mass is equal to the total external force acting on the system, which is $u\mathbf{e}_x$. (For this problem it is easier to consider the translational motion of the center of mass using Newton's second law rather than the angular momentum form.) Eq. (6.18) of Example 6.6 computed Newton's second law for the center of mass of the crane:

$$\frac{^\mathcal{I}d}{dt}\left(^\mathcal{T}\mathbf{p}_{G/O}\right) = m_G(\ddot{x}\mathbf{e}_x + \tilde{l}\ddot{\theta}\mathbf{e}_\theta - \tilde{l}\dot{\theta}^2\mathbf{e}_r) = (N - m_G g)\mathbf{e}_y + u\mathbf{e}_x.$$

The $\mathbf{e}_x$ component of this equation supplies the second equation of motion,

$$m_G\ddot{x} + m_G\tilde{l}\ddot{\theta}\cos\theta - m_G\tilde{l}\dot{\theta}^2\sin\theta = u. \tag{7.20}$$

Eqs. (7.19) and (7.20) are the two equations of motion, arrived at without considering the internal reaction forces. A small amount of algebra leads to the following separate equations of motion for $\ddot{x}$ and $\ddot{\theta}$ of Eqs. (6.5) and (6.6):

$$\left(1 - \frac{m_P}{m_G}\cos^2\theta\right)\ddot{x} - \frac{m_P}{2m_G}g\sin 2\theta - \frac{m_P}{m_G}l\dot{\theta}^2\sin\theta = \frac{u}{m_G}$$

$$\left(1 - \frac{m_P}{m_G}\cos^2\theta\right)\ddot{\theta} + \frac{m_P}{2m_G}\dot{\theta}^2\sin 2\theta + \frac{g}{l}\sin\theta = -\frac{u}{m_G l}\cos\theta.$$

As a check on the validity of these equations, it is interesting to examine what happens when m_Q becomes very large relative to m_P (and thus $m_G \gg m_P$). Our intuition tells us that, as the block becomes very heavy, it will look more and more like a fixed mass and the pendulum will behave according to the simple pendulum equations from Chapter 3. For large m_Q, $m_P/m_G \to 0$ and the equations of motion become approximately

$$\ddot{x} \approx \frac{u}{m_G}$$

$$\ddot{\theta} + \frac{g}{l}\sin\theta \approx -\frac{u}{m_G l}\cos\theta.$$

If $u = 0$, then we obtain the equations of motion of the simple pendulum. If $u \neq 0$ and constant, then we have recovered the equation of motion for a simple pendulum in an accelerating box (see Eq. (4.16) with $u = m_G a$).

7.4 Work and Energy of a Multiparticle System

Section 7.2 showed how the angular momentum of a collection of particles can be separated into the angular momentum of the center of mass about the origin and the angular momentum of the collection about its center of mass. We also derived separate equations of motion for each of these two angular momenta. This section performs a similar analysis to derive work and energy relationships for a system of particles.

7.4.1 Kinetic Energy

Just as we did for the total linear and angular momenta of the collection, we define the total kinetic energy of a system of particles as the sum of the kinetic energy of each particle.

Definition 7.2 The **total kinetic energy** of a system of particles about O is the sum of the individual kinetic energies:

$$
T_O \triangleq \sum_{i=1}^{N} T_{i/O} = \frac{1}{2} \sum_{i=1}^{N} m_i {}^{\mathcal{I}}\mathbf{v}_{i/O} \cdot {}^{\mathcal{I}}\mathbf{v}_{i/O} = \frac{1}{2} \sum_{i=1}^{N} m_i \| {}^{\mathcal{I}}\mathbf{v}_{i/O} \|^2. \qquad (7.21)
$$

We use the subscript O to indicate that this is the kinetic energy of the system relative to O (i.e., it is based on the velocity relative to origin O of the inertial frame). Substituting for ${}^{\mathcal{I}}\mathbf{v}_{i/O}$ using the triad

$$
{}^{\mathcal{I}}\mathbf{v}_{i/O} = {}^{\mathcal{I}}\mathbf{v}_{G/O} + {}^{\mathcal{I}}\mathbf{v}_{i/G}
$$

yields

$$
T_O = \frac{1}{2} \sum_{i=1}^{N} m_i {}^{\mathcal{I}}\mathbf{v}_{G/O} \cdot {}^{\mathcal{I}}\mathbf{v}_{G/O} + {}^{\mathcal{I}}\mathbf{v}_{G/O} \cdot \underbrace{\sum_{i=1}^{N} m_i {}^{\mathcal{I}}\mathbf{v}_{i/G}}_{=0} + \frac{1}{2} \sum_{i=1}^{N} m_i {}^{\mathcal{I}}\mathbf{v}_{i/G} \cdot {}^{\mathcal{I}}\mathbf{v}_{i/G}.
$$

The second term is zero according to (the derivative of) the center-of-mass corollary in Eq. (6.15). The result is another expression for the total kinetic energy,

$$
T_O = \underbrace{\frac{1}{2} \sum_{i=1}^{N} m_i \| {}^{\mathcal{I}}\mathbf{v}_{G/O} \|^2}_{\triangleq T_{G/O}} + \underbrace{\frac{1}{2} \sum_{i=1}^{N} m_i \| {}^{\mathcal{I}}\mathbf{v}_{i/G} \|^2}_{\triangleq T_G}. \qquad (7.22)
$$

Just as for the total angular momentum, the total kinetic energy separates into a kinetic energy $T_{G/O}$ of the center of mass and a kinetic energy T_G about the center of mass:

$$
T_O = T_{G/O} + T_G, \qquad (7.23)
$$

where

$$
T_{G/O} \triangleq \frac{1}{2} m_G \| {}^{\mathcal{I}}\mathbf{v}_{G/O} \|^2 \qquad (7.24)
$$

and

$$
T_G \triangleq \frac{1}{2} \sum_{i=1}^{N} m_i \|{}^{\mathcal{I}}\mathbf{v}_{i/G}\|^2.
\tag{7.25}
$$

Note that this notation might be a bit confusing, since, as we have discussed, the velocity is not measured relative to a point but to a frame. We are not measuring the velocity relative to a frame fixed to G; on the contrary, all of the velocities are inertial, as required by Definition 5.3 of kinetic energy in Chapter 5. T_G is simply that part of the total kinetic energy due to the change in time of the position of each point relative to the center of mass of the particle collection.

Example 7.8 Separation of Kinetic Energy of Two Hockey Pucks

This example computes the total kinetic energy of two sliding hockey pucks connected by a massless string, as shown in Figure 7.4. Recall from Example 7.4 that the position and velocity of the center of mass G are

$$
\mathbf{r}_{G/O} = \underbrace{\frac{m_P x_P + m_Q x_Q}{m_P + m_Q}}_{\triangleq x_G} \mathbf{e}_x + \underbrace{\frac{m_P y_P + m_Q y_Q}{m_P + m_Q}}_{\triangleq y_G} \mathbf{e}_y
$$

and

$$
{}^{\mathcal{I}}\mathbf{v}_{G/O} = \dot{x}_G \mathbf{e}_x + \dot{y}_G \mathbf{e}_y.
$$

Therefore the kinetic energy of the center of mass with respect to O is

$$
T_{G/O} = \frac{1}{2} m_G \|{}^{\mathcal{I}}\mathbf{v}_{G/O}\|^2 = \frac{1}{2} m_G (\dot{x}_G^2 + \dot{y}_G^2).
$$

To compute the kinetic energy about the center of mass, we use $\mathbf{r}_{P/G} = -l_P \mathbf{b}_1$ and $\mathbf{r}_{Q/G} = l_Q \mathbf{b}_1$, where $\mathbf{b}_1$ is the unit vector of the polar frame $\mathcal{B} = (G, \mathbf{b}_1, \mathbf{b}_2, \mathbf{b}_3)$ that is directed from G to Q at angle θ with respect to $\mathbf{e}_x$ (see Figure 7.4). We have

$$
T_G = \frac{1}{2} m_P \|l_P \dot{\theta} \mathbf{b}_2\|^2 + \frac{1}{2} m_Q \|l_Q \dot{\theta} \mathbf{b}_2\|^2 = \frac{1}{2}(m_P l_P^2 + m_Q l_Q^2) \dot{\theta}^2.
$$

The total kinetic energy of the two-particle system is, of course, $T_O = T_{G/O} + T_G$.

7.4.2 The Work–Kinetic-Energy Formula

What about the work–kinetic-energy relationship for a system of particles? Recall the work–kinetic-energy formula from Chapter 5 (Eq. (5.6)) that relates the total work $W_i^{(\text{tot})}$ done on particle i to its change in kinetic energy:

$$
W_i^{(\text{tot})}(\mathbf{r}_{i/O}(t); \gamma_i) = T_{i/O}(t_2) - T_{i/O}(t_1) \triangleq \Delta T_{i/O}.
$$

We introduce the Δ notation to concisely indicate the change in kinetic energy from time t_1 to t_2. The total work $W^{(\text{tot})}$ of a multiparticle system is defined as the sum of the work on each particle:

$$W^{(\text{tot})}(\{\mathbf{r}_{i/O}; \gamma_i\}_{i=1}^N) \triangleq \sum_{i=1}^N W_i^{(\text{tot})}(\mathbf{r}_{i/O}(t); \gamma_i)$$

$$= \sum_{i=1}^N \Delta T_{i/O} = \Delta T_O. \tag{7.26}$$

Note that we used the definition of the total kinetic energy. Also, the notation $\{\mathbf{r}_{i/O}; \gamma_i\}_{i=1}^N$ is short for $\{\mathbf{r}_{1/O}, \ldots, \mathbf{r}_{N/O}; \gamma_1, \ldots, \gamma_N\}$.

Let us examine the total-work term more carefully. Using Definition 5.1 for work from Chapter 5, we can write the total work as the sum of the work done by internal and external forces, which we call the *internal* and *external* work:

$$W^{(\text{tot})}(\{\mathbf{r}_{i/O}; \gamma_i\}_{i=1}^N) = \sum_{i=1}^N \underbrace{\int_{\gamma_i} \mathbf{F}_i^{(\text{ext})} \cdot {}^{\mathcal{I}}d\mathbf{r}_{i/O}}_{\triangleq W_i^{(\text{ext})}(\mathbf{r}_{i/O}; \gamma_i)} + \sum_{i=1}^N \underbrace{\int_{\gamma_i} \sum_{j=1}^N \mathbf{F}_{i,j} \cdot {}^{\mathcal{I}}d\mathbf{r}_{i/O}}_{\triangleq W_i^{(\text{int})}(\mathbf{r}_{i/O}; \gamma_i)}.$$

The total work is just the sum of the internal and external work on each particle over its trajectory γ_i. Sometimes it is convenient to make the same substitution as we did for the angular-momentum calculation (in going from Eq. (7.2) to Eq. (7.3)) and rewrite the second term in an alternate form:

$$W^{(\text{tot})}(\{\mathbf{r}_{i/O}; \gamma_i\}_{i=1}^N) = \underbrace{\sum_{i=1}^N \int_{\gamma_i} \mathbf{F}_i^{(\text{ext})} \cdot {}^{\mathcal{I}}d\mathbf{r}_{i/O}}_{\triangleq W^{(\text{ext})}(\{\mathbf{r}_{i/O}; \gamma_i\}_{i=1}^N)} + \underbrace{\frac{1}{2} \sum_{i=1}^N \sum_{j=1}^N \int_{\gamma_i} \mathbf{F}_{i,j} \cdot {}^{\mathcal{I}}d\mathbf{r}_{i/j}}_{\triangleq W^{(\text{int})}(\{\mathbf{r}_{i/O}; \gamma_i\}_{i=1}^N)}. \tag{7.27}$$

This form also illustrates an important fact about the work in multiparticle systems. When we calculated the total moment for the angular-momentum equation, the second term involving the double sum of internal forces was identically zero because of the internal-moment assumption. Here it is not zero because we have a dot product rather than a cross product. The second term in Eq. (7.27) represents the work done against internal forces when the particles move relative to one another. The existence of this term should make intuitive sense. Imagine a rubber sheet being deformed and stretched. Clearly, work is being done against the internal tension forces to stretch the sheet, that is, to move the particles apart. *A common mistake is to forget the internal work in a multiparticle system!*

We write the work–kinetic-energy formula for a system of particles as

$$\boxed{W^{(\text{tot})} = T_O(t_2) - T_O(t_1) = W^{(\text{ext})} + W^{(\text{int})}.} \tag{7.28}$$

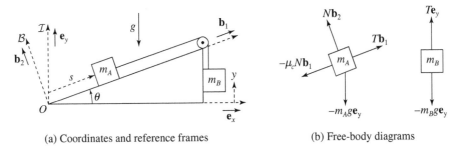

(a) Coordinates and reference frames   (b) Free-body diagrams

Figure 7.7 Pulley on a ramp.

To simplify the notation, we have dropped the arguments. The external and internal work are, respectively,

$$W^{(\text{ext})}(\{\mathbf{r}_{i/O}; \gamma_i\}_{i=1}^N) \triangleq \sum_{i=1}^N \int_{\gamma_i} \mathbf{F}_i^{(\text{ext})} \cdot {}^{\mathcal{I}}d\mathbf{r}_{i/O}$$

and

$$W^{(\text{int})}(\{\mathbf{r}_{i/O}; \gamma_i\}_{i=1}^N) \triangleq \frac{1}{2} \sum_{i=1}^N \sum_{j=1}^N \int_{\gamma_i} \mathbf{F}_{i,j} \cdot {}^{\mathcal{I}}d\mathbf{r}_{i/j}.$$

Eq. (7.28) is the work–kinetic-energy formula for a multiparticle system. We use it the same way we used the single-particle version: to make predictions about the state of the system at two different times. Note that we can also separate the total work into conservative and nonconservative parts. For example, the total work performed by conservative forces from time t_1 to time t_2 is $W^{(c)}(\{\mathbf{r}_{i/O}(t_1), \mathbf{r}_{i/O}(t_2)\}_{i=1}^N)$, or $W^{(c)}(t_1, t_2)$ for short.

Example 7.9 Pulley on a Ramp

Consider two blocks A and B connected by a rope of fixed length, as shown in Figure 7.7a. We assume that the blocks are initially at rest.

Let μ_c denote the coefficient of friction between block A and the ramp. Suppose the blocks are released from rest at time t_1 and $m_B > m_A$. In this example we use work-energy methods to find their speed at time t_2. The reference frames $\mathcal{I} = (O, \mathbf{e}_x, \mathbf{e}_y, \mathbf{e}_z)$ and $\mathcal{B} = (O, \mathbf{b}_1, \mathbf{b}_2, \mathbf{b}_3)$ are related by the transformation table

	$\mathbf{e}_x$	$\mathbf{e}_y$
$\mathbf{b}_1$	$\cos\theta$	$\sin\theta$
$\mathbf{b}_2$	$-\sin\theta$	$\cos\theta$.

Using the coordinates shown in Figure 7.7a, the kinematics of block A are

$$\mathbf{r}_{A/O} = s\mathbf{b}_1$$
$$^{\mathcal{I}}\mathbf{v}_{A/O} = \dot{s}\mathbf{b}_1$$
$$^{\mathcal{I}}\mathbf{a}_{A/O} = \ddot{s}\mathbf{b}_1$$

and, using the kinematic constraint $\dot{s} = -\dot{y}$ imposed by the rope, we have

$$\mathbf{r}_{B/O} = x_0\mathbf{e}_x + y\mathbf{e}_y$$
$$^{\mathcal{I}}\mathbf{v}_{B/O} = -\dot{s}\mathbf{e}_y$$
$$^{\mathcal{I}}\mathbf{a}_{B/O} = -\ddot{s}\mathbf{e}_y.$$

The total force on block A, from the free-body diagram in Figure 7.7b, is

$$\mathbf{F}_A = (T - \mu_c N)\mathbf{b}_1 + N\mathbf{b}_2 - m_A g\mathbf{e}_y,$$

and the total force on block B is

$$\mathbf{F}_B = (T - m_B g)\mathbf{e}_y.$$

Applying Newton's second law to block A yields

$$(T - \mu_c N)\mathbf{b}_1 + N\mathbf{b}_2 - m_A g\mathbf{e}_y = m_A \ddot{s}\mathbf{b}_1.$$

Using the transformation table and equating forces in the $\mathbf{b}_2$ direction gives the normal force:

$$N = m_A g \cos\theta.$$

Using $^{\mathcal{I}}d\mathbf{r}_{A/O} = {}^{\mathcal{I}}\mathbf{v}_{A/O}dt$ and $^{\mathcal{I}}d\mathbf{r}_{B/O} = {}^{\mathcal{I}}\mathbf{v}_{B/O}dt$, the total work on the system of two blocks during the period from t_1 to time t_2 is

$$W^{(\text{tot})}(\mathbf{r}_{A/O}, \mathbf{r}_{B/O}; \gamma_A, \gamma_B) = \int_{\gamma_A} \mathbf{F}_A \cdot {}^{\mathcal{I}}d\mathbf{r}_{A/O} + \int_{\gamma_B} \mathbf{F}_B \cdot {}^{\mathcal{I}}d\mathbf{r}_{B/O}$$

$$= \int_{s(t_1)}^{s(t_2)} ((T - \mu_c m_A g \cos\theta)\mathbf{b}_1 + m_A g \cos\theta\,\mathbf{b}_2 - m_A g\mathbf{e}_y)$$

$$\cdot (ds\,\mathbf{b}_1) + \int_{s(t_1)}^{s(t_2)} ((T - m_B g)\mathbf{e}_y) \cdot (-ds\,\mathbf{e}_y)$$

$$= \underbrace{(-\mu_c m_A g \cos\theta - m_A g \sin\theta + m_B g)\Delta s}_{=W^{(\text{ext})}(\mathbf{r}_{A/O}, \mathbf{r}_{B/O}; \gamma_A, \gamma_B)},$$

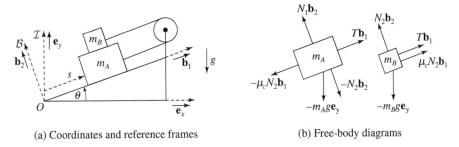

(a) Coordinates and reference frames   (b) Free-body diagrams

Figure 7.8 Pulley on a ramp, revisited.

where $\Delta s = s(t_2) - s(t_1)$. Note that, in this example, the total internal work done on the system is zero. By Eq. (7.28), the change in total work on the system equals the change in the total kinetic energy:

$$W^{(\mathrm{tot})}(\mathbf{r}_{A/O}, \mathbf{r}_{B/O}; \gamma_A, \gamma_B) = \frac{1}{2} m_A \dot{s}(t_2)^2 + \frac{1}{2} m_B \dot{s}(t_2)^2 = \frac{1}{2}(m_A + m_B)\dot{s}^2(t_2),$$

where we used the fact that the kinetic energy at t_1 is zero. If we also assume that $s(t_1) = 0$, then

$$\dot{s}(t_2) = \sqrt{\frac{2(-\mu_c m_A g \cos\theta - m_A g \sin\theta + m_B g)s(t_2)}{m_A + m_B}}.$$

Example 7.10 Pulley on a Ramp, Take 2

Now consider the system of two blocks A and B (masses m_A and m_B, respectively) connected by a rope of fixed length L, as shown in Figure 7.8a. Let μ_c denote the coefficient of friction between block A and block B, and assume there is no friction between A and the ramp. Suppose that the blocks are released from rest at time t_1 and $m_A > m_B$. In this example we use work-energy methods to find their speed at time t_2. Unlike in the previous example, the total internal work of the system is *not* zero.

The reference frames $\mathcal{I} = (O, \mathbf{e}_x, \mathbf{e}_y, \mathbf{e}_z)$ and $\mathcal{B} = (O, \mathbf{b}_1, \mathbf{b}_2, \mathbf{b}_3)$ are related by the same transformation table as in the previous example. Using the coordinates shown in Figure 7.8a, the kinematics of block A are

$$\mathbf{r}_{A/O} = s\mathbf{b}_1$$

$$^{\mathcal{I}}\mathbf{v}_{A/O} = \dot{s}\mathbf{b}_1$$

$$^{\mathcal{I}}\mathbf{a}_{A/O} = \ddot{s}\mathbf{b}_1,$$

and, assuming the pulley is located at $L\mathbf{b}_1$, we have

$$\mathbf{r}_{B/O} = (L - s)\mathbf{b}_1$$

$$^{\mathcal{I}}\mathbf{v}_{B/O} = -\dot{s}\mathbf{b}_1$$

$$^{\mathcal{I}}\mathbf{a}_{B/O} = -\ddot{s}\mathbf{b}_1.$$

From the free-body diagram in Figure 7.8b, the total force on block A is[2]

$$\mathbf{F}_A = (T - \mu_c N_2)\mathbf{b}_1 + (N_1 - N_2)\mathbf{b}_2 - m_A g \mathbf{e}_y,$$

and the total force on block B is

$$\mathbf{F}_B = (T + \mu_c N_2)\mathbf{b}_1 + N_2 \mathbf{b}_2 - m_B g \mathbf{e}_y.$$

Applying Newton's second law to block B yields

$$(T + \mu_c N_2)\mathbf{b}_1 + N_2 \mathbf{b}_2 - m_B g \mathbf{e}_y = -m_B \ddot{s} \mathbf{b}_1.$$

Using the transformation table and equating forces in the $\mathbf{b}_2$ direction gives the internal force between the blocks:

$$N_2 = m_B g \cos\theta.$$

Using $^I d\mathbf{r}_{A/O} = {}^I \mathbf{v}_{A/O} dt$ and $^I d\mathbf{r}_{B/O} = {}^I \mathbf{v}_{B/O} dt$, the total work on the system of two blocks during the period from t_1 to t_2 is

$$
\begin{aligned}
W^{(\text{tot})}(\mathbf{r}_{A/O}, \mathbf{r}_{B/O}; \gamma_A, \gamma_B) &= \int_{\gamma_A} \mathbf{F}_A \cdot {}^I d\mathbf{r}_{A/O} + \int_{\gamma_B} \mathbf{F}_B \cdot {}^I d\mathbf{r}_{B/O} \\
&= \int_{s(t_1)}^{s(t_2)} ((T - \mu_c m_B g \cos\theta)\mathbf{b}_1 + (N_1 - N_2)\mathbf{b}_2 - m_A g \mathbf{e}_y) \cdot (ds\,\mathbf{b}_1) \\
&\quad + \int_{s(t_1)}^{s(t_2)} ((T + \mu_c m_B g \cos\theta)\mathbf{b}_1 + N_2 \mathbf{b}_2 - m_B g \mathbf{e}_y) \cdot (-ds\,\mathbf{b}_1) \\
&= \underbrace{-2\mu_c m_B g \cos\theta \Delta s}_{=W^{(\text{int})}(\mathbf{r}_{A/O}, \mathbf{r}_{B/O}; \gamma_A, \gamma_B)} + \underbrace{(m_B - m_A) g \sin\theta \Delta s}_{=W^{(\text{ext})}(\mathbf{r}_{A/O}, \mathbf{r}_{B/O}; \gamma_A, \gamma_B)},
\end{aligned}
$$

where $\Delta s = s(t_2) - s(t_1)$. Note that, in this example, the total internal work on the system is not zero. By the work-energy formula in Eq. (7.28), the change in total work on the system equals the change in the total kinetic energy:

$$W^{(\text{tot})}(\mathbf{r}_{A/O}, \mathbf{r}_{B/O}; \gamma_A, \gamma_B) = \frac{1}{2} m_A \dot{s}(t_2)^2 + \frac{1}{2} m_B \dot{s}(t_2)^2 = \frac{1}{2}(m_A + m_B)\dot{s}^2(t_2).$$

Assuming $s(t_1) = 0$, we find that

$$\dot{s}(t_2) = \sqrt{\frac{(2(m_B - m_A)g \sin\theta - 4\mu_c m_B g \cos\theta)s(t_2)}{m_A + m_B}}.$$

Note that, if we had ignored the internal work in this example, we would have gotten the wrong answer!

We conclude this section by looking at the power expended by a collection of forces. Recall from Definition 5.2 that power is defined as the rate of change of work.

---

[2] The internal friction forces are oriented as though s increases; although we don't know this *a priori*, you get the same answer either way.

We showed that power is also equal to the rate of change of kinetic energy. Thus using the definition of total work in Eq. (7.26) and the work–kinetic-energy formula in Eq. (7.28), the total power acting on the collection of particles can be written as

$$P^{(\text{tot})} = \sum_{i=1}^{N} \mathbf{F}_i^{(\text{ext})} \cdot {}^{\mathcal{I}}\mathbf{v}_{i/O} + \frac{1}{2} \sum_{i=1}^{N} \sum_{j=1}^{N} \mathbf{F}_{i,j} \cdot {}^{\mathcal{I}}\mathbf{v}_{i/j} = \frac{d}{dt} T_O.$$

If we simply replace ${}^{\mathcal{I}}\mathbf{v}_{i/O}$ with ${}^{\mathcal{I}}\mathbf{v}_{G/O} + {}^{\mathcal{I}}\mathbf{v}_{i/G}$, we can write the power due to the external forces as a sum of the power due to the motion of the center of mass plus the power due to motion relative to the center of mass:

$$P^{(\text{tot})} = \mathbf{F}_G^{(\text{ext})} \cdot {}^{\mathcal{I}}\mathbf{v}_{G/O} + \sum_{i=1}^{N} \mathbf{F}_i^{(\text{ext})} \cdot {}^{\mathcal{I}}\mathbf{v}_{i/G} + \frac{1}{2} \sum_{i=1}^{N} \sum_{j=1}^{N} \mathbf{F}_{i,j} \cdot {}^{\mathcal{I}}\mathbf{v}_{i/j}. \quad (7.29)$$

The first term in Eq. (7.29) looks like the power associated with a single equivalent particle of mass m_G located at the center of mass of the collection. This leads us to ask whether the center of mass satisfies the same energy conservation laws as a single particle, just as it satisfies Newton's second law (Section 6.1) and the angular-momentum form (Section 7.3). Recall from Chapter 5 that we found the relationship between power and the change in kinetic energy by differentiating the kinetic-energy formula. We do that again here, using the separation property of kinetic energy and differentiating the kinetic energy of the center of mass in Eq. (7.24):

$$\frac{d}{dt}(T_{G/O}) = \frac{1}{2} m_G \left(\frac{{}^{\mathcal{I}}d}{dt} \left({}^{\mathcal{I}}\mathbf{v}_{G/O} \right) \cdot {}^{\mathcal{I}}\mathbf{v}_{G/O} + {}^{\mathcal{I}}\mathbf{v}_{G/O} \cdot \frac{{}^{\mathcal{I}}d}{dt} \left({}^{\mathcal{I}}\mathbf{v}_{G/O} \right) \right)$$

$$= \underbrace{m_G \frac{{}^{\mathcal{I}}d}{dt} \left({}^{\mathcal{I}}\mathbf{v}_{G/O} \right)}_{=\mathbf{F}_G^{(\text{ext})}} \cdot {}^{\mathcal{I}}\mathbf{v}_{G/O}.$$

Chapter 6 showed that we can apply Newton's second law to the center of mass, so we replace $m_G \frac{{}^{\mathcal{I}}d}{dt} \left({}^{\mathcal{I}}\mathbf{v}_{G/O} \right)$ with the net force $\mathbf{F}_G^{(\text{ext})}$ acting on the center of mass G to find

$$\boxed{\frac{d}{dt} T_{G/O}(t) = \mathbf{F}_G^{(\text{ext})} \cdot {}^{\mathcal{I}}\mathbf{v}_{G/O}(t) = P_{G/O},} \quad (7.30)$$

where $P_{G/O}$ is the power associated with total force acting on the center of mass.

This is a satisfying result. Eq. (7.30) shows that the first term of Eq. (7.29) is indeed the power associated with motion of just the center of mass. It also shows that, just as for linear and angular momentum, we can treat the kinetic energy of the center of mass of the collection as if it were the kinetic energy of a single equivalent particle, and it satisfies the same power relationship. In particular, if the total external force is zero, the kinetic energy of the center of mass is a constant (like the linear momentum), even if there are internal or external forces acting on the individual particles and even if each particle is accelerating. This result will prove very useful in the study of rigid bodies.

7.4.3 Total Energy

Just as in Chapter 5, when we discussed the energy of a single particle, we can now use the potential energy when either the external or internal work involves conservative forces. Thus we let $U_{i/O}^{(\text{ext})}(\mathbf{r}_{i/O})$ and $U_i^{(\text{int})}(\mathbf{r}_{i/1}, \ldots, \mathbf{r}_{i/N}) \triangleq \sum_{j=1}^{N} U_{i/j}^{(\mathbf{F}_{i,j})}(\mathbf{r}_{i/j})$ denote the scalar potentials corresponding to external and internal conservative forces, respectively, on particle i. Recalling the definition of potential energy, we have

$$\mathbf{F}_i^{(\text{c,ext})} = -\nabla U_{i/O}^{(\text{ext})}(\mathbf{r}_{i/O})$$

$$\mathbf{F}_{i,j}^{(\text{c})} = -\nabla U_{i/j}^{(\mathbf{F}_{i,j})}(\mathbf{r}_{i/j}).$$

Using Eq. (7.27), the total conservative work is

$$W^{(\text{c})}(\{\mathbf{r}_{i/O}; \gamma_i\}_{i=1}^N) = \underbrace{\sum_{i=1}^{N} \int_{\gamma_i} \mathbf{F}_i^{(\text{c,ext})} \cdot {}^{\mathcal{I}} d\mathbf{r}_{i/O}}_{\triangleq W^{(\text{c,ext})}(\{\mathbf{r}_{i/O}; \gamma_i\}_{i=1}^N)} + \underbrace{\frac{1}{2} \sum_{i=1}^{N} \sum_{j=1}^{N} \int_{\gamma_i} \mathbf{F}_{i,j}^{(\text{c})} \cdot {}^{\mathcal{I}} d\mathbf{r}_{i/j}}_{\triangleq W^{(\text{c,int})}(\{\mathbf{r}_{i/O}; \gamma_i\}_{i=1}^N)}. \quad (7.31)$$

Now we rewrite the total conservative work in Eq. (7.31) using the change in total potential energy for the conservative forces:

$$W^{(\text{c})}(t_1, t_2) = \underbrace{U_O^{(\text{ext})}(t_1) - U_O^{(\text{ext})}(t_2)}_{\triangleq -\Delta U_O^{(\text{ext})}} + \underbrace{U_O^{(\text{int})}(t_1) - U_O^{(\text{int})}(t_2)}_{\triangleq -\Delta U_O^{(\text{int})}}, \quad (7.32)$$

where

$$U_O^{(\text{ext})}(t) \triangleq \sum_{i=1}^{N} U_{i/O}^{(\text{ext})}(\mathbf{r}_{i/O}(t)) \quad (7.33)$$

$$U_O^{(\text{int})}(t) \triangleq \frac{1}{2} \sum_{i=1}^{N} \sum_{j=1}^{N} U_{i/j}^{(\mathbf{F}_{i,j})}(\mathbf{r}_{i/j}(t)). \quad (7.34)$$

Eq. (7.32) leads to our final work-energy formula for a multiparticle system. We define the total energy as the sum of the potential and kinetic energies, only now of the system of particles:

$$\boxed{E_O(t) \triangleq T_O(t) + U_O^{(\text{ext})}(t) + U_O^{(\text{int})}(t) = T_{G/O}(t) + T_G(t) + U_O^{(\text{ext})}(t) + U_O^{(\text{int})}(t).}$$

The work-energy formula then looks just like Eq. (5.16):

$$\boxed{E_O(t_2) = E_O(t_1) + W^{(\text{nc,ext})} + W^{(\text{nc,int})}.} \quad (7.35)$$

Thus Eq. (7.35) also necessitates an updated law of conservation of total energy.

Law 7.3 The law of **conservation of total energy of a multiparticle system** states that, if the total nonconservative force is zero, the total energy of a multiparticle system is a constant of the motion.

This law also suggests a power form, as in Chapter 5. Using the same line of reasoning, the power associated with the nonconservative forces is equal to the rate of change of the total energy:

$$\frac{d}{dt}E_O = P^{(nc)} = \sum_{i=1}^{N} \mathbf{F}_i^{(nc,ext)} \cdot {}^{\mathcal{I}}\mathbf{v}_{i/O} + \frac{1}{2}\sum_{i=1}^{N}\sum_{j=1}^{N} \mathbf{F}_{i,j}^{(nc)} \cdot {}^{\mathcal{I}}\mathbf{v}_{i/j}.$$

Example 7.11 Dissipation of Total Energy of a Pair of Sprung Masses

Consider a pair of identical particles of mass m on a horizontal surface connected by (a) a spring with constant k and rest length r_0 and (b) a viscous damper, as shown in Figure 7.9. There are no external (horizontal) forces acting on either mass, which implies that $T_{G/O}$ and $U_O^{(ext)}$ are constant. There is, however, nonconservative internal work due to the damper, which will eventually bring the distance between the masses to r_0, regardless of their initial conditions. The amount of work done by the damper is

$$W^{(nc,int)} = E_O(\infty) - E_O(0) = T_G(\infty) - T_G(0) + U_O^{(int)}(\infty) - U_O^{(int)}(0).$$

In this example we compute $W^{(nc,int)}$.

Consider a polar frame $\mathcal{B} = (G, \mathbf{e}_r, \mathbf{e}_\theta, \mathbf{e}_z)$ attached to the center of mass G, with unit vector $\mathbf{e}_r$ directed from G to Q at angle θ from $\mathbf{e}_1$. The distance between the masses is r. We have

$$\mathbf{r}_{P/G} = -\frac{r}{2}\mathbf{e}_r$$

$${}^{\mathcal{I}}\mathbf{v}_{P/G} = -\frac{\dot{r}}{2}\mathbf{e}_r - \frac{r}{2}\dot{\theta}\mathbf{e}_\theta$$

and similarly for mass Q for which $\mathbf{r}_{Q/G} = \frac{r}{2}\mathbf{e}_r$. The total internal potential energy is

$$U_O^{(int)} = \frac{1}{2}k(r - r_0)^2,$$

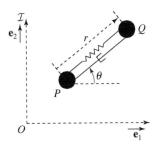

Figure 7.9 A pair of sprung masses.

and the kinetic energy about the center of mass is

$$T_G = \frac{1}{2}m\|^{\mathcal{I}}\mathbf{v}_{P/G}\|^2 + \frac{1}{2}m\|^{\mathcal{I}}\mathbf{v}_{Q/G}\|^2 = \frac{m}{4}\left(\dot{r}^2 + r^2\dot{\theta}^2\right).$$

The latter expression can be simplified using conservation of angular momentum about the center of mass:

$$^{\mathcal{I}}\mathbf{h}_G = \mathbf{r}_{P/G} \times m\,^{\mathcal{I}}\mathbf{v}_{P/G} + \mathbf{r}_{Q/G} \times m\,^{\mathcal{I}}\mathbf{v}_{Q/G} = \frac{mr^2\dot{\theta}}{2}\mathbf{e}_z \triangleq mh_G\mathbf{e}_z,$$

where $h_G = r(0)^2\dot{\theta}(0)/2$ is constant.

Using $r(\infty) = r_0$ and $\dot{r}(\infty) = 0$, we have $\dot{\theta}(\infty) = 2h_G/r_0^2$ and

$$W^{(\text{nc,int})} = \frac{mh_G^2}{r_0^2} - \frac{m}{4}\left(\dot{r}(0)^2 + \frac{4h_G^2}{r(0)^2}\right) - \frac{1}{2}k(r(0) - r_0),$$

which depends only on $\dot{\theta}(0)$, $r(0)$, and $\dot{r}(0)$.

7.5 Tutorials

Tutorial 7.1 The Two-Body Problem

Chapter 4 developed the equation of motion of a body acted on by a central gravitational force, where we assumed that the central attracting body was fixed in the inertial frame. In this chapter, we relax that assumption and treat the more realistic problem of two gravitating bodies (i.e., two very big particles) free to move in absolute space. This problem is depicted in Figure 7.10. For illustration, we have chosen $m_2 \gg m_1$, such as the sun and a planet or the earth and a satellite. Although the mass disparity is a common scenario, the result presented here is exact for two attracting bodies of any mass.

The key assumption is that there are no external forces acting on the bodies; the only force is the internal gravitational force acting along the line between them (parallel to the vector $\mathbf{r}_{1/2}$ in Figure 7.10). Therefore the center of mass G is either fixed or moving at a constant velocity in the inertial frame; the acceleration $^{\mathcal{I}}\mathbf{a}_{G/O}$ is zero. In addition, the total angular momentum $^{\mathcal{I}}\mathbf{h}_O$ from Definition 7.1 is constant because its inertial derivative in Eq. (7.5) is zero. In fact, the same is true for both the angular momentum $^{\mathcal{I}}\mathbf{h}_{G/O}$ of the center of mass G about the origin O and the angular momentum $^{\mathcal{I}}\mathbf{h}_G$ of the two bodies about the center of mass G, whose inertial derivatives are given in Eqs. (7.11) and (7.12), respectively. (The latter claim is true because the internal gravitational force is parallel to $\mathbf{r}_{1/2}$.) Because the total angular momentum $^{\mathcal{I}}\mathbf{h}_O$ is fixed, the bodies must move in a planar orbit that is orthogonal to $^{\mathcal{I}}\mathbf{h}_O$, just as in Example 4.9.

These observations indicate that the two bodies orbit about G such that the total angular momentum relative to G is constant. This result is slightly different from the picture given in Example 4.9, where the small body orbited the attracting origin. It is

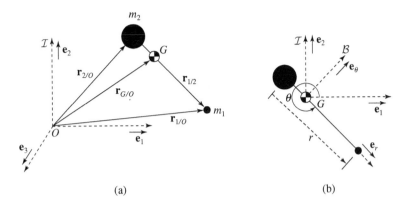

Figure 7.10 Two bodies of masses m_1 and m_2 in orbit about their common center of mass in the plane orthogonal to $\mathbf{e}_3$. (a) Inertial frame $\mathcal{I} = (O, \mathbf{e}_1, \mathbf{e}_2, \mathbf{e}_3)$. (b) Polar frame $\mathcal{B} = (G, \mathbf{e}_r, \mathbf{e}_\theta, \mathbf{e}_3)$.

also perhaps different from our intuitive image of the earth orbiting the sun. In fact, the earth and sun orbit their common center of mass; because the sun is so much larger than the earth, the center of mass of the two bodies is actually inside the sun. Consequently, the sun's motion is very small.[3] (The same is true of the earth's motion when it is orbited by a small satellite.) Nevertheless, it is very common and useful to find a description of the earth's motion relative to the sun (or, to be more general, body 1 relative to body 2) given by $\mathbf{r}_{1/2}(t)$.

Using the inertial frame $\mathcal{I} = (O, \mathbf{e}_1, \mathbf{e}_2, \mathbf{e}_3)$, we write the dynamics of body 1 relative to body 2. We start with the definition of the center of mass,

$$\mathbf{r}_{G/O} = \underbrace{\frac{m_1}{m_1 + m_2}}_{\triangleq \mu_1} \mathbf{r}_{1/O} + \underbrace{\frac{m_2}{m_1 + m_2}}_{\triangleq \mu_2} \mathbf{r}_{2/O},$$

where we have introduced the dimensionless masses $\mu_1 \triangleq m_1/(m_1 + m_2)$ and $\mu_2 \triangleq m_2/(m_1 + m_2)$. Note $\mu_1 + \mu_2 = 1$ and $m_1 + m_2 \triangleq m_G$. Rearranging and using the vector triad $\mathbf{r}_{2/O} = \mathbf{r}_{2/1} + \mathbf{r}_{1/O}$, we find

$$\mu_1 \mathbf{r}_{1/O} = \mathbf{r}_{G/O} - \mu_2 \mathbf{r}_{2/O} = \mathbf{r}_{G/O} - \mu_2 (\mathbf{r}_{2/1} + \mathbf{r}_{1/O})$$
$$= \mathbf{r}_{G/O} + \mu_2 \mathbf{r}_{1/2} - \mu_2 \mathbf{r}_{1/O},$$

which means

$$\mathbf{r}_{1/O} = \mathbf{r}_{G/O} + \mu_2 \mathbf{r}_{1/2}.$$

---

[3] Scientists searching for a planet about another star infer the presence of a planet by measuring the (small) motion of the star.

Consequently, the kinematics of body 1 are

$$^{\mathcal{I}}\mathbf{v}_{1/O} = {^{\mathcal{I}}\mathbf{v}_{G/O}} + \mu_2{^{\mathcal{I}}\mathbf{v}_{1/2}}$$

$$^{\mathcal{I}}\mathbf{a}_{1/O} = \underbrace{^{\mathcal{I}}\mathbf{a}_{G/O}}_{=0} + \mu_2{^{\mathcal{I}}\mathbf{a}_{1/2}},$$

where the absence of external forces implies that the acceleration of G is zero. Using Newton's second law, we find

$$\mathbf{F}_{1,2} = m_1{^{\mathcal{I}}\mathbf{a}_{1/O}} = m_1\mu_2{^{\mathcal{I}}\mathbf{a}_{1/2}}, \tag{7.36}$$

where $\mathbf{F}_{1,2}$ is the gravitational force on body 1 due to body 2. Using Newton's universal law of gravity for $\mathbf{F}_{1,2}$ (see Eq. (4.25)) we obtain the two-body vector equation of motion,

$$\frac{m_1 m_2}{m_1 + m_2}{^{\mathcal{I}}\mathbf{a}_{1/2}} = \frac{-Gm_1 m_2}{\|\mathbf{r}_{1/2}\|^3}\mathbf{r}_{1/2},$$

or more simply,

$$\frac{^{\mathcal{I}}d^2}{dt^2}(\mathbf{r}_{1/2}) + \frac{Gm_G}{\|\mathbf{r}_{1/2}\|^3}\mathbf{r}_{1/2} = 0. \tag{7.37}$$

Eq. (7.37) is the same equation of motion as in Eq. (4.26) except with m_O replaced by $m_G = m_1 + m_2$ and P/O replaced by $1/2$. The solution is thus the same (elliptical orbits of m_1 about m_2) but with slightly different periods.

As in Example 4.9, we can also examine the two-body problem in polar coordinates using components in polar frame $\mathcal{B} = (G, \mathbf{e}_r, \mathbf{e}_\theta, \mathbf{e}_3)$, which is fixed to G and rotating so that $\mathbf{e}_r$ is directed toward body 1. Let $r \triangleq \|\mathbf{r}_{1/2}\|$ denote the distance between body 1 and body 2, and let θ denote the angular separation between frame $\mathcal{B}$ and frame $\mathcal{I}$. (Assume θ increases in the direction of rotation of the bodies.) We have

$$\mathbf{r}_{1/2} = r\mathbf{e}_r$$

$$^{\mathcal{I}}\mathbf{v}_{1/2} = \dot{r}\mathbf{e}_r + r\dot{\theta}\mathbf{e}_\theta$$

$$^{\mathcal{I}}\mathbf{a}_{1/2} = (\ddot{r} - r\dot{\theta}^2)\mathbf{e}_r + (2\dot{r}\dot{\theta} + r\ddot{\theta})\mathbf{e}_\theta.$$

The gravitational force $\mathbf{F}_{1,2}$ is

$$\mathbf{F}_{1,2} = -\frac{Gm_1 m_2}{r^2}\mathbf{e}_r.$$

Substituting the expressions for $\mathbf{F}_{1,2}$ and $^{\mathcal{I}}\mathbf{a}_{1/2}$ into Eq. (7.36) yields

$$\ddot{\theta} = -\frac{2\dot{r}\dot{\theta}}{r} \tag{7.38}$$

$$\ddot{r} = -\frac{Gm_G}{r^2} + r\dot{\theta}^2, \tag{7.39}$$

which are the same equations of motion as in Example 4.9, again with m_O replaced by m_G.

Eqs. (7.38) and (7.39) are not the only results we have seen before. For example, computing the angular momentum ${}^{\mathcal{I}}\mathbf{h}_G$ of bodies 1 and 2 about G gives

$$
{}^{\mathcal{I}}\mathbf{h}_G = \sum_{i=1}^{2} m_i \mathbf{r}_{i/G} \times {}^{\mathcal{I}}\mathbf{v}_{i/G},
$$

where

$$
\mathbf{r}_{1/G} = \mu_2 r \mathbf{e}_r
$$

$$
{}^{\mathcal{I}}\mathbf{v}_{1/G} = \mu_2 (\dot{r}\mathbf{e}_r + r\dot{\theta}\mathbf{e}_\theta)
$$

$$
\mathbf{r}_{2/G} = -\mu_1 r \mathbf{e}_r
$$

$$
{}^{\mathcal{I}}\mathbf{v}_{2/G} = -\mu_1 (\dot{r}\mathbf{e}_r + r\dot{\theta}\mathbf{e}_\theta).
$$

Consequently, we have

$$
{}^{\mathcal{I}}\mathbf{h}_G = \frac{m_1 m_2}{m_1 + m_2} r^2 \dot{\theta} \mathbf{e}_3,
$$

which implies that the magnitude of the angular momentum is proportional to $r^2\dot{\theta}$—the same value as the specific angular momentum h_O in Tutorial 4.2!

In fact, all of the results from Example 4.9 and Tutorial 4.2 apply here, as well as those from Tutorial 5.2. The orbit of body 1 about body 2 is still a conic section. We can study orbits about the sun with no error or approximations, as long as we replace m_O in Example 4.9 with m_G, the sum of the two masses. This tutorial also shows why the analysis in Example 4.9, where we assumed the central body was fixed in absolute space, was a good approximation: if $m_1 \gg m_2$, we introduce little error by using m_1 instead of m_G.

Tutorial 7.2 The Energy of Two Falling Masses Connected by a Spring

This tutorial revisits the problem examined in Tutorial 6.1 of two falling masses connected by a spring. We showed that, even though the dynamics of the center of mass separates from the motion about the center of mass, the full equations of motion do not always separate. In Tutorial 6.1 we could treat the center-of-mass motion completely independently of the motion relative to the center of mass as long as the gravitational field was a constant; this was not the case when we treated gravity more precisely using the $1/r^2$ form. Here we look at the total energy of this nonrigid collection of particles.

First we write the kinetic energy as the sum of the individual kinetic energies,

$$
T_O = \frac{1}{2} m \dot{y}_m^2 + \frac{1}{2} M \dot{y}_M^2
$$

or equivalently, as the sum of the kinetic energy of the center of mass plus the kinetic energy due to motion relative to the center of mass,

$$T_O = \frac{1}{2}(m + M)\dot{y}_G^2 + \frac{1}{2}m\dot{y}_{m/G}^2 + \frac{1}{2}M\dot{y}_{M/G}^2.$$

What about the work done on the masses? The work done by the internal force (the spring) is equal to the change in the potential energy of the spring between the endpoints,

$$W^{(\text{int})} = \frac{1}{2}k\left[l - (y_m(t_f) - y_M(t_f))\right]^2 - \frac{1}{2}k\left[l - (y_m(t_0) - y_M(t_0))\right]^2,$$

which can also be written in terms of the motion relative to the center of mass,

$$W^{(\text{int})} = \frac{1}{2}k\left[l - (y_{m/G}(t_f) - y_{M/G}(t_f))\right]^2 - \frac{1}{2}k\left[l - (y_{m/G}(t_0) - y_{M/G}(t_0))\right]^2.$$

The external work is the work done in moving the masses down in the gravitational field. For the constant-gravity model, the work on each mass is

$$W_{m/O}^{(\text{ext})} = mg(y_m(t_f) - y_m(t_0))$$

$$W_{M/O}^{(\text{ext})} = Mg(y_M(t_f) - y_M(t_0)).$$

Adding these work sources together gives the total work,

$$W^{(\text{tot})} = W_{m/O}^{(\text{ext})} + W_{M/O}^{(\text{ext})} + W^{(\text{int})}$$

$$= \underbrace{(M + m)g(y_G(t_f) - y_G(t_0))}_{=W_{G/O}} + \frac{1}{2}k\left[l - (y_{m/G}(t_f) - y_{M/G}(t_f))\right]^2$$

$$- \frac{1}{2}k\left[l - (y_{m/G}(t_0) - y_{M/G}(t_0))\right]^2.$$

For the constant-gravity case, the total work is the work done on the center of mass by gravity ($W_{G/O}$) plus the internal work of the spring due to motion relative to the center of mass. In fact, because the force of gravity is a constant, we could have computed the work done on the center of mass directly. Since all of the forces, whether internal or external, are conservative, we can also write the following conservation law for the center of mass:

$$E_{G/O} = (M + m)gy_G + \frac{1}{2}(M + m)\dot{y}_G^2 = \text{constant}.$$

The situation is different for the $1/r^2$ gravity field. While the internal work is the same, the external work due to gravity, which equals the change in potential energy, is now equal to (see Table 5.1)

$$W^{(\text{ext})} = \mu m \left(\frac{1}{R_E + y_m(t_f)} - \frac{1}{R_E + y_m(t_0)} \right)$$
$$+ \mu M \left(\frac{1}{R_E + y_M(t_f)} - \frac{1}{R_E + y_M(t_0)} \right),$$

where $\mu = Gm_E$. This work expression cannot be separated into the work associated with the position of the center of mass and that associated with the motion relative to the center of mass (even though we can perform this separation for the kinetic energy). Consequently, we cannot apply a conservation-of-energy law solely to the center-of-mass motion. This failure to separate is due to the specific nonlinear nature of the gravitational force.

Nevertheless, we can obtain an expression for the conservation of total energy, since all the forces are conservative. The total energy of the system is

$$E_O = \frac{1}{2}m\dot{y}_m^2 + \frac{1}{2}M\dot{y}_M^2 + \frac{\mu m}{R_e + y_m} + \frac{\mu M}{R_e + y_M}$$
$$+ \frac{1}{2}k \left[l - (y_m - y_M) \right]^2 = \text{constant}.$$

Tutorial 7.3 Scattering

Chapter 6 examined in detail the collision problem of two solid particles colliding and rebounding with altered trajectories. In this tutorial we discuss a similar problem called *scattering*. Particles interact through field forces, such as gravity or electrostatics, and thus change their trajectories as they come into proximity of each other without actually colliding. We have already seen one example of scattering—hyperbolic orbits in the two-body problem. In that case a small body approaches from an infinite distance, interacts gravitationally with a large body, and moves off to infinity at some angle relative to the arrival trajectory. We derive the equations for that trajectory in more detail here.

Scattering is an important problem in physics and describes common dynamical behavior from the very small atomic scale to the large-scale motion of comets and other planetary bodies. In fact, the famous Rutherford experiment verified the atomic nature of matter by scattering charged particles off the nucleus.[4]

In this tutorial we consider two particles, P and Q, moving toward each other from an infinite separation on straight-line trajectories, each at the same constant speed v_∞ in an inertial frame $\mathcal{I}$, as shown in Figure 7.11a. Their trajectories are separated by a distance b, called the *impact parameter*. They interact through a

[4] Ernest Rutherford (1871–1937) was a New Zealand chemist and physicist. He is credited as the discoverer of the nucleus from his famous gold-foil experiment, where he measured the scattering of a beam of alpha particles directed at a thin gold foil. He was awarded the Nobel Prize in chemistry in 1908.

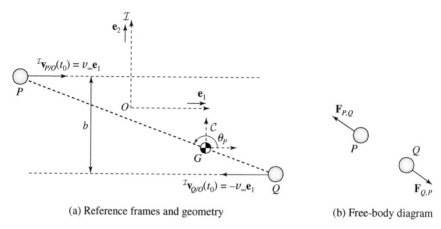

(a) Reference frames and geometry   (b) Free-body diagram

Figure 7.11 (a) Two charged particles traveling along straight-line paths toward each other with separation b. (b) The free-body diagram of each particle.

repulsive electrostatic force along the line between them. We are interested in the trajectory of each particle relative to the inertial frame $\mathcal{C}$ located at their center of mass. Since there are no external forces acting, the center of mass is either fixed or moving at constant velocity. The center of mass is located on the line connecting the particles, as shown in Figure 7.11a, and its position in $\mathcal{I}$ is given by

$$\mathbf{r}_{G/O} = \frac{1}{m_P + m_Q} \left(m_P \mathbf{r}_{P/O} + m_Q \mathbf{r}_{Q/O} \right).$$

The (constant) velocity of the center of mass in $\mathcal{I}$ can be found using the initial velocity of the two particles (their velocities at infinity):

$$
\begin{aligned}
{}^{\mathcal{I}}\mathbf{v}_{G/O}(t) &= \frac{1}{m_P + m_Q} \left(m_P {}^{\mathcal{I}}\mathbf{v}_{P/O}(t_0) + m_Q {}^{\mathcal{I}}\mathbf{v}_{Q/O}(t_0) \right) \\
&= - \left(\frac{m_Q - m_P}{m_P + m_Q} \right) v_\infty \mathbf{e}_1.
\end{aligned}
\tag{7.40}
$$

Let $m_Q \geq m_P$, which implies the center of mass moves in the $-\mathbf{e}_1$ direction unless $m_Q = m_P$, in which case G is fixed.

We use Newton's second law to find separately the equations of motion for the position of each particle relative to the center-of-mass frame $\mathcal{C}$. The repulsive force between the particles in the free-body diagram in Figure 7.11b is given by Coulomb's law,

$$\mathbf{F}_{P,Q} = -\frac{K \mathbf{r}_{Q/P}}{\|\mathbf{r}_{Q/P}\|^3} = -\mathbf{F}_{Q,P},$$

where $K > 0$ is a constant proportional to the charges on the particles. The position of Q relative to P is

$$\mathbf{r}_{Q/P} = \mathbf{r}_{Q/G} - \mathbf{r}_{P/G}.$$

The equations of motion for each particle in $\mathcal{C}$ are

$$m_P \frac{{}^{\mathcal{C}}d^2}{dt^2}\mathbf{r}_{P/G} = -\frac{K\mathbf{r}_{Q/P}}{\|\mathbf{r}_{Q/P}\|^3}$$

$$m_Q \frac{{}^{\mathcal{C}}d^2}{dt^2}\mathbf{r}_{Q/G} = \frac{K\mathbf{r}_{Q/P}}{\|\mathbf{r}_{Q/P}\|^3}.$$

We now use the fact that no external forces are acting on the system. The center of mass is thus moving at the constant velocity given in Eq. (7.40) and the relative positions of the two particles satisfy the center-of-mass corollary, which implies

$$\mathbf{r}_{Q/G} = -\frac{m_P}{m_Q}\mathbf{r}_{P/G}.$$

This equation allows us to rewrite the separation between the particles as

$$\mathbf{r}_{Q/P} = \mathbf{r}_{Q/G} - \mathbf{r}_{P/G} = -\frac{m_P}{m_Q}\mathbf{r}_{P/G} - \mathbf{r}_{P/G} = -\left(\frac{m_P + m_Q}{m_Q}\right)\mathbf{r}_{P/G}.$$

Performing a similar calculation for $\mathbf{r}_{Q/G}$ yields the following equations of motion in $\mathcal{C}$:

$$\frac{{}^{\mathcal{C}}d^2}{dt^2}\mathbf{r}_{P/G} - \frac{\mu_P \mathbf{r}_{P/G}}{\|\mathbf{r}_{P/G}\|^3} = 0 \tag{7.41}$$

$$\frac{{}^{\mathcal{C}}d^2}{dt^2}\mathbf{r}_{Q/G} - \frac{\mu_Q \mathbf{r}_{Q/G}}{\|\mathbf{r}_{Q/G}\|^3} = 0, \tag{7.42}$$

where

$$\mu_P = \frac{K}{m_P}\left(\frac{m_Q}{m_P + m_Q}\right)^2$$

$$\mu_Q = \frac{K}{m_Q}\left(\frac{m_P}{m_P + m_Q}\right)^2.$$

The absence of an external force allowed us to rewrite the equations of motion as two independent equations for each of the particles P and Q relative to the center of mass. Because of the symmetry of Eqs. (7.41) and (7.42), solving one generates the solution to the other. Let us solve for the motion of P relative to the center of mass. Eq. (7.41) has the same form as the orbit equation in Eq. (4.26), which means we can follow a similar procedure for solving it. Consider the position of P with respect to G in polar coordinates $(r_P, \theta_P)_{\mathcal{C}}$. This choice of coordinates allows us to separate the equation of motion into two scalar differential equations:

$$\ddot{\theta}_P = -\frac{2\dot{r}_P\dot{\theta}_P}{r_P} \tag{7.43}$$

$$\ddot{r}_P = \frac{\mu_P}{r_P^2} + r_P\dot{\theta}_P^2. \tag{7.44}$$

We now use the fact that the force between the particles acts along the line connecting them. Thus the internal-moment assumption holds and the total angular momentum about G is conserved:

$$^C\mathbf{h}_G = m_P\mathbf{r}_{P/G} \times {}^C\mathbf{v}_{P/G} + m_Q\mathbf{r}_{Q/G} \times {}^C\mathbf{v}_{Q/G} = \text{constant.} \tag{7.45}$$

We can calculate the total angular momentum using the initial conditions when the particles are infinitely far apart. Since in $\mathcal{I}$, the particles have equal and opposite velocities, ${}^{\mathcal{I}}\mathbf{v}_{P/O}(t_0) = v_\infty\mathbf{e}_1$ and ${}^{\mathcal{I}}\mathbf{v}_{Q/O}(t_0) = -v_\infty\mathbf{e}_1$, we can use Eq. (7.40) to find the initial velocities of each relative to G in $\mathcal{C}$:

$$^C\mathbf{v}_{P/G}(t_0) = v_\infty\mathbf{e}_1 + \left(\frac{m_Q - m_P}{m_P + m_Q}\right)v_\infty\mathbf{e}_1 = \frac{2m_Q}{m_P + m_Q}v_\infty\mathbf{e}_1 \tag{7.46}$$

$$^C\mathbf{v}_{Q/G}(t_0) = -v_\infty\mathbf{e}_1 + \left(\frac{m_Q - m_P}{m_P + m_Q}\right)v_\infty\mathbf{e}_1 = -\frac{2m_P}{m_P + m_Q}v_\infty\mathbf{e}_1. \tag{7.47}$$

Using the geometry shown in Figure 7.11a and the expressions for the two velocities in Eqs. (7.46) and (7.47), the total angular momentum is

$$^C\mathbf{h}_G(t) = {}^C\mathbf{h}_G(t_0) = \frac{2m_Pm_Q}{m_P + m_Q}v_\infty\left(\mathbf{r}_{P/G}(t_0) \times \mathbf{e}_1 - \mathbf{r}_{Q/G}(t_0) \times \mathbf{e}_1\right)$$

$$= -\frac{2m_Pm_Q}{m_P + m_Q}v_\infty b\mathbf{e}_3. \tag{7.48}$$

We can also use the center-of-mass corollary to substitute for $\mathbf{r}_{Q/G}$ and ${}^C\mathbf{v}_{Q/G}$ in the total angular momentum expression in Eq. (7.45) to rewrite it in terms of the angular momentum of P with respect to G only:

$$^C\mathbf{h}_G = m_P\mathbf{r}_{P/G} \times {}^C\mathbf{v}_{P/G} + m_Q\left(\frac{m_P}{m_Q}\right)^2 \mathbf{r}_{P/G} \times {}^C\mathbf{v}_{P/G}$$

$$= \left(\frac{m_P + m_Q}{m_Q}\right)m_P\mathbf{r}_{P/G} \times {}^C\mathbf{v}_{P/G}$$

$$= \left(\frac{m_P + m_Q}{m_Q}\right){}^C\mathbf{h}_{P/G}.$$

This equation states that the angular momentum of P with respect to G and the angular momentum of Q with respect to G are both conserved.

Equating this expression for the total angular momentum about G with Eq. (7.48) gives an expression for the specific angular momentum of P with respect to G in terms of the initial speeds:

$$\frac{^C\mathbf{h}_{P/G}}{m_P} = -\frac{2m_Q^2}{(m_P + m_Q)^2} v_\infty b \mathbf{e}_3 \triangleq h_P \mathbf{e}_3. \tag{7.49}$$

Eq. (7.49) allows us to perform the same substitution in the polar equations of motion as we did in Example 4.9 for the specific angular momentum in polar coordinates, $h_P = r_P^2 \dot{\theta}_P$, so Eqs. (7.43) and (7.44) become

$$\ddot{\theta}_P = \frac{-2\dot{r}_P h_P}{r_P^3} \tag{7.50}$$

$$\ddot{r}_P = \frac{\mu_P}{r_P^2} + \frac{h_P^2}{r_P^3}. \tag{7.51}$$

The result is the separation of the equations of motion so that we have a single differential equation for r_P, as in Eq. (4.47). We can thus follow the same procedure as in Tutorial 4.2, except we now make the change of variables $y = -1/r_P$ to account for the sign change due to the repulsive rather than attractive force. Substituting in Eq. (7.51), we find

$$\frac{d^2 y}{d\theta_P^2} + y = \frac{\mu_P}{h_P^2}.$$

This is the equation for simple harmonic motion, which, when put in terms of r_P, gives the conic equation

$$r_P(\theta_P) = \frac{-h_P^2/\mu_P}{1 + e \cos(\theta_P - \theta_0)},$$

where θ_0 is the initial angle and e is the eccentricity.

Assuming that P starts at infinity, as in Figure 7.11, $\theta_0 = \pi$ and the polar equation becomes

$$r_P(\theta_P) = \frac{h_P^2/\mu_P}{e \cos \theta_P - 1} > 0. \tag{7.52}$$

The radial position r_P in Eq. (7.52) is always positive since, as we'll show next, the eccentricity of the scattering encounter is greater than 1. The position of Q relative to G is then found from the center-of-mass corollary, $\mathbf{r}_{Q/G} = -(m_P/m_Q)\mathbf{r}_{P/G}$.

Computing the total angular momentum about G allowed us to find the equation for the orbit. To get the shape of the orbit we need the eccentricity e, which is found by examining the total energy, as in Tutorial 5.2. For the closed orbits considered there, the energy was negative and the eccentricity was thus less than one (corresponding to circular or elliptical orbits). Here the total energy is positive because of the nonzero velocity when the particles are at infinity. To see this, consider the constant total energy of the system, using $T_{G/O} = 0$,

$$E_O = \frac{1}{2} m_P \|^C\mathbf{v}_{P/G}\|^2 + \frac{1}{2} m_Q \|^C\mathbf{v}_{Q/G}\|^2 + U_{P/Q} = \text{constant},$$

where $U_{P/Q}$ is the potential energy of the two particles (computed by integrating the Coulomb force using $\|r_{Q/P}\| = \infty$ as the zero reference),

$$U_{P/Q} = \frac{K}{\|r_{Q/P}\|}.$$

Again using the center-of-mass corollary, we can rewrite the total energy as

$$E_O = \frac{1}{2}\frac{m_P}{m_Q}\left(m_P + m_Q\right)\|^{\mathcal{C}}\mathbf{v}_{P/G}\|^2 - \frac{Km_Q}{m_P + m_Q}\frac{1}{r_P}.$$

Following the same procedure as in Tutorial 5.2, we substitute the velocity in polar coordinates to find the specific energy

$$\varepsilon_O \triangleq \frac{1}{m_P + m_Q}\frac{m_Q}{m_P}E_O = \frac{1}{2}\dot{r}_P^2 + \frac{h_P^2}{2r_P^2} - \frac{\mu_P}{r_P}.$$

This equation is the same as Eq. (5.20). The eccentricity in terms of the specific energy and angular momentum is the same as Eq. (5.22):

$$e = \sqrt{1 + \frac{2\varepsilon_O h_P^2}{\mu_P^2}}. \tag{7.53}$$

As noted, for closed orbits the energy is negative and thus the eccentricity is less than one. Here the energy is positive, as seen by computing the total specific energy at the initial time when the particles are at infinity, so that $U_{P/Q} = 0$ and

$$\varepsilon_O = \frac{1}{2}\|^{\mathcal{C}}\mathbf{v}_{P/G}(t_0)\|^2 = \frac{2m_Q^2}{(m_P + m_Q)^2}v_\infty^2 > 0.$$

Substituting into Eq. (7.53) gives the eccentricity,

$$e = \sqrt{1 + \frac{16m_P^2 m_Q^2 v_\infty^4 b^2}{K^2(m_P + m_Q)^2}} > 1.$$

The particles thus follow a hyperbolic trajectory as they scatter off each other. We often describe such a hyperbola by the *asymptotes* or the straight-line trajectories at infinity. The angle between the asymptotes of a hyperbola,

$$\phi = 2\arcsin\left(\frac{1}{e}\right),$$

is called the *turning angle*. It indicates the angle by which the trajectory has rotated due to the scattering encounter.

Figure 7.12 shows an example of a pair of scattering trajectories. The two particles have equal charge and equal mass, making the center of mass fixed at the origin of the inertial frame.

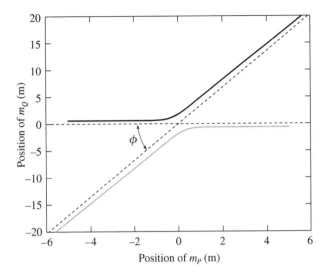

Figure 7.12 Trajectories of a scattering encounter between two particles of equal charge and mass, traveling in opposite directions at 5 m/s in the inertial frame. The dashed line shows an asymptote of the resulting hyperbolic trajectory. Turning angle ϕ is also shown.

7.6 Key Ideas

- The **total angular momentum** of a collection of particles is equal to the sum of the angular momentum of each individual particle:

$$^{\mathcal{I}}\mathbf{h}_O = \sum_{i=1}^{N} m_i \mathbf{r}_{i/O} \times {}^{\mathcal{I}}\mathbf{v}_{i/O}.$$

When the internal-moment assumption holds, the time derivative of $^{\mathcal{I}}\mathbf{h}_O$ is equal to the total external moment on the system:

$$\frac{^{\mathcal{I}}d}{dt}\left(^{\mathcal{I}}\mathbf{h}_O\right) = \sum_{i=1}^{N} \mathbf{r}_{i/O} \times \mathbf{F}_i^{(\text{ext})} = \mathbf{M}_O^{(\text{ext})}.$$

- When the internal-moment assumption holds and the total external moment acting on a system of particles about O is zero, then the **total angular momentum of the collection about O is conserved**.

- The total angular momentum of a collection of particles relative to point O can be separated into the angular momentum of the center of mass about O, $^{\mathcal{I}}\mathbf{h}_{G/O}$, and the angular momentum of the collection about the center of mass, $^{\mathcal{I}}\mathbf{h}_G$:

$$^{\mathcal{I}}\mathbf{h}_O = {}^{\mathcal{I}}\mathbf{h}_{G/O} + {}^{\mathcal{I}}\mathbf{h}_G,$$

where

$$^{\mathcal{I}}\mathbf{h}_{G/O} \triangleq m_G \mathbf{r}_{G/O} \times {}^{\mathcal{I}}\mathbf{v}_{G/O}$$

and

$$^{\mathcal{I}}\mathbf{h}_G \triangleq \sum_{i=1}^{N} m_i \mathbf{r}_{i/G} \times {}^{\mathcal{I}}\mathbf{v}_{i/G}.$$

- The dynamics of each of the angular momenta $^{\mathcal{I}}\mathbf{h}_{G/O}$ and $^{\mathcal{I}}\mathbf{h}_G$ are

$$\frac{^{\mathcal{I}}d}{dt}\left(^{\mathcal{I}}\mathbf{h}_{G/O}\right) = \mathbf{M}_{G/O}^{(\text{ext})}$$

and

$$\frac{^{\mathcal{I}}d}{dt}\left(^{\mathcal{I}}\mathbf{h}_G\right) = \mathbf{M}_{G}^{(\text{ext})},$$

where $\mathbf{M}_{G/O}^{(\text{ext})}$ is the external moment on the center mass relative to O and $\mathbf{M}_{G}^{(\text{ext})}$ is the external moment acting on all the particles relative to the center of mass.

- For some problems, it is convenient to consider the angular momentum of all the particles relative to an arbitrary point Q:

$$^{\mathcal{I}}\mathbf{h}_Q \triangleq \sum_{i=1}^{N} \mathbf{r}_{i/Q} \times m_i {}^{\mathcal{I}}\mathbf{v}_{i/Q}.$$

In terms of the angular momentum of the particles about their center of mass, this angular momentum can be written as

$$^{\mathcal{I}}\mathbf{h}_Q = {}^{\mathcal{I}}\mathbf{h}_G + m_G \mathbf{r}_{Q/G} \times {}^{\mathcal{I}}\mathbf{v}_{Q/G}.$$

- The dynamics of the angular momentum $^{\mathcal{I}}\mathbf{h}_Q$ are

$$\frac{^{\mathcal{I}}d}{dt}\left(^{\mathcal{I}}\mathbf{h}_Q\right) = \mathbf{M}_{Q}^{(\text{ext})} + \mathbf{r}_{Q/G} \times m_G {}^{\mathcal{I}}\mathbf{a}_{Q/O}.$$

The equation of motion for rotation relative to Q is thus coupled to the translational motion of Q.

- The **total kinetic energy** of a collection of particles is equal to the sum of the kinetic energy of each individual particle:

$$T_O \triangleq \sum_{i=1}^{N} T_{i/O} = \frac{1}{2}\sum_{i=1}^{N} m_i {}^{\mathcal{I}}\mathbf{v}_{i/O} \cdot {}^{\mathcal{I}}\mathbf{v}_{i/O} = \frac{1}{2}\sum_{i=1}^{N} m_i \|{}^{\mathcal{I}}\mathbf{v}_{i/O}\|^2.$$

- As with the total angular momentum, the **total kinetic energy** of the particles separates into the kinetic energy $T_{G/O}$ of the center of mass and the kinetic energy

T_G of the particles relative to the center of mass:

$$T_O = T_{G/O} + T_G,$$

where

$$T_{G/O} \triangleq \frac{1}{2} m_G \|{}^{\mathcal{I}}\mathbf{v}_{G/O}\|^2$$

and

$$T_G \triangleq \frac{1}{2} \sum_{i=1}^{N} m_i \|{}^{\mathcal{I}}\mathbf{v}_{i/G}\|^2.$$

- The **total work** on a system of particles is equal to the sum of the **external** and **internal** work:

$$W^{(\text{tot})} = W^{(\text{ext})} + W^{(\text{int})}.$$

- The **power on the center of mass** of the particles is related to the total external force acting on the center of mass and the change in kinetic energy of the center of mass:

$$\mathrm{P}_G = \mathbf{F}_G^{(\text{ext})} \cdot {}^{\mathcal{I}}\mathbf{v}_{G/O}(t) = \frac{d}{dt} T_{G/O}(t).$$

If the total external force is zero, the kinetic energy of the center of mass is a constant of the motion.

- Using $\Delta E_O \triangleq E_O(t_2) - E_O(t_1)$, the **work-energy formula** for a multiparticle system is

$$\Delta E_O = \Delta T_{G/O} + \Delta T_G + \Delta U_O^{(\text{ext})} + \Delta U_O^{(\text{int})}$$

$$= W^{(\text{nc,ext})}(\{\mathbf{r}_{i/O}; \gamma_i\}_{i=1}^N) + W^{(\text{nc,int})}(\{\mathbf{r}_{i/O}; \gamma_i\}_{i=1}^N).$$

If the total nonconservative internal and external forces are zero, the total energy of the system of particles is a constant of the motion.

7.7 Notes and Further Reading

To obtain one of the main ideas of this chapter, Eq. (7.5), which equates the rate of change of total angular momentum to the external moment on the collection, we had to make an assumption about the internal moments acting among the particles. Remarkably, a surprising number of texts pass over this issue quite quickly. They either ignore the internal-moment term in Eq. (7.3) (e.g., Greenwood 1988; Tongue and Sheppard 2005), incorrectly state that Newton's third law implies that the internal forces act along the line connecting the particles (e.g., Beer et al. 2007; Ginsberg 2008), or invoke Newton's third law to state that the sum of the internal moments in Eq. (7.3) must be zero (e.g., Meriam and Kraige 2001; Rao 2006). We have emphasized that Eq. (7.5) only holds for those cases where the internal-moment assumption applies (and we provide a counterexample in Problem 7.13). This subtle and historically contentious point dates back to Bernoulli and Euler and reminds us of the importance of always checking the assumptions that go into solving a problem. An excellent discussion of the history of the "law of moment of momentum" can be found in the engaging book of essays by Truesdell (1968). We revisit this point again in the notes for Chapter 9.

We also emphasized the importance of considering the internal forces and their contributions to the internal work on a system of particles. As in the derivation of the angular momentum equation, some texts ignore the internal forces when computing the work and energy of a collection. Among those that do include internal work is Bedford and Fowler (2002), where we found the material for Examples 7.9 and 7.10.

7.8 Problems

7.1 Consider the three-particle system shown in Figure 7.13. The mass, position, and velocity of each particle are $m_1 = m$, $\mathbf{r}_{1/O} = \mathbf{e}_x + 2\mathbf{e}_y$, ${}^{\mathcal{I}}\mathbf{v}_{1/O} = -\mathbf{e}_x - \mathbf{e}_y$; $m_2 = 2m$, $\mathbf{r}_{2/O} = 2.5\mathbf{e}_x + \mathbf{e}_y$, ${}^{\mathcal{I}}\mathbf{v}_{2/O} = 2\mathbf{e}_y$; and $m_3 = 3m$, $\mathbf{r}_{3/O} = \mathbf{e}_x + \mathbf{e}_y$, ${}^{\mathcal{I}}\mathbf{v}_{3/O} = \mathbf{e}_x$.

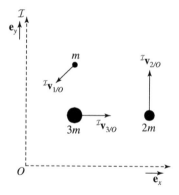

Figure 7.13 Problem 7.1.

    a. Find the total angular momentum ${}^{\mathcal{I}}\mathbf{h}_O$ about O.
    b. Find the angular momentum ${}^{\mathcal{I}}\mathbf{h}_{G/O}$ of the center of mass about O. [HINT: First find the center of mass $\mathbf{r}_{G/O}$.]
    c. Find the angular momentum ${}^{\mathcal{I}}\mathbf{h}_G$ about the center of mass.
    d. Show that ${}^{\mathcal{I}}\mathbf{h}_O = {}^{\mathcal{I}}\mathbf{h}_{G/O} + {}^{\mathcal{I}}\mathbf{h}_G$.

7.2 Repeat Problem 7.1 for kinetic energy:

    a. Find the total kinetic energy T_O with respect to O.
    b. Find the kinetic energy $T_{G/O}$ of the center of mass about O.
    c. Find the kinetic energy T_G about the center of mass.
    d. Show that $T_O = T_{G/O} + T_G$.

7.3 Find the equation of motion for the seesaw shown in Figure 7.14.

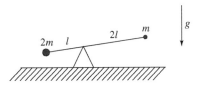

Figure 7.14 Problem 7.3.

7.4 A particle of mass m_1 slides without friction along a circular track of radius R cut from a block of mass m_2, as shown in Figure 7.15. The block slides without friction along a horizontal surface. Assume that x describes the horizontal displacement of the block and that θ describes the position of the particle relative to the vertical. Assume that $x(0) = 0$, $\dot{x}(0) = 0$, and $\dot{\theta}(0) = 0$. Be sure to explicitly point out any use of conserved quantities, such as energy or momentum.

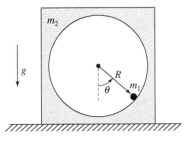

Figure 7.15 Problem 7.4.

a. If the particle starts at $\theta(0) = \theta_0$, what is the velocity of the particle and the velocity of the block when $\theta = 0$? [HINT: Start by carefully formulating the inertial velocity of the block and the particle in terms of x, $\dot{x}$, θ, and $\dot{\theta}$.]

b. If friction is introduced between the particle and the block (assuming no friction between the block and the horizontal surface) so that the particle oscillates until it lies at rest at $\theta = 0$, what is the steady-state position of the block, assuming $\theta(0) = \pi/2$?

7.5 The baton shown in Figure 7.16 is leaning against a vertical wall. Because there is no friction between the baton and the wall or between the baton and the floor, the baton slides along the floor. Find the angle θ when the baton falls, assuming it starts nearly upright. [HINT: The baton falls when the normal force from the wall is zero.]

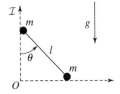

Figure 7.16 Problem 7.5.

7.6   Consider the two masses connected by a spring and damper from Example 7.11. Suppose they start moving with the initial conditions shown in Figure 7.17, with particle P moving to the right at a speed v_0. After reaching its steady-state motion, the spring has stretched to twice its original length. Assuming that the two particles have equal mass m, show that v_0 as a function of the other system parameters is given by $v_0 = 4r_0\sqrt{k/m}$, where r_0 is the unstretched length of the spring and k is its spring constant. Compute the work done by the damper as a fraction of the total initial energy.

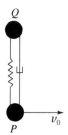

Figure 7.17 Problem 7.6.

7.7   Three identical masses lie on a frictionless horizontal surface. One mass is at rest and the other two masses, which are connected by a string of length l, slide with initial speed v_0, as shown in Figure 7.18a. The middle of the string meets the stationary mass, and all three masses slide together, as shown in Figure 7.18b. When the two masses at each end of the string collide, they stick together (Figure 7.18c).

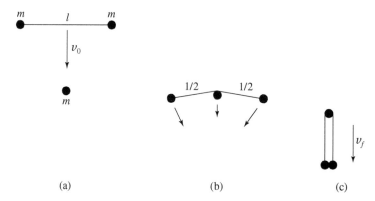

(a)   (b)   (c)

Figure 7.18 Problem 7.7. (a) Initial setup. (b) Intermediate configuration. (c) Final configuration.

      a. Find the final speed v_f of the three-mass system after the collision. [HINT: There are no external horizontal forces.]

      b. Find the velocity of each of the three masses at the instant of the collision.

7.8   Consider again the colliding rail cars in Problem 6.14. What is the maximum compression of the spring after the collision?

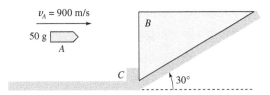

Figure 7.19 Problem 7.9.

7.9 The 10 kg wedge B shown in Figure 7.19 is held at rest on an incline of 30° by stop C when it is struck by the 50 g bullet A, which is traveling horizontally with speed $v_A = 900$ m/s. Assume that the bullet is stopped instantaneously by the block.

   a. Calculate the velocity with which the wedge starts up the incline.
   b. Assuming gravity acts vertically downward, how high does the block travel before stopping? (Assume zero friction.)

7.10 Suppose an apple of mass M is hanging motionless on a massless string of length l from a fixed point O when it is impaled by an arrow of mass m moving horizontally with speed v_0 (Figure 7.20). What is the maximum angle θ that the apple reaches after the collision?

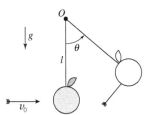

Figure 7.20 Problem 7.10.

7.11 Consider two masses m_1 and m_2 interacting gravitationally with no external forces, as in Tutorial 7.1. Find the equations of motion of each mass relative to the center of mass of the system. What do you think the resulting orbits look like?

7.12 James Bond is being chased by a group of S.P.E.C.T.R.E. agents and spots a ramp in front of a canyon. In an attempt to escape, he tries to jump his Aston Martin DB5 across the canyon. The ramp has an angle α and the canyon is L meters wide. Bond has mass m_P and the Aston Martin has mass m_C.

   a. Given that Bond drives off the ramp at constant speed v_o, what is the greatest possible width L the canyon can have so that Bond escapes safely?
   b. Find an expression for Bond's horizontal position, relative to the edge of the ramp, when he reaches the apex of the trajectory. What must this equal so that Bond knows he can cross the canyon safely at this point?
   c. At the peak of the trajectory Bond has either decided he won't make it, or he is simply taking some time out of his day to irritate Q by returning the car in less than pristine order. Regardless, he makes the decision to use the ejector seat. Assume that Bond and the car are two point

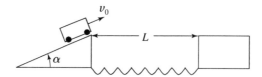

Figure 7.21 Problem 7.12.

masses. Model the ejector with a spring of constant k compressed by $\Delta\xi$. Assume that the spring is released quickly, so the resulting force is well modeled by an impulse. Also assume that the impulse due to the spring acts purely in the vertical direction between mass m_P and m_C. What is Bond's velocity ${}^{\mathcal{I}}\mathbf{v}_{P/O}(t^*)$ immediately after deployment (where t^* is the time he reaches the apex)?

d. Suppose that the canyon length is indeed too long to cross without ejection. Solve for the spring compression $\Delta\xi$ of Bond's ejector seat that ensures a safe crossing of the canyon. Express your answer in terms of L, v_0, α, k, m_P, and m_C.

7.13 Consider a slightly different scattering experiment than the one described in Tutorial 7.3. Rather than two charged particles interacting, here you analyze the encounter of a positively charged particle Q with a magnetic dipole P. Assume Q has charge $+q$ and P has a dipole moment $\mathbf{m} = m\mathbf{e}_3$, so that the dipole moment is always perpendicular to the plane of the encounter. The magnetic field $\mathbf{B}$ due to a dipole is

$$\mathbf{B} = \frac{\mu_0}{4\pi r^3}\left(3(\mathbf{m}\cdot\hat{\mathbf{r}})\hat{\mathbf{r}} - \mathbf{m}\right),$$

where r is the distance from the dipole, μ_0 is the permeability of free space, and $\hat{\mathbf{r}}$ is the unit vector in the direction of the dipole. Assume that both particles have mass m_Q and initial velocities ${}^{\mathcal{I}}\mathbf{v}_{Q/O}(t_0)$ and ${}^{\mathcal{I}}\mathbf{v}_{P/O}(t_0)$. The force on Q due to P is given by the force due to a dipole,

$$\mathbf{F}_{Q,P} = q\,{}^{\mathcal{I}}\mathbf{v}_{Q/O} \times \mathbf{B},$$

since we have confined the dipole to be perpendicular to the plane. By Newton's third law, the force on P must be equal and opposite.

a. What is the velocity of the center of mass? Is it conserved?

b. Find an expression for the total angular momentum of the two particles, relative to an inertially fixed point. Is it conserved? Is the motion still an orbit in the plane?

c. Find the equations of motion for the two particles relative to a frame with origin you chose in (b).

d. Simulate the trajectories of the two particles, and plot on the same set of axes. Also plot the angular momentum you computed in (a). Does it confirm your conclusion in (b)? Does the internal-moment assumption hold for this system? Assume the particles each have a mass of 1 kg, an impact parameter b of 0.1 m, and equal and opposite initial velocities relative to their center of mass of 5 m/s. Q has a charge of 1 C and the magnetic dipole moment of P is 10^5 A-m^2. In these units, $\mu_0 = 1$.

PART THREE

Relative Motion and Rigid-Body
Dynamics in Two Dimensions

CHAPTER EIGHT

Relative Motion in a Rotating Frame

This chapter is devoted to relative motion in a rotating reference frame. Of course, you have already been introduced to some rotating frames, namely, the polar and path frames. These frames were enormously useful for solving certain problems, simplifying the kinematics, and providing new insights into particle motion. This chapter examines problems where using multiple rotating frames leads to important new formulas for dynamical systems. In particular, we see how the kinematics of certain planar rigid bodies are simplified by attaching a reference frame to them and then applying our previous results. We then complete the study of planar kinematics by examining the relative motion of a particle in a translating and rotating frame of reference. This expands our treatment in Chapter 3, where we restricted the discussion of relative motion to a translating frame. This chapter introduces one of the most important formulas of the book and the precursor to our subsequent treatment of rigid-body dynamics—the *transport equation*.

8.1 Rotational Motion of a Planar Rigid Body

This section focuses on the kinematics of a rotating and translating rigid body. We have alluded to rigid bodies before without rigorously defining them. For now, you simply need to recognize that a rigid body is a collection of particles constrained so as not to move relative to one another. In many cases we consider a continuous collection that forms a solid body, such as a disk, rod, or sphere. Alternatively, we may consider a finite collection of particles connected by massless rigid links. What is important for our study of kinematics is that to every rigid body we can attach a *body frame*. Recall Definition 3.1: a reference frame is equivalent to a rigid body. Since all points of a rigid body are fixed with respect to one another, we can use them to define a reference frame. Thus as a rigid body translates and rotates in inertial space, so will a body frame attached to it. The distinction between the polar frame and the body frame here is subtle: a polar frame always points toward a moving point, whereas a body

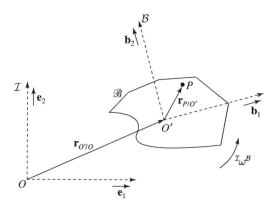

Figure 8.1 Planar rigid body $\mathscr{B}$ with attached body frame $\mathcal{B}$ translating and rotating in the plane with angular velocity ${}^{\mathcal{I}}\boldsymbol{\omega}^{\mathcal{B}}$ relative to inertial frame $\mathcal{I}$.

frame is fixed to a rigid body. As shown later, in many ways the body frame behaves just like a polar frame. This observation will allow us to reuse many previous results.

Figure 8.1 shows a rigid body $\mathscr{B}$ with an attached body frame $\mathcal{B} = (O', \mathbf{b}_1, \mathbf{b}_2, \mathbf{b}_3)$, where the origin O' is a point on the rigid body. O' is not necessarily the center of mass. The position of O' with respect to O is $\mathbf{r}_{O'/O}$. The rigid body (and thus frame $\mathcal{B}$) is rotating relative to the inertial frame with an angular velocity ${}^{\mathcal{I}}\boldsymbol{\omega}^{\mathcal{B}}$, as well as possibly translating with a velocity ${}^{\mathcal{I}}\mathbf{v}_{O'/O}$ and acceleration ${}^{\mathcal{I}}\mathbf{a}_{O'/O}$. Point P is an arbitrary point fixed to the rigid body, which implies that P does not move in frame $\mathcal{B}$.

Suppose we want to find the velocity of point P in the inertial frame. We start by writing the position of P relative to O using the vector triad,

$$\mathbf{r}_{P/O} = \mathbf{r}_{O'/O} + \mathbf{r}_{P/O'}.$$

We find the velocity of P by taking the time derivative of $\mathbf{r}_{P/O}$:

$$ {}^{\mathcal{I}}\mathbf{v}_{P/O} = \underbrace{\frac{{}^{\mathcal{I}}d}{dt}\left(\mathbf{r}_{O'/O}\right)}_{={}^{\mathcal{I}}\mathbf{v}_{O'/O}} + \frac{{}^{\mathcal{I}}d}{dt}\left(\mathbf{r}_{P/O'}\right). \tag{8.1}$$

The first term on the right-hand side of Eq. (8.1) is the translational velocity of $\mathscr{B}$, that is, the velocity ${}^{\mathcal{I}}\mathbf{v}_{O'/O}$ of the origin of frame $\mathcal{B}$. What about the second term? Although P has no velocity in $\mathcal{B}$, it is moving in $\mathcal{I}$ due to the rotation/translation of the body.

We approach this question the same way we approached the velocity in the polar frame in Chapter 3. Writing out the position of P as components in the body frame using the Cartesian coordinates $(x, y)_{\mathcal{B}}$ leads to

$$ {}^{\mathcal{I}}\mathbf{v}_{P/O'} = \frac{{}^{\mathcal{I}}d}{dt}\left(\mathbf{r}_{P/O'}\right) = \frac{{}^{\mathcal{I}}d}{dt}(x\mathbf{b}_1 + y\mathbf{b}_2) $$

$$ = x\frac{{}^{\mathcal{I}}d}{dt}\mathbf{b}_1 + y\frac{{}^{\mathcal{I}}d}{dt}\mathbf{b}_2, \tag{8.2}$$

where we have used the distributive property and product rule of vector differentiation and the fact that P is fixed in $\mathcal{B}$. The unit vector derivatives in Eq. (8.2) are the same as those treated in Chapter 3 when studying the polar and path reference frames. Since $\mathcal{B}$ is rotating with respect to $\mathcal{I}$ with angular velocity ${}^{\mathcal{I}}\boldsymbol{\omega}^{\mathcal{B}} \triangleq {}^{\mathcal{I}}\omega^{\mathcal{B}}\mathbf{b}_3$, the unit-vector derivatives are

$$\frac{{}^{\mathcal{I}}d}{dt}\mathbf{b}_i = {}^{\mathcal{I}}\boldsymbol{\omega}^{\mathcal{B}} \times \mathbf{b}_i \quad i = \{1, 2\}. \tag{8.3}$$

This is a good time to review Section 3.4.2. We are still restricting the motion to be planar, so the rotation of frame $\mathcal{B}$ with respect to $\mathcal{I}$ means a simple rotation by some angle θ about the $\mathbf{e}_3$ or, equivalently, $\mathbf{b}_3$ axis. The angular velocity is thus a simple angular velocity and is given by ${}^{\mathcal{I}}\boldsymbol{\omega}^{\mathcal{B}} = \dot{\theta}\mathbf{b}_3$. Note that the rotation rate $\dot{\theta}$ need not be constant! Although in Section 3.4.2 the development of Eq. (8.3) was for the polar (or path) frame, the derivation was completely general. *Eq. (8.3) describes the time derivative of an arbitrary unit vector $\mathbf{b}_i$ that is fixed in a frame $\mathcal{B}$ rotating with respect to frame $\mathcal{I}$ with angular velocity ${}^{\mathcal{I}}\boldsymbol{\omega}^{\mathcal{B}}$.*

Now we can substitute Eq. (8.3) into Eq. (8.2) to find

$$\mathbf{v}_{P/O'} = x({}^{\mathcal{I}}\boldsymbol{\omega}^{\mathcal{B}} \times \mathbf{b}_1) + y({}^{\mathcal{I}}\boldsymbol{\omega}^{\mathcal{B}} \times \mathbf{b}_2)$$

or, after using the distributive rule of the cross product,

$$\mathbf{v}_{P/O'} = {}^{\mathcal{I}}\boldsymbol{\omega}^{\mathcal{B}} \times \underbrace{(x\mathbf{b}_1 + y\mathbf{b}_2)}_{=\mathbf{r}_{P/O'}}. \tag{8.4}$$

The last term is the position of P in $\mathcal{B}$. This identification lets us write the final expression for the velocity of P in $\mathcal{I}$ by combining Eqs. (8.1) and (8.4):

$$\boxed{{}^{\mathcal{I}}\mathbf{v}_{P/O} = {}^{\mathcal{I}}\mathbf{v}_{O'/O} + {}^{\mathcal{I}}\boldsymbol{\omega}^{\mathcal{B}} \times \mathbf{r}_{P/O'}.} \tag{8.5}$$

Eq. (8.5) is a fundamental and important equation. It is our basic tool for finding the velocity of a point on a rigid body undergoing translational and rotational motion. Here we derived it under the restriction of planar rotation of the rigid body; later chapters show that it is completely general. It is the starting point for many applications in mechanics, some of which are shown in the following examples.

Example 8.1 Kinematics of a Compound Pendulum

A compound pendulum is a generalization of the simple pendulum we have been repeatedly studying. Rather than a particle attached to the end of a massless rod, a compound pendulum is an extended rigid body pinned at some point and allowed to swing, as shown in Figure 8.2. We restrict ourselves to planar motion, so this system has a single degree of freedom—the angle of the pendulum. We use the coordinate θ to measure the angle from the local vertical to a line connecting the pin to the center of mass. Figure 8.2b shows inertial frame $\mathcal{I} = (O, \mathbf{e}_1, \mathbf{e}_2, \mathbf{e}_3)$ located at the pin, labeled O, and body frame $\mathcal{B} = (O, \mathbf{b}_1, \mathbf{b}_2, \mathbf{b}_3)$ fixed to the pendulum at O. The angular

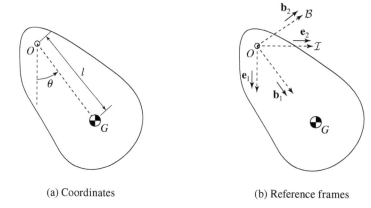

(a) Coordinates   (b) Reference frames

Figure 8.2 Compound pendulum consisting of a rigid body pinned at point O in the inertial frame.

velocity of the pendulum is thus $^{\mathcal{I}}\boldsymbol{\omega}^{\mathcal{B}} = \dot{\theta}\mathbf{b}_3$. In this example we calculate the inertial velocity $^{\mathcal{I}}\mathbf{v}_{G/O}$ of the center of mass of the pendulum.

Because the pendulum is pinned at O, there is no translational velocity ($^{\mathcal{I}}\mathbf{v}_{O'/O} = 0$), and the origin of $\mathcal{B}$ is the origin of $\mathcal{I}$. Using Eq. (8.5), the velocity of the center of mass is

$$^{\mathcal{I}}\mathbf{v}_{G/O} = {}^{\mathcal{I}}\boldsymbol{\omega}^{\mathcal{B}} \times \mathbf{r}_{G/O}.$$

Writing the vectors in the body frame, we obtain

$$^{\mathcal{I}}\mathbf{v}_{G/O} = \dot{\theta}\mathbf{b}_3 \times l\mathbf{b}_1 = l\dot{\theta}\mathbf{b}_2.$$

Example 8.2 The Velocity of a Rolling Wheel

This example computes the velocity of the center of a rolling wheel or disk, shown in Figure 8.3a. The rolling disk is an essential construct of mechanics and is used as one of the elements for building a dynamic model of many systems. The key idea is that the disk rolls without slipping. That means the disk actually only has one degree of freedom, even though it is both translating and rotating. Thus there must be a constraint. We call the constraint the *no-slip condition*. The no-slip condition states that *the point of the disk making contact with the ground has no instantaneous inertial velocity* (that is, it has the same velocity as the point it is touching on the ground). We use this constraint along with Eq. (8.5) to find the translational velocity of the center of the wheel.

Figure 8.3b shows the frames used for this problem. The no-slip condition states that $^{\mathcal{I}}\mathbf{v}_{P/O} = 0$, where P is the point of contact with the ground. In this case, Eq. (8.5) becomes

$$0 = {}^{\mathcal{I}}\mathbf{v}_{O'/O} + {}^{\mathcal{I}}\boldsymbol{\omega}^{\mathcal{B}} \times \mathbf{r}_{P/O'},$$

where $^{\mathcal{I}}\boldsymbol{\omega}^{\mathcal{B}} = \omega_B \mathbf{e}_3$, as drawn in Figure 8.3 (a positive rotation—the disk rolls to the left). Since the position of P relative to the center of the wheel is $\mathbf{r}_{P/O'} = -R\mathbf{e}_2$, we have

$$^{\mathcal{I}}\mathbf{v}_{O'/O} = -\omega_B \mathbf{e}_3 \times (-R\mathbf{e}_2) = -R\omega_B \mathbf{e}_1.$$

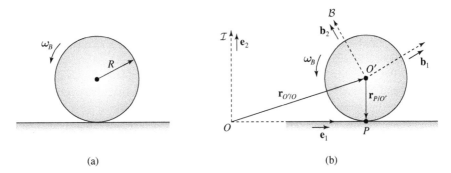

(a)   (b)

Figure 8.3 A disk of radius R rolling without slipping on a horizontal surface at angular rate ω_B. (a) Rolling disk. (b) Reference frames.

In Cartesian coordinates, where $\mathbf{r}_{O'/O} = x\mathbf{e}_1 + R\mathbf{e}_2$, the no-slip condition is

$$\dot{x} = -R\omega_B. \tag{8.6}$$

The no-slip condition in Eq. (8.6) is an example of a *motion constraint,* that is, a constraint on speeds rather than on coordinates. A motion constraint reduces the number of degrees of freedom in a problem but not necessarily the number of coordinates in terms of the others (see Section 2.4). In the case of the no-slip condition, Eq. (8.6) can be integrated if we know the initial position or angle of the wheel. Thus, we only need one coordinate for the one degree of freedom. We discuss general motion constraints in Chapter 13.

Example 8.3 The Gear Ratio

As you may know from riding a bike, gears are used to change an angular velocity. For example, a motor may produce an angular velocity that is faster than desired and a gear is used to reduce it. The *gear ratio* dictates the amount of reduction (or increase). Figure 8.4a shows two gears with centers fixed in inertial space. Gear $\mathcal{A}$ has radius r_A and is rotating at rate ω_A. If gear $\mathcal{B}$ has radius r_B, what is its angular rate ω_B?

Like the rolling disk discussed in the previous example, this problem has only one degree of freedom. That is, once the angle of gear $\mathcal{A}$ is set, the angle of gear $\mathcal{B}$ is

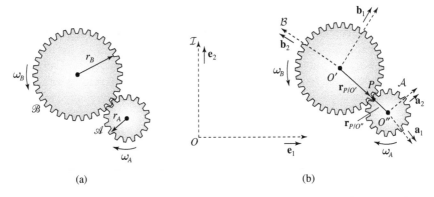

(a)   (b)

Figure 8.4 (a) Two fixed, interlocking gears of radius r_A and r_B. Gear $\mathcal{A}$ is rotating in the plane at the angular rate ω_A, and gear $\mathcal{B}$ is rotating at ω_B. (b) Reference frames.

determined. Since there are two coordinates (the angles of each gear), there must be a constraint. We call this constraint the *gear constraint*. The gear constraint requires that the points of contact between the two gears have the same inertial velocity. Otherwise, the gears would slip.

We solve this problem again using Eq. (8.5). Figure 8.4b shows the body frames $\mathcal{A}$ and $\mathcal{B}$ fixed to each gear. Point P is the point of contact between the two gears. Since the gears are fixed, the velocities of their centers are zero:

$$^{\mathcal{I}}\mathbf{v}_{O'/O} = {}^{\mathcal{I}}\mathbf{v}_{O''/O} = 0.$$

Using Eq. (8.5) on $\mathcal{A}$ to find the velocity of P, we have

$$^{\mathcal{I}}\mathbf{v}_{P/O} = {}^{\mathcal{I}}\boldsymbol{\omega}^{\mathcal{A}} \times \mathbf{r}_{P/O''},$$

where $^{\mathcal{I}}\boldsymbol{\omega}^{\mathcal{A}} = -\omega_A\mathbf{a}_3$. Likewise, for $\mathcal{B}$ we have

$$^{\mathcal{I}}\mathbf{v}_{P/O} = {}^{\mathcal{I}}\boldsymbol{\omega}^{\mathcal{B}} \times \mathbf{r}_{P/O'},$$

where $^{\mathcal{I}}\boldsymbol{\omega}^{\mathcal{B}} = \omega_B\mathbf{b}_3$. By the gear constraint, these two velocities must be the same:

$$\omega_B\mathbf{b}_3 \times (r_B\hat{\mathbf{r}}_{P/O'}) = \omega_A\mathbf{a}_3 \times (r_A\hat{\mathbf{r}}_{P/O''}) = \omega_A\mathbf{b}_3 \times (-r_A\hat{\mathbf{r}}_{P/O'})$$

which implies

$$\omega_B r_B = -\omega_A r_A. \tag{8.7}$$

Eq. (8.7) yields the following equation, called the *gear equation*:

$$\omega_B = -\left(\frac{r_A}{r_B}\right)\omega_A. \tag{8.8}$$

The ratio r_A/r_B is the gear ratio. Eq. (8.8) is another example of a motion constraint.

Example 8.4 A Two-Bar Linkage

A linkage consisting of two or more long, rigid bars is a device that converts rotational motion into translational motion or transfers rotational motion from one location to another (e.g., from the steering wheel of a car to the front wheels). In a three-link robot arm, for example, each link has one degree of freedom; the coordinate associated with each degree of freedom is given by the angle of the link (e.g., in the robot arm of Example 2.5, there are three angles: the shoulder, the elbow, and the wrist).

Linkages are ubiquitous in mechanical systems, yet all linkages rely on the same principle. If the rotation rate of each joint is known, the velocity of a point on the linkage can be found using Eq. (8.5). Usually linkages are combined with mechanical constraints to perform specific types of motion.

Figure 8.5 shows a simple two-bar linkage. This system has only one degree of freedom because the second link is constrained to always slide (without friction) on the horizontal surface. If the angular velocity of bar $\mathcal{A}$ is given as $^{\mathcal{I}}\boldsymbol{\omega}^{\mathcal{A}} = \omega_A\mathbf{a}_3$, what is the angular velocity $\omega_B\mathbf{b}_3$ of bar $\mathcal{B}$, and what is the velocity of its endpoint P?

Using Eq. (8.5), the velocity of P is

$$^{\mathcal{I}}\mathbf{v}_{P/O} = {}^{\mathcal{I}}\mathbf{v}_{O'/O} + {}^{\mathcal{I}}\boldsymbol{\omega}^{\mathcal{B}} \times \mathbf{r}_{P/O'}.$$

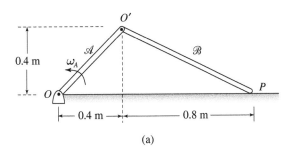

(a)

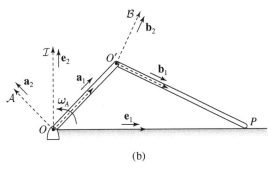

(b)

Figure 8.5 (a) Two-bar linkage constrained to move along a horizontal surface. (b) Reference frames.

Here the velocity $^{\mathcal{I}}\mathbf{v}_{O'/O}$ is also due to a rigid body rotation and is thus given by the same formula in reference frame $\mathcal{A}$:

$$^{\mathcal{I}}\mathbf{v}_{O'/O} = {}^{\mathcal{I}}\boldsymbol{\omega}^{\mathcal{A}} \times \mathbf{r}_{O'/O}.$$

Combining these two expressions yields

$$^{\mathcal{I}}\mathbf{v}_{P/O} = {}^{\mathcal{I}}\boldsymbol{\omega}^{\mathcal{A}} \times \mathbf{r}_{O'/O} + {}^{\mathcal{I}}\boldsymbol{\omega}^{\mathcal{B}} \times \mathbf{r}_{P/O'}.$$

From Figure 8.5a we can write down the instantaneous positions of P and O' in $\mathcal{I}$,

$$\mathbf{r}_{O'/O} = 0.4\mathbf{e}_1 + 0.4\mathbf{e}_2$$
$$\mathbf{r}_{P/O'} = 0.8\mathbf{e}_1 - 0.4\mathbf{e}_2.$$

Substituting for the positions as components in $\mathcal{I}$ gives the following expression for the velocity of point P:

$$^{\mathcal{I}}\mathbf{v}_{P/O} = \omega_A \mathbf{a}_3 \times (0.4\mathbf{e}_1 + 0.4\mathbf{e}_2) + \omega_B \mathbf{b}_3 \times (0.8\mathbf{e}_1 - 0.4\mathbf{e}_2)$$

$$= 0.4\omega_A \mathbf{e}_2 - 0.4\omega_A \mathbf{e}_1 + 0.8\omega_B \mathbf{e}_2 + 0.4\omega_B \mathbf{e}_1$$

$$= (-0.4\omega_A + 0.4\omega_B)\mathbf{e}_1 + (0.4\omega_A + 0.8\omega_B)\mathbf{e}_2. \qquad (8.9)$$

Now we use the constraint that point P moves only on the horizontal plane, which means its velocity is in the $\mathbf{e}_1$ direction:

$$^{\mathcal{I}}\mathbf{v}_{P/O} = v_P\mathbf{e}_1.$$

We can now solve for the two unknowns, v_P and ω_B. Since the $\mathbf{e}_2$ component in Eq. (8.9) is zero, we have

$$0 = 0.4\omega_A + 0.8\omega_B,$$

which results in

$$\omega_B = -\frac{\omega_A}{2} \text{ rad/s.}$$

The $\mathbf{e}_1$ component yields the speed of point P as a function of ω_A:

$$v_P = -0.4\omega_A + 0.4\omega_B = -0.4\omega_A - 0.2\omega_A = -0.6\omega_A \text{ m/s.}$$

8.2 Relative Motion in a Rotating Frame

Section 3.6 introduced the idea of relative motion, though restricted to translation only. We asked there how to develop the equations of motion of a particle in terms of a set of coordinates defined relative to a moving frame of reference. Since Newton's second law applies only to accelerations relative to an inertial frame, we had to be careful with the kinematics. We found a correction term to account for the translational acceleration of the frame. Now we generalize the treatment of relative motion to a frame that is both translating and rotating and develop the kinematics of a particle moving in a rotating frame. We then show how to use Newton's second law to develop the dynamics of a particle relative to a rotating frame. This problem is exceptionally important. For instance, every dynamics problem we solve on the surface of the earth is, in fact, relative to a translating and rotating frame!

8.2.1 The Transport Equation

Section 8.1 treated the problem of finding the inertial velocity of a particle P fixed to a translating and rotating frame of reference (a rigid body). In this section, we relax the restriction that P be fixed to the body and treat the more general problem of P moving relative to the body frame. In other words, suppose we know both the position and velocity of a particle in a reference frame $\mathcal{B}$ that is translating and rotating in the inertial frame. What is its inertial velocity? It turns out that answering this question requires only a slight extension of our earlier treatment.

The situation is illustrated in Figure 8.6, where body frame $\mathcal{B} = (O', \mathbf{b}_1, \mathbf{b}_2, \mathbf{b}_3)$ is translating and rotating relative to inertial frame $\mathcal{I} = (O, \mathbf{e}_1, \mathbf{e}_2, \mathbf{e}_3)$. Note the similarity to Figure 3.26. The important difference is that we now allow the body frame to also rotate in the plane. Particle P is free to move in the plane at any velocity relative to frame $\mathcal{B}$.

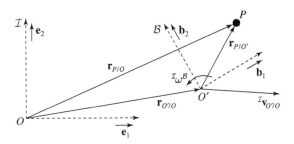

Figure 8.6 Body frame $\mathcal{B}$ translating and rotating with respect to inertial frame $\mathcal{I}$.

To find the inertial velocity of P, we begin as in Section 8.1, by taking the derivative of the vector triad locating P relative to O:

$$\mathcal{I}\mathbf{v}_{P/O} = \frac{\mathcal{I}d}{dt}\left(\mathbf{r}_{O'/O}\right) + \frac{\mathcal{I}d}{dt}\left(\mathbf{r}_{P/O'}\right) = \mathcal{I}\mathbf{v}_{O'/O} + \mathcal{I}\mathbf{v}_{P/O'}. \tag{8.10}$$

The first quantity on the right, $\mathcal{I}\mathbf{v}_{O'/O}$, is the translational velocity of frame $\mathcal{B}$. The second quantity is the inertial velocity of P with respect to O'. We write the components of $\mathbf{r}_{P/O'}$ using Cartesian coordinates $(x,y)_{\mathcal{B}}$ and take the derivative:

$$\mathcal{I}\mathbf{v}_{P/O'} = \frac{\mathcal{I}d}{dt}(x\mathbf{b}_1 + y\mathbf{b}_2) = \underbrace{\dot{x}\mathbf{b}_1 + \dot{y}\mathbf{b}_2}_{=\mathcal{B}\mathbf{v}_{P/O'}} + x\frac{\mathcal{I}d}{dt}\mathbf{b}_1 + y\frac{\mathcal{I}d}{dt}\mathbf{b}_2. \tag{8.11}$$

Since P can move in $\mathcal{B}$, we must include the rates of change of the Cartesian coordinates. This introduces the relative velocity of P with respect to frame $\mathcal{B}$, $\mathcal{B}\mathbf{v}_{P/O'}$. The last two terms in Eq. (8.11) result from the rotation of $\mathcal{B}$ in $\mathcal{I}$. We already solved for these two terms in the previous section. Eq. (8.11) thus simplifies to

$$\mathcal{I}\mathbf{v}_{P/O'} = \mathcal{B}\mathbf{v}_{P/O'} + \mathcal{I}\boldsymbol{\omega}^{\mathcal{B}} \times \mathbf{r}_{P/O'}, \tag{8.12}$$

or, explicitly writing the derivatives,

$$\boxed{\frac{\mathcal{I}d}{dt}\left(\mathbf{r}_{P/O'}\right) = \frac{\mathcal{B}d}{dt}\left(\mathbf{r}_{P/O'}\right) + \mathcal{I}\boldsymbol{\omega}^{\mathcal{B}} \times \mathbf{r}_{P/O'}.} \tag{8.13}$$

If $\mathcal{B}$ is not rotating, then $\mathcal{I}\boldsymbol{\omega}^{\mathcal{B}} = 0$ and Eq. (8.12) reproduces what we showed in Section 3.6, that is, $\mathcal{I}\mathbf{v}_{P/O'} = \mathcal{B}\mathbf{v}_{P/O'}$. However, the rotation of $\mathcal{B}$ adds the correction term $\mathcal{I}\boldsymbol{\omega}^{\mathcal{B}} \times \mathbf{r}_{P/O'}$. Eq. (8.12) is the most general equation for relating the inertial velocity of a particle to its motion in a translating and rotating frame.

Eq. (8.13) is an extraordinarily important equation, perhaps the most important one we encounter (short of Newton's second law). In fact, what is key about Eq. (8.13) is that it holds for *any* vector. It is not restricted to just the position. Thus the most general form of Eq. (8.13) is written in terms of an arbitrary vector $\mathbf{c}$ and two arbitrary reference frames $\mathcal{A}$ and $\mathcal{B}$. Since we can decompose $\mathbf{c}$ into components in the unit-vector directions of frame $\mathcal{B}$ and write the rotation of $\mathcal{B}$ in $\mathcal{A}$ using the angular velocity $^{\mathcal{A}}\boldsymbol{\omega}^{\mathcal{B}}$, the above development holds without modification. We call the resulting equation the *transport equation*. It is the backbone of all the kinematics to follow. Memorize it!

Definition 8.1 The **transport equation** is used to find the time derivative with respect to frame $\mathcal{A}$ of a vector $\mathbf{c}$ in terms of the time derivative of $\mathbf{c}$ with respect to frame $\mathcal{B}$ and the angular velocity $^{\mathcal{A}}\boldsymbol{\omega}^{\mathcal{B}}$ of $\mathcal{B}$ relative to $\mathcal{A}$:

$$\frac{^{\mathcal{A}}d}{dt}\mathbf{c} = \frac{^{\mathcal{B}}d}{dt}\mathbf{c} + {^{\mathcal{A}}\boldsymbol{\omega}^{\mathcal{B}}} \times \mathbf{c}.$$

Note that we can write the vectors on either side of the transport equation as components in either frame $\mathcal{A}$ or frame $\mathcal{B}$, whichever is most convenient for the problem at hand. We use Eqs. (8.10) and (8.12) to obtain the final expression for the inertial velocity of point P, given the translational velocity of frame $\mathcal{B}$, the angular velocity of $\mathcal{B}$ with respect to $\mathcal{I}$, and the relative velocity of P in $\mathcal{B}$:

$$^{\mathcal{I}}\mathbf{v}_{P/O} = {^{\mathcal{I}}\mathbf{v}_{O'/O}} + {^{\mathcal{B}}\mathbf{v}_{P/O'}} + {^{\mathcal{I}}\boldsymbol{\omega}^{\mathcal{B}}} \times \mathbf{r}_{P/O'}. \tag{8.14}$$

Example 8.5 Throwing the Brass Ring

You may have had the pleasure of riding a carousel, grabbing a brass ring, and trying to throw it into a clown's mouth while the carousel is rotating. The challenge is to figure out just the right speed and direction to throw the ring relative to the carousel. We assume that no (horizontal) forces are acting on the ring once it leaves your hand so that it travels in a straight line in an inertial frame with whatever initial velocity you give it.

A schematic of the situation is shown in Figure 8.7. The objective is for the ring to have the velocity $^{\mathcal{I}}\mathbf{v}_{P/O}$ in the inertial frame so that it will land in the clown's mouth. You are located at point O' on the carousel, which is rotating with angular velocity $^{\mathcal{I}}\boldsymbol{\omega}^{\mathcal{B}} = \omega_B \mathbf{e}_3$ (frame $\mathcal{B}$ is fixed to the carousel). We will find the velocity $^{\mathcal{B}}\mathbf{v}_{P/O'}$ that corresponds to $^{\mathcal{I}}\mathbf{v}_{P/O} = v_P\hat{\mathbf{r}}_{C/O'}$.

Solving Eq. (8.14) for the body-frame velocity yields

$$^{\mathcal{B}}\mathbf{v}_{P/O'} = {^{\mathcal{I}}\mathbf{v}_{P/O}} - {^{\mathcal{I}}\mathbf{v}_{O'/O}} - {^{\mathcal{I}}\boldsymbol{\omega}^{\mathcal{B}}} \times \mathbf{r}_{P/O'}.$$

At the instant you throw the ring it is at O', which implies $\mathbf{r}_{P/O'} = 0$. Since the carousel is going in a circle, you can use the transport equation to find the velocity of your horse (O'):

$$^{\mathcal{I}}\mathbf{v}_{O'/O} = \underbrace{^{\mathcal{B}}\mathbf{v}_{O'/O}}_{=0} + {^{\mathcal{I}}\boldsymbol{\omega}^{\mathcal{B}}} \times \mathbf{r}_{O'/O}.$$

Combining these two equations gives the desired velocity,

$$^{\mathcal{B}}\mathbf{v}_{P/O'} = {^{\mathcal{I}}\mathbf{v}_{P/O}} - \omega_B\mathbf{e}_3 \times \mathbf{r}_{O'/O} = v_P\hat{\mathbf{r}}_{C/O'} - \omega_B\mathbf{e}_3 \times \mathbf{r}_{O'/O}. \tag{8.15}$$

This equation shows what makes the game fun: in a split second you must estimate the angular velocity of the carousel, the position of your horse, and the direction to the clown. You then have to solve Eq. (8.15) (in your head) and throw.

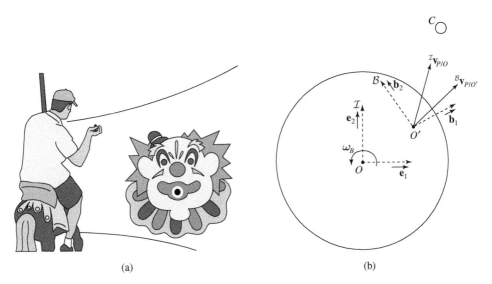

Figure 8.7 (a) Carousel with a body frame $\mathcal{B}$ rotating in inertial frame $\mathcal{I}$. (b) Carousel schematic (top view). $^{\mathcal{I}}\mathbf{v}_{P/O}$ is the velocity of a ring thrown at clown C.

Example 8.6 Linkage with a Sliding Contact

This example examines a linkage with a sliding contact, as shown in Figure 8.8. Unlike the linkages in Example 8.4, here the two links are not connected at a pinned joint, but rather a pin on one link is free to slide in a slot on the other link. This is similar to the transfer of translational motion to rotational motion in a steam engine. The goal is to find the angular velocity of the vertical bar and the velocity of the pin in the slot, given a fixed angular velocity of the slotted link.

As in Example 8.4, we begin by defining an inertial frame fixed at O and attaching body frames $\mathcal{A}$ and $\mathcal{B}$ to the two bars. We give bar $\mathcal{A}$ an angular velocity $^{\mathcal{I}}\boldsymbol{\omega}^{\mathcal{A}} = \omega_A \mathbf{a}_3$ of 2 rad/s and ask for the angular velocity of bar $\mathcal{B}$, $^{\mathcal{I}}\boldsymbol{\omega}^{\mathcal{B}} = \omega_B \mathbf{b}_3$, and the inertial velocity $^{\mathcal{I}}\mathbf{v}_{P/O}$ of pin P in the slot at the instant shown in Figure 8.8.

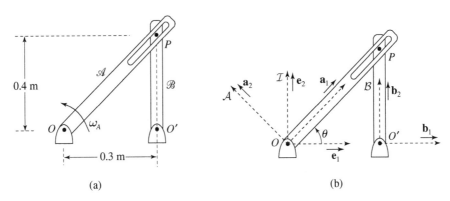

Figure 8.8 (a) A sliding linkage, where pin P is constrained to move along a slot in bar $\mathcal{A}$. (b) Reference frames.

We use the transport equation to write the inertial velocity of point P relative to O in terms of the body-frame velocity $^{\mathcal{A}}\mathbf{v}_{P/O}$:

$$^{\mathcal{I}}\mathbf{v}_{P/O} = {}^{\mathcal{A}}\mathbf{v}_{P/O} + {}^{\mathcal{I}}\boldsymbol{\omega}^{\mathcal{A}} \times \mathbf{r}_{P/O}. \tag{8.16}$$

Since $\mathbf{r}_{P/O} = 0.3\mathbf{e}_1 + 0.4\mathbf{e}_2$ and $^{\mathcal{I}}\boldsymbol{\omega}^{\mathcal{A}} = \omega_A\mathbf{e}_3 = 2\mathbf{e}_3$ rad/s, Eq. (8.16) simplifies to

$$^{\mathcal{I}}\mathbf{v}_{P/O} = {}^{\mathcal{A}}\mathbf{v}_{P/O} + 0.6\mathbf{e}_2 - 0.8\mathbf{e}_1. \tag{8.17}$$

Because of the slot, the velocity of P in $\mathcal{A}$ is always along $\mathbf{a}_1$, which implies

$$^{\mathcal{A}}\mathbf{v}_{P/O} = v_P\mathbf{a}_1 = v_P(\cos\theta\mathbf{e}_1 + \sin\theta\mathbf{e}_2),$$

where we have used the transformation array between frames $\mathcal{I}$ and $\mathcal{A}$,

	$\mathbf{e}_1$	$\mathbf{e}_2$
$\mathbf{a}_1$	$\cos\theta$	$\sin\theta$
$\mathbf{a}_2$	$-\sin\theta$	$\cos\theta$,

and $\theta = \arctan(0.4/0.3) = 53.1°$.

Substituting this result back into Eq. (8.17) yields the inertial velocity of P with respect to O as a function of v_P:

$$^{\mathcal{I}}\mathbf{v}_{P/O} = (v_P\cos\theta - 0.8)\mathbf{e}_1 + (v_P\sin\theta + 0.6)\mathbf{e}_2.$$

A second equation for $^{\mathcal{I}}\mathbf{v}_{P/O}$ is found by examining the velocity of P in reference frame $\mathcal{B}$ and again using the transport equation. As in Example 8.4, the velocity in $\mathcal{B}$ is zero (as the pin is fixed to the bar), so the transport equation becomes

$$^{\mathcal{I}}\mathbf{v}_{P/O} = \underbrace{{}^{\mathcal{I}}\mathbf{v}_{O'/O}}_{=0} + \underbrace{{}^{\mathcal{B}}\mathbf{v}_{P/O'}}_{=0} + {}^{\mathcal{I}}\boldsymbol{\omega}^{\mathcal{B}} \times \mathbf{r}_{P/O'}$$

$$= (\omega_B\mathbf{e}_3) \times (0.4\mathbf{e}_2) = -0.4\omega_B\mathbf{e}_1.$$

We now have two equations for $^{\mathcal{I}}\mathbf{v}_{P/O}$ that can be solved for v_P and ω_B. We have

$$(v_P\cos\theta - 0.8)\mathbf{e}_1 + (v_P\sin\theta + 0.6)\mathbf{e}_2 = -0.4\omega_B\mathbf{e}_1.$$

Equating components in the $\mathbf{e}_1$ and $\mathbf{e}_2$ directions yields the following two scalar equations in the unknowns v_P and ω_B:

$$v_P\cos\theta - 0.8 = -0.4\omega_B$$

$$v_P\sin\theta + 0.6 = 0,$$

which implies

$$v_P = -0.75 \text{ m/s}$$

$$\omega_B = 3.125 \text{ rad/s}.$$

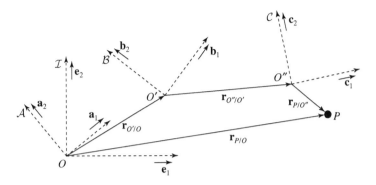

Figure 8.9 Multiple rotating body frames.

8.2.2 The Addition of Angular Velocities

Example 2.5 introduced a three-link robot arm as an example of a three-degree-of-freedom system. Figure 2.6b shows a simple planar model of the arm, where the three scalar angles of each arm relative to the previous one are used to describe its state. This is a sensible set of coordinates, since motors and sensors are often located at the joints and operate based on the relative rate between the two bodies. The question is, how do we combine the angular velocities of multiple frames to find the angular velocity with respect to an inertial frame? If we are going to study the dynamics of an individual arm, we need to know its inertial angular velocity. In this section we study how to go back and forth between the inertial and the relative angular velocities.

Figure 8.9 shows an arrangement of three connected frames, $\mathcal{A} = (O, \mathbf{a}_1, \mathbf{a}_2, \mathbf{a}_3)$, $\mathcal{B} = (O', \mathbf{b}_1, \mathbf{b}_2, \mathbf{b}_3)$, and $\mathcal{C} = (O'', \mathbf{c}_1, \mathbf{c}_2, \mathbf{c}_3)$. Each has some angular velocity, $^{\mathcal{I}}\boldsymbol{\omega}^{\mathcal{A}}$, $^{\mathcal{I}}\boldsymbol{\omega}^{\mathcal{B}}$, and $^{\mathcal{I}}\boldsymbol{\omega}^{\mathcal{C}}$, respectively, in the inertial frame. Suppose we want to find the inertial velocity of point P. We can use Eq. (8.14) to find

$$\frac{^{\mathcal{I}}d}{dt}\left(\mathbf{r}_{P/O}\right) = \frac{^{\mathcal{I}}d}{dt}\left(\mathbf{r}_{O''/O}\right) + {}^{\mathcal{C}}\mathbf{v}_{P/O''} + {}^{\mathcal{I}}\boldsymbol{\omega}^{\mathcal{C}} \times \mathbf{r}_{P/O''}.$$

We can use the vector triad and the transport equation on the first term to rewrite the derivative in terms of the translation and rotation of frame $\mathcal{B}$:

$$\frac{^{\mathcal{I}}d}{dt}\left(\mathbf{r}_{P/O}\right) = \frac{^{\mathcal{I}}d}{dt}\left(\mathbf{r}_{O'/O}\right) + {}^{\mathcal{B}}\mathbf{v}_{O''/O'} + {}^{\mathcal{I}}\boldsymbol{\omega}^{\mathcal{B}} \times \mathbf{r}_{O''/O'} + {}^{\mathcal{C}}\mathbf{v}_{P/O''} + {}^{\mathcal{I}}\boldsymbol{\omega}^{\mathcal{C}} \times \mathbf{r}_{P/O''}.$$

Finally, using Eq. (8.14) to replace the inertial velocity of point O' in terms of the angular velocity of reference frame $\mathcal{A}$ (which, in this setting, shares an origin with the inertial frame) yields

$$\begin{aligned}
^{\mathcal{I}}\mathbf{v}_{P/O} = {}^{\mathcal{A}}\mathbf{v}_{O'/O} + {}^{\mathcal{I}}\boldsymbol{\omega}^{\mathcal{A}} \times \mathbf{r}_{O'/O} + {}^{\mathcal{B}}\mathbf{v}_{O''/O'} \\
+ {}^{\mathcal{I}}\boldsymbol{\omega}^{\mathcal{B}} \times \mathbf{r}_{O''/O'} + {}^{\mathcal{C}}\mathbf{v}_{P/O''} + {}^{\mathcal{I}}\boldsymbol{\omega}^{\mathcal{C}} \times \mathbf{r}_{P/O''}.
\end{aligned} \tag{8.18}$$

This equation shows how to use the transport equation on multiple frames of reference to find the velocity of a point. However, as mentioned at the beginning of the section, we may not know the angular velocities with respect to an inertial frame but rather with respect to an intermediate frame that may also be rotating. We therefore rederive the velocity of P using a slightly different approach.

Rather than starting with the velocity of P in frame $\mathcal{C}$ and then working backward through frames $\mathcal{B}$ and $\mathcal{A}$, as we did above, we begin with the position of P as a vector sum, take the derivative, and work forward. We have

$$\mathbf{r}_{P/O} = \mathbf{r}_{O'/O} + \mathbf{r}_{O''/O'} + \mathbf{r}_{P/O''}.$$

The inertial velocity is found by taking the vector derivative to obtain

$$^{\mathcal{I}}\mathbf{v}_{P/O} = \frac{^{\mathcal{I}}d}{dt}\left(\mathbf{r}_{O'/O}\right) + \frac{^{\mathcal{I}}d}{dt}\left(\mathbf{r}_{O''/O'}\right) + \frac{^{\mathcal{I}}d}{dt}\left(\mathbf{r}_{P/O''}\right).$$

Now we use the transport equation separately on each term. We begin by transforming to derivatives in frame $\mathcal{A}$, using the angular velocity $^{\mathcal{I}}\boldsymbol{\omega}^{\mathcal{A}}$ of $\mathcal{A}$ relative to $\mathcal{I}$ (remember that, in this example, $\mathcal{A}$ is not translating in $\mathcal{I}$):

$$^{\mathcal{I}}\mathbf{v}_{P/O} = {}^{\mathcal{A}}\mathbf{v}_{O'/O} + {}^{\mathcal{I}}\boldsymbol{\omega}^{\mathcal{A}} \times \mathbf{r}_{O'/O} + \frac{^{\mathcal{A}}d}{dt}\left(\mathbf{r}_{O''/O'}\right) + {}^{\mathcal{I}}\boldsymbol{\omega}^{\mathcal{A}} \times \mathbf{r}_{O''/O'}$$
$$+ \frac{^{\mathcal{A}}d}{dt}\left(\mathbf{r}_{P/O''}\right) + {}^{\mathcal{I}}\boldsymbol{\omega}^{\mathcal{A}} \times \mathbf{r}_{P/O''}.$$

Next, we transform from derivatives in $\mathcal{A}$ to derivatives in $\mathcal{B}$, using the angular velocity $^{\mathcal{A}}\boldsymbol{\omega}^{\mathcal{B}}$ of $\mathcal{B}$ relative to $\mathcal{A}$:

$$^{\mathcal{I}}\mathbf{v}_{P/O} = {}^{\mathcal{A}}\mathbf{v}_{O'/O} + {}^{\mathcal{I}}\boldsymbol{\omega}^{\mathcal{A}} \times \mathbf{r}_{O'/O} + {}^{\mathcal{B}}\mathbf{v}_{O''/O'} + {}^{\mathcal{A}}\boldsymbol{\omega}^{\mathcal{B}} \times \mathbf{r}_{O''/O'}$$
$$+ {}^{\mathcal{I}}\boldsymbol{\omega}^{\mathcal{A}} \times \mathbf{r}_{O''/O'} + \frac{^{\mathcal{B}}d}{dt}\left(\mathbf{r}_{P/O''}\right) + {}^{\mathcal{A}}\boldsymbol{\omega}^{\mathcal{B}} \times \mathbf{r}_{P/O''} + {}^{\mathcal{I}}\boldsymbol{\omega}^{\mathcal{A}} \times \mathbf{r}_{P/O''}.$$

Finally, we use the transport equation one more time to get the velocity of P in frame $\mathcal{C}$, using the angular velocity $^{\mathcal{B}}\boldsymbol{\omega}^{\mathcal{C}}$ of $\mathcal{C}$ relative to $\mathcal{B}$:

$$^{\mathcal{I}}\mathbf{v}_{P/O} = {}^{\mathcal{A}}\mathbf{v}_{O'/O} + {}^{\mathcal{I}}\boldsymbol{\omega}^{\mathcal{A}} \times \mathbf{r}_{O'/O} + {}^{\mathcal{B}}\mathbf{v}_{O''/O'} + {}^{\mathcal{A}}\boldsymbol{\omega}^{\mathcal{B}} \times \mathbf{r}_{O''/O'} + {}^{\mathcal{I}}\boldsymbol{\omega}^{\mathcal{A}} \times \mathbf{r}_{O''/O'}$$
$$+ {}^{\mathcal{C}}\mathbf{v}_{P/O''} + {}^{\mathcal{B}}\boldsymbol{\omega}^{\mathcal{C}} \times \mathbf{r}_{P/O''} + {}^{\mathcal{A}}\boldsymbol{\omega}^{\mathcal{B}} \times \mathbf{r}_{P/O''} + {}^{\mathcal{I}}\boldsymbol{\omega}^{\mathcal{A}} \times \mathbf{r}_{P/O''}.$$

This expression is a bit cluttered, so we simplify it by factoring and combining the cross products:

$$^{\mathcal{I}}\mathbf{v}_{P/O} = {}^{\mathcal{A}}\mathbf{v}_{O'/O} + {}^{\mathcal{I}}\boldsymbol{\omega}^{\mathcal{A}} \times \mathbf{r}_{O'/O} + {}^{\mathcal{B}}\mathbf{v}_{O''/O'} + \left({}^{\mathcal{I}}\boldsymbol{\omega}^{\mathcal{A}} + {}^{\mathcal{A}}\boldsymbol{\omega}^{\mathcal{B}}\right) \times \mathbf{r}_{O''/O'}$$
$$+ {}^{\mathcal{C}}\mathbf{v}_{P/O''} + \left({}^{\mathcal{I}}\boldsymbol{\omega}^{\mathcal{A}} + {}^{\mathcal{A}}\boldsymbol{\omega}^{\mathcal{B}} + {}^{\mathcal{B}}\boldsymbol{\omega}^{\mathcal{C}}\right) \times \mathbf{r}_{P/O''}. \qquad (8.19)$$

Compare Eq. (8.19) to Eq. (8.18). Both are expressions for the inertial velocity of P, $^{\mathcal{I}}\mathbf{v}_{P/O}$. The two expressions must be the same. Comparing each term reveals an extremely important property of angular velocities, known as the *angular-velocity addition property:*

$$\boxed{{}^{\mathcal{I}}\boldsymbol{\omega}^{\mathcal{B}} = {}^{\mathcal{I}}\boldsymbol{\omega}^{\mathcal{A}} + {}^{\mathcal{A}}\boldsymbol{\omega}^{\mathcal{B}}}$$

and

$$ {}^{\mathcal{I}}\boldsymbol{\omega}^{\mathcal{C}} = {}^{\mathcal{I}}\boldsymbol{\omega}^{\mathcal{A}} + {}^{\mathcal{A}}\boldsymbol{\omega}^{\mathcal{B}} + {}^{\mathcal{B}}\boldsymbol{\omega}^{\mathcal{C}}.$$

In fact, we could have used any number of intermediate frames and found that the addition property holds for an arbitrary number of angular velocities.

We use the angular-velocity addition property a great deal in later chapters. It should be easy to remember, because it matches physical intuition. For example, consider a rotating disk on top of which is a second disk rotating with respect to the first; the top disk's rotation rate relative to absolute space is the sum of its rotation rate and the rotation rate of the first disk.

Finally, when the various frames in a problem are attached to rigid links, as in Section 8.1, the relative velocities of the origins are zero, and we have the multiple-frame generalization of Eq. (8.5):

$$ {}^{\mathcal{I}}\mathbf{v}_{P/O} = {}^{\mathcal{I}}\boldsymbol{\omega}^{\mathcal{A}} \times \mathbf{r}_{O'/O} + {}^{\mathcal{I}}\boldsymbol{\omega}^{\mathcal{B}} \times \mathbf{r}_{O''/O'} + {}^{\mathcal{I}}\boldsymbol{\omega}^{\mathcal{C}} \times \mathbf{r}_{P/O''}.$$

Example 8.7 A Three-Bar Linkage

Figure 8.10 shows three bars attached to one another and to two fixed points. The bars are allowed to pivot smoothly about their attachment points. We define three body frames: $\mathcal{A}$, which rotates with link AB; $\mathcal{B}$, which rotates with link BC; and $\mathcal{C}$, which rotates with link CD. All the frames are right handed, so the out-of-plane component is always in the direction of $\mathbf{e}_3 = \mathbf{e}_1 \times \mathbf{e}_2$. The links move only in the plane, so all angular velocities are in the $\mathbf{e}_3$ direction.

Suppose we know the initial orientation of all the links and the angular velocity ${}^{\mathcal{I}}\boldsymbol{\omega}^{\mathcal{A}}$ of link AB. We can then solve for the angular velocity ${}^{\mathcal{A}}\boldsymbol{\omega}^{\mathcal{B}}$ of link BC relative to link AB and for ${}^{\mathcal{B}}\boldsymbol{\omega}^{\mathcal{C}}$ of link CD relative to link BC. Since only one angular velocity is given, we need a constraint that will allow us to find the other two. This constraint comes from the fixed pivots A and D—namely, that the inertial

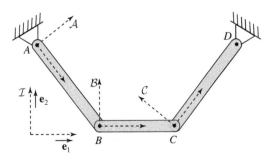

Figure 8.10 Three-bar linkage.

velocity $^{\mathcal{I}}\mathbf{v}_{D/A}$ must equal zero. Noting that $\mathbf{r}_{D/A} = \mathbf{r}_{D/C} + \mathbf{r}_{C/B} + \mathbf{r}_{B/A}$, we have $^{\mathcal{I}}\mathbf{v}_{D/A} = {}^{\mathcal{I}}\mathbf{v}_{D/C} + {}^{\mathcal{I}}\mathbf{v}_{C/B} + {}^{\mathcal{I}}\mathbf{v}_{B/A} = 0$. We use the body frames to write

$$^{\mathcal{I}}\mathbf{v}_{B/A} = {}^{\mathcal{A}}\mathbf{v}_{B/A} + {}^{\mathcal{I}}\boldsymbol{\omega}^{\mathcal{A}} \times \mathbf{r}_{B/A}$$

$$^{\mathcal{I}}\mathbf{v}_{C/B} = {}^{\mathcal{B}}\mathbf{v}_{C/B} + {}^{\mathcal{I}}\boldsymbol{\omega}^{\mathcal{B}} \times \mathbf{r}_{C/B}$$

$$^{\mathcal{I}}\mathbf{v}_{D/C} = {}^{\mathcal{C}}\mathbf{v}_{D/C} + {}^{\mathcal{I}}\boldsymbol{\omega}^{\mathcal{C}} \times \mathbf{r}_{D/C}.$$

Since these are rigid links, all body-frame velocities are zero:

$$^{\mathcal{I}}\mathbf{v}_{D/A} = {}^{\mathcal{I}}\boldsymbol{\omega}^{\mathcal{A}} \times \mathbf{r}_{B/A} + {}^{\mathcal{I}}\boldsymbol{\omega}^{\mathcal{B}} \times \mathbf{r}_{C/B} + {}^{\mathcal{I}}\boldsymbol{\omega}^{\mathcal{C}} \times \mathbf{r}_{D/C} = 0. \qquad (8.20)$$

From the angular-velocity addition property, we know that

$$^{\mathcal{I}}\boldsymbol{\omega}^{\mathcal{B}} = {}^{\mathcal{I}}\boldsymbol{\omega}^{\mathcal{A}} + {}^{\mathcal{A}}\boldsymbol{\omega}^{\mathcal{B}}$$

$$^{\mathcal{I}}\boldsymbol{\omega}^{\mathcal{C}} = {}^{\mathcal{I}}\boldsymbol{\omega}^{\mathcal{A}} + {}^{\mathcal{A}}\boldsymbol{\omega}^{\mathcal{B}} + {}^{\mathcal{B}}\boldsymbol{\omega}^{\mathcal{C}}.$$

Since the cross product is distributive over addition, Eq. (8.20) can be written as

$$0 = {}^{\mathcal{I}}\boldsymbol{\omega}^{\mathcal{A}} \times \left(\mathbf{r}_{B/A} + \mathbf{r}_{C/B} + \mathbf{r}_{D/C}\right) + {}^{\mathcal{A}}\boldsymbol{\omega}^{\mathcal{B}} \times \left(\mathbf{r}_{C/B} + \mathbf{r}_{D/C}\right) + {}^{\mathcal{B}}\boldsymbol{\omega}^{\mathcal{C}} \times \mathbf{r}_{D/C}$$

$$= {}^{\mathcal{I}}\boldsymbol{\omega}^{\mathcal{A}} \times \mathbf{r}_{D/A} + {}^{\mathcal{A}}\boldsymbol{\omega}^{\mathcal{B}} \times \mathbf{r}_{D/B} + {}^{\mathcal{B}}\boldsymbol{\omega}^{\mathcal{C}} \times \mathbf{r}_{D/C}.$$

We now have one equation with two unknowns, the angular rates $^{\mathcal{A}}\boldsymbol{\omega}^{\mathcal{B}}$ and $^{\mathcal{B}}\boldsymbol{\omega}^{\mathcal{C}}$. Fortunately, this is a vector equation, which means we can decompose it into multiple scalar equations. We do so by dotting with $\mathbf{e}_1$ and $\mathbf{e}_2$. Since the dot product is distributive over addition, we need only look at one of these terms, say, the first one. Using the scalar triple product identity (see Appendix B.4.2) yields

$$\mathbf{e}_1 \cdot \left(^{\mathcal{I}}\boldsymbol{\omega}^{\mathcal{A}} \times \mathbf{r}_{D/A}\right) = \mathbf{r}_{D/A} \cdot \left(\mathbf{e}_1 \times {}^{\mathcal{I}}\boldsymbol{\omega}^{\mathcal{A}}\right).$$

Recall that due to our frame definitions, all angular velocities are along the $\mathbf{e}_3 = \mathbf{e}_1 \times \mathbf{e}_2$ direction, so

$$\mathbf{r}_{D/A} \cdot \left(\mathbf{e}_1 \times {}^{\mathcal{I}}\boldsymbol{\omega}^{\mathcal{A}}\right) = -{}^{\mathcal{I}}\omega^{\mathcal{A}} \left(\mathbf{r}_{D/A} \cdot \mathbf{e}_2\right).$$

Applying this procedure to all terms for both the $\mathbf{e}_1$ and $\mathbf{e}_2$ directions produces two equations:

$$0 = {}^{\mathcal{I}}\omega^{\mathcal{A}} \left(\mathbf{r}_{D/A} \cdot \mathbf{e}_2\right) + {}^{\mathcal{A}}\omega^{\mathcal{B}} \left(\mathbf{r}_{D/B} \cdot \mathbf{e}_2\right) + {}^{\mathcal{B}}\omega^{\mathcal{C}} \left(\mathbf{r}_{D/C} \cdot \mathbf{e}_2\right)$$

$$0 = {}^{\mathcal{I}}\omega^{\mathcal{A}} \left(\mathbf{r}_{D/A} \cdot \mathbf{e}_1\right) + {}^{\mathcal{A}}\omega^{\mathcal{B}} \left(\mathbf{r}_{D/B} \cdot \mathbf{e}_1\right) + {}^{\mathcal{B}}\omega^{\mathcal{C}} \left(\mathbf{r}_{D/C} \cdot \mathbf{e}_1\right).$$

We now have two separate equations. Given the lengths and orientations of the three links, we can easily find $^{\mathcal{A}}\omega^{\mathcal{B}}$ and $^{\mathcal{B}}\omega^{\mathcal{C}}$, as well as the angular velocities in the inertial frame and the velocities of all points in the linkage.

8.3 Planar Kinetics in a Rotating Frame

This section completes our discussion of planar relative motion by examining how to use Newton's second law to find the equations of motion of a particle relative to a translating and rotating reference frame. We treated a special case of this in Section 3.6, where we examined only a frame in translation. Now that we have the transport equation and an understanding of the kinematics of a particle relative to a rotating frame, we can complete the picture by including rotation. Recall that the procedure was to find the inertial acceleration in terms of the relative acceleration and the motion of the frame and then to rewrite Newton's second law in terms of the acceleration relative to the moving frame. We follow a similar approach here.

We begin with Figure 8.11 and the transport equation for the inertial velocity of particle P in terms of its velocity in frame $\mathcal{B}$ and the motion of $\mathcal{B}$ from Eq. (8.14):

$$^{\mathcal{I}}\mathbf{v}_{P/O} = {}^{\mathcal{I}}\mathbf{v}_{O'/O} + {}^{\mathcal{B}}\mathbf{v}_{P/O'} + {}^{\mathcal{I}}\omega^{\mathcal{B}} \times \mathbf{r}_{P/O'}.$$

What is the inertial acceleration of P? With the transport equation, finding the acceleration of P in $\mathcal{I}$ in terms of its acceleration in $\mathcal{B}$ and the angular velocity of $\mathcal{B}$ in $\mathcal{I}$ is easy. We just apply the transport equation again to the velocity $^{\mathcal{I}}\mathbf{v}_{P/O}$:

$$^{\mathcal{I}}\mathbf{a}_{P/O} = \frac{^{\mathcal{I}}d}{dt}\left(^{\mathcal{I}}\mathbf{v}_{O'/O}\right) + \frac{^{\mathcal{I}}d}{dt}\left(^{\mathcal{B}}\mathbf{v}_{P/O'} + {}^{\mathcal{I}}\omega^{\mathcal{B}} \times \mathbf{r}_{P/O'}\right)$$

$$= {}^{\mathcal{I}}\mathbf{a}_{O'/O} + \frac{^{\mathcal{B}}d}{dt}\left(^{\mathcal{B}}\mathbf{v}_{P/O'} + {}^{\mathcal{I}}\omega^{\mathcal{B}} \times \mathbf{r}_{P/O'}\right)$$

$$+ {}^{\mathcal{I}}\omega^{\mathcal{B}} \times \left(^{\mathcal{B}}\mathbf{v}_{P/O'} + {}^{\mathcal{I}}\omega^{\mathcal{B}} \times \mathbf{r}_{P/O'}\right).$$

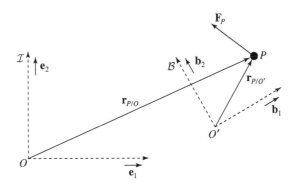

Figure 8.11 Particle P subject to force $\mathbf{F}_P$ moving relative to frame $\mathcal{B}$, which may rotate or translate relative to frame $\mathcal{I}$.

By distributing through the quantities in parentheses and using the product rule, we find the expression for the acceleration:

$$^{\mathcal{I}}\mathbf{a}_{P/O} = {}^{\mathcal{I}}\mathbf{a}_{O'/O} + {}^{\mathcal{B}}\mathbf{a}_{P/O'} + \frac{{}^{\mathcal{B}}d}{dt}\left({}^{\mathcal{I}}\boldsymbol{\omega}^{\mathcal{B}}\right) \times \mathbf{r}_{P/O'} + 2{}^{\mathcal{I}}\boldsymbol{\omega}^{\mathcal{B}} \times {}^{\mathcal{B}}\mathbf{v}_{P/O'}$$
$$+ {}^{\mathcal{I}}\boldsymbol{\omega}^{\mathcal{B}} \times \left({}^{\mathcal{I}}\boldsymbol{\omega}^{\mathcal{B}} \times \mathbf{r}_{P/O'}\right).$$

The quantity $\frac{{}^{\mathcal{B}}d}{dt}\left({}^{\mathcal{I}}\boldsymbol{\omega}^{\mathcal{B}}\right)$ is the *angular acceleration* and, for this planar case, is equal to $\ddot{\theta}\mathbf{e}_3$. The symbol ${}^{\mathcal{I}}\boldsymbol{\alpha}^{\mathcal{B}} \triangleq \frac{{}^{\mathcal{B}}d}{dt}\left({}^{\mathcal{I}}\boldsymbol{\omega}^{\mathcal{B}}\right) = \frac{{}^{\mathcal{I}}d}{dt}\left({}^{\mathcal{I}}\boldsymbol{\omega}^{\mathcal{B}}\right)$ is often used for the angular acceleration. This gives us our final expression for the inertial acceleration of P:

$$\boxed{\begin{aligned}^{\mathcal{I}}\mathbf{a}_{P/O} &= {}^{\mathcal{I}}\mathbf{a}_{O'/O} + {}^{\mathcal{B}}\mathbf{a}_{P/O'} + {}^{\mathcal{I}}\boldsymbol{\alpha}^{\mathcal{B}} \times \mathbf{r}_{P/O'} + 2{}^{\mathcal{I}}\boldsymbol{\omega}^{\mathcal{B}} \times {}^{\mathcal{B}}\mathbf{v}_{P/O'} \\ &\quad + {}^{\mathcal{I}}\boldsymbol{\omega}^{\mathcal{B}} \times \left({}^{\mathcal{I}}\boldsymbol{\omega}^{\mathcal{B}} \times \mathbf{r}_{P/O'}\right).\end{aligned}}$$

(8.21)

Eq. (8.21) is a general expression for the inertial acceleration of a point P in terms of its velocity and acceleration in a translating and rotating frame. It is this expression that we use when applying Newton's second law. Just as in the discussion of the polar frame in Section 3.5, the extra terms are purely kinematic. They arise because we are trying to describe motion relative to a noninertial frame. The fourth term, $2{}^{\mathcal{I}}\boldsymbol{\omega}^{\mathcal{B}} \times {}^{\mathcal{B}}\mathbf{v}_{P/O'}$, is the Coriolis acceleration and the last term, ${}^{\mathcal{I}}\boldsymbol{\omega}^{\mathcal{B}} \times \left({}^{\mathcal{I}}\boldsymbol{\omega}^{\mathcal{B}} \times \mathbf{r}_{P/O'}\right)$, is the centripetal acceleration. The polar frame equations in Section 3.5 are just special cases of Eq. (8.21). The difference here is that frame $\mathcal{B}$ can rotate at an arbitrary angular velocity and point P can move in $\mathcal{B}$.

As in Section 3.6, we can find a form of Newton's second law that allows us to write the equations of motion for the coordinates of P in frame $\mathcal{B}$. We just substitute from Eq. (8.21) into Newton's second law and solve for ${}^{\mathcal{B}}\mathbf{a}_{P/O'}$, which yields

$$m_P{}^{\mathcal{B}}\mathbf{a}_{P/O'} = \mathbf{F}_P - m_P{}^{\mathcal{I}}\mathbf{a}_{O'/O} - m_P{}^{\mathcal{I}}\boldsymbol{\alpha}^{\mathcal{B}} \times \mathbf{r}_{P/O'} - 2m_P{}^{\mathcal{I}}\boldsymbol{\omega}^{\mathcal{B}} \times {}^{\mathcal{B}}\mathbf{v}_{P/O'}$$
$$- m_P{}^{\mathcal{I}}\boldsymbol{\omega}^{\mathcal{B}} \times \left({}^{\mathcal{I}}\boldsymbol{\omega}^{\mathcal{B}} \times \mathbf{r}_{P/O'}\right).$$

(8.22)

This equation is similar to Eq. (3.57) and is, in fact, a more general version of it. In the absence of rotation of the body frame $\mathcal{B}$, Eq. (8.22) reduces exactly to Eq. (3.57). This equation tells us how to formulate the equations of motion for a particle relative to a translating and rotating frame. As before, the kinematic terms appear as fictional forces on the right-hand side. We don't include fictional forces on a free-body diagram; rather, we encourage using the kinematic expression for acceleration in Eq. (8.21) in Newton's second law.

Now we revisit the discussion in Section 3.5 and explore the physics a bit more. For example, we can find the force required to keep a particle P stationary in a translating

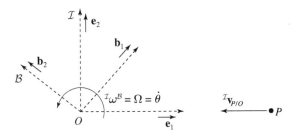

Figure 8.12 Point P traveling at a constant inertial velocity.

and rotating frame $\mathcal{B}$. Suppose, for instance, you are on a spinning platform at a playground. What force is required to stand on the platform? Since $\mathcal{B}$ is rotating, P is accelerating in $\mathcal{I}$ and thus a force is required. If $\mathcal{B}$ is neither translating nor rotationally accelerating, then $^{\mathcal{I}}\mathbf{a}_{O/O'} = 0$ and $^{\mathcal{I}}\boldsymbol{\alpha}^{\mathcal{B}} = 0$. Since P is fixed in $\mathcal{B}$, we have $^{\mathcal{B}}\mathbf{v}_{P/O'} = {}^{\mathcal{B}}\mathbf{a}_{P/O'} = 0$, and the Coriolis term is zero. Using Eq. (8.22), the force required to keep P fixed in $\mathcal{B}$ is

$$\mathbf{F}_P = m_P {}^{\mathcal{I}}\boldsymbol{\omega}^{\mathcal{B}} \times \left({}^{\mathcal{I}}\boldsymbol{\omega}^{\mathcal{B}} \times \mathbf{r}_{P/O'} \right).$$

This new, more general, expression for the kinematics in Eq. (8.22) also lets us more easily look at a slightly different scenario that lends additional insight into the geometric origins of the Coriolis and centripetal terms. Consider a particle P moving at a constant velocity in the $\mathbf{e}_1$ direction of the inertial frame, as shown in Figure 8.12. A body frame $\mathcal{B}$ with origin $O' = O$ is rotating at a constant rate Ω with angular velocity $^{\mathcal{I}}\boldsymbol{\omega}^{\mathcal{B}} = \Omega\mathbf{e}_3$. Just as in Figure 3.23, the particle, though moving at constant inertial velocity, is accelerating relative to frame $\mathcal{B}$.

Since there are no forces acting on the particle, we can find the acceleration relative to the body frame from Eq. (8.22):

$$^{\mathcal{B}}\mathbf{a}_{P/O} = -2{}^{\mathcal{I}}\boldsymbol{\omega}^{\mathcal{B}} \times {}^{\mathcal{B}}\mathbf{v}_{P/O} - {}^{\mathcal{I}}\boldsymbol{\omega}^{\mathcal{B}} \times \left({}^{\mathcal{I}}\boldsymbol{\omega}^{\mathcal{B}} \times \mathbf{r}_{P/O} \right),$$

where we have used the assumption that the angular velocity is constant. Again, the particle has zero inertial acceleration because there are no forces acting on it. But due to the kinematics of a rotating frame, P is accelerating relative to the body frame. We explore this idea more deeply in Example 8.8.

Finally, we can double-check that Eqs. (8.14) and (8.21) lead to the right answer for the polar frame. Let $\mathcal{B}$ be a polar frame with origin O. The angular velocity of $\mathcal{B}$ with respect to $\mathcal{I}$ is $^{\mathcal{I}}\boldsymbol{\omega}^{\mathcal{B}} = \dot{\theta}\mathbf{e}_3$ and the angular acceleration is $^{\mathcal{I}}\boldsymbol{\alpha}^{\mathcal{B}} = \ddot{\theta}\mathbf{e}_3$. The position of P in $\mathcal{B}$, written in terms of components in $\mathcal{B}$, is $\mathbf{r}_{P/O'} = r\mathbf{e}_r$. Plugging these equations into Eq. (8.14), the inertial velocity is

$$^{\mathcal{I}}\mathbf{v}_{P/O} = \dot{r}\mathbf{e}_r + r\dot{\theta}\mathbf{e}_\theta,$$

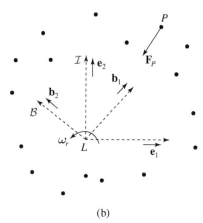

(a)   (b)

Figure 8.13 A simple planar model for a hurricane. (a) Hurricane. (b) Reference frames.
Image (a) courtesy of Shutterstock.

where we have used ${}^{\mathcal{B}}\mathbf{v}_{P/O'} = \dot{r}\mathbf{e}_r$ and ${}^{\mathcal{I}}\boldsymbol{\omega}^{\mathcal{B}} \times \mathbf{r}_{P/O} = r\dot{\theta}\mathbf{e}_\theta$. The acceleration is found
from Eq. (8.21) to be

$$\begin{aligned}
{}^{\mathcal{I}}\mathbf{a}_{P/O} &= \ddot{r}\mathbf{e}_r + r\ddot{\theta}\mathbf{e}_\theta + 2\dot{r}\dot{\theta}\mathbf{e}_\theta - r\dot{\theta}^2\mathbf{e}_r \\
&= (\ddot{r} - r\dot{\theta}^2)\mathbf{e}_r + (2\dot{r}\dot{\theta} + r\ddot{\theta})\mathbf{e}_\theta.
\end{aligned}$$

These are the same results as Eqs. (3.26) and (3.27).

Example 8.8 Hurricanes in Flatland[1]

This example explores a simple model of a hurricane by restricting the phenomenon to
a planar earth. The dynamics of a real hurricane are substantially more complicated
because the earth is a rotating sphere and the physics are more sophisticated than
we treat here. However, a simple planar model is an excellent representation of the
phenomenology of hurricanes and how the Coriolis effect acts to produce the classic
spiral pattern.

A hurricane, as shown in Figure 8.13a, forms when air in the upper atmosphere
is pushed toward a low-pressure system. Figure 8.13b shows our simple model of a
hurricane. The origin of the inertial frame $\mathcal{I} = (L, \mathbf{e}_1, \mathbf{e}_2, \mathbf{e}_3)$ is at the center of the
low pressure. The body frame $\mathcal{B} = (L, \mathbf{b}_1, \mathbf{b}_2, \mathbf{b}_3)$ is our planar model of the earth's
surface and is rotating at angular velocity ${}^{\mathcal{I}}\boldsymbol{\omega}^{\mathcal{B}} = \omega_r\mathbf{e}_3$ with respect to an inertial frame.
We assume that the upper-atmosphere air mass is not rotating with the earth.

Using a multiparticle model for the air, each air particle P is attracted to the center
of the low pressure (or pushed by the surrounding high-pressure air mass) by a central
force with magnitude f_L:

$$\mathbf{F}_P = -f_L\hat{\mathbf{r}}_{P/L},$$

---

[1] *Flatland: A Romance in Many Dimensions* is a satirical novel by Edwin Abbott Abbott written in 1884.
Originally written as social commentary, it has become popular among scientists and mathematicians for
its fanciful treatment of dimension.

where $\hat{\mathbf{r}}_{P/L}$ is a unit vector pointing from L to P. The particle thus accelerates toward L by Newton's second law:

$$^{\mathcal{I}}\mathbf{a}_{P/L} = -\frac{f_L}{m_P}\hat{\mathbf{r}}_{P/L}.$$

What is the equation of motion for particle P in frame $\mathcal{B}$? We use Eq. (8.22) to rewrite the equation of motion in terms of the acceleration in the earth-fixed frame:

$$^{\mathcal{B}}\mathbf{a}_{P/L} = -\frac{f_L}{m_P}\hat{\mathbf{r}}_{P/L} - 2\omega_r\mathbf{b}_3 \times {}^{\mathcal{B}}\mathbf{v}_{P/L} - \omega_r^2\mathbf{b}_3 \times (\mathbf{b}_3 \times \mathbf{r}_{P/L}).$$

To gain insight into the motion of the particle in $\mathcal{B}$, we use polar coordinates $(r_1, \theta)_{\mathcal{B}}$ and polar frame $\mathcal{C} = (L, \mathbf{e}_r, \mathbf{e}_\theta, \mathbf{e}_3)$. The resulting equation of motion is

$$(\ddot{r} - r\dot{\theta}^2)\mathbf{e}_r + (2\dot{r}\dot{\theta} + r\ddot{\theta})\mathbf{e}_\theta = -\frac{f_L}{m_P}\mathbf{e}_r - 2\omega_r\dot{r}\mathbf{e}_\theta + 2\omega_r r\dot{\theta}\mathbf{e}_r + \omega_r^2 r\mathbf{e}_r.$$

This vector equation corresponds to the following two scalar equations of motion:

$$\ddot{r} - r\dot{\theta}^2 - 2\omega_r r\dot{\theta} - \omega_r^2 r = -\frac{f_L}{m_P}$$

$$\ddot{\theta} + 2\frac{\dot{r}}{r}\dot{\theta} + 2\omega_r\frac{\dot{r}}{r} = 0.$$

A solution to these equations exists for $\ddot{\theta} = 0$. Solving the second equation for $\dot{\theta}$ yields $\dot{\theta} = -\omega_r$. Substituting this result into the first equation leaves the equation of motion

$$\ddot{r} = -\frac{f_L}{m_P}.$$

Thus each air particle rotates about the center of low pressure at a rate equal and opposite to the earth's rotation rate while accelerating inward toward the center. This motion forms a spiral pattern in the earth-fixed frame.

Example 8.9 The Simple Double Pendulum

This example uses the expressions developed for kinematics in a rotating frame to solve for the equations of motion of the simple double pendulum shown in Figure 8.14.[2] A simple double pendulum has two particles: the first, labeled P in Figure 8.14a, is suspended from a massless rod, as in the simple pendulum; the second particle, labeled Q in Figure 8.14a, is hanging from a massless rod attached to P. The double pendulum is a fascinating dynamical system capable of producing chaotic trajectories, which exhibit sensitive dependence on the initial conditions.

The simple double pendulum is a two-degree-of-freedom system. We choose as coordinates the two angles θ_1 and θ_2 shown in Figure 8.14a. These are the angles each rod makes with the vertical. The angle that the lower rod makes with the first rod

---

[2] A nonsimple double pendulum, or just the double pendulum, consists of two rigid rods hinged together. You are invited to solve the double pendulum in Problem 9.9.

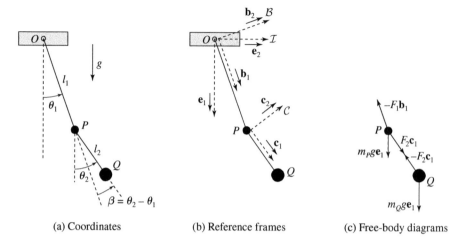

(a) Coordinates   (b) Reference frames   (c) Free-body diagrams

Figure 8.14 Solving the simple double pendulum.

is labeled $\beta = \theta_2 - \theta_1$ in Figure 8.14a. Our approach to solving this problem is very similar to the one taken for the simple pendulum in Example 3.8. Here we use the three reference frames shown in Figure 8.14b: the inertial frame $\mathcal{I} = (O, \mathbf{e}_1, \mathbf{e}_2, \mathbf{e}_3)$ and two rotating frames $\mathcal{B} = (O, \mathbf{b}_1, \mathbf{b}_2, \mathbf{b}_3)$ and $\mathcal{C} = (P, \mathbf{c}_1, \mathbf{c}_2, \mathbf{c}_3)$, where $\mathbf{e}_3 = \mathbf{b}_3 = \mathbf{c}_3$. Using the coordinates shown in Figure 8.14a, the transformation tables are

	$\mathbf{e}_1$	$\mathbf{e}_2$	$\mathbf{c}_1$	$\mathbf{c}_2$
$\mathbf{b}_1$	$\cos\theta_1$	$\sin\theta_1$	$\cos\beta$	$-\sin\beta$
$\mathbf{b}_2$	$-\sin\theta_1$	$\cos\theta_1$	$\sin\beta$	$\cos\beta$.

$$(8.23)$$

We use the angular-velocity addition property to find the inertial angular velocity of each rotating frame:

$$^{\mathcal{I}}\boldsymbol{\omega}^{\mathcal{B}} = \dot{\theta}_1 \mathbf{b}_3 \tag{8.24}$$

$$^{\mathcal{I}}\boldsymbol{\omega}^{\mathcal{C}} = {}^{\mathcal{I}}\boldsymbol{\omega}^{\mathcal{B}} + {}^{\mathcal{B}}\boldsymbol{\omega}^{\mathcal{C}}$$

$$= \dot{\theta}_1 \mathbf{b}_3 + \dot{\beta} \mathbf{c}_3 = \dot{\theta}_2 \mathbf{c}_3. \tag{8.25}$$

The kinematics of point P are exactly the same as in Example 3.11,

$$\mathbf{r}_{P/O} = l_1 \mathbf{b}_1$$

$$^{\mathcal{I}}\mathbf{v}_{P/O} = l_1 \dot{\theta}_1 \mathbf{b}_2$$

$$^{\mathcal{I}}\mathbf{a}_{P/O} = l_1 \ddot{\theta}_1 \mathbf{b}_2 - l_1 \dot{\theta}_1^2 \mathbf{b}_1. \tag{8.26}$$

These equations are found by taking the derivative of the polar-frame unit vectors using the angular velocity from Eq. (8.24).

To find the acceleration of mass Q, we could follow the same procedure, using the position $\mathbf{r}_{Q/O} = l_1 \mathbf{b}_1 + l_2 \mathbf{c}_1$. Instead we use Eq. (8.21), the general expression for the acceleration in rotating frames:

$$
{}^{\mathcal{I}}\mathbf{a}_{Q/O} = {}^{\mathcal{I}}\mathbf{a}_{P/O} + {}^{\mathcal{C}}\mathbf{a}_{Q/P} + {}^{\mathcal{I}}\boldsymbol{\alpha}^{\mathcal{C}} \times \mathbf{r}_{Q/P} + 2\,{}^{\mathcal{I}}\boldsymbol{\omega}^{\mathcal{C}} \times {}^{\mathcal{C}}\mathbf{v}_{Q/P}
$$

$$
+ {}^{\mathcal{I}}\boldsymbol{\omega}^{\mathcal{C}} \times \left({}^{\mathcal{I}}\boldsymbol{\omega}^{\mathcal{C}} \times \mathbf{r}_{Q/P} \right).
$$

Using the angular velocity from Eq. (8.25), the acceleration of P from Eq. (8.26), and the fact that ${}^{\mathcal{C}}\mathbf{a}_{Q/P} = {}^{\mathcal{C}}\mathbf{v}_{Q/P} = 0$, we have

$$
{}^{\mathcal{I}}\mathbf{a}_{Q/O} = l_1\ddot{\theta}_1\mathbf{b}_2 - l_1\dot{\theta}_1^2\mathbf{b}_1 + l_2\ddot{\theta}_2\mathbf{c}_2 - l_2\dot{\theta}_2^2\mathbf{c}_1.
$$

To find the equations of motion, we need the accelerations and forces expressed as components in a single frame. In this problem it is easiest to choose frame $\mathcal{B}$, making the inertial acceleration of Q equal to

$$
{}^{\mathcal{I}}\mathbf{a}_{Q/O} = -(l_1\dot{\theta}_1^2 + l_2\ddot{\theta}_2 \sin\beta + l_2\dot{\theta}_2^2 \cos\beta)\mathbf{b}_1 + (l_1\ddot{\theta}_1 + l_2\ddot{\theta}_2 \cos\beta - l_2\dot{\theta}_2^2 \sin\beta)\mathbf{b}_2.
$$

Now we use Newton's second law for each mass and the free-body diagrams in Figure 8.14c:

$$
-F_1\mathbf{b}_1 + F_2\mathbf{c}_1 + m_P g\mathbf{e}_1 = m_P(l_1\ddot{\theta}_1\mathbf{b}_2 - l_1\dot{\theta}_1^2\mathbf{b}_1)
$$

$$
-F_2\mathbf{c}_1 + m_Q g\mathbf{e}_1 = m_Q(-(l_1\dot{\theta}_1^2 + l_2\ddot{\theta}_2 \sin\beta + l_2\dot{\theta}_2^2 \cos\beta)\mathbf{b}_1
$$

$$
+ (l_1\ddot{\theta}_1 + l_2\ddot{\theta}_2 \cos\beta - l_2\dot{\theta}_2^2 \sin\beta)\mathbf{b}_2).
$$

Using the transformation tables in Eq. (8.23), we turn these two vector equations into the following four scalar equations in the four unknowns F_1, F_2, $\ddot{\theta}_1$, and $\ddot{\theta}_2$:

$$
-F_1 + F_2 \cos\beta + m_P g \cos\theta_1 = -m_P l_1 \dot{\theta}_1^2
$$

$$
F_2 \sin\beta - m_P g \sin\theta_1 = m_P l_1 \ddot{\theta}_1
$$

$$
-F_2 \cos\beta + m_Q g \cos\theta_1 = -m_Q(l_1\dot{\theta}_1^2 + l_2\ddot{\theta}_2 \sin\beta + l_2\dot{\theta}_2^2 \cos\beta)
$$

$$
-F_2 \sin\beta - m_Q g \sin\theta_1 = m_Q(l_1\ddot{\theta}_1 + l_2\ddot{\theta}_2 \cos\beta - l_2\dot{\theta}_2^2 \sin\beta).
$$

With some algebra, these equations can be solved for the equations of motion of the two degrees of freedom:

$$
\ddot{\theta}_1 = \frac{g\left(\mu_Q \cos\beta \sin\theta_2 - \sin\theta_1\right) + \mu_Q l_1 \sin\beta \cos\beta\dot{\theta}_1^2 + \mu_Q l_2 \sin\beta\dot{\theta}_2^2}{l_1\left(\mu_P + \mu_Q \sin^2\beta\right)}
$$

$$
\ddot{\theta}_2 = -\frac{l_1 \sin\beta\dot{\theta}_1^2 + l_2\mu_Q \sin\beta \cos\beta\dot{\theta}_2^2 + g \cos\theta_1 \sin\beta}{l_2\left(\mu_P + \mu_Q \sin^2\beta\right)},
$$

where $\mu_P = m_P/(m_Q + m_P)$ and $\mu_Q = m_Q/(m_Q + m_P)$. Problem 8.15 invites you to explore numerical solutions to these equations.

8.4 Tutorials

Tutorial 8.1 The Watt Linkage

A common dynamics problem is to convert rotation about an axis into the linear translation of a point. For example, a rotating motor may be used to translate the cylinder piston in an engine; another example is how a steering wheel is used to turn the wheels of a car. The first such device was invented and patented by James Watt[3] in 1784 and used linkages. It has come to be known as the *Watt linkage*. Watt's original design is shown in Figure 8.15. A simplified model of the Watt linkage is shown in Figure 8.16a. By rotating link $\mathcal{A}$ up and down about point A, the point P at the center of link $\mathcal{B}$ performs (almost) straight-line motion up and down for a large range of angles. We validate this claim using the results of this chapter.

We begin by defining three body frames: frame $\mathcal{A}$, which rotates with link $\mathcal{A}$ connecting points A and B; frame $\mathcal{B}$, which rotates with link $\mathcal{B}$ connecting points B and C; and frame $\mathcal{C}$, which rotates with link $\mathcal{C}$ connecting points C and D. The out-of-plane components of each frame are parallel to $\mathbf{e}_3$. The endpoints of the linkage are fixed in the inertial frame, so we have the constraints

$$^{\mathcal{I}}\mathbf{v}_{A/O} = 0 \tag{8.27}$$

$$^{\mathcal{I}}\mathbf{v}_{D/A} = 0. \tag{8.28}$$

Using the vector triad $\mathbf{r}_{D/A} = \mathbf{r}_{B/A} + \mathbf{r}_{C/B} + \mathbf{r}_{D/C}$, we can write the velocity of D with respect to A in the inertial frame as

$$^{\mathcal{I}}\mathbf{v}_{D/A} = {}^{\mathcal{I}}\mathbf{v}_{B/A} + {}^{\mathcal{I}}\mathbf{v}_{C/B} + {}^{\mathcal{I}}\mathbf{v}_{D/C}.$$

Each of these intermediate velocities can be rewritten using the transport equation:

$$^{\mathcal{I}}\mathbf{v}_{B/A} = {}^{\mathcal{A}}\mathbf{v}_{B/A} + {}^{\mathcal{I}}\boldsymbol{\omega}^{\mathcal{A}} \times \mathbf{r}_{B/A}$$

$$^{\mathcal{I}}\mathbf{v}_{C/B} = {}^{\mathcal{B}}\mathbf{v}_{C/B} + {}^{\mathcal{I}}\boldsymbol{\omega}^{\mathcal{B}} \times \mathbf{r}_{C/B}$$

$$^{\mathcal{I}}\mathbf{v}_{D/C} = {}^{\mathcal{C}}\mathbf{v}_{D/C} + {}^{\mathcal{I}}\boldsymbol{\omega}^{\mathcal{C}} \times \mathbf{r}_{D/C},$$

where $^{\mathcal{I}}\boldsymbol{\omega}^{\mathcal{A}} = \omega_A \mathbf{e}_3$, $^{\mathcal{I}}\boldsymbol{\omega}^{\mathcal{B}} = \omega_B \mathbf{e}_3$, and $^{\mathcal{I}}\boldsymbol{\omega}^{\mathcal{C}} = \omega_C \mathbf{e}_3$.

Assuming the links are rigid, all body velocities are zero. Using Eq. (8.28), this assumption gives

$$^{\mathcal{I}}\mathbf{v}_{D/A} = 0 = {}^{\mathcal{I}}\boldsymbol{\omega}^{\mathcal{A}} \times \mathbf{r}_{B/A} + {}^{\mathcal{I}}\boldsymbol{\omega}^{\mathcal{B}} \times \mathbf{r}_{C/B} + {}^{\mathcal{I}}\boldsymbol{\omega}^{\mathcal{C}} \times \mathbf{r}_{D/C}$$

$$= \omega_A l_1 \mathbf{a}_2 + \omega_B l_2 \mathbf{b}_1 + \omega_C l_3 \mathbf{c}_2$$

$$= \dot{\theta}_{AB} l_1 \mathbf{a}_2 + \dot{\theta}_{BC} l_2 \mathbf{b}_1 + \dot{\theta}_{CD} l_3 \mathbf{c}_2,$$

---

[3] James Watt (1736–1819) was a Scottish engineer known for his many innovative mechanical devices, including the Watt flyball governor and the Watt linkage.

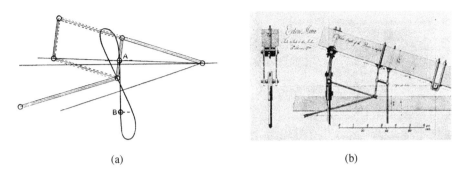

(a)   (b)

Figure 8.15 (a) The original Watt linkage for parallel motion. (Reprinted with the permission of Cambridge University Press.) (b) Its implementation in the Ecton mine engine.

where θ_{AB}, θ_{BC}, and θ_{CD} represent the angle each frame makes with respect to the inertial frame (assuming right-handed rotations). From this vector equation, we find two scalar equations by expressing all unit vectors as components in the inertial frame:

$$-\dot{\theta}_{AB}l_1 \sin \theta_{AB} + \dot{\theta}_{BC}l_2 \cos \theta_{BC} - \dot{\theta}_{CD}l_3 \sin \theta_{CD} = 0$$

$$\dot{\theta}_{AB}l_1 \cos \theta_{AB} + \dot{\theta}_{BC}l_2 \sin \theta_{BC} + \dot{\theta}_{CD}l_3 \cos \theta_{CD} = 0.$$

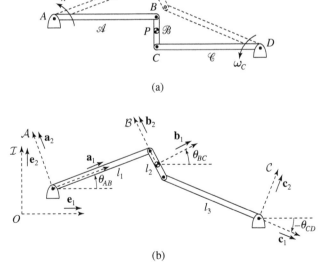

(a)

(b)

Figure 8.16 The Watt linkage converts rotational motion into translational motion. (a) Watt-linkage schematic. (b) Reference frames.

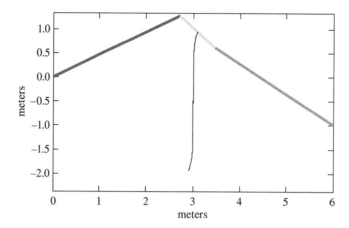

Figure 8.17 The trajectory of the center of the middle link of a Watt linkage as the left link is rotated between $\pm 30°$.

Now assuming we know the trajectory (or the constant angular velocity) of any of the three frames, we can integrate to find the others. Assuming that we control $\dot{\theta}_{AB} = \omega_A$, the equations of motion become

$$\dot{\theta}_{BC} = \omega_A \frac{l_1 \sin\left(\theta_{AB} - \theta_{CD}\right)}{l_2 \cos\left(\theta_{BC} - \theta_{CD}\right)} \tag{8.29}$$

$$\dot{\theta}_{CD} = -\omega_A \frac{l_1 \cos\left(\theta_{AB} - \theta_{BC}\right)}{l_3 \cos\left(\theta_{BC} - \theta_{CD}\right)}. \tag{8.30}$$

To integrate Eqs. (8.29) and (8.30), we must select a trajectory for ω_A that avoids singularities in these equations. A singularity occurs when:

$$\theta_{BC} - \theta_{CD} = \frac{\pi}{2}.$$

Physically this condition corresponds to when $\mathscr{B}$ is parallel to either one of the other two links. Figure 8.17 shows a trajectory of point P, the center of link $\mathscr{B}$, as it is rotated between roughly $\pi/2$ and $-\pi/2$. For this solution, link $\mathscr{A}$ is rotated at a rate of ± 0.5 rad/s between $\pm 30°$ and links $\mathscr{A}$, $\mathscr{B}$, and $\mathscr{C}$ have lengths of 3 m, 1 m, and 3 m, respectively.

Tutorial 8.2 Placing a Glass on a Table Using a Three-Link Robot Arm

This tutorial revisits the three-link robot arm introduced in Example 2.5. As described there, the arm is maneuvered by controlling the angle (and thus the angular rate) of each joint relative to the previous link. Our goal is to determine the angular velocities of each joint needed to place a glass on a table. The glass should follow a linear

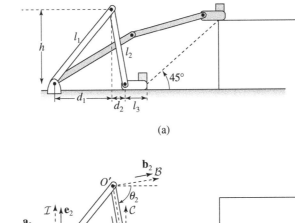

(a)

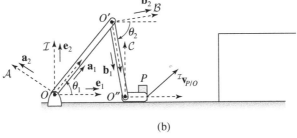

(b)

Figure 8.18 A three-link robot arm places a glass of water, labeled P, on a table at constant velocity. (a) Initial (white) and final (gray) configurations. (b) Reference frames.

trajectory at a constant velocity, as shown in Figure 8.18. We also require that the "wrist" link be level with the ground so that no water spills from the glass.

The geometry of the problem is shown in Figure 8.18a. Because the glass must travel an equal height and horizontal distance, we can write the components of its velocity in the inertial frame as

$$^{\mathcal{I}}\mathbf{v}_{P/O} = \frac{\sqrt{2}}{2} u_0 \mathbf{e}_1 + \frac{\sqrt{2}}{2} u_0 \mathbf{e}_2, \qquad (8.31)$$

where u_0 is the speed of the glass. We can use Eq. (8.18) to find $^{\mathcal{I}}\mathbf{v}_{P/O}$ again, this time in terms of the angular velocity of each arm relative to the inertial frame:

$$^{\mathcal{I}}\mathbf{v}_{P/O} = {}^{\mathcal{I}}\boldsymbol{\omega}^{\mathcal{A}} \times \mathbf{r}_{O'/O} + {}^{\mathcal{I}}\boldsymbol{\omega}^{\mathcal{B}} \times \mathbf{r}_{O''/O'} + \underbrace{{}^{\mathcal{I}}\boldsymbol{\omega}^{\mathcal{C}} \times \mathbf{r}_{P/O''}}_{=0},$$

where the body velocities are zero and the last term is zero because of the constraint that the glass remains level. From the geometry in Figure 8.18, we have

$$\mathbf{r}_{O'/O} = l_1 \mathbf{a}_1 = \sqrt{d_1^2 + h^2}\, \mathbf{a}_1$$

$$\mathbf{r}_{O''/O'} = l_2 \mathbf{b}_1 = \sqrt{d_2^2 + h^2}\, \mathbf{b}_1.$$

Because all motion in this system is planar, $^{\mathcal{I}}\boldsymbol{\omega}^{\mathcal{A}} = {}^{\mathcal{I}}\omega^{\mathcal{A}}\mathbf{a}_3$, and $^{\mathcal{I}}\boldsymbol{\omega}^{\mathcal{B}} = {}^{\mathcal{I}}\omega^{\mathcal{B}}\mathbf{b}_3$ so that the velocity of P is

$$^{\mathcal{I}}\mathbf{v}_{P/O} = l_1{}^{\mathcal{I}}\omega^{\mathcal{A}}\mathbf{a}_2 + l_2{}^{\mathcal{I}}\omega^{\mathcal{B}}\mathbf{b}_2. \tag{8.32}$$

If we define θ_1 as the angle of $\mathcal{A}$ with respect to $\mathcal{I}$ and θ_2 as the angle of $\mathcal{B}$ with respect to $\mathcal{I}$ as shown in Figure 8.18, the transformation arrays between $\mathcal{A}$ and $\mathcal{I}$ and $\mathcal{B}$ and $\mathcal{I}$ are, respectively,

	$\mathbf{e}_1$	$\mathbf{e}_2$
$\mathbf{a}_1$	$\cos\theta_1$	$\sin\theta_1$
$\mathbf{a}_2$	$-\sin\theta_1$	$\cos\theta_1$

and

	$\mathbf{e}_1$	$\mathbf{e}_2$
$\mathbf{b}_1$	$\cos\theta_2$	$-\sin\theta_2$
$\mathbf{b}_2$	$\sin\theta_2$	$\cos\theta_2$.

Note, $\dot{\theta}_1 = {}^{\mathcal{I}}\omega^{\mathcal{A}}$ and $\dot{\theta}_2 = -{}^{\mathcal{I}}\omega^{\mathcal{B}}$. At the initial time t_0 shown in Figure 8.18a, the transformation arrays are

	$\mathbf{e}_1$	$\mathbf{e}_2$
$\mathbf{a}_1$	d_1/l_1	h/l_1
$\mathbf{a}_2$	$-h/l_1$	d_1/l_1

and

	$\mathbf{e}_1$	$\mathbf{e}_2$
$\mathbf{b}_1$	d_2/l_2	$-h/l_2$
$\mathbf{b}_2$	h/l_2	d_2/l_2 ,

and the velocity in Eq. (8.32) expressed as components in the inertial frame is

$$^{\mathcal{I}}\mathbf{v}_{P/O}(t_0) = h\left(-{}^{\mathcal{I}}\omega^{\mathcal{A}}(t_0) + {}^{\mathcal{I}}\omega^{\mathcal{B}}(t_0)\right)\mathbf{e}_1 + \left(d_1{}^{\mathcal{I}}\omega^{\mathcal{A}}(t_0) + d_2{}^{\mathcal{I}}\omega^{\mathcal{B}}(t_0)\right)\mathbf{e}_2. \tag{8.33}$$

Comparing Eq. (8.33) to Eq. (8.31) gives two equations for the unknown inertial angular velocities at t_0:

$$h\left(-{}^{\mathcal{I}}\omega^{\mathcal{A}}(t_0) + {}^{\mathcal{I}}\omega^{\mathcal{B}}(t_0)\right) = \frac{\sqrt{2}}{2}u_0$$

$$d_1{}^{\mathcal{I}}\omega^{\mathcal{A}}(t_0) + d_2{}^{\mathcal{I}}\omega^{\mathcal{B}}(t_0) = \frac{\sqrt{2}}{2}u_0,$$

whose solutions are

$$^{\mathcal{I}}\omega^{\mathcal{A}}(t_0) = \frac{\sqrt{2}}{2}u_0\frac{1 - d_2/h}{d_1 + d_2} \tag{8.34}$$

$$^{\mathcal{I}}\omega^{\mathcal{B}}(t_0) = \frac{\sqrt{2}}{2}\frac{u_0}{h} + {}^{\mathcal{I}}\omega^{\mathcal{A}}(t_0). \tag{8.35}$$

Note that these angular velocities will only give a 45° trajectory at the very beginning of this system's motion (because d_1, d_2, and h change as the arms move). To find the angular velocities of the robot's links for $t > t_0$, we use the constraint that the inertial acceleration of the glass is zero. Returning to Eq. (8.32), we differentiate again to find

$$^{\mathcal{I}}\mathbf{a}_{P/O} = l_1\left({}^{\mathcal{I}}\alpha^{\mathcal{A}}\mathbf{a}_2 - ({}^{\mathcal{I}}\omega^{\mathcal{A}})^2\mathbf{a}_1\right) + l_2\left({}^{\mathcal{I}}\alpha^{\mathcal{B}}\mathbf{b}_2 - ({}^{\mathcal{I}}\omega^{\mathcal{B}})^2\mathbf{b}_1\right),$$

where $^{\mathcal{I}}\alpha^{\mathcal{A}}$ and $^{\mathcal{I}}\alpha^{\mathcal{B}}$ are the angular accelerations. Expressing the unit vectors as components in the inertial frame and using $^{\mathcal{I}}\mathbf{a}_{P/O} = 0$ produces

$$-l_1 {}^{\mathcal{I}}\alpha^A \sin\theta_1 - l_1({}^{\mathcal{I}}\omega^A)^2 \cos\theta_1 + l_2 {}^{\mathcal{I}}\alpha^B \sin\theta_2 - l_2({}^{\mathcal{I}}\omega^B)^2 \cos\theta_2 = 0$$
$$l_1 {}^{\mathcal{I}}\alpha^A \cos\theta_1 - l_1({}^{\mathcal{I}}\omega^A)^2 \sin\theta_1 + l_2 {}^{\mathcal{I}}\alpha^B \cos\theta_2 + l_2({}^{\mathcal{I}}\omega^B)^2 \sin\theta_2 = 0.$$

With some algebra, we can solve for the two angular accelerations:

$$^{\mathcal{I}}\alpha^A = \frac{-l_1({}^{\mathcal{I}}\omega^A)^2 \cos\beta - l_2({}^{\mathcal{I}}\omega^B)^2}{l_1 \sin\beta},$$

$$^{\mathcal{I}}\alpha^B = \frac{l_1({}^{\mathcal{I}}\omega^A)^2 + l_2({}^{\mathcal{I}}\omega^B)^2 \cos\beta}{l_2 \sin\beta},$$

where $\beta = \theta_1 + \theta_2$.

We now numerically integrate these first-order differential equations to find $^{\mathcal{I}}\omega^A(t)$, $^{\mathcal{I}}\omega^B(t)$, $\theta_1(t)$, and $\theta_2(t)$. The initial conditions for this integration are based on the initial geometry of the system and Eqs. (8.34) and (8.35).

Assuming that $d_1 = 0.5\,\text{m}$, $d_2 = 0.12\,\text{m}$, and $h = 1\,\text{m}$, suppose we want the velocity of the glass (u_0) to be $1\,\text{m/s}$. For these values, the initial angular velocities are $^{\mathcal{I}}\omega^A(t_0) = 1.00\,\text{rad/s}$ and $^{\mathcal{I}}\omega^B(t_0) = 1.71\,\text{rad/s}$. The resulting trajectories of the link angles and angular velocities are shown in Figure 8.19.

Finally, we can use the angular-velocity addition property to find the angular velocity of each arm relative to the previous arm:

$$^{\mathcal{I}}\omega^B = {}^{\mathcal{I}}\omega^A + {}^A\omega^B$$
$$^{\mathcal{I}}\omega^C = {}^{\mathcal{I}}\omega^A + {}^A\omega^B + {}^B\omega^C = 0.$$

Those two equations imply

$$^B\omega^C = -{}^{\mathcal{I}}\omega^A - {}^A\omega^B.$$

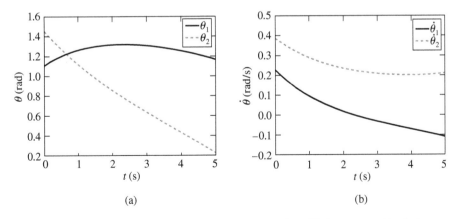

Figure 8.19 Trajectories of the angles and angular velocities of a three-link robot arm. The initial angular velocities have been scaled so that the entire trajectory takes exactly 5 s to execute. (a) Angles. (b) Angular velocities.

Tutorial 8.3 The Circular Restricted Three-Body Problem

Tutorial 7.1 outlined the solution to the two-body problem in orbital mechanics and showed that even when both bodies are free to move in absolute space, the solution for the motion of one body relative to the other is a conic section. The center of mass of the two bodies is either fixed or moves at constant inertial velocity.

Unfortunately, the more general problem of many bodies moving under their mutual gravitational attraction (the N-body problem) is not solvable: there is no simple way to write the trajectory of each body in terms of elementary functions. Even for just three bodies, the motion is extremely complex and unpredictable, except for some very specific arrangements (known as *choreographies*). In fact, it was Poincaré's[4] study of the three-body problem that led to the development of dynamical systems theory. Although a detailed discussion of the three-body problem and chaos is outside our scope, a simple analysis—called the *circular restricted three-body problem*—is an excellent example of the use of rotating frames for writing equations of motion.

The circular restricted three-body problem refers to the problem in which two large bodies (as in Tutorial 7.1) are in a circular orbit about each other and we wish to find the approximate equations of motion for a small third body of mass $m_3 \ll m_1, m_2$. The third body might be a spacecraft or an asteroid. In fact, several space missions have placed a satellite into an orbit in the earth/moon or sun/earth system where the gravitational attraction of both large bodies is important. Since we assume the third body has negligible mass, it does not affect the circular motion of the other two bodies (hence the adjective "restricted").

The situation is depicted in Figure 8.20. Since there are no external forces acting on bodies 1 and 2, we can consider the motion of body 3 relative to an inertial frame $\mathcal{I} = (G, \mathbf{e}_1, \mathbf{e}_2, \mathbf{e}_3)$ fixed to the center of mass G of bodies 1 and 2. We attach a rotating frame $\mathcal{B} = (G, \mathbf{b}_1, \mathbf{b}_2, \mathbf{b}_3)$ to G with $\mathbf{b}_1$ along the line connecting m_1 and m_2, and $\mathbf{b}_3$ parallel to $\mathbf{e}_3$ (and perpendicular to the plane of motion), so that $\mathbf{b}_2 = \mathbf{b}_3 \times \mathbf{b}_1$, as shown in Figure 8.20b. Because there are no external forces and body 1 and body 2 are in a circular orbit with fixed diameter, the total angular momentum of body 1 and body 2 about G is conserved and frame $\mathcal{B}$ rotates at a constant rate relative to $\mathcal{I}$ with angular velocity $^{\mathcal{I}}\boldsymbol{\omega}^{\mathcal{B}} = n\mathbf{e}_3$. (The common shorthand for the rotation rate of a circular orbit is the variable n.)

The free-body diagram for mass m_3 is shown in Figure 8.20c; the only two forces acting on mass m_3 are the gravitational attractions of masses m_1 and m_2. Also note that the center of mass G is not accelerating. This lets us write Newton's second law for the small third mass as

$$m_3\,{}^{\mathcal{I}}\mathbf{a}_{3/G} = -\frac{Gm_1m_3}{\|\mathbf{r}_{3/1}\|^3}\mathbf{r}_{3/1} - \frac{Gm_2m_3}{\|\mathbf{r}_{3/2}\|^3}\mathbf{r}_{3/2}. \tag{8.36}$$

We are interested, however, in the motion of m_3 relative to reference frame $\mathcal{B}$ rotating with the two larger bodies. We therefore replace the acceleration on the left-hand side of Eq. (8.36) with the expression in Eq. (8.21):

---

[4] Jules-Henri Poincaré (1854–1912) was a French mathematician, physicist, and philosopher. He made seminal contributions in topology, celestial mechanics, relativity, electromagnetism, and many other areas of mathematics and physics. He is considered the father of modern dynamical systems theory.

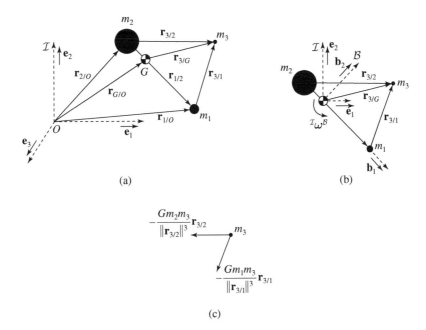

Figure 8.20 (a) Two bodies of masses m_1 and m_2 in a circular orbit about their common center of mass G in the plane orthogonal to $\mathbf{e}_3$. A third, small body of mass $m_3 \ll m_1$, m_2 orbits both of them. (b) Reference frames. (c) Free-body diagram for m_3.

$$^B\mathbf{a}_{3/G} + 2\,^{\mathcal{I}}\boldsymbol{\omega}^{B} \times \,^B\mathbf{v}_{3/G} + \,^{\mathcal{I}}\boldsymbol{\omega}^{B} \times \left(^{\mathcal{I}}\boldsymbol{\omega}^{B} \times \mathbf{r}_{3/G}\right) = -\frac{\mu_1}{\|\mathbf{r}_{3/1}\|^3}\mathbf{r}_{3/1} - \frac{\mu_2}{\|\mathbf{r}_{3/2}\|^3}\mathbf{r}_{3/2},$$

where we have substituted $\mu_1 = Gm_1$ and $\mu_2 = Gm_2$ and have divided by the mass m_3.

At this point it is easiest to write the equations of motion in terms of specific coordinates. We choose Cartesian coordinates in frame $\mathcal{B}$, $(x, y)_{\mathcal{B}}$, so that $\mathbf{r}_{3/G} = x\mathbf{b}_1 + y\mathbf{b}_2$. Note that the position of body 1 is $\mathbf{r}_{1/G} = x_1\mathbf{b}_1$ and that of body 2 is $\mathbf{r}_{2/G} = -x_2\mathbf{b}_1$. Only motion in the plane of the two-body circular orbit is considered here, so we ignore the z coordinate. Using $^{\mathcal{I}}\boldsymbol{\omega}^{B} = n\mathbf{e}_3$, $^B\mathbf{a}_{3/G} = \ddot{x}\mathbf{b}_1 + \ddot{y}\mathbf{b}_2$, $\mathbf{r}_{3/1} = (x - x_1)\mathbf{b}_1 + y\mathbf{b}_2$, and $\mathbf{r}_{3/2} = (x + x_2)\mathbf{b}_1 + y\mathbf{b}_2$, the equations of motion in the rotating frame are

$$\ddot{x} - 2n\dot{y} - n^2x = -\frac{\mu_1(x - x_1)}{\left((x - x_1)^2 + y^2\right)^{\frac{3}{2}}} - \frac{\mu_2(x + x_2)}{\left((x + x_2)^2 + y^2\right)^{\frac{3}{2}}} \qquad (8.37)$$

$$\ddot{y} + 2n\dot{x} - n^2y = -\frac{\mu_1 y}{\left((x - x_1)^2 + y^2\right)^{\frac{3}{2}}} - \frac{\mu_2 y}{\left((x + x_2)^2 + y^2\right)^{\frac{3}{2}}}. \qquad (8.38)$$

These equations have no general solution, and it is beyond our scope to study in detail various possible trajectories, either analytically or numerically, though very interesting and complex motions are possible. The important point is that we used the relative-motion results from Eq. (8.21) to develop equations of motion for the coordinates of a particle in a rotating frame of reference.

There are two calculations that lend great insight into the circular restricted three-body problem without solving for exact trajectories. First, we can find an integral of the motion in the same way we found the energy integral for many other problems. Multiplying Eq. (8.37) by $2\dot{x}$ and Eq. (8.38) by $2\dot{y}$ and then adding the two equations results in

$$2(\ddot{x}\dot{x} + \ddot{y}\dot{y} - n^2(x\dot{x} + y\dot{y})) = -\frac{2\mu_1((x-x_1)\dot{x} + y\dot{y})}{r_1^3}$$
$$-\frac{2\mu_2((x+x_2)\dot{x} + y\dot{y})}{r_2^3}, \tag{8.39}$$

where $r_1 \triangleq \|\mathbf{r}_{3/1}\| = \sqrt{(x-x_1)^2 + y^2}$ and $r_2 \triangleq \|\mathbf{r}_{3/2}\| = \sqrt{(x+x_2)^2 + y^2}$. Note that differentiating r_1^2 and r_2^2 yields $r_1\dot{r}_1 = 2(x-x_1)\dot{x} + y\dot{y}$ and $r_2\dot{r}_2 = (x+x_2)\dot{x} + y\dot{y}$, respectively. Consequently, we can multiply both sides of Eq. (8.39) by dt and integrate to get

$$2\int \dot{x}\,d\dot{x} + 2\int \dot{y}\,d\dot{y} - 2n^2 \int x\,dx - 2n^2 \int y\,dy = -2\mu_1 \int \frac{dr_1}{r_1^2} - 2\mu_2 \int \frac{dr_2}{r_2^2}$$

which equals

$$\dot{x}^2 + \dot{y}^2 - n^2(x^2 + y^2) = \frac{2\mu_1}{r_1} + \frac{2\mu_2}{r_2} - C, \tag{8.40}$$

where C is a constant of integration.

Eq. (8.40) is known as the *Jacobi integral,* and C is *Jacobi's constant. C* serves the same purpose as the total energy, since all trajectories of body 3 must satisfy Eq. (8.40). It is common practice to draw contours of constant C. These zero-velocity curves are used just as the ones in Tutorial 5.3. They determine the regions, for a given value of C, where trajectories must lie. For example, Figure 8.21 shows zero-velocity curves for the earth-moon system over a range of values of C from 2 to 18 m^2/s^2. A trajectory must lie entirely inside the curve determined by its initial condition. It is interesting that, for initial conditions close to either large mass, the constant-potential lines look like those for a single gravitating body and the orbits will be like those discussed for the two-body problem in Example 4.9 and Tutorial 4.2.

The second insightful calculation is to search for equilibrium solutions of Eqs. (8.37) and (8.38). These are points in frame $\mathcal{B}$ where body 3 remains stationary. We find these points by setting the speeds and accelerations equal to zero in Eqs. (8.37) and (8.38). There are five such points in this rotating system. Three of them lie on the line connecting masses m_1 and m_2 ($y = 0$). Eq. (8.38) is satisfied trivially there, and Eq. (8.37) reduces to the quintic polynomial,

$$n^2 x(x-x_1)^2(x+x_2)^2 - \mu_1(x+x_2)^2 - \mu_2(x-x_1)^2 = 0.$$

The three real roots of this polynomial correspond to the three equilibrium points on the x-axis, which we label L_1, L_2, and L_3.

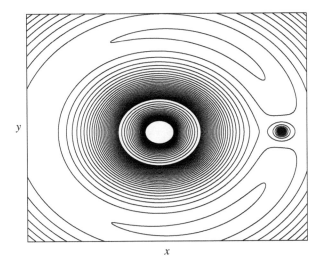

Figure 8.21 Zero-velocity curves for the earth-moon system. The earth is at the center and the moon to the right.

The other two equilibrium points lie on the vertices of an equilateral triangle such that $r_1 = r_2 = r$, the distance between body 1 and body 2. To see how these equilibrium points satisfy Eqs. (8.37) and (8.38), use the identity $n^2 = (\mu_1 + \mu_2)/r^3$ and the center-of-mass corollary, $m_1 x_1 = m_2 x_2$. The final two equilibrium points are then

$$L_4 = \left(\frac{r}{2} - \mu^* r, \frac{\sqrt{3}}{2} r \right)_{\mathcal{B}}$$

$$L_5 = \left(\frac{r}{2} - \mu^* r, -\frac{\sqrt{3}}{2} r \right)_{\mathcal{B}},$$

where $\mu^* \triangleq m_1/(m_1 + m_2)$ is the *mass ratio* of the large bodies.

These five points were originally found by Lagrange[5] in 1772 and are now often referred to as the *Lagrange points*. They are schematically shown in Figure 8.22. It can be shown that the collinear points L_1, L_2, and L_3 are all unstable, whereas, for certain ratios of m_1/m_2, the equilateral points L_4 and L_5 are stable.[6] For instance, the sun/Jupiter L_4 and L_5 points are stable equilibrium locations in the sun/Jupiter system. It was a great verification of the theory when asteroids were discovered to be residing at these points more than 100 years after Lagrange's first calculation.

---

[5] Joseph-Louis Lagrange (1736–1813) was an Italian-French mathematician and astronomer known for his many contributions to mechanics.

[6] Recall that, if a particle starts near a stable equilibrium point, then it stays near that point for all time, whereas if a particle starts near an unstable equilibrium point, it is not guaranteed to stay nearby.

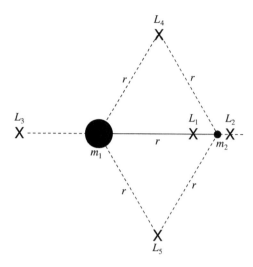

Figure 8.22 Schematic of the five Lagrange points in the circular restricted three-body problem.

8.5 Key Ideas

- If P is a point fixed to a rigid body $\mathscr{B}$, the inertial velocity of P with respect to O is

$$^{\mathcal{I}}\mathbf{v}_{P/O} = {}^{\mathcal{I}}\mathbf{v}_{O'/O} + {}^{\mathcal{I}}\boldsymbol{\omega}^{\mathcal{B}} \times \mathbf{r}_{P/O'},$$

where $\mathcal{B}$ is a reference frame fixed to $\mathscr{B}$ at point O', and $^{\mathcal{I}}\boldsymbol{\omega}^{\mathcal{B}}$ is the angular velocity of $\mathcal{B}$ with respect to $\mathcal{I}$.

- Under the **no-slip condition,** the point of a rolling disk or wheel in contact with the ground has zero velocity relative to the ground.

- The **transport equation** can be used to find the time derivative of a vector $\mathbf{c}$ with respect to frame $\mathcal{A}$:

$$\frac{^{\mathcal{A}}d}{dt}\mathbf{c} = \frac{^{\mathcal{B}}d}{dt}\mathbf{c} + {}^{\mathcal{A}}\boldsymbol{\omega}^{\mathcal{B}} \times \mathbf{c}.$$

For example, the inertial velocity of P with respect to the origin O' of a rotating frame $\mathcal{B}$ is

$$\frac{^{\mathcal{I}}d}{dt}\left(\mathbf{r}_{P/O'}\right) = \frac{^{\mathcal{B}}d}{dt}\left(\mathbf{r}_{P/O'}\right) + {}^{\mathcal{I}}\boldsymbol{\omega}^{\mathcal{B}} \times \mathbf{r}_{P/O'}.$$

- Angular velocities obey the **addition property**:

$$^{\mathcal{I}}\boldsymbol{\omega}^{\mathcal{B}} = {}^{\mathcal{I}}\boldsymbol{\omega}^{\mathcal{A}} + {}^{\mathcal{A}}\boldsymbol{\omega}^{\mathcal{B}},$$

where $\mathcal{A}$ and $\mathcal{B}$ are rotating reference frames.

- A general expression for the inertial acceleration of P with respect to O in terms of its velocity and acceleration in a rotating frame $\mathcal{B}$ is

$$^{\mathcal{I}}\mathbf{a}_{P/O} = {}^{\mathcal{I}}\mathbf{a}_{O'/O} + {}^{\mathcal{B}}\mathbf{a}_{P/O'} + {}^{\mathcal{I}}\boldsymbol{\alpha}^{\mathcal{B}} \times \mathbf{r}_{P/O'} + 2{}^{\mathcal{I}}\boldsymbol{\omega}^{\mathcal{B}} \times {}^{\mathcal{B}}\mathbf{v}_{P/O'}$$

$$+ {}^{\mathcal{I}}\boldsymbol{\omega}^{\mathcal{B}} \times \left({}^{\mathcal{I}}\boldsymbol{\omega}^{\mathcal{B}} \times \mathbf{r}_{P/O'} \right),$$

where O' is the origin of $\mathcal{B}$ and $^{\mathcal{I}}\boldsymbol{\alpha}^{\mathcal{B}} \triangleq \frac{{}^{\mathcal{B}}d}{dt}({}^{\mathcal{I}}\boldsymbol{\omega}^{\mathcal{B}}) = \frac{{}^{\mathcal{I}}d}{dt}({}^{\mathcal{I}}\boldsymbol{\omega}^{\mathcal{B}})$ is the **angular acceleration** of $\mathcal{B}$ with respect to $\mathcal{I}$.

8.6 Notes and Further Reading

Without question, the most important result of this chapter is the transport equation, relating, by means of the angular velocity, the derivative of a vector in one frame to the derivative of the same vector with respect to another frame. In one notation or another, the transport equation appears in every dynamics textbook, from introductory to advanced. It is the cornerstone of kinematics in rotating frames. Surprisingly, given its importance and ubiquitous application, there is no generally accepted nomenclature for referring to it. Most textbooks don't use a name at all (e.g., Greenwood 1988; Moon 1998; Meriam and Kraige 2001; Bedford and Fowler 2002; Hibbeler 2003; Tongue and Sheppard 2005; Beer et al. 2007). Kane (1978) discusses what he calls *kinematic theorems*, the transport equation being the first (the second refers to vector summation of velocities). The later book Kane and Levinson (1985) drops the reference and simply discusses "differentiation in two reference frames." Tenenbaum (2004) also refers to this equation as a kinematic theorem. Rao (2006) calls it the *transport theorem*, which unfortunately can cause confusion with the well-known Reynolds' transport theorem in fluid mechanics. It is very convenient to be able to reference this equation by name, so we refer to it as the "transport equation." We will often refer to it by name rather than by an equation number to get you used to its importance, just as we refer to Newton's second law.

This chapter also presented several important engineering examples that appear in most texts and arise frequently in practice, including linkages, gears, and wheels. We recommend many texts for further exploration of examples and problems, including Moon (1998), Meriam and Kraige (2001), Bedford and Fowler (2002), Hibbeler (2003), Tongue and Sheppard (2005), Beer et al. (2007), and Ginsberg (2008). Our particular examples are very similar to those in Bedford and Fowler (2002).

James Watt was a singularly important figure in the history of engineering. His inventions were transformational and were instrumental in ushering in the industrial revolution; many are still used today. We describe two in detail in this book (the Watt linkage and the Watt flyball governor). Readers interested in more detail on Watt and his work are directed to Dickinson (1936).

The three-body problem is a classic and important problem in orbital mechanics. It was first studied in detail by Poincaré and resulted in the discovery of chaos theory. An excellent historical discussion is in Diacu and Holmes (1996). An extremely thorough (albeit advanced) treatment, including the restricted problem, is in Szebehely's (1967) classic book. Vallado (2001) also has a good introductory discussion to the restricted problem. Those interested in a modern and much deeper treatment of the three- and four-body problems and motion about equilibrium points are directed to Gómez et al. (2001) and Koon et al. (2007).

8.7 Problems

8.1 Consider a body frame $\mathcal{B} = (O', \mathbf{b}_1, \mathbf{b}_2, \mathbf{b}_3)$ translating and rotating with respect to an inertial frame $\mathcal{I}$, as shown in Figure 8.23. The instantaneous position of particle P with respect to O' is $\mathbf{r}_{P/O'} = \mathbf{b}_1 + 2\mathbf{b}_2$. The velocity of O' with respect to O is $^{\mathcal{I}}\mathbf{v}_{O'/O} = 2\mathbf{e}_x$ m/s. The angular velocity of $\mathcal{B}$ with respect to $\mathcal{I}$ is $^{\mathcal{I}}\boldsymbol{\omega}^{\mathcal{B}} = 3\mathbf{b}_3$ rad/s.

    a. If P is fixed in $\mathcal{B}$, find the velocity $^{\mathcal{I}}\mathbf{v}_{P/O}$ of P with respect to O in $\mathcal{I}$.

    b. If P is moving with respect to O' in $\mathcal{B}$ at velocity $^{\mathcal{B}}\mathbf{v}_{P/O'} = 3\mathbf{b}_2$, find the velocity $^{\mathcal{I}}\mathbf{v}_{P/O}$ of P with respect to O in $\mathcal{I}$.

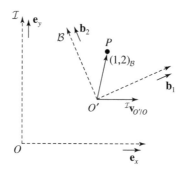

Figure 8.23 Problem 8.1.

8.2 Let $\mathcal{B} = (O', \mathbf{b}_1, \mathbf{b}_2, \mathbf{b}_3)$ be an arbitrary (planar) body frame with angular velocity $^{\mathcal{I}}\boldsymbol{\omega}^{\mathcal{B}}$ relative to an inertial frame $\mathcal{I} = (O, \mathbf{e}_1, \mathbf{e}_2, \mathbf{e}_3)$, where $\mathbf{e}_3 = \mathbf{b}_3$. Prove that $\frac{^{\mathcal{I}}d}{dt}(\mathbf{b}_i) = {^{\mathcal{I}}\boldsymbol{\omega}^{\mathcal{B}}} \times \mathbf{b}_i$, for $i = 1, 2, 3$.

8.3 Consider the bicycle model in Figure 8.24. Suppose crank shaft $\mathcal{A}$ is pedaled at $^{\mathcal{I}}\omega^{\mathcal{A}} = 2$ rpm and is connected to gear $\mathcal{B}$ by a chain of fixed length. The gear is fixed to wheel $\mathcal{C}$. If $r_A = 2r_B = 0.5r_C = 0.1$ m, how fast does the bike roll without slipping? If r_B increases, does the bike roll faster or slower?

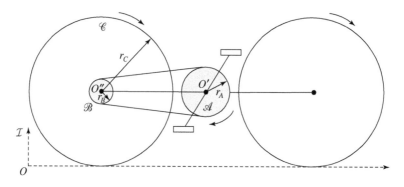

Figure 8.24 Problem 8.3.

8.4   An astronaut (white dot) is walking radially outward at v_0 relative to a space station, as shown in Figure 8.25. The space station is traveling with inertial velocity $^\mathcal{I}\mathbf{v}_{O'/O}$ and rotating with constant angular rate Ω. When the astronaut is at O' (when $r = 0$), she starts moving in the same direction as $^\mathcal{I}\mathbf{v}_{O'/O}$. Find the magnitude of the inertial velocity of the astronaut as a function of $^\mathcal{I}\mathbf{v}_{O'/O}$, Ω, v_0 and her radial distance r from O'.

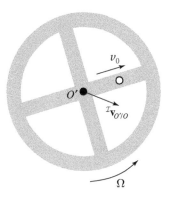

Figure 8.25 Problem 8.4.

8.5   Consider a carousel rotating at angular velocity $^\mathcal{I}\omega^\mathcal{B}$ relative to an inertial frame fixed to the ground with origin O. There is a clown on the nonrotating wall (into whose mouth you want to throw a brass ring) located at a vector position $\mathbf{r}_{C/O}$ relative to O, the center of the carousel. You are on a horse fixed to the carousel (ignore its up-and-down motion) at position $\mathbf{r}_{H/O}$. What is the velocity of the clown (rate of change of $\mathbf{r}_{C/H}$) relative to you (i.e., in the body frame of the carousel) in terms of $^\mathcal{I}\omega^\mathcal{B}$ and $\mathbf{r}_{C/O}$? (See Example 8.5.)

8.6   Suppose you are the gunner on a fighter jet traveling with inertial velocity $^\mathcal{I}\mathbf{v}_{J/O}$. The ground has radioed to you an incoming threat with inertial velocity $^\mathcal{I}\mathbf{v}_{T/O}$ (assume that the earth is an inertial frame). To fire and hit the threat, you need to know its velocity relative to your jet. Your jet is flying level, but it is executing an evasive turn, so it is rotating with angular velocity $^\mathcal{I}\omega^\mathcal{B}$. Based on knowledge of your position and information from the ground, you are able to compute the position of the target $\mathbf{r}_{T/J}$ relative to your jet. Write down the equation you would use to compute the velocity of the target relative to a frame fixed to your jet ($\mathcal{B}$ frame) in terms of $^\mathcal{I}\mathbf{v}_{T/O}$, $^\mathcal{I}\mathbf{v}_{J/O}$, $\mathbf{r}_{T/J}$, and $^\mathcal{I}\omega^\mathcal{B}$.

8.7   Consider a disk with a slot, as shown in Figure 8.26. The slot is offset from the diameter by distance l, and the disk is rotating with constant angular velocity Ω. Inside the slot is a particle P of mass m connected to a linear spring with spring constant k. At time $t = 0$ the spring is stretched an amount x_0 (assume that the unstretched length of the spring is zero). Note that the disk is in a plane and there is no gravity acting.

   a. Find the equation of motion for the mass position x inside the slot.
   b. For what Ω does $\ddot{x}$ go to zero? What physically is happening?

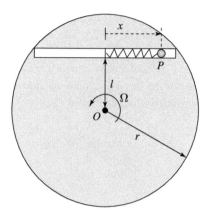

Figure 8.26 Problem 8.7.

    c. What is the normal force the disk applies to the mass in terms of m, x, $\dot{x}$, l, and Ω?

    d. Is energy conserved in this problem?

    e. What is the angular momentum $^{I}\mathbf{h}_{P/O}$ of the point mass relative to origin O? Is it conserved during the motion?

8.8   A particle of mass m slides inside a circular slot of radius r cut out of a massless disk of radius $R > r$, as shown in Figure 8.27. The particle is attached to a curvilinear spring with spring constant K and unstressed angle θ_0, where the angle θ describes the position of the particle relative to the attachment point A of the spring. Knowing that the disk rotates with constant angular velocity $\Omega\mathbf{e}_3$ about an axis through its center at O and assuming no gravity, determine the following:

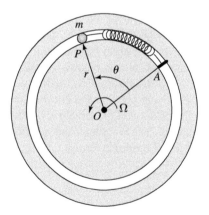

Figure 8.27 Problem 8.8.

    a. The velocity and acceleration of P relative to a fixed observer (you can pick the frame used to write the components).

    b. The differential equation of motion for the particle in terms of angle θ.

8.9   Sketch a planar model of a K-MAX helicopter delivering cargo to a moving
ship, as shown in Figure 8.28, using a point mass to represent the helicopter.
Derive the inertial kinematics of the cargo with respect to a point fixed to the
deck of the ship, assuming that the length of the cargo sling is fixed.

Figure 8.28 Problem 8.9.

8.10   Consider a disk of radius $R = 1$ m that is rotating at constant angular rate
$\Omega = 0.2$ rad/s about an inertially fixed point O, as shown in Figure 8.29.
Body frame $\mathcal{B} = (O, \mathbf{b}_1, \mathbf{b}_2, \mathbf{b}_3)$ is fixed to the disk. Particle P is launched
with respect to O in $\mathcal{B}$ at velocity ${}^{\mathcal{B}}\mathbf{v}_{P/O} = -v_0\mathbf{b}_1$, where $v_0 = 1$ m/s. There
are no external forces acting on P after it is launched.

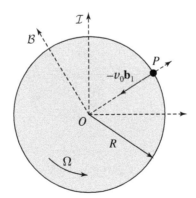

Figure 8.29 Problem 8.10.

a. Using the coordinates $(x, y)_{\mathcal{I}}$, find the equations of motion of P.

b. Integrate the equations of motion in MATLAB for 2 s and plot position $\mathbf{r}_{P/O}(t)$ from the perspective of an observer in $\mathcal{I}$. [HINT: Assume that $\mathcal{B}$ and $\mathcal{I}$ are aligned at the instant P is launched.]

c. Using the coordinates $(x, y)_{\mathcal{B}}$, find the equations of motion of P.

d. Integrate the equations of motion in MATLAB for 2 s, and plot the position $\mathbf{r}_{P/O}(t)$ from the perspective of an observer in $\mathcal{B}$.

8.11 Consider the locomotive wheel sketched in Figure 8.30. Assuming the wheel rolls without slipping and the piston moves only horizontally, derive the kinematic relationship between the velocity of a point on the piston and the velocity of the train.

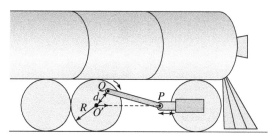

Figure 8.30 Problem 8.11.

8.12 An astronaut drops a wrench while on a space walk from her space station (based on a true story!) (Figure 8.31). Assume that there are no forces acting on the wrench and that the space station is translating at constant velocity $^{\mathcal{I}}\mathbf{v}_{O'/O}$ and rotating with respect to an inertial frame $\mathcal{I}$ at constant angular rate Ω. Let $\mathcal{B}$ be a body frame attached to the space station. Use the coordinates $(x, y)_{\mathcal{B}}$ to derive the kinematics of P with respect to O, and then find the equations of motion of x and y. If the astronaut releases the wrench from rest in $\mathcal{B}$, what is the initial velocity $^{\mathcal{I}}\mathbf{v}_{P/O}$?

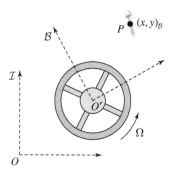

Figure 8.31 Problem 8.12.

8.13 Consider two particles, each of mass m, free to slide on rails forming a right angle and connected by a spring of spring constant k, as shown in Figure 8.32. The spring is unstretched when each mass is a distance x_0 from point O (so the unstretched length of the spring is $\sqrt{2}x_0$).

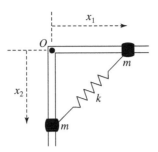

Figure 8.32 Problem 8.13.

a. How many degrees of freedom are there in this problem?
b. Is the linear momentum of the center of mass conserved during any motion of the masses?
c. Write down the equations of motion for the masses.
d. Now suppose the angle bracket rotates at constant angular velocity $\Omega\mathbf{e}_3$ perpendicular to the plane (with $\mathbf{e}_3$ going through O). What are the new equations of motion?
e. How many equilibrium arrangements are there for the masses while the bracket is rotating? Explain.

8.14 Consider a bead sliding down a frictionless wire under the influence of gravity as shown in Figure 8.33. Suppose the bead starts at rest at height $y = h$ at $x = 0$. Also assume that the shape of the wire is given by the function $y = f(x)$.

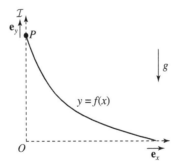

Figure 8.33 Problem 8.14.

a. Find an expression for the velocity $^\mathcal{I}\mathbf{v}_{P/O}(x)$ of the bead as a function of horizontal position x in terms of components in the inertial frame directions $\mathbf{e}_x$ and $\mathbf{e}_y$.
b. Now suppose the wire and bead are rotating together at angular rate Ω about the $\mathbf{e}_z$ axis. What is the inertial velocity of the bead now?

8.15 Recall the simple double pendulum considered in Example 8.9. We now explore some solutions to the equations of motion developed in the text.

a. Write a MATLAB function that accepts as inputs l_1, l_2, m_P, m_Q, a time span (initial and final times), and initial conditions for θ_1, θ_2, $\dot{\theta}_1$, and $\dot{\theta}_2$

and integrates the equations of motion of the simple double pendulum for these values over the specified time span.

b. Integrate the equations of motion for $m_P = m_Q = 4$ kg, $l_1 = l_2 = 1$ m, and initial conditions such that the pendulum starts from rest with both arms extended horizontally to the right of the attachment point (i.e., $\theta_1(0) = \theta_2(0) = \pi/2$). Plot the first 10 s of the resulting trajectories of points P and Q on one graph, using coordinates in the $D = (O, \mathbf{e}_x, \mathbf{e}_y, \mathbf{e}_z)$ frame, where $\mathbf{e}_x = \mathbf{e}_2$ and $\mathbf{e}_y = -\mathbf{e}_1$ (these would be the paths you observed the particles tracing out if you were observing the double pendulum in action). Repeat the integration for initial conditions with the first link as before and the second link starting pointing up, perpendicular to the first (i.e., $\theta_2(0) = -\pi$).

c. Perform integrations with unequal masses and link lengths. Explain how these differences affect the behavior of the double pendulum.

8.16 A particle P moves along a straight radial groove in a circular disk of radius a that is pivoted about a perpendicular axis through its center O. The particle moves relative to the disk so that

$$r = \frac{a}{2}(1 + \sin \omega t),$$

and the disk rotates according to

$$\theta = \theta_0 \sin \omega t.$$

Find the general expression for the absolute acceleration (i.e., in inertial space) of P in terms of a, θ_0, ω, and t.

8.17 Suppose you are the navigator on a ship in choppy seas so that the ship is rocking about its center of mass at rate ω_r (i.e., the ship's angle is given by $\phi = a \sin \omega_r t$), as shown in Figure 8.34.

Near the bow, a distance d from the center of mass, you have a pendulum clock. (You are in the early eighteenth century, and that is the only available clock.) Thus time is kept based on a calibrated period of the swinging pendulum. (You can ignore the height of the pendulum above the deck of the ship; it is exaggerated in the figure for clarity.)

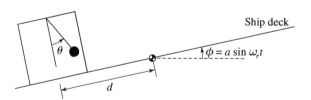

Figure 8.34 Problem 8.17.

a. Find the equation of motion for the angle of the clock's pendulum relative to its base (i.e., to a frame fixed to the ship).

b. What is the likelihood that the clock will keep accurate time while at sea?

CHAPTER NINE

Dynamics of a Planar Rigid Body

As discussed in Chapter 8, a rigid body is a collection of many particles that are constrained to be stationary relative to one another. In this chapter, we use the results of Chapter 7 to find equations for the translational and rotational motion of a rigid body. Our only restriction is that the motion is still confined to a plane. Despite this restriction, we will encounter many realistic and useful problems and lay the groundwork for the three-dimensional case studied in Part Four.

9.1 A Rigid Body Is a Multiparticle System

Chapters 6 and 7 discussed multiparticle systems in depth. In the most general, unrestricted, three-dimensional case, a system of N particles has $3N$ degrees of freedom. There is little we can do to simplify such a problem: a complete description requires $3N$ equations of motion. We did, however, introduce some useful facts that made solving many problems easier. For instance, the translational motion of the center of mass of the collection can be treated as if it were a single particle of mass m_G equal to the total mass of the collection. We then devoted our attention to understanding the motion of the collection relative to the center of mass. Even that was not a great simplification, however, as it still involved the same large number of equations. We thus restricted the discussion in Chapter 6 to either a small number of particles (collisions, orbits) or to a continuum of particles (variable-mass systems).

Chapter 7 introduced additional tools to study the motion of the particles relative to the center of mass (or some other fixed point), namely, angular momentum and energy. These tools supply concise formulas for writing the equations of motion relative to the center of mass. Nevertheless, the problem of scale did not necessarily go away, and many particles meant many equations. We thus limited the discussion in that chapter to systems with a small number of particles or ones with many constraints.

Now we turn to the special case of a planar rigid body. Figure 9.1 shows an example of a rigid and nonrigid body. Put simply, a rigid body is a collection of particles with

(a)   (b)

Figure 9.1 Illustration of (a) a rigid (frozen) fish and (b) a flexible (swimming) fish. Images courtesy of Shutterstock.

enough constraints to reduce the degrees of freedom to the motion of only a single object.

Definition 9.1 A **rigid body** is a collection of particles constrained to remain motionless relative to one another.

Mathematically, this definition is represented by the constraints

$$\frac{d}{dt}\|\mathbf{r}_{i/j}\| = 0 \qquad (9.1)$$

for all pairs i and j. Although this set of constraints is true for every rigid body, it is more than the minimum number of constraining links among particles needed to make a set of N particles rigid. (How many massless rods are necessary to connect a set of N particles so that the resulting collection is rigid? While this is an interesting question, its investigation is beyond the scope of this text.)

It is because of the many constraints holding the constituent particles fixed relative to one another that the number of degrees of freedom reduces dramatically. Recall that, in two dimensions, the number of degrees of freedom of N particles is $2N$. A two-dimensional rigid body composed of N particles, however, has only three degrees of freedom: two degrees of freedom describe the position of the center of mass (translation) and one describes the orientation of the body (rotation). This reduction results from the $2N - 3$ constraints of the form Eq. (9.1) that make the collection rigid (see Eq. (2.12)).

In three dimensions, a general collection of N particles has $3N$ degrees of freedom. A three-dimensional rigid body, however, has only six degrees of freedom: three

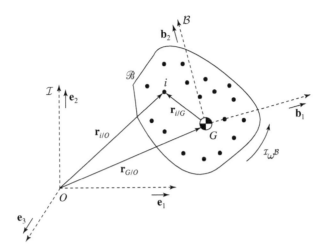

Figure 9.2 Planar rigid body $\mathscr{B}$ with attached body frame $\mathcal{B}$ translating and rotating with angular velocity $^{\mathcal{I}}\boldsymbol{\omega}^{\mathcal{B}}$ with respect to fixed inertial frame $\mathcal{I}$.

degrees of freedom describe the position of the center of mass and three describe the orientation of the body. Thus $3N - 6$ constraints of the form Eq. (9.1) are needed to make the collection rigid. This result holds even for a *continuous* rigid body, by which we mean a collection of N particles in the limit $N \to \infty$.

Figure 9.2 shows a multiparticle system collected together as a rigid body labeled $\mathscr{B}$. Because the particles do not move relative to one another, we can unambiguously attach a body frame to $\mathscr{B}$, which we label $\mathcal{B}$, and whose origin is located at the center of mass G of the rigid body. In the remainder of this chapter we develop a formalism for describing the translational motion of the body in the inertial frame and the rotational motion of the rigid body relative to $\mathcal{I}$, while still restricting ourselves to only two dimensions and thus three degrees of freedom (two of translation and one of rotation).

Before we move on, a word about models. This is a good time to review the discussion in Chapter 2 on Newton's laws. Recall that we emphasized that these three laws were not meant to explain matter and inertia but rather to provide predictive tools for describing motion. The concepts of point mass and force are abstractions—useful models that when used correctly with Newton's laws are remarkably effective at describing the motion of particles. The same is true here in this discussion of rigid bodies. The idea of constructing a rigid body as a continuous collection of point masses is a useful model that allows us to apply Newton's laws; its veracity is proven by its effectiveness. There is no reductionist theory that allows us to break down rigid bodies into actual point masses. At the atomic scale, quantum mechanics takes over and the Newtonian approach to forces and moments breaks down. Nevertheless, as long as we use our models consistently and correctly apply the laws of motion, the results are valid. The notes at the end of the chapter briefly discuss an alternative to the particle or *corpuscular* model for a rigid body and provide some references for further exploration.

9.2 Translation of the Center of Mass—Euler's First Law

Recall from Chapter 6 that the motion of the center of mass G of a collection of particles can be treated as if it were a single particle located at the center of mass with mass m_G equal to the total mass of the collection. This result applies unchanged to rigid bodies. In other words, we can extend Newton's second law and apply it to the center of mass of the rigid body. This has come to be known as *Euler's first law*, as it was Euler who first developed the equations of motion for a rigid body.

Law 9.1 Euler's first law states that the total mass m_G of a rigid body times the inertial acceleration of the center of mass equals the total external force on the rigid body:

$$\mathbf{F}_G = m_G{}^{\mathcal{I}}\mathbf{a}_{G/O}.$$

Note that we have dropped the (ext) superscript on the force that appears in Eq. (6.16) for multiparticle systems in this statement of Euler's first law for a rigid body. Since the internal forces in the rigid body will always cancel by Newton's third law, we simplify the notation for the translational equation of motion of the center of mass. It is implied that $\mathbf{F}_G$ represents the total external force acting on the body.

In many respects we have been using Euler's first law all along. Most of the objects for which we were finding trajectories were actually extended rigid bodies (e.g., a satellite or planet in orbit). It is this law that justifies our treatment of the translation of those bodies as if they were particles.

As discussed in Chapter 6, a consequence of Euler's first law is that, if the total external force is zero, then the linear momentum of the center of mass of a rigid body is conserved.

Since we most often consider continuous rigid bodies, it is necessary to modify Definition 6.2 for the center of mass of a continuous body. Instead of modeling the rigid body as a collection of particles of mass m_i and then summing over i, we assign to each point of the rigid body a differential mass dm and position $\mathbf{r}_{dm/O}$ relative to O. We then convert the sum in Eq. (6.13) to an integral over $\mathscr{B}$:

$$\mathbf{r}_{G/O} \overset{\triangle}{=} \frac{1}{m_G} \int_{\mathscr{B}} \mathbf{r}_{dm/O}dm.$$

Assuming the rigid body has density $\rho(\mathbf{r}_{dm/O})$, which may vary with position inside the body, we can replace the differential mass element dm by the density times the differential volume element dV to obtain the continuous formula for the center of mass:

$$\mathbf{r}_{G/O} = \frac{1}{m_G} \int_{\mathscr{B}} \mathbf{r}_{dm/O}\rho(\mathbf{r}_{dm/O})dV. \tag{9.2}$$

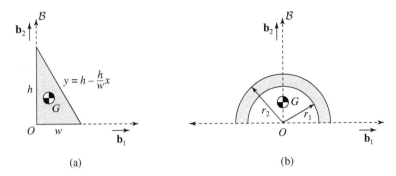

Figure 9.3 Computing the center of mass of a continuous rigid body. (a) Cartesian coordinates. (b) Polar coordinates.

(We discuss finding differential area and volume elements in different coordinate systems in Appendix A.) For example, in Cartesian coordinates, the density is $\rho(\mathbf{r}_{dm/O}) = \rho(x, y, z)$, and the differential volume element is $dV = dx\,dy\,dz$. (These coordinates can be expressed in either the body frame or instantaneously in an inertial frame, though the body frame is usually the simplest.)

For a continuum of particles, the center-of-mass corollary in Eq. (6.15) is

$$\int_{\mathcal{B}} \mathbf{r}_{dm/G}\,\rho(\mathbf{r}_{dm/G})dV = 0. \tag{9.3}$$

Example 9.1 Computing the Center of Mass of a Continuous Rigid Body (Cartesian Coordinates)

Consider the right-triangular rigid body shown in Figure 9.3a. To use Eq. (9.2) to find the center of mass, we start by writing down the position $\mathbf{r}_{dm/O} = x\mathbf{b}_1 + y\mathbf{b}_2$ and the volume element (actually, area element in two dimensions) $dV = dx\,dy$. Note that we express the position as vector components with respect to a body-fixed reference frame $\mathcal{B} = (O, \mathbf{b}_1, \mathbf{b}_2, \mathbf{b}_3)$. Assuming mass m and uniform density ρ, we have

$$\mathbf{r}_{G/O} = \frac{\rho}{m} \int_0^w \int_0^y (x\mathbf{b}_1 + y\mathbf{b}_2)dy\,dx$$

$$= \left(\frac{\rho}{m} \int_0^w \int_0^{h - \frac{h}{w}x} x\,dy\,dx \right) \mathbf{b}_1 + \left(\frac{\rho}{m} \int_0^w \int_0^{h - \frac{h}{w}x} y\,dy\,dx \right) \mathbf{b}_2$$

$$= \frac{\rho h w^2}{6m}\mathbf{b}_1 + \frac{\rho h^2 w}{6m}\mathbf{b}_2.$$

Using $m = \rho h w/2$ results in

$$\mathbf{r}_{G/O} = \frac{w}{3}\mathbf{b}_1 + \frac{h}{3}\mathbf{b}_2.$$

Example 9.2 Computing the Center of Mass of a Continuous Rigid Body (Polar Coordinates)

We now use polar coordinates in a Cartesian frame to compute the center of mass of the annular rigid body shown in Figure 9.3b. In polar coordinates, the position of a small mass element is $\mathbf{r}_{dm/O} = r\cos\theta\mathbf{b}_1 + r\sin\theta\mathbf{b}_2$ and the volume (area) element is $dV = rdrd\theta$. Assuming mass m and uniform density ρ, the center of mass integral in Eq. (9.2) becomes

$$\mathbf{r}_{G/O} = \frac{\rho}{m}\int_0^\pi\int_{r_1}^{r_2}(r\cos\theta\mathbf{b}_1 + r\sin\theta\mathbf{b}_2)rdrd\theta$$

$$= \frac{2\rho}{3m}(r_2^3 - r_1^3)\mathbf{b}_2.$$

Using $m = (\rho\pi r_2^2 - \rho\pi r_1^2)/2 = \rho\pi(r_2^2 - r_1^2)/2$, we obtain

$$\mathbf{r}_{G/O} = \frac{4(r_2^3 - r_1^3)}{3\pi(r_2^2 - r_1^2)}\mathbf{b}_2.$$

Example 9.3 A Falling Rigid Body

In Example 6.5 we found the equations of motion for a collection of particles falling in a uniform gravitational field and showed that they reduce to the single, simple equation for an equivalent falling particle of mass m_G located at the center of mass of the collection, G. Here we repeat that example for a rigid body $\mathcal{B}$. Figure 9.4a shows a rigid body of arbitrary shape falling in a uniform gravitational field with acceleration $-g\mathbf{e}_y$. Thus every mass element dm of the rigid body is acted on by a force $-dmg\mathbf{e}_y$, as shown in Figure 9.4b.

Using Euler's first law and integrating over the external force on each mass element dm yields:

$$m_G\frac{^\mathcal{I}d^2}{dt^2}(\mathbf{r}_{G/O}) = -\int_\mathcal{B}dmg\mathbf{e}_y = -m_Gg\mathbf{e}_y. \qquad (9.4)$$

Thus, as for the collection of free particles, the center-of-mass equation of motion is

$$\frac{^\mathcal{I}d^2}{dt^2}(\mathbf{r}_{G/O}) = -g\mathbf{e}_y,$$

which is independent of the orientation of the body. As before, this complete decoupling is due to the fact that, for a constant force field, the integral of the force in Eq. (9.4) is particularly simple.

If the force on each mass element instead depended on position (as for $1/r^2$ gravity), then the problem would be more complicated and the equation of motion

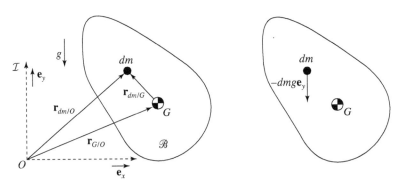

(a) Coordinates and reference frame   (b) External force on mass element dm

Figure 9.4 Planar rigid body falling freely in a uniform gravitational field.

for G would depend on the orientation of the rigid body. This situation is similar to that discussed in Tutorial 6.1. For a general rigid body, the center of gravity is not the same as the center of mass. Although the center of mass and center of gravity are the same for every rigid body in a uniform gravitational field, that is not the case in the more exact $1/r^2$ field. It is only for very special geometries—like a sphere—that they coincide. Such bodies are referred to as *centrobaric*.

9.3 Rotation about the Center of Mass— Euler's Second Law

In the last example of the previous section we explained that, for more complicated force laws, the equation of motion of the center of mass could depend on the orientation of the rigid body without carefully defining the term *orientation*. What, in fact, do we mean by orientation? And how might we find the rotational dynamics of a planar rigid body? In general, orientation refers to the arrangement of the particles relative to the center of mass in the inertial frame. Since the body is rigid, this is equivalent to describing the orientation, or rotation, of body frame $\mathcal{B}$ in the inertial frame. In two dimensions, the orientation of a rigid body in $\mathcal{I}$ has a single degree of freedom. We typically use the coordinate θ to describe the rotation of a planar rigid body about a single axis fixed in the inertial frame.

Again, for the planar, two-dimensional situation considered in this chapter, the orientation of $\mathcal{B}$ in $\mathcal{I}$ is given entirely by the single coordinate θ and the rotational motion is thus described by the angular velocity of $\mathcal{B}$ in $\mathcal{I}$. For now, the angular velocity is simple; that is, it is constrained to point in the $\mathbf{e}_3$ (and thus $\mathbf{b}_3$) direction $({}^{\mathcal{I}}\boldsymbol{\omega}^{\mathcal{B}} = {}^{\mathcal{I}}\omega^{\mathcal{B}}\mathbf{b}_3 = \dot{\theta}\mathbf{b}_3)$. In this section we search for the equation of motion for the angular velocity, in terms of the coordinate θ, by utilizing the separation of angular momentum described in Chapter 7. We begin as in that chapter with the equation of motion for the total angular momentum of a rigid body.

9.3.1 The Angular Momentum of a Rigid Body

Chapter 7 defined the total angular momentum of a collection of particles about an inertially fixed point O (see Definition 7.1):

$$^{\mathcal{I}}\mathbf{h}_O \triangleq \sum_{i=1}^{N} m_i \mathbf{r}_{i/O} \times {}^{\mathcal{I}}\mathbf{v}_{i/O}.$$

This expression holds unchanged for a discrete rigid body. For a continuous body, we extend it to integral form, just as for the definition of the center of mass:

$$^{\mathcal{I}}\mathbf{h}_O \triangleq \int_{\mathscr{B}} \mathbf{r}_{dm/O} \times {}^{\mathcal{I}}\mathbf{v}_{dm/O}\, dm.$$

To keep things simple, we often use the summation notation for a finite number of rigidly connected particles. Just remember that everything still applies if you instead use an integral over a continuous body.

Taking the derivative of the angular momentum, as in Eq. (7.1), once again yields

$$\frac{^{\mathcal{I}}d}{dt}\left(^{\mathcal{I}}\mathbf{h}_O\right) = \sum_{i=1}^{N} \mathbf{r}_{i/O} \times m_i \frac{^{\mathcal{I}}d}{dt}\left(^{\mathcal{I}}\mathbf{v}_{i/O}\right),$$

which, after substituting from Newton's second law and performing a bit of manipulation, reduces to

$$\frac{^{\mathcal{I}}d}{dt}\left(^{\mathcal{I}}\mathbf{h}_O\right) = \underbrace{\sum_{i=1}^{N} \mathbf{r}_{i/O} \times \mathbf{F}_i^{(\text{ext})}}_{=\mathbf{M}_O^{(\text{ext})}} + \frac{1}{2}\sum_{i=1}^{N}\sum_{j=1}^{N} (\mathbf{r}_{i/O} - \mathbf{r}_{j/O}) \times \mathbf{F}_{i,j}. \qquad (9.5)$$

At this point we are a bit stuck. We want eventually to obtain an expression just like Eq. (7.5): the rate of change of the angular momentum of a rigid body is equal solely to the external moments on it. In fact, as you will soon see, this is Euler's second law. However, to get there from our particle-dynamics model requires eliminating the internal-moment term in Eq. (9.5). In Chapter 7 we did this by invoking the internal-moment assumption (Eq. (7.4)). The result was a compact expression, but one that held only for multiparticle systems for which the internal-moment assumption applied.

It turns out that Newton's laws are not enough to find the equations of motion for a rigid body without making an additional axiomatic assumption. In fact, Euler, for this reason, abandoned the particle model for rigid bodies entirely and simply postulated new laws of motion for rigid bodies, just as Newton postulated his laws for particles. Although this approach is legitimate, and one taken by some authors, we instead postulate the internal-moment assumption as a property of rigid bodies. Recall the comments at the end of Section 9.1: our approach to dynamics is to develop models that, along with the laws of motion, effectively describe and predict motion. Part of our particle-based model of a rigid body is that the internal-moment assumption must hold. This assumption should not come as a surprise; in statics you may have analyzed the internal axial and shear forces and internal moments of a rigid body, always assuming that their sum was zero. The internal-moment assumption is part of the foundation of macroscopic material mechanics and will arise again in courses on vibration and elasticity.

Under the internal-moment assumption, the second term in Eq. (9.5) is zero, and we obtain Euler's second law.

Law 9.2 Euler's second law states that the rate of change of the inertial angular momentum of a rigid body about point O in the inertial frame is equal to the total external moment acting on the rigid body about O:

$$\boxed{\frac{^{\mathcal{I}}d}{dt}\left(^{\mathcal{I}}\mathbf{h}_O\right) = \mathbf{M}_O.} \tag{9.6}$$

Euler's second law simply states that Eq. (7.5) for the change in total angular momentum of a collection of particles applies to a rigid body. This fact is enormously important and is the foundation of everything we do in the remainder of the book. Note also that, as with Euler's first law, we have eliminated the (ext) superscript on the applied moment to simplify the notation, since, by our internal-moment assumption, Euler's second law applies only for external moments on the rigid body.

Also, as in Law 7.1, when the total external moment acting on the rigid body about O is zero, the total angular momentum of the rigid body about O is conserved. So, for instance, under a central force, the center of mass of a rigid body will orbit like a particle.

Eq. (9.6) is certainly useful for rigid-body problems. However, of even more use is the application of the angular momentum separation principle to rigid bodies. Look back over Section 7.2. Everything done there applies equally well to a rigid body. Thus we can write the total angular momentum of a rigid body as the sum of the angular momentum of the center of mass plus the angular momentum about the center of mass:

$$^{\mathcal{I}}\mathbf{h}_O = {}^{\mathcal{I}}\mathbf{h}_{G/O} + {}^{\mathcal{I}}\mathbf{h}_G,$$

where

$$^{\mathcal{I}}\mathbf{h}_{G/O} \triangleq m_G \mathbf{r}_{G/O} \times {}^{\mathcal{I}}\mathbf{v}_{G/O} \tag{9.7}$$

is the angular momentum of the center of mass relative to O, and

$$^{\mathcal{I}}\mathbf{h}_G \triangleq \sum_{i=1}^{N} m_i \mathbf{r}_{i/G} \times {}^{\mathcal{I}}\mathbf{v}_{i/G} \tag{9.8}$$

is the angular momentum about the center of mass. For a continuous rigid body,

$$^{\mathcal{I}}\mathbf{h}_G \triangleq \int_{\mathcal{B}} \mathbf{r}_{dm/G} \times {}^{\mathcal{I}}\mathbf{v}_{dm/G} \, dm.$$

Just as in Chapter 7, we can take the derivative of each of these angular momenta. Doing so for Eq. (9.7) results in

$$\frac{^{\mathcal{I}}d}{dt}\left(^{\mathcal{I}}\mathbf{h}_{G/O}\right) = \mathbf{r}_{G/O} \times \mathbf{F}_G \triangleq \mathbf{M}_{G/O}, \qquad (9.9)$$

where again we have removed the (ext) subscript for brevity. Eq. (9.9) is Euler's first law in angular-momentum form, analogous to that for a multiparticle system. Again, Eq. (9.9) states that the center-of-mass motion of a rigid body can be considered separately from its rotational motion; the center-of-mass motion satisfies Newton's second law, in either linear-momentum or angular-momentum form.

Taking the time derivative of the angular momentum about the center of mass in Eq. (9.8) follows the same process. To eliminate the internal moments, we again have to invoke the internal-moment assumption. Doing so results in Eq. (7.12) applied to a rigid body:

$$\boxed{\frac{^{\mathcal{I}}d}{dt}\left(^{\mathcal{I}}\mathbf{h}_G\right) = \mathbf{M}_G.} \qquad (9.10)$$

This equation is Euler's second law for the angular momentum and moments about the center of mass. The total moment $\mathbf{M}_G$ on the rigid body in Eq. (9.10) is

$$\boxed{\mathbf{M}_G \triangleq \sum_{i=1}^{N} \mathbf{r}_{i/G} \times \mathbf{F}_i^{(\text{ext})}.} \qquad (9.11)$$

When a continuous field force is acting on the body, let $\mathbf{f}_{dm}$ be the force per unit mass on the body (also called the *specific force*). In this case:

$$\boxed{\mathbf{M}_G \triangleq \int_{\mathcal{B}} \mathbf{r}_{dm/G} \times \mathbf{f}_{dm}^{(\text{ext})} dm.} \qquad (9.12)$$

When dealing with a collection of individual particles (in this case, rigidly attached), $\mathbf{F}_i$ is the total force on the ith particle and $\mathbf{r}_{i/G}$ is the position of that particle. For a continuous rigid body, there are an infinite number of particles. In that case, for the contact forces in Eq. (9.11), we interpret the moment equation slightly differently. $\mathbf{M}_G$ represents the net moment of some finite number of contact forces applied to the rigid body at locations $\mathbf{r}_{i/G}$. It should not be surprising by now that when we consider a rigid body (i.e., a system of particles), both the magnitude of the applied force and the location of its application are important when solving for the complete motion (translation and rotation). Likewise, the moment due to field forces, such as gravity, can depend on shape, location, and orientation. We highlight these observations in the next few examples.

Example 9.4 The Moment on a Rigid Body Due to Uniform Gravity

Here we reconsider the rigid body falling in a uniform gravitational field from Example 9.3 and shown in Figure 9.4. Example 9.3 showed that we could treat the rigid body as a single particle located at the center of mass with the force of gravity, $-m_G g \mathbf{e}_y$, acting on it. Here we ask whether there is a moment generated on the rigid body by the gravitational field, thus allowing us to separately write a rotational equation of motion. From Figure 9.4b, the force per unit mass on each mass element is just $-g \mathbf{e}_y$. The external moment from Eq. (9.12) is thus

$$\mathbf{M}_G = -\int_{\mathcal{B}} \mathbf{r}_{dm/G} \times g\mathbf{e}_y \, dm.$$

Since g and $\mathbf{e}_y$ are independent of position, we can factor them out of the integral to find

$$\mathbf{M}_G = -g \left(\int_{\mathcal{B}} \mathbf{r}_{dm/G} \, dm \right) \times \mathbf{e}_y.$$

However, the center-of-mass corollary requires that the integral in parentheses be zero (see Eq. (9.3)). Thus for any rigid body falling in a uniform gravitational field, the external moment about the center of mass must be zero:

$$\mathbf{M}_G = 0.$$

Note that this is not a general principle of dynamics but rather a consequence of the specific force field acting on the body. In fact, for the more general $1/r^2$ gravitational field, there is a moment about the center of mass of the falling body. We explore this fact in Tutorial 9.2.

Example 9.5 The Moment on a Vibration-Isolation Table

Figure 9.5 shows a side view of a planar model of a vibration-isolation system consisting of a heavy table of length L supported by two springy legs with spring constants k_A and k_B. (We assume the springs stay vertical when the table rotates.) This is a two-degree-of-freedom system described by the coordinates y (the height of the center of mass) and θ (the rotation angle of the table). Both coordinates are shown in Figure 9.5a. We use two frames, as shown in Figure 9.5b: inertial frame $\mathcal{I} = (O, \mathbf{e}_x, \mathbf{e}_y, \mathbf{e}_z)$ and body frame $\mathcal{B} = (G, \mathbf{b}_1, \mathbf{b}_2, \mathbf{b}_3)$. The transformation array between the inertial frame and the body frame has the usual form:

	$\mathbf{b}_1$	$\mathbf{b}_2$
$\mathbf{e}_x$	$\cos\theta$	$-\sin\theta$
$\mathbf{e}_y$	$\sin\theta$	$\cos\theta$.

Example 9.10 solves for the equations of motion of the table in translation and rotation using the separation principle. Here we find the moment on the table about its center of mass due to the discrete forces acting on it, as shown in Figure 9.5c.

The previous example showed that a uniform gravitational field can be modeled as a single force acting at the center of mass while producing no moment about the center of mass (independent of the shape of the rigid body). Thus the only moments

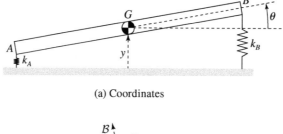

(a) Coordinates

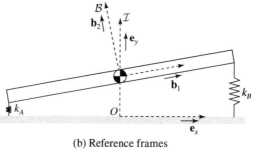

(b) Reference frames

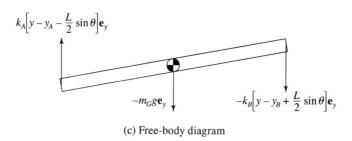

(c) Free-body diagram

Figure 9.5 A vibration-isolation system consists of a table of length L supported by two stiff springs.

about the center of mass of the table are produced by the spring forces $\mathbf{F}_A$ and $\mathbf{F}_B$ acting at each end. To find the spring forces we need to know the displacement of each spring as a function of the two coordinates, y and θ. Consider the position of point A relative to O:

$$\mathbf{r}_{A/O} = \mathbf{r}_{G/O} + \mathbf{r}_{A/G} = y\mathbf{e}_y - \frac{L}{2}\mathbf{b}_1$$

$$= y\mathbf{e}_y - \frac{L}{2}\cos\theta\mathbf{e}_x - \frac{L}{2}\sin\theta\mathbf{e}_y,$$

where we have used the transformation table between the unit vectors of $\mathcal{B}$ and $\mathcal{I}$. The spring force at A is given by the displacement of the spring in the $\mathbf{e}_y$ direction,

$$\mathbf{F}_A = -k_A(y - \frac{L}{2}\sin\theta - y_A)\mathbf{e}_y,$$

where y_A is the unstretched length of the spring at A. A similar analysis gives the spring force at B,

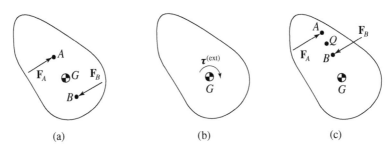

Figure 9.6 A couple and pure torque acting on a rigid body about the center of mass G and another point Q. (a) Couple about G. (b) Pure torque. (c) Couple about Q.

$$\mathbf{F}_B = -k_B(y + \frac{L}{2}\sin\theta - y_B)\mathbf{e}_y,$$

where y_B is the unstretched length of the spring at B.

The moment about G is found using Eq. (9.11):

$$\mathbf{M}_G = \mathbf{r}_{A/G} \times \mathbf{F}_A + \mathbf{r}_{B/G} \times \mathbf{F}_B$$

$$= -\frac{L}{2}\mathbf{b}_1 \times -k_A(y - \frac{L}{2}\sin\theta - y_A)\mathbf{e}_y + \frac{L}{2}\mathbf{b}_1 \times -k_B(y + \frac{L}{2}\sin\theta - y_B)\mathbf{e}_y$$

$$= k_A\frac{L}{2}\cos\theta(y - \frac{L}{2}\sin\theta - y_A)\mathbf{e}_z - k_B\frac{L}{2}\cos\theta(y + \frac{L}{2}\sin\theta - y_B)\mathbf{e}_z. \quad (9.13)$$

Example 9.6 The Moment of a Couple

As described in the text, the moment about the center of mass in Euler's second law arises either from a collection of contact forces acting on the body at discrete points or from a nonuniform field force. Among the set of contact forces we consider, there is a special set that warrants highlighting. These are forces that produce a moment on the rigid body, and thus rotational motion, but no net force and thus no translational acceleration. We call such a set of forces *a couple*. Figure 9.6a illustrates a couple acting on a rigid body. A couple consists of two forces, $\mathbf{F}_A$ and $\mathbf{F}_B = -\mathbf{F}_A$, of equal magnitude F and opposite directions acting on the rigid body at two collinear points equidistant from the center of mass. The net force on the rigid body is thus zero:

$$\mathbf{F}_G = \mathbf{F}_A + \mathbf{F}_B = 0.$$

Without loss of generality, assume the forces act perpendicular to the line connecting A and B.

The moment of the couple, however, is not zero. Let $\mathbf{b}$ be the unit vector directed from B to A, and let $\hat{\mathbf{F}}_A$ be the unit vector in the direction of $\mathbf{F}_A$. Using Eq. (9.11), the moment is

$$\mathbf{M}_G = \mathbf{r}_{A/G} \times \mathbf{F}_A + \mathbf{r}_{B/G} \times \mathbf{F}_B$$

$$= r\mathbf{b} \times F\hat{\mathbf{F}}_A + (-r\mathbf{b}) \times (-F\hat{\mathbf{F}}_A) = 2rF\mathbf{b} \times \hat{\mathbf{F}}_A = 2rF\mathbf{b}_3,$$

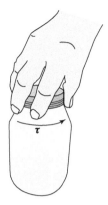

Figure 9.7 Pure torque.

where $\mathbf{b}_3$ points into the page and r is the distance from G to each of the two points of application, A and B. Thus a couple produces zero total force and a moment of magnitude $2rF$ about the center of mass. Note that a couple can consist of any number of forces equally distant from the center of mass, as long as their vector sum is zero. For n forces around the center of mass, the resulting moment has magnitude $M_G = nrF$.

It is common to consider the limiting case where the distances of the applied forces in a couple go to zero while the magnitude of the forces increases such that the product remains constant. The resulting moment about G is called a *pure torque* and written as $\boldsymbol{\tau}$. Figure 9.6b illustrates a pure torque. This is, of course, an abstraction, as all moments, if you examine them with enough fidelity, consist of a number of applied forces. However, the pure-torque concept can be very convenient for many problems. Examples of pure torques abound, including screwdrivers, electric drills, and motors of all types. Even twisting off the lid of a jar is an example of a pure torque (see Figure 9.7).

The most important property of couples and pure torques is that they are independent of the reference point. The moment on the rigid body is the same no matter where the couple is applied.[1] This property is actually fairly easy to show. Consider the same couple but now equidistant from some other point $Q \neq G$, as shown in Figure 9.6c. The moment about Q is (see Section 7.3)

$$\mathbf{M}_Q = \mathbf{r}_{A/Q} \times \mathbf{F}_A + \mathbf{r}_{B/Q} \times \mathbf{F}_B$$

$$= r\mathbf{b} \times F\hat{\mathbf{F}}_A + (-r\mathbf{b}) \times (-F\hat{\mathbf{F}}_A) = 2rF\mathbf{b} \times \hat{\mathbf{F}}_A.$$

What is the moment due to the couple about the center of mass? It is

$$\mathbf{M}_G = \mathbf{r}_{A/G} \times \mathbf{F}_A + \mathbf{r}_{B/G} \times \mathbf{F}_B,$$

where here the two vectors $\mathbf{r}_{A/G}$ and $\mathbf{r}_{B/G}$ are no longer equal and opposite. However, using the usual vector triads, we can rewrite the moment:

---

[1] This is in contrast to the moment due to a single force: the moment about an arbitrary point Q is generally different than the moment about the center of mass G.

$$\mathbf{M}_G = (\mathbf{r}_{Q/G} + \mathbf{r}_{A/Q}) \times \mathbf{F}_A + (\mathbf{r}_{Q/G} + \mathbf{r}_{B/Q}) \times \mathbf{F}_B$$

$$= (\mathbf{r}_{Q/G} + r\mathbf{b}) \times (F\hat{\mathbf{F}}_A) + (\mathbf{r}_{Q/G} - r\mathbf{b}) \times (-F\hat{\mathbf{F}}_A) = 2rF\mathbf{b} \times \hat{\mathbf{F}}_A = \mathbf{M}_Q.$$

The moment of the couple about the center of mass is exactly the same as the moment about Q. Thus *pure torques or couples can be applied anywhere on a rigid body and produce the same moment with no translational force.*

9.3.2 The Moment of Inertia

We use Eq. (9.10) in almost all our analysis of rigid-body motion, both here and in Chapter 11, when we study three-dimensional motion. To make it useful, however, we need an expression that relates the angular momentum about the center of mass to the scalar coordinate for orientation, θ, so we can convert Euler's second law to a scalar equation of motion. To do so requires some manipulation.

We begin with the expression for the angular momentum in Eq. (9.8),

$$^{\mathcal{I}}\mathbf{h}_G = \sum_{i=1}^{N} m_i \mathbf{r}_{i/G} \times {}^{\mathcal{I}}\mathbf{v}_{i/G}.$$

We again use the discrete form in this discussion for simplicity; the integral form for a continuous rigid body follows a similar set of steps. We next use the transport equation (Definition 8.1) to rewrite the inertial velocity of each particle in terms of the angular velocity of the rigid body:

$$^{\mathcal{I}}\mathbf{v}_{i/G} = \underbrace{{}^{\mathcal{B}}\mathbf{v}_{i/G}}_{=0} + {}^{\mathcal{I}}\boldsymbol{\omega}^{\mathcal{B}} \times \mathbf{r}_{i/G}. \tag{9.14}$$

The body-frame velocity is zero because the body is rigid. At this stage, the specialization to a rigid body comes into play. If the collection of particles is rigid, then by definition, the particles have no velocity in the body frame relative to the center of mass (see Section 8.1). The result is that the angular momentum about the center of mass is

$$^{\mathcal{I}}\mathbf{h}_G = \sum_{i=1}^{N} m_i \mathbf{r}_{i/G} \times ({}^{\mathcal{I}}\boldsymbol{\omega}^{\mathcal{B}} \times \mathbf{r}_{i/G}). \tag{9.15}$$

We simplify the following calculations by using the constraint that the rigid body only rotates in the plane. That is, $^{\mathcal{I}}\boldsymbol{\omega}^{\mathcal{B}} = {}^{\mathcal{I}}\omega^{\mathcal{B}}\mathbf{b}_3$, which yields

$$^{\mathcal{I}}\mathbf{h}_G = \sum_{i=1}^{N} m_i {}^{\mathcal{I}}\omega^{\mathcal{B}} \mathbf{r}_{i/G} \times (\mathbf{b}_3 \times \mathbf{r}_{i/G}). \tag{9.16}$$

We also assume that the body itself is confined to the plane, so that $\mathbf{r}_{i/G} = x_i\mathbf{b}_1 + y_i\mathbf{b}_2$ (in Cartesian coordinates). Thus the vector triple cross product in Eq. (9.16) results in a vector in the $\mathbf{b}_3$ direction (try it!), implying that the angular momentum is also in the $\pm\mathbf{b}_3$ direction. The angular momentum is thus $^{\mathcal{I}}\mathbf{h}_G = h_G\mathbf{b}_3$.

Given the direction of the angular momentum, we need to find an expression for its signed magnitude h_G, which can be positive or negative depending on the sign of the angular velocity. We do so by reversing the order of both cross products and then taking the dot product of Eq. (9.16) with $\mathbf{b}_3$:

$$h_G = {}^{\mathcal{I}}\mathbf{h}_G \cdot \mathbf{b}_3 = \sum_{i=1}^{N} m_i {}^{\mathcal{I}}\omega^{\mathcal{B}} \left[(\mathbf{r}_{i/G} \times \mathbf{b}_3) \times \mathbf{r}_{i/G} \right] \cdot \mathbf{b}_3.$$

Note that we have put parentheses around the first cross product $\mathbf{r}_{i/G} \times \mathbf{b}_3$ because, in the next step, we treat this cross product as a single vector. We now use the scalar triple product identity $(\mathbf{a} \times \mathbf{b}) \cdot \mathbf{c} = \mathbf{a} \cdot (\mathbf{b} \times \mathbf{c})$ (see Appendix B) to find

$$h_G = {}^{\mathcal{I}}\omega^{\mathcal{B}} \sum_{i=1}^{N} m_i \left(\mathbf{r}_{i/G} \times \mathbf{b}_3 \right) \cdot \left(\mathbf{r}_{i/G} \times \mathbf{b}_3 \right).$$

Since $\mathbf{b}_3$ is always perpendicular to $\mathbf{r}_{i/G}$ in this planar situation, we can use the definition of the cross product to show that

$$\left(\mathbf{r}_{i/G} \times \mathbf{b}_3 \right) \cdot \left(\mathbf{r}_{i/G} \times \mathbf{b}_3 \right) = \| (x_i \mathbf{b}_1 + y_i \mathbf{b}_2) \times \mathbf{b}_3 \|^2 = x_i^2 + y_i^2 = \| \mathbf{r}_{i/G} \|^2.$$

This expression makes the magnitude of the angular momentum of the rigid body about G equal to

$$h_G = {}^{\mathcal{I}}\omega^{\mathcal{B}} \underbrace{\sum_{i=1}^{N} m_i \| \mathbf{r}_{i/G} \|^2}_{\triangleq I_G}. \tag{9.17}$$

Eq. (9.17) leads us to the next definition.

Definition 9.2 The **moment of inertia** I_G of a planar rigid body about its center of mass G is the sum of the mass-weighted squared distance of each point on the body to the center of mass:

$$\boxed{I_G \triangleq \sum_{i=1}^{N} m_i \| \mathbf{r}_{i/G} \|^2.}$$

For a continuous rigid body $\mathcal{B}$ with density $\rho(\mathbf{r}_{dm/G})$, we have

$$\boxed{I_G \triangleq \int_{\mathcal{B}} \| \mathbf{r}_{dm/G} \|^2 dm = \int_{\mathcal{B}} \| \mathbf{r}_{dm/G} \|^2 \rho(\mathbf{r}_{dm/G}) dV.}$$

We practice calculating the planar moment of inertia in the examples below. Note that the planar moment of inertia is a scalar quantity.

With the introduction of the moment of inertia, Eq. (9.17) simplifies to $h_G = I_G {}^{\mathcal{I}}\omega^{\mathcal{B}}$. The angular momentum of the rigid body $\mathcal{B}$ with respect to absolute space can thus be compactly written as

$$\boxed{{}^{\mathcal{I}}\mathbf{h}_G = I_G {}^{\mathcal{I}}\boldsymbol{\omega}^{\mathcal{B}} = I_G {}^{\mathcal{I}}\boldsymbol{\omega}^{\mathcal{B}}\mathbf{b}_3.}$$ (9.18)

In the next section we use this equation to find the rotational equations of motion of a planar rigid body.

Example 9.7 Calculating the Planar Moment of Inertia (Cartesian Coordinates)

This example uses Cartesian coordinates to compute the planar moment of inertia about the center of mass of the square rigid body shown in Figure 9.8a. To use Definition 9.2, we first specify the position $\mathbf{r}_{dm/G} = x\mathbf{b}_1 + y\mathbf{b}_2$ and volume (actually area) element $dV = dxdy$. Note that we express the position as components in the body frame $\mathcal{B} = (G, \mathbf{b}_1, \mathbf{b}_2, \mathbf{b}_3)$. Assuming mass m and uniform density ρ, we have

$$I_G = \rho \int_{-l}^{l} \int_{-l}^{l} (x^2 + y^2)dxdy = \frac{8}{3}\rho l^4.$$

Using $\rho = m/(4l^2)$, we have $I_G = 2ml^2/3$.

Example 9.8 Calculating the Planar Moment of Inertia (Polar Coordinates)

We now use polar coordinates to compute the moment of inertia of the circular rigid body shown in Figure 9.8b. The position of a small mass element dm with respect to the center of mass G is $\mathbf{r}_{dm/G} = r\cos\theta\mathbf{b}_1 + r\sin\theta\mathbf{b}_2$ and the volume (area) element is $dV = rdrd\theta$. Assuming mass m and uniform density ρ, Definition 9.2 yields

$$I_G = \rho \int_{0}^{2\pi} \int_{0}^{R} (r^2\cos^2\theta + r^2\sin^2\theta)rdrd\theta = \frac{\pi}{2}\rho R^4.$$

Using $\rho = m/(\pi R^2)$, we have $I_G = mR^2/2$.

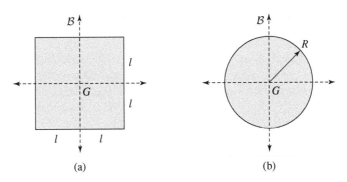

(a)   (b)

Figure 9.8 Calculating the planar moment of inertia. (a) Square. (b) Circle.

Appendix D supplies the moments of inertia for a variety of common rigid body shapes.

9.3.3 The Rotational Equation of Motion of a Planar Rigid Body

We can now use Eq. (9.18) for the angular momentum of a planar rigid body and Euler's second law to rewrite the rotational equation of motion in Eq. (9.10) in terms of the moment of inertia and angular velocity:

$$\frac{{}^{\mathcal{I}}d}{dt}\left({}^{\mathcal{I}}\mathbf{h}_G\right) = \frac{{}^{\mathcal{I}}d}{dt}\left(I_G{}^{\mathcal{I}}\boldsymbol{\omega}^{\mathcal{B}}\right) = \frac{d}{dt}\left(I_G\right){}^{\mathcal{I}}\boldsymbol{\omega}^{\mathcal{B}} + I_G{}^{\mathcal{I}}\boldsymbol{\alpha}^{\mathcal{B}}\mathbf{b}_3 = \mathbf{M}_G.$$

The beauty of this formulation of the equations of motion for a rigid body—using the separation principle and computing the moment of inertia as an integral of the mass distribution in the body—is that the time derivative of the moment of inertia is zero, which is seen from the definition:

$$\frac{d}{dt}\left(I_G\right) = \sum_{i=1}^{N} m_i \frac{d}{dt}\|\mathbf{r}_{i/G}\|^2 = 0.$$

The moment of inertia is constant because, if the body is rigid, the distance of any particle to the center of mass is constant. (Note that these derivatives have no frame identified because we are differentiating scalar quantities.)

Thus we can write the equation of motion for the angular momentum about the center of mass in its final form:

$$\boxed{\frac{{}^{\mathcal{I}}d}{dt}\left({}^{\mathcal{I}}\mathbf{h}_G\right) = I_G{}^{\mathcal{I}}\boldsymbol{\alpha}^{\mathcal{B}}\mathbf{b}_3 = \mathbf{M}_G,} \tag{9.19}$$

where we have also assumed that $\mathbf{M}_G$ is in the $\mathbf{b}_3$ direction and have used ${}^{\mathcal{I}}\alpha^{\mathcal{B}} \triangleq {}^{\mathcal{I}}\dot{\omega}^{\mathcal{B}} = \ddot{\theta}$. Note again that, as with a multiparticle system, when the total moment acting about the center of mass is zero, then the law of conservation of angular momentum holds and the angular momentum about the center of mass is constant. For the simple planar situations considered in this chapter, that means the rigid body rotates at a constant angular velocity ${}^{\mathcal{I}}\omega^{\mathcal{B}} = \dot{\theta}$.

Example 9.9 Disk Rolling down an Inclined Plane

This example determines the equation of motion for a rigid disk or wheel rolling down an inclined plane, as shown in Figure 9.9. Recall from Example 8.2 that this problem has only one degree of freedom, represented by coordinate θ in Figure 9.9a. This is because of the no-slip condition, which tells us that the distance the disk travels down the plane is entirely given by the amount the disk has rotated. This dependence is usually represented by the differential relationship derived in Example 8.2 (see Eq.(8.6)),

$$\dot{x} = -r\dot{\theta}, \tag{9.20}$$

where the coordinate x is the distance of the center of the disk from the bottom of the incline, r is the radius of the disk, and θ increases clockwise.

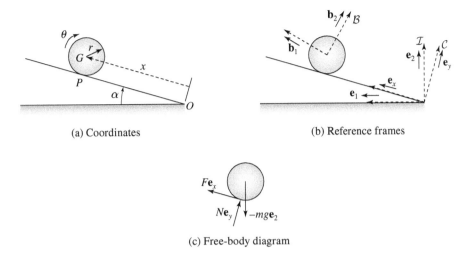

(a) Coordinates   (b) Reference frames

(c) Free-body diagram

Figure 9.9 Disk rolling down an incline of angle α.

We use the separation principle discussed in this section to derive the equation of motion for θ. Figure 9.9b identifies three reference frames: an inertial frame $\mathcal{I}$ located at O with its $\mathbf{e}_1$ direction along the ground, a second inertial frame $\mathcal{C}$ also located at O but inclined so that $\mathbf{e}_x$ is along the incline, and a body frame $\mathcal{B}$ fixed to the disk and located at its center of mass. The transformation array between the unit vectors of $\mathcal{I}$ and $\mathcal{C}$ is

	$\mathbf{e}_1$	$\mathbf{e}_2$
$\mathbf{e}_x$	$\cos\alpha$	$\sin\alpha$
$\mathbf{e}_y$	$-\sin\alpha$	$\cos\alpha$.

The free-body diagram for the disk is shown in Figure 9.9c.

We begin by finding the translational equation of motion for the center of mass G of the disk. From Euler's first law, using the free-body diagram in Figure 9.9c and Cartesian coordinates in $\mathcal{C}$, this equation becomes

$$m\ddot{x}\mathbf{e}_x = F\mathbf{e}_x + N\mathbf{e}_y - mg\mathbf{e}_2,$$

where we have assumed that the disk does not move in the $\mathbf{e}_y$ direction and F is the (static) friction force between the disk and the incline. Using the transformation array yields the scalar equation of motion in the $\mathbf{e}_x$ direction:

$$m\ddot{x} = -mg\sin\alpha + F.$$

By differentiating the no-slip condition in Eq. (9.20) we can substitute for $\ddot{x}$ to obtain

$$-mr\ddot{\theta} = -mg\sin\alpha + F. \tag{9.21}$$

To find F we use the rotational equation of motion for the disk in Eq. (9.19) with angular velocity $^{\mathcal{I}}\omega^{\mathcal{B}} = \dot{\theta}\mathbf{b}_3$ and moment about G given by $\mathbf{M}_G = rF\mathbf{b}_3$:

$$I_G\ddot{\theta} = rF, \tag{9.22}$$

where the moment of inertia of the disk about its center is given by $I_G = \frac{1}{2}mr^2$ (see Example 9.8). Eqs. (9.21) and (9.22) can be combined to eliminate F and obtain the final equation of motion for the rolling disk,

$$\frac{3}{2}mr\ddot{\theta} = mg \sin \alpha$$

or, when rearranged,

$$\ddot{\theta} = \frac{2g \sin \alpha}{3r}. \tag{9.23}$$

Example 9.10 Vibration Isolation

Example 9.5 introduced the vibration-isolation system consisting of a heavy table supported by two springy legs. The coordinates, reference frames, and free-body diagram are shown in Figure 9.5. In this example we find the equations of motion for the table's position and orientation.

Recall that the position of the table's center of mass is described by the height y and its orientation in the inertial frame by the rotation angle θ. The position of the center of mass in $\mathcal{I}$ is thus

$$\mathbf{r}_{G/O} = y\mathbf{e}_y,$$

and the angular velocity of the table in the inertial frame is $^{\mathcal{I}}\boldsymbol{\omega}^B = \dot{\theta}\mathbf{e}_z$.

To find the equations of motion we separately use Euler's first law on the center of mass and then Euler's second law in Eq. (9.19) for motion about the center of mass. There are three forces acting on the table: gravity acting at the center of mass,

$$\mathbf{F}_G = -m_G g\mathbf{e}_y,$$

and the two spring forces, $\mathbf{F}_A$ and $\mathbf{F}_B$, acting at endpoints A and B, respectively. Example 9.5 showed that

$$\mathbf{F}_A = -k_A \left(y - \frac{L}{2} \sin \theta - y_A \right) \mathbf{e}_y$$

$$\mathbf{F}_B = -k_B \left(y + \frac{L}{2} \sin \theta - y_B \right) \mathbf{e}_y,$$

where y_A and y_B are the unstretched lengths of the springs.

The translational equation of motion for the center of mass is found from Euler's first law in the $\mathbf{e}_y$ direction,

$$m_G\ddot{y} + m_G g + (k_A + k_B)y + (k_B - k_A)\frac{L}{2} \sin \theta = k_A y_A + k_B y_B. \tag{9.24}$$

The equation of motion for θ is found by substituting the moment expression in Eq. (9.13) into Euler's second law in Eq. (9.19):

$$I_G \ddot{\theta} + \frac{L}{2} \cos\theta (k_B - k_A) y + \left(\frac{L}{2}\right)^2 (k_A + k_B) \cos\theta \sin\theta$$

$$= \frac{L}{2} (k_B y_B - k_A y_A) \cos\theta. \tag{9.25}$$

Eqs. (9.24) and (9.25) are the equations of motion for the two degrees of freedom of the table—translation and rotation.

Eqs. (9.24) and (9.25) also indicate the existence of a nonzero equilibrium. When $\ddot{y} = \ddot{\theta} = 0$, the equilibrium position of the table is given by the solution of two equations:

$$y_0 + \frac{k_B - k_A}{k_B + k_A} \frac{L}{2} \sin\theta_0 = \frac{k_A y_A + k_B y_B - m_G g}{k_A + k_B}$$

$$y_0 + \frac{k_A + k_B}{k_B - k_A} \frac{L}{2} \sin\theta_0 = \frac{k_B y_B - k_A y_A}{k_B - k_A}.$$

The solution gives the equilibrium height and angle:

$$y_0 = \frac{y_B + y_A}{2} - \frac{m_G y (k_B + k_A)}{4 L k_A k_B}$$

$$\theta_0 = \arcsin\left(\frac{y_B - y_A}{L} + \frac{m_G g (k_B - k_A)}{2 L k_a k_B}\right).$$

As you might have expected, in equilibrium the springs are slightly compressed and the table is lower than the unstretched lengths of the springs due to gravity, while it has a slight tilt because of the unequal spring constants. If $k_A = k_B$, then the equilibrium reduces to a small vertical sag in y and no rotation ($\theta_0 = 0$). The equations of motion in y and θ also completely uncoupled. That is why it is important to find table legs with equal spring constants!

Finally, we can make the small-angle approximation and rewrite the equations of motion for the table for small motions about its equilibrium. Letting $y = \delta y + y_0$ and $\theta = \delta\theta + \theta_0$ in Eqs. (9.24) and (9.25) yields[2]

$$\delta\ddot{y} + \frac{(k_A + k_B)}{m_G} \delta y + \frac{(k_B - k_A)}{m_G} \frac{L}{2} \delta\theta = 0$$

$$\delta\ddot{\theta} + \frac{k_B - k_A}{I_G} \frac{L}{2} \delta y + \frac{k_A + k_B}{I_G} \frac{L^2}{4} \delta\theta = 0,$$

where δy and $\delta\theta$ are small deviations from the equilibrium values y_0, θ_0.

We discuss the solution of such coupled linear differential equations for vibrational problems in Section 12.1. Note again that if the spring constants are equal, $k_A = k_B$, these separate into two simple harmonic motions. Section 12.1 also discusses how such a system is used to isolate the table from the motion of the ground.

---

[2] We discuss linearization in Section 12.2.2.

9.3.4 Impulsive Moments on a Rigid Body

Chapter 4 (Definition 4.4) defined the angular impulse on a particle P associated with an applied moment to be

$$\overline{\mathbf{M}}_{P/O}(t_1, t_2) \triangleq \int_{t_1}^{t_2} \mathbf{M}_{P/O} \, dt.$$

We also solved for the integrated, impulsive form of the angular-momentum equation of motion for a particle (Eq. (4.17)). We can do the same for the equation of motion of the angular momentum of a rigid body in Eq. (9.10).

Suppose that an impulsive moment or torque is applied to a rigid body. Then integrating Eq. (9.10) gives

$$^{\mathcal{I}}\mathbf{h}_G(t_2) = {}^{\mathcal{I}}\mathbf{h}_G(t_1) + \overline{\mathbf{M}}_G(t_1, t_2). \tag{9.26}$$

In the planar case, Eq. (9.26) can be simplified nicely using Eq. (9.18) to substitute for the angular momentum:

$$^{\mathcal{I}}\boldsymbol{\omega}^{\mathcal{B}}(t_2) = {}^{\mathcal{I}}\boldsymbol{\omega}^{\mathcal{B}}(t_1) + \frac{1}{I_G}\overline{\mathbf{M}}_G(t_1, t_2), \tag{9.27}$$

where $^{\mathcal{I}}\boldsymbol{\omega}^{\mathcal{B}} = {}^{\mathcal{I}}\omega^{\mathcal{B}}\mathbf{b}_3$.

Thus the application of an impulsive moment to a planar rigid body instantaneously changes its angular velocity, just as an impulsive force instantaneously changes the velocity of a particle. In fact, the force associated with the impulsive moment will also cause the center of mass of the rigid body to instantaneously change its velocity, as described in Chapter 4.

Example 9.11 Cue Shot

Part of the art of playing pool is knowing just where to hit the cue ball with the cue. Hit it high and the cue ball rolls forward faster than the no-slip condition, making it follow another ball after contact. Strike it low, as in Figure 9.10, and the cue ball rolls too slowly and slides forward, making it bounce back after hitting another ball. This example computes the height above the centerline of the cue ball at which an impulsive strike by the cue makes it roll without slipping.

Figure 9.11 shows the model for the cue striking the cue ball a distance h above the centerline. The cue ball is a sphere with radius r, mass m, and moment of inertia about a central axis of $I_G = (2/5)mr^2$ (see Appendix D). It is struck by a force $\mathbf{F}_c = F_c\mathbf{e}_1$ over a short interval, resulting in an impulsive force $\overline{\mathbf{F}}_c(t_1, t_2)$ in the $\mathbf{e}_1$ direction. If the ball starts at rest at $t = t_1$, then from Eq. (4.2), its change in velocity will be

$$m{}^{\mathcal{I}}\mathbf{v}_{G/O}(t_2) = mv(t_2)\mathbf{e}_1 = \overline{\mathbf{F}}_c(t_1, t_2) = F_c\Delta t\mathbf{e}_1,$$

Figure 9.10 Pool cue hitting the cue ball. Image courtesy of Shutterstock.

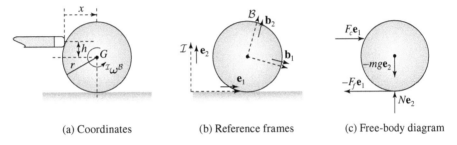

(a) Coordinates (b) Reference frames (c) Free-body diagram

Figure 9.11 A pool ball being struck a distance h above its centerline.

where t_2 is the time just after the impulse and $\Delta t = t_2 - t_1$. The impulse imparted by the cue also creates an impulsive moment on the cue ball about its center of mass:

$$\overline{\mathbf{M}}_G(t_1, t_2) = h\mathbf{e}_2 \times \overline{\mathbf{F}}_c = -hF_c\Delta t\mathbf{e}_3.$$

From Eq. (9.27) the angular velocity of the cue ball about its center of mass after impact satisfies

$$I_G{}^{\mathcal{I}}\omega^{\mathcal{B}}(t_2)\mathbf{e}_3 = -hF_c\Delta t\mathbf{e}_3.$$

Substituting for the linear impulse from the translational equation gives

$$I_G{}^{\mathcal{I}}\omega^{\mathcal{B}}(t_2) = \frac{2}{5}mr^2{}^{\mathcal{I}}\omega^{\mathcal{B}}(t_2) = -hmv(t_2). \qquad (9.28)$$

For the ball to roll without slipping, it must satisfy the no-slip condition at t_2,

$$v(t_2) = -r^{\mathcal{I}}\omega^{\mathcal{B}}(t_2).$$

Substituting this condition into Eq. (9.28) gives the required height above the center for the cue ball to start to roll without sliding:

$$h = \frac{2}{5}r.$$

If $h < (2/5)r$, then the ball slides faster than it rolls and friction will retard the horizontal motion, slowing it down. If $h > (2/5)r$, the ball rolls too fast for its translational motion and sliding friction acts forward, increasing the linear velocity while decreasing the angular velocity, so that the ball speeds up.

9.4 Rotation about an Arbitrary Body Point

There are many situations where we are interested in writing the angular momentum about an arbitrary point of the body rather than about the center of mass, as in Figure 9.12, even though angular momentum separation no longer holds. Section 7.3 discussed in detail the dynamics of a collection of particles relative to an arbitrary point Q. The resulting equation of motion, given in Eq. (7.17), was

$$\frac{^{\mathcal{I}}d}{dt}\left(^{\mathcal{I}}\mathbf{h}_Q\right) = \mathbf{M}_Q + \mathbf{r}_{Q/G} \times m_G{}^{\mathcal{I}}\mathbf{a}_{Q/O},$$

where the angular momentum about Q is

$$^{\mathcal{I}}\mathbf{h}_Q = \sum_{i=1}^{N} \mathbf{r}_{i/Q} \times m_i{}^{\mathcal{I}}\mathbf{v}_{i/Q}$$

and we have dropped the (ext) from the moment.

Fortunately, the derivation in the previous section holds exactly for motion relative to Q. If we also assume that Q is fixed to the body, so that $^{\mathcal{B}}\mathbf{v}_{i/Q} = 0$ for every mass particle, then use of the transport equation leads to

$$^{\mathcal{I}}\mathbf{h}_Q = {}^{\mathcal{I}}\omega^{\mathcal{B}}\mathbf{b}_3 \underbrace{\sum_{i=1}^{N} m_i \|\mathbf{r}_{i/Q}\|^2}_{\triangleq I_Q} = I_Q{}^{\mathcal{I}}\omega^{\mathcal{B}}\mathbf{b}_3. \tag{9.29}$$

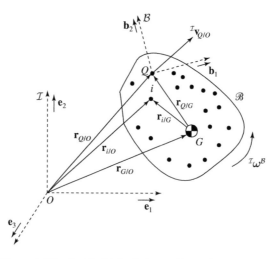

Figure 9.12 Planar rigid body $\mathscr{B}$ with body frame B attached to an arbitrary point Q.

The moment of inertia about Q for a discrete collection of particles is

$$I_Q \triangleq \sum_{i=1}^{N} m_i \|\mathbf{r}_{i/Q}\|^2,$$

and for a continuous rigid body $\mathscr{B}$ it is

$$I_Q = \int_{\mathscr{B}} \|\mathbf{r}_{dm/Q}\|^2 \rho(\mathbf{r}_{dm/O}) d\mathbf{r}_{dm/O}.$$

The equation of motion for the angular momentum about Q then reduces to

$$\frac{^\mathcal{I}d}{dt}\left(^\mathcal{I}\mathbf{h}_Q\right) = I_Q \, {}^\mathcal{I}\alpha^\mathcal{B}\mathbf{b}_3 = \mathbf{M}_Q + \mathbf{r}_{Q/G} \times m_G \, {}^\mathcal{I}\mathbf{a}_{Q/O}. \qquad (9.30)$$

Eq. (9.30) can be used to solve a rotational problem about an arbitrary point on the body. It is most useful for problems involving multiple rigid bodies where the attachment point is moving, thus making it easier to use Eq. (9.30) than to compute the angular momentum about the center of mass. Many problems have one point of the rigid body pinned in absolute space, so that Q is now an inertially fixed point, and Eq. (9.30) reduces to Euler's second law. The effort is in finding the moment of inertia and moments relative to Q. Usually we already have a description of the inertia of the rigid body and the moments acting on it relative to the center of mass, so it seems like extra work to start from the beginning to compute the moment of inertia about Q and total moment relative to Q. In fact, there are shortcuts that simplify finding these quantities. In the next two subsections we introduce two well-known theorems for making these calculations easier. A short discussion of the special case of a pinned rigid body then follows.

9.4.1 The Moment Transport Theorem

Eq. (9.30) requires an expression for the external moment $\mathbf{M}_Q$ on the rigid body about Q. This quantity is obtained from the definition of the moment:

$$\mathbf{M}_Q \triangleq \sum_{i=1}^{N} \mathbf{r}_{i/Q} \times \mathbf{F}_i^{(\text{ext})}. \qquad (9.31)$$

Suppose, however, that we already have an expression for the moment about the center of mass G from Eq. (9.11). It turns out that there is a simple formula for relating the moment about Q to the moment about G without having to recompute the sum in Eq. (9.31). We begin by replacing $\mathbf{r}_{i/Q}$ in Eq. (9.31) with the difference of vectors relative to O in Figure 9.12:

$$\mathbf{M}_Q \triangleq \sum_{i=1}^{N} (\mathbf{r}_{i/O} - \mathbf{r}_{Q/O}) \times \mathbf{F}_i^{(\text{ext})}.$$

We next substitute for $\mathbf{r}_{i/O}$ from the vector triad $\mathbf{r}_{i/O} = \mathbf{r}_{i/G} + \mathbf{r}_{G/O}$:

$$\mathbf{M}_Q = \sum_{i=1}^{N} (\mathbf{r}_{i/G} + \mathbf{r}_{G/O} - \mathbf{r}_{Q/O}) \times \mathbf{F}_i^{(\text{ext})}.$$

Distributing the terms in parentheses yields

$$\mathbf{M}_Q = \underbrace{\sum_{i=1}^{N} \mathbf{r}_{i/G} \times \mathbf{F}_i^{(\text{ext})}}_{\triangleq \mathbf{M}_G} + \sum_{i=1}^{N} (\mathbf{r}_{G/O} - \mathbf{r}_{Q/O}) \times \mathbf{F}_i^{(\text{ext})}.$$

The last term can be simplified using the vector triad $\mathbf{r}_{Q/G} + \mathbf{r}_{G/O} = \mathbf{r}_{Q/O}$ to obtain

$$\mathbf{M}_Q = \mathbf{M}_G - \mathbf{r}_{Q/G} \times \sum_{i=1}^{N} \mathbf{F}_i^{(\text{ext})}. \tag{9.32}$$

Eq. (9.32) is known as the *moment transport theorem*. It can be used to write the moment about point Q in terms of the moment about G and the vector between Q and G. We have chosen to use the center of mass here because it is a common reference point due to the separation of angular momenta, but it applies to any pair of points fixed to the body.

Example 9.12 The Gravitational Moment on a Compound Pendulum

Look back to Example 8.1 (and Figure 8.2), which considered the kinematics of a compound pendulum. Soon we will derive the full equations of motion for the compound pendulum. To do so, we need the moment acting on the rigid body relative to the attachment point O. To find the moment relative to O, we could compute the moment due to the gravity force on each mass element and integrate over the body. However, the moment transport theorem makes finding the moment much easier. We already showed (in Example 9.4) that the total moment about the center of mass of a rigid body in a uniform gravitational field is zero:

$$\mathbf{M}_G = 0.$$

Thus for the compound pendulum, the moment transport theorem (Eq. (9.32)) in continuous form gives (ignoring the force at O because it cancels)

$$\mathbf{M}_O = -\mathbf{r}_{O/G} \times \int_{\mathscr{B}} dm g \mathbf{e}_1$$

$$= -m_G g \mathbf{r}_{O/G} \times \mathbf{e}_1.$$

This moment is the same as if we treated the compound pendulum as a single particle at G and calculated the moment relative to O. As we discussed in Example 9.3, treating the moment on the compound pendulum as if it were equivalent to the moment on a simple pendulum located at the center of mass is not correct in a $1/r^2$ gravitational field because the center of mass and center of gravity are not the same for most rigid bodies.

9.4.2 The Parallel Axis Theorem

Recall that Eq. (7.16) in Section 7.3 expressed the angular momentum about point Q in terms of the angular momentum about the center of mass:

$$^{\mathcal{I}}\mathbf{h}_Q = {}^{\mathcal{I}}\mathbf{h}_G + m_G \mathbf{r}_{Q/G} \times {}^{\mathcal{I}}\mathbf{v}_{Q/G}. \tag{9.33}$$

This equation leads to an expression for the moment of inertia about Q in terms of the moment of inertia about the center of mass, which in turn eliminates the need to recalculate the moment of inertia for each new point. In Eq. (9.18) $^{\mathcal{I}}\mathbf{h}_G$ is given in terms of the moment of inertia about G. Look at the second term in Eq. (9.33) and use the transport equation:

$$m_G \mathbf{r}_{Q/G} \times {}^{\mathcal{I}}\mathbf{v}_{Q/G} = m_G \mathbf{r}_{Q/G} \times \bigg(\underbrace{{}^{\mathcal{B}}\mathbf{v}_{Q/G}}_{=0} + {}^{\mathcal{I}}\boldsymbol{\omega}^{\mathcal{B}} \times \mathbf{r}_{Q/G} \bigg),$$

where the velocity of Q in the body frame is zero, since we are assuming that Q is fixed to the rigid body. The result is the following triple cross product:

$$m_G \mathbf{r}_{Q/G} \times {}^{\mathcal{I}}\mathbf{v}_{Q/G} = m_G \mathbf{r}_{Q/G} \times \bigg({}^{\mathcal{I}}\boldsymbol{\omega}^{\mathcal{B}} \times \mathbf{r}_{Q/G} \bigg).$$

Just as in Section 9.3.2, we recognize that this triple cross product is always in the $\mathbf{b}_3$ direction for the planar case, leading to

$$m_G \mathbf{r}_{Q/G} \times \bigg({}^{\mathcal{I}}\boldsymbol{\omega}^{\mathcal{B}} \times \mathbf{r}_{Q/G} \bigg) = m_G {}^{\mathcal{I}}\omega^{\mathcal{B}} \| \mathbf{r}_{Q/G} \|^2 \mathbf{b}_3.$$

We now rewrite the expression for $^{\mathcal{I}}\mathbf{h}_Q$ in Eq. (9.33) as

$$^{\mathcal{I}}\mathbf{h}_Q = I_G {}^{\mathcal{I}}\omega^{\mathcal{B}} \mathbf{b}_3 + m_G {}^{\mathcal{I}}\omega^{\mathcal{B}} \| \mathbf{r}_{Q/G} \|^2 \mathbf{b}_3.$$

A comparison to Eq. (9.29) shows that the moment of inertia about Q can be written

$$\boxed{I_Q = I_G + m_G \| \mathbf{r}_{Q/G} \|^2.} \tag{9.34}$$

Eq. (9.34) is known as the *parallel axis theorem*. If the moment of inertia about an axis through the center of mass is known, then this formula can be used to find the moment of inertia about any other parallel axis in terms of the perpendicular distance between them.

Example 9.13 The Moment of Inertia of a Compound Pendulum

Examples 8.1 and 9.12 looked at a compound pendulum of arbitrary shape. In the next section we solve for the equation of motion of the compound pendulum and find that it depends on the moment of inertia about the pinned point O. Here we use the parallel axis theorem to find the moment of inertia for a specific pendulum geometry.

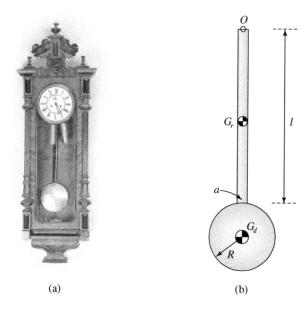

(a)   (b)

Figure 9.13 (a) Pendulum clock. (b) Compound pendulum model. Image (a) courtesy of Shutterstock.

A pendulum clock consists of a long rod of length l and width a with a circular disk of radius R at its end, as shown in Figure 9.13b. The center of mass G_r of the rod is located at its midpoint, and the center of mass G_d of the disk is at its center. From Example 9.8, the moment of inertia of the disk is

$$I_{G_d} = \frac{1}{2} m_d R^2.$$

From Appendix D, the moment of inertia of the rod about its center of mass is

$$I_{G_r} = \frac{m_r}{12}(l^2 + a^2).$$

Using the parallel axis theorem, the moment of inertia about the attachment point O is the sum $I_O = I_{O_d} + I_{O_r}$:

$$I_O = \frac{1}{2} m_d R^2 + m_d(l + R)^2 + \frac{m_r}{12}(l^2 + a^2) + m_r \frac{l^2}{4}$$

$$= m_d l^2 \left(1 + 2\left(\frac{R}{l}\right) + \frac{3}{2}\left(\frac{R}{l}\right)^2\right) + \frac{1}{3} m_r l^2 \left(1 + \frac{1}{4}\left(\frac{a}{l}\right)^2\right).$$

For a rod that is long and thin compared to the disk radius this is often approximated by

$$I_O \approx (m_d + \frac{1}{3} m_r) l^2.$$

This example also shows how the moment of inertia of a more complex object can be found by piecing together the moments of inertia of its parts and using the parallel axis theorem.

Example 9.14 Disk Rolling down Incline, Revisited

Example 9.9 computed the equation of motion for a disk or wheel rolling down an inclined plane by using the separation principle and finding equations for translation of the center of the disk and rotation of the disk. The no-slip condition was used to combine these into a single equation of motion for the angle θ of the disk. Here we recompute the equation of motion by using Eq. (9.30) about the contact point P.

The angular momentum of the disk about P is given by Eq. (9.29),

$$^{\mathcal{I}}\mathbf{h}_P = I_P \dot{\theta} \mathbf{b}_3,$$

where we used the fact that $^{\mathcal{I}}\omega^{\mathcal{B}} = \dot{\theta}$. The moment of inertia of the disk about P is found from the parallel axis theorem to be

$$I_P = \frac{1}{2}mr^2 + mr^2 = \frac{3}{2}mr^2.$$

From Example 8.2, the no-slip condition states that the point of contact of the disk and the ground is instantaneously fixed. Although $^{\mathcal{I}}\mathbf{a}_{P/O} \neq 0$, it is perpendicular to the ramp, implying $\mathbf{r}_{Q/G} \times {}^{\mathcal{I}}\mathbf{a}_{Q/O} = 0$ in Eq. (9.30), where Q is replaced by P here. Thus the moment equation is

$$\frac{^{\mathcal{I}}d}{dt}\left(^{\mathcal{I}}\mathbf{h}_P\right) = I_P\ddot{\theta}\mathbf{b}_3 = mgr\sin\alpha\mathbf{b}_3,$$

which results in the same equation of motion as in Eq. (9.23):

$$\ddot{\theta} = \frac{2g\sin\alpha}{3r}.$$

9.4.3 A Pinned Rigid Body

Example 9.14 was an example of the common situation where the point Q is fixed in inertial space. In this case, the acceleration of Q relative to the inertial frame in Eq. (9.30) is zero and the equation of motion for $^{\mathcal{I}}\mathbf{h}_Q$ reduces to Euler's second law,

$$\frac{^{\mathcal{I}}d}{dt}\left(^{\mathcal{I}}\mathbf{h}_Q\right) = I_Q\,{}^{\mathcal{I}}\alpha^{\mathcal{B}}\mathbf{b}_3 = \mathbf{M}_Q. \tag{9.35}$$

Thus if we use the parallel axis theorem to find the moment of inertia about Q, the equation of motion for the angular momentum about Q is the same as the equation of motion about the center of mass G. We just have to be sure to take the moment of the external forces about Q. This simplification is useful for many problems.

It is common practice, in particular for pinned problems, to introduce the definition of a new quantity, the *radius of gyration*, which is used to describe the moment of inertia of a body.

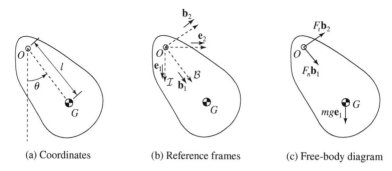

(a) Coordinates   (b) Reference frames   (c) Free-body diagram

Figure 9.14 Compound pendulum consisting of a rigid body with moment of inertia I_G pinned at point O in the inertial frame.

Definition 9.3 The **radius of gyration** k_Q relative to a point Q of a rigid body is the equivalent distance from Q of a particle of mass m_G needed to produce the same moment of inertia as the rigid body about Q:

$$m_G k_Q^2 \triangleq I_Q.$$

Thus the radius of gyration of a rigid body about its center of mass is $k_G \triangleq \sqrt{I_G/m_G}$. By the parallel axis theorem, the following also holds for an axis through a point Q a distance l away from the center of mass:

$$I_Q = I_G + m_G l^2 = m_G(k_G^2 + l^2).$$

Many handbooks list the radius of gyration rather than the moment of inertia; almost always, this quantity is the radius of gyration relative to the center of mass, in which case the subscript G is often dropped. The reason for using the radius of gyration rather than the moment of inertia is that it is easier to measure experimentally, as seen in the next example.

Example 9.15 The Compound Pendulum

This example returns to the compound pendulum of Examples 8.1 and 9.12. Our goal now is to find the equation of motion for θ.

Since the body is pinned at point O (Figure 9.14) the easiest approach to this problem is to balance the angular momentum and moment about O using Euler's second law for the total angular momentum in Eq. (9.35) with Q replaced by O. This approach allows us to ignore the reaction forces at the pin in Figure 9.14c. We could follow the same process as in Example 4.4 for the simple pendulum, but much of the work has already been done. The angular momentum of the compound pendulum is given by Eq. (9.29):

$$^\mathcal{I}\mathbf{h}_O = I_O\,^\mathcal{I}\omega^\mathcal{B}\mathbf{b}_3,$$

where I_O is given by the parallel axis theorem: $I_O = I_G + m_G l^2$. The moment about O was found in Example 9.12 using the moment transport theorem:

$$\mathbf{M}_O = -mg(-l\mathbf{b}_1) \times (\cos\theta\mathbf{b}_1 - \sin\theta\mathbf{b}_2) = -mgl\sin\theta\mathbf{b}_3,$$

where we still assume a uniform gravitational field. The resulting equation of motion from Eq. (9.35) is

$$\frac{{}^{\mathcal{I}}d}{dt}\left({}^{\mathcal{I}}\mathbf{h}_O\right) = I_O\ddot{\theta}\mathbf{b}_3 = -mgl\sin\theta\,\mathbf{b}_3,$$

or more simply,

$$\ddot{\theta} + \left(\frac{mgl}{I_O}\right)\sin\theta = 0. \qquad (9.36)$$

Eq. (9.36) is the same equation of motion as the simple pendulum but with frequency given in terms of the moment of inertia about the pinned point O. Thus for small angles, the compound pendulum oscillates sinusoidally just as a simple pendulum. Replacing the moment of inertia with the radius of gyration about O gives

$$\ddot{\theta} + \frac{g}{k_O^2/l}\sin\theta = 0. \qquad (9.37)$$

Eq. (9.37) looks just like the equation of motion for a simple pendulum of length k_O^2/l. The radius of gyration is thus related to the length of the simple pendulum, with mass m_G, of the same oscillation frequency. This is in fact how the moment of inertia is measured for most rigid bodies. Rather than trying to calculate I_G based on the geometry of what is often a complex mass distribution, the body is simply hung from a point and set oscillating. The oscillation frequency gives a direct measure of the radius of gyration from Eq. (9.37) in terms of the distance to the center of mass, l.

Example 9.16 Pulling without Slipping

This example examines the slip-free motion of a torqued wheel, as shown in Figure 9.15a. The torque comes from a string that is tied around a spool attached to the wheel's axle. As shown in this example, the direction of motion of the wheel (left or right) depends on the angle α of the rope.

Because the wheel is rolling without slipping, its translational dynamics are kinematically paired with its rotational dynamics; that is, we need only find one or the

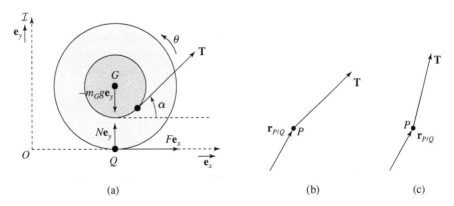

(a)   (b)   (c)

Figure 9.15 Pulling without slipping. (a) Free-body diagram. (b) $\mathbf{r}_{P/Q} \times \mathbf{T} < 0$. (c) $\mathbf{r}_{P/Q} \times \mathbf{T} > 0$.

other. We focus on deriving the rotational dynamics, for which we have at least two options. The first is to relate the rate of change of the angular momentum about the center of mass, $^{\mathcal{I}}\mathbf{h}_G = I_G{}^{\mathcal{I}}\boldsymbol{\omega}^B$, to the total moment about the center of mass, $\mathbf{M}_G = \sum_{i=1}^{N} \mathbf{r}_{i/G} \times \mathbf{F}_i^{(\text{ext})}$, using

$$\frac{^{\mathcal{I}}d}{dt}\left(^{\mathcal{I}}\mathbf{h}_G\right) = \mathbf{M}_G.$$

The second option is to consider instead the time rate of change of the angular momentum about the wheel's contact point Q with the ground. The angular momentum about Q is $^{\mathcal{I}}\mathbf{h}_Q = I_Q{}^{\mathcal{I}}\boldsymbol{\omega}^B$, and the total moment about Q is $\mathbf{M}_Q = \sum_{i=1}^{N} \mathbf{r}_{i/Q} \times \mathbf{F}_i^{(\text{ext})}$. Examination of the free-body diagram in Figure 9.15a helps make a decision. The first option has two torques and two unknowns, F and T (the tension in the string), whereas the second option has only one unknown T due to the tension in the string acting at point P. Pursuing the second option by using $^{\mathcal{I}}\mathbf{h}_Q = I_Q\dot{\theta}\mathbf{b}_3$ yields

$$\frac{^{\mathcal{I}}d}{dt}\left(^{\mathcal{I}}\mathbf{h}_Q\right) = I_Q\ddot{\theta}\mathbf{b}_3 = \mathbf{M}_Q = \mathbf{r}_{P/Q} \times \mathbf{T}.$$

At this point, we know enough to determine the direction of the wheel's motion given the angle α. Consider the arrangements of the vectors $\mathbf{r}_{P/Q}$ and $\mathbf{T}$ shown in Figure 9.15b and c. For small α, the cross product $\mathbf{r}_{P/Q} \times \mathbf{T}$ is negative, which implies that the wheel rolls right ($\ddot{\theta} < 0$); for large α, the cross product is positive, which implies that the wheel rolls left ($\ddot{\theta} > 0$). In the problems, you will have a chance to determine the critical rope angle, α^*, below which the wheel rolls right and above which it rolls left.

9.5 Work and Energy of a Rigid Body

The penultimate task of this chapter is to specialize the treatment of work and energy for multiparticle systems in Section 7.4 to a planar rigid body.

9.5.1 Kinetic Energy of a Rigid Body

As for a multiparticle system, we begin with the kinetic energy. Eq. (7.23) showed that the kinetic energy of a collection of particles separates into the kinetic energy due to motion of the center of mass and the kinetic energy due to motion about the center of mass:

$$\boxed{T_O = T_{G/O} + T_G.} \tag{9.38}$$

We can specialize the second term—the kinetic energy due to motion about the center of mass—to rigid bodies. It is given in Eq. (7.25) for a general collection of particles:

$$T_G \triangleq \frac{1}{2}\sum_{i=1}^{N} m_i{}^{\mathcal{I}}\mathbf{v}_{i/G} \cdot {}^{\mathcal{I}}\mathbf{v}_{i/G} = \frac{1}{2}\sum_{i=1}^{N} m_i\|{}^{\mathcal{I}}\mathbf{v}_{i/G}\|^2.$$

In a rigid body there is no movement of the individual particles relative to the center of mass in the body frame. We thus use the transport equation on $^{\mathcal{I}}\mathbf{v}_{i/G}$, as when developing angular momentum in Eq. (9.14). Letting the body-fixed velocity be zero allows us to rewrite the kinetic energy for motion about the center of mass:

$$T_G = \frac{1}{2}\sum_{i=1}^{N} m_i \|{}^{\mathcal{I}}\boldsymbol{\omega}^{\mathcal{B}} \times \mathbf{r}_{i/G}\|^2 = \frac{1}{2}({}^{\mathcal{I}}\omega^{\mathcal{B}})^2 \sum_{i=1}^{N} m_i \underbrace{\|\mathbf{b}_3 \times \mathbf{r}_{i/G}\|^2}_{=\|\mathbf{r}_{i/G}\|^2}.$$

This summation is just the definition of the moment of inertia about G! Thus we can write a simple expression for the kinetic energy of a rigid body rotating about its center of mass:

$$T_G = \frac{1}{2}I_G\left({}^{\mathcal{I}}\boldsymbol{\omega}^{\mathcal{B}} \cdot {}^{\mathcal{I}}\boldsymbol{\omega}^{\mathcal{B}}\right) = \frac{1}{2}I_G\|{}^{\mathcal{I}}\boldsymbol{\omega}^{\mathcal{B}}\|^2. \qquad (9.39)$$

The moment of inertia and angular velocity supply a very simple way for calculating the total kinetic energy of the rigid body about the center of mass. We can also substitute the expression for the angular momentum of a rigid body about its center of mass from Eq. (9.18) into Eq. (9.39) to formulate another commonly used expression for the kinetic energy of rotation:

$$T_G = \frac{1}{2}{}^{\mathcal{I}}\mathbf{h}_G \cdot {}^{\mathcal{I}}\boldsymbol{\omega}^{\mathcal{B}}. \qquad (9.40)$$

Example 9.17 A Flywheel

Flywheels are a way of storing energy using only mechanical components (unlike batteries, which use chemical storage). This is done by building up a high level of rotational kinetic energy in a rotor with a large moment of inertia, such as the one shown in Figure 9.16. The schematic shows a uniform disk of radius r rotating at rate $\dot{\theta}$. Eq. (9.39) allows us to write the kinetic energy of the flywheel simply as

$$T_G = \frac{1}{2}I_G\dot{\theta}^2.$$

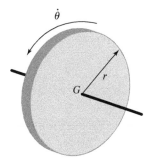

Figure 9.16 Flywheel schematic.

Since the moment of inertia about the center of mass of a uniform disk is given by $mr^2/2$, this expression becomes

$$T_G = \frac{1}{4}mr^2\dot{\theta}^2.$$

Applying this equation, we find that a 1 kg flywheel with a diameter of 1 m, rotating at 100 rpm (about 10.5 rad/s), stores almost 7 J of energy. For comparison, a typical flywheel used for electrical-power backup would have mass of 500 kg, be half a meter in diameter, and rotate at 30,000 rpm—that's more than 77 MJ!

We can also find the kinetic energy for rotation relative to an arbitrary point Q on the body, as done in Section 9.4 for angular momentum. We start by writing the total kinetic energy of the rigid body relative to the origin of the inertial frame:

$$T_O = T_{G/O} + T_G$$

$$= \frac{1}{2}m_G\|{}^{\mathcal{I}}\mathbf{v}_{G/O}\|^2 + \frac{1}{2}I_G\|{}^{\mathcal{I}}\boldsymbol{\omega}^{\mathcal{B}}\|^2.$$

We then replace ${}^{\mathcal{I}}\mathbf{v}_{G/O}$ in the first term with the usual vector triad, ${}^{\mathcal{I}}\mathbf{v}_{Q/O} = {}^{\mathcal{I}}\mathbf{v}_{G/O} + {}^{\mathcal{I}}\mathbf{v}_{Q/G}$, relating the velocity of Q to the velocity of G:

$$T_O = \underbrace{\frac{1}{2}m_G\|{}^{\mathcal{I}}\mathbf{v}_{Q/O}\|^2}_{\triangleq T_{Q/O}} - m_G\left({}^{\mathcal{I}}\mathbf{v}_{Q/O}\cdot{}^{\mathcal{I}}\mathbf{v}_{Q/G}\right) + \frac{1}{2}m_G\|{}^{\mathcal{I}}\mathbf{v}_{Q/G}\|^2$$

$$+ \frac{1}{2}I_G\|{}^{\mathcal{I}}\boldsymbol{\omega}^{\mathcal{B}}\|^2,$$

where $T_{Q/O} \triangleq \frac{1}{2}m_G\|{}^{\mathcal{I}}\mathbf{v}_{Q/O}\|^2$ is the kinetic energy of an equivalent particle of mass m_G located at point Q. Next we use the transport equation ${}^{\mathcal{I}}\mathbf{v}_{Q/G} = {}^{\mathcal{B}}\mathbf{v}_{Q/G} + {}^{\mathcal{I}}\boldsymbol{\omega}^{\mathcal{B}} \times \mathbf{r}_{Q/G} = {}^{\mathcal{I}}\boldsymbol{\omega}^{\mathcal{B}} \times \mathbf{r}_{Q/G}$ to replace ${}^{\mathcal{I}}\mathbf{v}_{Q/G}$ (Q is fixed to the rigid body):

$$T_O = T_{Q/O} - m_G{}^{\mathcal{I}}\mathbf{v}_{Q/O}\cdot\left({}^{\mathcal{I}}\boldsymbol{\omega}^{\mathcal{B}} \times \mathbf{r}_{Q/G}\right) + \underbrace{\frac{1}{2}m_G\|{}^{\mathcal{I}}\boldsymbol{\omega}^{\mathcal{B}} \times \mathbf{r}_{Q/G}\|^2}_{=\frac{1}{2}m_G({}^{\mathcal{I}}\boldsymbol{\omega}^{\mathcal{B}})^2\|\mathbf{r}_{Q/G}\|^2} + \frac{1}{2}I_G\|{}^{\mathcal{I}}\boldsymbol{\omega}^{\mathcal{B}}\|^2.$$

The last two terms should look familiar from our discussion of the parallel axis theorem. Using the same planar rotation assumptions as in Section 9.4.2, we can rewrite this equation in terms of the moment of inertia about Q and the kinetic energy about Q:

$$T_O = T_{Q/O} + T_Q - m_G{}^{\mathcal{I}}\mathbf{v}_{Q/O}\cdot\left({}^{\mathcal{I}}\boldsymbol{\omega}^{\mathcal{B}} \times \mathbf{r}_{Q/G}\right), \qquad (9.41)$$

where

$$T_Q = \frac{1}{2}(m_G\|\mathbf{r}_{Q/G}\|^2 + I_G)\|{}^{\mathcal{I}}\boldsymbol{\omega}^{\mathcal{B}}\|^2 = \frac{1}{2}I_Q\|{}^{\mathcal{I}}\boldsymbol{\omega}^{\mathcal{B}}\|^2$$

is the kinetic energy associated with rotation of the rigid body about Q. I_Q is the moment of inertia about point Q given by the parallel axis theorem in Eq. (9.34).

Note that Eq. (9.41) looks a lot like the separated expression for the kinetic energy in Eq. (9.38) except for a cross term between the rotation of the rigid body and the translation of point Q. As expected, complete separation only occurs when considering rotation about the center of mass. However, for the special case of a pinned rigid body (such as the compound pendulum), the inertial velocity of Q is zero, and the rotational kinetic energy about Q is easily calculated using the angular velocity of the rigid body and the parallel axis theorem.

9.5.2 Rigid Body Work

As in Chapter 7, we turn next to the work–kinetic-energy formula for a rigid body. Since a rigid body is a collection of particles (a fact used throughout this chapter), the work–kinetic-energy formula in Eq. (7.26) still holds:

$$W^{(\text{tot})} = T_O(t_2) - T_O(t_1),$$

where $W^{(\text{tot})}$ is short for $W^{(\text{tot})}(\{\mathbf{r}_{i/O}; \gamma_i\}_{i=1}^N)$, the total external work on the body.

How might we further simplify this formula when specializing it to a rigid collection of particles? The first obvious simplification is to eliminate the internal work term. Since the body is rigid, the constituent particles cannot move relative to one another, and thus no internal work can be done over a trajectory. The work–kinetic-energy formula thus immediately reduces to

$$W = \Delta T_{G/O} + \Delta T_G, \tag{9.42}$$

where we have used the separation property developed previously to separate the change in total kinetic energy $\Delta T_O \triangleq T_O(t_2) - T_O(t_1)$ into the change of kinetic energy of the center of mass $\Delta T_{G/O} \triangleq T_{G/O}(t_2) - T_{G/O}(t_1)$ plus the change in kinetic energy about the center of mass $\Delta T_G \triangleq T_G(t_2) - T_G(t_1)$. We have also dropped the (tot) superscript because there is no internal work.

Recall that Chapter 7 showed that the power associated with motion of the center of mass was equal to the change in kinetic energy of the center of mass and that this kinetic energy is therefore conserved if the total external force on the system is zero (see Eq. (7.30)). This result remains very useful and important even for rigid bodies. However, for a rigid body we can go a bit further. It is helpful to separate the total work into the work associated with the motion of the center of mass and that associated with the motion of the particles relative to the center of mass.

To begin, recall the definition of the external work on a collection of particles. In the absence of internal work (i.e., for a rigid body), we have

$$W(\{\mathbf{r}_{i/O}; \gamma_i\}_{i=1}^N) \triangleq \sum_{i=1}^N \int_{\gamma_i} \mathbf{F}_i^{(\text{ext})} \cdot {}^{\mathcal{I}} d\mathbf{r}_{i/O}.$$

Next we rewrite the work in the velocity form of Eq. (5.1):

$$W(\{\mathbf{r}_{i/O}; \gamma_i\}_{i=1}^N) = \sum_{i=1}^N \int_{t_1}^{t_2} \mathbf{F}_i^{(ext)} \cdot {}^{\mathcal{I}}\mathbf{v}_{i/O} dt.$$

We now use a vector triad to replace the velocity with respect to the origin with motion relative to the center of mass:

$$W(\{\mathbf{r}_{i/O}; \gamma_i\}_{i=1}^N) = \underbrace{\sum_{i=1}^N \int_{t_1}^{t_2} \mathbf{F}_i^{(ext)} \cdot {}^{\mathcal{I}}\mathbf{v}_{G/O} dt}_{\triangleq W_{G/O}(\mathbf{r}_{G/O}, \theta; \gamma_G, \gamma_\theta)} + \sum_{i=1}^N \int_{t_1}^{t_2} \mathbf{F}_i^{(ext)} \cdot {}^{\mathcal{I}}\mathbf{v}_{i/G} dt. \quad (9.43)$$

$W_{G/O}(\mathbf{r}_{G/O}, \theta; \gamma_G, \gamma_\theta)$ is the work done by the vector sum of all forces on the body to move the center of mass along the trajectory γ_G while the rigid body rotates along a rotational trajectory γ_θ. Exchanging the integration and summation operators yields

$$W_{G/O}(\mathbf{r}_{G/O}, \theta; \gamma_G, \gamma_\theta) \triangleq \int_{t_1}^{t_2} \mathbf{F}_G \cdot {}^{\mathcal{I}}\mathbf{v}_{G/O} dt. \quad (9.44)$$

Recall again how Chapter 7 showed that the rate of change of the kinetic energy of the center of mass of a collection of particles was equal to the power acting on the center of mass, or the derivative of the work in Eq. (9.44). Thus we can identify the work in Eq. (9.44) with the change in kinetic energy associated with motion of the center of mass:

$$\boxed{W_{G/O} = \Delta T_{G/O}.} \quad (9.45)$$

Again, if the total external force on the center of mass is zero, the translational kinetic energy of the center of mass of a rigid body is conserved.

Also of interest is the simplification of the second (relative-motion) term for rigid bodies in Eq. (9.43). Using the transport equation and recognizing that the particles do not move relative to the center of mass in the body frame allows us to make the following substitution:

$$W(\{\mathbf{r}_{i/O}; \gamma_i\}_{i=1}^N) = W_{G/O}(\mathbf{r}_{G/O}, \theta; \gamma_G, \gamma_\theta) + \sum_{i=1}^N \int_{t_1}^{t_2} \mathbf{F}_i^{(ext)} \cdot ({}^{\mathcal{I}}\boldsymbol{\omega}^{\mathcal{B}} \times \mathbf{r}_{i/G}) dt$$

$$= W_{G/O}(\mathbf{r}_{G/O}, \theta; \gamma_G, \gamma_\theta) + \int_{t_1}^{t_2} \underbrace{\sum_{i=1}^N (\mathbf{r}_{i/G} \times \mathbf{F}_i^{(ext)})}_{\triangleq \mathbf{M}_G} \cdot {}^{\mathcal{I}}\boldsymbol{\omega}^{\mathcal{B}} dt,$$

where we used the scalar triple product to rearrange terms (see Section B.4.2). The summation over the cross product in the second term is the total external moment $\mathbf{M}_G$ acting on the rigid body! Thus we can write the total work done over the trajectory of the rigid body in the simple form

$$W = W_{G/O} + W_G,$$

where the work $W_G = W_G(\mathbf{r}_{G/O}, \theta; \gamma_G, \gamma_\theta)$ corresponding to motion of the rigid body about the center of mass is

$$W_G(\mathbf{r}_{G/O}, \theta; \gamma_G, \gamma_\theta) \triangleq \int_{t_1}^{t_2} \mathbf{M}_G \cdot {}^{\mathcal{I}}\boldsymbol{\omega}^{\mathcal{B}} dt. \tag{9.46}$$

The corresponding power associated with rotation about the center of mass is

$$\mathbf{P}_G = \mathbf{M}_G \cdot {}^{\mathcal{I}}\boldsymbol{\omega}^{\mathcal{B}}.$$

Note that just as the work on the center of mass can depend on orientation, the work of motion about the center of mass possibly depends on the position of the center of mass, since the moment $\mathbf{M}_G$ can in general depend on both position and orientation. (Tutorial 9.2 illustrates this dependence.)

Our previous result identifying the change in kinetic energy of the center of mass to the work done on the center of mass lets us identify this second term as the change in kinetic energy of rotation about the center of mass,

$$W_G = \Delta T_G. \tag{9.47}$$

Thus, as for the kinetic energy of the center of mass, if the total external moment about G is zero (even if there are external forces), the rotational kinetic energy of a rigid body is conserved. Note that the moment $\mathbf{M}_G$ in Eq. (9.46) includes all external moments applied to the rigid body, even couples and pure torques.

9.5.3 Potential Energy and Total Energy of a Planar Rigid Body

Our final step in this section is to replace the work term by a change in potential energy when the applied forces (and torques) are conservative. Recall that Chapter 7 replaced the internal and external work terms (when the forces on the various particles were conservative) by a change in internal and external potentials for each particle. Since for rigid bodies there is no change in internal potential energy, we can define the total energy just as in Chapter 7:

$$E_O(t) \triangleq T_{G/O}(t) + T_G(t) + U_O(t).$$

The total potential is due to the (external) conservative forces acting on the rigid body:

$$U_O(t) = \sum_{i=1}^{N} U_{i/O}(\mathbf{r}_{i/O}(t)).$$

In this equation, $U_{i/O}(\mathbf{r}_{i/O}(t))$ is the potential energy associated with each discrete force acting at point i of the rigid body. For field forces, where every mass element of the body contributes to the potential, we add an integral over the body $\mathscr{B}$ and replace the potential energy with the potential per unit mass $U_{dm/O}(\mathbf{r}_{dm/O})$:

$$U_O(t) = \int_{\mathscr{B}} U_{dm/O}(\mathbf{r}_{dm/O})dm. \tag{9.48}$$

As in Chapter 7, the total energy can be used in the work-energy formula leading to Eq. (7.35), where the change in total energy is equal to the total nonconservative work:

$$E_O(t_2) = E_O(t_1) + W_{G/O}^{(nc)} + W_G^{(nc)}.$$

Of course, conservation of energy holds for a rigid body just as for a multiparticle system when no nonconservative work is being done.

Example 9.18 The Energy of a Compound Pendulum

Recall that Example 5.12 solved for the motion of a simple pendulum using energy considerations. This example does the same for the compound pendulum of Example 9.15. Because the compound pendulum is pinned at O, it is easiest to find the kinetic energy using the results for an arbitrary point (Eq. (9.41)) with $Q = O$:

$$T_O = \frac{1}{2}I_O\dot{\theta}^2,$$

where we have used the fact that the pinned point is not moving. As in Example 9.15, the moment of inertia about O is equal to $I_G + m_G l^2$ by the parallel axis theorem.

From our result in Example 9.3 we know that the potential energy of the compound pendulum in a uniform gravitational field must be $m_G g$ times the height of the center of mass (see Tutorial 9.2 for a rigid body in a $1/r^2$ gravitational field):

$$U_O = -m_G g l \cos\theta,$$

where the zero of potential energy is where $\theta = \pi/2$.

The total energy of the compound pendulum relative to the attachment point O is thus

$$E_O = \frac{1}{2}I_O\dot{\theta}^2 - m_G g l \cos\theta, \tag{9.49}$$

where we used the fact that no nonconservative forces are acting on the pendulum. (Remember that the constraint forces at the attachment point do no work.)

At this point, the problem is identical to Example 5.12 with $m_p l^2$ replaced by I_O. For instance, we can use Eq. (9.49) to find the angular rate of the swing at any angle given an initial height (or, equivalently, $E_O(0)$). We can also use Eq. (9.49) to integrate and solve for the trajectory in terms of the initial energy, as in Eq. (5.18).

Example 9.19 The Energy of a Disk Rolling down an Inclined Plane

This example revisits the disk rolling down the inclined plane of Example 9.9 and Figure 9.9a. Rather than find the equations of motion, however, we instead solve for the total energy of the rolling disk. We determine how fast the disk is traveling at the bottom of the plane, assuming it starts rolling some distance $x(t_0)$ up the plane.

Since we are assuming a uniform gravitational field, there is no potential energy associated with the orientation of the disk, only the height of the center of mass:

$$U_O = m_G g x \sin \alpha.$$

The kinetic energy of the disk is given by the sum of the translational kinetic energy of the center of mass and the rotational kinetic energy. Again using the inertial frame $\mathcal{C}$ as in Example 9.9, the velocity of the center of mass is

$${}^{\mathcal{C}}\mathbf{v}_{G/O} = -\dot{x}\mathbf{e}_x.$$

The total kinetic energy is then

$$T_O = T_{G/O} + T_G = \frac{1}{2} m_G \dot{x}^2 + \frac{1}{2} I_G \dot{\theta}^2.$$

Using the no-slip condition, $\dot{x} = -r\dot{\theta}$, and the disk inertia, $I_G = \frac{1}{2} m_G r^2$, we can write the total energy of the disk as

$$E_O = \frac{3}{4} m_G \dot{x}^2 + m_G g x \sin \alpha.$$

Is energy conserved in this problem? There are three forces acting on the disk: gravity, the normal force due to the incline, and the friction between the disk and the incline. We have included the conservative gravitational force in the potential above. The normal force is a constraint force and does no work. What about friction? Because the disk is rolling and not slipping, the friction force is always what it needs to be to make the velocity of the contact point relative to the incline zero. In other words, it is a constraint force; it reduces the degrees of freedom from two (rotation and translation) to one, where the constraint equation is the no-slip condition. It thus does no work, which implies the total energy is conserved.

We can thus find the final velocity of the disk, given an initial distance up the incline, by simply equating the energies at the two points:

$$m_G g x(t_0) \sin \alpha = \frac{3}{4} m_G \dot{x}(t_f)^2$$

or

$$\dot{x}(t_f) = \sqrt{\frac{4}{3} g x(t_0) \sin \alpha}.$$

Note the difference from the result for a sliding particle, which would be $\dot{x}(t_f) = \sqrt{2gx(t_0)\sin \alpha}$. The key idea here is that we need to include the kinetic energy of both translation and rotation of the disk in the total energy.

9.6 A Collection of Rigid Bodies and Particles

We now turn to a general system in two dimensions consisting of interconnected particles and rigid bodies. Remarkably, we have all the tools in place to solve for the motion of an arbitrary system. Recall that Section 2.3 discussed the art of modeling a physical system. Every dynamical system analyzed should be reduced to a collection of particles, rigid bodies, and perhaps massless rods or strings (or other constraining elements). The previous chapters of the book showed how to find the equations of motion for a particle and for a multiparticle system. This chapter demonstrated how to find the equations of motion for the translation and orientation (in the plane) of a rigid body. Combining these methods enables us to solve for the motion of any system.

The steps involved are actually rather straightforward. Once the system under study is modeled as a collection of connected particles and rigid bodies (actually, this is the hardest part), we simply break each connection and add constraint forces or torques between the elements. We then find the equations of motion separately for each of the constituent elements (particles or rigid bodies) using Newton's or Euler's laws. Remember, particles moving in the plane have two degrees of freedom and thus two equations of motion and rigid bodies have three degrees of freedom and thus three equations of motion. Of course, we must include the constraint forces in each of the equations of motion. (Don't forget to draw the free-body diagrams!) Just as we have done throughout the book, we then manipulate the vector equations of motion (and constraint equations) to eliminate the constraint forces and find a scalar equation of motion, in each coordinate, corresponding to each degree of freedom.[3]

To see that this method works, consider N_p particles and N_r rigid bodies. They may be connected through constraints or force-inducing elements, such as springs. We know immediately from the results of Chapter 6 and this chapter that the center of mass of the entire collection obeys Newton's second law using the total external force and that, if the total external force is zero, the center of mass is stationary or moves at constant velocity. That is, the total linear momentum, which is the same as the linear momentum of the center of mass of the entire collection, obeys Eq. (6.16):

$$^{\mathcal{I}}\mathbf{p}_{G/O} = \sum_{i=1}^{N_p} m_i\,^{\mathcal{I}}\mathbf{v}_{i/O} + \sum_{j=1}^{N_r} m_j\,^{\mathcal{I}}\mathbf{v}_{G_j/O}$$

and

$$\frac{^{\mathcal{I}}d}{dt}\left(^{\mathcal{I}}\mathbf{p}_{G/O}\right) = \mathbf{F}_G^{(\text{ext})},$$

where $^{\mathcal{I}}\mathbf{v}_{i/O}$ is the inertial velocity of particle i and $^{\mathcal{I}}\mathbf{v}_{G_j/O}$ is the inertial velocity of the center of mass G_j of rigid body j.

---

[3] It is worth noting here that the added complexity associated with introducing constraint forces and then eliminating them by additional constraint equations is one of the motivations behind the Lagrangian method described in Chapter 13. This advanced approach develops equations of motion for each coordinate (degree of freedom) directly, without the need to introduce constraint forces.

We can also look at the total angular momentum of the collection relative to some point Q that may or may not be accelerating. Following the definition of total angular momentum about an arbitrary point Q, we sum the individual angular momenta of each particle and rigid body:

$$^{\mathcal{I}}\mathbf{h}_Q = \sum_{i=1}^{N_p} {}^{\mathcal{I}}\mathbf{h}_{i/Q} + \sum_{j=1}^{N_r} {}^{\mathcal{I}}\mathbf{h}_{j/Q},$$

where $^{\mathcal{I}}\mathbf{h}_{i/Q}$ is the angular momentum of particle i relative to Q and $^{\mathcal{I}}\mathbf{h}_{j/Q} = {}^{\mathcal{I}}\mathbf{h}_{G_j/Q} + {}^{\mathcal{I}}\mathbf{h}_{G_j}$ is the total angular momentum of rigid body j relative to Q. Note that we may pick from a variety of choices for Q. It might be inertially fixed, it might be the center of mass of the entire collection, it might be the center of mass of one of the rigid bodies, or it might be one of the contact points.

We can now take the derivative of $^{\mathcal{I}}\mathbf{h}_Q$ and distribute the derivative through each sum. The derivative of the total angular momentum of the collection of particles is equal to the external moment on it plus any (internal) moment from the rigid bodies. Likewise, by Euler's second law, the derivative of each of the rigid-body angular momenta is equal to the external moment plus the (internal) moment from the particles and other rigid bodies. However, if we again make the assumption that the internal moments cancel (which will always be true if the only forces and moments acting among the particles and rigid bodies are constraint forces), then we have Euler's second law for the entire collection:

$$\frac{^{\mathcal{I}}d}{dt}\left(^{\mathcal{I}}\mathbf{h}_Q\right) = \mathbf{M}_Q + \mathbf{r}_{Q/G_p} \times m_{G_p}{}^{\mathcal{I}}\mathbf{a}_{Q/O} + m_{G_r}\sum_{j=1}^{N_r}\mathbf{r}_{Q/j} \times {}^{\mathcal{I}}\mathbf{a}_{Q/O}, \tag{9.50}$$

where G_p is the center of mass of the collection of particles, m_{G_p} is the total mass of the particles, and m_{G_r} is the total mass of all the rigid bodies. While we can certainly use Eq. (9.50) to try and find equations of motion, it is only slightly helpful for problems with many degrees of freedom because it does not provide enough information to find all the equations of motion. Its most important consequence is that, for an inertially fixed point Q and zero external moment, the total angular momentum of a collection of particles and rigid bodies is conserved. A classic example that uses conservation of angular momentum of a system of rigid bodies is the satellite momentum wheel. If a satellite encounters an impulsive disturbance that causes it to start to rotate (thus gaining angular momentum), an internal wheel can be spun up such that the total angular momentum stays fixed (only internal moments are generated), but the angular momentum of the satellite goes to zero. Almost all low-earth orbiting satellites use momentum wheels.

Finally, we can use the separation principle on each of the rigid bodies to rewrite the total angular momentum as

$$^{\mathcal{I}}\mathbf{h}_Q = \sum_{i=1}^{N_p} {}^{\mathcal{I}}\mathbf{h}_{i/Q} + \sum_{j=1}^{N_r} {}^{\mathcal{I}}\mathbf{h}_{G_j/Q} + \sum_{j=1}^{N_r} {}^{\mathcal{I}}\mathbf{h}_{G_j}. \tag{9.51}$$

We can then examine the derivatives of each term in Eq. (9.51) just as in Chapter 7. We would once again find Euler's second law for the motion of each rigid body about its center of mass (and the motion of its center of mass relative to Q), only we would have to be sure to include the internal constraint forces and moments acting among the elements. That is exactly the process for finding equations of motion as described at the beginning of the section.

We can perform a similar analysis for the energy of the collection of particles and rigid bodies. Using the separation principle, the total energy is the sum of the kinetic energy of all particles and rigid bodies plus the internal and external potential energies of the system:

$$E_O(t) = T_{G_p/O}(t) + T_{G_p}(t) + \sum_{j=1}^{N_r} \left(T_{j/O}(t) + T_j(t) \right) + U_O^{(\text{ext})}(t) + U_O^{(\text{int})}(t), \quad (9.52)$$

where we have simplified a bit and considered only the energy relative to an inertially fixed point O. The change in total energy across a trajectory is then given exactly by Eq. (7.35); it is equal to the sum of the external and internal work done on the system. Just as in Chapter 7, care must be taken to properly account for the internal forces and internal work. Energy is not like angular momentum! Nevertheless, if the total nonconservative work is zero, the total energy of the collection of rigid bodies and particles is conserved.

Admittedly, this all sounds a bit complicated. It really isn't. Studying the following examples is a good way to see how to approach these more complex dynamics problems.

Example 9.20 The Crane, Take 3

This example reexamines the overhead crane first discussed in Example 6.2. The difference is that here instead of a simple pendulum hanging from the block (a point mass at the end of a massless rod), there is a rigid arm, as in Figure 9.17a. Note that this problem still has two degrees of freedom: the configuration is completely determined by the horizontal position of the block and the angle of the arm. As usual, this implies the presence of constraints and thus constraint forces. Since the arm is now a rigid body, and thus nominally has three planar degrees of freedom (two translation and one rotation), there must now be two constraint forces between the arm and the block rather than the single tension force found for a simple pendulum. Figure 9.17c shows these constraint forces, where the normal force N constraining the block to motion only in x is still there. But now there is both a tangential reaction force and a normal reaction force where the arm joins the block rather than just a tension force as with a simple pendulum. (To solve this problem, we choose to express these forces as components in the $\mathcal{I}$ frame.)

As described above, this problem can be solved by breaking apart the rigid bodies and solving for the dynamics for each. The constraint forces can then be eliminated to find the two equations of motion for x and θ. As in Example 6.2, $\mathcal{I} = (O, \mathbf{e}_x, \mathbf{e}_y, \mathbf{e}_z)$ denotes the inertial frame, and $\mathcal{B} = (Q, \mathbf{e}_r, \mathbf{e}_\theta, \mathbf{e}_z)$ denotes a body frame fixed to the pendulum arm. The frames $\mathcal{B}$ and $\mathcal{I}$ are related by the transformation

	$\mathbf{e}_x$	$\mathbf{e}_y$
$\mathbf{e}_r$	$\sin\theta$	$-\cos\theta$
$\mathbf{e}_\theta$	$\cos\theta$	$\sin\theta$.

(9.53)

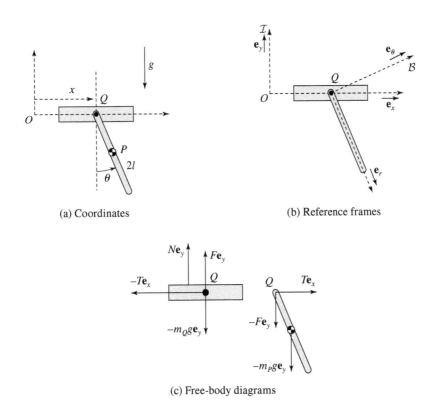

(a) Coordinates

(b) Reference frames

(c) Free-body diagrams

Figure 9.17 The crane, revisited.

The kinematics of block Q are the same as in Example 6.2:

$$\mathbf{r}_{Q/O} = x\mathbf{e}_x$$

$$^{\mathcal{I}}\mathbf{v}_{Q/O} = \dot{x}\mathbf{e}_x$$

$$^{\mathcal{I}}\mathbf{a}_{Q/O} = \ddot{x}\mathbf{e}_x.$$

Likewise, the kinematics of P, in this case the center of mass of the rigid rod, are similar to those in Example 6.2:

$$\mathbf{r}_{P/O} = x\mathbf{e}_x + l\mathbf{e}_r$$

$$^{\mathcal{I}}\mathbf{v}_{P/O} = \dot{x}\mathbf{e}_x + l\dot{\theta}\mathbf{e}_\theta$$

$$^{\mathcal{I}}\mathbf{a}_{P/O} = \ddot{x}\mathbf{e}_x + l\ddot{\theta}\mathbf{e}_\theta - l\dot{\theta}^2\mathbf{e}_r,$$

where l is half the length of the pendulum. Note that we have implicitly used a constraint equation by changing to polar coordinates and setting $r = l$.

We can now use Figure 9.17c and Euler's first law to write the translational equations of motion for the block Q and the center of mass P of the rigid rod:

$$(N + F - m_Q g)\mathbf{e}_y - T\mathbf{e}_x = m_Q \ddot{x}\mathbf{e}_x \tag{9.54}$$

$$(-F - m_P g)\mathbf{e}_y + T\mathbf{e}_x = m_P(\ddot{x}\mathbf{e}_x + l\ddot{\theta}\mathbf{e}_\theta - l\dot{\theta}^2 \mathbf{e}_r). \tag{9.55}$$

Eqs. (9.54) and (9.55) represent four scalar equations, but there are now five unknown quantities, $\ddot{x}, \ddot{\theta}, N, F$, and T. The fifth equation comes from considering the rotational equation of motion for the rigid arm about its center of mass P:

$$\frac{^{\mathcal{I}}d}{dt}\left(^{\mathcal{I}}\mathbf{h}_P\right) = I_P\ddot{\theta}\mathbf{e}_z = (-l\mathbf{e}_r) \times (-F\mathbf{e}_y) - (l\mathbf{e}_r \times T\mathbf{e}_x)$$

$$= Fl\sin\theta\mathbf{e}_z - Tl\cos\theta\mathbf{e}_z.$$

These five equations can be manipulated to eliminate the constraint forces:

$$(m_Q + m_P)\ddot{x} + m_P l\ddot{\theta}\cos\theta - m_P l\dot{\theta}^2\sin\theta = 0 \tag{9.56}$$

$$(I_P + m_P l^2)\ddot{\theta} + m_P g l\sin\theta + m_P l\ddot{x}\cos\theta = 0. \tag{9.57}$$

Finally, as done in Example 6.2, we can algebraically solve Eqs. (9.56) and (9.57) for the equations of motion in x and θ:

$$\left(1 - \frac{3m_P}{4m_G}\cos^2\theta\right)\ddot{x} - \frac{3}{8}\frac{m_P}{m_G}g\sin 2\theta - \frac{m_P}{m_G}l\dot{\theta}^2\sin\theta = 0$$

$$\left(1 - \frac{3}{4}\frac{m_P}{m_G}\cos^2\theta\right)\ddot{\theta} + \frac{3}{8}\frac{m_P}{m_G}\dot{\theta}^2\sin 2\theta + \frac{3}{4}\frac{g}{l}\sin\theta = 0,$$

where $m_G = m_P + m_Q$ and we used the fact that, for a thin rod, $I_P = \frac{1}{3}m_P l^2$. In the limit of $I_P \to 0$, the equations of motion are identical to the equations found for the point-mass crane.

Example 9.21 Particle on a Beam

A classic example of a combination rigid-body–particle system is a point mass sliding on a pinned beam (or seesaw). This system is a common example that you might explore in a course on automatic control. As shown in Figure 9.18a, the system consists of a thin rigid beam, pinned to a support at its center of mass, and a particle, which slides on the beam without friction. Note that this problem only has two degrees of freedom and thus can be completely described using only two coordinates: θ, the angle of the beam from horizontal, and x, the distance of the particle from the center of the beam. We define an inertial frame $\mathcal{I} = (O, \mathbf{e}_x, \mathbf{e}_y, \mathbf{e}_z)$ and body frame $\mathcal{B} = (O, \mathbf{e}_r, \mathbf{e}_\theta, \mathbf{e}_z)$, both attached to the center of the beam O, with the body frame fixed to the beam (Figure 9.18b). The two frames are related by the transformation array

	$\mathbf{e}_x$	$\mathbf{e}_y$
$\mathbf{e}_r$	$\cos\theta$	$\sin\theta$
$\mathbf{e}_\theta$	$-\sin\theta$	$\cos\theta$.

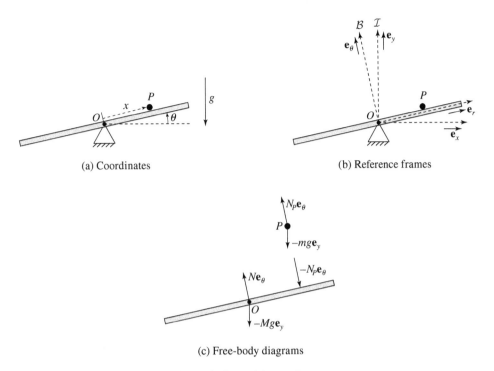

(a) Coordinates   (b) Reference frames

(c) Free-body diagrams

Figure 9.18 Particle on a beam.

Figure 9.18c shows the free-body diagrams of the two components of the system. Note that there are two normal forces involved here: N, the force of the beam support on the beam, and N_P, the force of the particle on the beam. Due to Newton's third law, N_P must appear with opposite signs in both free-body diagrams. We start with the particle P, whose kinematics are

$$\mathbf{r}_{P/O} = x\mathbf{e}_r$$

$$^\mathcal{I}\mathbf{v}_{P/O} = \dot{x}\mathbf{e}_r + x\dot{\theta}\mathbf{e}_\theta$$

$$^\mathcal{I}\mathbf{a}_{P/O} = (\ddot{x} - x\dot{\theta}^2)\mathbf{e}_r + \left(2\dot{x}\dot{\theta} + x\ddot{\theta}\right)\mathbf{e}_\theta.$$

Note that we have carried out the inertial derivatives using the body-frame coordinates.

Applying Newton's second law and separating components in the $\mathbf{e}_r$ and $\mathbf{e}_\theta$ directions produces the following two equations for P:

$$m(\ddot{x} - x\dot{\theta}^2) = -mg\sin\theta \qquad (9.58)$$

$$m(2\dot{x}\dot{\theta} + x\ddot{\theta}) = N_P - mg\cos\theta. \qquad (9.59)$$

We now have two equations of motion for the two degrees of freedom, but the problem is not yet solved, since N_P is unknown. To find it, we must consider the rotational equation of motion of the beam. The beam is rotating about its center of mass O and the only force producing a moment about O is N_P. All other forces in the free-body diagram pass through O and contribute no moment. Thus the rotational equation of motion is

$$\frac{^{\mathcal{I}}d}{dt}\left(^{\mathcal{I}}\mathbf{h}_O\right) = I_O\ddot{\theta}\mathbf{e}_z = x\mathbf{e}_r \times -N_P\mathbf{e}_\theta = -xN_P\mathbf{e}_z, \qquad (9.60)$$

where I_O is the moment of inertia of the beam about O. From Eq. (9.60), we find

$$N_P = -\frac{\ddot{\theta}I_O}{x},$$

which is valid as long as $x \neq 0$.

Substituting the expression for N_P into Eq. (9.59) produces the equations of motion for this system:

$$\ddot{x} - x\dot{\theta}^2 = -g\sin\theta$$

$$\ddot{\theta}\left(x^2 + \frac{I_O}{m}\right) = -gx\cos\theta - 2x\dot{x}\dot{\theta}.$$

Example 9.22 Yo-Yo De-Spin

In the early days of the space program a technique was developed to de-spin satellites, that is, remove the satellite's angular momentum, without the use of thrusters or fuel. Dubbed "yo-yo de-spin," it involved two masses connected to cords and wrapped symmetrically around a cylindrical satellite. When released, the masses would unwind and carry away the angular momentum due to the spin of the satellite. This device is an excellent example of how a designer can use the conservation of total angular momentum and total energy of a system of particles and rigid bodies. It is still in use today.

Our planar model for the yo-yo device is shown in Figure 9.19. It consists of a circular satellite of radius R and moment of inertia I_G about its center of mass G. It has an attached body frame $\mathcal{B} = (G, \mathbf{b}_1, \mathbf{b}_2, \mathbf{b}_3)$ and is rotating about its center of mass with angular velocity

$$^{\mathcal{I}}\boldsymbol{\omega}^{\mathcal{B}} = {}^{\mathcal{I}}\omega^{\mathcal{B}}\mathbf{b}_3.$$

There are two particles, P and Q, each of mass m, attached to cords that unwind from the satellite. In Figure 9.19 they are shown after the cords have unwrapped a distance l. Because of the symmetry, and equal masses, the center of mass of the entire system is also at G, and the two particles unwind an equal length. We have also introduced a polar frame $\mathcal{C} = (G, \mathbf{e}_r, \mathbf{e}_\theta, \mathbf{c}_3)$, where the radial unit vector $\mathbf{e}_r$ is directed toward the contact point of the cord connected to P and the satellite. The angular velocity of $\mathcal{C}$ in $\mathcal{B}$ is thus the usual:

$$^{\mathcal{B}}\boldsymbol{\omega}^{\mathcal{C}} = \dot{\theta}\mathbf{c}_3.$$

We also assume that, at $t = 0$, the cords are completely wrapped around the satellite and $\mathcal{C}$ is aligned with $\mathcal{B}$. That is, $\theta(0) = 0$. Thus the instantaneous cord length is $l = R\theta$.

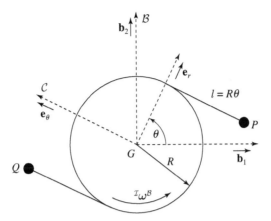

Figure 9.19 A yo-yo de-spin system consists of a cylindrical satellite and two masses, P and Q, attached to cords wrapped around the satellite.

Because there are no external forces or moments on the system, G is the origin of an inertial frame and the total angular momentum about G is conserved. The total angular momentum consists of the sum of the angular momenta of the satellite and the particles. From the kinematics of the problem, we have for particle P:

$$\mathbf{r}_{P/G} = R\mathbf{e}_r - R\theta\mathbf{e}_\theta$$

$$^\mathcal{I}\mathbf{v}_{P/G} = R^\mathcal{I}\boldsymbol{\omega}^\mathcal{C} \times \mathbf{e}_r - R\dot{\theta}\mathbf{e}_\theta - R\theta^\mathcal{I}\boldsymbol{\omega}^\mathcal{C} \times \mathbf{e}_\theta.$$

We use the addition property of angular velocities to find

$$^\mathcal{I}\boldsymbol{\omega}^\mathcal{C} = {}^\mathcal{I}\boldsymbol{\omega}^\mathcal{B} + {}^\mathcal{B}\boldsymbol{\omega}^\mathcal{C}$$

$$= ({}^\mathcal{I}\omega^\mathcal{B} + \dot{\theta})\mathbf{c}_3.$$

Substituting this expression into the previous one yields the inertial velocity of P:

$$^\mathcal{I}\mathbf{v}_{P/G} = R\theta({}^\mathcal{I}\omega^\mathcal{B} + \dot{\theta})\mathbf{e}_r + R^\mathcal{I}\omega^\mathcal{B}\mathbf{e}_\theta.$$

Similarly, we have for Q:

$$\mathbf{r}_{Q/G} = -R\mathbf{e}_r + R\theta\mathbf{e}_\theta$$

$$^\mathcal{I}\mathbf{v}_{Q/G} = -R\theta({}^\mathcal{I}\omega^\mathcal{B} + \dot{\theta})\mathbf{e}_r - R^\mathcal{I}\omega^\mathcal{B}\mathbf{e}_\theta.$$

The (constant) total angular momentum of the system about G is

$$^\mathcal{I}\mathbf{h}_G = I_G{}^\mathcal{I}\omega^\mathcal{B}\mathbf{c}_3 + m\mathbf{r}_{P/G} \times {}^\mathcal{I}\mathbf{v}_{P/G} + m\mathbf{r}_{Q/G} \times {}^\mathcal{I}\mathbf{v}_{Q/G},$$

which, after substituting for the position and velocities of P and Q, gives

$$^\mathcal{I}\mathbf{h}_G = \left(I_G{}^\mathcal{I}\omega^\mathcal{B} + 2mR^2\left[{}^\mathcal{I}\omega^\mathcal{B} + \theta^2({}^\mathcal{I}\omega^\mathcal{B} + \dot{\theta})\right]\right)\mathbf{c}_3.$$

Likewise, because there are no external or internal forces or moments, the total energy about G is equal to the total kinetic energy and is also a constant of the motion.

(The tension forces in the cords are constraint forces and therefore do no work.) The total kinetic energy is found from Eq. (9.52) to be

$$E_O = T_G = \frac{1}{2}I_G({}^{\mathcal{I}}\omega^B)^2 + \frac{1}{2}m\|{}^{\mathcal{I}}\mathbf{v}_{P/G}\|^2 + \frac{1}{2}m\|{}^{\mathcal{I}}\mathbf{v}_{Q/G}\|^2$$

$$= \frac{1}{2}I_G({}^{\mathcal{I}}\omega^B)^2 + mR^2\left[({}^{\mathcal{I}}\omega^B)^2 + \theta^2({}^{\mathcal{I}}\omega^B + \dot{\theta})^2\right].$$

Note that, because the total energy and angular momentum are conserved, they are equal to their values at $t = 0$, which implies

$${}^{\mathcal{I}}\mathbf{h}_G = (I_G + 2mR^2){}^{\mathcal{I}}\omega_0^B\mathbf{c}_3$$

$$E_O = \frac{1}{2}(I_G + 2mR^2)({}^{\mathcal{I}}\omega_0^B)^2,$$

where ${}^{\mathcal{I}}\omega_0^B = {}^{\mathcal{I}}\omega^B(t = 0)$.

The expressions for the angular momentum and energy at t and $t = 0$ can be equated to give the following two differential equations in the two unknowns $\dot{\theta}$ and ${}^{\mathcal{I}}\omega^B$:

$$c({}^{\mathcal{I}}\omega_0^B - {}^{\mathcal{I}}\omega^B) = \theta^2({}^{\mathcal{I}}\omega^B + \dot{\theta}) \tag{9.61}$$

$$c[({}^{\mathcal{I}}\omega_0^B)^2 - ({}^{\mathcal{I}}\omega^B)^2] = \theta^2({}^{\mathcal{I}}\omega^B + \dot{\theta})^2, \tag{9.62}$$

where

$$c = \frac{I_G}{2mR^2} + 1.$$

Dividing Eq. (9.62) by Eq. (9.61) gives

$${}^{\mathcal{I}}\omega_0^B + {}^{\mathcal{I}}\omega^B = {}^{\mathcal{I}}\omega^B + \dot{\theta},$$

which tells us that $\dot{\theta} = {}^{\mathcal{I}}\omega_0^B = $ constant. The cords unwrap at a constant rate equal to the initial spin rate of the satellite. Letting $\theta(t) = {}^{\mathcal{I}}\omega_0^B t$ and substituting into Eq. (9.61) gives the expression for the spin rate of the satellite as a function of time:

$${}^{\mathcal{I}}\omega^B(t) = {}^{\mathcal{I}}\omega_0^B\left(\frac{c - ({}^{\mathcal{I}}\omega_0^B)^2t^2}{c + ({}^{\mathcal{I}}\omega_0^B)^2t^2}\right). \tag{9.63}$$

If we wish to completely de-spin the satellite, so that ${}^{\mathcal{I}}\omega^B(t_f) = 0$, then we can use Eq. (9.63) to find the final time t_f for zero spin rate as a function of the initial spin rate and mass properties of the satellite and particles:

$$t_f = \frac{\sqrt{c}}{{}^{\mathcal{I}}\omega_0^B}.$$

This expression can be used to find the length of cord l_f necessary for complete de-spin. Using $l(t) = R\theta(t)$, we have

$$l_f = R\theta(t_f) = R^{\mathcal{I}}\omega_0^{\mathcal{B}} t_f = R\sqrt{\frac{I_G}{2mR^2} + 1}.$$

9.7 Tutorials

Tutorial 9.1 The Center of Percussion

This tutorial revisits the compound pendulum of Example 9.15. We solve it here using the separation of translational and angular momentum. Although this approach is more complicated than that applied in Example 9.15 (because it involves introducing and eliminating the reaction forces), it introduces the important concept of the *center of percussion*.

An important observation is that, for a simple pendulum with only a point mass at the end of a massless rod or string, the moment of inertia about the center of mass is zero (it has no extent, after all), and *the transverse reaction force is zero*. This is the case, though, only for the simple pendulum. For the more realistic physical pendulum, there is always a transverse reaction force that must be accounted for in the dynamics.

Consider again the compound pendulum in Figure 9.14. Here we find the equation of motion for θ by writing Euler's first law for the center of mass and then balancing angular momentum about the center of mass. The position of the center of mass, expressed as components in the body frame $\mathcal{B}$, is

$$\mathbf{r}_{G/O} = l\mathbf{b}_1.$$

The velocity is

$$^{\mathcal{I}}\mathbf{v}_{G/O} = \frac{^{\mathcal{I}}d}{dt}(l\mathbf{b}_1) = l^{\mathcal{I}}\boldsymbol{\omega}^{\mathcal{B}} \times \mathbf{b}_1 = l\dot{\theta}\mathbf{b}_2$$

and the acceleration is

$$^{\mathcal{I}}\mathbf{a}_{G/O} = \frac{^{\mathcal{I}}d}{dt}(l\dot{\theta}\mathbf{b}_2) = l\ddot{\theta}\mathbf{b}_2 - l\dot{\theta}^2\mathbf{b}_1.$$

Using the free-body diagram in Figure 9.14c, we can write Euler's first law for the center of mass,

$$F_n\mathbf{b}_1 + F_t\mathbf{b}_2 + m_G g \cos\theta\mathbf{b}_1 - m_G g \sin\theta\mathbf{b}_2 = m_G(l\ddot{\theta}\mathbf{b}_2 - l\dot{\theta}^2\mathbf{b}_1), \quad (9.64)$$

where the necessary reaction forces at the pin have been included and we have used the transformation array to express the gravitational force $m_G g\mathbf{e}_1$ in the pendulum body frame. Eq. (9.64) is equivalent to the two scalar equations:

$$F_n + m_G g \cos\theta = -m_G l\dot{\theta}^2 \quad (9.65)$$

$$F_t - m_G g \sin\theta = m_G l\ddot{\theta}. \quad (9.66)$$

Since there are three unknowns—$\ddot{\theta}$, F_n, and F_t—we need three equations. For the third equation we use the time derivative of the angular momentum about the center of mass, $^I\mathbf{h}_G = I_G\dot{\theta}\mathbf{b}_3$. The equation of motion for rotation about the center of mass is given by Eq. (9.19). The moment about the center of mass comes from the transverse reaction force at the pin,

$$\mathbf{M}_G = \mathbf{r}_{O/G} \times F_t\mathbf{b}_2 = -F_t l\mathbf{b}_3.$$

The rotational equation of motion is

$$\frac{^I d}{dt}\left(^I\mathbf{h}_G\right) = I_G\ddot{\theta}\mathbf{b}_3 = -F_t l\mathbf{b}_3$$

or, in scalar form,

$$I_G\ddot{\theta} = -F_t l. \tag{9.67}$$

Eqs. (9.65)–(9.67) provide three equations in the three unknowns F_t, F_n, and $\ddot{\theta}$. The equation of motion for $\ddot{\theta}$ is found by solving Eq. (9.67) for F_t and substituting into Eq. (9.66),

$$\ddot{\theta} + \frac{m_G g l}{I_G + m_G l^2}\sin\theta = 0.$$

This equation is, of course, identical to the equation of motion found in Example 9.15 (Eq. (9.36)). However, this approach has given us an expression for each of the two reaction forces:

$$F_n = -m_G(g\cos\theta + l\dot{\theta}^2)$$

$$F_t = -\frac{I_G\ddot{\theta}}{l} = \frac{m_G g I_G}{I_G + m_G l^2}\sin\theta.$$

We next explore a slight modification to the compound-pendulum problem. Suppose that an external force $F_A\mathbf{b}_2$ is applied to the pendulum at point P a distance l' from the attachment point O (below the center of mass G), as shown in Figure 9.20. This force is usually an impulsive force, but it need not be. In this case, the $\mathbf{b}_2$ translational equation in Eq. (9.66) is modified to become

$$F_t + F_A - m_G g\sin\theta = m_G l\ddot{\theta}. \tag{9.68}$$

The rotational equation is similarly modified. Here we find it easiest to return to the angular momentum equation of motion about the pinned point O, as in Example 9.15, but with an added moment due to the applied force (see Figure 9.20):

$$I_O\ddot{\theta} = m_G k_O^2\ddot{\theta} = F_A l' - m_G g\sin\theta. \tag{9.69}$$

We make one further simplification and drop gravity from the problem. For instance, we may consider the pendulum to be rotating horizontally or that the force F_A is very large compared to gravity and short enough that gravity has little effect on the

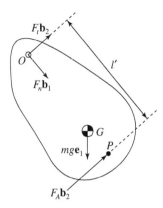

Figure 9.20 Free-body diagram of a compound pendulum with an applied force at point P a distance l' from attachment point O.

motion. In that case, Eqs. (9.68) and (9.69) can be combined to eliminate $\ddot{\theta}$, resulting in the ratio of transverse reaction force F_t to applied force F_A:

$$\frac{F_t}{F_A} = \frac{l'}{k_O^2/l} - 1 = \frac{l' - k_O^2/l}{k_O^2/l}, \tag{9.70}$$

where k_O is the radius of gyration defined in Definition 9.3. Recall that k_O^2/l is the length of the simple pendulum of mass m_G with equal oscillation frequency. Eq. (9.70) tells us something profound. If the external force is applied at a point P a distance $l' = k_O^2/l$ from the attachment point, there is no transverse reaction force! This point, the center of percussion, is easily found for any rigid body from the radius of gyration and the mounting point. A good batter in baseball knows where to find the center of percussion of a baseball bat; hitting a ball there feels effortless. Likewise, tennis rackets are designed with a "sweet spot" at the center of the face so that a proper swing will avoid a painful transverse reaction force.

Tutorial 9.2 A Dumbbell Satellite in a Circular Orbit

Chapters 4 and 7 presented examples that found the orbital motion of a satellite about a large gravitating body, such as the earth. In them we treated the satellite as a point mass acted on by a gravitational force given by Newton's universal law of gravity. A satellite is not, however, a point mass but an extended rigid body. It is thus acted on not only by a gravitational force at its center of mass but also by a gravitational moment about its center of mass due to the gradient of the gravitational field. In this tutorial we calculate the gravitational moment on a simple model of a satellite consisting of two point masses separated by a massless rod, as shown in Figure 9.21, and find its equation of motion. We often call this system a *dumbbell satellite*. It is a reasonable model of a satellite with a large central mass and an extended smaller mass, such as an antenna or sensor. In fact, many satellites are intentionally designed this way, often by including an extended mass on a long boom.

As in Example 4.9, we locate the center of mass G of the satellite using polar coordinates $(r, \theta)_{\mathcal{I}}$. We also introduce a polar frame $\mathcal{C} = (O, \mathbf{e}_r, \mathbf{e}_\theta, \mathbf{e}_z)$, where $\mathbf{e}_r$

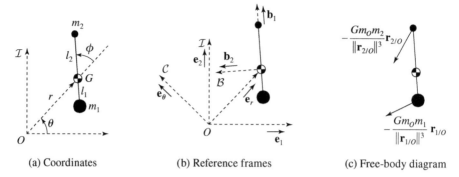

(a) Coordinates   (b) Reference frames   (c) Free-body diagram

Figure 9.21 Dumbbell satellite subject to a moment stemming from a gravitational gradient.

points from O to G. Thus the position of the center of mass in $\mathcal{I}$ can be expressed in terms of components in $\mathcal{C}$ as

$$\mathbf{r}_{G/O} = r\mathbf{e}_r.$$

In this example we are only interested in finding an expression for the external moment about G acting on the satellite. In particular, we find an approximate formula for the moment as a function of the satellite's orientation in $\mathcal{C}$. Figure 9.21a introduces the rotational angle ϕ describing the rotation of the satellite relative to the $\mathbf{e}_r$ direction. The transformation array between the body frame $\mathcal{B} = (G, \mathbf{b}_1, \mathbf{b}_2, \mathbf{b}_3)$ and polar frame $\mathcal{C}$ is

	$\mathbf{e}_r$	$\mathbf{e}_\theta$
$\mathbf{b}_1$	$\cos\phi$	$\sin\phi$
$\mathbf{b}_2$	$-\sin\phi$	$\cos\phi$.

We now write the moment on the satellite about G using the definition in Eq. (9.11), the free-body diagram in Figure 9.21c, and the universal law of gravity:

$$\mathbf{M}_G = \mathbf{r}_{1/G} \times \left(-\frac{Gm_O m_1}{\|\mathbf{r}_{1/O}\|^3} \mathbf{r}_{1/O} \right) + \mathbf{r}_{2/G} \times \left(-\frac{Gm_O m_2}{\|\mathbf{r}_{2/O}\|^3} \mathbf{r}_{2/O} \right). \quad (9.71)$$

We next substitute into Eq. (9.71) the two vector triads $\mathbf{r}_{1/O} = \mathbf{r}_{G/O} + \mathbf{r}_{1/G}$ and $\mathbf{r}_{2/O} = \mathbf{r}_{G/O} + \mathbf{r}_{2/G}$ and use the fact that a vector crossed into itself is zero to get

$$\mathbf{M}_G = -\frac{Gm_O m_1}{\|\mathbf{r}_{G/O} + \mathbf{r}_{1/G}\|^3} \mathbf{r}_{1/G} \times \mathbf{r}_{G/O} - \frac{Gm_O m_2}{\|\mathbf{r}_{G/O} + \mathbf{r}_{2/G}\|^3} \mathbf{r}_{2/G} \times \mathbf{r}_{G/O}. \quad (9.72)$$

From Figure 9.21b and the transformation array, we compute the two cross products,

$$\mathbf{r}_{1/G} \times \mathbf{r}_{G/O} = -l_1 \mathbf{b}_1 \times r\mathbf{e}_r = rl_1 \sin\phi \mathbf{b}_3$$

$$\mathbf{r}_{2/G} \times \mathbf{r}_{G/O} = l_2 \mathbf{b}_1 \times r\mathbf{e}_r = -rl_2 \sin\phi \mathbf{b}_3.$$

The expressions in the denominator are a bit more complicated. To simplify them we use the fact that the satellite is much smaller than its distance from the center of the earth: $l_1 \ll r$ and $l_2 \ll r$. This fact allows us to approximate the moment by dropping any terms of order $(l/r)^2$ and higher. We can expand the cubed magnitude in the denominator in Eq. (9.72) as

$$
\|\mathbf{r}_{G/O} + \mathbf{r}_{1/G}\|^3 = \left((\mathbf{r}_{G/O} + \mathbf{r}_{1/G}) \cdot (\mathbf{r}_{G/O} + \mathbf{r}_{1/G})\right)^{3/2}
$$

$$
= \left(r^2 + 2\mathbf{r}_{G/O} \cdot \mathbf{r}_{1/G} + l_1^2\right)^{3/2}
$$

$$
= r^3 \left(1 + 2\frac{\mathbf{r}_{G/O} \cdot \mathbf{r}_{1/G}}{r^2} + \frac{l_1^2}{r^2}\right)^{3/2}.
$$

We find a similar expression for the term involving $\mathbf{r}_{2/G}$. Since we are assuming that l_1/r and l_2/r are small, we can use the binomial expansion in Appendix A (Example A.1) to simplify the moment expression:

$$
\mathbf{M}_G \approx - \frac{Gm_O m_1}{r^2}(l_1 \sin \phi) \left(1 - 3\frac{\mathbf{r}_{G/O} \cdot \mathbf{r}_{1/G}}{r^2}\right) \mathbf{b}_3
$$

$$
+ \frac{Gm_O m_2}{r^2}(l_2 \sin \phi) \left(1 - 3\frac{\mathbf{r}_{G/O} \cdot \mathbf{r}_{2/G}}{r^2}\right) \mathbf{b}_3.
$$

The two leading terms cancel by the center-of-mass corollary (Eq. (6.15)). Using the transformation array, the two dot products are

$$
\mathbf{r}_{G/O} \cdot \mathbf{r}_{1/G} = -rl_1 \cos \phi
$$

$$
\mathbf{r}_{G/O} \cdot \mathbf{r}_{2/G} = rl_2 \cos \phi.
$$

The approximate moment about G is

$$
\mathbf{M}_G \approx -\frac{3}{2}\frac{Gm_O}{r^3}\left(m_1 l_1^2 + m_2 l_2^2\right) \sin 2\phi \mathbf{b}_3.
$$

We can simplify this result a bit further by recognizing that the quantity in parentheses is just the moment of inertia of the satellite about its center of mass: $I_G = m_1 l_1^2 + m_2 l_2^2$. The simplified expression is

$$
\mathbf{M}_G \approx -\frac{3}{2}\frac{Gm_O}{r^3}I_G \sin 2\phi \mathbf{b}_3. \tag{9.73}
$$

The relationship of the external gravitational moment to the moment of inertia turns out to hold for any symmetric satellite. You can explore this relation in Problem 9.3.

Note that the moment in Eq. (9.73) depends on the satellite's position coordinate r. Thus the rotational equations of motion do not completely separate from the translational equations of motion; the external moment depends on the position in orbit. However, recall that Example 4.9 considered a satellite in a circular orbit of radius r_0 and determined that the angular rate was constant and equal to $\dot{\theta}_0 = \sqrt{Gm_O/r_0^3}$. In that case, the moment on a two-mass satellite in a circular orbit reduces to

$$
\mathbf{M}_G \approx -\frac{3}{2}\dot{\theta}_0^2 I_G \sin 2\phi \mathbf{b}_3,
$$

which is independent of the translational motion.

To find the equation of motion for the satellite in a circular orbit, we first calculate its angular momentum about the center of mass:

$$^{\mathcal{I}}\mathbf{h}_G = I_G{}^{\mathcal{I}}\boldsymbol{\omega}^{\mathcal{B}} = I_G\left(^{\mathcal{I}}\boldsymbol{\omega}^{\mathcal{C}} + {}^{\mathcal{C}}\boldsymbol{\omega}^{\mathcal{B}}\right) = I_G\left(\dot{\theta}_0 + \dot{\phi}\right)\mathbf{b}_3.$$

Using Euler's second law in Eq. (9.19), we take the inertial derivative of $^{\mathcal{I}}\mathbf{h}_G$ and set it equal to the external moment about G to find the approximate equation of motion:

$$I_G\ddot{\phi} = -\frac{3}{2}\dot{\theta}_0^2 I_G \sin 2\phi,$$

where we have used the fact that, for a circular orbit, $\ddot{\theta} = 0$. Dividing both sides by I_G gives the final equation of motion:

$$\ddot{\phi} + \frac{3}{2}\dot{\theta}_0^2 \sin 2\phi = 0. \tag{9.74}$$

Eq. (9.74) looks just like the equation of motion for a simple pendulum! In fact, if we assume that the satellite starts aligned with $\mathbf{e}_r$ (the *local vertical*) and has only a small deviation ϕ, the equation of motion can be approximated using the small-angle formula:

$$\ddot{\phi} + 3\dot{\theta}_0^2\phi = 0.$$

This is the familiar equation for simple harmonic motion. The dumbbell satellite oscillates about the local vertical at a frequency of $\sqrt{3}$ times its orbit rate. This is why satellites are often designed to be long and thin or are equipped with a long boom and a mass at its end. If the goal is to keep it pointed toward the earth (e.g., to take scientific data or relay communications), the moment from the gravitational gradient will keep it roughly pointed in that direction.

Tutorial 9.3 The Falling Chimney

The falling chimney is a classic problem in dynamics. Figure 9.22a is a photograph of a famous industrial smokestack falling over in Glasgow, Scotland. Notice that it breaks as it falls. Our goal is to find the distance from the bottom of the chimney where it will break (if at all) and at what angle of fall.

We approach this problem by first solving for the equations of motion of the entire chimney falling as a rigid body pinned at its base. This problem is an inversion of that of the compound pendulum already solved but is otherwise identical. Figure 9.22b and c show the coordinate θ and the reference frames used. We "break" the chimney at some distance x from the bottom and treat each segment as a linked rigid body. We include the constraint forces—which in this case are the internal shear stress, normal stress, and moment in the chimney—and solve for the motion of each segment. By using the constraint that, before the break, the two segments must be moving together, we can solve for the internal forces. The chimney will break at the location of maximum stress.

We begin by treating the entire chimney as an inverted compound pendulum. (Note that we are ignoring the very slight error introduced because the chimney actually tips on its corner rather than on an attachment point through the centerline.) As in

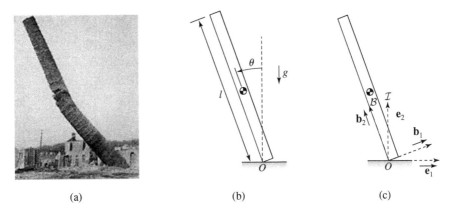

Figure 9.22 (a) A falling chimney. (b) Coordinates. (c) Reference frames.

Example 9.15, we sum moments about the attachment point O and set them equal to the change in angular momentum:

$$I_O \ddot{\theta} \mathbf{b}_3 = \mathbf{M}_O = mg \frac{l}{2} \sin \theta,$$

where we used the following transformation array between body frame $\mathcal{B}$ and inertial frame $\mathcal{I}$:

	$\mathbf{e}_1$	$\mathbf{e}_2$
$\mathbf{b}_1$	$\cos \theta$	$\sin \theta$
$\mathbf{b}_2$	$-\sin \theta$	$\cos \theta$.

As in the compound pendulum, the reaction forces at the base do not contribute to the moment about O.

The moment of inertia of a long rod about an axis through its center of mass is $I_G = ml^2/12$ (see Appendix D). Using the parallel axis theorem (Eq. (9.34)), the inertia about O is $I_O = ml^2/3$. We can then solve for the angular acceleration of the chimney:

$$\ddot{\theta} = \frac{3}{2} \frac{g}{l} \sin \theta. \tag{9.75}$$

Just as for the simple pendulum, inverting the compound pendulum changes the sign on the moment term, resulting in an unstable solution (which may be why it tips!).

Next we find the equations of motion for the upper segment of the chimney using the free-body diagram shown in Figure 9.23b. We apply the separation principle and solve for the motion of the center of mass and the motion about the center of mass. The density of the chimney is assumed to be uniform, so the mass of the upper segment is a linear fraction of the total mass: $(1 - x/l)m$. The translational equations of motion for the center of mass G of the upper segment are

$$\left(1 - \frac{x}{l}\right) m {}^{\mathcal{I}}\mathbf{a}_{G/O} = V\mathbf{b}_1 + N\mathbf{b}_2 - \left(1 - \frac{x}{l}\right) mg(\sin \theta \mathbf{b}_1 + \cos \theta \mathbf{b}_2), \tag{9.76}$$

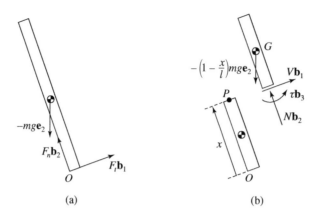

Figure 9.23 (a) Free-body diagram of the rigid rod. (b) Free-body diagram of the segment above the cut.

where V is the shear force on the segment, and N is the axial normal force. The rotational equation of motion about the center of mass due to the internal bending moment[4] $\boldsymbol{\tau} = \tau\mathbf{b}_3$ and the moment about G from the shear force is found using Eq. (9.19):

$$I_G\ddot{\theta} = \tau + V\frac{l}{2}\left(1 - \frac{x}{l}\right),\tag{9.77}$$

where the moment of inertia I_G of the segment about its center of mass is given by

$$I_G = \frac{1}{12}\left(1 - \frac{x}{l}\right)m(l - x)^2 = \frac{1}{12}ml^2\left(1 - \frac{x}{l}\right)^3.$$

We have also used the kinematic constraint that, before the chimney breaks, the rotation angle of the upper segment must be the same as the rotation angle of the entire chimney (modeled as a compound pendulum). In other words, the angular acceleration $\ddot{\theta}$ in Eq. (9.77) is the same as that in Eq. (9.75).

To solve for the shear and internal bending moment requires determining $^{\mathcal{I}}\mathbf{a}_{G/O}$ in Eq. (9.76). We find this quantity by using the kinematic constraint that the top segment must travel with the rest of the rod (as they are a solid piece). Thus point P at the top of the lower segment must have the same acceleration as the matching point on the upper segment. We thus use the results of Chapter 8. Starting with the lower segment, since P is fixed in the body frame attached to this segment and frame $\mathcal{B}$ is not translating, the inertial acceleration of P is found from Eq. (8.21) to be

$$^{\mathcal{I}}\mathbf{a}_{P/O} = {}^{\mathcal{I}}\boldsymbol{\alpha}^{\mathcal{B}} \times \mathbf{r}_{P/O} + {}^{\mathcal{I}}\boldsymbol{\omega}^{\mathcal{B}} \times \left({}^{\mathcal{I}}\boldsymbol{\omega}^{\mathcal{B}} \times \mathbf{r}_{P/O}\right).$$

Using the position of P, $\mathbf{r}_{P/O} = x\mathbf{b}_2$, the acceleration of P is

$$^{\mathcal{I}}\mathbf{a}_{P/O} = -x\ddot{\theta}\mathbf{b}_1 - x\dot{\theta}^2\mathbf{b}_2.\tag{9.78}$$

---

[4] The internal bending moment in a rod is an example of a pure torque. We thus use τ to represent it.

Next we find the acceleration of P on the upper segment. To do that, we take the derivative of the vector triad,

$$\mathbf{r}_{P/O} = \mathbf{r}_{G/O} + \mathbf{r}_{P/G},$$

and use the fact that $\mathbf{r}_{P/G} = -\frac{1}{2}(l - x)\mathbf{b}_2$ to find

$$^{\mathcal{I}}\mathbf{v}_{P/O} = {}^{\mathcal{I}}\mathbf{v}_{G/O} - \frac{l}{2}\left(1 - \frac{x}{l}\right)\frac{^{\mathcal{I}}d}{dt}\mathbf{b}_2$$

$$= {}^{\mathcal{I}}\mathbf{v}_{G/O} + \frac{l}{2}\left(1 - \frac{x}{l}\right)\dot{\theta}\mathbf{b}_1.$$

Taking the inertial derivative a second time gives

$$^{\mathcal{I}}\mathbf{a}_{P/O} = {}^{\mathcal{I}}\mathbf{a}_{G/O} + \frac{l}{2}\left(1 - \frac{x}{l}\right)\ddot{\theta}\mathbf{b}_1 + \frac{l}{2}\left(1 - \frac{x}{l}\right)\dot{\theta}^2\mathbf{b}_2. \tag{9.79}$$

Setting Eqs. (9.78) and (9.79) equal to each other gives a kinematic expression for the center-of-mass acceleration of the top segment:

$$^{\mathcal{I}}\mathbf{a}_{G/O} = -\frac{l}{2}\left(1 + \frac{x}{l}\right)\left(\ddot{\theta}\mathbf{b}_1 + \dot{\theta}^2\mathbf{b}_2\right).$$

We can now use this expression for the inertial acceleration of G in Eq. (9.76) to find the shear force. From the $\mathbf{b}_1$ component we have

$$V - \left(1 - \frac{x}{l}\right)mg\sin\theta = -\frac{l}{2}\left(1 + \frac{x}{l}\right)\left(1 - \frac{x}{l}\right)m\ddot{\theta}.$$

Substituting from the equation of motion for $\ddot{\theta}$ in Eq. (9.75) gives an expression for the shear force as a function of x and the angle of the chimney:

$$V = \frac{1}{4}\left(1 - \frac{x}{l}\right)\left(1 - 3\frac{x}{l}\right)mg\sin\theta. \tag{9.80}$$

Eqs. (9.76) and (9.77) and the shear force in Eq. (9.80) can be used to solve for the internal bending moment of the chimney:

$$\tau = \frac{x}{4}\left(1 - \frac{x}{l}\right)^2 mg\sin\theta. \tag{9.81}$$

Figure 9.24 plots the internal bending moment and shear force on the upper segment from Eqs. (9.80) and (9.81). From Newton's third law, the corresponding moment and shear on the lower segment are equal and opposite. A falling chimney is likely to fail in two possible ways, depending on the details of construction. As the maximum shear occurs at the base (and is equal to the tangent reaction force), many chimneys will fail by shearing off at the bottom. More likely, the chimney will fail in bending. Figure 9.24 shows that the maximum bending moment occurs at exactly

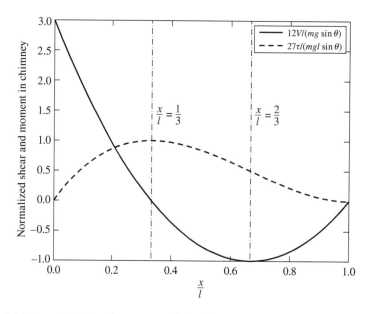

Figure 9.24 Normalized bending moment (dashed line) and shear force (solid line) on upper segment as functions of distance along chimney. (Note: the moment has been scaled so that the maximum occurs at 1 in the plot.)

1/3 the distance up from the base. The angle at which it fails depends on the strength of the chimney, as the moment at x increases sinusoidally with θ. A fun experiment is to build a tower of blocks and pull out the bottom; the tower should split apart at exactly 1/3 of the total number of blocks (it helps to use a multiple of three blocks).

It is also interesting to note that the moment on the lower segment is negative (because it is equal and opposite to that on the top segment). Thus a cross section of the chimney is subject to tension at its lower edge and compression at its upper edge as it falls. Since brick or cement chimneys are constructed to support compression, they will always fail at the lower edge first, causing the top segment to fall slightly backward, as seen in Figure 9.22a.

Finally, it is interesting to perform a somewhat more detailed analysis of the stress in the chimney on a cross section. The stress depends on the moment, the axial force, and the angle of fall. It turns out that, depending on the cross-section size, the length of the chimney, and the angle of rupture, failure occurs somewhere between 1/3 and 1/2 height of the chimney. That is why the failure in Figure 9.22a occurs closer to 1/2 than 1/3. You will explore this phenomenon in more depth in Problem 9.7.

9.8 Key Ideas

- A **rigid body** is a collection of particles rigidly attached to one another. That is, each particle is constrained to remain motionless relative to every other particle.

- The position of the **center of mass** of a continuous rigid body $\mathscr{B}$ with density $\rho(\mathbf{r}_{dm/O})$ is

$$\mathbf{r}_{G/O} = \frac{1}{m_G} \int_{\mathscr{B}} \mathbf{r}_{dm/O} \rho(\mathbf{r}_{dm/O}) dV.$$

- The center of mass of a rigid body obeys **Euler's first law**,

$$\mathbf{F}_G = m_G{}^{\mathcal{I}}\mathbf{a}_{G/O}.$$

- The angular momentum of a rigid body about an inertially fixed point O obeys **Euler's second law,**

$$\frac{{}^{\mathcal{I}}d}{dt}\left({}^{\mathcal{I}}\mathbf{h}_O\right) = \mathbf{M}_O.$$

Likewise, the angular momentum of a rigid body about its center of mass also obeys Euler's second law,

$$\frac{{}^{\mathcal{I}}d}{dt}\left({}^{\mathcal{I}}\mathbf{h}_G\right) = \mathbf{M}_G.$$

- The **moment of inertia** I_G of a planar rigid body about its center of mass G is the sum of the mass-weighted squared distance of each point on the body to the center of mass. For a rigid collection of N particles,

$$I_G \triangleq \sum_{i=1}^{N} m_i \|\mathbf{r}_{i/G}\|^2$$

and, for a continuous rigid body $\mathscr{B}$,

$$I_G \triangleq \int_{\mathscr{B}} \|\mathbf{r}_{dm/G}\|^2 dm.$$

- The **angular momentum** with respect to inertial frame $\mathcal{I}$ of a planar rigid body $\mathscr{B}$ about its center of mass G is

$$^{\mathcal{I}}\mathbf{h}_G = I_G{}^{\mathcal{I}}\boldsymbol{\omega}^{\mathcal{B}} = I_G{}^{\mathcal{I}}\omega^{\mathcal{B}}\mathbf{b}_3,$$

where $\mathcal{B}$ is a frame attached to $\mathscr{B}$.

- A **pure torque** $\boldsymbol{\tau}$ is a moment acting on a rigid body that does not arise from a specific, unique force acting at a known location.

- The **rotational dynamics of a planar rigid body** with constant inertia are

$$\frac{{}^{\mathcal{I}}d}{dt}\left({}^{\mathcal{I}}\mathbf{h}_G\right) = I_G{}^{\mathcal{I}}\alpha^{\mathcal{B}}\mathbf{b}_3 = \mathbf{M}_G.$$

For a set of discrete forces, the total moment is

$$\mathbf{M}_G \triangleq \sum_{i=1}^{N} \mathbf{r}_{i/G} \times \mathbf{F}_i^{(\text{ext})},$$

where $\mathbf{F}_i^{(\text{ext})}$ is the total external force acting on particle i. For a continuous field force acting on the body, we have

$$\mathbf{M}_G \triangleq \int_{\mathscr{B}} \mathbf{r}_{dm/G} \times \mathbf{f}_{dm}^{(\text{ext})} dm,$$

where $\mathbf{f}_{dm}$ is the force per unit mass on the body.

- The dynamics of a rigid body undergoing an **angular impulse** are

$$^{\mathcal{I}}\mathbf{h}_G(t_2) = {}^{\mathcal{I}}\mathbf{h}_G(t_1) + \overline{\mathbf{M}}_G(t_1, t_2),$$

where

$$\overline{\mathbf{M}}_{P/O}(t_1, t_2) \triangleq \int_{t_1}^{t_2} \mathbf{M}_{P/O} dt.$$

- The dynamics of a planar rigid body **about an arbitrary body point** Q are

$$\frac{^{\mathcal{I}}d}{dt}\left(^{\mathcal{I}}\mathbf{h}_Q\right) = I_Q{}^{\mathcal{I}}\alpha^{\mathcal{B}}\mathbf{b}_3 = \mathbf{M}_Q + \mathbf{r}_{Q/G} \times m_G{}^{\mathcal{I}}\mathbf{a}_{Q/O},$$

where

$$\mathbf{M}_Q = \mathbf{M}_G - \mathbf{r}_{Q/G} \times \sum_{i=1}^{N} \mathbf{F}_i^{(\text{ext})}.$$

- The **parallel axis theorem** relates the moment of inertia of a rigid body about an arbitrary body point Q to the moment of inertia about the center of mass G:

$$I_Q = I_G + m_G \|\mathbf{r}_{Q/G}\|^2.$$

- The **radius of gyration** k_Q relative to a point Q of a rigid body is the equivalent distance from Q of a particle of mass m_G needed to produce the same moment of inertia as the rigid body about Q:

$$m_G k_Q^2 \triangleq I_Q.$$

- The total **kinetic energy** of a rigid body both translating relative to an inertial origin O and rotating about its center of mass G is

$$T_O = T_{G/O} + T_G,$$

where the kinetic energy about the center of mass is

$$T_G = \frac{1}{2}I_G\|^{\mathcal{I}}\boldsymbol{\omega}^{\mathcal{B}}\|^2 = \frac{1}{2}{}^{\mathcal{I}}\mathbf{h}_G \cdot {}^{\mathcal{I}}\boldsymbol{\omega}^{\mathcal{B}}.$$

- The **total work** on a rigid body is

$$W = W_{G/O} + W_G = T_O(t_2) - T_O(t_1),$$

where $W_{G/O} = W_{G/O}(\mathbf{r}_{G/O}, \theta; \gamma_G, \gamma_\theta)$ is the work due to the total force acting on the rigid body along paths γ_G and γ_θ, and

$$W_G(\mathbf{r}_{G/O}, {}^{\mathcal{I}}\boldsymbol{\omega}^{\mathcal{B}}; \gamma_G, \gamma_\theta) = \int_{t_1}^{t_2} \mathbf{M}_G \cdot {}^{\mathcal{I}}\boldsymbol{\omega}^{\mathcal{B}} dt$$

is the work due to moments acting about the center of mass.

- External work satisfies the **conservation law** for kinetic energy of the center of mass and kinetic energy about the center of mass:

$$W_{G/O} = \Delta T_{G/O}$$

$$W_G = \Delta T_G.$$

- The **total energy** of a rigid body $\mathcal{B}$ is

$$E_O(t) \triangleq T_{G/O}(t) + T_G(t) + U_O(t),$$

where, for a set of discrete forces,

$$U_O(t) = \sum_{i=1}^{N} U_{i/O}(\mathbf{r}_{i/O}(t)),$$

and, for a continuous field force acting on the body,

$$U_O(t) = \int_{\mathcal{B}} U_{dm/O}(\mathbf{r}_{dm/O}) dm.$$

- The **work-energy formula** for a rigid body relates the change in total energy to the nonconservative work:

$$E_O(t_2) = E_O(t_1) + W_{G/O}^{(\mathrm{nc})} + W_G^{(\mathrm{nc})}.$$

9.9 Notes and Further Reading

We have taken a decidedly Newtonian approach in our development in this book. We have done so primarily for pedagogical reasons, as the material is more approachable and easier to learn by starting with particle mechanics, moving to multiparticle systems, and then on to rigid bodies. We have thus followed most contemporary physics and engineering texts in treating a rigid body as a collection of particles and then deriving Euler's laws. As pointed out in the text, this approach requires making an additional axiomatic assumption about internal moments of a rigid body that does not follow from Newton's third law. This difficulty caused Euler to abandon the particle or "corpuscular" model for rigid bodies entirely and postulate his laws of motion as axiomatic for rigid bodies, just as Newton's laws are for particles. This approach eliminates the need for any assumptions about the internal forces in a rigid body and makes the development of continuum mechanics more natural. It also has some logical appeal, since all bodies studied in dynamics are in fact rigid; particles are merely an abstraction. Thus Euler's laws can be considered more fundamental and general than Newton's. An excellent discussion of the history of Euler's treatment of rigid bodies and the nature of Euler's laws can be found in Truesdell (1968). Rao (2006) is an example of a recent text that presents Euler's laws as axiomatic rather than derivable from Newton's.

Only a few other texts make the distinction between moments and pure torques. Our notation came from Rao (2006) and Tenenbaum (2004). The dynamics of pool is a fascinating and complex subject, incorporating collisions, rigid-body dynamics, and complex models of friction. Fetter and Walecka (1980) has an insightful discussion (our pool ball example came from there). The yo-yo de-spin problem is a classic example in spacecraft attitude dynamics and control. It appears in many texts, including Kaplan (1976) and Wertz (1978). Our example closely follows that in Kaplan (1976). The falling chimney problem is also a classic problem in introductory courses, combining topics from both statics and dynamics. Variations can be found in many texts, including Bedford and Fowler (2002). There is a particularly interesting short paper by Varieschi and Kamiya (2003) that we used as a reference for both the tutorial and problem.

9.10 Problems

9.1   Find the position of the center of mass $\mathbf{r}_{G/O}$ of the planar rigid body in Figure 9.25. The rigid body has uniform density, with $m = 2$ kg, $r_1 = 0.3$ m, and $r_2 = 0.4$ m.

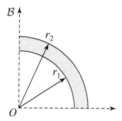

Figure 9.25 Problem 9.1.

9.2   Find the moment of inertia I_G of the planar rigid body in Figure 9.26. The rigid body has uniform density, with $m = 2$ kg, $w = 0.3$ m, and $h = 0.6$ m.

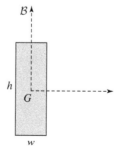

Figure 9.26 Problem 9.2.

9.3 Show, using the moment definition in Eq. (9.12), that the gravitational gradient moment on any long, axisymmetric satellite is given by Eq. (9.73).

9.4 Following Example 9.11, suppose the cue ball is hit dead center ($h = 0$). If the coefficient of sliding friction between the cue ball and the felt is μ, how long will the cue ball slide before it starts rolling without slipping?

9.5 Recall the sliding collar in Problem 3.20 in Chapter 3. It is shown again in Figure 9.27 with an added twist. Rather than being fixed, the shaft, with dimensions shown and mass M, is connected to the pivot point O by two massless rods and is allowed to swing in the plane like a pendulum. The collar has mass m (and can be treated as a point mass) and the shaft is a distance l away from O. The spring has spring constant k and its unstretched length is l. The position of the mass along the shaft is x and the angle of the perpendicular to the shaft with the local vertical is θ.

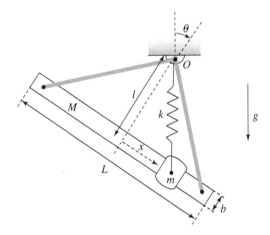

Figure 9.27 Problem 9.5.

a. How many degrees of freedom are there?
b. Show that the total angular momentum of the system is

$$^I\mathbf{h}_O = \left(\frac{M(L^2 + b^2)}{12} + (M + m)l^2 + mx^2\right)\dot{\theta}\mathbf{e}_z - ml\dot{x}\mathbf{e}_z.$$

Is $^I\mathbf{h}_O$ conserved?
c. Show that the equations of motion for the system are

$$\left(\frac{M(L^2 + b^2)}{12} + Ml^2 + mx^2\right)\ddot{\theta} + 2mx\dot{x}\dot{\theta} - mlx\dot{\theta}^2 + Mgl\sin\theta$$

$$-mgx\cos\theta + kxl\left(1 - \frac{l}{\sqrt{x^2 + l^2}}\right) = 0$$

$$\ddot{x} - l\ddot{\theta} - x\dot{\theta}^2 + \frac{k}{m}x\left(1 - \frac{l}{\sqrt{x^2 + l^2}}\right) - g\sin\theta = 0.$$

d. Is total energy conserved?

e. Suppose we start the system with the shaft at a positive angle of $\theta(0) = 30°$ and the point mass/collar at a positive stretch of $x(0) = 60$ cm. Setting $L = 2$ m, $l = 2$ m, $b = 10$ cm, $m = 0.25$ kg, $M = 10$ kg, $k = 1$ N/m, and $g = 9.8$ m/s^2, use MATLAB to simulate the motion of the shaft and the collar over the time interval $[0, 30]$ s. Plot x versus t, θ versus t, and $\|^{\mathcal{I}}\mathbf{h}_O\|$ versus t.

9.6 Consider the rotating disk with an offset slot, as shown in Figure 9.28. Inside the slot is a particle P of mass m connected to a spring with spring constant k. The disk has mass M and moment of inertia I about the center axis (not including the small mass m). The disk is free to rotate about its axis by an angle θ. You can assume that the unstretched length of the spring is zero and at $t = 0$ the spring is stretched by an amount x_0.

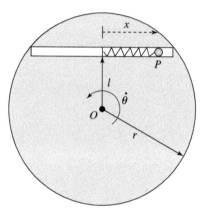

Figure 9.28 Problem 9.6.

a. Show that the equations of motion for the mass in the slot and the angle θ of the disk are

$$\ddot{\theta} = \frac{-2mx\dot{x}\dot{\theta} + mlx\dot{\theta}^2 - klx}{I + mx^2}$$

$$\ddot{x} = l\ddot{\theta} + x\dot{\theta}^2 - (k/m)x.$$

b. Is energy conserved?

c. What is the total angular momentum $^{\mathcal{I}}\mathbf{h}_O$ of the system? Is it conserved?

d. Suppose we start the system by offsetting the mass an amount $x_0 = 60$ cm. If the disk has a radius of 1 m, $I = 2.5$ kg-m^2, $l = 75$ cm, $m = 0.25$ kg, and $k = 1$ N/m, use MATLAB to simulate the motion of the mass and the disk over the time interval $[0, 20]$ s. Plot x versus t, θ versus t, and $\|^{\mathcal{I}}\mathbf{h}_O\|$ versus t.

9.7 In this problem you examine a slightly more realistic model for breakage of the falling chimney in Tutorial 9.3. You may recall from a statics class that the stress in a cross section can be written as a function of the axial force and bending moment. In particular, the maximum stress at the edges of the cross section (in tension and compression) is

$$\sigma = \frac{N}{A} \mp \frac{a\tau}{2J},$$

where A is the cross sectional area, a is the side length of a square cross section, and J is the area moment of inertia, $J = a^4/12$. Our calculation of where the chimney will break in Tutorial 9.3 was a bit simplified. Rather than failing at the maximum shear or bending moment point, it will in fact fail at the point of maximum internal stress (in particular, tension on the lower edge).

    a. As we did for shear and moment in Tutorial 9.3, find an expression for the axial force in the chimney as a function of the position x along the chimney.

    b. Find an expression for the internal stress of the chimney as a function of position x along the chimney, angle of fall θ, and aspect ratio l/a.

    c. Plot the stress in the chimney versus distance along the chimney for various fall angles and aspect ratios. Show that, depending on the aspect ratio (tall thin chimneys versus short fat ones), it will fail either at 1/3 or 1/2 the distance along the chimney.

9.8   Using the results of Problem 9.2, consider the compound crane illustrated in Figure 9.29. A compound pendulum is attached to mass M at point Q. Mass $M = 2$ kg slides horizontally without friction. The distance between Q and G is 0.25 m.

    a. Find the equations of motion in x and θ.

    b. Integrate the equations of motion in MATLAB, and create a MATLAB file that animates the solution.

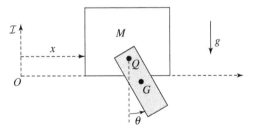

Figure 9.29 Problem 9.8.

9.9   Recall the simple double pendulum explored in Example 8.9. Repeat the derivation of the equations of motion for the double pendulum, but now model it as two connected rigid bodies with moments of inertia (about their centers of mass) I_1 and I_2, masses m_1 and m_2, and lengths l_1 and l_2. Solve for the equations of motion, assuming the links are both thin rods.

9.10   A thin rigid cylindrical rod is leaned against a wall as in Figure 9.30, with its other end resting on the floor. The rod slides (without friction) under the force

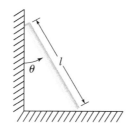

Figure 9.30 Problem 9.10.

of gravity. Find the equations of motion for the angle θ that the rod makes with the wall.

9.11   For the sliding rod in Problem 9.10, find the angle θ at which the rod loses contact with the wall, assuming the rod starts sliding with $\theta(0) = 0$ and $\dot{\theta}(0)$ small.

9.12   In Figure 9.31 a flatbed truck of mass M, containing a crane modeled as a long thin rod pivoted at its end, suddenly accelerates forward. Assume the rod is 2 m long, weighs 50 kg, and leans at an angle θ of 60° from the flatbed.

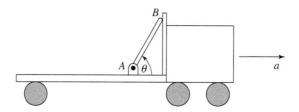

Figure 9.31 Problem 9.12.

   a. If the acceleration is 5 m/s², what is the normal force on the crane rod at B?

   b. What is the largest possible acceleration a before the rod will lose contact with the truck at B?

9.13   Consider a wheel of radius R_2, illustrated in Figure 9.32. String is wrapped around a concentric spool of radius R_1, making an angle of α with the horizon-

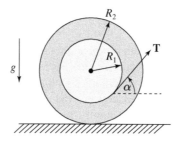

Figure 9.32 Problem 9.13.

tal. When tension **T** is applied to the string, the wheel rolls without slipping in a direction (left or right) determined by the angle α. Find the critical angle α^* above which the wheel rolls to the left and below which it rolls to the right.

9.14 A uniformly dense marble of mass $m = 0.05$ kg and radius $R = 0.01$ m is released from rest at the top of a ramp and rolls without slipping down the ramp and off a table, as shown in Figure 9.33. Find the distance d from the foot of the table where the marble lands on the floor. The ramp height is $h_1 = 0.2$ m, and its width is $w_1 = 0.4$ m. The marble rolls a distance of $w_2 = 0.15$ m on the table, which has height $h_2 = 1$ m. [HINT: Consider both the rotational and translational motion of the rolling marble.]

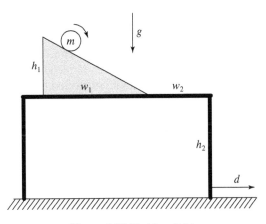

Figure 9.33 Problem 9.14.

9.15 A marble rolls without slipping on a walkway that accelerates horizontally with constant acceleration a relative to an inertially fixed point O, as shown in Figure 9.34. Find, in terms of a, the acceleration of the marble's center of mass G relative to O. For the moment of inertia of the marble, assume $I_G = (2/5)mR^2$, where m and R are the mass and radius of the marble, respectively. [HINT: Consider both the translational and rotational dynamics of the marble.]

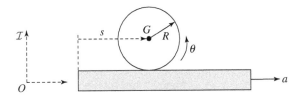

Figure 9.34 Problem 9.15.

9.16 A wheelchair of mass m rolls without slipping up a ramp of angle ϕ, as shown in Figure 9.35. Find the horizontal force **T** necessary to roll the wheelchair up the ramp at a constant speed. Assume that force acts at a distance r from the center of the wheel, which has radius R. Also assume that the center of mass G of the wheelchair is located in line with the center of the wheels.

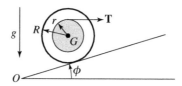

Figure 9.35 Problem 9.16.

9.17 Returning to the particle on a beam system of Example 9.21, replace the particle with a flat disk that rolls on the beam without slipping.

    a. How many degrees of freedom are there in this modified system?
    b. Find the equations of motion of the disk and beam.

9.18 Sketch a model of two trapeze artists on a trapeze using two coupled rigid rods (see Figure 9.36). How many degrees of freedom does this system have? Select coordinates and find the equations of motion for each artist. [HINT: Consider a double-pendulum model.] Choosing reasonable values for the parameters in your model, integrate the equations of motion in MATLAB and plot the solution for several different initial conditions. Be sure to label your axes and add a legend to each plot.

Figure 9.36 Problem 9.18.

9.19 A uniform disk rotor with moment of inertia I_G, similar to the flywheel in Figure 9.16, is connected to a DC motor with input voltage V. The current i in the armature circuit is governed by Kirchhoff's current law and can be expressed as

$$L\frac{di}{dt} + Ri + k\dot{\theta} - V = 0,$$

where L is the electric inductance, R the resistance, k the electromotive force constant, and $\dot{\theta}$ is the rotation rate of the rotor. The torque applied to the rotor by the motor is

$$\tau = ki,$$

and the mechanical linkage between the motor and rotor has a damping ratio b.

a. Show that the equation governing the motion of the rotor is

$$I_G \ddot{\theta} = -b\dot{\theta} + ki.$$

b. Using MATLAB, integrate the two governing equations of this system. Set the rotor moment of inertia I_G to 0.01 kg-m^2, the damping ratio b to 0.2 N-m-s, the resistance R to 1 ohm, the inductance L to 0.75 H, and the electromotive force constant k to 0.03 N-m/amp. Plot the rotor torque as a function of time for 5 s after a constant voltage of 1 V is applied to a previously turned off motor (i.e., all initial conditions are zero).

c. Repeat the previous part, but now plot the torque as a function of time when the voltage is ramped from 0 to 1 V over a 5 s period (integrate for at least 10 s).

9.20 Two identical particles of mass m are suspended on a rigid rod of length l to form a dynamic pendulum, as shown in Figure 9.37. Particle P is fixed to one end of the rod; the other end is attached to a fixed pivot. Particle Q, which is free to slide up and down the rod, is connected to P by a spring of constant k and rest length l_0.

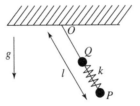

Figure 9.37 Problem 9.20.

a. Find the internal potential energy $U_O^{(\text{int})}$ of the system.
b. Find the external potential energy $U_O^{(\text{ext})}$ of the system.
c. Find the total energy $E_O^{(\text{tot})}$ of the system.
d. Is the total energy conserved?

9.21 The Cub Scouts has an annual competition called the Pinewood Derby. Each scout builds a wooden car that will race down an inclined track (inclined at some angle θ) powered only by gravity. The goal is to build a car that will make it down the slope as fast as possible. The total mass of the car (including the wheels) must be $\leq M$, and the radius of the wheels of the car must be R, though the wheel weight can be any fraction of the total you like. In this problem you will investigate whether it is better to make the wheels heavy or light (with ballast added to the car body).

Derive an expression for the time it takes the car to reach a certain position x down the incline in terms of R, θ, g, M, and the mass m_w of one of the four wheels. From this expression, make a prediction as to which type of car (one with heavier or lighter wheels) will get to the bottom of the track fastest. Ignore air resistance and assume that each of the four wheels rolls without slipping while going down the track.

9.22 A thin uniform rod of mass m_r and length l is fitted with a small collar of mass m_c that can slide freely (without friction), as shown in Figure 9.38. A motor at the rod's attachment point O rotates the rod with a constant torque τ. Assume the system is acting in plane (so that gravity can be neglected).

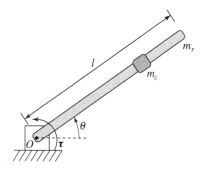

Figure 9.38 Problem 9.22.

a. Find the equations of motion of the collar.
b. Using MATLAB, integrate the equations of motion found in part (a). Let $\tau = 1$ N-m, $l = 1$ m, $m_c = 1$ kg, and $m_r = 1/3$ kg. Start the system from rest with $\theta = 0$ and with the collar 0.1 m from O and integrate for 1 s. Plot the collar's position along the length of rod and the rod's angle θ versus time.
c. Integrate the system in part (b) until the collar leaves the rod and find the value of θ and the collar's velocity at this time.
d. Repeat your integrations for at least two other values of τ between 1 and 3 N-m, and find the value of θ at which the collar leaves the rod in each instance. Explain your findings.

PART FOUR

Dynamics in Three Dimensions

CHAPTER TEN

Particle Kinematics and Kinetics in Three Dimensions

In this part we reach the final stage of our discussion of Newtonian dynamics. At last we study the full three-dimensional motion of particles and rigid bodies. This chapter reconsiders the kinematics of a particle, allowing motion in all three dimensions. While many of our two-dimensional results apply with only minor modifications or some new understanding, we will need to develop new and more sophisticated approaches to describing orientation in three dimensions. The next chapter undertakes the more challenging task of generalizing the motion of rigid bodies to rotation in three dimensions.

10.1 Two New Coordinate Systems

Chapter 3 introduced the concepts of vectors, coordinate systems, and reference frames and described how to find the velocity and acceleration of a particle in terms of different coordinate systems and in different frames. Chapter 8 introduced the study of the relative motion of a particle in a translating and rotating frame. Both chapters treated motion in only two dimensions. We now relax that restriction. This chapter repeats our entire treatment of particle dynamics, allowing motion in all three dimensions. We start by introducing two new coordinate systems—cylindrical and spherical coordinates[1]—that can be used to describe the configuration of a particle. But first we reprise our fundamental coordinate system—Cartesian coordinates—to set the framework for the three-dimensional analysis.

10.1.1 Cartesian Coordinates in Three Dimensions

Recall that we can locate a particle in an inertial reference frame most directly by using Cartesian coordinates $(x, y, z)_{\mathcal{I}}$:

$$\boxed{\mathbf{r}_{P/O} = x\mathbf{e}_x + y\mathbf{e}_y + z\mathbf{e}_z,}$$

---

[1] We do not address three-dimensional path coordinates, as this is a more advanced subject.

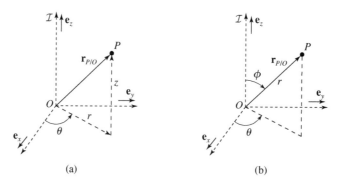

Figure 10.1 (a) Cylindrical $(r, \theta, z)_{\mathcal{I}}$ and (b) spherical $(r, \theta, \phi)_{\mathcal{I}}$ coordinates locate point P in reference frame $\mathcal{I}$ in three dimensions.

where $\mathcal{I} = (O, \mathbf{e}_x, \mathbf{e}_y, \mathbf{e}_z)$. The inertial velocity and acceleration of P are found from vector differentiation and Definition 3.3:

$$^{\mathcal{I}}\mathbf{v}_{P/O} = \frac{^{\mathcal{I}}d}{dt}\left(\mathbf{r}_{P/O}\right) = \dot{x}\mathbf{e}_x + \dot{y}\mathbf{e}_y + \dot{z}\mathbf{e}_z$$

$$^{\mathcal{I}}\mathbf{a}_{P/O} = \frac{^{\mathcal{I}}d}{dt}\left(^{\mathcal{I}}\mathbf{v}_{P/O}\right) = \ddot{x}\mathbf{e}_x + \ddot{y}\mathbf{e}_y + \ddot{z}\mathbf{e}_z.$$

Finding the equations of motion in Cartesian coordinates involves applying Newton's second law to each of the three directions, just as in Chapter 3.

What about other coordinate systems? As in the planar case, it is often convenient to use another coordinate system to describe the position of P in $\mathcal{I}$ (usually because of constraints in the problem). The two most common three-dimensional coordinate systems other than Cartesian coordinates are the cylindrical coordinates shown in Figure 10.1a and the spherical coordinates shown in Figure 10.1b.

10.1.2 Cylindrical Coordinates

Cylindrical coordinates $(r, \theta, z)_{\mathcal{I}}$ are a simple generalization of polar coordinates. All previous results for polar coordinates apply to the components of motion in the plane defined by $\mathbf{e}_x$ and $\mathbf{e}_y$; we simply add the third Cartesian coordinate z in the $\mathbf{e}_z$ direction and apply Newton's second law in that direction. Thus the position of P with respect to O in cylindrical coordinates is

$$\mathbf{r}_{P/O} = r\cos\theta\mathbf{e}_x + r\sin\theta\mathbf{e}_y + z\mathbf{e}_z \tag{10.1}$$

and the inertial kinematics expressed as components in $\mathcal{I}$ are

$$^{\mathcal{I}}\mathbf{v}_{P/O} = (\dot{r}\cos\theta - r\dot{\theta}\sin\theta)\mathbf{e}_x + (\dot{r}\sin\theta + r\dot{\theta}\cos\theta)\mathbf{e}_y + \dot{z}\mathbf{e}_z \tag{10.2}$$

$$^{\mathcal{I}}\mathbf{a}_{P/O} = (\ddot{r}\cos\theta - 2\dot{r}\dot{\theta}\sin\theta - r\ddot{\theta}\sin\theta - r\dot{\theta}^2\cos\theta)\mathbf{e}_x$$

$$+ (\ddot{r}\sin\theta + 2\dot{r}\dot{\theta}\cos\theta + r\ddot{\theta}\cos\theta - r\dot{\theta}^2\sin\theta)\mathbf{e}_y + \ddot{z}\mathbf{e}_z. \tag{10.3}$$

Compare these equations to Eqs. (3.18) and (3.19). They are identical except for the addition of the rates in the z-direction; we thus solve dynamics problems in cylindrical coordinates just as in polar coordinates and all previous results apply. The complexity of these equations, however, leads us again to ask whether a different frame would further simplify the kinematics. The answer is yes, but we defer that discussion until after introducing spherical coordinates.

10.1.3 Spherical Coordinates

The spherical coordinates $(r, \theta, \phi)_{\mathcal{I}}$ illustrated in Figure 10.1b are the three-dimensional analogue of polar coordinates. They consist of the radial distance r of point P from O, the *azimuth angle* θ, and the *polar angle* ϕ. Simple geometric considerations give the components of the vector $\mathbf{r}_{P/O}$ expressed in terms of spherical coordinates:

$$\mathbf{r}_{P/O} = r \cos \theta \sin \phi \mathbf{e}_x + r \sin \theta \sin \phi \mathbf{e}_y + r \cos \phi \mathbf{e}_z. \tag{10.4}$$

Just as for Cartesian and cylindrical coordinates, the velocity and acceleration of P relative to $\mathcal{I}$ can be found by differentiating Eq. (10.4) and using the definition of the vector derivative and the chain rule:

$$
\begin{aligned}
{}^{\mathcal{I}}\mathbf{v}_{P/O} = {}&(\dot{r} \sin \phi \cos \theta + r\dot{\phi} \cos \phi \cos \theta - r\dot{\theta} \sin \phi \sin \theta)\mathbf{e}_x \\
&+ (\dot{r} \sin \phi \sin \theta + r\dot{\phi} \cos \phi \sin \theta + r\dot{\theta} \sin \phi \cos \theta)\mathbf{e}_y \\
&+ (\dot{r} \cos \phi - r\dot{\phi} \sin \phi)\mathbf{e}_z
\end{aligned}
\tag{10.5}
$$

$$
\begin{aligned}
{}^{\mathcal{I}}\mathbf{a}_{P/O} = {}&(\ddot{r} \sin \phi \cos \theta + r\ddot{\phi} \cos \phi \cos \theta - r\ddot{\theta} \sin \phi \sin \theta + 2\dot{r}\dot{\phi} \cos \phi \cos \theta \\
&- 2\dot{r}\dot{\theta} \sin \phi \sin \theta - r\dot{\phi}^2 \sin \phi \cos \theta - r\dot{\theta}^2 \sin \phi \cos \theta - 2r\dot{\phi}\dot{\theta} \cos \phi \sin \theta)\mathbf{e}_x \\
&+ (\ddot{r} \sin \phi \sin \theta + r\ddot{\phi} \cos \phi \sin \theta + r\ddot{\theta} \sin \phi \cos \theta + 2\dot{r}\dot{\phi} \cos \phi \sin \theta \\
&+ 2\dot{r}\dot{\theta} \sin \phi \cos \theta - r\dot{\phi}^2 \sin \phi \sin \theta - r\dot{\theta}^2 \sin \phi \sin \theta + 2r\dot{\phi}\dot{\theta} \cos \phi \cos \theta)\mathbf{e}_y \\
&+ (\ddot{r} \cos \phi - r\ddot{\phi} \sin \phi - 2\dot{r}\dot{\phi} \sin \phi - r\dot{\phi}^2 \cos \phi)\mathbf{e}_z.
\end{aligned}
\tag{10.6}
$$

These equations can be used in Newton's second law to find equations of motion for the three spherical coordinates of a particle. We rarely do so, however, since these expressions are even more complicated and algebraically messy than their polar-coordinate counterparts. This strongly motivates us, again, to examine whether the introduction of a new reference frame would simplify things. We do so after a couple of simple examples.

Example 10.1 The Spherical Pendulum

Chapters 3 and 4 introduced the concepts of coordinate systems, vector derivatives, and equations of motion by examining the simple pendulum with varying levels of sophistication. We do the same here using the *spherical pendulum,* shown in Figure 10.2. The pendulum mass is a distance l away from the attachment point, just as in the simple pendulum, only here it is free to swing in two directions. It is thus a two-degree-of-freedom problem, which requires two coordinates to describe the configuration.

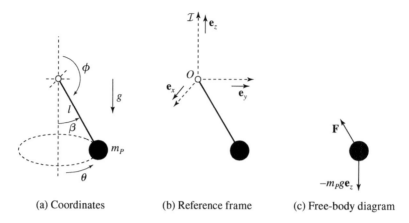

(a) Coordinates   (b) Reference frame   (c) Free-body diagram

Figure 10.2 Spherical pendulum using spherical coordinates $(r, \theta, \phi)_{\mathcal{I}}$.

The geometry of the pendulum is such that it is naturally described in terms of the spherical coordinates $(r = l, \theta, \phi = \pi - \beta)_{\mathcal{I}}$, as in Figure 10.2a. It is thus sensible to find equations of motion for the two angles θ and ϕ. We could do so by combining the free-body diagram with the acceleration from Eq. (10.6), as done in Chapter 3 for the simple pendulum. However, as stated above, this is rarely done because of the complexity of isolating the two coordinates. We wait, therefore, until after we have introduced some three-dimensional rotating frames.

Example 10.2 Latitude and Longitude

A good example of applying spherical coordinates is the use of latitude and longitude to locate a point on the surface of the earth. Figure 10.3 shows a point P located at coordinates $(\lambda, L)_{\mathcal{G}}$ on a spherical earth. The longitude λ corresponds to the spherical angle θ. The latitude L is the complement of the spherical angle ϕ, so that $L = \pi/2 - \phi$ (ϕ is sometimes referred to as the co-latitude).

The frame $\mathcal{G} = (O, \mathbf{g}_1, \mathbf{g}_2, \mathbf{g}_3)$ in which these spherical coordinates are defined is located at the center of the earth O and fixed to the rotating surface. $\mathcal{G}$ is called a *geographic reference frame.* This frame is defined by directions on the earth that provide the zero reference for latitude and longitude. By international agreement, $\mathbf{g}_3$ is along the spin-axis direction and $\mathbf{g}_1$ is in the equatorial plane and directed toward the *zero meridian,* which is the great circle that passes through Greenwich, England. The unit vector $\mathbf{g}_2$ completes the right-handed set.

Because $(\lambda, L)_{\mathcal{G}}$ are not exactly spherical coordinates, it is worthwhile to use Eq. (10.4) to find the position of P with respect to O in terms of the latitude and longitude. Using R for the radius of the earth, we have

$$\mathbf{r}_{P/O} = R \cos \lambda \cos L \mathbf{g}_1 + R \sin \lambda \cos L \mathbf{g}_2 + R \sin L \mathbf{g}_3.$$

Of course, the geographic frame is not an inertial frame because the earth is rotating in absolute space. It has a simple angular velocity with respect to inertial frame $\mathcal{I}$, also located at O and fixed to the distant stars, which implies

$$^{\mathcal{I}}\boldsymbol{\omega}^{\mathcal{G}} = \Omega_E \mathbf{g}_3.$$

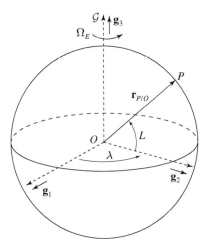

Figure 10.3 Spherical earth with latitude and longitude locating point P on the surface.

The angular velocity $^{\mathcal{I}}\boldsymbol{\omega}^{\mathcal{G}}$ of $\mathcal{G}$ with respect to $\mathcal{I}$ is directed along an axis through the north pole. We call the inertial frame centered at O the *geocentric inertial frame*. The fact that $\mathcal{G}$ is not inertial has interesting consequences when trying to solve for the equations of motion of particles traveling with respect to it (e.g., airplanes or ships). Tutorial 10.1 explores these consequences.

10.2 The Cylindrical and Spherical Reference Frames

This section introduces two reference frames that are three-dimensional equivalents of the polar frame in two dimensions. These frames rotate in $\mathcal{I}$ to stay aligned with a particle and result in compact and convenient vector expressions for the position, velocity, and acceleration, just as with the polar frame in two dimensions. Look back over Section 3.4. We had two objectives there: (a) to write the velocity and acceleration as components in a new, more convenient frame and (b) to use that frame to find a streamlined approach to calculating the velocity and acceleration via the angular velocity. The same motivations hold here. We just need to be much more careful in treating the angular velocity.

10.2.1 The Cylindrical Frame

Cylindrical frame $\mathcal{B}_c = (O, \mathbf{e}_r, \mathbf{e}_\theta, \mathbf{e}_z)$ (Figure 10.4a) shares much in common with the polar frame. Here, however, the unit vector $\mathbf{e}_r$ does not always point at particle P but rather at the projection of the position of P onto the plane spanned by $\mathbf{e}_x$ and $\mathbf{e}_y$. Thus the unit vectors $\mathbf{e}_r$ and $\mathbf{e}_\theta$ coincide with the polar-frame unit vectors. The $\mathbf{e}_z$ unit vector coincides with the corresponding inertial frame unit vector.

Just as in the planar discussion, we can write down a transformation table that allows us to write the unit vectors of one frame in terms of the unit vectors of another

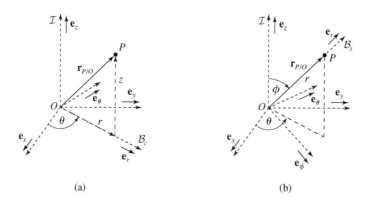

Figure 10.4 Cylindrical and spherical reference frames, $\mathcal{B}_c = (O, \mathbf{e}_r, \mathbf{e}_\theta, \mathbf{e}_z)$ and $\mathcal{B}_s = (O, \mathbf{e}_\phi, \mathbf{e}_\theta, \mathbf{e}_r)$. (a) Cylindrical frame. (b) Spherical frame.

frame. Since the $\mathbf{e}_z$ unit vectors of the two frames coincide, the three-dimensional generalization of the transformation table is

	$\mathbf{e}_x$	$\mathbf{e}_y$	$\mathbf{e}_z$
$\mathbf{e}_r$	$\cos\theta$	$\sin\theta$	0
$\mathbf{e}_\theta$	$-\sin\theta$	$\cos\theta$	0
$\mathbf{e}_z$	0	0	1 .

$$(10.7)$$

This is a three-dimensional version of our vector transformation table, in which one unit-vector direction of the two frames coincides. *Every transformation table that results from a simple rotation about a single axis resembles Eq. (10.7);* that is, it looks like a two-dimensional table with a new row and a new column containing zeros and one.

Eq. (10.7) can be used to express the kinematics of P in Eqs. (10.1)–(10.3) as components in the cylindrical frame rather than the inertial one. Using Eqs. (10.1)–(10.3) and Eq. (10.7), we obtain the following expressions:

$$\mathbf{r}_{P/O} = r\mathbf{e}_r + z\mathbf{e}_z \qquad (10.8)$$

$$^{\mathcal{I}}\mathbf{v}_{P/O} = \dot{r}\mathbf{e}_r + r\dot{\theta}\mathbf{e}_\theta + \dot{z}\mathbf{e}_z \qquad (10.9)$$

$$^{\mathcal{I}}\mathbf{a}_{P/O} = (\ddot{r} - r\dot{\theta}^2)\mathbf{e}_r + (2\dot{r}\dot{\theta} + r\ddot{\theta})\mathbf{e}_\theta + \ddot{z}\mathbf{e}_z. \qquad (10.10)$$

These expressions should come as no surprise, given the relationship between cylindrical and polar coordinates. Note that, as in Section 3.4.2, Eqs. (10.8)–(10.10) could also have been derived using $^{\mathcal{I}}\boldsymbol{\omega}^{\mathcal{B}_c} = \dot{\theta}\mathbf{e}_z$ and the transport equation (Definition 8.1) to differentiate $\mathbf{r}_{P/O}$ and $^{\mathcal{I}}\mathbf{v}_{P/O}$.

10.2.2 The Spherical Frame

The spherical frame $\mathcal{B}_s = (O, \mathbf{e}_\phi, \mathbf{e}_\theta, \mathbf{e}_r)$ is shown in Figure 10.4b. It consists of a unit vector $\mathbf{e}_r$ directed along $\mathbf{r}_{P/O}$ toward P, a unit vector $\mathbf{e}_\phi$ orthogonal to $\mathbf{e}_r$ and in the plane defined by $\mathbf{e}_r$ and $\mathbf{e}_z$, and a unit vector $\mathbf{e}_\theta = \mathbf{e}_r \times \mathbf{e}_\phi$ completing the right-

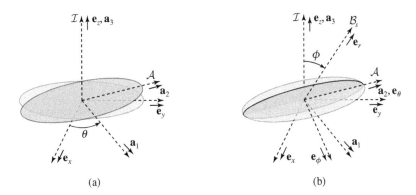

(a)   (b)

Figure 10.5 Relating spherical frame $\mathcal{B}_s = (O, \mathbf{e}_\phi, \mathbf{e}_\theta, \mathbf{e}_r)$ to inertial frame $\mathcal{I} = (O, \mathbf{e}_x, \mathbf{e}_y, \mathbf{e}_z)$ by means of two rotations by θ and ϕ, respectively, and intermediate frame $\mathcal{A} = (O, \mathbf{a}_1, \mathbf{a}_2, \mathbf{a}_3)$. (a) θ rotation about $\mathbf{e}_x$. (b) ϕ rotation about $\mathbf{a}_2$.

handed set (which implies that $\mathbf{e}_\phi = \mathbf{e}_\theta \times \mathbf{e}_r$ and $\mathbf{e}_r = \mathbf{e}_\phi \times \mathbf{e}_\theta$). How do we describe the orientation of $\mathcal{B}_s$ in $\mathcal{I}$ and relate the unit vectors defining $\mathcal{B}_s$ to those defining $\mathcal{I}$? None of the unit vectors are aligned anymore, so it is no longer as simple as just adding a new row and column to the two-dimensional transformation table. The answer is to use an *intermediate* frame.

Consider a frame $\mathcal{A}$ initially aligned with $\mathcal{I}$ that is rotated about the $\mathbf{e}_z$ axis by angle θ, as shown in Figure 10.5a. This frame is identical to the cylindrical frame in Figure 10.4a. The relationship between $\mathcal{I} = (O, \mathbf{e}_x, \mathbf{e}_y, \mathbf{e}_z)$ and $\mathcal{A} = (O, \mathbf{a}_1, \mathbf{a}_2, \mathbf{a}_3)$ is the transformation table for the cylindrical frame:

	$\mathbf{e}_x$	$\mathbf{e}_y$	$\mathbf{e}_z$
$\mathbf{a}_1$	$\cos\theta$	$\sin\theta$	0
$\mathbf{a}_2$	$-\sin\theta$	$\cos\theta$	0
$\mathbf{a}_3$	0	0	1 .

(10.11)

We can also describe the relative orientation between spherical frame $\mathcal{B}_s = (O, \mathbf{e}_\phi, \mathbf{e}_\theta, \mathbf{e}_r)$ in terms of the $\mathcal{A}$ unit vectors because it differs from $\mathcal{A}$ by only a simple rotation of ϕ about the $\mathbf{a}_2 = \mathbf{e}_\theta$ axis, as shown in Figure 10.5b. This transformation table is

	$\mathbf{a}_1$	$\mathbf{a}_2$	$\mathbf{a}_3$
$\mathbf{e}_\phi$	$\cos\phi$	0	$-\sin\phi$
$\mathbf{e}_\theta$	0	1	0
$\mathbf{e}_r$	$\sin\phi$	0	$\cos\phi$.

(10.12)

Successively using the two transformation tables in Eqs. (10.11) and (10.12) allows us to change between inertial frame $\mathcal{I}$ and spherical frame $\mathcal{B}_s$. For example, we can transform the position of P from components in $\mathcal{B}_s$ to components in $\mathcal{I}$. Written as components in the spherical frame $\mathcal{B}_s$, the position of P is

$$\boxed{\mathbf{r}_{P/O} = r\mathbf{e}_r.}$$

Using the transformation table in Eq. (10.12), $\mathbf{r}_{P/O}$ is written as components in $\mathcal{A}$:

$$\mathbf{r}_{P/O} = r \sin \phi \mathbf{a}_1 + r \cos \phi \mathbf{a}_3.$$

We then use the second transformation table in Eq. (10.11) to write $\mathbf{r}_{P/O}$ as components in $\mathcal{I}$:

$$\mathbf{r}_{P/O} = r \cos \theta \sin \phi \mathbf{e}_x + r \sin \theta \sin \phi \mathbf{e}_y + r \cos \phi \mathbf{e}_z,$$

which is the same as the spherical-coordinate position in Eq. (10.4).

We can also use these transformation tables successively on the unit vectors to form a single array for transforming unit vectors between $\mathcal{I}$ and $\mathcal{B}_s$. Writing out the relationship between the unit vectors of $\mathcal{I}$ and $\mathcal{A}$ gives

$$\mathbf{e}_x = \cos \theta \mathbf{a}_1 - \sin \theta \mathbf{a}_2$$
$$\mathbf{e}_y = \sin \theta \mathbf{a}_1 + \cos \theta \mathbf{a}_2$$
$$\mathbf{e}_z = \mathbf{a}_3.$$

Then, using the transformation table in Eq. (10.12), we have

$$\mathbf{e}_x = \cos \theta \cos \phi \mathbf{e}_\phi - \sin \theta \mathbf{e}_\theta + \cos \theta \sin \phi \mathbf{e}_r$$
$$\mathbf{e}_y = \sin \theta \cos \phi \mathbf{e}_\phi + \cos \theta \mathbf{e}_\theta + \sin \theta \sin \phi \mathbf{e}_r$$
$$\mathbf{e}_z = - \sin \phi \mathbf{e}_\phi + \cos \phi \mathbf{e}_r,$$

which can be written as a single transformation table

	$\mathbf{e}_x$	$\mathbf{e}_y$	$\mathbf{e}_z$
$\mathbf{e}_\phi$	$\cos \theta \cos \phi$	$\sin \theta \cos \phi$	$- \sin \phi$
$\mathbf{e}_\theta$	$- \sin \theta$	$\cos \theta$	0
$\mathbf{e}_r$	$\cos \theta \sin \phi$	$\sin \theta \sin \phi$	$\cos \phi$.

(10.13)

Eq. (10.13) is the three-dimensional version of the relative-orientation transformation between the inertial frame and the spherical frame. Remember, the trick was to use an intermediate frame to derive this transformation.

Eq. (10.13) can now be used to change Eqs. (10.5) and (10.6), expressed as components in the inertial frame, to ones expressed as components in the spherical frame, analogous to what we did in Chapter 3 for the polar frame. Making the proper substitutions and performing a bit of algebra (actually, a lot) results in

$$^{\mathcal{I}}\mathbf{v}_{P/O} = r\dot{\phi}\mathbf{e}_{\phi} + r\dot{\theta}\sin\phi\,\mathbf{e}_{\theta} + \dot{r}\mathbf{e}_{r} \tag{10.14}$$

$$^{\mathcal{I}}\mathbf{a}_{P/O} = (2\dot{r}\dot{\phi} + r\ddot{\phi} - r\dot{\theta}^2\cos\phi\sin\phi)\mathbf{e}_{\phi}$$
$$+ (2\dot{r}\dot{\theta}\sin\phi + 2r\dot{\theta}\dot{\phi}\cos\phi + r\ddot{\theta}\sin\phi)\mathbf{e}_{\theta}$$
$$+ (\ddot{r} - r\dot{\phi}^2 - r\dot{\theta}^2\sin^2\phi)\mathbf{e}_{r}. \tag{10.15}$$

Just as with the polar and cylindrical frames, the spherical frame expressions are simpler and easier to work with than their inertial frame counterparts. Now it is much easier to find equations of motion in terms of spherical coordinates.

Example 10.3 The Spherical Pendulum, Take 2

Example 10.1 introduced the spherical pendulum—the three-dimensional version of the simple pendulum studied in Chapter 3. However, we postponed finding its equations of motion (in terms of the spherical coordinates θ and ϕ) until after we introduced the spherical frame. Let us find them now. For reference, Figure 10.6 again shows the geometry, with the addition of the spherical frame $\mathcal{B}_s$ and the reaction force in terms of the $\mathbf{e}_r$ unit vector in the free-body diagram.

The inertial kinematics of the pendulum mass relative to the inertial frame and expressed in spherical coordinates in the spherical frame are given by Eqs. (10.14)

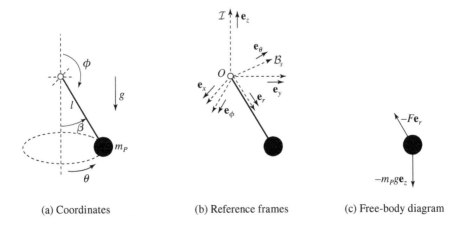

(a) Coordinates   (b) Reference frames   (c) Free-body diagram

Figure 10.6 Spherical pendulum using spherical coordinates $(r, \theta, \phi)_{\mathcal{I}}$ and spherical frame $\mathcal{B}_s = (O, \mathbf{e}_{\phi}, \mathbf{e}_{\theta}, \mathbf{e}_r)$. Also shown is the angle from the vertical, $\beta = \pi - \phi$.

and (10.15) with $r = l$ and $\dot{r} = \ddot{r} = 0$. The kinematics of P are

$$\mathbf{r}_{P/O} = l\mathbf{e}_r$$

$$^{\mathcal{I}}\mathbf{v}_{P/O} = l\dot{\phi}\mathbf{e}_\phi + l\dot{\theta}\sin\phi\,\mathbf{e}_\theta$$

$$^{\mathcal{I}}\mathbf{a}_{P/O} = (l\ddot{\phi} - l\dot{\theta}^2\cos\phi\sin\phi)\mathbf{e}_\phi + (2l\dot{\theta}\dot{\phi}\cos\phi + l\ddot{\theta}\sin\phi)\mathbf{e}_\theta$$

$$- (l\dot{\phi}^2 + l\dot{\theta}^2\sin^2\phi)\mathbf{e}_r.$$

The equations of motion for the pendulum bob are found from Newton's second law by equating the forces from the free-body diagram in Figure 10.6c with the acceleration times the mass, which yields

$$-F\mathbf{e}_r - m_P g\mathbf{e}_z = m_P(l\ddot{\phi} - l\dot{\theta}^2\cos\phi\sin\phi)\mathbf{e}_\phi + m_P(2l\dot{\theta}\dot{\phi}\cos\phi + l\ddot{\theta}\sin\phi)\mathbf{e}_\theta$$

$$- m_P(l\dot{\phi}^2 + l\dot{\theta}^2\sin^2\phi)\mathbf{e}_r. \tag{10.16}$$

Since it is easiest to solve Eq. (10.16) in the spherical frame, we write $\mathbf{e}_z$ in terms of $\mathbf{e}_\phi$ and $\mathbf{e}_r$ using Eq. (10.13):

$$-F\mathbf{e}_r - m_P g(-\sin\phi\mathbf{e}_\phi + \cos\phi\mathbf{e}_r) = m_P(l\ddot{\phi} - l\dot{\theta}^2\cos\phi\sin\phi)\mathbf{e}_\phi$$

$$+ m_P(2l\dot{\theta}\dot{\phi}\cos\phi + l\ddot{\theta}\sin\phi)\mathbf{e}_\theta$$

$$- m_P(l\dot{\phi}^2 + l\dot{\theta}^2\sin^2\phi)\mathbf{e}_r. \tag{10.17}$$

Eq. (10.17) corresponds to the three scalar equations in the $\mathbf{e}_r$, $\mathbf{e}_\theta$, and $\mathbf{e}_\phi$ directions,

$$-F - m_P g\cos\phi = -m_P l\dot{\phi}^2 - m_P l\dot{\theta}^2\sin^2\phi \tag{10.18}$$

$$0 = 2m_P l\dot{\theta}\dot{\phi}\cos\phi + m_P l\ddot{\theta}\sin\phi \tag{10.19}$$

$$m_P g\sin\phi = m_P l\ddot{\phi} - m_P l\dot{\theta}^2\cos\phi\sin\phi. \tag{10.20}$$

Eqs. (10.19) and (10.20) provide the two equations of motion,

$$\ddot{\phi} - \dot{\theta}^2\sin\phi\cos\phi - \frac{g}{l}\sin\phi = 0 \tag{10.21}$$

$$\ddot{\theta} + 2\dot{\theta}\dot{\phi}\cot\phi = 0, \tag{10.22}$$

while Eq. (10.18) provides an expression for the reaction force in the rod.

At first these may seem a bit surprising, since the sign on the gravity term seems to have switched from the simple pendulum. However, this is only because we have defined ϕ as the angle down from the vertical $\mathbf{e}_z$ axis for spherical coordinates. If we write the equations of motion in terms of the angle $\beta = \pi - \phi$ (see Figure 10.6a), then the equations of motion become

$$\ddot{\beta} - \dot{\theta}^2\sin\beta\cos\beta + \frac{g}{l}\sin\beta = 0$$

$$\ddot{\theta} + 2\dot{\theta}\dot{\beta}\cot\beta = 0.$$

Spherical coordinates allowed for a straightforward derivation of the equations, as well as lent insight to the problem. Additionally, we can use this formulation to study equilibrium points and steady-state solutions of the spherical pendulum. As with the planar pendulum, there are two equilibrium solutions, a stable one where the pendulum hangs down vertically ($\beta = 0$) and an unstable one where the pendulum is suspended upright ($\beta = \pi$). Unlike the planar pendulum, however, there is a third steady-state solution where the pendulum hangs at a fixed angle while rotating in a circle. You are invited to explore this phenomenon in Problem 10.8.

Although Eqs. (10.14) and (10.15) are simpler than Eqs. (10.5) and (10.6), it took quite a bit of algebra (much of which we spared you) to get there. Can we use the angular velocity to find an easier route to the kinematics, as we did for polar coordinates? The answer is yes, and the process is almost identical, but the concept of the angular velocity needs to be revisited.

10.2.3 Angular Velocity of the Spherical Frame

To find a formula for computing the velocity and acceleration directly in the spherical frame, we follow the same procedure as for polar coordinates in Chapter 3. We start by taking the derivative of the position of P expressed in the spherical frame:

$$\frac{^{\mathcal{I}}d}{dt}\left(\mathbf{r}_{P/O}\right) = \dot{r}\mathbf{e}_r + r\frac{^{\mathcal{I}}d}{dt}(\mathbf{e}_r). \tag{10.23}$$

What is the unit-vector derivative $\frac{^{\mathcal{I}}d}{dt}(\mathbf{e}_r)$? This question is more difficult to answer than before because $\mathcal{B}_s$ may not simply be rotating about a fixed axis along one of its unit-vector directions. To find an expression, we turn again to the intermediate frame $\mathcal{A}$. Because frame $\mathcal{A}$ is changing orientation with respect to $\mathcal{B}_s$ by rotating only about the $\mathbf{a}_2 = \mathbf{e}_\theta$ axis (a simple rotation), we can use our previous results and the unit-vector form of the transport equation to write the derivative of $\mathbf{e}_r$ with respect to $\mathcal{A}$:

$$\frac{^{\mathcal{A}}d}{dt}(\mathbf{e}_r) = {}^{\mathcal{A}}\boldsymbol{\omega}^{\mathcal{B}_s} \times \mathbf{e}_r, \tag{10.24}$$

where ${}^{\mathcal{A}}\boldsymbol{\omega}^{\mathcal{B}_s} = \dot{\phi}\mathbf{a}_2 = \dot{\phi}\mathbf{e}_\theta$.

Likewise, the transport equation can be used to relate the derivatives between $\mathcal{A}$ and $\mathcal{I}$ because the relative orientation between $\mathcal{A}$ and $\mathcal{I}$ is also only a simple rotation, this time about $\mathbf{e}_z = \mathbf{a}_3$:

$$\frac{^{\mathcal{I}}d}{dt}(\mathbf{e}_r) = \frac{^{\mathcal{A}}d}{dt}(\mathbf{e}_r) + {}^{\mathcal{I}}\boldsymbol{\omega}^{\mathcal{A}} \times \mathbf{e}_r, \tag{10.25}$$

where ${}^{\mathcal{I}}\boldsymbol{\omega}^{\mathcal{A}} = \dot{\theta}\mathbf{e}_z = \dot{\theta}\mathbf{a}_3$. Combining Eqs. (10.24) and (10.25) yields

$$\frac{^{\mathcal{I}}d}{dt}(\mathbf{e}_r) = {}^{\mathcal{I}}\boldsymbol{\omega}^{\mathcal{A}} \times \mathbf{e}_r + {}^{\mathcal{A}}\boldsymbol{\omega}^{\mathcal{B}_s} \times \mathbf{e}_r = \left({}^{\mathcal{I}}\boldsymbol{\omega}^{\mathcal{A}} + {}^{\mathcal{A}}\boldsymbol{\omega}^{\mathcal{B}_s}\right) \times \mathbf{e}_r.$$

This result looks just like the unit-vector derivative in Chapter 3. Recall that we first introduced the angular velocity in Chapter 3 as an operator that computes unit-vector derivatives. The same is true here, only now it is the sum of two simple angular velocities. In fact, we have partially verified the addition property of angular velocities in three dimensions! Introducing the definition

$$^{\mathcal{I}}\boldsymbol{\omega}^{\mathcal{B}_s} \triangleq {}^{\mathcal{I}}\boldsymbol{\omega}^{\mathcal{A}} + {}^{\mathcal{A}}\boldsymbol{\omega}^{\mathcal{B}_s} \tag{10.26}$$

allows us to write the vector derivative of $\mathbf{e}_r$ as

$$\frac{{}^{\mathcal{I}}d}{dt}(\mathbf{e}_r) = {}^{\mathcal{I}}\boldsymbol{\omega}^{\mathcal{B}_s} \times \mathbf{e}_r. \tag{10.27}$$

A similar development leads to the derivatives of $\mathbf{e}_\theta$ and $\mathbf{e}_\phi$.

Eq. (10.27) is the same as the planar equation in Eq. (3.32). The only difference is in the definition of the angular velocity ${}^{\mathcal{I}}\boldsymbol{\omega}^{\mathcal{B}_s}$. In Chapter 3, it was easy to relate the angular velocity to the geometry of motion, as it corresponded to a simple rotation about an axis perpendicular to the plane. Here the angular velocity results from the sum of two simple angular velocities associated with the rotation relative to an intermediate frame. Nevertheless, it still acts as the appropriate operator to provide the unit-vector derivative. What we have not (yet) shown is the physical and geometric interpretations of the three-dimensional angular velocity.

When dealing only with planar rotations, it was clear that the angular velocity corresponded to the rate of rotation about a particular axis. That is no longer obvious here. What, then, does the angular velocity defined in Eq. (10.26) represent? You might have guessed by now that it represents the instantaneous rate of rotation about an axis aligned with ${}^{\mathcal{I}}\boldsymbol{\omega}^{\mathcal{B}_s}$ (we call this the *instantaneous axis of rotation*). That is, in fact, true. However, we postpone validating this claim until we treat general frame rotations. For now, it is sufficient to remember that the angular-velocity addition property applied to the intermediate frame gives the operator needed to perform unit-vector derivatives of the spherical frame.

To perform the cross product in Eq. (10.27), it is convenient to express the angular velocity entirely as components in the $\mathcal{B}_s$ frame. From the two simple angular velocities and Eq. (10.26), we have

$$^{\mathcal{I}}\boldsymbol{\omega}^{\mathcal{B}_s} = \dot{\phi}\mathbf{e}_\theta + \dot{\theta}\mathbf{a}_3,$$

which, using Eq. (10.12), becomes

$$^{\mathcal{I}}\boldsymbol{\omega}^{\mathcal{B}_s} = -\dot{\theta}\sin\phi\mathbf{e}_\phi + \dot{\phi}\mathbf{e}_\theta + \dot{\theta}\cos\phi\mathbf{e}_r. \tag{10.28}$$

This is the spherical-frame angular velocity. We can substitute Eq. (10.28) into Eq. (10.27) to get the unit-vector derivatives in terms of the spherical coordinate angular rates of change (remember the order of terms matters when taking the cross product). For each of the unit vectors in $\mathcal{B}_s$, we have

$$\frac{{}^{\mathcal{I}}d}{dt}(\mathbf{e}_r) = \dot{\phi}\mathbf{e}_\phi + \dot{\theta}\sin\phi\mathbf{e}_\theta$$

$$\frac{{}^{\mathcal{I}}d}{dt}(\mathbf{e}_\theta) = -\dot{\theta}\sin\phi\mathbf{e}_r - \dot{\theta}\cos\phi\mathbf{e}_\phi$$

$$\frac{{}^{\mathcal{I}}d}{dt}(\mathbf{e}_\phi) = -\dot{\phi}\mathbf{e}_r + \dot{\theta}\cos\phi\mathbf{e}_\theta.$$

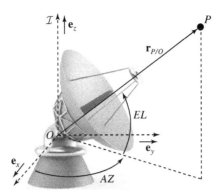

Figure 10.7 Radar tracking antenna. Image courtesy of Shutterstock.

Substituting the $\mathbf{e}_r$ derivative back into Eq. (10.23) gives the velocity

$$^{\mathcal{I}}\mathbf{v}_{P/O} = r\dot{\phi}\mathbf{e}_\phi + r\dot{\theta}\sin\phi\mathbf{e}_\theta + \dot{r}\mathbf{e}_r,$$

which matches Eq. (10.14). We can take the derivative again and use the other two angular-velocity cross products to find the acceleration expressed as components in spherical frame $\mathcal{B}_s$. The result, of course, is the same as in Eq. (10.15).

Example 10.4 Target Tracking and Intercept

Figure 10.7 shows a radar dish antenna. This device is used to track such targets as an airplane or a missile by sending out radar pulses and timing their return. To point at the target, it turns toward the maximum strength of the return signal. To find the range to the target, it computes the round-trip time of each radar pulse. The Doppler shift of the return pulses provides the range rate.

The dish antenna rotates about two axes to stay pointed at the target: the vertical axis, with azimuth angle AZ, and the horizontal axis, with elevation angle EL. These angles are related to the spherical coordinate angles, with EL corresponding to the co-latitude (i.e., $EL = \pi/2 - \phi$). We can imagine a spherical frame fixed to the dish; the mechanisms that rotate and tilt the dish are physical manifestations of the intermediate frame. Thus, if encoders on the motors provide the angles, the exact position of the target relative to the antenna can be found using the spherical-coordinate formula:

$$\mathbf{r}_{P/O} = r\cos(AZ)\cos(EL)\mathbf{e}_x + r\sin(AZ)\cos(EL)\mathbf{e}_y + r\sin(EL)\mathbf{e}_z.$$

Likewise, if the range rate and angular rates are provided, then the target velocity can be found from Eq. (10.14).

Alternatively, we may know the target position and velocity (and acceleration) from on-board instruments, such as a Global Positioning System (GPS) receiver, and may want to use that information to command the antenna. Suppose the position and inertial velocity in Cartesian coordinates $(x, y, z)_{\mathcal{I}}$ have been provided. We can then find the modified spherical coordinates $(AZ, EL, r)_{\mathcal{I}}$ by inverting the spherical coordinate equations to obtain

$$r = \sqrt{x^2 + y^2 + z^2}$$

$$EL = \arcsin\left(\frac{z}{r}\right)$$

$$AZ = \arcsin\left(\frac{y}{\sqrt{x^2 + y^2}}\right) = \arccos\left(\frac{x}{\sqrt{x^2 + y^2}}\right).$$

The rates of change of the spherical coordinates in this case are most easily found by inverting the velocity equation in Eq. (10.14), using $^{\mathcal{I}}\mathbf{v}_{P/O} = \dot{x}\mathbf{e}_x + \dot{y}\mathbf{e}_y + \dot{z}\mathbf{e}_z$ and Eq. (10.13) with $\theta = AZ$ and $\phi = \pi/2 - EL$:

$$\dot{r} = \dot{x}\cos(AZ)\cos(EL) + \dot{y}\sin(AZ)\cos(EL) + \dot{z}\sin(EL)$$

$$\dot{AZ} = (-\dot{x}\sin(AZ) + \dot{y}\cos(AZ))/(r\cos(EL))$$

$$\dot{EL} = (-\dot{x}\cos(AZ)\sin(EL) - \dot{y}\sin(AZ)\sin(EL) + \dot{z}\cos(EL))/r.$$

The angular velocity of the frame fixed to the dish is given by Eq. (10.28).

10.3 Linear Momentum, Angular Momentum, and Energy

Now that we know how to write the velocity and acceleration of a particle in three dimensions, and in cylindrical and spherical coordinates, we need to return to the material in Chapters 4–7 and generalize it, too, to three dimensions. In those chapters we always used a vector treatment, specializing to two dimensions only when necessary. As a consequence, our results can be used almost unchanged. For completeness, however, we review the important ideas.

10.3.1 Linear Momentum and Angular Momentum

Chapter 4 introduced the concept of linear impulse and the resulting momentum balance in Eq. (4.2), which we repeat here:

$$m_P{}^{\mathcal{I}}\mathbf{v}_{P/O}(t_2) = m_P{}^{\mathcal{I}}\mathbf{v}_{P/O}(t_1) + \int_{t_1}^{t_2} \mathbf{F}_P dt.$$

That treatment was completely general and thus translates unchanged to three dimensions. In practice, of course, when writing the velocities in whatever coordinate system is convenient for the problem at hand, we should be careful to use the new three-dimensional results.

Likewise for angular momentum and angular impulse. The definition of angular momentum in Definition 4.2 is unchanged in three dimensions:

$$^{\mathcal{I}}\mathbf{h}_{P/O} \triangleq \mathbf{r}_{P/O} \times {}^{\mathcal{I}}\mathbf{p}_{P/O} = \mathbf{r}_{P/O} \times m_P{}^{\mathcal{I}}\mathbf{v}_{P/O}.$$

The resulting formulation of Newton's second law is also unchanged, as the definition of the moment was also a general vector equation:

$$\frac{^{\mathcal{I}}d}{dt}\left(^{\mathcal{I}}\mathbf{h}_{P/O}\right) = \mathbf{M}_{P/O},$$

where $\mathbf{M}_{P/O} \triangleq \mathbf{r}_{P/O} \times \mathbf{F}_P$. The impulse form of the angular momentum equation in Eq. (4.17) also follows.

Example 10.5 Solving the Spherical Pendulum Using Angular Momentum

We now re-solve the spherical pendulum using angular momentum of a particle. This reduces the number of equations and avoids having to eliminate the reaction force in the rod, which produces no moment about the attachment point. Using the kinematics from Example 10.3, the angular momentum of the pendulum bob about the attachment point O becomes

$$^{\mathcal{I}}\mathbf{h}_{P/O} = \mathbf{r}_{P/O} \times m_P{}^{\mathcal{I}}\mathbf{v}_{P/O} = l\mathbf{e}_r \times m_P(l\dot{\phi}\mathbf{e}_\phi + l\dot{\theta}\sin\phi\mathbf{e}_\theta)$$

$$= m_P l^2 \dot{\phi}\mathbf{e}_\theta - m_P l^2 \dot{\theta}\sin\phi\mathbf{e}_\phi. \tag{10.29}$$

We now take the inertial derivative of the angular momentum,

$$\frac{^{\mathcal{I}}d}{dt}\left(^{\mathcal{I}}\mathbf{h}_{P/O}\right) = m_P l^2 \left(\ddot{\phi}\mathbf{e}_\theta + \dot{\phi}\frac{^{\mathcal{I}}d}{dt}\mathbf{e}_\theta - \ddot{\theta}\sin\phi\mathbf{e}_\phi - \dot{\theta}\dot{\phi}\cos\phi\mathbf{e}_\phi - \dot{\theta}\sin\phi\frac{^{\mathcal{I}}d}{dt}\mathbf{e}_\phi\right),$$

which, after substituting for the unit-vector derivatives and simplifying, gives

$$\frac{^{\mathcal{I}}d}{dt}\left(^{\mathcal{I}}\mathbf{h}_{P/O}\right) = m_P l^2 \left(\ddot{\phi} - \dot{\theta}^2 \sin\phi\cos\phi\right)\mathbf{e}_\theta - m_P l^2\left(\ddot{\theta}\sin\phi + 2\dot{\theta}\dot{\phi}\cos\phi\right)\mathbf{e}_\phi.$$

This equation equals the moment acting on the pendulum bob relative to origin O. The force that produces a moment about O is gravity, which implies (see Eq. (10.13))

$$\mathbf{M}_{P/O} = \mathbf{r}_{P/O} \times \mathbf{F}_{P/O} = l\mathbf{e}_r \times -m_P g\mathbf{e}_z$$

$$= -m_P gl\mathbf{e}_r \times (-\sin\phi\mathbf{e}_\phi + \cos\phi\mathbf{e}_r)$$

$$= m_P gl\sin\phi\mathbf{e}_\theta.$$

Equating the moment on the pendulum bob with the rate of change of angular momentum yields two scalar equations:

$$0 = 2m_P l^2 \dot{\theta}\dot{\phi}\cos\phi + m_P l^2 \ddot{\theta}\sin\phi$$

$$m_P gl\sin\phi = m_P l^2 \ddot{\phi} - m_P l^2 \dot{\theta}^2 \sin\phi\cos\phi,$$

which, after simplifying, result in the same equations of motion as Eqs. (10.21) and (10.22):

$$\ddot{\phi} - \dot{\theta}^2 \sin\phi \cos\phi - \frac{g}{l}\sin\phi = 0 \tag{10.30}$$

$$\ddot{\theta} + 2\dot{\theta}\dot{\phi}\cot\phi = 0. \tag{10.31}$$

Now we explore the equations of motion a bit more. At first glance they seem less useful than you might expect because Eq. (10.31) goes to infinity when ϕ is 0 or π. This is because, at those angles (the pendulum straight up or straight down), the other spherical angle, θ, is undefined. However, if you examine Eq. (10.31) closely you will see it is integrable. Multiplying by $\sin^2\phi$ gives

$$\ddot{\theta}\sin^2\phi + 2\dot{\theta}\dot{\phi}\sin\phi\cos\phi = 0.$$

This equation is equal to

$$\frac{d}{dt}(m_P l^2\dot{\theta}\sin^2\phi) = 0, \tag{10.32}$$

where we have multiplied by $m_P l^2$. Using the transformation table for spherical coordinates in Eq. (10.13) to rewrite the angular momentum in Eq. (10.29) as components in the inertial frame, we see that the quantity in Eq. (10.32) is the $\mathbf{e}_z$ component of the angular momentum of the pendulum bob relative to O. Eq. (10.32) states that the z-component of the angular momentum is a constant of the motion. This is because gravity cannot produce a moment about the z-axis.

We can use the fact that $h_z = {}^{\mathcal{I}}\mathbf{h}_{P/O} \cdot \mathbf{e}_z = m_P l^2\dot{\theta}\sin^2\phi$ is conserved to rewrite Eq. (10.30) as

$$\ddot{\phi} - \frac{h_z^2\cos\phi}{m_P^2 l^4\sin^3\phi} - \frac{g}{l}\sin\phi = 0.$$

Using conservation of one component of the angular momentum, we have again been able to reduce this two-degree-of-freedom problem to a single differential equation of motion. A reduction of this sort is often possible in problems that exhibit *symmetry*—in this case, the spherical pendulum is symmetric about the $\mathbf{e}_z$ axis.

10.3.2 Energy of a Particle in Three Dimensions

We now generalize our treatment of work and energy of a particle to three dimensions. Definition 5.1 is a vector equation, so it still applies in three dimensions; we just need to be careful about our choice of coordinates. The definition of power remains unchanged and the work–kinetic-energy formula in Eq. (5.6) remains

$$T_{P/O}(t_2) = T_{P/O}(t_1) + W_P^{(\mathrm{tot})}(\mathbf{r}_{P/O}; \gamma_P),$$

where the kinetic energy is

$$T_{P/O} = \frac{1}{2}m_P\left({}^{\mathcal{I}}\mathbf{v}_{P/O} \cdot {}^{\mathcal{I}}\mathbf{v}_{P/O}\right) = \frac{1}{2}m_P\|{}^{\mathcal{I}}\mathbf{v}_{P/O}\|^2.$$

The vector definition of a conservative force still holds, as does the treatment that led us to introduce the potential energy. Thus the final work energy formula in Eq. (5.16) remains

$$E_{P/O}(t_2) = E_{P/O}(t_1) + W_P^{(nc)}(\mathbf{r}_{P/O}; \gamma_P),$$

where the total energy is given by $E_{P/O}(t) = T_{P/O}(t) + U_{P/O}(t)$, and the potential energy, for conservative forces, satisfies $\mathbf{F}_P^{(c)} = -\nabla U_{P/O}^{(\mathbf{F}_P)}(\mathbf{r}_{P/O})$. The important thing to remember here is that we are now considering the potential energy to be a function of three spatial coordinates and the gradient operator must be treated accordingly. (Appendix A shows how to write the gradient of a scalar function in terms of the various three-dimensional coordinate systems.)

Example 10.6 The Energy of a Spherical Pendulum

We now look at the energy of a spherical pendulum. Since this example contains a single particle, albeit with two degrees of freedom, the kinetic energy is calculated using the velocity from Example 10.3:

$$T_{P/O} = \frac{1}{2} m_P \|{}^{\mathcal{I}}\mathbf{v}_{P/O}\|^2 = \frac{1}{2} m_P l^2 (\dot{\phi}^2 + \dot{\theta}^2 \sin^2 \phi).$$

The potential energy is a function of the height of the pendulum:

$$U_{P/O} = m_P g \mathbf{r}_{P/O} \cdot \mathbf{e}_z,$$

which, using spherical coordinates and the transformation table (Eq. (10.13)), gives

$$U_{P/O} = m_P g l \cos \phi.$$

The total energy is the sum of the kinetic and potential energies:

$$E_{P/O} = \frac{1}{2} m_P l^2 (\dot{\phi}^2 + \dot{\theta}^2 \sin^2 \phi) + m_P g l \cos \phi.$$

Because the only force doing work here is gravity, which is conservative, the total energy of the pendulum is conserved.

We can further simplify the expression for the energy. Recall that Example 10.5 showed that the component of the angular momentum of the pendulum bob about the $\mathbf{e}_z$ axis is a constant of the motion:

$$h_z = m_P l^2 \dot{\theta} \sin^2 \phi = \text{constant}.$$

This result can be used to solve for $\dot{\theta}$, which we then substitute into the energy expression. After a bit of algebra, we find

$$E_{P/O} = \frac{1}{2} m_P l^2 \dot{\phi}^2 + \frac{h_z^2}{2 m_P l^2 \sin^2 \phi} + m_P g l \cos \phi,$$

which is a function of ϕ and $\dot{\phi}$ only.

10.4 Relative Motion in Three Dimensions

This section generalizes to three dimensions our treatment of relative motion in Chapter 8. Unlike in the previous section, where the treatment of particles in two dimensions transferred directly, here the situation becomes quite a bit more complicated. Recall what we set out to do in Chapter 8. Our goal was to consider a body frame $\mathcal{B}$ rotating and translating with respect to an inertial frame $\mathcal{I}$. Point P was allowed to move arbitrarily in $\mathcal{B}$, and we described its inertial velocity and inertial acceleration in terms of its motion in $\mathcal{B}$. The only restriction was that $\mathcal{B}$ changed its orientation in $\mathcal{I}$ by means of a simple rotation about the $\mathbf{b}_3 = \mathbf{e}_3$ axis; the motion of P was confined to a plane. The result was the transport equation and the acceleration equation in Eq. (8.21), which we have used frequently since.

The restriction to a simple rotation between $\mathcal{B}$ and $\mathcal{I}$ was important. It allowed us to derive, using the transformation table, the formula for the time derivative of a body-frame unit vector in absolute space. Now we need to generalize this derivation to an arbitrary rotation between $\mathcal{B}$ and $\mathcal{I}$. The basic approach is the same as in Section 10.2.3—we will use intermediate frames. In fact, we just need one more rotation and one more intermediate frame.

10.4.1 Orientation Angles

If you review Chapter 3, particularly Section 3.4, you will see that before introducing the polar and path frames and using them to develop the kinematics of a particle moving in the plane, we first introduced the idea of relative orientation and the transformation table. We have used that idea frequently since. In particular, it was essential for the development of the transport equation (Definition 8.1). There we considered only three-degree-of-freedom problems, two for translation and one for rotation, so a single angle was sufficient to describe the orientation of the body frame in absolute space.

Now things get a bit more complicated, but we follow the same general approach. In three dimensions, a rigid body has, in general, six degrees of freedom—three for translation and three for rotation. It takes three scalar coordinates to completely describe the orientation of a rigid body (or body frame) in absolute space. Our first task, then, before turning to the kinematics of a particle or rigid body, is to find a way of describing the general three-dimensional orientation of a body frame (or a rigid body) relative to another frame using three scalar coordinates.

There are, as might be expected, many ways to describe the orientation of a frame, say $\mathcal{B}$, with respect to a second frame, say $\mathcal{I}$ (i.e., there are many sets of three scalar coordinates), just as there are many scalar coordinates that can be used to describe the position of a point in a reference frame. Our discussion of the spherical frame leads us to think of a particular approach—using simple rotations about the axes of intermediate frames. For the spherical frame, we had two angles, θ and ϕ, that oriented the frame. For a general body frame $\mathcal{B}$, three angles are needed to describe the orientation of $\mathcal{B}$ in $\mathcal{I}$. The most common set (and notation) are angles ψ, θ, and ϕ, which involve two intermediate frames. In similar fashion to the spherical frame,

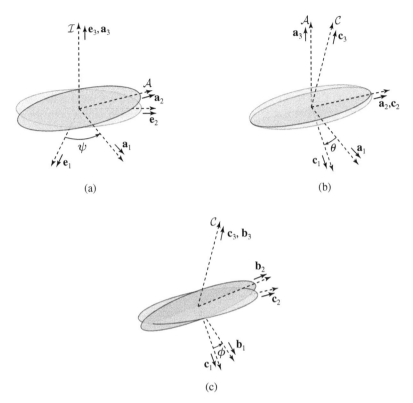

Figure 10.8 A 3-2-3 ordered rotation via the Euler angles $(\psi, \theta, \phi)_{\mathcal{B}}^{\mathcal{I}}$ of frame $\mathcal{B}$ in frame $\mathcal{I}$ using intermediate frames $\mathcal{A}$ and $\mathcal{C}$. (a) ψ rotation about $\mathbf{e}_3$. (b) θ rotation about $\mathbf{a}_2$. (c) ϕ rotation about $\mathbf{c}_3$.

we first rotate about the $\mathbf{e}_3$ axis of frame $\mathcal{I}$ by an angle ψ to get to the intermediate frame $\mathcal{A}$. We then rotate by θ about the $\mathbf{a}_2$ axis of frame $\mathcal{A}$ to reach the intermediate frame $\mathcal{C}$. (This frame is the same as the spherical frame!) Finally, we rotate about the $\mathbf{c}_3$ axis by ϕ to get to the $\mathcal{B}$ frame. Note that all are right-handed rotations.

The set of angles $(\psi, \theta, \phi)_{\mathcal{B}}^{\mathcal{I}}$ are known as *orientation angles* or *Euler angles*. (Note the slight change in notation when the scalar coordinates describe the orientation of $\mathcal{B}$ in $\mathcal{I}$ rather than the location of a particle relative to an origin; we need to indicate both frames.) The particular ordered set of rotations described above and shown in Figure 10.8 is known as a 3-2-3 rotation, as it involves an ordered sequence of rotations about the 3-axis, the 2-axis, and the 3-axis of a series of intermediate reference frames. It is possible to describe any orientation of reference frame $\mathcal{B}$ in $\mathcal{I}$ by means of these three angles. Thus they constitute a set of scalar coordinates for describing the orientation of a reference frame. In fact, you have already seen an example of using Euler angles in Example 10.4, where we used azimuth (AZ) and elevation (EL) to describe the orientation of a radar tracking dish.

As you might have guessed, there are also many other possible sets of Euler angles (e.g., a 3-1-3 rotation or a 1-2-3 rotation). In all, there are 12 possible sets of three

Euler-angle rotations that can be used to describe the orientation of one frame with respect to another. Which set is best depends on the specific mechanical system being analyzed. We use a few different sets in the examples that follow.

We begin by examining how to transform a vector from one set of unit vectors to another set in terms of the Euler angles, just as in the planar case. That is, how do we write the unit vectors of frame $\mathcal{B}$ in terms of those in $\mathcal{I}$? We use the same approach as for the spherical frame and write the transformation tables between each pair of intermediate frames related by a simple rotation. Thus the transformation between the inertial frame $\mathcal{I}$ and the first intermediate frame $\mathcal{A}$ is

	$\mathbf{e}_1$	$\mathbf{e}_2$	$\mathbf{e}_3$
$\mathbf{a}_1$	$\cos\psi$	$\sin\psi$	0
$\mathbf{a}_2$	$-\sin\psi$	$\cos\psi$	0
$\mathbf{a}_3$	0	0	1 .

The second transformation between $\mathcal{A}$ and $\mathcal{C}$ is

	$\mathbf{a}_1$	$\mathbf{a}_2$	$\mathbf{a}_3$
$\mathbf{c}_1$	$\cos\theta$	0	$-\sin\theta$
$\mathbf{c}_2$	0	1	0
$\mathbf{c}_3$	$\sin\theta$	0	$\cos\theta$,

and the transformation between $\mathcal{C}$ and $\mathcal{B}$ is

	$\mathbf{c}_1$	$\mathbf{c}_2$	$\mathbf{c}_3$
$\mathbf{b}_1$	$\cos\phi$	$\sin\phi$	0
$\mathbf{b}_2$	$-\sin\phi$	$\cos\phi$	0
$\mathbf{b}_3$	0	0	1 .

These transformation tables allow us to write down the unit vectors of $\mathcal{B}$ as a sum of components in each frame, and eventually in frame $\mathcal{I}$. Thus $\mathbf{b}_1$ is

$$\mathbf{b}_1 = \cos\phi\mathbf{c}_1 + \sin\phi\mathbf{c}_2$$
$$= \cos\phi\cos\theta\mathbf{a}_1 + \sin\phi\mathbf{a}_2 - \cos\phi\sin\theta\mathbf{a}_3$$
$$= (\cos\phi\cos\theta\cos\psi - \sin\phi\sin\psi)\mathbf{e}_1 + (\cos\phi\cos\theta\sin\psi + \sin\phi\cos\psi)\mathbf{e}_2$$
$$- \cos\phi\sin\theta\mathbf{e}_3.$$

Likewise, solving for $\mathbf{b}_2$ and $\mathbf{b}_3$ as components in $\mathcal{I}$ yields

$$\mathbf{b}_2 = -(\cos\psi\cos\theta\sin\phi + \cos\phi\sin\psi)\mathbf{e}_1 - (\sin\psi\cos\theta\sin\phi - \cos\phi\cos\psi)\mathbf{e}_2$$
$$+ \sin\theta\sin\phi\mathbf{e}_3$$
$$\mathbf{b}_3 = \cos\psi\sin\theta\mathbf{e}_1 + \sin\psi\sin\theta\mathbf{e}_2 + \cos\theta\mathbf{e}_3.$$

As for the spherical frame, we can consolidate the transformations into a table relating frame $\mathcal{I}$ and frame $\mathcal{B}$:

	$\mathbf{e}_1$	$\mathbf{e}_2$	$\mathbf{e}_3$
$\mathbf{b}_1$	$\cos\phi\cos\theta\cos\psi - \sin\phi\sin\psi$	$\cos\phi\cos\theta\sin\psi + \sin\phi\cos\psi$	$-\cos\phi\sin\theta$
$\mathbf{b}_2$	$-\cos\psi\cos\theta\sin\phi - \cos\phi\sin\psi$	$-\sin\psi\cos\theta\sin\phi + \cos\phi\cos\psi$	$\sin\theta\sin\phi$
$\mathbf{b}_3$	$\cos\psi\sin\theta$	$\sin\psi\sin\theta$	$\cos\theta$

$$(10.33)$$

Eq. (10.33) contains a three-dimensional transformation that relates the unit vectors of the inertial frame to those of an arbitrary body frame $\mathcal{B}$ whose orientation is described by a set of 3-2-3 Euler angles $(\psi, \theta, \phi)_{\mathcal{B}}^{\mathcal{I}}$. As in the planar case, the row i, column j element is given by $\mathbf{b}_i \cdot \mathbf{e}_j$.

It is also convenient to describe these unit-vector transformations by means of matrix operations. We thus conclude this section with a rederivation using matrix notation. We have not used this notation very much, but it becomes very helpful here and in the next chapter.

Recall that Chapter 3 introduced a matrix notation for writing the components of a vector in a frame, in which the magnitudes of the components are the elements of a column matrix and the unit vectors are implied by the subscript. Thus a vector $\mathbf{r}$ written as components in frame $\mathcal{B}$ as $\mathbf{r} = x\mathbf{b}_1 + y\mathbf{b}_2 + z\mathbf{b}_3$ is written in matrix notation as

$$[\mathbf{r}]_{\mathcal{B}} = \begin{bmatrix} x \\ y \\ z \end{bmatrix}_{\mathcal{B}} . \tag{10.34}$$

Suppose we use the transformation table to express $\mathbf{r}$ in frame $\mathcal{C}$. Its components are then given by $\mathbf{r} = (x\cos\phi - y\sin\phi)\mathbf{c}_1 + (x\sin\phi + y\cos\phi)\mathbf{c}_2 + z\mathbf{c}_3$, which in matrix form is

$$[\mathbf{r}]_{\mathcal{C}} = \begin{bmatrix} r_1 \\ r_2 \\ r_3 \end{bmatrix}_{\mathcal{C}} = \begin{bmatrix} x\cos\phi - y\sin\phi \\ x\sin\phi + y\cos\phi \\ z \end{bmatrix}_{\mathcal{C}} . \tag{10.35}$$

Comparing Eq. (10.35) with Eq. (10.34) shows that $[\mathbf{r}]_{\mathcal{B}}$ and $[\mathbf{r}]_{\mathcal{C}}$ are related by a simple matrix multiplication:

$$\begin{bmatrix} r_1 \\ r_2 \\ r_3 \end{bmatrix}_{\mathcal{C}} = \underbrace{\begin{bmatrix} \cos\phi & -\sin\phi & 0 \\ \sin\phi & \cos\phi & 0 \\ 0 & 0 & 1 \end{bmatrix}}_{\triangleq\, {}^{\mathcal{C}}C^{\mathcal{B}}} \begin{bmatrix} x \\ y \\ z \end{bmatrix}_{\mathcal{B}} . \tag{10.36}$$

The matrix ${}^{\mathcal{C}}C^{\mathcal{B}}$ defined in Eq. (10.36) is the transformation matrix in three dimensions, also called the *direction-cosine matrix*. It acts to transform the components of a vector from one frame to another using matrix multiplication.

Note that we haven't really done anything new here. The elements of ${}^{\mathcal{C}}C^{\mathcal{B}}$ are just the elements of the corresponding transformation table. The value of writing the transformation table as a matrix is that it converts many vector operations to linear algebra. For instance, if we write the other two intermediate transformations as matrices,

$$^{\mathcal{A}}C^{\mathcal{C}} = \begin{bmatrix} \cos\theta & 0 & \sin\theta \\ 0 & 1 & 0 \\ -\sin\theta & 0 & \cos\theta \end{bmatrix}$$

and

$$^{\mathcal{I}}C^{\mathcal{A}} = \begin{bmatrix} \cos\psi & -\sin\psi & 0 \\ \sin\psi & \cos\psi & 0 \\ 0 & 0 & 1 \end{bmatrix},$$

then the net transformation between frames $\mathcal{I}$ and $\mathcal{B}$ is given by the matrix multiplication $^{\mathcal{I}}C^{\mathcal{B}} = {}^{\mathcal{I}}C^{\mathcal{A}\mathcal{A}}C^{\mathcal{C}\mathcal{C}}C^{\mathcal{B}}$, which yields

$$^{\mathcal{I}}C^{\mathcal{B}} = \begin{bmatrix} \cos\phi\cos\theta\cos\psi - \sin\phi\sin\psi & -\cos\psi\cos\theta\sin\phi - \cos\phi\sin\psi & \cos\psi\sin\theta \\ \cos\phi\cos\theta\sin\psi + \sin\phi\cos\psi & -\sin\psi\cos\theta\sin\phi + \cos\phi\cos\psi & \sin\psi\sin\theta \\ -\cos\phi\sin\theta & \sin\theta\sin\phi & \cos\theta \end{bmatrix}.$$

$$(10.37)$$

Turning successive rotations of frames into a series of matrix multiplications is a general property of the transformation matrix. The notation we have chosen, which explicitly notes the starting and ending frames, helps remind you of the proper order of multiplications. It is also still true that the elements of $^{\mathcal{I}}C^{\mathcal{B}}$ are given by $^{\mathcal{I}}C^{\mathcal{B}}_{ij} = \mathbf{b}_j \cdot \mathbf{e}_i$, $i, j = 1, 2, 3$. This explains why it is called the direction-cosine matrix: each element is the cosine of the angle between the corresponding pair of unit vectors.

Compare the expression for $^{\mathcal{I}}C^{\mathcal{B}}$ to our old notation using the transformation table in Eq. (10.33). This should convince you that, even for this more complicated matrix including all three rotations, it is still true that $^{\mathcal{I}}C^{\mathcal{B}} = (^{\mathcal{B}}C^{\mathcal{I}})^T$. In other words, we have

$$\boxed{[\mathbf{r}]_{\mathcal{I}} = {}^{\mathcal{I}}C^{\mathcal{B}}[\mathbf{r}]_{\mathcal{B}}} \qquad (10.38)$$

and

$$[\mathbf{r}]_{\mathcal{B}} = {}^{\mathcal{B}}C^{\mathcal{I}}[\mathbf{r}]_{\mathcal{I}} = (^{\mathcal{I}}C^{\mathcal{B}})^T[\mathbf{r}]_{\mathcal{I}} \qquad (10.39)$$

for any transformation matrix $^{\mathcal{I}}C^{\mathcal{B}}$. Of course, because these equations are simple matrix equations, we should have been able to get from Eq. (10.38) to Eq. (10.39) simply by multiplying on the left by the inverse of $^{\mathcal{I}}C^{\mathcal{B}}$. This is true, and it implies that $(^{\mathcal{I}}C^{\mathcal{B}})^{-1} = (^{\mathcal{I}}C^{\mathcal{B}})^T$ or, equivalently, that $^{\mathcal{I}}C^{\mathcal{B}}(^{\mathcal{I}}C^{\mathcal{B}})^T = I$, where I is the 3×3 identity matrix. Matrices that satisfy this property are called *orthogonal*. Orthogonality is a general property of all transformation matrices.

Finally, though we won't prove it, a transformation matrix also has the property $|^{\mathcal{I}}C^{\mathcal{B}}| = 1$ and $|^{\mathcal{I}}C^{\mathcal{B}} - I| = 0$, where the notation $|\cdot|$ represents the matrix determinant. These two properties imply that one of the eigenvalues of the transformation matrix is always unity. Thus there always exists a vector (an eigenvector of $^{\mathcal{I}}C^{\mathcal{B}}$) such that

$$\begin{bmatrix} c_1 \\ c_2 \\ c_3 \end{bmatrix}_{\mathcal{I}} = {}^{\mathcal{I}}C^{\mathcal{B}} \begin{bmatrix} c_1 \\ c_2 \\ c_3 \end{bmatrix}_{\mathcal{B}}.$$

We use this property later when discussing angular velocity in three dimensions.

Example 10.7 Gimbals

Example 10.4 described the orientation of the body frame attached to a radar dish by what we now know is a 3-2-3 orientation set (though the third angle in the set was not used). Even though we did not yet have a transformation table, we were nonetheless able to describe the position of a target fixed in the body frame in terms of components in the inertial frame using two Euler angles (AZ and EL) and its distance from the origin. The mechanism pointing the dish is a physical representation of the intermediate frame used to define the spherical body frame.

This example describes *gimbals*, another common mechanical system used to manifest a set of sequential Euler angles. You may be familiar with gimbals from a visit to an amusement park. A very popular ride, shown in Figure 10.9, involves strapping the patron into the center gimbal and allowing her to rotate freely to any arbitrary orientation. Gimbals date back to ancient Greece and have been used for millennia to keep objects oriented with the vertical. In modern times they have found their most common use in inertial navigation systems, where they support a *gimbaled gyroscope.* Chapter 11 covers much more about gyroscopes; for now, you only need to know that, in the absence of torques, the orientation of a spinning gyroscope stays fixed in absolute space.

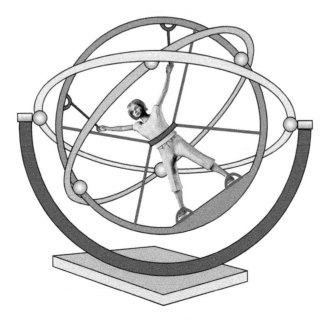

Figure 10.9 Amusement park ride using three gimbals.

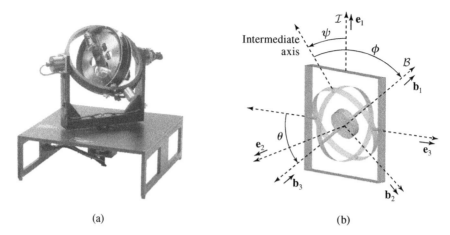

Figure 10.10 (a) Gimbaled gyroscope. (Made by Educational Control Products-ECP.) (b) Gimbal schematic.

Figure 10.10a shows a picture of a gimbaled gyroscope. By attaching the outer gimbal to a vehicle and making the joints free of friction, the orientation of the gyroscope stays fixed in absolute space while the vehicle (e.g., airplane, submarine, or spacecraft) moves around it. Thus using encoders to measure the angles of each gimbal directly provides the Euler angles that describe the orientation of a vehicle body frame relative to the gyro body frame (which is an inertial frame).

A schematic drawing of a typical set of gimbals is shown in Figure 10.10b. The orientation of the central disk relative to absolute space is represented by a 3-1-3 rotation set. We again use the Euler angles $(\psi, \theta, \phi)^{\mathcal{I}}_{\mathcal{B}}$ to represent the orientation of $\mathcal{B}$ in $\mathcal{I}$, only now ψ represents a rotation about the inertial 3-axis, θ represents a rotation about the intermediate 1-axis, and ϕ represents a rotation about the final 3-axis. Vectors in the body frame (e.g., the location of the rider's head) can be transformed to components in the inertial frame by means of a transformation table (or direction-cosine matrix) found in a similar way to the 3-2-3 one in Eq. (10.33). The resulting 3-1-3 transformation table is

	$\mathbf{e}_1$	$\mathbf{e}_2$	$\mathbf{e}_3$
$\mathbf{b}_1$	$-\sin\phi\cos\theta\sin\psi + \cos\phi\cos\psi$	$\sin\phi\cos\theta\cos\psi + \cos\phi\sin\psi$	$\sin\phi\sin\theta$
$\mathbf{b}_2$	$-\sin\psi\cos\theta\cos\phi - \sin\phi\cos\psi$	$\cos\psi\cos\theta\cos\phi - \sin\phi\sin\psi$	$\sin\theta\cos\phi$
$\mathbf{b}_3$	$\sin\psi\sin\theta$	$-\cos\psi\sin\theta$	$\cos\theta$

Example 10.8 The Orientation of an Airplane

The airplane is a classic example of a three-dimensional rigid body undergoing translation and rotation as it flies from point to point and changes orientation (perhaps pitching to increase angle of attack or rolling and yawing to enter a turn). It is thus an excellent example of rigid-body motion, as there are no constraints; it has the full six degrees of freedom. As shown in the next chapter, the separation principle still holds in three dimensions, and we can examine the translational motion of the airplane

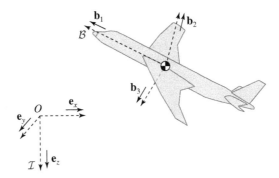

Figure 10.11 The orientation of an airplane body frame relative to the inertial frame is described by a 3-2-1 Euler angle set $(\psi, \theta, \phi)_{\mathcal{B}}^{\mathcal{I}}$.

as if it were a point at its center of mass and separately solve for its orientation. We postpone the treatment of airplane dynamics until later, but here we show the common description of the orientation of an airplane in the inertial frame.

Figure 10.11 shows a diagram of a typical aircraft with an attached body frame located at the center of mass of the plane and a fixed inertial frame. We assume the inertial frame is on the surface of the earth and ignore the curvature and rotation of the earth. (Yes, the inertial z-axis points down so that an airplane at altitude has a negative $\mathbf{e}_z$ component.) By convention the $\mathbf{b}_1$ axis of the body frame is along the centerline of the fuselage and points through the nose of the aircraft. The $\mathbf{b}_3$ axis points downward, so that the plane maintains symmetry on either side, and the $\mathbf{b}_2$ axis points to the right (looking forward) through the wing to complete the right-handed set.

Also by convention, the orientation of the airplane body frame relative to the fixed inertial frame is described by the ordered 3-2-1 set of Euler angles $(\psi, \theta, \phi)_{\mathcal{B}}^{\mathcal{I}}$, where the first rotation, ψ, is a *yaw* rotation about the inertial 3-axis, the second rotation, θ, is a *pitch* rotation about the intermediate body 2-axis, and the third rotation, ϕ, is a *roll* rotation about the body 1-axis. We can again find the transformation table (or direction-cosine matrix) to convert a vector in the body frame into components in absolute space by means of the same procedure using intermediate frames. The result is the following transformation table:

	$\mathbf{e}_x$	$\mathbf{e}_y$	$\mathbf{e}_z$
$\mathbf{b}_1$	$\cos\psi\cos\theta$	$\sin\psi\cos\theta$	$-\sin\theta$
$\mathbf{b}_2$	$\cos\psi\sin\theta\sin\phi - \cos\phi\sin\psi$	$\sin\psi\sin\theta\sin\phi + \cos\phi\cos\psi$	$\cos\theta\sin\phi$
$\mathbf{b}_3$	$\cos\phi\sin\theta\cos\psi + \sin\phi\sin\psi$	$\sin\psi\sin\theta\cos\phi - \sin\phi\cos\psi$	$\cos\theta\cos\phi$.

10.4.2 The Angular Velocity and the Transport Equation in Three Dimensions

We now have the tools to generalize the concept of angular velocity and re-derive the transport equation in three dimensions, thus completing our description of particle kinematics and kinetics. We repeat the derivation allowing for an arbitrary change of orientation of the body frame $\mathcal{B}$ in $\mathcal{I}$. The procedure is very similar to the treatment in

Section 10.2.3, where we calculated the rate of change of the unit vectors associated with the spherical frame. The consequence, as before, is the introduction of the angular velocity as an operator used to find the vector derivative.

We begin by taking the inertial derivative of the position $\mathbf{r}_{P/O} = x\mathbf{b}_1 + y\mathbf{b}_2 + z\mathbf{b}_3$, written as components in body frame $\mathcal{B}$:

$$\frac{{}^{\mathcal{I}}d}{dt}(\mathbf{r}_{P/O}) = \underbrace{\dot{x}\mathbf{b}_1 + \dot{y}\mathbf{b}_2 + \dot{z}\mathbf{b}_3}_{=\frac{{}^{\mathcal{B}}d}{dt}(\mathbf{r}_{P/O})} + x\frac{{}^{\mathcal{I}}d}{dt}(\mathbf{b}_1) + y\frac{{}^{\mathcal{I}}d}{dt}(\mathbf{b}_2) + z\frac{{}^{\mathcal{I}}d}{dt}(\mathbf{b}_3). \quad (10.40)$$

Once again, we need an expression for the unit-vector derivatives. As for the spherical frame, the trick is to use intermediate frames so that each incremental rotation is planar. Let us consider the 3-2-3 rotation with Euler angles $(\psi, \theta, \phi)_{\mathcal{B}}^{\mathcal{I}}$ described in the previous section. We can then use the transport equation already derived for a simple angular velocity (planar rotation) to relate the derivative of the unit vector $\mathbf{b}_i$ in $\mathcal{I}$ to its derivative in the intermediate frame $\mathcal{A}$:

$$\frac{{}^{\mathcal{I}}d}{dt}(\mathbf{b}_i) = \frac{{}^{\mathcal{A}}d}{dt}(\mathbf{b}_i) + {}^{\mathcal{I}}\boldsymbol{\omega}^{\mathcal{A}} \times \mathbf{b}_i, \quad (10.41)$$

where ${}^{\mathcal{I}}\boldsymbol{\omega}^{\mathcal{A}} = \dot{\psi}\mathbf{a}_3$, and $i = 1, 2, 3$. We can do the same for frames $\mathcal{A}$ and $\mathcal{C}$:

$$\frac{{}^{\mathcal{A}}d}{dt}(\mathbf{b}_i) = \frac{{}^{\mathcal{C}}d}{dt}(\mathbf{b}_i) + {}^{\mathcal{A}}\boldsymbol{\omega}^{\mathcal{C}} \times \mathbf{b}_i, \quad (10.42)$$

where ${}^{\mathcal{A}}\boldsymbol{\omega}^{\mathcal{C}} = \dot{\theta}\mathbf{c}_2$. Finally, we use the transport equation one last time between frames $\mathcal{C}$ and $\mathcal{B}$ to obtain

$$\frac{{}^{\mathcal{C}}d}{dt}(\mathbf{b}_i) = \underbrace{\frac{{}^{\mathcal{B}}d}{dt}(\mathbf{b}_i)}_{=0} + {}^{\mathcal{C}}\boldsymbol{\omega}^{\mathcal{B}} \times \mathbf{b}_i, \quad (10.43)$$

where ${}^{\mathcal{C}}\boldsymbol{\omega}^{\mathcal{B}} = \dot{\phi}\mathbf{b}_3$. The derivative in $\mathcal{B}$ is zero because $\mathbf{b}_i$ is a unit vector fixed in $\mathcal{B}$. Combining Eqs. (10.41)–(10.43) gives the following expression for the inertial derivative of $\mathbf{b}_i$:

$$\begin{aligned}
\frac{{}^{\mathcal{I}}d}{dt}(\mathbf{b}_i) &= {}^{\mathcal{I}}\boldsymbol{\omega}^{\mathcal{A}} \times \mathbf{b}_i + {}^{\mathcal{A}}\boldsymbol{\omega}^{\mathcal{C}} \times \mathbf{b}_i + {}^{\mathcal{C}}\boldsymbol{\omega}^{\mathcal{B}} \times \mathbf{b}_i \\
&= \underbrace{\left({}^{\mathcal{I}}\boldsymbol{\omega}^{\mathcal{A}} + {}^{\mathcal{A}}\boldsymbol{\omega}^{\mathcal{C}} + {}^{\mathcal{C}}\boldsymbol{\omega}^{\mathcal{B}}\right)}_{\triangleq {}^{\mathcal{I}}\boldsymbol{\omega}^{\mathcal{B}}} \times \mathbf{b}_i.
\end{aligned} \quad (10.44)$$

We define the angular velocity ${}^{\mathcal{I}}\boldsymbol{\omega}^{\mathcal{B}}$ as the sum of the simple angular velocities associated with the intermediate frames, just as for the spherical frame. As in the planar case, the vector quantity we call the angular velocity is an operator that provides unit-vector derivatives. We have also shown that the angular velocity can be found by summing the simple angular velocities of successive rotations. This is a generalization of the addition property established in Section 8.2.2, but here the successive rotations need not be co-axial. We still have not discussed what the angular velocity represents physically; because the three intermediate rotations are not co-axial, the physical interpretation of $^{\mathcal{I}}\boldsymbol{\omega}^{\mathcal{B}}$ may not be obvious. Nevertheless, as noted in the discussion of the spherical frame, it is a fact that this angular velocity represents an instantaneous rate of rotation about an axis along the vector direction.

We can now substitute the unit-vector derivative contained in Eq. (10.44) back into Eq. (10.40) to find

$$\frac{^{\mathcal{I}}d}{dt}\left(\mathbf{r}_{P/O}\right) = \frac{^{\mathcal{B}}d}{dt}\left(\mathbf{r}_{P/O}\right) + x\,{}^{\mathcal{I}}\boldsymbol{\omega}^{\mathcal{B}} \times \mathbf{b}_1 + y\,{}^{\mathcal{I}}\boldsymbol{\omega}^{\mathcal{B}} \times \mathbf{b}_2 + z\,{}^{\mathcal{I}}\boldsymbol{\omega}^{\mathcal{B}} \times \mathbf{b}_3,$$

or

$$\boxed{\frac{^{\mathcal{I}}d}{dt}\left(\mathbf{r}_{P/O}\right) = \frac{^{\mathcal{B}}d}{dt}\left(\mathbf{r}_{P/O}\right) + {}^{\mathcal{I}}\boldsymbol{\omega}^{\mathcal{B}} \times \mathbf{r}_{P/O},} \qquad (10.45)$$

where

$$^{\mathcal{I}}\boldsymbol{\omega}^{\mathcal{B}} = {}^{\mathcal{I}}\boldsymbol{\omega}^{\mathcal{A}} + {}^{\mathcal{A}}\boldsymbol{\omega}^{\mathcal{C}} + {}^{\mathcal{C}}\boldsymbol{\omega}^{\mathcal{B}} = \dot{\psi}\mathbf{a}_3 + \dot{\theta}\mathbf{c}_2 + \dot{\phi}\mathbf{b}_3. \qquad (10.46)$$

Thus *the transport equation is valid in three dimensions,* as long as we understand how to find $^{\mathcal{I}}\boldsymbol{\omega}^{\mathcal{B}}$!

Example 10.9 Airplane Kinematics

This example continues our study of the airplane. At first glance, describing the translational motion of the airplane in the inertial frame (see Figure 10.12) and finding the equations of motion using Newton's second law are easy. We treat the airplane as a point mass located at its center of mass G and write its position and velocity as components in $\mathcal{I}$ in terms of Cartesian coordinates:

$$\mathbf{r}_{G/O} = x\mathbf{e}_1 + y\mathbf{e}_2 + z\mathbf{e}_3$$

$$^{\mathcal{I}}\mathbf{v}_{G/O} = \dot{x}\mathbf{e}_1 + \dot{y}\mathbf{e}_2 + \dot{z}\mathbf{e}_3.$$

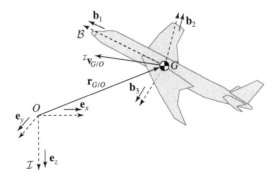

Figure 10.12 Position and velocity of an airplane with body frame $\mathcal{B}$ in inertial frame $\mathcal{I}$.

The equations of motion are three second-order differential equations in terms of the forces on the aircraft. However, the translational motion of an airplane is almost never described this way. We do use Cartesian coordinates to describe its position, but typically the velocity is expressed in terms of components in the body frame rather than the inertial frame. This is because most of the forces and moments on the plane are due to aerodynamics and are functions of the orientation of the velocity in the body frame of the aircraft. For instance, the angle that the velocity makes with the body $\mathbf{b}_1$ axis (through the nose) is called the *angle of attack* and the lift on the plane is directly proportional to it. It is thus much more convenient to calculate the lift by working with the velocity expressed in terms of components in $\mathcal{B}$.

There are two approaches for doing this. The first is by means of the transport equation. That approach involves first expressing the position of the plane's center of mass as components in the body frame, which is most easily done in matrix notation. We have

$$[\mathbf{r}_{G/O}]_\mathcal{B} = {}^\mathcal{B}C^\mathcal{I} \begin{bmatrix} x \\ y \\ z \end{bmatrix}_\mathcal{I} ,$$

where the direction-cosine matrix ${}^\mathcal{B}C^\mathcal{I}$ is derived from the transformation table for the 3-2-1 Euler angles in Example 10.8. We then use the transport equation (in matrix form) to find the components of the inertial velocity in the body frame,

$$[{}^\mathcal{I}\mathbf{v}_{G/O}]_\mathcal{B} = \frac{{}^\mathcal{B}d}{dt}\left({}^\mathcal{B}C^\mathcal{I} \begin{bmatrix} x \\ y \\ z \end{bmatrix}_\mathcal{I} \right) + \left[{}^\mathcal{I}\boldsymbol{\omega}^\mathcal{B} \times \mathbf{r}_{G/O} \right]_\mathcal{B} ,$$

where an airplane's angular-velocity components in body axes are commonly given by the letters p, q, and r:

$$ {}^\mathcal{I}\boldsymbol{\omega}^\mathcal{B} = p\mathbf{b}_1 + q\mathbf{b}_2 + r\mathbf{b}_3. $$

Although this approach certainly works to provide the velocity in terms of components in the body frame and has the added advantage of expressing the velocity components as functions of the rate of change of Cartesian coordinates, the process is tedious, and the resulting expression is extremely complex, particularly if we try to replace the rates of change of the Euler angles with the angular-velocity components. Instead, standard practice is to simply introduce new variables for the component magnitudes of the velocity along body axes:

$$ {}^\mathcal{I}\mathbf{v}_{G/O} = u\mathbf{b}_1 + v\mathbf{b}_2 + w\mathbf{b}_3. $$

When we find the dynamical equations of motion later using Newton's second law, we do so in terms of these three scalars (u, v, and w). This approach greatly reduces the amount of algebra needed to derive the airplane equations of motion. (Remember, we need the velocity expressed in the body frame to determine the forces and moments.) The disadvantage is that we are no longer describing the translational state of the airplane center of mass by coordinates and their rates but rather by the coordinates and the component magnitudes of the velocity in the body frame, which are functions of the coordinate rates. In fact, the kinematic equations relating the rate of change of the coordinates to the new velocity variables are given by the transformation

$$[{}^{\mathcal{I}}\mathbf{v}_{G/O}]_{\mathcal{I}} = \begin{bmatrix} \dot{x} \\ \dot{y} \\ \dot{z} \end{bmatrix}_{\mathcal{I}} = {}^{\mathcal{I}}C^{\mathcal{B}} \begin{bmatrix} u \\ v \\ w \end{bmatrix}_{\mathcal{B}}, \tag{10.47}$$

where

$${}^{\mathcal{I}}C^{\mathcal{B}} = \begin{bmatrix} \cos\theta\cos\psi & \cos\psi\sin\theta\sin\phi - \cos\phi\sin\psi & \cos\psi\sin\theta\cos\phi + \sin\phi\sin\psi \\ \sin\psi\cos\theta & \sin\psi\sin\theta\sin\phi + \cos\phi\cos\psi & \sin\psi\sin\theta\cos\phi - \cos\psi\sin\phi \\ -\sin\theta & \cos\theta\sin\phi & \cos\theta\cos\phi \end{bmatrix}.$$

Thus, given trajectories $u(t)$, $v(t)$, and $w(t)$ for the velocity and Euler angles $\psi(t)$, $\theta(t)$, and $\phi(t)$ (found from solving the equations of motion, done later in this chapter and the next), Eq. (10.47) can be integrated to find the inertial position of the plane. We further discuss the implications of this change of variables later in the chapter.

10.4.3 Angular Velocity and Euler-Angle Rates

Our last step in the kinematics of three-dimensional rotation is to relate the components of the angular velocity of the body frame $\mathcal{B}$ in $\mathcal{I}$ to the rates of change of the Euler angles, as these are the scalar coordinates used to describe the orientation of $\mathcal{B}$ in $\mathcal{I}$. Recall that, for the planar case, where the angular velocity is simple, this calculation was straightforward. The angular-velocity magnitude was simply $\dot{\theta}$, the rate of change of rotation of the body frame. In three dimensions, however, it gets more complicated.

From Eq. (10.46), we already have

$$\boxed{{}^{\mathcal{I}}\boldsymbol{\omega}^{\mathcal{B}} = \dot{\psi}\mathbf{a}_3 + \dot{\theta}\mathbf{c}_2 + \dot{\phi}\mathbf{b}_3}$$

for the 3-2-3 rotation. The problem is that this expression is written in terms of unit vectors of three different frames. It is most convenient to have an expression for ${}^{\mathcal{I}}\boldsymbol{\omega}^{\mathcal{B}}$ in terms of components in $\mathcal{B}$ when applying the transport equation. This is accomplished using the 3-2-3 transformation tables in Section 10.4.1, which yield

$${}^{\mathcal{I}}\boldsymbol{\omega}^{\mathcal{B}} = -\dot{\psi}\sin\theta\mathbf{c}_1 + \dot{\psi}\cos\theta\mathbf{c}_3 + \dot{\theta}\mathbf{c}_2 + \dot{\phi}\mathbf{b}_3$$
$$= (\dot{\theta}\sin\phi - \dot{\psi}\sin\theta\cos\phi)\mathbf{b}_1 + (\dot{\theta}\cos\phi + \dot{\psi}\sin\theta\sin\phi)\mathbf{b}_2 + (\dot{\phi} + \dot{\psi}\cos\theta)\mathbf{b}_3.$$

If we write the components of the angular velocity ${}^{\mathcal{I}}\boldsymbol{\omega}^{\mathcal{B}} = \omega_1\mathbf{b}_1 + \omega_2\mathbf{b}_2 + \omega_3\mathbf{b}_3$ then the expression for ${}^{\mathcal{I}}\boldsymbol{\omega}^{\mathcal{B}}$ is equivalent to the three scalar equations

$$\boxed{\begin{aligned} \omega_1 &= \dot{\theta}\sin\phi - \dot{\psi}\sin\theta\cos\phi \\ \omega_2 &= \dot{\theta}\cos\phi + \dot{\psi}\sin\theta\sin\phi \\ \omega_3 &= \dot{\phi} + \dot{\psi}\cos\theta. \end{aligned}} \tag{10.48}$$

These are called the *kinematic equations of rotation*. They relate the components of the angular velocity to the rate of change of the scalar coordinates describing orientation, just as there are equations that relate the components of the velocity of a particle to the rate of change of the scalar coordinates describing the position of the particle (e.g., Cartesian, cylindrical, or spherical coordinates). We thus have a parallel representation of the kinematics of three-dimensional orientation to that of three-dimensional translation. Our goal in the next chapter is to develop the dynamics of a rigid body, represented by a body frame $\mathcal{B}$. The resulting equations of motion are differential equations for the components of the angular velocity that we can solve simultaneously with the kinematic equations of rotation to find complete rotational trajectories. Just as with translation, each of the three degrees of freedom associated with rotation has a position-like coordinate (the Euler angles) and a speed-like coordinate (the angular-velocity components). Alternatively, we can substitute from Eq. (10.48) into the equations of motion for $^{\mathcal{I}}\boldsymbol{\omega}^{\mathcal{B}}$ to find three second-order differential equations for the Euler angles, as we did for the other coordinates earlier in the book.

You probably noticed that, if we have integrable equations of motion for the rates of change of ω_1, ω_2, and ω_3, then we also need the inverse of the kinematic equations of rotation to solve for the Euler angles $(\psi, \theta, \phi)_{\mathcal{B}}^{\mathcal{I}}$. Fortunately, these equations are not difficult to find. Simple algebra yields

$$\dot{\psi} = (-\omega_1 \cos \phi + \omega_2 \sin \phi) \csc \theta$$

$$\dot{\theta} = \omega_1 \sin \phi + \omega_2 \cos \phi \qquad\qquad (10.49)$$

$$\dot{\phi} = (\omega_1 \cos \phi - \omega_2 \sin \phi) \cot \theta + \omega_3.$$

You may also notice that these equations have a problem: at $\theta = 0$ or π they are singular (the right-hand side goes to infinity). This is a well-known problem with the Euler angle description of orientation and is referred to as *gimbal lock*. It occurs because there is an ambiguity between ψ and ϕ for describing the orientation of $\mathcal{B}$ in $\mathcal{I}$ when $\theta = 0$. (This problem came up in Example 10.3 for the spherical pendulum.) We call it gimbal lock because the singular value of θ corresponds to alignment of two of the gimbals (see Example 10.7), which creates a measurement ambiguity. Gimbal lock is one of the reasons that another set of Euler angles might be used for a particular problem. It is beyond our scope to explore this issue in more detail. The notes to this chapter suggest a number of texts for further reading.

Example 10.10 Kinematic Equations of Rotation for 3-1-3 and 3-2-1 Euler Angles

Examples 10.7 and 10.8 introduced two alternative sets of Euler angles convenient for specific geometries: the 3-1-3 angles for a set of gimbals and the 3-2-1 angles describing the orientation of an airplane. We also defined the transformation table (or direction-cosine matrix) for each. Just as they have different transformation tables, they also have different kinematic equations of rotation to relate the rates of change of the Euler angles to the angular-velocity components in the body frame. Thus, following the procedure above, the angular velocity for the 3-1-3 set is

$$^{\mathcal{I}}\boldsymbol{\omega}^{\mathcal{B}} = \dot{\psi}\mathbf{a}_3 + \dot{\theta}\mathbf{c}_1 + \dot{\phi}\mathbf{b}_3.$$

Using the appropriate frame transformations gives the 3-1-3 kinematic equations of rotation:

$$\omega_1 = \dot{\theta} \cos\phi + \dot{\psi} \sin\theta \sin\phi$$

$$\omega_2 = -\dot{\theta} \sin\phi + \dot{\psi} \sin\theta \cos\phi$$

$$\omega_3 = \dot{\phi} + \dot{\psi} \cos\theta.$$

Likewise, for the 3-2-1 set we have

$$^{\mathcal{I}}\boldsymbol{\omega}^{\mathcal{B}} = \dot{\psi}\mathbf{a}_3 + \dot{\theta}\mathbf{c}_2 + \dot{\phi}\mathbf{b}_1.$$

The frame transformations lead us to the 3-2-1 kinematic equations of rotation,

$$\omega_1 = p = \dot{\phi} - \dot{\psi} \sin\theta$$

$$\omega_2 = q = \dot{\theta} \cos\phi + \dot{\psi} \cos\theta \sin\phi$$

$$\omega_3 = r = -\dot{\theta} \sin\phi + \dot{\psi} \cos\theta \cos\phi,$$

where we have added the airplane notation for the angular-velocity components as a reminder (see Example 10.9).

Example 10.11 The Barrel Roll and Other Maneuvers

This example examines how to use the inverse relationship between the angular-velocity components (p, q, and r) of the airplane and the Euler angle rates ($\dot{\psi}$, $\dot{\theta}$, and $\dot{\phi}$). Inverting the 3-2-1 kinematic equations from the previous example yields

$$\dot{\psi} = (q \sin\phi + r \cos\phi) \sec\theta$$

$$\dot{\theta} = q \cos\phi - r \sin\phi \qquad\qquad (10.50)$$

$$\dot{\phi} = (q \sin\phi + r \cos\phi) \tan\theta + p.$$

Imagine the airplane performing a barrel roll about the $\mathbf{b}_3$ axis, as shown in Figure 10.13. During such a maneuver, $q = r = 0$, and $p \neq 0$. Eq. (10.50) tells us that $\dot{\psi} = \dot{\theta} = 0$ and $\dot{\phi} = p$, perhaps not a surprise. However, if the airplane performs a loop de loop maneuver, during which $p = r = 0$ and $q \neq 0$, observe that

$$\dot{\psi} = q \sin\phi \sec\theta$$

$$\dot{\theta} = q \cos\phi$$

$$\dot{\phi} = q \sin\phi \tan\theta.$$

That is, the loop de loop maneuver generates nonzero rates of change of all three Euler angles! In fact, it is only when the airplane loops with exactly zero roll ($\phi = 0$) that the expected result of $\dot{\psi} = \dot{\phi} = 0$ and $\dot{\theta} = q$ holds. Therefore, it is in general a mistake to assume that simple three-dimensional rotations of a rigid body are associated with simple angular rates. However, simple rotations are associated with simple angular velocities, provided that the angular velocity is expressed as components in the body frame.

Figure 10.13 Barrel roll. Image courtesy of Shutterstock.

Example 10.12 Target Tracking, Part 2

This example revisits the target-tracking problem explored in Example 10.4. Here we study the kinematics using Euler angles and the angular velocity rather than spherical coordinates. Figure 10.14 shows the dish antenna in the inertial frame, now with a body frame added and $\mathbf{b}_2$ pointing along the symmetry axis of the dish.

In this problem we solve for the commanded rates of change of the Euler angles so that the dish stays pointed at target P, traveling at constant velocity in the $\mathbf{e}_x$ direction, $^{\mathcal{I}}\mathbf{v}_{P/O} = v\mathbf{e}_x$. For example, P may be a ship with known position and velocity that needs to maintain communication with a land base via the dish.

The rotation taking the dish from the inertial frame to the arbitrary body-frame orientation is given by the 3-1-3 Euler angle set with $\psi = AZ$, $\theta = EL$, and $\phi = 0$. (No final rotation is needed because of the symmetry of the dish.) The goal is to find first-order differential equations for the Euler angles that can be solved to determine the angular-rate commands for the dish motors.

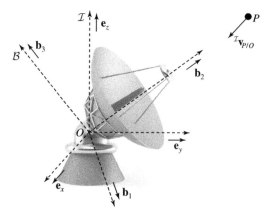

Figure 10.14 Dish antenna with attached body frame $\mathcal{B} = (O, \mathbf{b}_1, \mathbf{b}_2, \mathbf{b}_3)$ tracking point P traveling at velocity $^{\mathcal{I}}\mathbf{v}_{P/O}$. Image courtesy of Shutterstock.

We begin by assuming that the target has position $\mathbf{r}_{P/O}$. If the distance to the target is d, then

$$\mathbf{r}_{P/O} = d\mathbf{b}_2.$$

We use the transport equation to rewrite the target inertial velocity,

$$^{\mathcal{I}}\mathbf{v}_{P/O} = {}^{\mathcal{B}}\mathbf{v}_{P/O} + {}^{\mathcal{I}}\boldsymbol{\omega}^{\mathcal{B}} \times \mathbf{r}_{P/O} = v\mathbf{e}_x,$$

where $^{\mathcal{B}}\mathbf{v}_{P/O} = \dot{d}\mathbf{b}_2$, the range rate in body axes, and $^{\mathcal{I}}\boldsymbol{\omega}^{\mathcal{B}} = \omega_1\mathbf{b}_1 + \omega_2\mathbf{b}_2 + \omega_3\mathbf{b}_3$. Writing this out as components in $\mathcal{B}$ gives

$$v\mathbf{e}_x = \dot{d}\mathbf{b}_2 + \omega_1 d\mathbf{b}_3 - \omega_3 d\mathbf{b}_1.$$

Using the transformation table derived in Example 10.7 for the 3-1-3 Euler angles allows us to convert this to the following three scalar equations (by writing $\mathbf{e}_x$ in body axes):

$$v\cos\psi = -\omega_3 d$$

$$-v\sin\psi\cos\theta = \dot{d}$$

$$v\sin\psi\sin\theta = \omega_1 d,$$

where we have set $\phi = 0$. These equations are solved for ω_1 and ω_3, which are then substituted into the kinematic equations for a 3-1-3 Euler angle set from Example 10.10. Inverting the angular-velocity relationship for the 3-1-3 system gives

$$\dot{\psi} = (\omega_1\sin\phi + \omega_2\cos\phi)\sec\theta$$

$$\dot{\theta} = \omega_1\cos\phi - \omega_2\sin\phi$$

$$\dot{\phi} = -(\omega_1\sin\phi + \omega_2\cos\phi)\cot\theta + \omega_3.$$

Setting $\dot{\phi} = \phi = 0$ allows us to use the third equation to solve for $\omega_2 = \omega_3\tan\theta$. The three angular velocities are then substituted into the first two equations to give the following set of differential equations:

$$\dot{\psi} = -\frac{v\cos\psi}{d\cos\theta}$$

$$\dot{\theta} = \frac{v\sin\psi\sin\theta}{d}$$

$$\dot{d} = -v\sin\psi\cos\theta.$$

Integrating these equations from a known set of initial conditions (e.g., when the antenna locks onto a target) provides the Euler-angle commands to track the target.

10.4.4 Multiple Frames and the Addition Property

Although it seems like we have already demonstrated the addition property of angular velocities in three dimensions in Section 10.4.2, that is not quite true, as each angular velocity in the sum was a simple angular velocity associated with an intermediate frame. Nevertheless, the general addition property of angular velocities in three dimensions is in fact true. That is, suppose we have three frames undergoing arbitrary rotations with respect to one another. In other words, suppose that frame $\mathcal{A}$ is rotating in frame $\mathcal{I}$ with angular velocity $^{\mathcal{I}}\boldsymbol{\omega}^{\mathcal{A}}$. Suppose also that frame $\mathcal{B}$ is rotating in frame $\mathcal{A}$ with angular velocity $^{\mathcal{A}}\boldsymbol{\omega}^{\mathcal{B}}$, and suppose frame $\mathcal{C}$ is rotating in frame $\mathcal{B}$ with angular velocity $^{\mathcal{B}}\boldsymbol{\omega}^{\mathcal{C}}$. Then it is a fact that

$$^{\mathcal{I}}\boldsymbol{\omega}^{\mathcal{C}} = {}^{\mathcal{I}}\boldsymbol{\omega}^{\mathcal{A}} + {}^{\mathcal{A}}\boldsymbol{\omega}^{\mathcal{B}} + {}^{\mathcal{B}}\boldsymbol{\omega}^{\mathcal{C}}. \tag{10.51}$$

This addition property, of course, applies to any number of frames.

Fortunately, we have already derived Eq. (10.51)! Review Section 8.2.2 and Figure 8.9. The derivation there is entirely in terms of vectors and uses only the transport equation. Because we have now shown that the transport equation is true in three dimensions, the treatment in Section 8.2.2 is completely general and applies in three dimensions as well. Thus not only is the addition property true for general angular velocities in three dimensions, but the equation for the velocity of point P in $\mathcal{I}$ involving multiple frames of reference is still

$$^{\mathcal{I}}\mathbf{v}_{P/O} = {}^{\mathcal{A}}\mathbf{v}_{O'/O} + {}^{\mathcal{I}}\boldsymbol{\omega}^{\mathcal{A}} \times \mathbf{r}_{O'/O} + {}^{\mathcal{B}}\mathbf{v}_{O''/O'} + {}^{\mathcal{I}}\boldsymbol{\omega}^{\mathcal{B}} \times \mathbf{r}_{O''/O'}$$
$$+ {}^{\mathcal{C}}\mathbf{v}_{P/O''} + {}^{\mathcal{I}}\boldsymbol{\omega}^{\mathcal{C}} \times \mathbf{r}_{P/O''}.$$

Note the origin of frames $\mathcal{I}$ and $\mathcal{A}$ is O, the origin of frame $\mathcal{B}$ is O', and frame $\mathcal{C}$ has origin O''.

10.4.5 Three-Dimensional Particle Kinetics in a Rotating Frame

If you look back at Section 8.3, where the acceleration of a particle relative to a translating and rotating frame is discussed, you will see that we never used the planar assumption. Because we have just shown that the transport equation applies in three dimensions, the entire development in Section 8.3 holds. Nevertheless, for completeness, and to remind you yet again of their importance, we restate the fundamental equations for velocity and acceleration involving translating and rotating frames. That is, if frame $\mathcal{B}$ is translating and rotating with respect to frame $\mathcal{I}$, as shown in Figure 10.15, then the transport equation can be used to write the velocity of point P in $\mathcal{I}$ in terms of its velocity in $\mathcal{B}$ and the velocity of the origin O' of frame $\mathcal{B}$:

$$^{\mathcal{I}}\mathbf{v}_{P/O} = {}^{\mathcal{I}}\mathbf{v}_{O'/O} + {}^{\mathcal{B}}\mathbf{v}_{P/O'} + {}^{\mathcal{I}}\boldsymbol{\omega}^{\mathcal{B}} \times \mathbf{r}_{P/O'}.$$

Likewise, using the same derivation as in Section 8.3, the acceleration of P is

$$^{\mathcal{I}}\mathbf{a}_{P/O} = {}^{\mathcal{I}}\mathbf{a}_{O'/O} + {}^{\mathcal{B}}\mathbf{a}_{P/O'} + {}^{\mathcal{I}}\boldsymbol{\alpha}^{\mathcal{B}} \times \mathbf{r}_{P/O'} + 2{}^{\mathcal{I}}\boldsymbol{\omega}^{\mathcal{B}} \times {}^{\mathcal{B}}\mathbf{v}_{P/O'}$$
$$+ {}^{\mathcal{I}}\boldsymbol{\omega}^{\mathcal{B}} \times \left({}^{\mathcal{I}}\boldsymbol{\omega}^{\mathcal{B}} \times \mathbf{r}_{P/O'}\right). \tag{10.52}$$

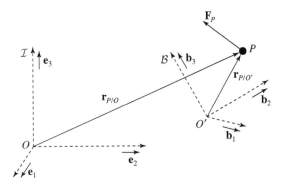

Figure 10.15 Particle P subject to force $\mathbf{F}_P$ moving relative to frame $\mathcal{B}$, which may rotate or translate in three dimensions relative to frame $\mathcal{I}$.

Just as we did earlier in the book, we can now apply Newton's second law to a particle (or collection of particles), given only their velocities and accelerations in a translating and rotating frame. Our discussion of the centripetal and Coriolis accelerations in Section 8.3 is unchanged, though we can now treat slightly more complex examples.

Example 10.13 Tool Bag in Space

This example uses Eq. (10.52) to study the motion of an object relative to a moving reference frame. Imagine that an astronaut on a space walk outside a spacecraft accidentally drops her tool bag and it drifts away from the spacecraft.[2] In this example, P is the tool bag, O' is the spacecraft, and O is fixed relative to the stars. Frame $\mathcal{B}$ is attached to the spacecraft. We would like to understand the motion of the tool bag relative to frame $\mathcal{B}$, since this is the perspective of the astronaut. Suppose that during the space walk the spacecraft thrusters are off, so $^{\mathcal{I}}\mathbf{a}_{O'/O} = 0$, $^{\mathcal{I}}\boldsymbol{\alpha}^{\mathcal{B}} = 0$, and $^{\mathcal{I}}\boldsymbol{\omega}^{\mathcal{B}}$ is constant. Furthermore, since there are no forces acting on the tool bag, $^{\mathcal{I}}\mathbf{a}_{P/O} = 0$. Rearranging Eq. (10.52) gives

$$^{\mathcal{B}}\mathbf{a}_{P/O'} = -2\,^{\mathcal{I}}\boldsymbol{\omega}^{\mathcal{B}} \times \,^{\mathcal{B}}\mathbf{v}_{P/O'} - \,^{\mathcal{I}}\boldsymbol{\omega}^{\mathcal{B}} \times \left(^{\mathcal{I}}\boldsymbol{\omega}^{\mathcal{B}} \times \mathbf{r}_{P/O'}\right). \qquad (10.53)$$

Using the Cartesian coordinates $(x, y, z)_{\mathcal{B}}$ to describe the position of the tool bag relative to O' and writing Eq. (10.53) in matrix notation with respect to frame $\mathcal{B}$ yields

$$\begin{bmatrix} \ddot{x} \\ \ddot{y} \\ \ddot{z} \end{bmatrix}_{\mathcal{B}} = -2 \begin{bmatrix} \omega_1 \\ \omega_2 \\ \omega_3 \end{bmatrix}_{\mathcal{B}} \times \begin{bmatrix} \dot{x} \\ \dot{y} \\ \dot{z} \end{bmatrix}_{\mathcal{B}} - \begin{bmatrix} \omega_1 \\ \omega_2 \\ \omega_3 \end{bmatrix}_{\mathcal{B}} \times \begin{bmatrix} \omega_1 \\ \omega_2 \\ \omega_3 \end{bmatrix}_{\mathcal{B}} \times \begin{bmatrix} x \\ y \\ z \end{bmatrix}_{\mathcal{B}}, \qquad (10.54)$$

where ω_1, ω_2, and ω_3 are the components of $^{\mathcal{I}}\boldsymbol{\omega}^{\mathcal{B}}$ expressed in frame $\mathcal{B}$.

---

[2] This incident happened outside the International Space Station on November 18, 2008.

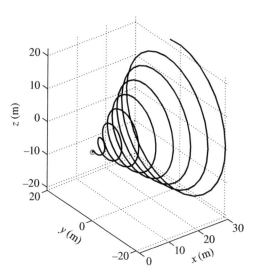

Figure 10.16 Tool bag in space.

Now, without loss of generality, suppose that the (constant) angular velocity of the spacecraft is aligned with the first axis of frame $\mathcal{B}$, so that $\omega_1 \neq 0$, and $\omega_2 = \omega_3 = 0$. Eq. (10.54) is equivalent to

$$\ddot{x} = 0$$
$$\ddot{y} = 2\omega_1\dot{z} + \omega_1^2 y \qquad (10.55)$$
$$\ddot{z} = -2\omega_1\dot{y} + \omega_1^2 z.$$

Integrating Eq. (10.55) using $\omega_1 = 0.5$ m/s, $x(0) = y(0) = z(0) = 0$ m, $\dot{x}(0) = 0.3$ m/s, $\dot{y}(0) = 0.1$ m/s, and $\dot{z}(0) = 0.2$ m/s yields the trajectory shown in Figure 10.16. From the perspective of the spacecraft (small circle), the tool bag spirals away at a constant rotational rate in the y-z plane and a constant (linear) rate in the x direction.

Example 10.14 The Translational Equations of Motion for an Airplane

This problem uses the results from Example 10.9 to find the translational equations of motion for an airplane, that is, the dynamic equations that let us solve for the trajectory of the plane in Cartesian coordinates in the inertial frame. Recall that we can use our most basic results from the beginning of the book (and this chapter) to write three simple equations of motion in Cartesian coordinates,

$$\ddot{x} = f_x/m_G$$
$$\ddot{y} = f_y/m_G$$
$$\ddot{z} = f_z/m_G,$$

where f_x, f_y, and f_z are the component magnitudes of the applied force in the inertial frame. Note that here we have assumed that the separation principle for rigid bodies still holds in three dimensions and that the airplane can be treated as a point mass located at its center of mass, something not verified until the next chapter.

In the absence of any other information about the forces on the plane, these equations seem like a reasonable approach. However, remember that we don't usually write the velocity of an airplane in terms of components in the inertial frame but rather as components in the body frame,

$$^{\mathcal{I}}\mathbf{v}_{G/O} = u\mathbf{b}_1 + v\mathbf{b}_2 + w\mathbf{b}_3.$$

It is thus convenient to have equations of motion for the body-frame components of the velocity. This is easily done using the transport equation and Newton's second law applied to the center of mass:[3]

$$\frac{\mathbf{F}_G}{m_G} = \frac{^{\mathcal{B}}d}{dt}\left(^{\mathcal{I}}\mathbf{v}_{G/O}\right) + {}^{\mathcal{I}}\boldsymbol{\omega}^{\mathcal{B}} \times {}^{\mathcal{I}}\mathbf{v}_{G/O}. \tag{10.56}$$

This equation looks a bit odd, as we are taking the derivative of the inertial velocity relative to the body frame; we have never done that before. This approach is fine, though. Remember, the transport equation works on any vector. Here we are just applying it to the inertial velocity. This is a variation of the acceleration equation in Eq. (10.52).

Performing the operations on the components of the vectors in Eq. (10.56) results in the three scalar equations,

$$\dot{u} = rv - qw + F_1/m_G \tag{10.57}$$

$$\dot{v} = pw - ru + F_2/m_G \tag{10.58}$$

$$\dot{w} = qu - pv + F_3/m_G, \tag{10.59}$$

where F_1, F_2, and F_3 are the three component magnitudes of the total force on the plane in the body frame. Eqs. (10.57)–(10.59) are combined with the three kinematic equations in Eq. (10.47) to get the six first-order equations of motion for airplane translation. Of course, they can't be integrated without explicit expressions for the forces. We discuss the applied forces and the solution trajectories in Chapter 12.

10.5 Derivations—Euler's Theorem and the Angular Velocity

In this section we show that the angular velocity introduced above as an operator corresponds to the instantaneous axis of rotation of body frame $\mathcal{B}$ in inertial frame $\mathcal{I}$. Note that the material in this section is not essential for the continuity of the rest of the book. You can simply take our word that the angular velocity represents a time-varying rotation about an instantaneous axis aligned with it—the so-called *Euler axis*. However, we encourage you to read on, as a familiarity with Euler's theorem is extremely helpful in understanding the rotation of rigid bodies.

---

[3] We justify this approach in the next chapter.

Recall that, when angular velocity was introduced in Chapter 3, it was obviously aligned with the axis of rotation of the body. The first three parts of the book were confined to considering planar motion and to body frames that rotated in absolute space about a single axis only, usually denoted the 3-axis and represented by a unit vector such as $\mathbf{b}_3$. The angular velocity was thus given by $^{\mathcal{I}}\boldsymbol{\omega}^{\mathcal{B}} = \dot{\theta}\mathbf{b}_3$. We called this a *simple* angular velocity. It was directed along the axis of rotation, and its magnitude was given by the rate of rotation about that axis.

This chapter introduced the more general three-dimensional rotation of reference frame $\mathcal{B}$. We used our earlier result for the simple angular velocity, namely, that it is an operator that provides unit-vector derivatives (by the transport equation) to show that the transport equation still holds in three dimensions, where the angular velocity is given by the sum of simple angular velocities associated with intermediate frames, as in Eq. (10.44). We now show that this angular velocity is directed along the instantaneous axis of rotation and that its magnitude is equal to the instantaneous rate of rotation. To begin, we return to the relative orientation of two reference frames.

10.5.1 Euler's Theorem

Section 10.4.1 showed that the relative orientation of frame $\mathcal{B}$ in $\mathcal{I}$ can be described by three angles (the Euler angles) and that the magnitude of vector components can be transformed from one frame to the other using the transformation matrix $^{\mathcal{I}}C^{\mathcal{B}}$. We now introduce another way to describe the orientation of a rigid body (or reference frame) relative to another frame: *Euler's Theorem of Rotation.*

Theorem 10.1 Euler's Theorem of Rotation The orientation of a reference frame $\mathcal{B}$ relative to another frame $\mathcal{I}$ can be described by a single, simple rotation of $\mathcal{B}$, initially aligned with $\mathcal{I}$, about an axis fixed in both $\mathcal{B}$ and $\mathcal{I}$.

Figure 10.17 depicts the geometry of Euler's theorem, where the two frames $\mathcal{I}$ and $\mathcal{B}$ are initially aligned and then $\mathcal{B}$ rotates by an angle θ about the axis of rotation L, whose direction is given by the unit vector $\mathbf{k}$. Specifying $\mathbf{k}$ and the rotation angle θ is another way of describing the orientation of $\mathcal{B}$ in $\mathcal{I}$. Observe that there are still three independent quantities describing the orientation (not four), because $\mathbf{k} = k_1\mathbf{e}_1 + k_2\mathbf{e}_2 + k_3\mathbf{e}_3$ is a unit vector, thus implying the constraint $\|\mathbf{k}\|^2 = k_1^2 + k_2^2 + k_3^2 = 1$.

Recall also that the transformation matrix $^{\mathcal{I}}C^{\mathcal{B}}$ always has one eigenvalue equal to 1 (as discussed in Section 10.4.1). The eigenvector associated with that eigenvalue is the unit vector $\mathbf{k}$! This is because if $\mathbf{k}$ is along the axis of rotation, it is necessarily the same in both frames:

$$[\mathbf{k}]_{\mathcal{I}} = {}^{\mathcal{I}}C^{\mathcal{B}}[\mathbf{k}]_{\mathcal{B}} = [\mathbf{k}]_{\mathcal{B}}.$$

This expression is an eigenvalue equation for matrix $^{\mathcal{I}}C^{\mathcal{B}}$ with unity eigenvalue. Remember that $\mathbf{k}$ is fixed in both $\mathcal{I}$ and $\mathcal{B}$. Its components in $\mathcal{I}$ and $\mathcal{B}$ are the same before and after the rotation, which implies

$$\mathbf{k} = k_1\mathbf{e}_1 + k_2\mathbf{e}_2 + k_3\mathbf{e}_3 = k_1\mathbf{b}_1 + k_2\mathbf{b}_2 + k_3\mathbf{b}_3.$$

Euler's theorem is a profound and important result that is useful for understanding angular velocity. Before we prove Euler's theorem, however, it is helpful to introduce a corollary[4] that is used both in our proof and in the discussion of angular velocity.

---

[4] This corollary follows the discussion of a simple rotation in Kane et al. (1983).

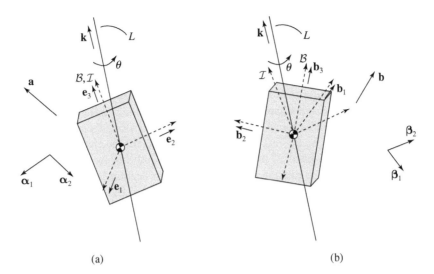

Figure 10.17 Change of vector **a** in absolute space to vector **b** after an Euler rotation of angle θ about axis L. (a) Before rotation. (b) After rotation.

Corollary 10.1 (Kane et al. 1983) Consider an arbitrary vector **a** fixed in frame $\mathcal{I}$ and a vector **b** fixed in $\mathcal{B}$. If $\mathcal{B}$ is aligned with $\mathcal{I}$ prior to the rotation, such that **b** = **a**, then after $\mathcal{B}$ undergoes a rotation by an amount θ about the line L (the Euler axis), the vector **b** can be written in terms of **a**, the unit vector **k** along the axis of rotation, and the angle θ by means of

$$\mathbf{b} = \mathbf{a} \cos \theta - (\mathbf{a} \times \mathbf{k}) \sin \theta + (\mathbf{a} \cdot \mathbf{k})\mathbf{k}(1 - \cos \theta). \tag{10.60}$$

To prove this corollary, we introduce two unit vectors fixed in $\mathcal{I}$, $\boldsymbol{\alpha}_1$ and $\boldsymbol{\alpha}_2$, both perpendicular to the rotation axis L, so that the three vectors $\boldsymbol{\alpha}_1$, $\boldsymbol{\alpha}_2$, and **k** form a right-handed orthogonal set, as in Figure 10.17a. Thus we can write vector **a** as components in these unit-vector directions fixed in $\mathcal{I}$,

$$\mathbf{a} = a_1 \boldsymbol{\alpha}_1 + a_2 \boldsymbol{\alpha}_2 + a_3 \mathbf{k}. \tag{10.61}$$

We next introduce two unit vectors fixed in frame $\mathcal{B}$, $\boldsymbol{\beta}_1$ and $\boldsymbol{\beta}_2$, also both perpendicular to L, and such that before the rotation $\boldsymbol{\alpha}_1 = \boldsymbol{\beta}_1$ and $\boldsymbol{\alpha}_2 = \boldsymbol{\beta}_2$. Likewise, the three vectors $\boldsymbol{\beta}_1$, $\boldsymbol{\beta}_2$, and **k** form a right-handed orthogonal set fixed in $\mathcal{B}$. Since **b** is aligned with **a** before the rotation, its components in these directions are the same:

$$\mathbf{b} = a_1 \boldsymbol{\beta}_1 + a_2 \boldsymbol{\beta}_2 + a_3 \mathbf{k}. \tag{10.62}$$

Since **b** is fixed in $\mathcal{B}$, these are its components after the rotation as well.

Now we express **b** as components in terms of $\boldsymbol{\alpha}_1$, $\boldsymbol{\alpha}_2$, and **k** after the rotation, as shown in Figure 10.17b. This is done by writing the two unit vectors $\boldsymbol{\beta}_1$ and $\boldsymbol{\beta}_2$ in terms of unit vectors $\boldsymbol{\alpha}_1$ and $\boldsymbol{\alpha}_2$. Since these unit vectors differ only by a simple rotation of angle θ about a mutually orthogonal direction, **k**, we can use the transformation table for a simple planar rotation to find

$$\boldsymbol{\beta}_1 = \cos \theta \boldsymbol{\alpha}_1 + \sin \theta \boldsymbol{\alpha}_2$$
$$\boldsymbol{\beta}_2 = -\sin \theta \boldsymbol{\alpha}_1 + \cos \theta \boldsymbol{\alpha}_2$$

after the rotation. Substituting these into Eq. (10.62) gives

$$\mathbf{b} = (a_1 \cos\theta - a_2 \sin\theta)\boldsymbol{\alpha}_1 + (a_1 \sin\theta + a_2 \cos\theta)\boldsymbol{\alpha}_2 + a_3\mathbf{k}.$$

This equation is the same result obtained by substituting the expression for $\mathbf{a}$ in Eq. (10.61) into Eq. (10.60), thus verifying Corollary 10.1. Another nice check of Eq. (10.60) is to substitute $\mathbf{k}$ for $\mathbf{a}$ and show that the result is $\mathbf{b} = \mathbf{k}$, as expected, because $\mathbf{k}$ does not change with the rotation, as it is along the axis of rotation.

We can now use Eq. (10.60) to find expressions for the elements of the direction-cosine matrix in terms of the component magnitudes of the unit vector $\mathbf{k} = k_1\mathbf{e}_1 + k_2\mathbf{e}_2 + k_3\mathbf{e}_3$ and the rotation angle θ. Recall from Section 10.4.1 that the elements of the direction-cosine matrix, $^{\mathcal{I}}C^{\mathcal{B}}$, between reference frame $\mathcal{I} = (O, \mathbf{e}_1, \mathbf{e}_2, \mathbf{e}_3)$ and $\mathcal{B} = (O, \mathbf{b}_1, \mathbf{b}_2, \mathbf{b}_3)$ are given by

$$^{\mathcal{I}}C^{\mathcal{B}}_{ij} = \mathbf{e}_i \cdot \mathbf{b}_j \quad i, j = 1, 2, 3.$$

Using the fact that $\mathbf{b}_j = \mathbf{e}_j$ prior to the rotation, we can substitute for $\mathbf{b}_j$ from Eq. (10.60) to obtain

$$^{\mathcal{I}}C^{\mathcal{B}}_{ij} = \mathbf{e}_i \cdot \left(\mathbf{e}_j \cos\theta - (\mathbf{e}_j \times \mathbf{k})\sin\theta + (\mathbf{e}_j \cdot \mathbf{k})\mathbf{k}(1 - \cos\theta)\right).$$

Using component magnitudes of $\mathbf{k}$ in $\mathcal{I}$ (or $\mathcal{B}$) results in the following expressions for each element of the transformation matrix:

$$^{\mathcal{I}}C^{\mathcal{B}}_{11} = \cos\theta + k_1^2(1 - \cos\theta)$$

$$^{\mathcal{I}}C^{\mathcal{B}}_{12} = -k_3 \sin\theta + k_1 k_2(1 - \cos\theta)$$

$$^{\mathcal{I}}C^{\mathcal{B}}_{13} = k_2 \sin\theta + k_1 k_3(1 - \cos\theta)$$

$$^{\mathcal{I}}C^{\mathcal{B}}_{21} = k_3 \sin\theta + k_1 k_2(1 - \cos\theta)$$

$$^{\mathcal{I}}C^{\mathcal{B}}_{22} = \cos\theta + k_2^2(1 - \cos\theta)$$

$$^{\mathcal{I}}C^{\mathcal{B}}_{23} = -k_1 \sin\theta + k_2 k_3(1 - \cos\theta)$$

$$^{\mathcal{I}}C^{\mathcal{B}}_{31} = -k_2 \sin\theta + k_3 k_1(1 - \cos\theta)$$

$$^{\mathcal{I}}C^{\mathcal{B}}_{32} = k_1 \sin\theta + k_3 k_2(1 - \cos\theta)$$

$$^{\mathcal{I}}C^{\mathcal{B}}_{33} = \cos\theta + k_3^2(1 - \cos\theta).$$

As expected, we can write the transformation matrix either in terms of the three Euler angles describing the orientation of $\mathcal{B}$ in $\mathcal{I}$ (as in Eq. (10.37)) or in terms of the components of the unit vector $\mathbf{k}$ along the Euler axis of rotation and the rotation angle θ. Either set is a valid description of the orientation of $\mathcal{B}$ in $\mathcal{I}$.

There is something even more important about these relationships. We can use them to prove Euler's theorem. Suppose that the rigid body (reference frame $\mathcal{B}$) has some arbitrary orientation in $\mathcal{I}$. This orientation could be described by the Euler angles $(\psi, \theta, \phi)^{\mathcal{I}}_{\mathcal{B}}$, as in the previous section. And as in Section 10.4.1, we can write the transformation matrix in terms of these angles. The inverse of the above relationships then provides expressions for the components of the unit vector $\mathbf{k}$ along the Euler axis of rotation and the rotation angle in terms of the components of this matrix:

$$\theta = 2 \arccos \left(\frac{\sqrt{1 + {}^{\mathcal{I}}C_{11}^{\mathcal{B}} + {}^{\mathcal{I}}C_{22}^{\mathcal{B}} + {}^{\mathcal{I}}C_{33}^{\mathcal{B}}}}{2} \right) \tag{10.63}$$

$$k_1 = \frac{{}^{\mathcal{I}}C_{32}^{\mathcal{B}} - {}^{\mathcal{I}}C_{23}^{\mathcal{B}}}{2 \sin \theta} \tag{10.64}$$

$$k_2 = \frac{{}^{\mathcal{I}}C_{13}^{\mathcal{B}} - {}^{\mathcal{I}}C_{31}^{\mathcal{B}}}{2 \sin \theta} \tag{10.65}$$

$$k_3 = \frac{{}^{\mathcal{I}}C_{21}^{\mathcal{B}} - {}^{\mathcal{I}}C_{12}^{\mathcal{B}}}{2 \sin \theta}. \tag{10.66}$$

Eqs. (10.63)–(10.66) constitute a proof of Euler's theorem! In other words, given any arbitrary orientation and the transformation matrix (or Euler angles) describing it, we can always find a single axis of rotation and an angle about that axis corresponding to the orientation by using Eqs. (10.63)–(10.66).

10.5.2 The Angular Velocity

Finally, we turn our attention back to the angular velocity. Suppose now that the rigid body (frame $\mathcal{B}$) is changing orientation with time in $\mathcal{I}$. The rate of change of vectors is given by the transport equation, but what does the angular velocity physically represent? If the orientation of $\mathcal{B}$ is changing with time, then over any small time interval, that change could be represented by an Euler axis rotation by using Euler's theorem. Thus the Euler axis represents the instantaneous axis of rotation. If we again consider two vectors $\mathbf{a}$ and $\mathbf{b}$, with $\mathbf{a}$ fixed in $\mathcal{I}$ and $\mathbf{b}$ fixed in $\mathcal{B}$ and initially aligned, then a very short time later the vector $\mathbf{b}$ is given by Eq. (10.60) in Corollary 10.1 (where θ is small). What, then, is the rate of change of $\mathbf{b}$? Taking the inertial derivative of Eq. (10.60) yields

$$\frac{{}^{\mathcal{I}}d}{dt}\mathbf{b} = -\dot{\theta}(\mathbf{a} \sin \theta + (\mathbf{a} \times \mathbf{k}) \cos \theta - (\mathbf{a} \cdot \mathbf{k})\mathbf{k} \sin \theta), \tag{10.67}$$

where $\mathbf{k}$ is along the instantaneous axis of rotation and we have used the fact that $\mathbf{a}$ and $\mathbf{k}$ are fixed in $\mathcal{I}$. We also know the derivative of $\mathbf{b}$ from the transport equation:

$$\frac{{}^{\mathcal{I}}d}{dt}\mathbf{b} = \underbrace{\frac{{}^{\mathcal{B}}d}{dt}\mathbf{b}}_{=0} + {}^{\mathcal{I}}\boldsymbol{\omega}^{\mathcal{B}} \times \mathbf{b}. \tag{10.68}$$

Eqs. (10.67) and (10.68) for the inertial velocity of $\mathbf{b}$ should, of course, be the same. Let us suppose that ${}^{\mathcal{I}}\boldsymbol{\omega}^{\mathcal{B}} = \dot{\theta}\mathbf{k}$, that is, that the angular velocity is a simple rotation of rate $\dot{\theta}$ about the Euler axis. Then substituting this expression for ${}^{\mathcal{I}}\boldsymbol{\omega}^{\mathcal{B}}$ into Eq. (10.68) and using the expression for $\mathbf{b}$ in Eq. (10.60) gives

$${}^{\mathcal{I}}\boldsymbol{\omega}^{\mathcal{B}} \times \mathbf{b} = \dot{\theta}[(\mathbf{k} \times \mathbf{a}) \cos \theta - \mathbf{k} \times (\mathbf{a} \times \mathbf{k}) \sin \theta + (\mathbf{a} \cdot \mathbf{k}) \underbrace{\mathbf{k} \times \mathbf{k}}_{=0}(1 - \cos \theta)]$$

$$= -\dot{\theta} \left[(\mathbf{a} \times \mathbf{k}) \cos \theta + \mathbf{a} \sin \theta - (\mathbf{a} \cdot \mathbf{k})\mathbf{k} \sin \theta \right], \tag{10.69}$$

where we have used the triple vector cross product identity from Appendix B: $\mathbf{k} \times (\mathbf{a} \times \mathbf{k}) = \mathbf{a}(\mathbf{k} \cdot \mathbf{k}) - \mathbf{k}(\mathbf{a} \cdot \mathbf{k})$. Eq. (10.69) is identical to Eq. (10.67), as required. This shows that the angular velocity is indeed given by $^{\mathcal{I}}\boldsymbol{\omega}^{\mathcal{B}} = \dot{\theta}\mathbf{k}$; it is directed along the instantaneous Euler axis and is equal in magnitude to the instantaneous rate of rotation about that axis. *The angular velocity represents the rotation rate about an instantaneous axis of rotation parallel to it.*

10.6 Tutorials

Tutorial 10.1 The Local Vertical

Example 10.2 introduced the geocentric inertial frame $\mathcal{I} = (O, \mathbf{e}_1, \mathbf{e}_2, \mathbf{e}_3)$, and the geographic frame $\mathcal{G} = (O, \mathbf{g}_1, \mathbf{g}_2, \mathbf{g}_3)$; the former is an inertial frame located at the center of the earth and the latter is located at the earth's center and fixed to the earth. The earth rotates in absolute space at a rate Ω_E (i.e., once per day). The angular velocity is

$$^{\mathcal{I}}\boldsymbol{\omega}^{\mathcal{G}} = \Omega_E \mathbf{g}_3.$$

In this tutorial we are interested in examining the equations of motion of particles relative to a "local" frame on the surface of the earth. That is, we consider a point O' somewhere on the surface of the earth. Figure 10.18a locates that point with the spherical coordinates $(R_E, \theta, \phi)_{\mathcal{G}}$, where R_E is the earth's radius, θ is the longitude, and ϕ is the co-latitude ($90°$ minus the latitude). We then place frame $\mathcal{B} = (O', \mathbf{b}_1, \mathbf{b}_2, \mathbf{b}_3)$ at O' with $\mathbf{b}_3$ along the radius direction to the center of the earth, $\mathbf{b}_1$ pointed south (i.e., perpendicular to $\mathbf{b}_3$ and in the plane formed by $\mathbf{b}_3$ and $\mathbf{g}_3$), and $\mathbf{b}_2 = \mathbf{b}_3 \times \mathbf{b}_1$ (as shown in Figure 10.18b).

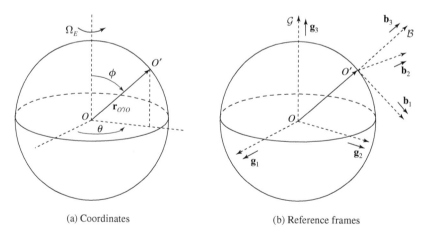

(a) Coordinates   (b) Reference frames

Figure 10.18 A spherical earth with point O' located on the surface by spherical coordinates ϕ and θ. Reference frame $\mathcal{G}$ is located at the center O of the earth fixed to the earth's surface and rotating in absolute space with angular velocity $^{\mathcal{I}}\boldsymbol{\omega}^{\mathcal{G}} = \Omega_E \mathbf{g}_3$. Local frame $\mathcal{B}$ is also fixed to the earth at O'.

Because the earth is very large, for most problems considered on the surface of the earth, $\mathcal{B}$ can be treated as inertial (this assumption holds for almost every problem we have considered in the book). For problems involving gravity near the surface of the earth, for instance, we usually assume that the force of gravity is constant and is always directed downward in frame $\mathcal{B}$ (i.e., in the $\mathbf{b}_3$ direction). This tutorial tests how good an assumption that is.

To consider the dynamics in frame $\mathcal{B}$ of particle P, located at $\mathbf{r}_{P/O'}$, we begin with the expression for the inertial acceleration of P in Eq. (10.52) and Newton's second law,

$$
^{\mathcal{I}}\mathbf{a}_{P/O} = {}^{\mathcal{I}}\mathbf{a}_{O'/O} + {}^{\mathcal{B}}\mathbf{a}_{P/O'} + 2^{\mathcal{I}}\boldsymbol{\omega}^{\mathcal{B}} \times {}^{\mathcal{B}}\mathbf{v}_{P/O'} + {}^{\mathcal{I}}\boldsymbol{\omega}^{\mathcal{B}} \times \left({}^{\mathcal{I}}\boldsymbol{\omega}^{\mathcal{B}} \times \mathbf{r}_{P/O'}\right) = \frac{\mathbf{F}_P}{m_P},
$$

(10.70)

where the angular acceleration has been set to zero. (Because the local frame $\mathcal{B}$ is fixed to the earth, it has zero angular velocity relative to the geographic frame $\mathcal{G}$ and thus $^{\mathcal{I}}\boldsymbol{\omega}^{\mathcal{B}} = {}^{\mathcal{I}}\boldsymbol{\omega}^{\mathcal{G}}$.)

We are interested in finding equations of motion for P in the local frame. We rearrange Eq. (10.70) slightly to solve for the acceleration of P in $\mathcal{B}$:

$$
^{\mathcal{B}}\mathbf{a}_{P/O'} = \frac{\mathbf{F}_P}{m_P} - {}^{\mathcal{I}}\mathbf{a}_{O'/O} - 2^{\mathcal{I}}\boldsymbol{\omega}^{\mathcal{G}} \times {}^{\mathcal{B}}\mathbf{v}_{P/O'} - {}^{\mathcal{I}}\boldsymbol{\omega}^{\mathcal{G}} \times \left({}^{\mathcal{I}}\boldsymbol{\omega}^{\mathcal{G}} \times \mathbf{r}_{P/O'}\right). \tag{10.71}
$$

We immediately see that P behaves slightly differently than we would have expected had the local frame been inertial. Its motion is affected by the Coriolis and centripetal terms as well as the acceleration of the origin O' of the frame. To complete the description, we need to solve for that acceleration. That is done using the transport equation, as in Chapter 8. The inertial velocity of O' is

$$
^{\mathcal{I}}\mathbf{v}_{O'/O} = \underbrace{\frac{^{\mathcal{G}}d}{dt}\left(\mathbf{r}_{O'/O}\right)}_{=0} + {}^{\mathcal{I}}\boldsymbol{\omega}^{\mathcal{G}} \times \mathbf{r}_{O'/O},
$$

where the derivative in the $\mathcal{G}$ frame is zero because point O' is fixed to the surface of the earth. Taking the inertial derivative again gives the acceleration of O':

$$
^{\mathcal{I}}\mathbf{a}_{O'/O} = {}^{\mathcal{I}}\boldsymbol{\omega}^{\mathcal{G}} \times ({}^{\mathcal{I}}\boldsymbol{\omega}^{\mathcal{G}} \times \mathbf{r}_{O'/O}).
$$

Next we find the components of this acceleration in the geographic frame. Using spherical coordinates for the position (Eq. (10.4)), we have

$$
\mathbf{r}_{O'/O} = R_E \cos\theta \sin\phi\,\mathbf{g}_1 + R_E \sin\theta \sin\phi\,\mathbf{g}_2 + R_E \cos\phi\,\mathbf{g}_3.
$$

The acceleration of O' is

$$
^{\mathcal{I}}\mathbf{a}_{O'/O} = -R_E \Omega_E^2(\cos\theta \sin\phi\,\mathbf{g}_1 + \sin\theta \sin\phi\,\mathbf{g}_2).
$$

Finally, it is more convenient to express the acceleration of O' in terms of components in the local frame, as that is where we want to find equations of motion of P.

Because the local frame $\mathcal{B}$ is the same as a spherical frame, we use the transformation table in Eq. (10.13):

	$\mathbf{g}_1$	$\mathbf{g}_2$	$\mathbf{g}_3$
$\mathbf{b}_1$	$\cos\theta\cos\phi$	$\sin\theta\cos\phi$	$-\sin\phi$
$\mathbf{b}_2$	$-\sin\theta$	$\cos\theta$	0
$\mathbf{b}_3$	$\cos\theta\sin\phi$	$\sin\theta\sin\phi$	$\cos\phi$.

Transforming the acceleration gives (after a bit of algebra)

$$^{\mathcal{I}}\mathbf{a}_{O'/O} = -R_E\Omega_E^2(\cos\phi\sin\phi\,\mathbf{b}_1 + \sin^2\phi\,\mathbf{b}_3). \tag{10.72}$$

It should not come as a surprise that this acceleration is independent of longitude due to the rotational symmetry about $\mathbf{g}_3$.

We now use Eq. (10.71) and the acceleration in Eq. (10.72) to solve the following problem. We consider a particle P at rest at the origin of the local frame. Because gravity is acting, there must be a reaction force to keep the particle stationary (applied by the earth's surface). Because the particle is at the origin and at rest, $\mathbf{r}_{P/O'}$ and $^{\mathcal{B}}\mathbf{v}_{P/O'}$ are both zero, leaving us, from Newton's second law, with

$$0 = -g\mathbf{b}_3 + \frac{\mathbf{N}}{m_P} + R_E\Omega_E^2(\cos\phi\sin\phi\,\mathbf{b}_1 + \sin^2\phi\,\mathbf{b}_3),$$

where $\mathbf{N}$ is the reaction force from the ground. Solving for $\mathbf{N}$ yields

$$\mathbf{N} = m_P(g - R_E\Omega_E^2\sin^2\phi)\mathbf{b}_3 - m_P R_E\Omega_E^2\cos\phi\sin\phi\,\mathbf{b}_1.$$

$\mathbf{N}$ is not vertical! That is, the reaction force keeping the particle stationary is not directed along the line through the center of the earth, but is offset due to the rotation of the earth. If we were to drop an object, it would not fall toward the center of the earth but rather along the direction $-\mathbf{N}$. We call this direction the *local vertical*. For instance, if we were to hang a pendulum, its equilibrium would be along the local vertical rather than the geographic one. A pendulum that identifies the local vertical is called a *plumb bob*. In fact, there is no local measurement we can take to differentiate between the local gravity vector and the one due only to the earth's mass.

How big is this effect? If we define the small unitless quantity ϵ by

$$\epsilon = \frac{R_E\Omega_E^2}{g},$$

then we can find the angle between $\mathbf{N}$ and the $\mathbf{b}_3$ direction by taking their dot product and dividing by $\|\mathbf{N}\|$. Letting β be the angle between the local vertical and $\mathbf{b}_3$, we have

$$\cos\beta = \frac{1 - \epsilon\sin^2\phi}{\sqrt{\epsilon^2\sin^2\phi\cos^2\phi + (1 - \epsilon\sin^2\phi)^2}}. \tag{10.73}$$

The radius of the earth is approximately 6,378 km, and its angular rotation rate is roughly 7.3×10^{-5} rad/s. Assuming the acceleration due to gravity at the surface is $g = 9.8$ m/s^2, we find that $\epsilon = 3.4 \times 10^{-3}$. Suppose we consider the local vertical at

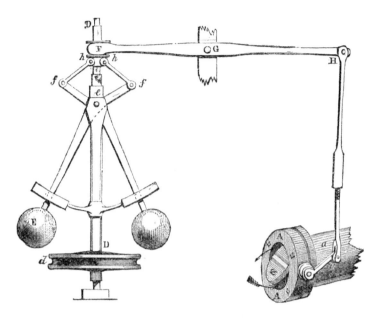

Figure 10.19 Watt flyball governor. Drawing from Routledge (1900).

a latitude of $\phi = 45°$ (somewhere in Canada, say). Then, from Eq. (10.73), it is offset from the earth's radial direction by an angle of roughly $0.1°$.

As a final note, Eq. (10.71) can be used to estimate the error made when assuming that a frame fixed to the surface of the earth is inertial. In most of our examples, we have treated earth-fixed frames as inertial. That is only an approximation, as the earth is rotating. The three terms in Eq. (10.71) provide the correction to account for the earth's rotation.

Tutorial 10.2 The Flyball Governor

The flyball governor is one of the earliest examples of an automatic control system. Its original purpose was to regulate the speed of the grinding stone in a flour mill. However, it was made famous by James Watt in the late eighteenth century when he adapted it to maintain the speed of a steam engine under load. An illustration of the Watt flyball governor is shown in Figure 10.19. The vertical motion of the balls opens or closes a steam valve that is connected to a rotary engine, the spin rate of which determines the motion of the balls.

This tutorial derives the equation of motion of a simple flyball-governor model using point masses and massless rods. Assume $\Omega > 0$ is the rotational speed of the shaft (D in Figure 10.19) and ignore the lever control system. We find that the angle the balls make with the rotating shaft is determined by the shaft rotation rate. This is an excellent example of solving a particle-dynamics problem in three dimensions with multiple rotating frames.

Our model of the governor is shown in Figure 10.20. There are three masses: the two balls on the swinging massless rods (they are simple pendula) and the bottom mass that moves up and down the shaft. The position of the lower mass M is given by y and the angle of the rods holding the balls is given by θ. Nevertheless, this

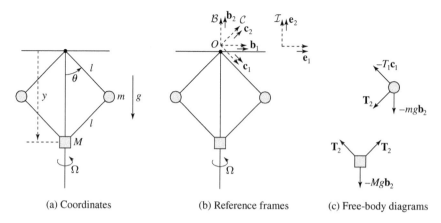

(a) Coordinates   (b) Reference frames   (c) Free-body diagrams

Figure 10.20 Simple model of the flyball governor.

is a single-degree-of-freedom problem because of the linked rods: the geometric constraint between y and θ is

$$y = 2l \cos \theta. \tag{10.74}$$

By symmetry we need only find the equation of motion for one of the balls; the other ball, because of the constraint, behaves exactly the same way.

We introduce three frames as shown in Figure 10.20b. As usual, $\mathcal{I} = (O, \mathbf{e}_1, \mathbf{e}_2, \mathbf{e}_3)$ is an inertial frame fixed in absolute space. Frame $\mathcal{B} = (O, \mathbf{b}_1, \mathbf{b}_2, \mathbf{b}_3)$ is an intermediate frame rotating with the shaft at $^{\mathcal{I}}\boldsymbol{\omega}^{\mathcal{B}} = \Omega \mathbf{b}_2$. Frame $\mathcal{C} = (O, \mathbf{c}_1, \mathbf{c}_2, \mathbf{c}_3)$ is a body frame fixed to the pendulum rod and rotating in frame $\mathcal{B}$ with angular velocity $^{\mathcal{B}}\boldsymbol{\omega}^{\mathcal{C}} = \dot{\theta}\mathbf{b}_3$. We use the addition property of angular velocities to find the angular velocity of the body frame $\mathcal{C}$ in $\mathcal{I}$:

$$^{\mathcal{I}}\boldsymbol{\omega}^{\mathcal{C}} = {}^{\mathcal{I}}\boldsymbol{\omega}^{\mathcal{B}} + {}^{\mathcal{B}}\boldsymbol{\omega}^{\mathcal{C}}$$

$$= \Omega \mathbf{b}_2 + \dot{\theta}\mathbf{b}_3.$$

Frame $\mathcal{C}$ is related to frame $\mathcal{B}$ by the transformation table

	$\mathbf{b}_1$	$\mathbf{b}_2$,	$\mathbf{b}_3$
$\mathbf{c}_1$	$\sin \theta$	$-\cos \theta$	0
$\mathbf{c}_2$	$\cos \theta$	$\sin \theta$	0
$\mathbf{c}_3$	0	0	1 .

We proceed by finding the equation of motion for θ, using the free-body diagrams in Figure 10.20c. The kinematics of the lower mass are

$$\mathbf{r}_{M/O} = -y\mathbf{b}_2$$

$$^{\mathcal{I}}\mathbf{v}_{M/O} = -\dot{y}\mathbf{b}_2$$

$$^{\mathcal{I}}\mathbf{a}_{M/O} = -\ddot{y}\mathbf{b}_2.$$

From the free-body diagram, the force on M is

$$\mathbf{F}_M = (2T_2 \cos \theta - Mg)\mathbf{b}_2.$$

Newton's second law then gives

$$-M\ddot{y} = 2T_2 \cos\theta - Mg.$$

Solving for T_2 yields

$$T_2 = \frac{M}{2\cos\theta}(g - \ddot{y}). \tag{10.75}$$

We can eliminate $\ddot{y}$ by differentiating Eq. (10.74) twice to obtain

$$\ddot{y} = -2l(\dot{\theta}^2 \cos\theta + \ddot{\theta}\sin\theta). \tag{10.76}$$

Substituting Eq. (10.76) into Eq. (10.75) yields

$$T_2 = \frac{Mg}{2\cos\theta} + Ml\dot{\theta}^2 + Ml\ddot{\theta}\tan\theta. \tag{10.77}$$

The equation of motion for the ball is found by using angular momentum, which avoids having to solve for the internal force T_1. The position expressed in the body frame is

$$\mathbf{r}_{m/O} = l\mathbf{c}_1.$$

The velocity is found from the transport equation to be

$$^{\mathcal{I}}\mathbf{v}_{m/O} = {}^{\mathcal{I}}\boldsymbol{\omega}^{\mathcal{C}} \times l\mathbf{c}_1$$

$$= (\Omega\mathbf{b}_2 + \dot{\theta}\mathbf{b}_3) \times l(\sin\theta\mathbf{b}_1 - \cos\theta\mathbf{b}_2)$$

$$= \dot{\theta}l\cos\theta\mathbf{b}_1 + \dot{\theta}l\sin\theta\mathbf{b}_2 - \Omega l\sin\theta\mathbf{b}_3.$$

We use this expression to compute the angular momentum of a ball in absolute space:

$$^{\mathcal{I}}\mathbf{h}_{m/O} = \mathbf{r}_{m/O} \times m^{\mathcal{I}}\mathbf{v}_{m/O}$$

$$= l(\sin\theta\mathbf{b}_1 - \cos\theta\mathbf{b}_2) \times m(\dot{\theta}l\cos\theta\mathbf{b}_1 + \dot{\theta}l\sin\theta\mathbf{b}_2 - \Omega l\sin\theta\mathbf{b}_3)$$

$$= m\Omega l^2\sin\theta\cos\theta\mathbf{b}_1 + m\Omega l^2\sin^2\theta\mathbf{b}_2 + m\dot{\theta}l^2\mathbf{b}_3. \tag{10.78}$$

The equation of motion comes from setting the inertial derivative of the angular momentum equal to the moment about O. The moment is found from the forces on m in the free-body diagram:

$$\mathbf{M}_{m/O} = \mathbf{r}_{m/O} \times \mathbf{F}_m$$

$$= l(\sin\theta\mathbf{b}_1 - \cos\theta\mathbf{b}_2)$$

$$\times (-T_1(\sin\theta\mathbf{b}_1 - \cos\theta\mathbf{b}_2) - T_2(\sin\theta\mathbf{b}_1 + \cos\theta\mathbf{b}_2) - mg\mathbf{b}_2)$$

$$= (-2T_2l\sin\theta\cos\theta - mgl\sin\theta)\mathbf{b}_3.$$

Perhaps not surprisingly, the moment on the ball is only in the $\mathbf{b}_3$ direction, implying that the components of the angular momentum in the $\mathbf{b}_1$ and $\mathbf{b}_2$ directions are conserved. Thus, to find the equation of motion for the ball, we use the transport equation on the angular momentum and set the $\mathbf{b}_3$ component equal to the moment. Taking the inertial derivative of Eq. (10.78) gives

$$\frac{{}^I d}{dt}({}^I\mathbf{h}_{m/O}) = \frac{{}^B d}{dt}({}^I\mathbf{h}_{m/O}) + {}^I\boldsymbol{\omega}^B \times {}^I\mathbf{h}_{m/O}$$

$$= m\Omega\dot{\theta}l^2(\cos^2\theta - \sin^2\theta)\mathbf{b}_1 + 2m\Omega\dot{\theta}l^2\sin\theta\cos\theta\mathbf{b}_2 + m\ddot{\theta}l^2\mathbf{b}_3$$

$$+ \Omega\mathbf{b}_2 \times m(\Omega l^2\sin\theta\cos\theta\mathbf{b}_1 + \Omega l^2\sin^2\theta\mathbf{b}_2 + \dot{\theta}l^2\mathbf{b}_3).$$

Setting the $\mathbf{b}_3$ component equal to the moment yields the second scalar equation of motion:

$$ml^2(\ddot{\theta} - \Omega^2\sin\theta\cos\theta) = -2T_2 l\sin\theta\cos\theta - mgl\sin\theta.$$

Our final step is to substitute for T_2 from Eq. (10.77) and solve for $\ddot{\theta}$ to yield our desired equation of motion for the flyball governor:

$$\ddot{\theta} = \frac{\left[-(m+M)g + (m\Omega^2 - 2M\dot{\theta}^2)l\cos\theta\right]\sin\theta}{(m + 2M\sin^2\theta)l}.$$

The equation of motion is often simplified for the case $M \ll m$ by dividing through by m and dropping terms of order $\frac{M}{m} \ll 1$. The approximate equation of motion is

$$\ddot{\theta} \approx \left(-\frac{g}{l} + \Omega^2\cos\theta\right)\sin\theta.$$

Our interest in the flyball governor is motivated by its equilibrium solutions. Since it is used as a feedback controller, we are interested in seeing whether there is a steady-state solution with nonzero θ that keeps Ω fixed. Setting $\ddot{\theta} = 0$ gives the following condition for an equilibrium:

$$\left(-\frac{g}{l} + \Omega^2\cos\theta\right)\sin\theta = 0.$$

The equilibrium condition is satisfied by the equilibrium angles $\theta_1^* = 0$, which correspond to the balls hanging vertically downward. There is also an additional pair of equilibrium points at

$$\theta_2^* = \pm\arccos\left(\frac{g}{l\Omega^2}\right),$$

as long as

$$\frac{g}{l\Omega^2} \leq 1.$$

The latter condition is satisfied for $\Omega \geq \Omega_0$, where

$$\Omega_0 = \sqrt{\frac{g}{l}}. \tag{10.79}$$

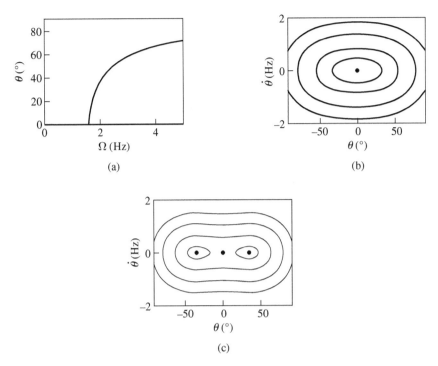

Figure 10.21 Flyball governor with $M \ll m$ and $l = 0.1$ m. (a) Bifurcation diagram showing bifurcation value $\Omega_0 = 1.58$ Hz. (b) Phase portrait for $\Omega < \Omega_0$ showing one equilibrium plotted as a dot. (c) Phase portrait for $\Omega > \Omega_0$ showing three equilibria.

This analysis shows that the flyball governor exhibits a fascinating behavior known as a *bifurcation*. Depending on the value of Ω, the system has either one equilibrium point or three equilibrium points. We call Ω the *bifurcation parameter,* and the critical value where the number of equilibrium points changes is called the *bifurcation value.* In this case, the bifurcation value is given by Eq. (10.79).

We can graphically represent this behavior in a *bifurcation diagram,* a plot of the equilibrium values as a function of the bifurcation parameter. Such a diagram for the flyball governor is given in Figure 10.21a. We can also often graph the trajectories of the system in a *phase portrait.* In a phase portrait, rather than plotting the state variables versus time (here, θ and $\dot{\theta}$), we instead plot them versus each other for various trajectories. This plot gives a picture of the overall behavior of the system for different initial conditions. For example, the phase portrait of a spring-mass system consists of just circles of varying sizes. An isolated dot on the phase portrait is an equilibrium solution: a location where the system remains for all time. Phase portraits are excellent tools for visually displaying equilibria. Figure 10.21b shows a phase portrait for the flyball governor for $\Omega < \Omega_0$. Note that there is only one equilibrium at $\theta = 0$ (represented by the dot). Figure 10.21c shows a phase portrait of the flyball governor for $\Omega > \Omega_0$. Note that the equilibria have bifurcated into three possible values, corresponding to $\theta_1^* = 0$ and $\theta_2^* = \pm \arccos\left(\frac{g}{l\Omega^2}\right)$.

10.7 Key Ideas

- This chapter introduced three three-dimensional coordinate systems: **Cartesian coordinates**, $(x, y, z)_\mathcal{I}$, **cylindrical coordinates**, $(r, \theta, z)_\mathcal{I}$, and **spherical coordinates**, $(r, \theta, \phi)_\mathcal{I}$. The position of P with respect to O expressed in terms of vector components in frame $\mathcal{I} = (O, \mathbf{e}_x, \mathbf{e}_y, \mathbf{e}_z)$ using each of these coordinate systems is

$$\mathbf{r}_{P/O} = x\mathbf{e}_x + y\mathbf{e}_y + z\mathbf{e}_z \qquad \text{(Cartesian)}$$
$$= r\cos\theta\,\mathbf{e}_x + r\sin\theta\,\mathbf{e}_y + z\mathbf{e}_z \qquad \text{(cylindrical)}$$
$$= r\cos\theta\sin\phi\,\mathbf{e}_x + r\sin\theta\sin\phi\,\mathbf{e}_y + r\cos\phi\,\mathbf{e}_z \qquad \text{(spherical)}.$$

- The **cylindrical frame** $\mathcal{B}_c = (O, \mathbf{e}_r, \mathbf{e}_\theta, \mathbf{e}_z)$ is like a polar frame, except that the unit vector $\mathbf{e}_r$ points at the projection of P onto the x-y plane, instead of at P directly. The kinematics of P using cylindrical coordinates in the cylindrical frame are

$$\mathbf{r}_{P/O} = r\mathbf{e}_r + z\mathbf{e}_z$$
$${}^\mathcal{I}\mathbf{v}_{P/O} = \dot{r}\mathbf{e}_r + r\dot{\theta}\mathbf{e}_\theta + \dot{z}\mathbf{e}_z$$
$${}^\mathcal{I}\mathbf{a}_{P/O} = (\ddot{r} - r\dot{\theta}^2)\mathbf{e}_r + (2\dot{r}\dot{\theta} + r\ddot{\theta})\mathbf{e}_\theta + \ddot{z}\mathbf{e}_z.$$

- The **spherical frame** $\mathcal{B}_s = (O, \mathbf{e}_\phi, \mathbf{e}_\theta, \mathbf{e}_r)$ is defined with respect to an inertial frame $\mathcal{I}$ using two simple rotations and one intermediate frame. The unit vector $\mathbf{e}_r$ is directed from O to P. The kinematics of P using spherical coordinates in the spherical frame are

$$\mathbf{r}_{P/O} = r\mathbf{e}_r$$
$${}^\mathcal{I}\mathbf{v}_{P/O} = r\dot{\phi}\mathbf{e}_\phi + r\dot{\theta}\sin\phi\,\mathbf{e}_\theta + \dot{r}\mathbf{e}_r$$
$${}^\mathcal{I}\mathbf{a}_{P/O} = (2\dot{r}\dot{\phi} + r\ddot{\phi} - r\dot{\theta}^2\cos\phi\sin\phi)\mathbf{e}_\phi$$
$$\qquad\qquad + (2\dot{r}\dot{\theta}\sin\phi + 2r\dot{\theta}\dot{\phi}\cos\phi + r\ddot{\theta}\sin\phi)\mathbf{e}_\theta$$
$$\qquad\qquad + (\ddot{r} - r\dot{\phi}^2 - r\dot{\theta}^2\sin^2\phi)\mathbf{e}_r.$$

- The expressions for three-dimensional **linear momentum, angular momentum,** and **energy** of a particle are identical to the two-dimensional expressions.

- The **Euler angles** $(\psi, \theta, \phi)_\mathcal{B}^\mathcal{I}$ are a set of coordinates describing the orientation of a three-dimensional body frame $\mathcal{B}$ with respect to frame $\mathcal{I}$. A set of Euler angles is defined by three simple rotations and two intermediate frames.

- The entries in the three-dimensional transformation table that relates the unit vectors of frame $\mathcal{B}$ to frame $\mathcal{I}$ form a 3×3 **transformation matrix** ${}^\mathcal{I}C^\mathcal{B}$ that is used to transform the components of a vector from $\mathcal{B}$ to $\mathcal{I}$ or vice versa. In matrix notation we have

$$[\mathbf{r}_{P/O}]_\mathcal{I} = {}^\mathcal{I}C^\mathcal{B}[\mathbf{r}_{P/O}]_\mathcal{B}.$$

- The **transport equation** is valid in three dimensions:

$$\frac{^{\mathcal{A}}d}{dt}\mathbf{a} = \frac{^{\mathcal{B}}d}{dt}\mathbf{a} + {}^{\mathcal{A}}\boldsymbol{\omega}^{\mathcal{B}} \times \mathbf{a},$$

where $\mathbf{a}$ is an arbitrary vector, and ${}^{\mathcal{A}}\boldsymbol{\omega}^{\mathcal{B}}$ is the angular velocity of frame $\mathcal{B}$ with respect to frame $\mathcal{A}$.

- The **angular-velocity addition property** holds in three dimensions.
- The **three-dimensional kinematics** of a particle in a rotating frame $\mathcal{B}$ with origin O' are

$$^{\mathcal{I}}\mathbf{v}_{P/O} = {}^{\mathcal{I}}\mathbf{v}_{O'/O} + {}^{\mathcal{B}}\mathbf{v}_{P/O'} + {}^{\mathcal{I}}\boldsymbol{\omega}^{\mathcal{B}} \times \mathbf{r}_{P/O'}$$

$$^{\mathcal{I}}\mathbf{a}_{P/O} = {}^{\mathcal{I}}\mathbf{a}_{O'/O} + {}^{\mathcal{B}}\mathbf{a}_{P/O'} + {}^{\mathcal{I}}\boldsymbol{\alpha}^{\mathcal{B}} \times \mathbf{r}_{P/O'} + 2{}^{\mathcal{I}}\boldsymbol{\omega}^{\mathcal{B}} \times {}^{\mathcal{B}}\mathbf{v}_{P/O'}$$

$$+ {}^{\mathcal{I}}\boldsymbol{\omega}^{\mathcal{B}} \times \left({}^{\mathcal{I}}\boldsymbol{\omega}^{\mathcal{B}} \times \mathbf{r}_{P/O'}\right),$$

where ${}^{\mathcal{I}}\boldsymbol{\alpha}^{\mathcal{B}} = \frac{^{\mathcal{B}}d}{dt}\left({}^{\mathcal{I}}\boldsymbol{\omega}^{\mathcal{B}}\right) = \frac{^{\mathcal{I}}d}{dt}\left({}^{\mathcal{I}}\boldsymbol{\omega}^{\mathcal{B}}\right)$ is the angular acceleration.

- The **angular velocity** in terms of the rates of change of the 3-2-3 Euler angles $(\psi, \theta, \phi)^{\mathcal{I}}_{\mathcal{B}}$ is

$$^{\mathcal{I}}\boldsymbol{\omega}^{\mathcal{B}} = \dot{\psi}\mathbf{a}_3 + \dot{\theta}\mathbf{c}_2 + \dot{\phi}\mathbf{b}_3,$$

which yields the **kinematic equations of rotation**

$$\omega_1 = \dot{\theta}\sin\phi - \dot{\psi}\sin\theta\cos\phi$$

$$\omega_2 = \dot{\theta}\cos\phi + \dot{\psi}\sin\theta\sin\phi$$

$$\omega_3 = \dot{\phi} + \dot{\psi}\cos\theta.$$

- The angular velocity represents the rate of rotation about an instantaneous axis of rotation aligned with it.
- **Euler's theorem of rotation** states that any change in the orientation of frame $\mathcal{B}$ in frame $\mathcal{I}$ can be produced by a single, simple rotation about an axis fixed in both $\mathcal{B}$ and $\mathcal{I}$.

10.8 Notes and Further Reading

Our primary reference for much of the material on the kinematics of rotation in this chapter is the book on spacecraft dynamics by Kane et al. (1983). This book has an excellent and thorough treatment of several different descriptions of orientation (including quaternions, Rodriguez parameters, and both body and space angles) as well as the kinematics of rigid bodies. It is an excellent source for delving further. It is also the only text we know of that discusses all of the different sets of Euler angles, including two appendices with every direction-cosine matrix and corresponding set of kinematic equations of rotation. Although all textbooks discuss Euler angles, not all use the same set, and some may switch for different applications (e.g., the standard set used for spacecraft dynamics is different from that used for airplane dynamics).

Our discussion of the physical interpretation of the velocity and Euler's theorem is also motivated by Kane's books (Kane 1978; Kane and Levinson 1985; Kane et al. 1983). Many introductory texts, in both physics and engineering, use infinitesimal rotations and qualitative arguments to explain the angular velocity. We prefer to treat it as an operator and show, using Euler's theorem, that it is along the instantaneous axis of rotation. Eq. (10.60) in Corollary 10.1 and its derivation come directly from Kane et al. (1983), as does our statement of Euler's theorem.

10.9 Problems

10.1 Write the position of P with respect to O in frame $\mathcal{I}$ as shown in Figure 10.22 using

  a. Cartesian coordinates,
  b. cylindrical coordinates,
  c. spherical coordinates.

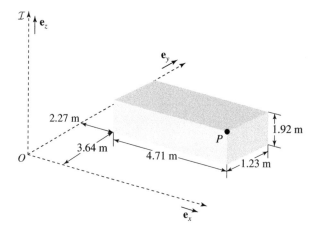

Figure 10.22 Problem 10.1.

10.2 Let $\mathcal{I} = (O, \mathbf{e}_1, \mathbf{e}_2, \mathbf{e}_3)$ be an inertial reference frame, $\mathcal{A} = (O, \mathbf{a}_1, \mathbf{a}_2, \mathbf{a}_3)$ be a cylindrical frame, and $\mathcal{B} = (O, \mathbf{e}_\phi, \mathbf{e}_\theta, \mathbf{e}_r)$ be a spherical frame, as shown in Figure 10.23. Derive the following:

  a. The transformation tables between $\mathcal{A}$ and $\mathcal{I}$, and between $\mathcal{B}$ and $\mathcal{A}$.
  b. The angular velocity of $\mathcal{A}$ with respect to $\mathcal{I}$, and of $\mathcal{B}$ with respect to $\mathcal{I}$.
  c. The position $\mathbf{r}_{P/O}$ and velocity $^{\mathcal{I}}\mathbf{v}_{P/O}$ of P with respect to O, expressed as components in $\mathcal{A}$ using cylindrical coordinates $(r, \theta, z)_{\mathcal{I}}$.
  d. The position $\mathbf{r}_{P/O}$ and velocity $^{\mathcal{I}}\mathbf{v}_{P/O}$ of P with respect to O, expressed as components in $\mathcal{B}$ using spherical coordinates $(r, \theta, \phi)_{\mathcal{I}}$.

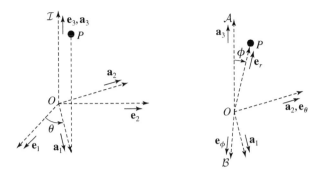

Figure 10.23 Problem 10.2.

10.3 Let $C = (O, \mathbf{e}_\phi, \mathbf{e}_\theta, \mathbf{e}_r)$ be a spherical frame and $\mathcal{I} = (O, \mathbf{e}_1, \mathbf{e}_2, \mathbf{e}_3)$ be an inertial frame such that $^{\mathcal{I}}\boldsymbol{\omega}^C = -\dot\theta \sin\phi \mathbf{e}_\phi + \dot\phi \mathbf{e}_\theta + \dot\theta \cos\phi \mathbf{e}_r$. Derive $\frac{^{\mathcal{I}}d}{dt}(\mathbf{e}_\theta)$ and $\frac{^{\mathcal{I}}d}{dt}(\mathbf{e}_r)$.

10.4 Solve for the equation of motion of the spherical-pendulum angle ϕ using angular momentum. Assume it rotates at a constant angular speed Ω. See Figure 10.24.

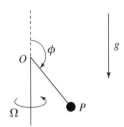

Figure 10.24 Problem 10.4.

10.5 Find the angular velocity of a football spinning at Ω about its long axis and traveling at a constant speed v_0 tangent to a semicircle of radius R, as shown in Figure 10.25.

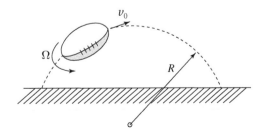

Figure 10.25 Problem 10.5.

10.6 Repeat Problem 10.5 where the football travels along a parabolic trajectory.

10.7 A particle of mass m slides without friction inside a three-dimensional conical funnel, whose walls have unit slope and whose center axis is vertical (Figure 10.26). If the particle is launched inside the funnel at height h and horizontal speed s_0, find the vertical speed of the particle when it reaches a height of $h/2$. (Note that a solution exists only for sufficiently small s_0.) [HINT: Use a cylindrical frame and cylindrical coordinates to write the kinematics of P with respect to O.]

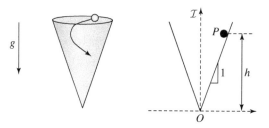

Figure 10.26 Problem 10.7.

10.8 Find the conditions for the steady-state solution of a spherical pendulum at a nonzero angle other than π.

10.9 Integrate the equations of motion for the flyball governor for different Ω and show the bifurcation discussed in Tutorial 10.2.

10.10 Example 10.3 derived the equations of motion for a spherical pendulum. When dealing with a real pendulum, however, we are actually operating in a rotating frame. Repeat the derivation of the equations of motion, but this time, place the pendulum in a local vertical reference frame at latitude β (see Tutorial 10.1). You may use whatever reference frames and coordinate systems you consider to be appropriate, but remember that your equations of motion must reflect the motion of the pendulum in all three dimensions relative to the local vertical.

    a. Using your new equations of motion, explain what effect the rotation of the earth has on the plane of the pendulum's swing.

    b. French physicist Jean-Bernard-Léon Foucault (1819–1868) predicted that a pendulum could be used to prove the rotation of the earth. In 1851, a 67 m pendulum with a 28 kg bob was constructed in Paris (latitude 48° N). Estimate the rate of rotation of the plane of this pendulum's swing.

    c. What happens if a Foucault pendulum is constructed at the north (or south) pole? At the equator?

10.11 Suppose you are operating an unmanned surveillance aircraft (UAV, or unmanned aerial vehicle) from a ground station at point O, as shown in Figure 10.27. Your job is to search for and take a picture of a target ground facility over the next ridge. Assume a flat earth. Suppose the UAV flies level at constant speed v_0 (its velocity always points through its nose, and its wings are

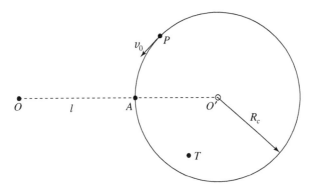

Figure 10.27 Bird's-eye view of system in Problem 10.11. Point P represents the UAV. Point T represents the target facility, but the target is not necessarily located as shown in the figure.

parallel to the ground), counter-clockwise in a circle of radius R_C centered at a surveyed point O' a distance l away from your station and at altitude h. Let B be a frame fixed to the aircraft and C be a frame fixed to a camera mounted under the aircraft, which can rotate with respect to B on gimbals. The camera lens points in the $\mathbf{c}_3$ direction.

a. Suppose the camera sees the facility when the UAV is at a position on its circular trajectory on the line between O and O' (point A in Figure 10.27). At the time, the orientation of the camera with respect to the UAV is given by the 3-2-3 Euler angle set $(\psi = \psi_A, \theta = \theta_A, \phi = 0)^B_C$, and the radar ranging gives its distance along the line of sight to the camera as $d = d_A$ m. What is the position of the target from your station in Cartesian coordinates (useful for future targeting)?

b. The UAV now must keep the target in the camera's sights as it continues to fly in its circle. Find expressions for the rate of change of the Euler angles so that the camera stays pointed at the target. Assume that this can be done with ϕ kept at zero for all time (since the camera is symmetric about the axis through the lens). Find a system of differential equations that would allow you to solve for ψ, θ, and d for all time.

10.12 Consider an experimental high-altitude vehicle dropped from a large C-130 transport plane. The plane and vehicle are beyond the visible range of the ground system. Sensors on the plane measure the position and velocity of the released vehicle relative to the C-130, $\mathbf{r}_{P/O'}$ and $^B\mathbf{v}_{P/O'}$. The C-130 telemeters this information to the ground observers as well as its own inertial velocity $^I\mathbf{v}_{O'/O}$, its angular velocity in inertial space $^I\boldsymbol{\omega}^B$, and its inertial position $\mathbf{r}_{O'/O}$, obtained from its inertial navigation system and GPS sensors. The ground observers must point their cameras and other equipment to pick up the experimental vehicle; thus they must know the vehicle's velocity in inertial space. Write down the vector expression they use to find the vehicle's velocity from the telemetered data.

10.13 Consider the spinning disk in a single gimbal shown in Figure 10.28. The disk has a diameter of 50 mm, and the gimbal has a diameter of 100 mm. Find the inertial velocities of points A and B if the rotor is spinning at 10 rad/s while the gimbal rotates about the vertical axis at 2 rad/s. Express your answer as components in an intermediate frame fixed to the gimbal but not spinning with the rotor.

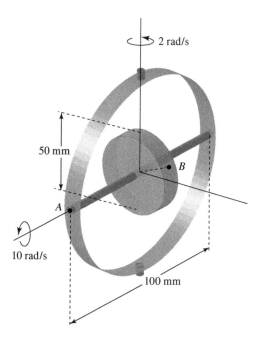

Figure 10.28 Problem 10.13.

CHAPTER ELEVEN

Multiparticle and Rigid-Body Dynamics in Three Dimensions

This chapter concludes our study of Newtonian mechanics with an examination of the motion of multiparticle systems and rigid bodies in three dimensions. By the end of the chapter you will have all the tools and skills you need to tackle any dynamics problem. Although it may seem like a fairly minor extension to move from the simple planar rotations of Chapter 9 to the general three-dimensional rotations studied in this chapter, you will see that a rich set of behavior emerges. The complex motion of spinning bodies has allowed for an amazing array of engineered devices from toys to spacecraft.

11.1 Euler's Laws in Three Dimensions

We begin this chapter with the simple observation that the basic ideas of Chapters 6 and 7 translate unchanged to three dimensions. We were careful in those chapters to develop our ideas using a vector treatment so that the fundamental results were independent of dimension. Thus the law of conservation of total linear momentum is the same in three dimensions, and Definition 6.2 of the center of mass still applies, including the center-of-mass corollary in Eq. (6.15). The discussion of collisions extends to three dimensions because we considered two objects whose impact velocities are merely confined to a plane, as is the case in three dimensions in the absence of external forces. The idea that a multiparticle system can be treated by applying Newton's second law to each particle still holds—except we now have three coordinates for each particle rather than two. Chapter 10 covered the three-dimensional equations of motion of a particle.

Also still true is the all-important result from Chapter 6—that the center of mass of a particle collection follows a trajectory determined by solving Newton's second law independently of the relative motion of the particles. That is, the three-dimensional

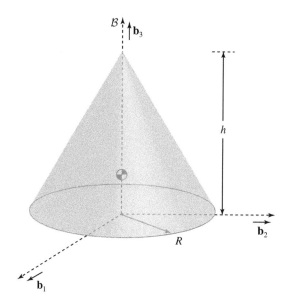

Figure 11.1 Computing the center of mass of a continuous rigid body.

translational dynamics of the center of mass of a collection of particles is

$$\mathbf{F}_G^{(\text{ext})} = m_G{}^{\mathcal{I}}\mathbf{a}_{G/O},$$

(11.1)

where $\mathbf{F}_G^{(\text{ext})}$ represents the total external force acting on the system. When applied to a rigid body, Eq. (11.1) is equivalent to Euler's first law (Law 9.1) in three dimensions.

Example 11.1 Computing the Center of Mass of a Continuous Rigid Body in Three Dimensions

Example 9.1 computed the center of mass of a uniform planar triangle. The same equations allow us to find the center of mass of a continuous three-dimensional body $\mathcal{B}$. Consider the cone in Figure 11.1, which has base radius R and height h. Because the cone can be easily described as a surface of revolution about the $\mathbf{b}_3$ axis, it makes sense to use cylindrical coordinates here. As before, we start by writing the position $\mathbf{r}_{dm/O} = r\cos\theta\mathbf{b}_1 + r\sin\theta\mathbf{b}_2 + z\mathbf{b}_3$ and the volume element $dV = r\,dr\,d\theta\,dz$. We also assume a mass m and uniform density ρ. We apply Eq. (9.2) to write the center of mass as

$$\mathbf{r}_{G/O} = \frac{\rho}{m}\int_{\mathcal{B}}(r\cos\theta\mathbf{b}_1 + r\sin\theta\mathbf{b}_2 + z\mathbf{b}_3)r\,dr\,d\theta\,dz.$$

To find the limits of integration for the three integrals, we must consider the shape of the cone. The limits on θ and z are simple—θ goes from 0 to 2π, and z goes from 0 to h. The upper limit on r, however, depends on the slope of the cone and is given by $R(h-z)/h$. Finally, before performing the integration, we note that, if we integrate over θ first, the $\mathbf{b}_1$ and $\mathbf{b}_2$ terms go to zero, thereby simplifying the remaining integrations:

$$
\begin{aligned}
\mathbf{r}_{G/O} &= \frac{\rho}{m} \int_0^h \int_0^{R(h-z)/h} \int_0^{2\pi} (r\cos\theta\mathbf{b}_1 + r\sin\theta\mathbf{b}_2 + z\mathbf{b}_3)r\,d\theta\,dr\,dz \\
&= \frac{\rho}{m} \int_0^h \int_0^{R(h-z)/h} 2\pi rz\mathbf{b}_3 dr\,dz \\
&= \frac{\rho}{m} \int_0^h \frac{\pi R^2 z(h-z)^2}{h^2}\mathbf{b}_3 dz \\
&= \frac{\rho}{m} \frac{h^2\pi R^2}{12}\mathbf{b}_3.
\end{aligned}
$$

Because $m = \rho \int_{\mathscr{B}} dV = \rho\pi R^2 h/3$, the center-of-mass position is

$$
\mathbf{r}_{G/O} = \frac{h}{4}\mathbf{b}_3.
$$

This result is very useful because it states that the center of mass of any right circular cone will be one-fourth of the way up its central axis.

Our treatment of the angular momentum of multiparticle systems and rigid bodies also extends to three dimensions. Recall that the total angular momentum of a collection of particles (Definition 7.1) is

$$
{}^{\mathcal{I}}\mathbf{h}_O \triangleq \sum_{i=1}^N {}^{\mathcal{I}}\mathbf{h}_{i/O} = \sum_{i=1}^N m_i \mathbf{r}_{i/O} \times {}^{\mathcal{I}}\mathbf{v}_{i/O}.
$$

For a continuous rigid body,

$$
{}^{\mathcal{I}}\mathbf{h}_O \triangleq \int_{\mathscr{B}} \mathbf{r}_{dm/O} \times {}^{\mathcal{I}}\mathbf{v}_{dm/O} dm.
$$

When the internal-moment assumption holds, the angular-momentum form of Newton's second law in Eq. (7.5) applies for the total angular momentum of the collection:

$$
\boxed{\frac{{}^{\mathcal{I}}d}{dt}\left({}^{\mathcal{I}}\mathbf{h}_O\right) = \sum_{i=1}^N \mathbf{r}_{i/O} \times \mathbf{F}_i^{(\text{ext})} = \mathbf{M}_O^{(\text{ext})}.}
\tag{11.2}
$$

These equations are still true in three dimensions, so that the derivations in Section 7.1 still hold. For rigid bodies, everything in Section 9.1 through Section 9.2 and Section 9.3.1 applies unchanged in three dimensions. We did not use the planar assumption until Section 9.3.2. Thus the internal-moment assumption still applies as a property of a rigid body and Euler's second law in Eq. (9.6) holds unchanged for the motion of a rigid body relative to an inertially fixed point O:

$$\frac{^{\mathcal{I}}d}{dt}\left(^{\mathcal{I}}\mathbf{h}_O\right) = \mathbf{M}_O,$$

where we have dropped the superscript (ext) for a rigid body.

The other important result from Chapter 7 is the separation principle in Section 7.2: the total angular momentum of the collection separates into the angular momentum of the center of mass G relative to O and the angular momentum about the center of mass. Those results were completely general and, consequently, still apply in three dimensions. We have

$$^{\mathcal{I}}\mathbf{h}_O = {}^{\mathcal{I}}\mathbf{h}_{G/O} + {}^{\mathcal{I}}\mathbf{h}_G,$$

where

$$^{\mathcal{I}}\mathbf{h}_{G/O} \triangleq m_G \mathbf{r}_{G/O} \times {}^{\mathcal{I}}\mathbf{v}_{G/O}$$

and

$$^{\mathcal{I}}\mathbf{h}_G \triangleq \sum_{i=1}^{N} m_i \mathbf{r}_{i/G} \times {}^{\mathcal{I}}\mathbf{v}_{i/G}.$$

For a continuous rigid body,

$$^{\mathcal{I}}\mathbf{h}_G \triangleq \int_{\mathscr{B}} \mathbf{r}_{dm/G} \times {}^{\mathcal{I}}\mathbf{v}_{dm/G}\, dm.$$

Again, the center of mass obeys Newton's second law (Euler's first law) as if the collection were a single particle of mass m_G located at G. When the internal-moment assumption holds, the angular momentum $\mathbf{h}_G$ about the center of mass satisfies the equation of motion

$$\frac{^{\mathcal{I}}d}{dt}\left(^{\mathcal{I}}\mathbf{h}_G\right) = \sum_{i=1}^{N} \mathbf{r}_{i/G} \times \mathbf{F}_i^{(\text{ext})} = \mathbf{M}_G^{(\text{ext})}.$$

For a rigid body, the internal-moment assumption always holds, and this equation is Euler's second law about the center of mass:

$$\frac{^{\mathcal{I}}d}{dt}\left(^{\mathcal{I}}\mathbf{h}_G\right) = \mathbf{M}_G. \tag{11.3}$$

Even before developing three-dimensional expressions for the angular momentum in terms of the angular velocity and mass properties of a rigid body, we can already solve interesting and useful problems just by using the angular-momentum expressions in Euler's first and second laws. This observation is very important; in many

Figure 11.2 Toy gyroscope on its stand. Image courtesy of Shutterstock.

problems it is useful to go back to these basic principles rather than jump right to the more sophisticated tools you will soon learn. You will be surprised how much can be done just by examining angular momentum.

Example 11.2 The Gyropendulum

This example derives the equations of motion for an inverted *spherical gyropendulum.* The device is very similar to the spherical pendulum studied in Example 10.3 except that, instead of the mass particle at the end of the massless rod, there is a spinning rotor. This is an excellent model of the toy gyroscope depicted in Figure 11.2; when spinning, the gyro doesn't tip over but *precesses* around the vertical. The result derived here qualitatively predicts this motion.

Our simple model of the gyropendulum, including the reference frames and free-body diagram, is shown in Figure 11.3. If the rotor radius is small compared to the length of the pendulum, we can very effectively model it as a single particle with a constant angular momentum of magnitude h about its symmetry axis. This approximation is common in three-dimensional dynamics, when the motion is dominated by the effect of the large-spin angular momentum.

To begin, we review the results of Example 10.3 for the nonspinning case. We still locate P using spherical coordinates (Figure 11.3a) and introduce the spherical frame $\mathcal{B}_s$ (Figure 11.3b). The only difference here is that we are considering motion for

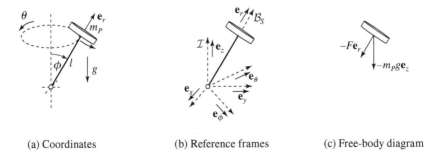

(a) Coordinates   (b) Reference frames   (c) Free-body diagram

Figure 11.3 Spherical gyropendulum using spherical coordinates and the spherical frame.

$0 < \phi < \pi/2$. The equations of motion are thus the same as for the inverted spherical pendulum:

$$\ddot{\phi} - \dot{\theta}^2 \sin\phi \cos\phi - \frac{g}{l} \sin\phi = 0$$

$$\ddot{\theta} + 2\dot{\theta}\dot{\phi} \cot\phi = 0.$$

Recall that Example 10.3 discussed the existence of a steady-state solution where the pendulum hangs at a constant offset angle while circling around the vertical. Here we show that for certain values of the angular momentum, an inverted steady-state solution with a constant offset also exists.[1]

For the spinning case, we begin as for the spherical pendulum in Example 10.5. The position and velocity of the rotor in spherical coordinates is

$$\mathbf{r}_{P/O} = l\mathbf{e}_r$$

$$^{\mathcal{I}}\mathbf{v}_{P/O} = l\dot{\phi}\mathbf{e}_\phi + l\dot{\theta}\sin\phi\mathbf{e}_\theta.$$

To find the angular momentum, we use the separation principle and write the total angular momentum about O as the sum of the angular momentum of the rotor center of mass P plus the angular momentum, $h\mathbf{e}_r$, about the rotor center of mass

$$^{\mathcal{I}}\mathbf{h}_O = \mathbf{r}_{P/O} \times m_P{}^{\mathcal{I}}\mathbf{v}_{P/O} + h\mathbf{e}_r$$

$$= m_P l^2 \dot{\phi}\mathbf{e}_\theta - m_P l^2 \dot{\theta}\sin\phi\mathbf{e}_\phi + h\mathbf{e}_r. \qquad (11.4)$$

We can now take the derivative of the total angular momentum as done in Example 10.5, using the angular velocity for the unit-vector derivatives, and set it equal to the total moment about O, as in Eq. (11.2). The angular velocity of the spherical frame, $\mathcal{B}_S$ (Eq. (10.28)), is

$$^{\mathcal{I}}\boldsymbol{\omega}^{\mathcal{B}_S} = -\dot{\theta}\sin\phi\mathbf{e}_\phi + \dot{\phi}\mathbf{e}_\theta + \dot{\theta}\cos\phi\mathbf{e}_r,$$

which we use to find the rate of change of the unit vectors in the angular momentum equation in Eq. (11.4). The rotational kinematics are

$$\frac{^{\mathcal{I}}d}{dt}\left(^{\mathcal{I}}\mathbf{h}_O\right) = m_P l^2 (\ddot{\phi} - \dot{\theta}^2 \sin\phi \cos\phi)\mathbf{e}_\theta - m_P l^2 \left(\ddot{\theta}\sin\phi + 2\dot{\theta}\dot{\phi}\cos\phi\right)\mathbf{e}_\phi$$

$$+ h\dot{\theta}\sin\phi\mathbf{e}_\theta + h\dot{\phi}\mathbf{e}_\phi.$$

As in Example 10.5, the only moment about O is due to gravity:

$$\mathbf{M}_O = \mathbf{r}_{P/O} \times \mathbf{F}_{P/O} = l\mathbf{e}_r \times (-m_P g\mathbf{e}_z)$$

$$= -m_P g l\mathbf{e}_r \times (-\sin\phi\mathbf{e}_\phi + \cos\phi\mathbf{e}_r)$$

$$= m_P g l \sin\phi\mathbf{e}_\theta.$$

---

[1] Note that the stability of these solutions is a separate question. Stability refers to the tendency of the pendulum to return to the steady state if slightly perturbed away from it. Clearly the downward pendulum equilibrium is stable but the inverted one is not. Examining stability is a more complicated question (though the inverted gyropendulum is, in fact, stable); Chapter 12 presents an introduction to this subject.

Equating the moment on the pendulum rotor about O with the rate of change of angular momentum about O yields the following two scalar equations:

$$0 = -2m_p l^2 \dot{\theta} \dot{\phi} \cos \phi - m_p l^2 \ddot{\theta} \sin \phi + h\dot{\phi}$$

$$m_p g l \sin \phi = m_p l^2 \ddot{\phi} - m_p l^2 \dot{\theta}^2 \sin \phi \cos \phi + h\dot{\theta} \sin \phi.$$

Solving for $\ddot{\theta}$ and $\ddot{\phi}$ produces the two equations of motion for the simplified gyro-pendulum:

$$\ddot{\theta} + 2\dot{\theta}\dot{\phi} \cot \phi - \frac{h\dot{\phi}}{m_p l^2 \sin \phi} = 0 \qquad (11.5)$$

$$\ddot{\phi} - \dot{\theta}^2 \sin \phi \cos \phi - \frac{g}{l} \sin \phi + \frac{h\dot{\theta} \sin \phi}{m_p l^2} = 0. \qquad (11.6)$$

We now look for a constant-offset solution. Is there a solution with constant $\dot{\theta} = \dot{\theta}_0 > 0$ and constant $\phi = \phi_0$, where $0 < \phi_0 < \pi/2$? Setting $\ddot{\phi} = \dot{\phi} = 0$ in Eq. (11.6) yields

$$\dot{\theta}^2 \cos \phi_0 - \frac{h}{m_p l^2}\dot{\theta} + \frac{g}{l} = 0. \qquad (11.7)$$

Thus for $\dot{\theta} = \dot{\theta}_0$, the steady-state tilt of the pendulum is

$$\phi_0 = \arccos \left(\frac{h}{m_p l^2 \dot{\theta}_0} - \frac{g}{l\dot{\theta}_0^2} \right).$$

Such a solution exists as long as

$$0 \leq \frac{g}{l\dot{\theta}_0^2} - \frac{h}{m_p l^2 \dot{\theta}_0} \leq 1.$$

Alternatively, we can ask what the constant precession rate $\dot{\theta}_0 > 0$ of the pendulum is about the vertical for a given h and tilt angle ϕ_0. This problem involves solving the quadratic equation in Eq. (11.7) to obtain

$$\dot{\theta}_0 = \frac{\frac{h}{m_p l^2} \pm \sqrt{\frac{h^2}{m_p^2 l^4} - \frac{4\cos\phi_0 g}{l}}}{2\cos\phi_0}.$$

Such a solution exists as long as

$$\frac{h^2}{m_p^2 l^4} - \frac{4\cos\phi_0 g}{l} \geq 0.$$

Thus we can state a condition on the angular momentum of the rotor for which an inverted steady-state angle ϕ_0 exists:

$$h \geq \sqrt{4\cos\phi_0 g m_p^2 l^3}.$$

We conclude this section with a reminder regarding momentum conservation. Just as in Chapters 7 and 9, Euler's second law leads to the law of conservation of angular momentum.

Law 11.1 Under the internal-moment assumption, the law of **conservation of total angular momentum** states that, if the total external moment acting on a system of particles about the origin O is zero, then the total inertial angular momentum of the collection about O is a constant of the motion. Likewise, if the total external moment about the center of mass G of the collection is zero, then the inertial angular momentum of the collection about G is a constant of the motion.

11.2 Three-Dimensional Rotational Equations of Motion of a Rigid Body

The primary objective in this section is to study the rotational motion of a rigid body about its center of mass, employing the separation of angular momentum. Up to now it may seem like there is not much new. The general expressions for the dynamics of the angular momentum (Euler's first and second laws) are exactly the same in three dimensions. In fact, as described at the beginning of the chapter, in translation there is nothing new. Newton's second law (or Euler's first law) is simply applied to the center of mass in three dimensions, just as we did in two, and all results from Chapter 10 apply. The changes come when considering rotation.

Although the analysis of angular momentum in Chapters 7 and 9 used vectors and thus was completely general, we only found the planar equations of motion. This simplification is significant: by confining motion to a plane we constrained the angular momentum to be always perpendicular to the plane. In other words, the angular-momentum direction was fixed and, in fact, parallel to the angular velocity. The number of rotational degrees of freedom collapsed from three to one and our rotational equation of motion reduced to a single scalar equation. Here we allow the particles or rigid body to move arbitrarily. The total angular momentum can thus change both its direction and magnitude and is not necessarily parallel to the angular velocity. Although the resulting equations are concise, these new degrees of freedom result in amazingly complex and interesting behavior.

Our main observation is that Eq. (11.3), Euler's second law, is not the most convenient form of the vector equation of motion, as it differentiates the angular momentum with respect to absolute space. Now that there are additional degrees of freedom, the angular momentum can change in more complicated ways and, for most problems, we are more interested in describing that change with respect to a body frame. This is because (a) we are considering the motion of particles relative to a moving body frame; (b) we are solving for the motion of a rigid body and it is easier to express the angular momentum in its body frame; or (c) the applied moment is most easily expressed in a body frame.

The first step, then, is to find a differential equation for the angular momentum with respect to the body frame. To do so, we apply the transport equation to Eq. (11.3). Because the angular momentum is a vector, we can use the transport equation and rewrite Eq. (11.3) to find the following alternative form of Euler's second law in three dimensions:

$$\frac{{}^{\mathcal{B}}d}{dt}\left({}^{\mathcal{I}}\mathbf{h}_G\right) + {}^{\mathcal{I}}\boldsymbol{\omega}^{\mathcal{B}} \times {}^{\mathcal{I}}\mathbf{h}_G = \mathbf{M}_G.$$

(11.8)

Eq. (11.8) is the key vector equation for all rotational rigid-body dynamics. It is almost always the best place to start when solving three-dimensional dynamics problems. Note that when we were confined to planar motion this extra step was unnecessary, as the angular momentum was, by design, always aligned with the angular velocity and thus the cross product between the two was zero. Now the use of the transport equation has given us a much more useful form of Euler's second law for motion about the center of mass.

However, we can't yet use Eq. (11.8) to solve problems exactly, as the three-dimensional relationship between the angular momentum and the angular velocity has not yet been discussed. We will do that soon, including how to incorporate the Euler angles. Nevertheless, it is important not to lose sight, amidst all the mathematics, that Eq. (11.8) is always the starting point. We are solving for the time evolution of the angular momentum of a multiparticle system or a rigid body. The next subsection discusses how this equation is approximated to *qualitatively* understand the motion of a spinning body. The following two subsections modify Eq. (11.8) to provide exact scalar equations of motion for the rotational trajectories in three dimensions.

11.2.1 Qualitative Analysis of a Spinning Body

Just as for the gyropendulum in Example 11.2, many physical systems involve a rapidly spinning body or a system with a connected spinning rotor. It is often very useful in these systems to develop a qualitative understanding of the expected behavior. By this we mean an approximate prediction of the dominant motion without an exact solution of the detailed trajectories. For instance, in the gyropendulum, if the rotor is spinning clockwise, is the pendulum going to precess to the right or to the left? What direction does the earth's spin axis move due to the torques exerted on it by the sun and moon? If you hold a spinning bicycle wheel and tilt it to the left, do you feel a clockwise or counterclockwise torque? Fortunately, all these questions can be easily answered using Eq. (11.8) without resorting to finding and solving exact equations of motion. The ability to qualitatively assess three-dimensional behavior is an important engineering skill to develop.

The key point is that the angular momentum of a spinning body is dominated by the contribution from the rapidly spinning part. All other motions make only a very small contribution to the total angular momentum (or its change with time). Thus as a first approximation, it is usually adequate to ignore all but the spin angular momentum. If the spin rate is roughly constant, this observation suggests that we assume

$$\frac{{}^{\mathcal{B}}d}{dt}\left({}^{\mathcal{I}}\mathbf{h}_G\right) \approx 0.$$

The result is that the equation of motion reduces to the simple vector equation

$$\boxed{{}^{\mathcal{I}}\boldsymbol{\omega}^{\mathcal{B}} \times {}^{\mathcal{I}}\mathbf{h}_G \approx \mathbf{M}_G.}$$

(11.9)

Eq. (11.9) is not a differential equation! Our qualitative analysis says simply that the angular velocity ${}^{\mathcal{I}}\boldsymbol{\omega}^{\mathcal{B}}$ of the body frame $\mathcal{B}$ is such that it satisfies the cross product in Eq. (11.9). The only requirement is that we choose a body frame $\mathcal{B}$ where the spin angular momentum is roughly constant and is large compared to the contributions from the other motions. Getting comfortable with this sort of qualitative analysis of problems is enormously helpful. It gives you a sense of what to expect when you set out on the more complicated exact solution, which helps avoid mistakes. It also provides an intuition that is indispensable when you move from analyzing a system to synthesizing a design. Some examples will help clarify what we mean.

Example 11.3 Qualitative Analysis of the Gyropendulum

This example reconsiders the gyropendulum. In Example 11.2 we found the exact equations of motion and then were able to work out several equilibrium solutions. What if instead we had begun by just using Eq. (11.9) to determine the general characteristics of the motion before solving for it exactly?

Figure 11.3a shows that the angular momentum is dominated by the spin of the rotor:

$${}^{\mathcal{I}}\mathbf{h}_O \approx h\mathbf{e}_r.$$

The moment on the gyropendulum about O is again, from the free-body diagram in Figure 11.3c,

$$\mathbf{M}_O = m_P g l \sin \phi \mathbf{e}_\theta.$$

Thus a good estimate of the rotational dynamics is found from Eq. (11.9). We have

$${}^{\mathcal{I}}\boldsymbol{\omega}^{\mathcal{B}} \times h\mathbf{e}_r \approx m_P g l \sin \phi \mathbf{e}_\theta.$$

(11.10)

For this cross product to be satisfied, the angular velocity must be perpendicular to $\mathbf{e}_\theta$ and have a component perpendicular to $\mathbf{e}_r$. Thus there must be a component of ${}^{\mathcal{I}}\boldsymbol{\omega}^{\mathcal{B}}$ that is in the $-\mathbf{e}_\phi$ direction (watch the sign!). This observation implies

$${}^{\mathcal{I}}\boldsymbol{\omega}^{\mathcal{B}} \approx -\frac{m_P g l}{h} \sin \phi \mathbf{e}_\phi = \frac{m_P g l}{h}(\mathbf{e}_z - \cos \phi \mathbf{e}_r).$$

This qualitative analysis shows that, for a positive spin of the rotor, the body frame fixed to the gyropendulum precesses about the $\mathbf{e}_z$ axis in the counterclockwise direction when viewed from above.

Figure 11.4 Holding a spinning bicycle wheel on a swivel chair.

Example 11.4 The Spinning Bicycle Wheel

One of our favorite science-museum exhibits is the spinning bicycle wheel (Figure 11.4). In this exhibit, you sit in a swivel chair and pick up a large (usually somewhat heavy) bicycle wheel. You then have a friend spin it up as fast as he or she can. Holding the wheel out in front of you, you tilt it either to the right or to the left. As you do, you and the swivel chair will turn. It is an amazing yet simple demonstration of gyroscopic motion. Using Eq. (11.9) we can quite easily determine the direction of rotation of the chair.

We begin by connecting a body frame $\mathcal{B} = (G, \mathbf{b}_1, \mathbf{b}_2, \mathbf{b}_3)$ to the bicycle wheel, with $\mathbf{b}_1$ along the axle and directed to your left, $\mathbf{b}_2$ pointed toward you, and $\mathbf{b}_3$ pointed upward from your lap to complete the right-handed set. Note that we have not attached the body frame to the spinning wheel, just to the axle. This practice is important and is something done quite often in rigid-body analysis. For a symmetric body that is spinning, we often define the body frame as an intermediate frame. That is because we are not necessarily interested in the angle that the spinning rotor makes in absolute space, but rather with the orientation of the rotor's spin axis. This is perfectly fine (as long as the rotor is symmetric). The angular velocity in Eq. (11.9) is that of the wheel axle with respect to $\mathcal{I}$.

Figure 11.5 Spinning top.

Once spun up, the bicycle wheel has a rather hefty angular momentum directed in the $\mathbf{b}_1$ direction: $^{\mathcal{I}}\mathbf{h}_G = h\mathbf{b}_1$. When you tilt the bicycle wheel to the left (in the direction of the angular momentum), you are applying a positive torque to it along the $\mathbf{b}_2$ direction: $\mathbf{M}_G = M\mathbf{b}_2$.[2] An angular velocity of the body frame in absolute space must be induced to satisfy Eq. (11.9). In other words, we have

$$^{\mathcal{I}}\boldsymbol{\omega}^{\mathcal{B}} \times h\mathbf{b}_1 = M\mathbf{b}_2.$$

For this equation to be satisfied, the angular velocity must have a component $^{\mathcal{I}}\omega^{\mathcal{B}}\mathbf{b}_3$, where $^{\mathcal{I}}\omega^{\mathcal{B}} > 0$. Thus there is a positive angular velocity pointed upward; the wheel and the chair (with you in it) rotate counterclockwise when viewed from above. If you turn the wheel to the right, which requires a negative moment, then the chair will rotate clockwise.

Remember, this analysis is only approximate. A detailed treatment would use the full vector equation of motion, employ multiple frames, and include reaction forces in the chair. Nevertheless, if all we are interested in is predicting the direction the chair rotates, the qualitative analysis works amazingly well!

Example 11.5 A Spinning Top

This example takes a qualitative look at the impact of friction on a top that spins on a rounded (as opposed to pointed) tip (see Figure 11.5). Let $\mathcal{C} = (O, \mathbf{c}_1, \mathbf{c}_2, \mathbf{c}_3)$ denote the body frame fixed to the top. Figure 11.6 illustrates frame $\mathcal{C}$ and the two intermediate frames $\mathcal{B}$ and $\mathcal{A}$ that allow us to relate the orientation of $\mathcal{C}$ to an inertial frame $\mathcal{I}$ using Euler angles ψ and θ. The third Euler angle ϕ describes the top's primary angle of rotation about $\mathbf{b}_3 = \mathbf{c}_3$; as for the bicycle wheel in the previous example, we assume $\dot{\phi} \gg 0$. Thus we can approximate the top's angular momentum about its center of mass as $^{\mathcal{I}}\mathbf{h}_G \approx h_G\mathbf{b}_3$, where h_G is a positive constant. Under this approximation, the rate of change of $^{\mathcal{I}}\mathbf{h}_G$ with respect to frame $\mathcal{B}$ is zero, and we can gain insight into the behavior of the top using Eq. (11.9).

---

[2] Note that this torque is an external moment on the wheel because, for you to be able to apply this moment without counterrotating, you must push against the chair.

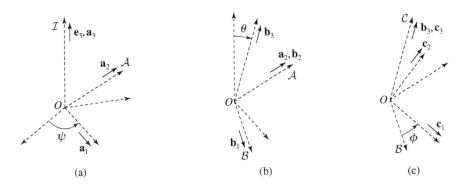

Figure 11.6 Reference frames for the spinning top. (a) Intermediate frame $\mathcal{A}$. (b) Intermediate frame $\mathcal{B}$. (c) Body frame $\mathcal{C}$.

To apply Eq. (11.9), we must first compute the (external) moment $\mathbf{M}_G$ about the center of mass. Since the weight of the top acts at its center of mass and therefore does not exert a moment about it, the only forces that produce moments about G are the contact forces from the ground. Let us assume that the top makes contact with the ground at point O' (as shown in Figure 11.7a) and d is the radius of curvature of the bottom. In this case, the moment about G is

$$\mathbf{M}_G = \mathbf{r}_{O'/G} \times \mathbf{F}_{O'},$$

where

$$\mathbf{r}_{O'/G} = -l\mathbf{b}_3 + d\mathbf{b}_1.$$

Using the free-body diagram in Figure 11.7b yields

$$\mathbf{F}_{O'} = N\mathbf{a}_3 - \mu N\mathbf{b}_2,$$

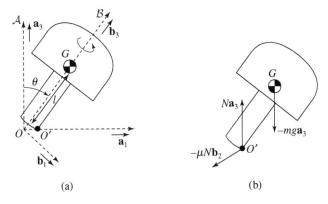

Figure 11.7 (a) Side view of spinning top in inertial frame $\mathcal{A}$. (b) Free-body diagram used for the spinning top.

where μ is the coefficient of friction. If, in addition, we assume that the bottom radius d is significantly smaller than the height l of the center of mass relative to the bottom, then the position of O' relative to G is roughly $\mathbf{r}_{O'/G} \approx -l\mathbf{b}_3$. The transformation table relating the unit vectors of frames $\mathcal{A}$ and $\mathcal{B}$ is

	$\mathbf{b}_1$	$\mathbf{b}_2$	$\mathbf{b}_3$
$\mathbf{a}_1$	$\cos\theta$	0	$\sin\theta$
$\mathbf{a}_2$	0	1	0
$\mathbf{a}_3$	$-\sin\theta$	0	$\cos\theta$.

Therefore the moment about G is approximately

$$\mathbf{M}_G \approx (-l\mathbf{b}_3) \times (-N\sin\theta\mathbf{b}_1 - \mu N\mathbf{b}_2 + N\cos\theta\mathbf{b}_3)$$
$$= lN\sin\theta\mathbf{b}_2 - l\mu N\mathbf{b}_1. \tag{11.11}$$

Using $^{\mathcal{I}}\mathbf{h}_G \approx h_G\mathbf{b}_3$ and Eq. (11.11) in Eq. (11.9) to find $^{\mathcal{I}}\boldsymbol{\omega}^{\mathcal{B}}$ yields

$$^{\mathcal{I}}\boldsymbol{\omega}^{\mathcal{B}} \approx -\frac{lN\sin\theta}{h_G}\mathbf{b}_1 - \frac{l\mu N}{h_G}\mathbf{b}_2. \tag{11.12}$$

Now consider the angular velocity $^{\mathcal{I}}\boldsymbol{\omega}^{\mathcal{B}}$ in terms of the Euler angle rates:

$$^{\mathcal{I}}\boldsymbol{\omega}^{\mathcal{B}} = \dot{\psi}\mathbf{a}_3 + \dot{\theta}\mathbf{b}_2 = -\dot{\psi}\sin\theta\mathbf{b}_1 + \dot{\theta}\mathbf{b}_2 + \dot{\psi}\cos\theta\mathbf{b}_3. \tag{11.13}$$

Comparing Eqs. (11.12) and (11.13) reveals that $\dot{\psi} \approx lN/h_G > 0$ and $\dot{\theta} \approx -l\mu N/h_G < 0$. The first result implies that the top precesses counterclockwise at a constant rate about the $\mathbf{a}_3$ axis. The second result implies that the angle θ that the top is tipped away from vertical decreases at a rate proportional to the coefficient of friction, μ. That is, friction causes the top to right itself! Although we are unable to predict the $\mathbf{b}_3$ component of the angular velocity, we can predict the basic motion of the top.

11.2.2 The Moment of Inertia Tensor

Although a qualitative understanding of motion is tremendously useful, eventually we do want to find scalar differential equations of motion that can be used to describe rigid-body trajectories—that is, the evolution of its orientation—in terms of angular-velocity components and Euler angles. To do so, we follow a process similar to that in Chapter 9 for planar motion: we find an expression for the angular momentum in terms of the angular velocity. This then allows us to write the scalar component magnitudes of Eq. (11.3) as differential equations in the coefficients of the angular velocity expressed as components in the body frame (i.e., ω_1, ω_2, and ω_3).

Chapter 9 used the planar assumption in several places. We now relax that assumption. The result is a bit more complicated, but it fully describes the three-dimensional angular momentum of a rigid body. We begin, as before, with the equation for the angular momentum of a rigid body (viewed as a collection of N particles) about its center of mass G:

$$^{\mathcal{I}}\mathbf{h}_G = \sum_{i=1}^{N} m_i \mathbf{r}_{i/G} \times {}^{\mathcal{I}}\mathbf{v}_{i/G}.$$

We use the transport equation on the inertial velocity of each particle, recognizing that the body is rigid (so that there is no body-frame velocity) to find

$$^{\mathcal{I}}\mathbf{v}_{i/G} = \underbrace{{}^{\mathcal{B}}\mathbf{v}_{i/G}}_{=0} + {}^{\mathcal{I}}\boldsymbol{\omega}^{\mathcal{B}} \times \mathbf{r}_{i/G}.$$

This result is substituted back into the formula for $^{\mathcal{I}}\mathbf{h}_G$ to again find Eq. (9.15),

$$^{\mathcal{I}}\mathbf{h}_G = \sum_{i=1}^{N} m_i \mathbf{r}_{i/G} \times \left({}^{\mathcal{I}}\boldsymbol{\omega}^{\mathcal{B}} \times \mathbf{r}_{i/G} \right).$$

We now deviate from our planar approach. Since motion is allowed in all three directions, we can no longer simply substitute for the angular velocity and assume it is fixed in the $\mathbf{b}_3$ direction. Instead, applying the vector triple cross-product rule (see Appendix B) gives

$$^{\mathcal{I}}\mathbf{h}_G = \sum_{i=1}^{N} m_i \left[(\mathbf{r}_{i/G} \cdot \mathbf{r}_{i/G}) {}^{\mathcal{I}}\boldsymbol{\omega}^{\mathcal{B}} - \mathbf{r}_{i/G} (\mathbf{r}_{i/G} \cdot {}^{\mathcal{I}}\boldsymbol{\omega}^{\mathcal{B}}) \right]. \qquad (11.14)$$

Now things get a bit more sophisticated. How do we factor out the angular velocity as was done in Chapter 9 for the planar case? We do it by introducing the *tensor product* $\otimes$. The tensor product is described in Appendix B. Unlike the cross product, which produces a vector described by three component magnitudes, the tensor product of two vectors results in a second-rank tensor described by nine scalars. We use the scalar tensor triple product described in Appendix B to rewrite the second term in Eq. (11.14):

$$^{\mathcal{I}}\mathbf{h}_G = \sum_{i=1}^{N} m_i \left[(\mathbf{r}_{i/G} \cdot \mathbf{r}_{i/G}) {}^{\mathcal{I}}\boldsymbol{\omega}^{\mathcal{B}} - (\mathbf{r}_{i/G} \otimes \mathbf{r}_{i/G}) \cdot {}^{\mathcal{I}}\boldsymbol{\omega}^{\mathcal{B}} \right].$$

Don't be intimidated by this new notation. It seems much more complicated than it really is. Soon we will rewrite the results with matrices so that we can work with the scalar component magnitudes directly. The tensor notation is convenient, however, because it permits compact writing of the equations of motion in a vector form independent of a specific reference frame, as has been done throughout the book.

We now factor out the angular velocity, as for the planar case, to obtain

$$^{\mathcal{I}}\mathbf{h}_G = \underbrace{\left(\sum_{i=1}^{N} m_i \left[(\mathbf{r}_{i/G} \cdot \mathbf{r}_{i/G}) \mathbb{U} - (\mathbf{r}_{i/G} \otimes \mathbf{r}_{i/G}) \right] \right)}_{\triangleq \mathbb{I}_G} \cdot {}^{\mathcal{I}}\boldsymbol{\omega}^{\mathcal{B}},$$

where $\mathbb{U}$ is the unity tensor and $\mathbb{I}_G$ is the moment of inertia tensor. This equation is the three-dimensional analog to Eq. (9.17), except here the moment of inertia is a tensor rather than a scalar.

Definition 11.1 The **moment of inertia tensor** $\mathbb{I}_G$ of a rigid body about its center of mass G is the mass-weighted sum

$$\mathbb{I}_G \triangleq \sum_{i=1}^{N} m_i \left[(\mathbf{r}_{i/G} \cdot \mathbf{r}_{i/G})\mathbb{U} - (\mathbf{r}_{i/G} \otimes \mathbf{r}_{i/G}) \right].$$

For a continuous rigid body $\mathcal{B}$,

$$\mathbb{I}_G \triangleq \int_{\mathcal{B}} \left[(\mathbf{r}_{dm/G} \cdot \mathbf{r}_{dm/G})\mathbb{U} - (\mathbf{r}_{dm/G} \otimes \mathbf{r}_{dm/G}) \right] \, dm.$$

Using Definition 11.1, the angular momentum of the body about its center of mass can be written in the compact form

$$^{\mathcal{I}}\mathbf{h}_G = \mathbb{I}_G \cdot {}^{\mathcal{I}}\boldsymbol{\omega}^{\mathcal{B}}. \tag{11.15}$$

Note, again, that the value of tensor notation is that it is frame independent. The equation for the angular momentum in terms of the moment of inertia tensor in Eq. (11.15) is the same no matter which frame we elect to express its components. However, at some point we may need to choose a frame in which to express the vector components. Appendix B describes how to find the components I_{ij} of tensor $\mathbb{I}_G$ in frame $\mathcal{B} = (G, \mathbf{b}_1, \mathbf{b}_2, \mathbf{b}_3)$:

$$\mathbb{I}_G = \sum_{i=1}^{3} \sum_{j=1}^{3} I_{ij} \mathbf{b}_i \otimes \mathbf{b}_j.$$

For instance, suppose that the position of particle k with respect to G has Cartesian coordinates $(x_k, y_k, z_k)_{\mathcal{B}}$. Then the discrete form in Definition 11.1 yields the following expressions for the components of the rigid body's moment of inertia tensor in that frame. When $i = j$,

$$I_{11} = \sum_{k=1}^{N} m_k \left(y_k^2 + z_k^2 \right) \tag{11.16}$$

$$I_{22} = \sum_{k=1}^{N} m_k \left(x_k^2 + z_k^2 \right) \tag{11.17}$$

$$I_{33} = \sum_{k=1}^{N} m_k \left(x_k^2 + y_k^2 \right), \tag{11.18}$$

where we used $\mathbf{r}_{k/G} = x_k\mathbf{b}_1 + y_k\mathbf{b}_2 + z_k\mathbf{b}_3$. The quantities I_{11}, I_{22}, and I_{33} are called the *moments of inertia*. When $i \neq j$,

$$I_{12} = I_{21} = -\sum_{k=1}^{N} m_k x_k y_k \tag{11.19}$$

$$I_{13} = I_{31} = -\sum_{k=1}^{N} m_k x_k z_k \tag{11.20}$$

$$I_{23} = I_{32} = -\sum_{k=1}^{N} m_k y_k z_k. \tag{11.21}$$

These are called the *products of inertia*.

As an alternative to tensor notation, we now reprise this analysis using matrix notation. In fact, it is in three-dimensional rigid-body analysis that matrix notation comes in particularly handy. Recall that we can use matrix notation to write the component magnitudes of a vector in a frame:

$$[\mathbf{r}_{k/G}]_\mathcal{B} = \begin{bmatrix} x_k \\ y_k \\ z_k \end{bmatrix}_\mathcal{B},$$

where, again, Cartesian coordinates locate particle k with respect to G in $\mathcal{B}$. Appendix B similarly shows how to write the components of a tensor in matrix notation:

$$[\mathbb{I}_G]_\mathcal{B} = \begin{bmatrix} I_{11} & I_{12} & I_{13} \\ I_{21} & I_{22} & I_{23} \\ I_{31} & I_{32} & I_{33} \end{bmatrix}_\mathcal{B}.$$

This is known as the *moment of inertia matrix*. $[\mathbb{I}_G]_\mathcal{B}$ is a 3×3 symmetric matrix with elements calculated in a frame $\mathcal{B}$ fixed to the rigid body. The diagonal elements are the moments of inertia of the rigid body (given by Eqs. (11.16)–(11.18)) and the off-diagonal elements are the products of inertia (given by Eqs. (11.19)–(11.21)).

Appendix B shows how various vector and tensor operations can be written as matrix products, allowing us to find expressions for the moment of inertia matrix in terms of simple matrix operations. Thus we can use Definition 11.1 to write out the discrete form of the moment of inertia tensor components in matrix notation:

$$[\mathbb{I}_G]_\mathcal{B} = \sum_{i=1}^{N} m_i \left(([\mathbf{r}_{i/G}]_\mathcal{B}^T [\mathbf{r}_{i/G}]_\mathcal{B}) I - [\mathbf{r}_{i/G}]_\mathcal{B} [\mathbf{r}_{i/G}]_\mathcal{B}^T \right),$$

where I is the 3×3 identity matrix.

For a continuous rigid body,

$$[\mathbb{I}_G]_\mathcal{B} = \int_\mathscr{B} \left(\|\mathbf{r}_{dm/G}\|^2 I - [\mathbf{r}_{dm/G}]_\mathcal{B} [\mathbf{r}_{dm/G}]_B^T \right) dm.$$

As with the scalar moment of inertia in Chapter 9, the moment of inertia matrix is usually computed using a position-dependent density $\rho(\mathbf{r}_{dm/G})$ for the rigid body:

$$[\mathbb{I}_G]_{\mathcal{B}} = \int_{\mathscr{B}} \left(\|\mathbf{r}_{dm/G}\|^2 I - [\mathbf{r}_{dm/G}]_{\mathcal{B}} [\mathbf{r}_{dm/G}]_{\mathcal{B}}^T \right) \rho(\mathbf{r}_{dm/G}) dV,$$

where dV is the differential volume element (see Appendix A).

As done earlier with the components of the tensor, let's see what is going on by writing out the matrix integral explicitly using the Cartesian coordinates $(x,\ y,\ z)_{\mathcal{B}}$ of mass element dm in frame $\mathcal{B}$. Using these coordinates, the position $[\mathbf{r}_{dm/G}]_{\mathcal{B}}$ is $[x\ y\ z]_{\mathcal{B}}^T$. Assuming a constant density ρ, the moment of inertia matrix is

$$[\mathbb{I}_G]_{\mathcal{B}} = \rho \int_{\mathscr{B}} \begin{bmatrix} y^2 + z^2 & -xy & -xz \\ -yx & x^2 + z^2 & -yz \\ -zx & -zy & x^2 + y^2 \end{bmatrix}_{\mathcal{B}} dx\,dy\,dz. \qquad (11.22)$$

The body-frame components of the angular momentum from Eq. (11.15) in matrix notation are

$$\boxed{[^{\mathcal{I}}\mathbf{h}_G]_{\mathcal{B}} = [\mathbb{I}_G]_{\mathcal{B}} [^{\mathcal{I}}\boldsymbol{\omega}^{\mathcal{B}}]_{\mathcal{B}}.} \qquad (11.23)$$

Example 11.6 Computing the Moment of Inertia of a Cube

This example uses the formulation of the moment of inertia matrix in Cartesian coordinates to find the moment of inertia matrix of a constant-density cube about its center. Consider a body frame $\mathcal{B}$ aligned with the edges of a cube and whose origin is coincident with the center of the cube, as shown in Figure 11.8. Assume the length of each side of the cube is $2l$. Also assume that the mass of the cube is m, which implies that $\rho = m/(8l^3)$. Then the matrix in Eq. (11.22) becomes

$$[\mathbb{I}_G]_{\mathcal{B}} = \frac{m}{8l^3} \int_{-l}^{l} \int_{-l}^{l} \int_{-l}^{l} \begin{bmatrix} y^2 + z^2 & -xy & -xz \\ -yx & x^2 + z^2 & -yz \\ -zx & -zy & x^2 + y^2 \end{bmatrix}_{\mathcal{B}} dx\,dy\,dz. \qquad (11.24)$$

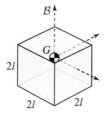

Figure 11.8 Computing the moment of inertia of a cube.

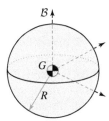

Figure 11.9 Computing the moment of inertia of a sphere.

Eq. (11.24) is a 3×3 matrix of integrals. Let I_{ij} denote the entry in the ith row and jth column of this matrix, where $i, j = 1, 2, 3$. Evaluating the first integral yields

$$I_{11} = \frac{m}{8l^3} \int_{-l}^{l} \int_{-l}^{l} \int_{-l}^{l} (y^2 + z^2) dx\, dy\, dz = \frac{2}{3} ml^2.$$

In fact, we have $I_{22} = I_{33} = (2/3)ml^2$ as well. Evaluating an off-diagonal integral yields

$$I_{12} = \frac{m}{8l^3} \int_{-l}^{l} \int_{-l}^{l} \int_{-l}^{l} (-xy) dx\, dy\, dz = 0.$$

In fact, *all* the off-diagonal integrals evaluate to zero. Therefore the moment of inertia matrix relative to a frame perpendicular to the faces of a constant-density cube is simply $[\mathbb{I}_G]_{\mathcal{B}} = (2/3)ml^2 I$, where I is the 3×3 identity matrix.

Example 11.7 Computing the Moment of Inertia of a Sphere

The same procedure can be used to compute the moment of inertia matrix of a constant-density sphere of radius R. We again introduce body frame $\mathcal{B}$ with origin at the center of the sphere, as shown in Figure 11.9. Assuming again that the mass of the sphere is m, then the density is $\rho = 3m/(4\pi R^3)$. For this geometry, the moment of inertia matrix is most easily computed using spherical coordinates. Using the relationships from Eq. (10.4),

$$x = r \cos \theta \sin \phi$$
$$y = r \sin \theta \sin \phi$$
$$z = r \cos \phi,$$

we can replace the Cartesian coordinates in Eq. (11.22). The result is an integral expression for the moment of inertia matrix:

$$[\mathbb{I}_G]_{\mathcal{B}} = \frac{3m}{4\pi R^3} \int_0^{\pi} \sin \phi \, d\phi \int_0^R r^4 dr \tag{11.25}$$

$$\times \int_0^{2\pi} \begin{bmatrix} (\sin^2 \theta \sin^2 \phi + \cos^2 \phi) & -\frac{1}{2} \sin 2\theta \sin^2 \phi & -\frac{1}{2} \cos \theta \sin 2\phi \\ -\frac{1}{2} \sin 2\theta \sin^2 \phi & (\cos^2 \theta \sin^2 \phi + \cos^2 \phi) & -\frac{1}{2} \sin \theta \sin 2\phi \\ -\frac{1}{2} \cos \theta \sin 2\phi & -\frac{1}{2} \sin \theta \sin 2\phi & \sin^2 \phi \end{bmatrix}_{\mathcal{B}} d\theta,$$

where we used the volume element in spherical coordinates, $dV = r^2 \sin\phi\, dr\, d\phi\, d\theta$. All products of inertia are zero because each includes an integral in θ over a full period of sine or cosine. This is true no matter how we orient the frame chosen relative to the sphere. Unlike a cube, where we can choose a specific geometry for the reference frame $\mathcal{B}$, for a sphere the moment of inertia matrix is always diagonal, no matter what body frame is chosen. We'll revisit this important observation later.

Completing the integrals over dr and $d\theta$ leaves (it helps to consult an integral table)

$$[\mathbb{I}_G]_{\mathcal{B}} = \frac{3mR^2}{40} \int_0^\pi \begin{bmatrix} (3+\cos 2\phi) & 0 & 0 \\ 0 & (3+\cos 2\phi) & 0 \\ 0 & 0 & 4\sin^2\phi \end{bmatrix}_{\mathcal{B}} \sin\phi\, d\phi.$$

Completing the remaining integrals over $d\phi$ results in the moment of inertia matrix of a constant-density sphere: $[\mathbb{I}_G]_{\mathcal{B}} = (2/5)mR^2 I$, where I is the 3×3 identity matrix.

The sphere is an example of an *isoinertial* body. An isoinertial body has identical moment of inertia about any line through the center of mass and zero products of inertia, no matter which body frame is chosen. It is an interesting fact that all centrobaric[3] bodies are also isoinertial. One of the consequences of a body being isoinertial is that the angular momentum is always aligned with the angular velocity in the body. This is not true in general for a three-dimensional rigid body, as we show next.

Example 11.8 The Moment of Inertia and Angular Momentum of a Rotating Rod

This example computes the moment of inertia tensor and angular momentum of the long cylindrical rod shown in Figure 11.10. The rod has a circular cross section of radius R and length l. We again introduce a body frame $\mathcal{B}$ with origin at the center of mass G of the rod and with $\mathbf{b}_3$ aligned along the long axis of the rod. Assuming uniform density and mass m, the density is $\rho = m/(\pi R^2 l)$. For this body, cylindrical coordinates are best for computing the moments of inertia. Using our usual relationship between cylindrical and Cartesian coordinates and $dV = r\, dr\, d\theta\, dz$, the moment of inertia matrix is

$$[\mathbb{I}_G]_{\mathcal{B}} = \frac{m}{\pi R^2 l} \int_0^R r\, dr \int_0^{2\pi} d\theta$$

$$\times \int_{-l/2}^{l/2} \begin{bmatrix} r^2\sin^2\theta + z^2 & -r^2\cos\theta\sin\theta & -rz\cos\theta \\ -r^2\cos\theta\sin\theta & r^2\cos^2\theta + z^2 & -rz\sin\theta \\ -rz\cos\theta & -rz\sin\theta & r^2 \end{bmatrix}_{\mathcal{B}} dz.$$

Completing the integrals gives the moment of inertia matrix for the rod:

$$[\mathbb{I}_G]_{\mathcal{B}} = \begin{bmatrix} \frac{mR^2}{4} + \frac{ml^2}{12} & 0 & 0 \\ 0 & \frac{mR^2}{4} + \frac{ml^2}{12} & 0 \\ 0 & 0 & \frac{mR^2}{2} \end{bmatrix}_{\mathcal{B}}.$$

---

[3] Recall that for a centrobaric body the center of mass and the center of gravity coincide.

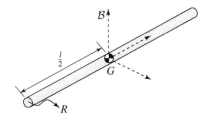

Figure 11.10 Computing the moment of inertia of a long cylindrical rod.

Unlike for the cube and sphere, the cylindrical rod has a different moment of inertia about the z axis than it does about the x and y axes. Thus if the rod is rotating with angular velocity $[^\mathcal{I}\boldsymbol{\omega}^\mathcal{B}]_\mathcal{B} = [\omega_1\ \omega_2\ \omega_3]_\mathcal{B}^T$, the angular momentum of the rod is

$$[^\mathcal{I}\mathbf{h}_G]_\mathcal{B} = \begin{bmatrix} (\frac{mR^2}{4} + \frac{ml^2}{12})\omega_1 \\ (\frac{mR^2}{4} + \frac{ml^2}{12})\omega_2 \\ \frac{mR^2}{2}\omega_3 \end{bmatrix}_\mathcal{B},$$

which is *not* aligned with the angular velocity.

Often we will solve problems incorporating a "long thin rod". This description refers to a cylindrical rod with $R \ll l$. In this case, we usually ignore the inertia about the z-axis and approximate the moment of inertia matrix by

$$[\mathbb{I}_G]_\mathcal{B} = \begin{bmatrix} \frac{ml^2}{12} & 0 & 0 \\ 0 & \frac{ml^2}{12} & 0 \\ 0 & 0 & 0 \end{bmatrix}_\mathcal{B}.$$

11.2.3 The Rotational Equations of Motion

As for the planar case in Chapter 9, we can now take the expression for the angular momentum of the rigid body in terms of the moment of inertia tensor and the angular velocity in Eq. (11.15) and substitute into the body-frame vector equation of motion in Eq. (11.8):

$$\underbrace{\frac{^\mathcal{B}d}{dt}(\mathbb{I}_G)}_{=0} \cdot {}^\mathcal{I}\boldsymbol{\omega}^\mathcal{B} + \mathbb{I}_G \cdot \frac{^\mathcal{B}d}{dt}({}^\mathcal{I}\boldsymbol{\omega}^\mathcal{B}) + {}^\mathcal{I}\boldsymbol{\omega}^\mathcal{B} \times (\mathbb{I}_G \cdot {}^\mathcal{I}\boldsymbol{\omega}^\mathcal{B}) = \mathbf{M}_G.$$

Note that we assumed the product rule applies to the scalar tensor triple product and took the derivative in the frame of a tensor. This process highlights that we are repeating the result found in Chapter 9, which is that the derivative of the moment of inertia is zero, since it is defined in the body frame of the rigid body. That means the rotational equation of motion simplifies to the important form:

$$\boxed{\mathbb{I}_G \cdot \frac{^\mathcal{B}d}{dt}({}^\mathcal{I}\boldsymbol{\omega}^\mathcal{B}) + {}^\mathcal{I}\boldsymbol{\omega}^\mathcal{B} \times (\mathbb{I}_G \cdot {}^\mathcal{I}\boldsymbol{\omega}^\mathcal{B}) = \mathbf{M}_G.} \qquad (11.26)$$

Eq. (11.26) is the complete equation of motion for a rigid body rotating in absolute space. Its components provide three differential equations for the angular speeds. Note that it is a first-order differential equation in the angular-velocity component magnitudes. Sometimes we may wish to substitute the kinematic equations of rotation to obtain second-order differential equations in the Euler angles. Most of the time we will leave them as first-order differential equations in the angular velocities and combine them with the kinematic equations of rotation to get six first-order equations that completely describe the rotational trajectory. Note that this is the same first-order form used to numerically integrate equations of motion.

When converting Eq. (11.26) into scalar equations of motion, it is most convenient to use matrix notation. We could just differentiate Eq. (11.23) directly and repeat our derivation in matrix form. In fact, if you are finding tensor notation confusing, you can always solve three-dimensional rigid-body rotation problems in matrix form; just be sure you are careful to keep track of which frame of reference the components are written in. We could alternatively use the matrix form of tensor operations in Appendix B to directly convert Eq. (11.26). In either case, the matrix form of the rotational equations of motion, expressed in a frame $\mathcal{B}$ fixed to the rigid body, is

$$[\mathbb{I}_G]_\mathcal{B} \frac{^\mathcal{B}d}{dt}\left([^\mathcal{I}\boldsymbol{\omega}^\mathcal{B}]_\mathcal{B}\right) + [^\mathcal{I}\boldsymbol{\omega}^\mathcal{B}\times]_\mathcal{B}[\mathbb{I}_G]_\mathcal{B}[^\mathcal{I}\boldsymbol{\omega}^\mathcal{B}]_\mathcal{B} = [\mathbf{M}_G]_\mathcal{B}, \qquad (11.27)$$

where we have dropped the moment of inertia matrix derivative, since we assume the moments and products of inertia are constant in the body frame. We have also used the *cross-product equivalent matrix* to write the cross product with the angular velocity in matrix notation as

$$[^\mathcal{I}\boldsymbol{\omega}^\mathcal{B}\times]_\mathcal{B} = \begin{bmatrix} 0 & -\omega_3 & \omega_2 \\ \omega_3 & 0 & -\omega_1 \\ -\omega_2 & \omega_1 & 0 \end{bmatrix}_\mathcal{B}.$$

We describe this convenient tool in Appendix B.

Example 11.9 A Spinning Symmetric Rigid Body

This example uses the matrix form of the rotational dynamics in Eq. (11.27) to derive the exact equations of motion of the spinning rigid body shown in Figure 11.11a. The body is symmetric and is rapidly spinning; it has a small moment of inertia about the spin axis and equal moments of inertia about the other axes. We consider the case where the body is subject to a constant external torque about its center of mass. Our qualitative analysis (see Examples 11.3 and 11.4) using Eq. (11.9) predicts a steady rotation of the spin-axis direction (check this!); here we verify this prediction using a quantitative analysis. Observe that this is an excellent model of the earth's *geoid* experiencing small moments due to the gravitational attraction of the sun and moon. We discuss this model a bit more at the end of the example.

Let $\mathcal{C}$ denote the body frame fixed to the rigid body with origin at its center of mass. The intermediate frames $\mathcal{A}$ and $\mathcal{B}$ allow us to describe the orientation of $\mathcal{C}$ relative to an inertial frame $\mathcal{I}$ using the Euler angles ψ and θ, as depicted in Figure 11.11b

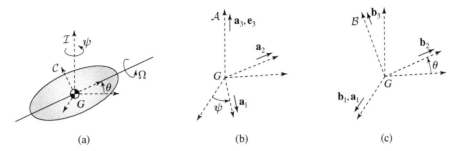

Figure 11.11 Spinning symmetric rigid body undergoing constant torque. (a) Coordinates. (b) Frame $\mathcal{A}$. (c) Frame $\mathcal{B}$.

and c. The rigid body spins rapidly at angular rate Ω about the $\mathbf{b}_2 = \mathbf{c}_2$ axis. (This is a set of 3-1-2 Euler angles.) The angular velocity of $\mathcal{C}$ relative to $\mathcal{B}$ is therefore $^{\mathcal{B}}\boldsymbol{\omega}^{\mathcal{C}} = \Omega\mathbf{b}_2$. The angular velocity of $\mathcal{B}$ relative to $\mathcal{I}$ is

$$^{\mathcal{I}}\boldsymbol{\omega}^{\mathcal{B}} = \dot{\psi}\mathbf{a}_3 + \dot{\theta}\mathbf{b}_1 = \dot{\theta}\mathbf{b}_1 + \dot{\psi}\sin\theta\,\mathbf{b}_2 + \dot{\psi}\cos\theta\,\mathbf{b}_3, \qquad (11.28)$$

where we expressed $^{\mathcal{I}}\boldsymbol{\omega}^{\mathcal{B}}$ as components in frame $\mathcal{B}$ using the transformation table between frames $\mathcal{A}$ and $\mathcal{B}$:

	$\mathbf{b}_1$	$\mathbf{b}_2$	$\mathbf{b}_3$
$\mathbf{a}_1$	1	0	0
$\mathbf{a}_2$	0	$\cos\theta$	$-\sin\theta$
$\mathbf{a}_3$	0	$\sin\theta$	$\cos\theta$.

The angular velocity of $\mathcal{C}$ relative to $\mathcal{I}$ is

$$^{\mathcal{I}}\boldsymbol{\omega}^{\mathcal{C}} = \dot{\theta}\mathbf{b}_1 + (\dot{\psi}\sin\theta + \Omega)\mathbf{b}_2 + \dot{\psi}\cos\theta\,\mathbf{b}_3.$$

We now compute the angular momentum $^{\mathcal{I}}\mathbf{h}_G$ of the rigid body about its center of mass and express it in the intermediate frame $\mathcal{B}$ fixed to the body symmetry axis. (As in our qualitative analysis, choosing an intermediate frame to perform rotational analysis of a spinning body often helps to simplify the calculations.) Because of the physical symmetry, the body frame $\mathcal{C}$ yields a diagonal moment of inertia matrix,

$$[\mathbb{I}_G]_{\mathcal{C}} = \begin{bmatrix} I_1 & 0 & 0 \\ 0 & I_2 & 0 \\ 0 & 0 & I_1 \end{bmatrix}_{\mathcal{C}},$$

with identical first and third entries. Note that, because of the rotational symmetry about $\mathbf{b}_2$, *the moment of inertia is unaffected by the spin,* which implies that $[\mathbb{I}_G]_B = [\mathbb{I}_G]_C$.[4] In matrix notation, the angular momentum expressed in frame B is thus

$$[^\mathcal{I}\mathbf{h}_G]_B = [\mathbb{I}_G]_B \left[^\mathcal{I}\boldsymbol{\omega}^\mathcal{C}\right]_B = \begin{bmatrix} I_1\dot{\theta} \\ I_2(\dot{\psi}\sin\theta + \Omega) \\ I_1\dot{\psi}\cos\theta \end{bmatrix}_B . \tag{11.29}$$

We next assume a constant applied moment about the $\mathbf{b}_1$ axis of $\mathbf{M}_G = -M_0\mathbf{b}_1$. Depending on the specific application, this moment could arise from a number of sources (e.g., aerodynamic loading on an airplane or gravitational torque on a satellite). The important thing here is that it is a constant in the B frame.

Now we are ready to compute the rotational equations of motion using Eq. (11.27). The only caveat is that we will produce the equations of motion using the transport equation by considering the rate of change of the angular momentum with respect to the intermediate frame B instead of the body frame C. (Either choice is valid; this choice simplifies the calculation.) Also note that we will perform the matrix operations with respect to frame B; the important thing is that we are consistent. (This is why we use a frame subscript in our matrix notation!) The rotational dynamics with respect to frame B are

$$\frac{^B d}{dt}[^\mathcal{I}\mathbf{h}_G]_B + [^\mathcal{I}\boldsymbol{\omega}^B\times]_B[^\mathcal{I}\mathbf{h}_G]_B = [\mathbf{M}_G]_B.$$

Using Eqs. (11.28) and (11.29) yields the matrix equation

$$\begin{bmatrix} I_1\ddot{\theta} \\ I_2\frac{d}{dt}(\dot{\psi}\sin\theta + \Omega) \\ I_1(\ddot{\psi}\cos\theta - \dot{\psi}\dot{\theta}\sin\theta) \end{bmatrix}_B$$

$$+ \begin{bmatrix} 0 & -\dot{\psi}\cos\theta & \dot{\psi}\sin\theta \\ \dot{\psi}\cos\theta & 0 & -\dot{\theta} \\ -\dot{\psi}\sin\theta & \dot{\theta} & 0 \end{bmatrix}_B \begin{bmatrix} I_1\dot{\theta} \\ I_2(\dot{\psi}\sin\theta + \Omega) \\ I_1\dot{\psi}\cos\theta \end{bmatrix}_B = \begin{bmatrix} -M_0 \\ 0 \\ 0 \end{bmatrix}_B ,$$

which, in turn, yields the following rotational equations of motion:

$$I_1\ddot{\theta} + (I_1 - I_2)\dot{\psi}^2 \sin\theta\cos\theta - I_2\Omega\dot{\psi}\cos\theta = -M_0 \tag{11.30}$$

$$\frac{d}{dt}\left(\dot{\psi}\sin\theta + \Omega\right) = 0 \tag{11.31}$$

$$I_1\ddot{\psi}\cos\theta + (I_2 - 2I_1)\dot{\psi}\dot{\theta}\sin\theta + I_2\Omega\dot{\theta} = 0. \tag{11.32}$$

Eq. (11.31) is a perfect differential and thus integrates to yield a constant of the motion,

$$\dot{\psi}\sin\theta + \Omega = C, \tag{11.33}$$

---

[4] It is very common in the analysis of spinning symmetric rigid bodies to analyze only the orientation of an intermediate body frame that is not spinning with the body, since we are typically not interested in the spin orientation. As long as the body is symmetric with respect to this frame, so that the moments of inertia are the same, this approach is valid.

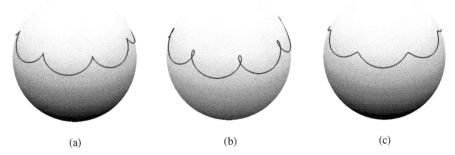

(a)   (b)   (c)

Figure 11.12 Motion of the rotating axis of a spinning body projected onto a sphere. Qualitative analysis correctly predicts precession about the vertical axis, but fails to predict small oscillations (nutation) that are evident in the exact solution. (a) $\dot{\psi}(0) = 0$. (b) $\dot{\psi}(0) = -0.1$ rad/s. (c) $\dot{\psi}(0) = 0.1$ rad/s.

where C is a constant that depends on the initial conditions. In fact, Eq. (11.33) implies that the component of the angular momentum about the spin axis is conserved. We use this observation to simplify Eqs. (11.30) and (11.32):

$$I_1\ddot{\theta} + I_1\dot{\psi}^2 \sin\theta \cos\theta - I_2 C\dot{\psi} \cos\theta = -M_0 \tag{11.34}$$

$$I_1\ddot{\psi} \cos\theta + I_2 C\dot{\theta} - 2I_1\dot{\psi}\dot{\theta} \sin\theta = 0. \tag{11.35}$$

Eqs. (11.34) and (11.35) are two differential equations of motion that can be integrated to yield trajectories for ψ and θ. The resulting solution exhibits *precession* (the slow rotation of the symmetry axis in ψ) and *nutation* (the wobble of the symmetry axis in θ). Figure 11.12 illustrates how precession and nutation depend on the initial value of $\dot{\psi}$ by plotting the projection of the tip of the symmetry axis $\mathbf{b}_2$ onto a sphere. The plots were generated using MATLAB to integrate Eqs. (11.34) and (11.35) with $\Omega(0) = 20$ rad/s, $\theta(0) = 30°$, $I_1 = 10$ kg-m^2, $I_2 = 1$ kg-m^2, and $M_0 = 5$ N-m. Note how the shape of the projected trajectory depends on $\dot{\psi}(0)$.

Precession and nutation are exhibited by any spinning body undergoing a constant torque. For example, as we alluded to at the beginning of the example, the earth can be modeled as a spinning rigid body that is being torqued by the sun and moon (due to slight asphericities in its shape). Its symmetry axis precesses with a period of approximately 26,000 years, known as the *precession of the equinoxes*. There is also a small nutation about the precession of roughly 9 arcseconds, known as the *Chandler wobble*.

Although the exact solution to these equations of motion predicts behavior that is significantly more complicated than our qualitative analysis, the qualitative solution was approximately correct. Recall that we predicted the rigid body would precess at a roughly constant rate. Our simulations support this prediction. In fact, we can find an approximate solution for precession and nutation by considering small deviations of the spin axis. Assuming that θ and $\dot{\theta}$ are both small, the equations of motion reduce to

$$I_1\ddot{\theta} + I_1\dot{\psi}^2\theta - I_2 C\dot{\psi} = -M_0 \tag{11.36}$$

$$I_1\ddot{\psi} + I_2 C\dot{\theta} = 0. \tag{11.37}$$

Eq. (11.37) is again a perfect differential that can be integrated to yield

$$\dot{\psi} = -\frac{I_2}{I_1}C\theta + k,$$

where k is a constant that depends on the initial conditions. Substituting into Eq. (11.36) yields

$$\ddot{\theta} + \left(k^2 + \frac{I_2^2}{I_1^2}C^2\right)\theta = -\frac{M_0}{I_1} + \frac{I_2}{I_1}Ck,$$

which is the equation for simple harmonic motion! Thus θ oscillates (nutates) about a constant offset, while ψ precesses at a roughly constant rate with small oscillations. Eq. (11.33) shows that the spin rate will oscillate by a small amount about the nominal value as well.

11.2.4 Principal Axes and Euler's Equations

Although the scalar equations represented by the matrix form of the equation of motion in Eq. (11.27) work for every problem, in their most general form (that is, for an arbitrary body frame and moment of inertia matrix), they can be quite complicated. The moment of inertia matrix has nine elements. Because it is a symmetric matrix, only six elements are independent. Nonetheless, the various matrix multiplications can result in rather cumbersome equations. Fortunately, there is an important and common simplification that makes the resulting equations easier to write down and easier to solve: we can assume that the moment of inertia matrix is diagonal. In fact, we did that in the previous example. Finding a general rule involves the concept of *principal axes.*

For any rigid body, we can always find a body frame $\mathcal{B}_p$ within which the products of inertia are zero and the moment of inertia matrix is diagonal. These special axes are called principal axes. Many bodies have obvious symmetries, such as in Example 11.9; in these cases the principal axes correspond to the symmetry axes of the body. For situations where it is easier to calculate the moments and products of inertia in other axes, it is always possible to transform to principal axes through an ordered rotation. We discuss how to do that in Section 11.5. As shown there, the rotation matrix consists of the eigenvectors of the moment of inertia matrix and the principal moments of inertia are the eigenvalues. For now, it is important just to remember that, for most problems, it is wise to choose a principal-axes body frame. This frame almost always corresponds to symmetry axes of the body. Appendix D tabulates the principal moments of inertia and principal axes for a number of simple rigid bodies.

Choosing a principal-axes body frame greatly simplifies the equations of motion for the rigid body. Since the moment of inertia matrix takes the simple diagonal form

$$[\mathbb{I}_G]_{\mathcal{B}_p} = \begin{bmatrix} I_1 & 0 & 0 \\ 0 & I_2 & 0 \\ 0 & 0 & I_3 \end{bmatrix}_{\mathcal{B}_p},$$

substituting into the matrix form of the rotational equation of motion gives

$$
\begin{bmatrix} I_1 & 0 & 0 \\ 0 & I_2 & 0 \\ 0 & 0 & I_3 \end{bmatrix}_{\mathcal{B}_p} \begin{bmatrix} \dot{\omega}_1 \\ \dot{\omega}_2 \\ \dot{\omega}_3 \end{bmatrix}_{\mathcal{B}_p}
$$

$$
+ \begin{bmatrix} 0 & -\omega_3 & \omega_2 \\ \omega_3 & 0 & -\omega_1 \\ -\omega_2 & \omega_1 & 0 \end{bmatrix}_{\mathcal{B}_p} \begin{bmatrix} I_1 & 0 & 0 \\ 0 & I_2 & 0 \\ 0 & 0 & I_3 \end{bmatrix}_{\mathcal{B}_p} \begin{bmatrix} \omega_1 \\ \omega_2 \\ \omega_3 \end{bmatrix}_{\mathcal{B}_p} = \begin{bmatrix} M_1 \\ M_2 \\ M_3 \end{bmatrix}_{\mathcal{B}_p} , \quad (11.38)
$$

where $M_i \mathbf{b}_i$, $i = 1, 2, 3$ are the components of $\mathbf{M}_G$ in $\mathcal{B}_p$.

The matrix equation in Eq. (11.38) is equivalent to the following three scalar equations for the angular velocity components in $\mathcal{B}_p$:

$$
\begin{array}{rcl}
I_1\dot{\omega}_1 + (I_3 - I_2)\omega_2\omega_3 & = & M_1 \\
I_2\dot{\omega}_2 + (I_1 - I_3)\omega_1\omega_3 & = & M_2 \\
I_3\dot{\omega}_3 + (I_2 - I_1)\omega_1\omega_2 & = & M_3.
\end{array}
\qquad (11.39)
$$

These three scalar equations are called *Euler's equations.* They are the cornerstone of most rigid-body rotation problems. There is no new physics here—Euler's equations are just another way to write the angular-momentum equation from Eq. (11.8) in terms of component magnitudes. Collectively these equations are the scalar version of Euler's second law.

Example 11.10 Torque-Free Motion

One of the most important examples of rigid-body motion and the use of Euler's equations is the study of torque-free motion of a spinning rigid body—the rotational motion of a rigid body in free space in the absence of any torques on it. For the planar situations studied in Chapter 9, this example is completely trivial; the rigid body either will not rotate or will spin at a constant angular velocity about the fixed axis. However, in three dimensions, the motion results in some surprising effects. Even though the angular momentum must be fixed in magnitude and direction in absolute space (by conservation of angular momentum), *the angular velocity (and the body frame) can change orientation with respect to the angular momentum.* Note that this problem well approximates a wide variety of physical systems, such as rotating satellites, falling objects, and planetary rotation.

We begin by writing Euler's equations (Eq. (11.39)) in the absence of applied moments or torques:

$$
\begin{array}{rcl}
I_1\dot{\omega}_1 + (I_3 - I_2)\omega_2\omega_3 & = & 0 \\
I_2\dot{\omega}_2 + (I_1 - I_3)\omega_1\omega_3 & = & 0 \\
I_3\dot{\omega}_3 + (I_2 - I_1)\omega_1\omega_2 & = & 0.
\end{array}
$$

Without loss of generality, we consider a rigid body spinning about its $\mathbf{b}_3$ axis. Thus we label the angular velocity about the 3-axis as $\omega_s + \omega_3$, where ω_s is a (large) constant spin rate and ω_3 is a small perturbation. We also assume that the other two angular velocity component magnitudes, ω_1 and ω_2, are small compared to ω_s. Dropping the small second-order products $\omega_1\omega_2$, $\omega_1\omega_3$, and $\omega_2\omega_3$, Euler's equations become

$$I_1\dot{\omega}_1 + (I_3 - I_2)\omega_s\omega_2 \approx 0 \qquad (11.40)$$

$$I_2\dot{\omega}_2 + (I_1 - I_3)\omega_s\omega_1 \approx 0 \qquad (11.41)$$

$$I_3\dot{\omega}_3 \approx 0. \qquad (11.42)$$

These equations (with $\omega_s = \omega_3$) are exact for a symmetric rigid body (where $I_2 = I_1$). That is, for a symmetric torque-free rigid body, the motion we find in this example is true for all values of ω_1 and ω_2, not just for small deviations.

Eq. (11.42) shows that the spin of the body about the 3-axis stays roughly constant (and exactly so for a symmetric body). For the angular rates about the other two axes, we differentiate Eqs. (11.40) and (11.41) and combine the result to find the separate equations of motion,

$$\ddot{\omega}_1 + \omega_n^2\omega_1 = 0$$

$$\ddot{\omega}_2 + \omega_n^2\omega_2 = 0,$$

where

$$\omega_n^2 = \frac{(I_3 - I_2)(I_3 - I_1)\omega_s^2}{I_1 I_2}. \qquad (11.43)$$

As long as $\omega_n^2 > 0$, these two equations of motion represent simple harmonic motion in ω_1 and ω_2! Thus the angular rates about the 1- and 2-axes remain small and oscillate about their small initial values at the same frequency ω_n. The spin axis, rather than being directly along the body 3-axis, is slightly offset from the 3-axis and rotates about it. From Eq. (11.43) we see that this occurs when $I_3 > I_1$ and $I_3 > I_2$, or $I_3 < I_1$ and $I_3 < I_2$. In other words, when the body spins about either the principal axis of maximum moment of inertia (the *major axis*) or the principal axis of minimum moment of inertia (the *minor axis*) its motion is stable (Figure 11.13). This motion of the angular velocity about the 3-axis of the body (the symmetry axis for symmetric rigid bodies) is often called *free-body precession.*

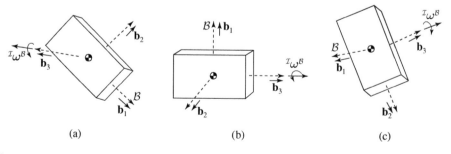

Figure 11.13 Torque-free motion. (a) Spin about major axis. (b) Spin about minor axis. (c) Spin about intermediate axis.

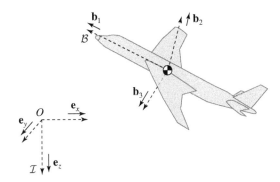

Figure 11.14 Inertial frame and body frame of an airplane.

When the spin occurs about the intermediate axis of inertia, the equations of motion become unstable and produce exponential growth of the angular-velocity components over time. The motion rapidly deviates from a simple spin about the 2-axis. You should remember this fundamental result of rigid-body analysis: *torque-free spin about the intermediate axis of inertia of a rigid body is unstable.*

Example 11.11 The Rotational Equations of Motion of an Airplane

The previous chapter (Examples 10.4, 10.10, and 10.14) solved for the translational and rotational kinematics of an airplane in flight but only for the translational equations of motion. We complete the treatment here with the rotational equations of motion. This turns out to be straightforward, using the same matrix treatment that led us to Euler's equations. Unfortunately, we can't simply substitute into Euler's equations because the airplane's body frame $\mathcal{B}$, shown in Figure 11.14, does not coincide with principal axes. Rather, the moment of inertia matrix in the body frame is

$$[\mathbb{I}_G]_\mathcal{B} = \begin{bmatrix} I_1 & 0 & -I_{13} \\ 0 & I_2 & 0 \\ -I_{13} & 0 & I_3 \end{bmatrix}_\mathcal{B}. \tag{11.44}$$

Most aircraft are symmetric about the vertical plane (containing $\mathbf{b}_1$ and $\mathbf{b}_3$), thus making $I_{12} = I_{23} = 0$. However, they don't have front-to-back symmetry, which implies nonzero products of inertia. Of course, it is certainly possible to find the principal axes for any aircraft, but they would differ for each plane and would not coincide with the obviously convenient geometric directions of the conventional body frame. Because the body frame in Figure 11.14 is so convenient for expressing the forces and moments and describing the configuration of the plane, it is universally used at the expense of having slightly more complicated rotational equations of motion.

Using the conventional notation for the angular-velocity components, $[{}^{\mathcal{I}}\boldsymbol{\omega}^{\mathcal{B}}]_{\mathcal{B}} = [p \; q \; r]_{\mathcal{B}}^{T}$ (see Example 10.9), the equations of motion in matrix form for the symmetric airplane are

$$
\begin{bmatrix} I_1 & 0 & -I_{13} \\ 0 & I_2 & 0 \\ -I_{13} & 0 & I_3 \end{bmatrix}_{\mathcal{B}} \begin{bmatrix} \dot{p} \\ \dot{q} \\ \dot{r} \end{bmatrix}_{\mathcal{B}}
$$

$$
+ \begin{bmatrix} 0 & -r & q \\ r & 0 & -p \\ -q & p & 0 \end{bmatrix}_{\mathcal{B}} \begin{bmatrix} I_1 & 0 & -I_{13} \\ 0 & I_2 & 0 \\ -I_{13} & 0 & I_3 \end{bmatrix}_{\mathcal{B}} \begin{bmatrix} p \\ q \\ r \end{bmatrix}_{\mathcal{B}} = \begin{bmatrix} M_1 \\ M_2 \\ M_3 \end{bmatrix}_{\mathcal{B}},
$$

where M_i is the ith component of the applied moment in the body frame. Solving for $\dot{p}, \dot{q}$, and $\dot{r}$ requires multiplying both sides from the left by the inverse of the moment of inertia matrix. We compute

$$
[\mathbb{I}_G]_{\mathcal{B}}^{-1} = \begin{bmatrix} \dfrac{1}{I_1 - I_{13}^2/I_3} & 0 & \dfrac{I_{13}}{I_1 I_3 - I_{13}^2} \\ 0 & \dfrac{1}{I_2} & 0 \\ \dfrac{I_{13}}{I_1 I_3 - I_{13}^2} & 0 & \dfrac{1}{I_3 - I_{13}^2/I_1} \end{bmatrix}_{\mathcal{B}}.
$$

The rotational equations of motion for the symmetric airplane are

$$
\dot{p} = \frac{1}{I_1 I_3 - I_{13}^2} \left[I_3(I_3 - I_2)qr - I_3 I_{13} pq + I_{13}(I_2 - I_1)pq + I_{13}^2 qr \right]
$$

$$
+ \frac{1}{I_1 I_3 - I_{13}^2} \left(I_3 M_1 + I_{13} M_3 \right)
$$

$$
\dot{q} = \frac{1}{I_2} \left[(I_1 - I_3)pr + I_{13}(p^2 - r^2) \right] + \frac{M_2}{I_2}
$$

$$
\dot{r} = \frac{1}{I_1 I_3 - I_{13}^2} \left[I_1(I_2 - I_1)pq + I_1 I_{13} qr + I_{13}(I_3 - I_2)qr - I_{13}^2 pq \right]
$$

$$
+ \frac{1}{I_1 I_3 - I_{13}^2} \left(I_{13} M_1 + I_1 M_3 \right).
$$

These three equations are combined with the inverse kinematic equations from Example 10.11 (Eq. (10.50)) to obtain the set of six first-order differential equations that describe airplane orientation. The inverse kinematic equations are

$$
\dot{\psi} = (q \sin \phi + r \cos \phi) \sec \theta \tag{11.45}
$$

$$
\dot{\theta} = q \cos \phi - r \sin \phi \tag{11.46}
$$

$$
\dot{\phi} = (q \sin \phi + r \cos \phi) \tan \theta + p. \tag{11.47}
$$

Note that these six equations differ from most equations of motion we have found in that they do not consist of second-order differential equations in the coordinates

$((\psi, \theta, \phi)_{\mathcal{B}}^{\mathcal{I}}$ in this case). Rather than use the coordinates and their rates to describe the rotational state, we use the coordinates and the angular-velocity components. Nevertheless, these six first-order differential equations can be integrated using the same numerical approach as we have used throughout the book.

For completeness, we repeat the six translational equations (kinematic and kinetic) from Example 10.14 (Eqs. (10.57)–(10.59)):

$$\dot{u} = rv - qw + F_1/m_G$$
$$\dot{v} = pw - ru + F_2/m_G$$
$$\dot{w} = qu - pv + F_3/m_G$$

and

$$\begin{bmatrix} \dot{x} \\ \dot{y} \\ \dot{z} \end{bmatrix}_{\mathcal{I}} = {}^{\mathcal{I}}C^{\mathcal{B}} \begin{bmatrix} u \\ v \\ w \end{bmatrix}_{\mathcal{B}}.$$

These 12 differential equations completely describe the six degrees of freedom of an airplane in flight. Of course, we can't yet integrate them (even numerically), as we have not yet found the forces and moments on the airplane. We discuss these further in Chapter 12.

11.3 The Moment Transport Theorem and the Parallel Axis Theorem in Three Dimensions

We now reexamine the motion of a rigid body relative to an arbitrary point Q other than the center of mass G, as done for the planar case in Chapter 9. This means revisiting the discussions in Sections 9.4.1 and 9.4.2, where we derived the moment transport theorem and the parallel axis theorem. The treatment in Section 9.4.1 was again completely general in terms of vectors in three dimensions. Thus our final result for the moment transport theorem in Eq. (9.32) holds without change:

$$\mathbf{M}_Q = \mathbf{M}_G - \mathbf{r}_{Q/G} \times \sum_{i=1}^{N} \mathbf{F}_i^{(\text{ext})}. \tag{11.48}$$

In words, given the moment on a rigid body about its center of mass, Eq. (11.48) is used to find the resulting moment about some other point Q of the rigid body.

For the parallel axis theorem we are not so fortunate; treating that in three dimensions is more involved and requires a bit of care. In particular, we need to go back to the beginning of the derivation in Section 9.4 and recompute the angular momentum. Nevertheless, we once again take advantage of the fact that the treatment in Chapter 7 was entirely general. Thus, just as in Chapter 9 for planar rigid bodies, the equation of motion for the angular momentum of the rigid body about Q is given by Eq. (7.17):

$$\frac{{}^{\mathcal{I}}d}{dt} \left({}^{\mathcal{I}}\mathbf{h}_Q \right) = \mathbf{M}_Q + \mathbf{r}_{Q/G} \times m_G {}^{\mathcal{I}}\mathbf{a}_{Q/O}, \tag{11.49}$$

where we omitted the superscript (ext), as we are treating a rigid body. As before, Eq. (11.49) is most useful when the rigid body is pinned at Q and the resulting equation reduces to Euler's second law for the total angular momentum. Also, just as for the angular momentum about the center of mass, we can rewrite this equation of motion using the transport equation to be in terms of the body-frame derivative:

$$\frac{^{B}d}{dt}\left(^{I}\mathbf{h}_{Q}\right) + {}^{I}\boldsymbol{\omega}^{B} \times {}^{I}\mathbf{h}_{Q} - \mathbf{r}_{Q/G} \times m_{G}{}^{I}\mathbf{a}_{Q/O} = \mathbf{M}_{Q}. \qquad (11.50)$$

The final step is to find an expression for the angular momentum $^{I}\mathbf{h}_{Q}$, so we can develop a modified form of Euler's equations. What we would like to find is a relationship between $^{I}\mathbf{h}_{Q}$ and the angular velocity, somehow incorporating the moment of inertia matrix. When treating motion in two dimensions in Section 9.4, this relationship turned out to be relatively straightforward. We could write down $^{I}\mathbf{h}_{Q}$ in terms of a scalar moment of inertia about the point Q and relate this moment of inertia to the one about the center of mass G using the parallel axis theorem. Our objective here is to generalize this to three dimensions.

The most obvious approach is to start with the definition of $^{I}\mathbf{h}_{Q}$ and manipulate it in the same way done for $^{I}\mathbf{h}_{G}$ earlier. Thus, as in Section 9.4, we start with

$$^{I}\mathbf{h}_{Q} = \sum_{i=1}^{N} \mathbf{r}_{i/Q} \times m_{i}{}^{I}\mathbf{v}_{i/Q}$$

and go through the same series of substitutions (using the transport equation) and manipulations as done in the last section. That is, we use the transport equation to substitute $^{I}\boldsymbol{\omega}^{B} \times \mathbf{r}_{i/Q}$ for the inertial velocity and use the triple vector cross product rule to obtain the analog of Eq. (11.14). Using the scalar tensor triple product to find the angular momentum in terms of a moment of inertia about Q results in

$$^{I}\mathbf{h}_{Q} = \mathbb{I}_{Q} \cdot {}^{I}\boldsymbol{\omega}^{B}, \qquad (11.51)$$

where the moment of inertia tensor about Q is similar to that about G:

$$\mathbb{I}_{Q} \triangleq \sum_{i=1}^{N} m_{i}\left[(\mathbf{r}_{i/Q} \cdot \mathbf{r}_{i/Q}))\mathbb{U} - (\mathbf{r}_{i/Q} \otimes \mathbf{r}_{i/Q})\right]. \qquad (11.52)$$

For a continuous body,

$$\mathbb{I}_{Q} \triangleq \int_{\mathscr{B}}\left[(\mathbf{r}_{dm/Q} \cdot \mathbf{r}_{dm/Q})\mathbb{U} - (\mathbf{r}_{dm/Q} \otimes \mathbf{r}_{dm/Q})\right] \, dm.$$

Just as for the center-of-mass version, we substitute Eq. (11.51) into Eq. (11.50) for the equations of motion of the rigid body, but with the added term due to the inertial motion of Q. In tensor form, we have

$$\mathbb{I}_{Q} \cdot \frac{^{B}d}{dt}\left(^{I}\boldsymbol{\omega}^{B}\right) + {}^{I}\boldsymbol{\omega}^{B} \times \left(\mathbb{I}_{Q} \cdot {}^{I}\boldsymbol{\omega}^{B}\right) - \mathbf{r}_{Q/G} \times m_{G}{}^{I}\mathbf{a}_{Q/O} = \mathbf{M}_{Q}. \qquad (11.53)$$

In matrix form, Eq. (11.53) becomes

$$[\mathbb{I}_Q]_\mathcal{B} \frac{^\mathcal{B}d}{dt}\left([^\mathcal{I}\boldsymbol{\omega}^\mathcal{B}]_\mathcal{B}\right) + [^\mathcal{I}\boldsymbol{\omega}^\mathcal{B}\times]_\mathcal{B}[\mathbb{I}_Q]_\mathcal{B}[^\mathcal{I}\boldsymbol{\omega}^\mathcal{B}]_\mathcal{B}$$

$$- m_G[\mathbf{r}_{Q/G}\times]_\mathcal{B}[^\mathcal{I}\mathbf{a}_{Q/o}]_\mathcal{B} = [\mathbf{M}_Q]_\mathcal{B}. \qquad (11.54)$$

As in the planar case, if the body is pinned in absolute space at Q, then the correction term is zero and we have the same equations of motion as before, except that the moment of inertia matrix is taken relative to Q (which is now inertially fixed) rather than G. If the inertia matrix is also diagonal, then we again have Euler's equations. It is important to remember that, in this derivation, we assumed Q is fixed to the body, which implies $\frac{^\mathcal{B}d}{dt}(\mathbf{r}_{i/Q}) = 0$ for all points i on the rigid body. This assumption implies the moment of inertia matrix about Q is fixed in $\mathcal{B}$ and thus factors out of the time derivative.

To find an expression for $\mathbb{I}_Q$, we begin once again by recognizing that the developments in Chapter 7 were independent of dimension. Our vector equations apply equally well in three dimensions. Thus we can use the same equation as in Section 9.4.2 when discussing the (planar) parallel axis theorem; that is, the expression for $^\mathcal{I}\mathbf{h}_Q$ in terms of $^\mathcal{I}\mathbf{h}_G$ is

$$^\mathcal{I}\mathbf{h}_Q = {}^\mathcal{I}\mathbf{h}_G + m_G\mathbf{r}_{Q/G} \times {}^\mathcal{I}\mathbf{v}_{Q/G}. \qquad (11.55)$$

We replace the inertial velocity of point Q by means of the transport equation, recognizing that the body-frame velocity of Q is zero (because Q is attached to the body) to obtain

$$^\mathcal{I}\mathbf{h}_Q = {}^\mathcal{I}\mathbf{h}_G + m_G\mathbf{r}_{Q/G} \times \left({}^\mathcal{I}\boldsymbol{\omega}^\mathcal{B} \times \mathbf{r}_{Q/G}\right).$$

We again use the triple vector cross product rule to transform this expression:

$$^\mathcal{I}\mathbf{h}_Q = {}^\mathcal{I}\mathbf{h}_G + m_G\left[(\mathbf{r}_{Q/G}\cdot\mathbf{r}_{Q/G})^\mathcal{I}\boldsymbol{\omega}^\mathcal{B} - (\mathbf{r}_{Q/G}\cdot{}^\mathcal{I}\boldsymbol{\omega}^\mathcal{B})\mathbf{r}_{Q/G}\right].$$

This equation looks familiar. It is very similar to the expression in the summation over mass particles in our original derivation of the moment of inertia tensor (prior to Definition 11.1). We can thus use the same tensor product identity on the second term (see Appendix B) to rewrite it as

$$^\mathcal{I}\mathbf{h}_Q = \mathbb{I}_G \cdot {}^\mathcal{I}\boldsymbol{\omega}^\mathcal{B} + m_G\left[(\mathbf{r}_{Q/G}\cdot\mathbf{r}_{Q/G})^\mathcal{I}\boldsymbol{\omega}^\mathcal{B} - (\mathbf{r}_{Q/G}\otimes\mathbf{r}_{Q/G})\cdot{}^\mathcal{I}\boldsymbol{\omega}^\mathcal{B}\right],$$

where we have substituted from Eq. (11.15) for the angular momentum about the center of mass.

Finally, we can factor out the angular velocity to get our final expression for the angular momentum about Q,

$$^\mathcal{I}\mathbf{h}_Q = \left(\mathbb{I}_G + m_G\left[(\mathbf{r}_{Q/G}\cdot\mathbf{r}_{Q/G})\mathbb{U} - (\mathbf{r}_{Q/G}\otimes\mathbf{r}_{Q/G})\right]\right)\cdot{}^\mathcal{I}\boldsymbol{\omega}^\mathcal{B},$$

where $\mathbb{U}$ is the unity tensor. Comparing this result to Eq. (11.51) for the angular momentum about Q shows that the moment of inertia tensor about Q is

$$\mathbb{I}_Q = \mathbb{I}_G + m_G \left[(\mathbf{r}_{Q/G} \cdot \mathbf{r}_{Q/G})\mathbb{U} - (\mathbf{r}_{Q/G} \otimes \mathbf{r}_{Q/G}) \right]. \qquad (11.56)$$

Eq. (11.56) is the three-dimensional version of the parallel axis theorem. It provides a formula for computing the moment of inertia tensor about an arbitrary point Q fixed to a rigid body, given the moment of inertia tensor about the center of mass G and the position of Q with respect to G. It is used the same way as the planar form was used in Section 9.4.2.

We can also write the parallel axis theorem in matrix form, which is useful when actually doing computations:

$$[\mathbb{I}_Q]_{\mathcal{B}} = [\mathbb{I}_G]_{\mathcal{B}} + m_G \left(\|\mathbf{r}_{Q/G}\|^2 I - [\mathbf{r}_{Q/G}]_{\mathcal{B}}[\mathbf{r}_{Q/G}]_{\mathcal{B}}^T \right), \qquad (11.57)$$

where I is the 3×3 identity matrix.

Example 11.12 Transporting the Space Shuttle atop a 747

The space shuttle was frequently transported atop a specially equipped Boeing 747 aircraft as shown in Figure 11.15. To readjust the airplane's control systems, the total moment of inertia of the combined system is needed. Rather than compute that quantity directly, it is easier to use the known inertia tensors of the two vehicles. Each has a moment of inertia matrix relative to their individual centers of mass, as in Eq. (11.44). Let $[\mathbb{I}_{G_p}]_{\mathcal{B}_p}$ be the moment of inertia matrix of the plane in frame $\mathcal{B}_p$ relative to its center of mass, and $[\mathbb{I}'_{G_s}]_{\mathcal{B}_s}$ be the moment of inertia matrix of the shuttle in frame $\mathcal{B}_s$ relative to its center of mass. We have

$$[\mathbb{I}_{G_p}]_{\mathcal{B}_p} = \begin{bmatrix} I_1 & 0 & -I_{13} \\ 0 & I_2 & 0 \\ -I_{13} & 0 & I_3 \end{bmatrix}_{\mathcal{B}_p} \quad \text{and} \quad [\mathbb{I}'_{G_s}]_{\mathcal{B}_s} = \begin{bmatrix} I'_1 & 0 & -I'_{13} \\ 0 & I'_2 & 0 \\ -I'_{13} & 0 & I'_3 \end{bmatrix}_{\mathcal{B}_s}.$$

Because of the symmetries of both the plane and the shuttle, we assume that the center of mass of the shuttle is offset from that of the aircraft along the plane of symmetry and that the two body frames are not rotated with respect to each other. What we seek is the composite moment of inertia matrix expressed in frame $\mathcal{B}_p$ relative to the center of mass of the aircraft. To find it, we first need to find the shuttle's moment of inertia matrix relative to the plane's center of mass and then add it to the plane's moment of inertia matrix.

We begin by writing the position of the plane's center of mass relative to the shuttle's:

$$[\mathbf{r}_{G_p/G_s}]_{\mathcal{B}_s} = \begin{bmatrix} d_1 \\ 0 \\ d_3 \end{bmatrix}_{\mathcal{B}_s}.$$

Figure 11.15 Space shuttle being transported on a Boeing 747. Image courtesy of Shutterstock.

We then find the moment of inertia of the shuttle relative to G_p using Eq. (11.57):

$$
[\mathbb{I}'_{G_p}]_{\mathcal{B}_s} = \begin{bmatrix} I'_1 & 0 & -I'_{13} \\ 0 & I'_2 & 0 \\ -I'_{13} & 0 & I'_3 \end{bmatrix}_{\mathcal{B}_s} + m_s \begin{bmatrix} d_3^2 & 0 & -d_1 d_3 \\ 0 & d_1^2 + d_3^2 & 0 \\ -d_1 d_3 & 0 & d_1^2 \end{bmatrix}_{\mathcal{B}_s}.
$$

Since the two body frames align, the shuttle's moment of inertia matrix is the same expressed in $\mathcal{B}_p$. Adding it to the plane's moment of inertia matrix gives the total moment of inertia matrix of the plane plus shuttle relative to the center of mass of the plane:

$$
[\mathbb{I}^{(\text{tot})}_{G_p}]_{\mathcal{B}_p} = \begin{bmatrix} I_1 + I'_1 + m_s d_3^2 & 0 & -I_{13} - I'_{13} - m_s d_1 d_3 \\ 0 & I_2 + I'_2 + m_s(d_1^2 + d_3^2) & 0 \\ -I_{13} - I'_{13} - m_s d_1 d_3 & 0 & I_3 + I'_3 + m_s d_1^2 \end{bmatrix}.
$$

Example 11.13 Exact Solution of a Demonstration Gyroscope

This example uses the matrix form of the rotational dynamics in Eq. (11.54) and the parallel axis theorem in Eq. (11.57) to derive the exact equations of motion of a demonstration gyroscope depicted in Figure 11.16a. The gyro consists of a large spinning rotor at one end of a (massless) rod, able to spin on its axis, and a small

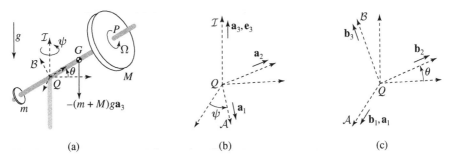

(a)   (b)   (c)

Figure 11.16 Exact solution of a demonstration gyroscope. (a) Coordinates. (b) Frame $\mathcal{A}$. (c) Frame $\mathcal{B}$.

mass whose position is adjustable at the other end of the rod. We treat the entire arrangement as a single rigid body. When the small mass is placed so that the support is exactly at the center of mass G, there are no moments or torques on the system and it balances. However, when the small mass is moved to misalign the center of mass, gravity produces a moment about the pinned point Q, resulting in interesting and complex motions of the gyro. Using a qualitative analysis as in Section 11.2.1, we predict a steady counterclockwise rotation of the gyro axis (check this!). In what follows, we perform a more careful quantitative treatment to verify this prediction and to study other aspects of the motion.

Figure 11.16a defines two frames: an inertial frame $\mathcal{I}$ located at the top of the support post at point Q and an intermediate body frame $\mathcal{B}$ fixed to the gyro shaft but not spinning with the wheel. We are interested in the orientation of $\mathcal{B}$ in $\mathcal{I}$ and the resulting trajectory. Since we are not concerned with the rotation of the rotor, there are only two degrees of freedom in this problem and the orientation of $\mathcal{B}$ in $\mathcal{I}$ is described by the two Euler angles, ψ and θ, shown in Figure 11.16a. These are the first two angles in a 3-1-3 Euler rotation set. The intermediate frame $\mathcal{A}$ shown in Figure 11.16b allows us to find the angular velocity of $\mathcal{B}$ relative to $\mathcal{I}$,

$$^{\mathcal{I}}\boldsymbol{\omega}^{\mathcal{B}} = \dot{\psi}\mathbf{a}_3 + \dot{\theta}\mathbf{b}_1 = \dot{\theta}\mathbf{b}_1 + \dot{\psi}\sin\theta\mathbf{b}_2 + \dot{\psi}\cos\theta\mathbf{b}_3, \qquad (11.58)$$

where we have expressed $^{\mathcal{I}}\boldsymbol{\omega}^{\mathcal{B}}$ in terms of components in frame $\mathcal{B}$ using the transformation table between frames $\mathcal{A}$ and $\mathcal{B}$:

	$\mathbf{b}_1$	$\mathbf{b}_2$	$\mathbf{b}_3$
$\mathbf{a}_1$	1	0	0
$\mathbf{a}_2$	0	$\cos\theta$	$-\sin\theta$
$\mathbf{a}_3$	0	$\sin\theta$	$\cos\theta$.

To find the angular momentum of the gyro relative to Q, we use the same approach as in Examples 11.4 and 11.5 and add the spin angular momentum of the rotor, $h\mathbf{b}_2$, to the angular momentum due to the motion of the rigid body about Q:

$$[^{\mathcal{I}}\mathbf{h}_Q]_{\mathcal{B}} = [\mathbb{I}_Q]_{\mathcal{B}}\,[\,^{\mathcal{I}}\boldsymbol{\omega}^{\mathcal{B}}\,]_{\mathcal{B}} + [h\mathbf{b}_2]_{\mathcal{B}},$$

where $[\mathbb{I}_Q]_{\mathcal{B}}$ is the moment of inertia matrix relative to Q given by the parallel axis theorem in Eq. (11.57). As in Example 11.9, we are ignoring the spin angle of the rotor and only accounting for its angular-momentum contribution. We also are solving for motion relative to a point that is not the center of mass, which necessitates the use of the parallel axis theorem.

Again, using the same approach followed in Example 11.9, we are only interested in the orientation of the intermediate body frame. We showed there that this approach was valid as long as the body was symmetric about the spin axis. That is the case here too, which allows us to write the moment of inertia matrix of the rotor about its center of mass P as the following diagonal matrix in frame $\mathcal{B}$:

$$[\mathbb{I}_P]_{\mathcal{B}} = \begin{bmatrix} I_1 & 0 & 0 \\ 0 & I_2 & 0 \\ 0 & 0 & I_1 \end{bmatrix}_{\mathcal{B}},$$

which has identical first and third entries because of the symmetry of the rotor. We now use the parallel axis theorem in Eq. (11.57) to find the moment of inertia of the rotor about point Q. Adding the contribution to the rigid-body inertia of the small mass, we obtain

$$[\mathbb{I}_Q]_B = \begin{bmatrix} I_1 & 0 & 0 \\ 0 & I_2 & 0 \\ 0 & 0 & I_1 \end{bmatrix}_B + M \begin{bmatrix} l^2 & 0 & 0 \\ 0 & 0 & 0 \\ 0 & 0 & l^2 \end{bmatrix}_B + m \begin{bmatrix} l'^2 & 0 & 0 \\ 0 & 0 & 0 \\ 0 & 0 & l'^2 \end{bmatrix}_B, \quad (11.59)$$

where $l = \|\mathbf{r}_{P/Q}\|$ is the distance of the rotor from the attachment point and $l' = \|\mathbf{r}_{m/Q}\|$ is the distance of the small mass from the attachment point.

In matrix notation, the angular momentum expressed in frame B is

$$[^\mathcal{I}\mathbf{h}_Q]_B = [\mathbb{I}_Q]_B [^\mathcal{I}\boldsymbol{\omega}^B]_B + [h\mathbf{b}_2]_B = \begin{bmatrix} I_1'\dot\theta \\ I_2\dot\psi \sin\theta + h \\ I_1'\dot\psi \cos\theta \end{bmatrix}_B, \quad (11.60)$$

where $I_1' = I_1 + Ml^2 + ml'^2$.

The next step is to compute the moment about Q. Here we refer to Examples 9.4 and 9.12, which showed that the moment about the center of mass of a rigid body due to a uniform gravity field is zero and thus that the moment on a compound pendulum is just the moment about Q due to the total force at the center of mass acting about Q. In those examples we never used the planar restriction, so they apply unchanged here. The moment is thus easily computed using the moment transport theorem in Eq. (11.48), where we ignore the force at Q because it cancels:

$$\mathbf{M}_Q = \mathbf{r}_{G/Q} \times \mathbf{F}_G = l_G\mathbf{b}_2 \times (-m_G g \sin\theta \mathbf{b}_2 - m_G g \cos\theta \mathbf{b}_3)$$

$$= -m_G g l_G \cos\theta \mathbf{b}_1, \quad (11.61)$$

where $m_G = m + M$ and $l_G = \|\mathbf{r}_{G/Q}\|$.

Finally, we find the rotational equations of motion by using Eq. (11.50) with $^\mathcal{I}\mathbf{a}_{Q/O} = 0$ and expressed in matrix notation:

$$\frac{^B d}{dt}[^\mathcal{I}\mathbf{h}_Q]_B + [^\mathcal{I}\boldsymbol{\omega}^B\times]_B[^\mathcal{I}\mathbf{h}_Q]_B = [\mathbf{M}_Q]_B.$$

Using Eqs. (11.58)–(11.61) yields the matrix equation

$$\begin{bmatrix} I_1'\ddot\theta \\ \frac{d}{dt}(I_2\dot\psi \sin\theta + h) \\ I_1'(\ddot\psi \cos\theta - \dot\psi\dot\theta \sin\theta) \end{bmatrix}_B$$

$$+ \begin{bmatrix} 0 & -\dot\psi \cos\theta & \dot\psi \sin\theta \\ \dot\psi \cos\theta & 0 & -\dot\theta \\ -\dot\psi \sin\theta & \dot\theta & 0 \end{bmatrix}_B \begin{bmatrix} I_1'\dot\theta \\ I_2\dot\psi \sin\theta + h \\ I_1'\dot\psi \cos\theta \end{bmatrix}_B = \begin{bmatrix} -m_G g l_G \cos\theta \\ 0 \\ 0 \end{bmatrix}_B,$$

which, in turn, yields the following rotational equations of motion:

$$I_1'\ddot{\theta} + (I_1' - I_2)\dot{\psi}^2 \sin\theta \cos\theta - h\dot{\psi}\cos\theta = -m_G g l_G \cos\theta \qquad (11.62)$$

$$\frac{d}{dt}(I_2\dot{\psi}\sin\theta + h) = 0 \qquad (11.63)$$

$$I_1'\ddot{\psi}\cos\theta + (I_2 - 2I_1')\dot{\psi}\dot{\theta}\sin\theta + h\dot{\theta} = 0. \qquad (11.64)$$

Observe the similarity between this result and the one we obtained in Example 11.9 (Eqs. (11.30)–(11.32)). In particular, Eq. (11.63) is a perfect differential, indicating that $I_2\dot{\psi}\sin\theta + h$ is a constant of the motion. The demo gyro exhibits the same kind of precession and nutation characteristics that are illustrated in Figure 11.12. We defer further analysis of this observation to Tutorial 11.2.

To conclude, we note that the parallel axis theorem in Eq. (11.56) is only valid for transporting the moment of inertia tensor from about the center of mass to about an arbitrary point fixed in the body. It cannot be used to transport the moment of inertia from one arbitrary point to another. That is because to obtain ${}^I\mathbf{h}_Q$ in terms of ${}^I\mathbf{h}_G$ in Eq. (11.55), we used the center of mass corollary. Nevertheless, there are occasions when it is useful to be able to find $\mathbb{I}_Q$, the moment of inertia tensor about an arbitrary point Q, given $\mathbb{I}_P$, the moment of inertia tensor about another arbitrary point P. You can derive that expression in Problem 11.18.

11.4 Dynamics of Multibody Systems in Three Dimensions

Section 9.6 discussed how to extend our results for the dynamics of rigid bodies and particles to a collection of connected bodies and particles. Our conclusion was quite simple: just break apart all connections, introduce constraint forces, and solve each part separately. The constraint forces can be eliminated algebraically by properly combining the equations of motion and constraints. In some cases we can avoid the extra algebraic step of introducing constraint forces by using the parallel axis theorem and the angular-momentum equation relative to a moving point. We also showed that the total angular momentum and total energy of the system are conserved if the total external moment is zero.

This idea is completely general and applies equally well to three-dimensional rigid-body problems. Simply break the components apart and use the equations of motion introduced in this chapter or use the new three-dimensional version of the parallel axis theorem (Eq. (11.56)). We can even use conservation of three-dimensional angular momentum (or total energy) to study the behavior of the entire system. We specifically highlight these ideas again in three dimensions because of a special class of connected systems consisting of a rigid body with an attached spinning rotor (like the spinning gyropendulum). Such a system is often used to stabilize a large moving mass, such as a camera or ship, or to control the attitude of a rigid body, such as a spacecraft or submarine. A rigid body with an attached spinning rotor is known as a *gyrostat*. It takes advantage of what we discovered earlier in our qualitative analysis: when subject to an external moment, a system with significant angular momentum precesses rather than spinning out of control. We illustrate how conservation of angular momentum can be used in the example below. Tutorial 11.2 shows a more complex example using gyroscopic stabilization.

Example 11.14 The Control Moment Gyroscope

The control moment gyroscope (CMG) is a device that utilizes conservation of total angular momentum to control the rotation and orientation of a rigid body. It is commonly used on large geosynchronous satellites. To a very good approximation such a satellite is free of external torques and thus the total angular momentum is conserved. Periodically, though, it may encounter an impulsive disturbance that causes it to gain angular momentum. Rather than use a thruster (which applies an external moment but requires fuel), the satellite applies an internal torque to one or more large spinning disks that have significant angular momentum about their centers of mass. By rotating the CMG, an angular-momentum component can be produced in the direction opposite the satellite's angular momentum. Since the total angular momentum must be conserved (only internal moments are used to rotate the CMG), the effect of the disturbance on the satellite's rotation can be canceled out.

A fairly simple model of a satellite with a control moment gyroscope shows how this works in principle. Let $^{\mathcal{I}}\mathbf{h}_s = \mathbb{I}_s \cdot {}^{\mathcal{I}}\boldsymbol{\omega}^{\mathcal{B}}$ be the angular momentum of the satellite, where $\mathbb{I}_s$ is the moment of inertia tensor of the satellite and $^{\mathcal{I}}\boldsymbol{\omega}^{\mathcal{B}}$ is the angular velocity of a body-fixed frame $\mathcal{B}$ attached to the satellite. Likewise, $^{\mathcal{I}}\mathbf{h}_r$ is the angular momentum of the control moment gyroscope's rotor. The total angular momentum of the system is the sum of the satellite and gyroscope angular momenta and is a constant of the motion:

$$^{\mathcal{I}}\mathbf{h}_T = \mathbb{I}_s \cdot {}^{\mathcal{I}}\boldsymbol{\omega}^{\mathcal{B}} + {}^{\mathcal{I}}\mathbf{h}_r = \text{constant}.$$

A CMG operates by reorienting the rotor spin in the body-frame so as to eliminate the angular velocity of $\mathcal{B}$ with respect to $\mathcal{I}$. The simplest model assumes the CMG consists of a spinning rotor on a three-axis gimbal as in Example 10.7. If we introduce a frame $\mathcal{C}$ fixed to the gyroscope rotor, then expressed as components in this frame the rotor angular momentum is $h\mathbf{c}_3$. The objective is to set the CMG angular momentum equal to the total angular momentum and thus force $^{\mathcal{I}}\boldsymbol{\omega}^{\mathcal{B}} = 0$. Introducing matrix notation and the direction-cosine matrix for the gimbals between frame $\mathcal{C}$ and frame $\mathcal{B}$ yields

$$^{\mathcal{B}}C^{\mathcal{C}} \begin{bmatrix} 0 \\ 0 \\ h \end{bmatrix}_{\mathcal{C}} = \left[{}^{\mathcal{I}}\mathbf{h}_T \right]_{\mathcal{B}}.$$

These three equations can be solved for the 3-1-3 Euler angles $(\psi, \theta, \phi)^{\mathcal{B}}_{\mathcal{C}}$.

A real CMG system is considerably more complicated (Figure 11.17). Typically, for reasons of redundancy and ease of assembly, CMGs are only mounted on a single or double gimbal. It is thus common for up to four rotors to be employed on a satellite and the solution for the gimbal motion involves more complicated control laws. An excellent description of such a system can be found in Bryson (1994) and Wie (1998).

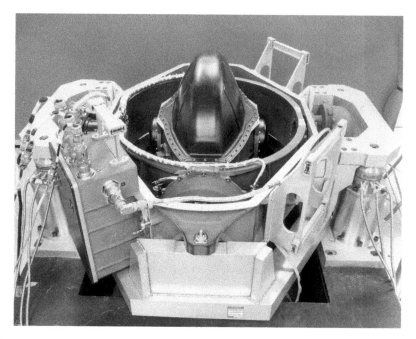

Figure 11.17 A control moment gyroscope of the type used on the International Space Station. Image courtesy of L-3 Communications, Space, and Navigation.

11.5 Rotating the Moment of Inertia Tensor

Suppose you know the moment of inertia tensor of a rigid body $\mathscr{B}$ in some body frame $\mathcal{B}_1$. How do you find the moment of inertia written in some other frame $\mathcal{B}_2$? The answer is very similar to how we rotate the component magnitudes of a vector from one frame to another. Recall from Eq. (10.38) that, if we have the direction-cosine matrix describing the orientation of $\mathcal{B}_1$ in $\mathcal{B}_2$, then we can transform the component magnitudes of a vector $\mathbf{r}$ written in $\mathcal{B}_1$ to those written in $\mathcal{B}_2$ using

$$[\mathbf{r}]_{\mathcal{B}_2} = {}^{\mathcal{B}_2}C^{\mathcal{B}_1}[\mathbf{r}]_{\mathcal{B}_1}.$$

A similar expression holds for the moment of inertia tensor. If we have the magnitudes of the moments and products of inertia in $\mathcal{B}_1$, then we can also find the moment of inertia tensor in $\mathcal{B}_2$ by using the direction-cosine matrix ${}^{\mathcal{B}_2}C^{\mathcal{B}_1}$. We just have to use it twice:

$$[\mathbb{I}_G]_{\mathcal{B}_2} = {}^{\mathcal{B}_2}C^{\mathcal{B}_1}[\mathbb{I}_G]_{\mathcal{B}_1}\left({}^{\mathcal{B}_2}C^{\mathcal{B}_1}\right)^T. \tag{11.65}$$

We have already seen one example where this expression would have been helpful. Recall the discussion of principal axes in Section 11.2.4. We showed there the simplifications that arise in the equations of motion by choosing a principal-axes frame but also noted that it sometimes is computationally (or experimentally) easier to find the moments and products of inertia in some other frame. To find the principal-axes frame, you then need to find the direction-cosine matrix such that Eq. (11.65) results in a diagonal matrix whose components are the principal moments of inertia. This is an example of *diagonalization,* which you may remember from a linear algebra class. For the matrix $[\mathbb{I}_G]_{\mathcal{B}_2}$ to be diagonal, the direction-cosine matrix in Eq. (11.65) must be the matrix of eigenvectors of $[\mathbb{I}_G]_{\mathcal{B}_1}$ and the resulting principal moments of inertia are the eigenvalues of $[\mathbb{I}_G]_{\mathcal{B}_1}$. This is one of the most important reasons for discussing rotation of the moment of inertia tensor—to transform to and from the principal axes.

One way to derive Eq. (11.65) is through the angular-momentum expression. Suppose we have the angular velocity $^{\mathcal{I}}\boldsymbol{\omega}^{\mathcal{B}_1}$ of $\mathcal{B}_1$ in $\mathcal{I}$. The angular momentum of the rigid body expressed as components in $\mathcal{B}_1$ is

$$[^{\mathcal{I}}\mathbf{h}_G]_{\mathcal{B}_1} = [\mathbb{I}_G]_{\mathcal{B}_1}[^{\mathcal{I}}\boldsymbol{\omega}^{\mathcal{B}_1}]_{\mathcal{B}_1}. \tag{11.66}$$

Suppose we would instead like the angular momentum expressed as components in frame $\mathcal{B}_2$. We can transform it by multiplying both sides of Eq. (11.66) by the direction-cosine matrix:

$$[^{\mathcal{I}}\mathbf{h}_G]_{\mathcal{B}_2} = {}^{\mathcal{B}_2}C^{\mathcal{B}_1}[^{\mathcal{I}}\mathbf{h}_G]_{\mathcal{B}_1} = {}^{\mathcal{B}_2}C^{\mathcal{B}_1}[\mathbb{I}_G]_{\mathcal{B}_1}[^{\mathcal{I}}\boldsymbol{\omega}^{\mathcal{B}_1}]_{\mathcal{B}_1}. \tag{11.67}$$

The left-hand side is just $[^{\mathcal{I}}\mathbf{h}_G]_{\mathcal{B}_2}$, according to Eq. (10.38). For the right-hand side, we also use Eq. (10.38), but on the angular velocity:

$$[^{\mathcal{I}}\boldsymbol{\omega}^{\mathcal{B}_1}]_{\mathcal{B}_2} = {}^{\mathcal{B}_2}C^{\mathcal{B}_1}[^{\mathcal{I}}\boldsymbol{\omega}^{\mathcal{B}_1}]_{\mathcal{B}_1}. \tag{11.68}$$

Using the orthogonality property of the direction cosine (i.e., that its inverse is also its transpose), we can multiply both sides of Eq. (11.68) by $({}^{\mathcal{B}_2}C^{\mathcal{B}_1})^T$ from the left and then substitute the result into Eq. (11.67) to find

$$[^{\mathcal{I}}\mathbf{h}_G]_{\mathcal{B}_2} = {}^{\mathcal{B}_2}C^{\mathcal{B}_1}[\mathbb{I}_G]_{\mathcal{B}_1}({}^{\mathcal{B}_2}C^{\mathcal{B}_1})^T[^{\mathcal{I}}\boldsymbol{\omega}^{\mathcal{B}_1}]_{\mathcal{B}_2}.$$

Finally, we observe that $^{\mathcal{I}}\boldsymbol{\omega}^{\mathcal{B}_1} = {}^{\mathcal{I}}\boldsymbol{\omega}^{\mathcal{B}_2}$ (since both frames are fixed to the rigid body), which allows us to write the final form of the transformed angular momentum:

$$[^{\mathcal{I}}\mathbf{h}_G]_{\mathcal{B}_2} = {}^{\mathcal{B}_2}C^{\mathcal{B}_1}[\mathbb{I}_G]_{\mathcal{B}_1}({}^{\mathcal{B}_2}C^{\mathcal{B}_1})^T[^{\mathcal{I}}\boldsymbol{\omega}^{\mathcal{B}_2}]_{\mathcal{B}_2}. \tag{11.69}$$

We now compare Eq. (11.69) to the definition of the matrix form of the angular momentum in terms of the moment of inertia matrix:

$$[^{\mathcal{I}}\mathbf{h}_G]_{\mathcal{B}_2} = [\mathbb{I}_G]_{\mathcal{B}_2}[^{\mathcal{I}}\boldsymbol{\omega}^{\mathcal{B}_2}]_{\mathcal{B}_2}.$$

Therefore it must be true that

$$[\mathbb{I}_G]_{\mathcal{B}_2} = {}^{\mathcal{B}_2}C^{\mathcal{B}_1}[\mathbb{I}_G]_{\mathcal{B}_1}({}^{\mathcal{B}_2}C^{\mathcal{B}_1})^T.$$

Example 11.15 A Long Thin Rod in a Gimbal

Consider the long thin rod connected to an outer gimbal by means of two rotations, as shown in Figure 11.18. We wish to find the moment of inertia matrix for the rod in frame $\mathcal{A}$ fixed to the outer gimbal. The rod has undergone a 3-1-3 rotation with respect

Figure 11.18 Gimbaled rod.

to the outer gimbal of $\psi = 30°$, $\theta = 60°$, and $\phi = 0$. The direction-cosine matrix $^{\mathcal{A}}C^{\mathcal{B}_p}$ for the principal-axes frame $\mathcal{B}_p$ fixed to the rod relative to the outer gimbal is thus found from Example 10.7:

$$
^{\mathcal{A}}C^{\mathcal{B}_p} =
\begin{bmatrix}
\cos 30° & \sin 30° & 0 \\
-\sin 30° \cos 60° & \cos 30° \cos 60° & \sin 60° \\
\sin 30° \sin 60° & -\cos 30° \sin 60° & \cos 60°
\end{bmatrix}
=
\begin{bmatrix}
\frac{\sqrt{3}}{2} & \frac{1}{2} & 0 \\
-\frac{1}{4} & \frac{\sqrt{3}}{4} & \frac{\sqrt{3}}{2} \\
\frac{\sqrt{3}}{4} & -\frac{3}{4} & \frac{1}{2}
\end{bmatrix}
$$

Using Eq. (11.65) and the moment of inertia matrix found in Example 11.8, the moment of inertia of the gimbaled rod in $\mathcal{A}$ is

$$
[\mathbb{I}_G]_{\mathcal{A}} =
\underbrace{\begin{bmatrix}
\frac{\sqrt{3}}{2} & \frac{1}{2} & 0 \\
-\frac{1}{4} & \frac{\sqrt{3}}{4} & \frac{\sqrt{3}}{2} \\
\frac{\sqrt{3}}{4} & -\frac{3}{4} & \frac{1}{2}
\end{bmatrix}}_{=\,^{\mathcal{A}}C^{\mathcal{B}_p}}
\underbrace{\begin{bmatrix}
\frac{ml^2}{12} & 0 & 0 \\
0 & \frac{ml^2}{12} & 0 \\
0 & 0 & 0
\end{bmatrix}}_{=[\mathbb{I}_G]_{\mathcal{B}_p}}
\underbrace{\begin{bmatrix}
\frac{\sqrt{3}}{2} & -\frac{1}{4} & \frac{\sqrt{3}}{4} \\
\frac{1}{2} & \frac{\sqrt{3}}{4} & -\frac{3}{4} \\
0 & \frac{\sqrt{3}}{2} & \frac{1}{2}
\end{bmatrix}}_{=\left(^{\mathcal{A}}C^{\mathcal{B}_p}\right)^T}.
$$

Multiplying these matrices gives the moment of inertia matrix of the rotated rod in frame $\mathcal{A}$:

$$
[\mathbb{I}_G]_{\mathcal{A}} = \frac{ml^2}{12}
\begin{bmatrix}
1 & 0 & 0 \\
0 & \frac{1}{4} & -\frac{\sqrt{3}}{4} \\
0 & -\frac{\sqrt{3}}{4} & \frac{3}{4}
\end{bmatrix}.
$$

Why would we ever want to rotate the moment of inertia tensor to another frame? In what situations would this be useful? As it turns out, there are many such situations.

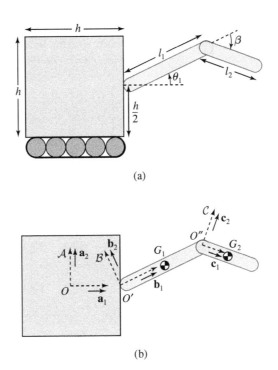

Figure 11.19 A model of a crane. (a) Crane schematic. (b) Reference frames.

A common reason for using Eq. (11.65) is to find the moment of inertia tensor of a rigid body composed of many connected rigid bodies of different shapes. The easiest way to find the composite moment of inertia tensor is to find the inertia tensor of each individual rigid body in its own principal-axes frame, rotate it into the body frame of the composite rigid body, and then use the parallel axis theorem to sum the constituent moment of inertia matrices, as demonstrated in the next example.

Example 11.16 The Moment of Inertia of a Crane

A crane, shown schematically in Figure 11.19, is composed of a cab (modeled as a cube of mass M) and two rigid links that can rotate freely in a plane (modeled as long thin rods of masses m_1 and m_2). We wish to find an expression for the moment of inertia tensor $\mathbb{I}_O$ of the whole system about the center of the cab (point O) and expressed as components in frame $\mathcal{A} = (O, \mathbf{a}_1, \mathbf{a}_2, \mathbf{a}_3)$ fixed to the cab. This might be necessary, for instance, to find the torque needed to rotate the cab around the $\mathbf{a}_2$ axis. To calculate $\mathbb{I}_O$, we find the moment of inertia tensor for each of the three components (cab and two arms), express each as a matrix in $\mathcal{A}$, and sum them.

Example 11.6 determined the moment of inertia of a constant-density cube. Using this result, we write the moment of inertia of the cab about point O in frame $\mathcal{A}$ as

$$[\mathbb{I}_O^{\text{cab}}]_{\mathcal{A}} = \frac{2}{3}M\frac{h^2}{4}I = \frac{Mh^2}{6}I, \tag{11.70}$$

where M is the mass of the cab and I is the identity matrix.

Example 11.8 determined the moment of inertia of a long thin rod aligned with the $\mathbf{b}_3$ axis. Here we have a long thin rod aligned with the $\mathbf{b}_1$ axis. The moment of inertia of the first arm about its center of mass (G_1, located halfway along the arm) in frame $\mathcal{B}$ is

$$[\mathbb{I}^1_{G_1}]_{\mathcal{B}} = \frac{m_1}{12} \begin{bmatrix} 0 & 0 & 0 \\ 0 & l^2_1 & 0 \\ 0 & 0 & l^2_1 \end{bmatrix}_{\mathcal{B}},$$

where m_1 is the mass of the first arm. Following Eq. (11.65),

$$[\mathbb{I}^1_{G_1}]_{\mathcal{A}} = {}^{\mathcal{A}}C^{\mathcal{B}}[\mathbb{I}^1_{G_1}]_{\mathcal{B}}\left({}^{\mathcal{A}}C^{\mathcal{B}}\right)^T,$$

where

$$^{\mathcal{A}}C^{\mathcal{B}} = \begin{bmatrix} \cos\theta_1 & -\sin\theta_1 & 0 \\ \sin\theta_1 & \cos\theta_1 & 0 \\ 0 & 0 & 1 \end{bmatrix}.$$

We can use the parallel axis theorem (Eq. (11.57)) to calculate $[\mathbb{I}^1_O]$ by noting that

$$\mathbf{r}_{O/G_1} = -\frac{l_1}{2}\mathbf{b}_1 - \frac{h}{2}\mathbf{a}_1 = -\left(\frac{h}{2} + \frac{l_1}{2}\cos\theta_1\right)\mathbf{a}_1 - \frac{l_1}{2}\sin\theta_1\mathbf{a}_2.$$

Putting this all together, we have

$$[\mathbb{I}^1_O]_{\mathcal{A}} = {}^{\mathcal{A}}C^{\mathcal{B}}[\mathbb{I}^1_{G_1}]_{\mathcal{B}}\left({}^{\mathcal{A}}C^{\mathcal{B}}\right)^T + m_1\left(\|\mathbf{r}_{O/G_1}\|^2 I - [\mathbf{r}_{O/G_1}]_{\mathcal{A}}[\mathbf{r}_{O/G_1}]^T_{\mathcal{A}}\right) \quad (11.71)$$

$$= \frac{m_1}{12} \begin{bmatrix} 4l^2_1\sin^2\theta_1 & I_{12} & 0 \\ I_{21} & 3h^2 + 4l^2_1\cos^2\theta_1 + 6hl_1\cos\theta_1 & 0 \\ 0 & 0 & 3h^2 + 4l^2_1 + 6hl_1\cos\theta_1 \end{bmatrix}_{\mathcal{A}},$$

where $I_{12} = I_{21} = -3hl_1\sin\theta_1 - 2l^2_1\sin 2\theta_1$.

The moment of inertia of the second arm in frame $\mathcal{C}$ about its center of mass G_2 is the same as that of arm 1 in frame $\mathcal{B}$ about G_1 (except with l_2 and m_2 instead of l_1 and m_1), so we perform the same calculations as before, with

$$^{\mathcal{A}}C^{\mathcal{C}} = {}^{\mathcal{A}}C^{\mathcal{B}\,\mathcal{B}}C^{\mathcal{C}} = \begin{bmatrix} \cos\theta_1 & -\sin\theta_1 & 0 \\ \sin\theta_1 & \cos\theta_1 & 0 \\ 0 & 0 & 1 \end{bmatrix} \begin{bmatrix} \cos\beta & \sin\beta & 0 \\ -\sin\beta & \cos\beta & 0 \\ 0 & 0 & 1 \end{bmatrix}$$

$$= \begin{bmatrix} \cos\theta_2 & \sin\theta_2 & 0 \\ -\sin\theta_2 & \cos\theta_2 & 0 \\ 0 & 0 & 1 \end{bmatrix},$$

where $\theta_2 = \beta - \theta_1$. It is important to note that in Figure 11.19, β is defined as increasing in the opposite direction from θ_1, making it a left-handed, rather than a right-handed, rotation. (There's nothing wrong with doing this, but one must be careful to keep track of the direction of each rotation—which is why the direction-cosine matrix $^{\mathcal{B}}C^{\mathcal{C}}$ has opposite signs from $^{\mathcal{A}}C^{\mathcal{B}}$ on the sine terms.) The angle θ_2 increases in the same direction as θ_1.

Using $^{A}C^{B}$ and $^{A}C^{C}$, we can write

$$\mathbf{r}_{O/G_2} = -\frac{l_2}{2}\mathbf{c}_1 - l_1\mathbf{b}_1 - \frac{h}{2}\mathbf{a}_1$$

$$= -\left(\frac{h}{2} + l_1\cos\theta_1 + \frac{l_2}{2}\cos\theta_2\right)\mathbf{a}_1 + \left(-l_1\sin\theta_1 + \frac{l_2}{2}\sin\theta_2\right)\mathbf{a}_2$$

and thus, using $[\mathbb{I}_O^2]_A = {}^{A}C^{C}[\mathbb{I}_{G_2}^2]({}^{A}C^{C})^T + m_2\left(\|\mathbf{r}_{O/G_2}\|^2 I - [\mathbf{r}_{O/G_2}]_A[\mathbf{r}_{O/G_2}]_A^T\right)$,

$$[\mathbb{I}_O^2]_A = \frac{m_2}{12}\begin{bmatrix} I_{11} & I_{12} & 0 \\ I_{21} & I_{22} & 0 \\ 0 & 0 & I_{33} \end{bmatrix}_A, \tag{11.72}$$

where

$$I_{11} = 12l_1^2\sin^2\theta_1 - 12l_1l_2\sin\theta_1\sin\theta_2 + 4l_2^2\sin^2\theta_2$$

$$I_{12} = I_{21} = 2l_2^2\sin 2\theta_2 + 3hl_2\sin\theta_2 - 6l_1l_2\sin(\theta_1 - \theta_2) - 6hl_1\sin\theta_1 - 6l_1^2\sin 2\theta_1$$

$$I_{22} = 3\left(h + 2l_1\cos\theta_1\right)^2 + 4l_2^2\cos^2\theta_2 + 6l_2(h + 2l_1\cos\theta_1)\cos\theta_2$$

$$I_{33} = 3h^2 + 12l_1\left(h\cos\theta_1 + l_1\right) + 4l_2^2 + 6hl_2\cos\theta_2 + 12l_1l_2\cos\beta.$$

The total moment of inertia about O in frame A is the sum of the inertia matrices in Eqs. (11.70)–(11.72):

$$[\mathbb{I}_O]_A = [\mathbb{I}_O^{\text{cab}}]_A + [\mathbb{I}_O^1]_A + [\mathbb{I}_O^2]_A = \frac{1}{12}\begin{bmatrix} I_{11}' & I_{12}' & 0 \\ I_{21}' & I_{22}' & 0 \\ 0 & 0 & I_{33}' \end{bmatrix}_A,$$

where

$$I_{11}' = 2Mh^2 + 4(m_1 + 3m_2)l_1^2\sin^2\theta_1 - 12l_1l_2m_2\sin\theta_1\sin\theta_2 + 4m_2l_2^2\sin^2\theta_2$$

$$I_{12}' = I_{21}' = -2(m_1 + 3m_2)l_1^2\sin 2\theta_1 + 2m_2l_2^2\sin 2\theta_2$$

$$- 3\left(hl_1(m_1 + 2m_2)\sin\theta_1 + l_2m_2(2l_1\sin(\theta_1 - \theta_2) - h\sin\theta_2)\right)$$

$$I_{22}' = 2Mh^2 + 3(m_1 + m_2)h^2 + 4(m_1 + 3m_2)l_1^2\cos^2\theta_1$$

$$+ 6(m_1 + 2m_2)hl_1\cos\theta_1 + 6m_2l_2(h + 2l_1\cos\theta_1)\cos\theta_2 + 4m_2l_2^2\cos^2\theta_2$$

$$I_{33}' = 2Mh^2 + 3(m_1 + m_2)h^2 + 4(m_1 + 3m_2)l_1^2 + 4m_2l_2^2 + 6(m_1 + 2m_2)hl_1\cos\theta_1$$

$$+ 6m_2l_2(h\cos\theta_2 + 2l_1\cos\beta).$$

11.6 Angular Impulse in Three Dimensions

Section 9.3.4 discussed how to treat a rigid body after an impulsive moment $\overline{\mathbf{M}}_G(t_1, t_2)$ is applied. The solution to the vector equation of motion for the angular momentum in Eq. (11.3) does not change for an impulsive moment because we did not use the planar assumption in Section 9.3.4. We have

$$^{\mathcal{I}}\mathbf{h}_G(t_2) = {}^{\mathcal{I}}\mathbf{h}_G(t_1) + \overline{\mathbf{M}}_G(t_1, t_2). \tag{11.73}$$

While Eq. (11.73) is true, it is not as useful for solving problems as one might think because we usually are looking for the angular velocity at $t = t_2$ so we can integrate to find the orientation of the rigid body (or describe the trajectory after the impulse). For a planar rigid body, this was accomplished by dividing by the moment of inertia. It is slightly more complicated in three dimensions but not excessively so. Here it is easiest to turn to matrix notation. We rewrite Eq. (11.73), using Eq. (11.23), to obtain

$$[\mathbb{I}_G]_{\mathcal{B}}[^{\mathcal{I}}\boldsymbol{\omega}^{\mathcal{B}}(t_2)]_{\mathcal{B}} = [\mathbb{I}_G]_{\mathcal{B}}[^{\mathcal{I}}\boldsymbol{\omega}^{\mathcal{B}}(t_1)]_{\mathcal{B}} + [\overline{\mathbf{M}}_G(t_1, t_2)]_{\mathcal{B}}. \tag{11.74}$$

Eq. (11.74) allows us to solve for the instantaneous change in the angular velocity due to an impulsive moment by inverting the moment of inertia matrix and multiplying from the left:

$$[^{\mathcal{I}}\boldsymbol{\omega}^{\mathcal{B}}(t_2)]_{\mathcal{B}} = [^{\mathcal{I}}\boldsymbol{\omega}^{\mathcal{B}}(t_1)]_{\mathcal{B}} + [\mathbb{I}_G]_{\mathcal{B}}^{-1}[\overline{\mathbf{M}}_G(t_1, t_2)]_{\mathcal{B}}.$$

Just as in the planar case, the application of an impulsive moment instantaneously changes the angular velocity. If the moment impulse is a result of a force (rather than a pure torque), then the velocity of the center of mass will also be changed, just as in Example 9.11. We are able to find the angular velocity at t_2 because the moment of inertia matrix is always invertible.

11.7 Work and Energy of a Rigid Body in Three Dimensions

Our final task in this chapter is to extend the discussion in Chapters 7 and 9 of work and energy for planar multiparticle systems and rigid bodies to three dimensions. Although all the basic principles still hold, in some cases we need to use a bit of care, just as when discussing the angular momentum.

11.7.1 Kinetic Energy of a Rigid Body in Three Dimensions

The earlier derivation in Chapters 7 and 9 of the separation of motion of the center of mass and motion about the center of mass of a collection of particles was completely general. The result that the kinetic energy can be written as the sum of the kinetic energy due to motion of the center of mass and the kinetic energy due to motion about the center of mass applies unchanged in three dimensions:

$$\boxed{T_O = T_{G/O} + T_G.}$$

As before, $T_{G/O}$ is the translational kinetic energy of the center of mass:

$$\boxed{T_{G/O} \triangleq \frac{1}{2} m_G \|{}^{\mathcal{I}}\mathbf{v}_{G/O}\|^2.}$$

The derivation of T_G—the kinetic energy due to motion about the center of mass—in terms of the moment of inertia tensor for a rigid body requires a bit of care. By definition, for a system of N particles we have

$$T_G \triangleq \frac{1}{2} \sum_{i=1}^{N} m_i {}^{\mathcal{I}}\mathbf{v}_{i/G} \cdot {}^{\mathcal{I}}\mathbf{v}_{i/G} = \frac{1}{2} \sum_{i=1}^{N} m_i \| {}^{\mathcal{I}}\mathbf{v}_{i/G} \|^2.$$

Applying the transport equation to ${}^{\mathcal{I}}\mathbf{v}_{i/G}$ and making use of the fact that we are studying a rigid collection of N particles yields

$$T_G = \frac{1}{2} \sum_{i=1}^{N} m_i \| \underbrace{{}^{\mathcal{B}}\mathbf{v}_{i/G}}_{=0} + {}^{\mathcal{I}}\boldsymbol{\omega}^{\mathcal{B}} \times \mathbf{r}_{i/G} \|^2 = \frac{1}{2} \sum_{i=1}^{N} m_i ({}^{\mathcal{I}}\boldsymbol{\omega}^{\mathcal{B}} \times \mathbf{r}_{i/G}) \cdot ({}^{\mathcal{I}}\boldsymbol{\omega}^{\mathcal{B}} \times \mathbf{r}_{i/G}).$$

Next we apply two cross-product identities from Appendix B—the scalar triple product and the vector triple product. Sequential application of these two identities results in

$$T_G = \frac{1}{2} \sum_{i=1}^{N} m_i {}^{\mathcal{I}}\boldsymbol{\omega}^{\mathcal{B}} \cdot (\mathbf{r}_{i/G} \times ({}^{\mathcal{I}}\boldsymbol{\omega}^{\mathcal{B}} \times \mathbf{r}_{i/G}))$$

$$= \frac{1}{2} {}^{\mathcal{I}}\boldsymbol{\omega}^{\mathcal{B}} \cdot \sum_{i=1}^{N} m_i \left[(\mathbf{r}_{i/G} \cdot \mathbf{r}_{i/G}) {}^{\mathcal{I}}\boldsymbol{\omega}^{\mathcal{B}} - \mathbf{r}_{i/G}(\mathbf{r}_{i/G} \cdot {}^{\mathcal{I}}\boldsymbol{\omega}^{\mathcal{B}}) \right], \qquad (11.75)$$

where we pulled the angular velocity out of the sum over i because it does not depend on i.

Note the similarity of Eq. (11.75) to Eq. (11.14). As in Section 11.2.2, we now invoke tensor analysis to complete the derivation. Factoring out the angular velocity to the right yields

$$T_G = \frac{1}{2} {}^{\mathcal{I}}\boldsymbol{\omega}^{\mathcal{B}} \cdot \underbrace{\left(\sum_{i=1}^{N} m_i \left[(\mathbf{r}_{i/G} \cdot \mathbf{r}_{i/G}) \mathbb{U} - (\mathbf{r}_{i/G} \otimes \mathbf{r}_{i/G}) \right] \right)}_{=\mathbb{I}_G} \cdot {}^{\mathcal{I}}\boldsymbol{\omega}^{\mathcal{B}},$$

where $\mathbb{U}$ is the unity tensor and $\mathbb{I}_G$ is the moment of inertia tensor about the center of mass. The result is that the rotational kinetic energy of a general rigid body about its center of mass has the compact form

$$\boxed{T_G = \frac{1}{2} {}^{\mathcal{I}}\boldsymbol{\omega}^{\mathcal{B}} \cdot \mathbb{I}_G \cdot {}^{\mathcal{I}}\boldsymbol{\omega}^{\mathcal{B}}.}$$

In matrix notation, the rotational kinetic energy is

$$T_G = \frac{1}{2} [{}^{\mathcal{I}}\boldsymbol{\omega}^{\mathcal{B}}]_{\mathcal{B}}^{T} [\mathbb{I}_G]_{\mathcal{B}} [{}^{\mathcal{I}}\boldsymbol{\omega}^{\mathcal{B}}]_{\mathcal{B}}.$$

Although not quite as simple as for the planar case, we still have an elegant expression for the kinetic energy of rotation that depends only on the angular velocity of the rigid body and its mass properties. For an arbitrary body frame, the resulting expression for T_G can be rather cumbersome. However, as long as we work in principal axes, the following expression:

$$T_G = \frac{1}{2}(I_1\omega_1^2 + I_2\omega_2^2 + I_3\omega_3^2) \tag{11.76}$$

holds, where ω_1, ω_2, and ω_3 are the coefficients of ${}^{\mathcal{I}}\boldsymbol{\omega}^{\mathcal{B}}$ expressed as components in $\mathcal{B}$.

Our result for T_G also shows that, as in Eq. (9.40), we can write the kinetic energy of rotation in terms of the angular momentum in three dimensions:

$$T_G = \frac{1}{2}{}^{\mathcal{I}}\mathbf{h}_G \cdot {}^{\mathcal{I}}\boldsymbol{\omega}^{\mathcal{B}}.$$

Finally, although we will not rederive it here (you can try it in Problem 11.4), we can write the total kinetic energy in terms of rotation relative to an arbitrary point Q on the rigid body exactly as in Eq. (9.41):

$$T_O = T_{Q/O} + T_Q - m_G{}^{\mathcal{I}}\mathbf{v}_{Q/O} \cdot ({}^{\mathcal{I}}\boldsymbol{\omega}^{\mathcal{B}} \times \mathbf{r}_{Q/G}), \tag{11.77}$$

where the kinetic energy due to rotation relative to Q is

$$T_Q = \frac{1}{2}{}^{\mathcal{I}}\boldsymbol{\omega}^{\mathcal{B}} \cdot \mathbb{I}_Q \cdot {}^{\mathcal{I}}\boldsymbol{\omega}^{\mathcal{B}}$$

and the moment of inertia tensor relative to Q is given by the parallel axis theorem in Eq. (11.56).

11.7.2 Three-Dimensional Rigid-Body Work, Potential Energy, and Total Energy

Our next task is to generalize the work–kinetic-energy formulas for three-dimensional rigid bodies. Fortunately, the task is again greatly simplified by the fact that the development in Chapter 9 was completely general; it never actually invoked the planar restriction. Thus we can simply jump right to the work–kinetic-energy formula in Eq. (9.42),

$$W = \Delta T_{G/O} + \Delta T_G,$$

where $W = W(\{\mathbf{r}_{i/O}; \gamma_i\}_{i=1}^N)$ refers to the total external work on all of the particles along their three-dimensional trajectories. For a rigid body, we again separate the external work into the work due to motion of the center of mass $W_{G/O} = W_{G/O}(\mathbf{r}_{G/O}, \theta, \mathbf{k}; \gamma_G, \gamma_{\theta,\mathbf{k}})$ and the work due to rotation about the center of mass $W_G = W_G(\mathbf{r}_{G/O}, \theta, \mathbf{k}; \gamma_G, \gamma_{\theta,\mathbf{k}})$. The former satisfies the three-dimensional version of Eq. (9.45),

$$W_{G/O} = \Delta T_{G/O},$$

whereas the latter satisfies the three-dimensional version of Eq. (9.47):

$$W_G = \Delta T_G.$$

Remember, though, that for the most general cases, the work W_G can depend on the trajectory of the center of mass embedded in $W_{G/O}$ and vice versa.

As in Chapter 9, γ_G represents the trajectory of the center of mass. However, we have had to modify our notation a bit for motion about the center of mass. When we were considering a planar rigid body, the rotational trajectory was entirely described by the single angle θ. Thus the work was the integral of the moment times the angular displacement over the trajectory. Here, things are somewhat more complicated, as the rigid body can rotate arbitrarily in three dimensions. Rather than parameterize the work by the three Euler angles (which vary, depending on the set of angles chosen), we have selected to parameterize it by the unit vector $\mathbf{k}$ along the Euler axis and the instantaneous rotation angle θ about that axis. Thus the instantaneous work is the product of the moment about the Euler axis times the rotation angle θ. As shown below, this choice is sensible, as the work is more often computed by means of the velocity form, integrating the inner product of the external moment and the angular velocity. It also highlights the fact that only moments along the Euler axis (that is, in the same direction as the angular velocity) do work. Moments perpendicular to that axis do no work, just as forces perpendicular to the translational direction of motion (like constraint forces) do no work. As shown in Section 10.5, the angular velocity is along the instantaneous axis of rotation; it is always aligned with the Euler axis. Thus we describe the rotational trajectory by the parameter $\gamma_{\theta,\mathbf{k}}$.

As in Section 9.5.2, we write the work due to motion about the center of mass in terms of the applied moments:

$$W_G(\mathbf{r}_{G/O}, \theta, \mathbf{k}; \gamma_G, \gamma_{\theta,\mathbf{k}}) = \int_{t_1}^{t_2} \mathbf{M}_G \cdot {}^{\mathcal{I}}\boldsymbol{\omega}^{\mathcal{B}} \, dt.$$

The corresponding power is

$$\mathbf{P}_G = \mathbf{M}_G \cdot {}^{\mathcal{I}}\boldsymbol{\omega}^{\mathcal{B}}.$$

As in Chapter 9, our final step is to replace the external work by a potential energy when the corresponding forces or torques are conservative. There was no restriction made in Section 9.5.3 to planar motion other than in computing the pure torque from its corresponding potential. Thus the total energy is still defined as

$$\boxed{E_O(t) \triangleq T_{G/O}(t) + T_G(t) + U_O(t),}$$

where the total potential energy is

$$U_O(t) = \sum_{i=1}^{N} U_i(\mathbf{r}_{i/O}(t)).$$

Likewise, if we have a rigid body $\mathcal{B}$ rather than a collection of N particles, the external potential is given by Eq. (9.48):

$$U_O(t) = \int_{\mathcal{B}} U_{dm/O}(\mathbf{r}_{dm/O}) \, dm.$$

The result is that the same work-energy formula with conservation of energy holds in three dimensions:

$$\boxed{E_O(t_2) = E_O(t_1) + W_{G/O}^{(nc)} + W_G^{(nc)}.}$$

Example 11.17 Torque-Free Motion, Revisited

Example 11.10 examined the torque-free motion of a spinning rigid body, but restricted to small deviations of the angular velocity from the nominal spin. For spin about the major or minor axis of inertia, the angular velocity rotated about the inertia axis; spin about the intermediate axis was unstable. In this example we take a different approach—using energy conservation—to find the complete set of trajectories corresponding to the possible motion of the angular velocity in the body frame.

We begin by noting that, if the total external force and total external moment are zero, then the total energy of the rigid body is equal to its kinetic energy and the kinetic energy is a constant of the motion. Thus in a principal-axes body frame, the kinetic energy about the center of mass is given by Eq. (11.76):

$$T_G = \frac{1}{2}(I_1\omega_1^2 + I_2\omega_2^2 + I_3\omega_3^2).$$

Dividing through by T_G yields

$$\frac{\omega_1^2}{2T_G/I_1} + \frac{\omega_2^2}{2T_G/I_2} + \frac{\omega_3^2}{2T_G/I_3} = 1. \tag{11.78}$$

Eq. (11.78) is the equation for an ellipsoid in the body frame with semi-axes $\sqrt{2T_G/I_1}$, $\sqrt{2T_G/I_2}$, and $\sqrt{2T_G/I_3}$. Recall that the Cartesian equation for an ellipse in two dimensions is

$$\frac{x^2}{a^2} + \frac{y^2}{b^2} = 1,$$

where a is the semi-major axis and b is the semi-minor axis. Eq. (11.78) simply adds a third dimension. We call this form the *inertia ellipsoid*, as its shape is determined by the three principal moments of inertia. Typically, the axes are numbered such that $I_3 > I_2 > I_1$. Eq. (11.78) states that, for a given set of initial conditions on the angular velocity, which establishes the kinetic energy T_G, *the tip of the angular velocity will move in the body frame in such a way as to always be on the surface of the inertia ellipsoid.*

However, we have another constraint. Since the total external moment is zero, the angular momentum is also a constant of the motion; it is fixed in both magnitude and direction (in the inertial frame). Expressing the inertia tensor and angular velocity as components in a principal-axes body frame, we can find the squared magnitude of the angular momentum using Eq. (11.15) or Eq. (11.23):

$$\|^\mathcal{I}\mathbf{h}_G\|^2 = h_G^2 = I_1^2\omega_1^2 + I_2^2\omega_2^2 + I_3^2\omega_3^2.$$

After a manipulation similar to that done on Eq. (11.78) we find

$$\frac{\omega_1^2}{h_G^2/I_1^2} + \frac{\omega_2^2}{h_G^2/I_2^2} + \frac{\omega_3^2}{h_G^2/I_3^2} = 1. \tag{11.79}$$

This is the equation for a second ellipsoid, whose size depends upon the angular momentum, that the angular velocity must also trace. Thus the path of the angular velocity in the body frame is given by the curve representing the intersection of the first ellipsoid in Eq. (11.78) and the second ellipsoid in Eq. (11.79). These curves are called *polhodes*. Figure 11.20 shows various polhode curves on the surface of the first ellipsoid. Note that the curves collapse to a point at each of the three principal

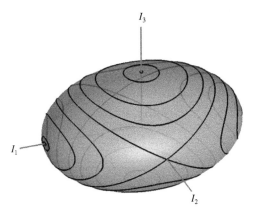

Figure 11.20 Inertia ellipsoid of a rigid body showing the polhode curves that trace the path of the angular velocity for torque-free motion.

axes, representing spin about each of these axes, as discussed in Example 11.10. For slight deviations from the axes of maximum and minimum moments of inertia (I_3 and I_1, respectively) the angular velocity encircles the principal axis, as discussed in Example 11.10. For slight deviations from the intermediate axis, however, the angular velocity travels away from the axis along one of two unstable trajectories called *separatrices*. This graphical approach gives a complete picture of all curves traced by the angular velocity, not just those for small motions about a spin axis.

We conclude this example with an important observation. Suppose that the rigid body is spinning about a single principal axis with moment of inertia I_i, where $i = 1$, 2, or 3. Then Eqs. (11.78) and (11.79) can be combined to find

$$T_G = \frac{h_G^2}{2I_i}.$$

This equation tells us that, for torque-free motion where h_G is constant, the kinetic energy is a maximum for spin about the minimum axis of inertia and a minimum for spin about the maximum axis of inertia. Suppose that there is some nonconservative work in the rigid body, causing dissipation. This work will always cause the angular velocity to change from spin about the minimum to that about the maximum axis to minimize the energy. This is admittedly a qualitative argument and ignores the specific mechanisms for such motion, but it was observed in the early days of the space program, when satellites that were set spinning about the minimum axis of inertia began to tumble out of control due to internal dissipation in antenna structures. So despite the theoretical stability of the polhode paths about the minimum axis, such spin is often unstable anyway!

11.8 Tutorials

Tutorial 11.1 Torque-Free Motion of a Symmetric Rigid Body

Example 11.10 used Euler's equations to approximately solve for the motion of the angular velocity in the body frame of a torque-free spinning rigid body. Example 11.17

used energy to find the set of permissible trajectories of the angular velocity in the body frame. This tutorial returns again to the torque-free rigid body and solves for the trajectory of the body in the inertial frame. To simplify, we only examine the special case of a symmetric rigid body. In this case $I_1 = I_2$, and there is no intermediate axis. Recall that spin about the I_3-axis (maximum or minimum moment of inertia) is stable. If you've ever seen someone juggling pins, you have seen torque-free motion of a spinning symmetric rigid body. (In that case, the rotation is about the major axis.)

We begin again with Euler's equations, with $\omega_3 = \omega_s$ the (large) spin about the 3-axis. From Example 11.10 we have

$$\ddot{\omega}_1 + \omega_n^2 \omega_1 = 0$$

$$\ddot{\omega}_2 + \omega_n^2 \omega_2 = 0,$$

where

$$\omega_n^2 = \frac{(I_O - J)^2}{I_O^2} \omega_s^2, \tag{11.80}$$

$I_1 = I_2 = I_O$, and $J = I_3$. Thus the variation of each angular velocity component is simple harmonic with solution

$$\omega_1 = -\omega_0 \cos \omega_n t$$

$$\omega_2 = \omega_0 \sin \omega_n t,$$

where we arbitrarily chose $\omega_1(t_0) = -\omega_0$ and $\omega_2(t_0) = 0$. It is possible to verify that the oscillatory motion of ω_1 and ω_2 is exactly $90°$ out of phase by substituting back into Eqs. (11.40) and (11.41).

How does the motion of the body in the inertial frame deviate from simple spin due to the small ω_1 and ω_2 perturbations? Examining the angular momentum will answer this question. Since there are no torques on the rigid body, the angular momentum $^I\mathbf{h}_G$ has constant magnitude and direction. Using the three-dimensional expression for the angular momentum of a rigid body in principal axes (Eq. (11.15) or Eq. (11.23)), we can write the components of $^I\mathbf{h}_G$ in the body frame:

$$^I\mathbf{h}_G = -I_O\omega_0 \cos \omega_n t\, \mathbf{b}_1 + I_O\omega_0 \sin \omega_n t\, \mathbf{b}_2 + J\omega_s \mathbf{b}_3. \tag{11.81}$$

The $\mathbf{b}_3$ symmetry axis of the rigid body *cones* about the inertial direction defined by $^I\mathbf{h}_G$. We find the angle θ between the body symmetry axis and the angular momentum by dotting $^I\mathbf{h}_G$ in Eq. (11.81) with $\mathbf{b}_3$ and dividing by $\|^I\mathbf{h}_G\|$:

$$\cos \theta = \frac{^I\mathbf{h}_G \cdot \mathbf{b}_3}{\|^I\mathbf{h}_G\|} = \frac{J\omega_s}{\sqrt{I_O^2\omega_0^2 + J^2\omega_s^2}}. \tag{11.82}$$

We call this coning motion free-body precession; the angle θ is the nutation angle. This effect is also often referred to as *spin stabilization* because spinning the rigid body stabilizes the orientation of its spin axis against impulsive disturbances. Suppose we want the symmetry axis of the body to point in a particular inertial direction. Inevitably, it will be perturbed by a small impulse. If the body is not spinning, it will develop a small angular velocity in response to the impulse and drift away. However, by spinning it, the small angular velocity due to the impulse results in a precession about the desired inertial direction at an angle given by Eq. (11.82). For this reason all satellites are spun before being transferred from a low orbit to a high one (see Tutorial 4.3). Any small alignment errors in the applied impulse induce a small coning motion rather than causing the satellite to tumble out of control.

We finish this tutorial by examining a 3-2-3 Euler-angle description of the rigid body. Since the angular momentum defines a fixed direction in absolute space, we can use it to define one of the axes of the inertial frame. We thus let $e_3 = {}^{\mathcal{I}}\mathbf{h}_G/\|{}^{\mathcal{I}}\mathbf{h}_G\|$. Eq. (11.82) determines the fixed tilt angle between ${}^{\mathcal{I}}\mathbf{h}_G$ (and thus e_3) and the body $\mathbf{b}_3$-axis, implying that $\dot{\theta} = 0$.[5] The kinematic equations of rotation in Eq. (10.48) become

$$\omega_1 = -\omega_0 \cos \omega_n t = -\dot{\psi} \sin \theta \cos \phi \qquad (11.83)$$

$$\omega_2 = \omega_0 \sin \omega_n t = \dot{\psi} \sin \theta \sin \phi \qquad (11.84)$$

$$\omega_3 = \omega_s = \dot{\phi} + \dot{\psi} \cos \theta. \qquad (11.85)$$

Squaring and adding Eqs. (11.83) and (11.84) leaves

$$\omega_0^2 = \dot{\psi}^2 \sin^2 \theta.$$

Since θ and ω_0 are fixed, $\dot{\psi}$ is a constant and represents the precession rate of the $\mathbf{b}_3$ axis about the angular-momentum direction.

We now use the transformation array from Eq. (10.33) to solve for the motion of the body frame and the angular velocity in the inertial frame. First, we can see again that the body symmetry axis cones about the inertial e_3 direction (aligned with the angular momentum) by writing $\mathbf{b}_3$ as components in $\mathcal{I}$ using the third row of the transformation array,

$$\mathbf{b}_3 = \cos \psi \sin \theta e_1 + \sin \psi \sin \theta e_2 + \cos \theta e_3. \qquad (11.86)$$

Since θ and $\dot{\psi}$ are constant, the e_1 and e_2 components vary sinusoidally 90° out of phase; $\mathbf{b}_3$ rotates or *cones* about e_3 at a precession rate $\dot{\psi}$.

To find the precession rate, we can use the transformation array in Eq. (10.33) to write the angular momentum, ${}^{\mathcal{I}}\mathbf{h}_G = h_G e_3$, in the body frame,

$${}^{\mathcal{I}}\mathbf{h}_G = h_G e_3 = -h_G \sin \theta \cos \phi \mathbf{b}_1 + h_G \sin \theta \sin \phi \mathbf{b}_2 + h_G \cos \theta \mathbf{b}_3.$$

Since ${}^{\mathcal{I}}\mathbf{h}_G = \mathbb{I}_G \cdot {}^{\mathcal{I}}\boldsymbol{\omega}^{\mathcal{B}} = I_O \omega_1 \mathbf{b}_1 + I_O \omega_2 \mathbf{b}_2 + J \omega_s \mathbf{b}_3$, we can equate components and use the kinematical equations in Eqs. (11.83)–(11.85) to find that $\dot{\psi} = h_G/I_O$. The body symmetry axis thus precesses in a positive direction about e_3 at a rate given by the magnitude of the angular momentum.

We likewise use the transformation array in Eq. (10.33) to write the angular velocity, ${}^{\mathcal{I}}\boldsymbol{\omega}^{\mathcal{B}} = \omega_1 \mathbf{b}_1 + \omega_2 \mathbf{b}_2 + \omega_3 \mathbf{b}_3$, as components in the inertial frame,

$${}^{\mathcal{I}}\boldsymbol{\omega}^{\mathcal{B}} = (\omega_s \sin \theta - \omega_0 \cos \theta) \cos \psi e_1 + (\omega_s \sin \theta - \omega_0 \cos \theta) \sin \psi e_2$$
$$+ (\omega_s \cos \theta + \omega_0 \sin \theta) e_3,$$

---

[5] Unlike that for the symmetric rigid body undergoing torque, the nutation angle of the torque-free rigid body does not vary with time.

where we also used the fact that $\omega_3 = \omega_s$ and substituted for ω_1 and ω_2 from Eqs. (11.83) and (11.84), using the fact that $\dot{\psi} > 0$. Comparing to Eq. (11.86) shows that the angular velocity and the symmetry axis are in the same plane as the angular momentum and rotate around it either in phase or $180°$ out of phase, depending upon the sign of the constant amplitudes of the $\mathbf{e}_1$ and $\mathbf{e}_2$ components, $\omega_s \sin\theta - \omega_0 \cos\theta$. The sign of this term can be found in terms of the moments of inertia by substituting for $\cos\theta$ (from Eq. (11.82)) and $\sin\theta$. We find an expression for $\sin\theta$ by taking the cross product of $\mathbf{b}_3$ with the angular momentum and using the definition of the cross product (Definition B.3),

$$\sin\theta = \frac{\|\mathbf{b}_3 \times {}^{\mathcal{I}}\mathbf{h}_G\|}{\|{}^{\mathcal{I}}\mathbf{h}_G\|} = \frac{I_O \omega_0}{\sqrt{I_O^2 \omega_0^2 + J^2 \omega_s^2}}.$$

The amplitude of the $\mathbf{e}_1$ and $\mathbf{e}_2$ components of the angular velocity is thus

$$\omega_s \sin\theta - \omega_0 \cos\theta = \frac{(I_O - J)\omega_0 \omega_s}{\sqrt{I_O^2 \omega_0^2 + J^2 \omega_s^2}}.$$

Last, we can find an expression for the *relative spin*, $\dot{\phi}$, from Eq. (11.85) and the $\mathbf{b}_3$ component of the angular momentum,

$$\dot{\phi} = \left(\frac{I_O - J}{I_O}\right)\frac{h_G}{J}\cos\theta = \frac{I_O - J}{I_O}\omega_s,$$

where we also used $\dot{\psi} = h_G/I_O$.

If $I_O > J$, then $(\omega_s \sin\theta - \omega_0 \cos\theta) > 0$ and $\dot{\phi} > 0$, implying that $\mathbf{b}_3$ and ${}^{\mathcal{I}}\boldsymbol{\omega}^B$ are on the same side of ${}^{\mathcal{I}}\mathbf{h}_G$ and rotate around it in the same direction as the relative spin. This motion is called *direct* or *prograde* precession and the rigid body with $I_O > J$ is called *prolate*. It is long and thin, like a pencil.

If $I_O < J$, then $(\omega_s \sin\theta - \omega_0 \cos\theta) < 0$ and $\dot{\phi} < 0$, implying that $\mathbf{b}_3$ and ${}^{\mathcal{I}}\boldsymbol{\omega}^B$ are on opposite sides of ${}^{\mathcal{I}}\mathbf{h}_G$ and rotate about it in the opposite direction of the relative spin. This motion is called *retrograde* precession and the rigid body with $I_O < J$ is called *oblate*. It is flat and broad, like a pancake.

The torque-free precession of a rigid body is often geometrically described by two right circular cones rolling on each other, as shown in Figure 11.21. The *space cone* surrounds the fixed angular momentum in absolute space and is defined by the angle between ${}^{\mathcal{I}}\mathbf{h}_G$ and ${}^{\mathcal{I}}\boldsymbol{\omega}^B$. The *body cone* surrounds the body-frame 3-axis and is defined by the angle between ${}^{\mathcal{I}}\boldsymbol{\omega}^B$ and $\mathbf{b}_3$. (The angular velocity is directed along the line of contact between the two cones.) For a prolate rigid body, precession of $\mathbf{b}_3$ about ${}^{\mathcal{I}}\mathbf{h}_G$ is described by the body cone rolling around the outside of the space cone, as in Figure 11.21a. For an oblate rigid body, the inside of the body cone rolls around the space cone, as in Figure 11.21b. Note the different arrangements of the three vectors, ${}^{\mathcal{I}}\mathbf{h}_G$, ${}^{\mathcal{I}}\boldsymbol{\omega}^B$, and $\mathbf{b}_3$, in each case.

Tutorial 11.2 Ship Stabilization

At one time, ship designers considered using a large gyroscope mounted to the hull to damp out the rocking motion. While this approach proved less practical for very

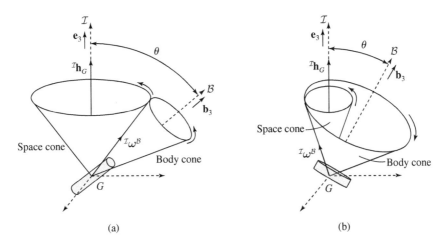

Figure 11.21 Space and body cones for the precession of a rigid body. (a) Prolate case $I_O > J$. (b) Oblate case $I_O < J$.

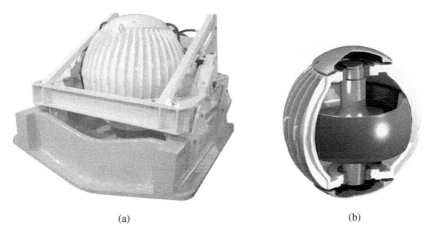

Figure 11.22 Example of a commercially available ship-stabilizing gyroscope for private yachts. (a) The gyroscope and gimbal mounted to the hull. (b) Cutaway view of the rotor. Images courtesy of Seakeeper, Inc.

large ships because of the large torques imposed on the superstructure, such a system is available for small yachts (see Figure 11.22). It is an excellent example of using a gyrostat to stabilize a vehicle. In this tutorial, we explore how it works.

Figure 11.23 shows the front view of a ship in inertial frame $\mathcal{I} = (O, \mathbf{e}_1, \mathbf{e}_2, \mathbf{e}_3)$, whose orientation is described by the single Euler angle ϕ about the $\mathbf{e}_1$ axis. Attached to the hull of the ship is a spinning disk (the rotor) that is allowed to pitch forward and backward about pivots attached to the rotor and the ship. The angle that the plane of the rotor makes with the hull is given by coordinate ϵ.

We consider the simple situation where the ship is traveling straight ahead and rolling back and forth because of wave action. We solve this problem by separating the ship and the rotor and including constraint torques at the pivot. Figure 11.24 shows the two body frames, frame $\mathcal{B} = (G, \mathbf{b}_1, \mathbf{b}_2, \mathbf{b}_3)$ attached to the ship at its center of

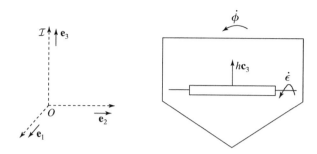

Figure 11.23 Front view of a rocking ship with gyro stabilization.

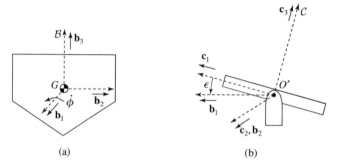

Figure 11.24 Ship and rotor body frames and coordinates. (a) Ship frame (front view). (b) Rotor frame (side view).

mass, with $\mathbf{b}_1 = \mathbf{e}_1$, and a frame $\mathcal{C} = (O', \mathbf{c}_1, \mathbf{c}_2, \mathbf{c}_3)$ attached to the rotor, with $\mathbf{c}_2 = \mathbf{b}_2$. The transformation table between the ship's body frame and the inertial frame is

	$\mathbf{b}_1$	$\mathbf{b}_2$	$\mathbf{b}_3$
$\mathbf{e}_1$	1	0	0
$\mathbf{e}_2$	0	$\cos\phi$	$-\sin\phi$
$\mathbf{e}_3$	0	$\sin\phi$	$\cos\phi$.

The angle that the rotor makes with the ship's hull is given by angle ϵ about the $\mathbf{b}_2 = \mathbf{c}_2$ axis, resulting in the transformation table

	$\mathbf{c}_1$	$\mathbf{c}_2$	$\mathbf{c}_3$
$\mathbf{b}_1$	$\cos\epsilon$	0	$-\sin\epsilon$
$\mathbf{b}_2$	0	1	0
$\mathbf{b}_3$	$\sin\epsilon$	0	$\cos\epsilon$.

Our goal is to find equations of motion for ϕ and ϵ. To do so, we split apart the two rigid bodies at the pivot and introduce the constraint moment acting between them to keep the rotor and ship aligned, given by pure torques about the $\mathbf{b}_1$ and $\mathbf{b}_3$ axes, as shown in the free-body diagrams in Figure 11.25. The moment on the ship is $\mathbf{M}_s = \tau_1\mathbf{b}_1 + \tau_3\mathbf{b}_3$, and the moment on the rotor is equal and opposite: $\mathbf{M}_r = -\tau_1\mathbf{b}_1 - \tau_3\mathbf{b}_3$.

We find the rigid-body equations for the ship and rotor separately, starting with the ship. The angular velocity of the ship in the inertial frame is

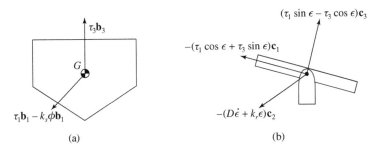

Figure 11.25 (a) Ship and (b) rotor moment diagrams.

$$^{\mathcal{I}}\boldsymbol{\omega}^{\mathcal{B}} = \dot{\phi}\mathbf{b}_1.$$

If we assume the ship has a moment of inertia about $\mathbf{b}_1$ of I_1, then the angular momentum of the ship about its center of mass is

$$^{\mathcal{I}}\mathbf{h}_G = I_1\dot{\phi}\mathbf{b}_1.$$

We now use Eq. (11.26) or Eq. (11.27) and the moment diagram in Figure 11.25a to find the equation of motion of the ship. Since the angular velocity and angular momentum are aligned, the cross product from the transport equation is zero, and we have simply

$$I_1\ddot{\phi} = -k_s\phi + \tau_1, \tag{11.87}$$

where we have introduced a *restoring moment* due to the water with spring constant k_s.

We next find the equations of motion for the rotor. The angular velocity of frame $\mathcal{C}$ in $\mathcal{B}$ is

$$^{\mathcal{B}}\boldsymbol{\omega}^{\mathcal{C}} = \dot{\epsilon}\mathbf{c}_2.$$

Combining this expression with the angular velocity of $\mathcal{B}$ in $\mathcal{I}$ by means of the addition property and using the transformation array between $\mathcal{B}$ and $\mathcal{C}$ gives the angular velocity of the rotor in absolute space:

$$^{\mathcal{I}}\boldsymbol{\omega}^{\mathcal{C}} = \dot{\phi}\cos\epsilon\mathbf{c}_1 + \dot{\epsilon}\mathbf{c}_2 - \dot{\phi}\sin\epsilon\mathbf{c}_3. \tag{11.88}$$

We assume that the symmetric rotor has a large inertia I_r about its spin axis and a smaller moment of inertia J about each of the transverse axes, giving it a moment of inertia tensor:

$$[\mathbb{I}_{O'}]_{\mathcal{C}} = \begin{bmatrix} J & 0 & 0 \\ 0 & J & 0 \\ 0 & 0 & I_r \end{bmatrix}_{\mathcal{C}}.$$

Multiplying the moment of inertia tensor and the angular velocity yields the angular momentum of the rotor as components in frame $\mathcal{C}$,

$$^{\mathcal{I}}\mathbf{h}_{O'} = \mathbb{I}_{O'} \cdot {}^{\mathcal{I}}\boldsymbol{\omega}^{\mathcal{C}} + h\mathbf{c}_3 = J\dot{\phi}\cos\epsilon\mathbf{c}_1 + J\dot{\epsilon}\mathbf{c}_2 + (h - I_r\dot{\phi}\sin\epsilon)\mathbf{c}_3, \tag{11.89}$$

where we have included the constant spin angular momentum h about the $\mathbf{c}_3$ axis.

We now use Eq. (11.26) or Eq. (11.27) to find the equations of motion, but this time in the $\mathcal{C}$ frame. The moment on the rotor from the moment diagram in Figure 11.25b expressed as components in frame $\mathcal{C}$ is

$$\mathbf{M}_r = -(\tau_1 \cos \epsilon + \tau_3 \sin \epsilon)\mathbf{c}_1 + (\tau_1 \sin \epsilon - \tau_3 \cos \epsilon)\mathbf{c}_3 - (D\dot{\epsilon} + k_r \epsilon)\mathbf{c}_2,$$

where we have introduced a restoring torque k_r and a viscous damper D on the pivot between the rotor and the ship. In most systems the torque and damping would be supplied by a feedback-control system on the rotor, but it also works with passive components. (Note that we did not include the equal-and-opposite pivot forces when examining the equations of motion of the ship, as we only consider motion of the ship about the $\mathbf{b}_1 = \mathbf{e}_1$ direction.)

Taking the inertial derivative of Eq. (11.89) using the transport equation and the angular velocity in Eq. (11.88) and setting the result equal to the moment gives three scalar equations:

$$J\ddot{\phi} \cos \epsilon + \dot{\epsilon}(h - I_r \dot{\phi} \sin \epsilon) = -\tau_1 \cos \epsilon - \tau_3 \sin \epsilon \tag{11.90}$$

$$J\ddot{\epsilon} - J\dot{\phi}^2 \sin \epsilon \cos \epsilon - \dot{\phi} \cos \epsilon(h - I_r \dot{\phi} \sin \epsilon) + D\dot{\epsilon} + k_r \epsilon = 0 \tag{11.91}$$

$$I_r \ddot{\phi} \sin \epsilon + I_r \dot{\phi}\dot{\epsilon} \cos \epsilon = -\tau_1 \sin \epsilon + \tau_3 \cos \epsilon. \tag{11.92}$$

We can reduce these to two equations for ϕ and ϵ by multiplying Eq. (11.90) by $\cos \epsilon$ and Eq. (11.92) by $\sin \epsilon$, adding, and solving for τ_1. Then τ_1 is substituted into Eq. (11.87) to obtain the equation of motion for $\ddot{\phi}$. Combined with Eq. (11.91), this result yields our two equations of motion. Let $\omega_s^2 = k_s/I_1$ be the natural frequency of the ship's rocking, $\omega_r^2 = k_r/J$ be the natural frequency of the rotor oscillation, $2\zeta_r \omega_r = D/J$ be the rotor damping, $K_1 = h/I_1$, and $K_2 = h/J$. We have

$$\left(1 + \frac{J \cos^2 \epsilon + I_r \sin^2 \epsilon}{I_1}\right) \ddot{\phi} + \omega_s^2 \phi + K_1 \dot{\epsilon} \cos \epsilon = 0$$

$$\ddot{\epsilon} + 2\zeta_r \omega_r \dot{\epsilon} + \omega_r^2 \epsilon - \dot{\phi}^2 \sin \epsilon \cos \epsilon - \cos \epsilon \dot{\phi}\left(K_2 - \frac{I_r \dot{\phi} \sin \epsilon}{J}\right) = 0.$$

Although these equations are fairly complicated, they are easily integrated to show that adding the gyroscope stabilizes the ship. Figure 11.26 shows a numerical simulation for typical parameter values and three different damping constants. For these simulations, we chose a natural frequency for the ship's rocking of 1 Hz and for the rotor of 10 Hz. As the damping constant in the rotor attachment increases, the ship's rocking damps out more quickly.

Analysis of the ship stabilization system can be simplified by *linearization*. If we assume that both the ship and the rotor motions are small ($\epsilon, \dot{\epsilon}, \phi, \dot{\phi} \ll 1$), then we can drop all second-order terms using the small-angle approximation to find the following linear equations of motion for the ship and rotor:

$$\ddot{\phi} + \bar{\omega}_s^2 \phi + \bar{K}_1 \dot{\epsilon} = 0$$

$$\ddot{\epsilon} + 2\zeta_r \omega_r \dot{\epsilon} + \omega_r^2 \epsilon - K_2 \dot{\phi} = 0,$$

where $\bar{\omega}_s^2 = \omega_s^2/(1 + J/I_1)$ and $\bar{K}_1 = K_1/(1 + J/I_1)$. Chapter 12 explains how solutions of coupled linear equations of motion are always harmonic, as in Figure 11.26.

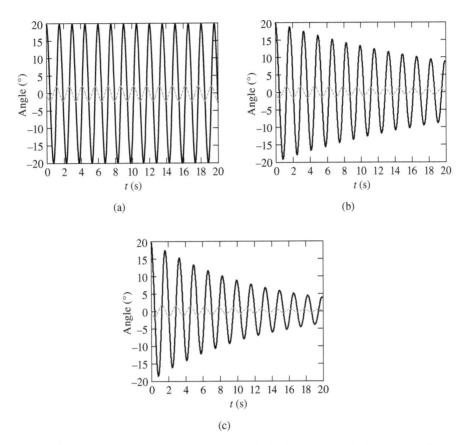

Figure 11.26 Roll of the ship (solid curves) and pitch of the rotor (dashed curves) for three different damping constants. Here $\omega_s = 1$ Hz, $\omega_r = 10$ Hz, $K_1 = 50$ s^{-1}, and $K_2 = 100$ s^{-1}. (a) $\zeta_r = 0$. (b) $\zeta_r = 0.4$. (c) $\zeta_r = 0.8$.

11.9 Key Ideas

- The **translational dynamics** of the center of mass of a multiparticle system in three dimensions are

$$\mathbf{F}_G^{(\text{ext})} = m_G{}^{\mathcal{I}}\mathbf{a}_{G/O}.$$

- The **total angular momentum** of a three-dimensional N-particle system about point O has the dynamics

$$\frac{{}^{\mathcal{I}}d}{dt}\left({}^{\mathcal{I}}\mathbf{h}_O\right) = \sum_{i=1}^{N} \mathbf{r}_{i/O} \times \mathbf{F}_i^{(\text{ext})} \triangleq \mathbf{M}_O^{(\text{ext})}.$$

- The **total angular momentum** of a three-dimensional multiparticle system about point O separates into the angular momentum of the center of mass G about O and the angular momentum about G:

$${}^{\mathcal{I}}\mathbf{h}_O = {}^{\mathcal{I}}\mathbf{h}_{G/O} + {}^{\mathcal{I}}\mathbf{h}_G,$$

where

$$^{\mathcal{I}}\mathbf{h}_{G/O} \triangleq \mathbf{r}_{G/O} \times m_G{}^{\mathcal{I}}\mathbf{v}_{G/O}$$

and

$$^{\mathcal{I}}\mathbf{h}_{G} \triangleq \sum_{i=1}^{N} m_i \mathbf{r}_{i/G} \times {}^{\mathcal{I}}\mathbf{v}_{i/G}.$$

- The **rotational dynamics** of a three-dimensional rigid body about its center of mass G are

$$\frac{^{\mathcal{I}}d}{dt}\left(^{\mathcal{I}}\mathbf{h}_{G}\right) = \frac{^{\mathcal{B}}d}{dt}\left(^{\mathcal{I}}\mathbf{h}_{G}\right) + {}^{\mathcal{I}}\boldsymbol{\omega}^{\mathcal{B}} \times {}^{\mathcal{I}}\mathbf{h}_{G} = \mathbf{M}_{G},$$

where $\mathcal{B}$ is a body frame and $\mathbf{M}_G \triangleq \sum_{i=1}^{N} \mathbf{r}_{i/G} \times \mathbf{F}_i^{(\text{ext})}$ is the total external moment on the body.

- **Qualitative analysis** of a rapidly spinning rigid body is based on the approximation

$$\frac{^{\mathcal{B}}d}{dt}\left(^{\mathcal{I}}\mathbf{h}_{G}\right) \approx 0,$$

which simplifies the rotational dynamics to the vector cross product

$$^{\mathcal{I}}\boldsymbol{\omega}^{\mathcal{B}} \times {}^{\mathcal{I}}\mathbf{h}_{G} \approx \mathbf{M}_{G}.$$

- The **moment of inertia tensor** of a rigid body about its center of mass is

$$\mathbb{I}_{G} \triangleq \sum_{i=1}^{N} m_i \left[(\mathbf{r}_{i/G} \cdot \mathbf{r}_{i/G})\mathbb{U} - (\mathbf{r}_{i/G} \otimes \mathbf{r}_{i/G}) \right],$$

where $\mathbb{U}$ is the unity tensor, and the rigid body is made up of N particles. For a continuous rigid body $\mathcal{B}$, the moment of inertia tensor is

$$\mathbb{I}_{G} \triangleq \int_{\mathcal{B}} \left[(\mathbf{r}_{dm/G} \cdot \mathbf{r}_{dm/G})\mathbb{U} - (\mathbf{r}_{dm/G} \otimes \mathbf{r}_{dm/G}) \right] \, dm.$$

- The **moment of inertia matrix** of a rigid body about its center of mass is

$$[\mathbb{I}_{G}]_{\mathcal{B}} = \sum_{i=1}^{N} m_i \left(([\mathbf{r}_{i/G}]_{\mathcal{B}}^{T}[\mathbf{r}_{i/G}]_{\mathcal{B}})I - [\mathbf{r}_{i/G}]_{\mathcal{B}}[\mathbf{r}_{i/G}]_{\mathcal{B}}^{T} \right).$$

where I is the 3×3 identity matrix. For a continuous rigid body $\mathcal{B}$, the moment of inertia matrix is

$$[\mathbb{I}_{G}]_{\mathcal{B}} = \int_{\mathcal{B}} \left(([\mathbf{r}_{dm/G}]_{\mathcal{B}}^{T}[\mathbf{r}_{dm/G}]_{\mathcal{B}})I - [\mathbf{r}_{dm/G}]_{\mathcal{B}}[\mathbf{r}_{dm/G}]_{\mathcal{B}}^{T} \right) dm.$$

- A three-dimensional **body frame** can always be chosen so that the moment of inertia matrix is diagonal. In this case, the axes of the body frame are called the **principal axes,** and the nonzero entries in the moment of inertia matrix are called the **principal moments of inertia.**

- The **angular momentum** of a rigid body about its center of mass is

$$^{\mathcal{I}}\mathbf{h}_G = \mathbb{I}_G \cdot {}^{\mathcal{I}}\boldsymbol{\omega}^{\mathcal{B}}$$

or, in matrix form,

$$[^{\mathcal{I}}\mathbf{h}_G]_{\mathcal{B}} = [\mathbb{I}_G]_{\mathcal{B}}[^{\mathcal{I}}\boldsymbol{\omega}^{\mathcal{B}}]_{\mathcal{B}}.$$

- The **rotational equations of motion** of a rigid body are written in terms of the angular velocity of the body in absolute space:

$$\mathbb{I}_G \cdot \frac{{}^{\mathcal{B}}d}{dt}\left({}^{\mathcal{I}}\boldsymbol{\omega}^{\mathcal{B}}\right) + {}^{\mathcal{I}}\boldsymbol{\omega}^{\mathcal{B}} \times \left(\mathbb{I}_G \cdot {}^{\mathcal{I}}\boldsymbol{\omega}^{\mathcal{B}}\right) = \mathbf{M}_G.$$

- **Euler's equations** are the three scalar rotational equations of motion of a rigid body expressed in the principal axes of a body frame,

$$\begin{array}{rcl}
I_1\dot{\omega}_1 + (I_3 - I_2)\omega_2\omega_3 & = & M_1 \\
I_2\dot{\omega}_2 + (I_1 - I_3)\omega_1\omega_3 & = & M_2 \\
I_3\dot{\omega}_3 + (I_2 - I_1)\omega_1\omega_2 & = & M_3,
\end{array}$$

where I_1, I_2, and I_3 are the **principal moments of inertia**.

- The **parallel axis theorem** in three dimensions provides a formula for computing the moment of inertia tensor about an arbitrary point Q fixed to a rigid body:

$$\mathbb{I}_Q = \mathbb{I}_G + m_G\left[(\mathbf{r}_{Q/G} \cdot \mathbf{r}_{Q/G})\mathbb{U} - (\mathbf{r}_{Q/G} \otimes \mathbf{r}_{Q/G})\right]$$

or, in matrix form,

$$[\mathbb{I}_Q]_{\mathcal{B}} = [\mathbb{I}_G]_{\mathcal{B}} + m_G\left(\|\mathbf{r}_{Q/G}\|^2 I - [\mathbf{r}_{Q/G}]_{\mathcal{B}}[\mathbf{r}_{Q/G}]_{\mathcal{B}}^T\right).$$

- The moment of inertia matrix expressed as components in a body-fixed frame $\mathcal{B}_1$ can be rotated so that it is expressed as components in a different frame $\mathcal{B}_2$ by means of

$$[\mathbb{I}_G]_{\mathcal{B}_2} = {}^{\mathcal{B}_2}C^{\mathcal{B}_1}[\mathbb{I}_G]_{\mathcal{B}_1}\left({}^{\mathcal{B}_2}C^{\mathcal{B}_1}\right)^T,$$

where ${}^{\mathcal{B}_2}C^{\mathcal{B}_1}$ is the direction-cosine matrix from $\mathcal{B}_1$ to $\mathcal{B}_2$.

- The **total kinetic energy** of a three-dimensional multiparticle system about point O separates into the kinetic energy of the center of mass G about O and the kinetic energy about G:

$$T_O = T_{G/O} + T_G,$$

where

$$T_{G/O} = \frac{1}{2}m_G\|{}^{\mathcal{I}}\mathbf{v}_{G/O}\|^2$$

and

$$T_G = \frac{1}{2}{}^{\mathcal{I}}\boldsymbol{\omega}^{\mathcal{B}} \cdot \mathbb{I}_G \cdot {}^{\mathcal{I}}\boldsymbol{\omega}^{\mathcal{B}} = \frac{1}{2}{}^{\mathcal{I}}\mathbf{h}_G \cdot {}^{\mathcal{I}}\boldsymbol{\omega}^{\mathcal{B}}.$$

- The **total energy** of a three-dimensional rigid body is

$$E_O \triangleq T_{G/O} + T_G + U_O$$

and the three-dimensional **work-energy formula** is

$$E_O(t_2) = E_O(t_1) + W_{G/O}^{(\text{nc})} + W_G^{(\text{nc})}.$$

11.10 Notes and Further Reading

Most introductory texts rely solely on a matrix approach when treating three-dimensional rigid bodies and thus use only scalar equations of motion in the body frame. We have chosen to use a tensor-based approach to be consistent with our earlier vector treatment and to be able to formulate the equations of motion in frame-independent terms. Two other books that use a similar notation are Rao (2006) and Tenenbaum (2004). A third notational approach, that of *dyadics,* is used throughout Kane's books (Kane 1978; Kane and Levinson 1985; Kane et al. 1983). Goldstein (1980) also has a good discussion of tensors and dyadics. Readers interested in exploring more mathematically advanced approaches employing differential geometry and group theory can consult O'Reilly (2008), Arnold (1989), or Marsden and Ratiu (1999). (These texts also primarily use the energy-based methods that are introduced in Chapter 13.)

As you might have surmised, we have only touched the surface here on the variety of examples demonstrating rigid-body motion. For a slightly more advanced treatment of torque-free motion (including the Poinsot construction for large-angle paths in space) readers can consult Goldstein (1980), Greenwood (1988), or Hand and Finch (1998). Goldstein (1980) and Hand and Finch (1998) also have excellent discussions of the spinning Lagrange top. An excellent collection of aerospace examples (with some applications to automatic control) can be found in Bryson (1994). Our ship stabilization example is similar to one in that book. Three other references with good discussions of space-vehicle dynamics are Kaplan (1976), Kane et al. (1983), and Wie (1998). A good discussion of the rigid-body dynamics of an airplane can be found in Stengel (2004). We did not discuss the topic of semi-rigid bodies, that is, bodies with moving parts that change the moment of inertia. This greatly complicates the analysis, but is important for some systems, such as a spacecraft with moving masses for attitude control. One reference that touches on the topic is Kaplan (1976).

There is some confusion in the literature regarding the use of the terms *precession* and *nutation*. Most texts follow the traditional approach of using *precession* to describe the circular motion of the body symmetry axis about an inertial direction (e.g., Fetter and Walecka 1980; Goldstein 1980; Greenwood 1988; Hand and Finch 1998). To highlight the differences between torque-free and torqued bodies, we have chosen to use the expression *free-body precession* to refer to this motion of a torque-free rigid body. The *nutation angle* refers to the offset of the symmetry axis from an inertial direction and *nutation* refers to the oscillatory motion of this angle for spinning bodies subject to torque. There are a number of books on spacecraft dynamics, however, that use *nutation* to refer to the motion of the angular velocity about the angular momentum in torque-free motion (e.g., Wertz 1978; Bryson 1994). Thus *nutation dampers,* such as the one described in Problem 11.19, damp out this motion of the angular velocity. We find this nomenclature confusing and have tried to consistently call this motion *precession;* a nutation damper thus brings the nutation angle to zero.

11.11 Problems

11.1 How many degrees of freedom does a rigid airplane have while flying in the air? While rolling on a runway?

11.2 Find the equations of motion for a gyropendulum where the spinning rotor is a disk. Assume the rotor has fixed angular momentum h about its axis and has a transverse moment of inertia J. That is, treat the spinning mass not as a particle but as a rigid body.

11.3 Show that the three-dimensional parallel axis theorem in Eq. (11.56) reduces to the planar theorem for rigid-body rotation in two dimensions (Eq. (9.34)).

11.4 Verify Eq. (11.77) for the kinetic energy about an inertially fixed point O (T_O) in terms of the kinetic energy, T_Q, about an arbitrary point of the body Q and the kinetic energy, $T_{Q/O}$, due to translation of Q.

11.5 Find the angular velocity of the rigid two-wheel system shown in Figure 11.27, assuming that both wheels roll without slipping at rate Ω and $R_2 > R_1$. (Assume the wheels are fixed to the axle.)

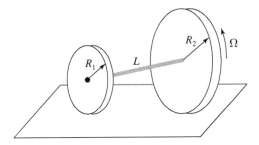

Figure 11.27 Problem 11.5.

11.6 Recall the right circular cone of height h, radius R, and mass m of Example 11.1. Let O be the tip of the cone and assume it has uniform density.

    a. Find the moment of inertia matrix about the center of mass G.
    b. Find the moment of inertia matrix about O.

11.7 Example 11.6 determined the moment of inertia of a solid cube. Now you will find the moment of inertia of a hollow cube, where the wall thickness is much smaller than the cube height.

    a. Consider a thin square plate of height h and mass m_p. Show that the moment of inertia of such a plate, relative to its center of mass G and expressed in a body frame $\mathcal{B}$ attached to the plate, is

$$[\mathbb{I}_G]_\mathcal{B} = \begin{bmatrix} \dfrac{m_p h^2}{12} & 0 & 0 \\[2ex] 0 & \dfrac{m_p h^2}{12} & 0 \\[2ex] 0 & 0 & \dfrac{m_p h^2}{6} \end{bmatrix}_\mathcal{B}.$$

    b. A hollow cube can be modeled as a structure built of six thin plates. For one of the plates, find the moment of inertia relative to the center of the cube O.

    c. Sum the contributions of all the plates to find the moment of inertia of the cube about its center of mass. Be careful to consider the rotation of the plates with respect to one another.

11.8 Use a qualitative analysis to predict the motion of a frisbee that is inclined at angle θ to the oncoming wind, as shown from the side in Figure 11.28. How does this motion differ from what would happen if the frisbee were not spinning?

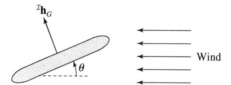

Figure 11.28 Problem 11.8.

11.9 A three-dimensional top with a pointed tip fixed at O on a frictionless surface spins rapidly about its long axis, as shown in Figure 11.29. Angle ψ is determined by the projection of the center of mass G onto the horizontal plane. The top is inclined at angle θ and spins about its long axis at rate $\dot{\phi} \gg 0$. Use a qualitative description of the rotational dynamics to explain the gyroscopic motion of the top. [HINT: Consider the rotational dynamics about O (i.e., $\frac{^\mathcal{I}d}{dt}(^\mathcal{I}\mathbf{h}_O) = \mathbf{M}_O$), and use the approximation $^\mathcal{I}\mathbf{h}_O = {}^\mathcal{I}\mathbf{h}_{G/O} + {}^\mathcal{I}\mathbf{h}_G \approx {}^\mathcal{I}\mathbf{h}_G.$]

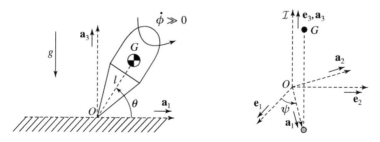

Figure 11.29 Problem 11.9.

11.10 Use MATLAB to integrate the torque-free rigid-body rotational equations of motion in Example 11.10 with $I_3 > I_2 > I_1$ (i.e., Euler's equations with $M_1 = M_2 = M_3 = 0$). For each of the following three sets of initial conditions, briefly describe in words the resulting motion of the rigid body:

    a. $\omega_1 > 1$ and $\omega_2 \approx \omega_3 \ll 1$,

    b. $\omega_2 > 1$ and $\omega_3 \approx \omega_1 \ll 1$, and

    c. $\omega_3 > 1$ and $\omega_1 \approx \omega_2 \ll 1$.

11.11 A rotary pendulum, shown in Figure 11.30, is composed of a uniform disk of radius R and a thin uniform rod of length l and mass m_r, pinned by its end to the edge of the disk. The rod swings in the plane tangent to the edge of the disk. Assuming the disk has a moment of inertia J about its center, find the equations of motion of this system.

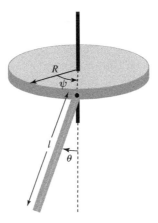

Figure 11.30 Problem 11.11.

11.12 Find the equations of motion of a passenger on the amusement park ride shown in Figure 11.31 by developing a physical model and analyzing it. (How many degrees of freedom are in your model?)

Figure 11.31 Problem 11.12. Image courtesy of Shutterstock.

11.13 Consider again the demonstration gyro from Example 11.13. Instead of the small mass having a fixed position, suppose it is connected to a spring of spring constant k and unstretched length ℓ_0 that is attached to the post.

    a. How many degrees of freedom does the system have?

    b. Find the equations of motion for the entire system.

    c. Suppose the gyro is spinning but otherwise balanced and an impulse is applied to the small mass. What is the total angular momentum before and after the impulse?

    d. Let the mass of the gyro be 5 kg, the small mass be 2 kg, and set $l = 0.5$ m and $l' = 0.25$ m. The gyro is spinning at a rate of 1 rad/s, and has the moment of inertia of a thin circular disk with a radius of 10 cm. The spring has a constant of 10 N/m and a rest length of zero. Numerically integrate the equations of motion after an impulse is applied (you can choose the size of the impulse). Qualitatively describe the resulting motion. Remember that the system is at rest before the impulse is applied when choosing your initial conditions.

    e. Suppose a damper is added in addition to the spring. What is the steady-state configuration of the system?

11.14 Figure 11.32 shows the basin of a typical washing machine filled with clothes. The basin spins about its axis and rotates about the pivot point at its center. When balanced, it stays level and simply spins about its axis. However, you may have experienced imbalance during the spin cycle, when it sounds like the basin will break through the machine. Figure 11.32b shows such a situation.

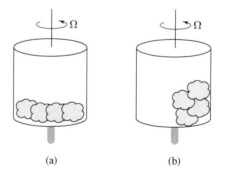

                    (a)   (b)

Figure 11.32 Problem 11.14. (a) Balanced load. (b) Unbalanced load.

    a. How many degrees of freedom does the system have? Select a set of scalar coordinates to describe the degrees of freedom.

    b. Describe qualitatively the expected motion when the clothes are imbalanced.

11.15 Consider the upper stage of a long narrow rocket used to transfer a satellite from a low-earth orbit to geosynchronous orbit (thus it is out of the atmosphere). Assume a small misalignment of the thrust vector $\mathbf{F}_T$ from the center line, as shown in Figure 11.33. For large high-thrust solid rocket motors, the rockets are always spun about the long axis. Explain the reason for spinning

Figure 11.33 Problem 11.15.

the rocket stage. [HINT: Consider the rotational and translational motion of the rocket when spinning and when not spinning.]

11.16 Suppose you take your new all-terrain vehicle (ATV) out for a spin, literally. Assume you are traveling at a constant speed v, and you take a sharp circular turn (either because it is fun or because you have to suddenly avoid an obstacle). Unfortunately, you take the turn too fast and roll over (a common problem).

    a. Find the maximum speed that you can take a turn of radius of curvature R_c before rolling over (the inside wheels[6] leaving the ground) in terms of L, the distance between the wheels, and d, the height above the ground of the vehicle's center of mass. Assume that the ATV travels the turn *without slipping;* that is, the radius of curvature is constant.

    b. Find the maximum speed for a typical ATV (with a wheelbase L of 3.5 ft and d of about 4 ft), assuming a sharp turn with radius of curvature of 5 ft.

    c. What changes might you make to the ATV design to prevent rollover?

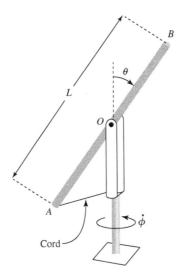

Figure 11.34 Problem 11.17.

11.17 Consider a slender rod AB of mass m and length L pinned at its midpoint O to a vertical shaft, as shown in Figure 11.34. The rod can rotate by θ about the pin, and the shaft can rotate by ϕ about its long axis. The mass of the shaft

---

[6] The inside wheels are the two wheels on the side of the car inside the turn, that is, toward the center of the circular arc.

and collar may be neglected. The cord keeps the rod at an initial angle of 30°
with the vertical while the shaft is initially rotating freely at angular velocity
$\dot{\phi} = 200$ rad/s. The cord suddenly breaks.

    a. Find an expression for the tension in the cord before it breaks.
    b. Find the differential equations of motion for the rod ($\ddot{\theta}$ and $\ddot{\phi}$) after the
       break.

11.18 Show that the most general parallel axis theorem for transporting the moment
of inertia tensor from an arbitrary point P of the body to a different arbitrary
point Q of the body is

$$\mathbb{I}_Q = \mathbb{I}_P + m_G \left[\begin{array}{c} (\mathbf{r}_{Q/P} \cdot \mathbf{r}_{Q/P})\mathbb{U} + 2(\mathbf{r}_{P/G} \cdot \mathbf{r}_{Q/P})\mathbb{U} \\ -\mathbf{r}_{Q/P} \otimes \mathbf{r}_{Q/P} - \mathbf{r}_{Q/P} \otimes \mathbf{r}_{P/G} - \mathbf{r}_{P/G} \otimes \mathbf{r}_{Q/P} \end{array} \right],$$

where, as usual, G refers to the center of mass of the rigid body.

11.19 Consider a torque-free cylindrical satellite spinning about its major axis,
as shown in Figure 11.35. If the angular momentum, angular velocity, and
symmetry (spin) axis are all aligned, it will spin uniformly about the $\mathbf{b}_2$ axis,
ensuring that all instruments stay pointed. However, should it be disturbed (by
a micrometeorite impact, perhaps) so that the angular momentum is moved,
the satellite body frame will precess, as seen in Tutorial 11.1. Many satellites
are equipped with what is called a *nutation damper* to remove the small
nonspin-axis angular-velocity components. This device consists of a small
wheel free to spin about its symmetry axis at rotation rate Ω but fixed to the
satellite about the other two axes, as shown in Figure 11.35. The wheel is
equipped with a viscous damper, so that it is subjected to a moment about its
spin axis of magnitude $-D\Omega$.

    Suppose the satellite (including the wheel) has principal moments of inertia
(I_T, I_S, I_T), where $I_T < I_S$, and the wheel has a moment of inertia about its

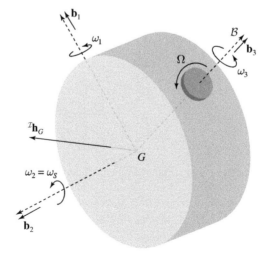

Figure 11.35 Problem 11.19.

spin axis of I_W. Find the equations of motion for the three angular-velocity components of the satellite, ω_1, ω_2, and ω_3, and the spin rate Ω of the wheel. Using initial conditions $\Omega(0) = 0$, $\omega_1(0) = 0.2$ rad/s, $\omega_2(0) = 2$ rad/s, $\omega_3(0) = 0$, $I_S/I_T = 1.5$, $I_W/I_T = 0.06$, and $D/I_T = 0.5$, show, numerically, that the system does indeed damp the initial precession.

11.20 Consider the spinning, torque-free cylindrical satellite in Figure 11.36. Similar to Problem 11.19, the cylinder is symmetric, with moment of inertia J about $\mathbf{b}_2$ and moment of inertia I about $\mathbf{b}_1$ and $\mathbf{b}_3$. Note that $J > I$. The cylinder is shown with an arbitrary angular velocity in the body frame. We would like the cylinder to be spinning about the $\mathbf{b}_2$ axis, its symmetry axis (the axis of maximum moment of inertia). As you know, if the cylinder has angular velocity components in the other directions ($\mathbf{b}_1$ and $\mathbf{b}_3$), it will precess. As in Problem 11.19, we have added a nutation damper.

The nutation damper here consists of a particle of mass m connected to a spring of spring constant k and a viscous damper of damping constant b. The particle is free to move only in the $\mathbf{b}_2$ direction. Let y represent the displacement of the particle from the center of mass in the $\mathbf{b}_2$ direction. The spring is unstretched when $y = 0$. The mass-spring-damper system is displaced a distance l in the $\mathbf{b}_3$ direction from the center of mass of the cylinder. Note that the mass properties of the cylinder do not include the nutation damper.

For all parts you may assume that the center of mass of the cylinder is either stationary in inertial space or moving at constant velocity.

    a. Draw free-body diagrams for the cylinder and the particle. Include all forces and moments (including constraint forces!).

    b. Find the equations of motion for the angular velocity components of the cylinder and the displacement of the particle. You may assume that the cylinder is nominally spinning about $\mathbf{b}_2$ at ω_s and that ω_1, ω_3, and

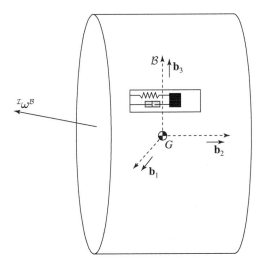

Figure 11.36 Problem 11.20.

y, the displacement of the mass, can all be treated as small quantities (thus higher powers and products among them can all be dropped). Show that the resulting equations of motion are

$$\dot{\omega}_1 - \Omega\omega_3 = -\frac{ml}{I}\left(\omega_n^2 - \omega_s^2\right) y - 2\frac{ml}{I}\zeta\omega_n\dot{y}$$

$$\dot{\omega}_3 + \Omega\omega_1 = 0$$

$$\ddot{y} + 2\zeta\omega_n\dot{y} + \omega_n^2 y + l\omega_s\omega_3 = l\dot{\omega}_1,$$

where $\Omega = \frac{J-I}{I}\omega_s$, $\omega_n^2 = k/m$, and $2\zeta\omega_n = b/m$.

c. Using MATLAB, numerically integrate the equations of motion over the interval $[0, 40]$ s and plot $\omega_1(t)$ and $\omega_3(t)$ on a single plot and $y(t)$ on a second plot. Explain your results. (That is, does the design work?) Use $J = 1000$ kg-m^2, $I = 200$ kg-m^2, $m = 25$ kg, $l = 1$ m, $\omega_n = \sqrt{k/m}_P = 1$ rad/s, and $\zeta = 0.9$. Assume that particle starts at rest $(y(0) = \dot{y}(0) = 0)$ and the cylinder has initial angular velocity components $\omega_1(0) = 0.1$ rad/s, $\omega_2(0) = \omega_s = 1$ rad/s, and $\omega_3(0) = -0.2$ rad/s.

d. What do you think would happen if the cylinder were long and thin rather than short and fat (that is, $J < I$)?

11.21 Repeat Problem 11.19 with $I_T > I_S$. That is, suppose the satellite is a minor-axis spinner with a nutation damper. Repeat the simulation with $I_T/I_S = 1.5$. What do you notice? Does this result make sense?

PART FIVE

Advanced Topics

CHAPTER TWELVE

Some Important Examples

This chapter applies the tools and methods developed in the first four parts of the book to examine three important examples in a bit more depth. In the process, we introduce some key concepts in dynamical systems that may be important if you continue in more advanced studies. The first section explores simple harmonic motion and vibrations, introducing you to the solution of multiple-degree-of-freedom linear systems and mode shapes. The second section explores the ideas of equilibrium and linearization to show how the behavior of a complex dynamical system can often be predicted by a set of linear equations of motion. As an example, we return to the airplane and present the linearized motion about a particular steady-state solution. The final section returns to the topic of collisions introduced in Chapter 6 and applies it to the collision of finite-sized particles.

12.1 An Introduction to Vibrations and Linear Systems

Chapter 2 introduced simple harmonic motion because of its importance and ubiquity. Although the mechanical system solved in Tutorials 2.3 and 2.4 was a simple one-dimensional model of a mass on a spring, by now you may have noticed how often simple harmonic motion appears in dynamics. For example, the blades of a wind turbine may vibrate due to wind gusts (as in Figure 12.1). When we talk about vibrations, we simply mean the response of a dynamical system described by either Eq. (2.20) or Eq. (2.24). For example, when we considered the solution of the pendulum for small angles, we were able to convert its nonlinear equation of motion to simple harmonic form. The importance of the solution to this differential equation lies in its application to so many systems!

This section reviews the solution of simple harmonic motion and expands it to the case of forced oscillations. We then show how this solution can be applied to any linear dynamic system by generalizing the solution approach. We conclude by introducing higher degree-of-freedom systems incorporating coupled oscillators.

Figure 12.1 Vibrations of the blades of a wind turbine exhibit simple harmonic motion.

12.1.1 A Review of Simple Harmonic Motion

Here we briefly review the *homogeneous* solution from Chapter 2 for the undamped and damped simple harmonic oscillator. By homogenous we mean the solution of the equations of motion for certain initial conditions with no other forces acting. We are looking for the solution of

$$\ddot{z} + \omega_0^2 z = 0 \tag{12.1}$$

for the undamped case, and of

$$\boxed{\ddot{z} + 2\zeta\omega_0\dot{z} + \omega_0^2 z = 0} \tag{12.2}$$

for the damped case (ω_0 is the natural frequency and ζ is the damping ratio). Recall that we found the solution by recognizing that the general solution must be the sum of two complex exponentials,

$$z(t) = Az_1(t) + Bz_2(t),$$

where $z_1(t) = e^{\lambda_1 t}$ and $z_2(t) = e^{\lambda_2 t}$ for some λ_1, λ_2, and A and B are constants given by the initial conditions. Substituting $z(t)$ into Eq. (12.2) shows that λ_1 and λ_2 must be solutions to the *characteristic equation:*

$$\lambda^2 + 2\zeta\omega_0\lambda + \omega_0^2 = 0, \tag{12.3}$$

which has the two roots

$$\lambda_{1,2} = -\zeta\omega_0 \pm \omega_0\sqrt{\zeta^2 - 1}.$$

For the undamped case ($\zeta = 0$), the roots are $\lambda_{1,2} = \pm\omega_0 i$ and the response is sinusoidal. Using $e^{i\omega_0 t} = \cos\omega_0 t + i\sin\omega_0 t$ and the initial conditions yields

$$z(t) = z(0)\cos(\omega_0 t) + \frac{\dot{z}(0)}{\omega_0}\sin(\omega_0 t)$$

$$\dot{z}(t) = -z(0)\omega_0\sin(\omega_0 t) + \dot{z}(0)\cos(\omega_0 t).$$

The important point here is that the constant ω_0 is the *natural frequency* of oscillation of the solution. Figure 2.8 shows this sinusoidal response.

For the damped case ($\zeta > 0$), we found three types of solutions: (a) an *under-damped* solution, where $0 < \zeta < 1$; (b) a *critically damped* solution, where $\zeta = 1$; and (c) an *overdamped* solution, where $\zeta > 1$. The underdamped solution is a decaying sinusoidal oscillation with *damped frequency* $\omega_d \triangleq \omega_0\sqrt{1 - \zeta^2}$:

$$z(t) = e^{-\zeta\omega_0 t}\left(c_1\cos(\omega_d t) + c_2\sin(\omega_d t)\right),$$

where c_1 and c_2 are constants that depend upon the initial conditions:

$$c_1 = z(0)$$

$$c_2 = \frac{\dot{z}(0) + z(0)\zeta\omega_0}{\omega_0\sqrt{1 - \zeta^2}}.$$

The amplitude of the underdamped oscillations remains inside a decaying *envelope* given by $e^{-\zeta\omega_0 t}$. The underdamped solution is plotted in Figure 2.9. Note also that, if $\zeta = 0$, we recover the undamped solution.

The overdamped solution is characterized by an exponential decay with no oscillation,

$$z(t) = Ae^{-\omega_0\left(\zeta+\sqrt{\zeta^2-1}\right)t} + Be^{-\omega_0\left(\zeta-\sqrt{\zeta^2-1}\right)t},$$

where A and B are found from the initial conditions to be

$$A = -\frac{\dot{z}(0) + z(0)\omega_0\left(\zeta - \sqrt{\zeta^2 - 1}\right)}{2\omega_0\sqrt{\zeta^2 - 1}}$$

$$B = \frac{\dot{z}(0) + z(0)\omega_0\left(\zeta + \sqrt{\zeta^2 - 1}\right)}{2\omega_0\sqrt{\zeta^2 - 1}}.$$

The overdamped solution is also plotted in Figure 2.9.

Finally, the critically damped solution ($\zeta = 1$) is the boundary between the over-damped and underdamped solutions. It is also characterized by an exponential decay, though the general solution is $z(t) = Ae^{-i\zeta\omega_0 t} + Bte^{-i\zeta\omega_0 t}$:

$$z(t) = z(0)e^{-\zeta\omega_0 t} + (\dot{z}(0) + z(0)\zeta\omega_0)te^{-\zeta\omega_0 t}.$$

It is also shown in Figure 2.9.

It is also interesting to look at the total energy E_O of the simple harmonic oscillator. Using the solution for the underdamped case and $E_0 = \frac{1}{2}m\dot{z}^2 + \frac{1}{2}kz^2$, we find

$$E_O = \frac{1}{2}m\left[-\zeta\omega_0 e^{-\zeta\omega_0 t}(c_1\cos(\omega_d t) + c_2\sin(\omega_d t)) + \omega_d e^{-\zeta\omega_0 t}(-c_1\sin(\omega_d t) + c_2\cos(\omega_d t))\right]^2$$
$$+ \frac{1}{2}ke^{-2\zeta\omega_0 t}(c_1\cos(\omega_d t) + c_2\sin(\omega_d t))^2.$$

Using $\omega_0^2 = k/m$, this expression can be simplified to

$$E_O = \frac{1}{2}e^{-2\zeta\omega_0 t}k(c_1^2 + c_2^2) + \frac{1}{2}\zeta^2 ke^{-2\zeta\omega_0 t}\left[(c_1^2 - c_2^2)\cos(2\omega_d t) + 2c_1 c_2\sin(2\omega_d t)\right]$$
$$+ \frac{1}{2}m\zeta\omega_d e^{-2\zeta\omega_0 t}\left[(c_1^2 - c_2^2)\sin(2\omega_d t) - 2c_1 c_2\cos(2\omega_d t)\right].$$

As expected, the total energy is a constant of the motion for $\zeta = 0$. For the damped case, the damping coefficient gives the rate at which energy is removed from the system, that is, the rate of nonconservative work done by the damper.

Remember, the solution of any dynamical system for which the equation of motion can be reduced to either Eq. (12.1) or Eq. (12.2) will show this behavior. Of course, any system that we can model by means of a mass on a spring will respond this way!

12.1.2 Forced Response and Resonance

This section examines *forced vibrations*. Consider a mass on a spring with rest length x_0. What form does the solution take when we allow an arbitrary force to act on the mass, as shown in Figure 12.2? The resulting equation of motion is

$$\boxed{\ddot{x} + 2\zeta\omega_0\dot{x} + \omega_0^2 x = \frac{-F_P(t)}{m_P},} \qquad (12.4)$$

where x is the displacement of the mass from the rest position.

The general solution of Eq. (12.4) is beyond the scope of this text. However, there are three specific types of forces considered here that have important, and ubiquitous, solutions: (a) an impulsive force, (b) a constant force, and (c) a sinusoidal driving

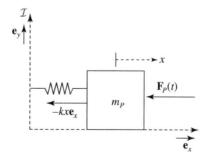

Figure 12.2 Mass on a spring with an arbitrary applied force in the $-\mathbf{e}_x$ direction.

force. Example 4.1 examined the impulsive force. We found that considering an impulsive force on the mass is equivalent to finding the unforced, or homogenous, solution with initial conditions given by Eqs. (4.3) and (4.4). We also looked at the constant-force solution toward the end of Tutorial 2.3. Nevertheless, let us revisit that case, as it introduces the process we will use for more complicated forces.

In general, when searching for the solution of a driven or *inhomogeneous* equation of motion, such as Eq. (12.4), we write it as a sum of two parts, the homogenous solution and the particular solution:

$$x(t) = x_h(t) + x_p(t).$$

The homogeneous solution is the sinusoidal response already found for $F_P(t) = 0$. This general solution must be correct, since it satisfies the equation of motion. At this point the linearity of the differential equation becomes relevant. It is only because the equation of motion is linear in x that we can write the general solution as the sum of these two parts. Thus the response due to initial conditions is entirely contained in $x_h(t)$. To find the forced response, we just have to guess the right $x_p(t)$ due to the forcing function. For a constant force, that is something we have already seen— $x_p(t) = x_p^{(c)}$ is a constant. Plugging into the equation of motion with $F_P = F_0$ gives

$$x_p = \frac{-F_0}{m_P \omega_0^2}.$$

Thus the general solution of the underdamped case with a constant driving force becomes

$$x(t) = \frac{-F_0}{m_P \omega_0^2} + e^{-\zeta \omega_0 t} \left(c_1 \cos(\omega_d t) + c_2 \sin(\omega_d t) \right). \tag{12.5}$$

Example 12.1 Bungee Jump

This example looks at the oscillatory solution of a mass on a spring-damper under the weight of gravity. Assume the initial position is zero (the spring is unstretched) but that there is some initial velocity at $t = 0$. This is an excellent model of a bungee jumper at the moment the bungee cord starts to stretch (Figure 12.3). The bungee jumper falls the length of the cord l before it begins to stretch. Let x be the stretch of the cord. From conservation of total energy, when the cord begins to stretch the initial velocity is

$$\dot{x}(0) = \sqrt{2gh} = \sqrt{2gl}.$$

The equation of motion of the mass-spring-damper is

$$\ddot{x} + 2\zeta \omega_0 \dot{x} + \omega_0^2 x = g.$$

The solution for the cord stretch over time is found from Eq. (12.5):

$$x(t) = \frac{g}{\omega_0^2} + e^{-\zeta \omega_0 t} \left(\frac{\sqrt{2gl\omega_0^2} - g\zeta}{\omega_0^2 \sqrt{1 - \zeta^2}} \sin(\omega_d t) - \frac{g}{\omega_0^2} \cos(\omega_d t) \right).$$

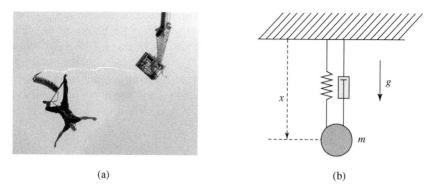

(a)   (b)

Figure 12.3 A model for a bungee jumper is a mass falling on a spring and damper. (a) Bungee jumper. (b) Mass on a spring-damper system. Image (a) courtesy of Shutterstock.

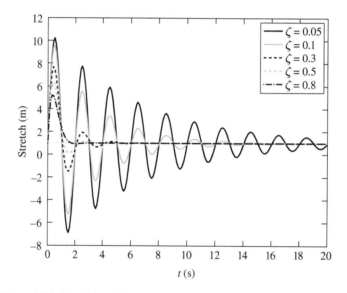

Figure 12.4 Stretch in a 50 m bungee cord for five different damping ratios.

Figure 12.4 plots the stretch for a 50 m bungee cord and five different damping ratios, assuming that a typical cord with an adult hanging on it has a natural frequency of roughly 0.5 Hz. Observe that smaller damping results in a much larger initial overshoot of the steady-state position, making for a much more exciting experience.

The next, and final, forcing function we consider is a sinusoidal one, $F(t) = F_0 \sin(\omega_f t + \phi_f)$, where ω_f is the *driving frequency* and ϕ_f is the phase of the forcing function. This function is important not only because of the ubiquity of harmonic forces but also because it can be used as the building block of a broader class of forcing functions. You may recall from a math course that any periodic function can be represented as the sum of sines and cosines of multiples of the base period (a Fourier series). Here again linearity comes into play. Since the equation of motion is linear, if the force is actually the sum of many simple functions, then the solution will just be the sum of the particular solution for each one. Thus by finding the solution for a sinusoidal forcing function, we are actually finding the response to any periodic function!

The equation of motion for a sinusoidal applied force is

$$\ddot{x} + 2\zeta\omega_0\dot{x} + \omega_0^2 x = \frac{F_0}{m_P}\sin(\omega_f t + \phi_f). \tag{12.6}$$

As with a constant force, the general solution is the sum of the homogeneous and particular solutions. For a particular solution, in this case we guess a sinusoidal form:

$$x_p(t) = a\cos(\omega_f t) + b\sin(\omega_f t).$$

Plugging this guess into Eq. (12.6) and collecting terms yields

$$\left[(\omega_0^2 - \omega_f^2)a + 2\zeta\omega_0\omega_f b\right]\cos(\omega_f t) - \left[2\zeta\omega_0\omega_f a - (\omega_0^2 - \omega_f^2)b\right]\sin(\omega_f t)$$

$$= \frac{F_0}{m_P}\left[\sin\phi_f\cos(\omega_f t) + \cos\phi_f\sin(\omega_f t)\right].$$

Matching terms on each side that multiply $\cos(\omega_f t)$ and $\sin(\omega_f t)$ results in two linear equations in the two unknowns, a and b:

$$(\omega_0^2 - \omega_f^2)a + 2\zeta\omega_0\omega_f b = \frac{F_0}{m_P}\sin\phi_f$$

$$-2\zeta\omega_0\omega_f a + (\omega_0^2 - \omega_f^2)b = \frac{F_0}{m_P}\cos\phi_f.$$

These equations are solved for the coefficients of the particular solution:

$$a = \frac{(F_0/m_P)\left[(\omega_0^2 - \omega_f^2)\sin\phi_f - 2\zeta\omega_0\omega_f\cos\phi_f\right]}{\left[(\omega_0^2 - \omega_f^2)^2 + 4\zeta^2\omega_0^2\omega_f^2\right]} \tag{12.7}$$

$$b = \frac{(F_0/m_P)\left[(\omega_0^2 - \omega_f^2)\cos\phi_f + 2\zeta\omega_0\omega_f\sin\phi_f\right]}{\left[(\omega_0^2 - \omega_f^2)^2 + 4\zeta^2\omega_0^2\omega_f^2\right]}. \tag{12.8}$$

Thus our guess for the particular solution is correct. In the presence of a sinusoidal forcing function with frequency ω_f, the mass-spring-damper system will respond by oscillating at frequency ω_f. However, the vibration response will be shifted in phase by an amount depending on the difference between the forcing frequency and natural frequency. (This can be seen by setting $\phi_f = 0$ in Eqs. (12.6)–(12.8). Although the input forcing function is then purely sinusoidal, the system response has both sine and cosine components.)

More important is what happens to the amplitude of the system response as a function of ω_f. Suppose that the forcing function is very slow compared to the natural frequency of the system: $\omega_f \ll \omega_0$. Then we can approximate the coefficients in Eqs. (12.7) and (12.8) by

$$a \approx \frac{-2\zeta(\omega_f/\omega_0)F_0}{m_P\omega_0^2}$$

$$b \approx \frac{F_0}{m_P\omega_0^2},$$

where, for simplicity, we have set $\phi_f = 0$. The amplitude of the response thus looks just like the constant-force case with a very small phase shift due to the small cosine component. Thus mass-spring-damper systems driven by a very slow forcing function will follow the force almost exactly. For instance, if you are traveling in a car over rolling hills, the car will follow the hills with very little compression in the shock absorbers.

Suppose instead that the forcing function is very fast compared to the natural frequency of the system: $\omega_f \gg \omega_0$. In that case, the response coefficients are approximately

$$a \approx \frac{-2\zeta(\omega_0/\omega_f)F_0}{m_P\omega_f^2}$$

$$b \approx \frac{-F_0}{m_P\omega_f^2}.$$

In this case the response of the system is attenuated by $1/\omega_f^2$. The higher the applied frequency is relative to the natural frequency, the smaller the response. This is an example of *vibration isolation*. For a very low natural frequency, the mass does not move in response to high-frequency force disturbances. This is like your car traveling over a bumpy road. If the shocks are working correctly, they will absorb the high-frequency forces and the car won't vibrate. All vibration-isolation systems work on this basic concept (e.g., Example 9.10).

What if the forcing function is close to the natural frequency, $\omega_f \approx \omega_0$? For undamped systems ($\zeta = 0$) we immediately see a problem. Both of the coefficients in the particular solution go to infinity! The response of the system completely blows up. Even for a damped system (ζ small), the amplitude response can be highly amplified:

$$a \approx \frac{-F_0}{2\zeta\omega_0^2 m_P}$$

$$b \approx 0.$$

This amplification when the driving force is near the natural frequency is called *resonance*. It is a real effect and is responsible for many amazing (and catastrophic) situations. As a designer it is extremely important to understand the environment to which the oscillatory system will be exposed in order to avoid a resonant situation.

These three types of responses are summarized in Figure 12.5, which shows the normalized amplitude and phase of the particular solution as a function of the

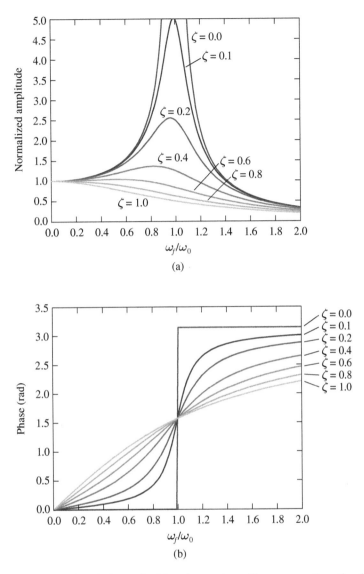

Figure 12.5 Amplitude and phase of a driven harmonic oscillator versus the ratio of driving frequency over natural frequency for various damping ratios. (a) Amplitude response. (b) Phase response.

ratio of driving frequency to natural frequency for various damping coefficients (the amplitude is given by $\sqrt{a^2 + b^2}$ and the phase by $\arctan(b/a)$). These curves are often referred to as the *transfer function* of the mass-spring-damper system. It shows how input forces are transferred to output response. Note that, at low frequencies, the forces are directly transferred to oscillations, whereas at high frequencies, the force-driven oscillations are attenuated.

12.1.3 Simple Harmonic Motion of a Linear System

Before discussing our final topic of multiple-degree-of-freedom vibrations, it is help-ful to rederive the simple harmonic oscillator solution using a slightly different ap-proach. This approach will be essential to the rest of our analysis. In the process, we will also show how the solution for simple harmonic motion applies to all linear equations of motion, as alluded to earlier.

The modified approach involves solving the equations of motion in first-order form. Thus, instead of considering the damped simple harmonic oscillator given by the second-order differential equation in Eq. (12.2), we instead write it in first-order form as

$$
\begin{bmatrix} \dot{z} \\ \ddot{z} \end{bmatrix} = \underbrace{\begin{bmatrix} 0 & 1 \\ -\omega_0^2 & -2\zeta\omega_0 \end{bmatrix}}_{\triangleq A} \begin{bmatrix} z \\ \dot{z} \end{bmatrix}.
\tag{12.9}
$$

We assume an exponential solution as before, only now the constant is a matrix:

$$
\begin{bmatrix} z(t) \\ \dot{z}(t) \end{bmatrix} = \begin{bmatrix} a_1 \\ a_2 \end{bmatrix} e^{\lambda t}.
\tag{12.10}
$$

Substituting Eq. (12.10) into the matrix equation of motion in Eq. (12.9) gives

$$
\lambda \begin{bmatrix} a_1 \\ a_2 \end{bmatrix} e^{\lambda t} = \begin{bmatrix} 0 & 1 \\ -\omega_0^2 & -2\zeta\omega_0 \end{bmatrix} \begin{bmatrix} a_1 \\ a_2 \end{bmatrix} e^{\lambda t}.
$$

Combining the right and left sides leaves

$$
\begin{bmatrix} \lambda & -1 \\ \omega_0^2 & \lambda + 2\zeta\omega_0 \end{bmatrix} \begin{bmatrix} a_1 \\ a_2 \end{bmatrix} e^{\lambda t} = 0.
\tag{12.11}
$$

Eq. (12.11) is just an eigenvalue problem! Since $e^{\lambda t}$ can never be zero, this equation can only be true if the matrix product is zero. If $a_1 \neq 0$ and $a_2 \neq 0$, then the matrix on the left must be *singular.* That means its determinant must be zero, which, in this case, provides the same characteristic equation for the eigenvalues as in Eq. (12.3):

$$
\lambda^2 + 2\zeta\omega_0\lambda + \omega_0^2 = 0.
\tag{12.12}
$$

Thus our solution approach in Section 12.1.1 can be reformulated as an eigenvalue problem for the matrix A. The eigenvalues are the roots of the characteristic equation in Eq. (12.3), and the oscillatory response is identical. What about $[a_1 \ a_2]^T$? This is an eigenvector of A. You may recall from linear algebra that the eigenvectors of a matrix have arbitrary length. For the simple harmonic oscillator in Eq. (12.11), a convenient choice for the two eigenvectors is

$$
\begin{bmatrix} a_1 \\ a_2 \end{bmatrix} = \begin{bmatrix} 1 \\ \lambda_1 \end{bmatrix} \quad \text{and} \quad \begin{bmatrix} 1 \\ \lambda_2 \end{bmatrix},
$$

where λ_1 and λ_2 are the two roots of Eq. (12.12).

The general homogeneous solution is

$$\begin{bmatrix} z(t) \\ \dot{z}(t) \end{bmatrix} = c_1 \begin{bmatrix} 1 \\ \lambda_1 \end{bmatrix} e^{\lambda_1 t} + c_2 \begin{bmatrix} 1 \\ \lambda_2 \end{bmatrix} e^{\lambda_2 t},$$

where, as before, the constants c_1 and c_2 are found from initial conditions.

As stated before, this analysis of simple harmonic motion applies to any linear system. Before we move on to multiple-degree-of-freedom vibrations, it is helpful to look at the general treatment. Suppose we have a single-degree-of-freedom system that is described in first-order form by a set of two linear differential equations. We can write them as the matrix equation

$$\begin{bmatrix} \dot{z} \\ \ddot{z} \end{bmatrix} = A \begin{bmatrix} z \\ \dot{z} \end{bmatrix}. \tag{12.13}$$

Guessing a solution of the same form as Eq. (12.10) and plugging into Eq. (12.13) yields

$$\lambda \begin{bmatrix} a_1 \\ a_2 \end{bmatrix} e^{\lambda t} = A \begin{bmatrix} a_1 \\ a_2 \end{bmatrix} e^{\lambda t},$$

which results in the same eigenvalue problem:

$$(\lambda I - A) \begin{bmatrix} a_1 \\ a_2 \end{bmatrix} e^{\lambda t} = 0.$$

Thus the solution of any single-degree-of-freedom system described by a linear equation of motion is harmonic (a complex exponential) and the natural frequencies are given by the eigenvalues of the matrix A. The next section generalizes this result to systems with many degrees of freedom.

12.1.4 Natural Frequencies and Mode Shapes of Coupled Oscillators

Until now, the only vibration problems we have examined have had a single degree of freedom, like a mass on a spring. Though this is a good model for a remarkable number of systems, it still falls short in many cases. This section expands the discussion to a multiple-degree-of-freedom system. In particular, we look at a model of a collection of *coupled* simple harmonic oscillators. This description is an excellent one for the vibration of a large structure or a continuous material. As you may have learned in statics, many structural elements respond to a small deformation with an internal force that is linearly proportional to the deformation: exactly our model for a linear spring. Thus we can break up these members into many small masses connected to one another by springs (and dampers). Two very simple examples are shown in Figure 12.6. Figure 12.6a shows a model of the axial compression and extension of a vertical beam or tower by a collection of N stacked masses. Figure 12.6b shows a model of the bending of a horizontal beam (or perhaps a bridge) by the displacement of a collection of N masses.

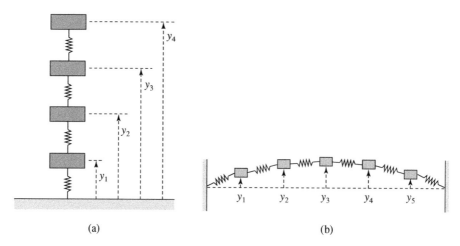

(a)   (b)

Figure 12.6 Two vibration models using coupled oscillators with identical springs and masses. (a) Axial beam model. (b) Bending beam model.

In each of these situations we are still considering one-dimensional motion for simplicity; the tower only moves up and down in the y-direction and the beam-bending model assumes that each mass element only moves vertically up and down in the y-direction. Thus the total number of degrees of freedom is N, the number of masses in the model. Thus in either case, we can write the equation of motion for the ith mass as

$$m_i \ddot{y}_i = -k(y_i - y_{i+1}) - k(y_i - y_{i-1}),$$

which can be simplified to

$$m_i \ddot{y}_i + 2ky_i - k(y_{i+1} + y_{i-1}) = 0.$$

Note that, to simplify things, we have assumed that every spring constant is k and have ignored damping. Breaking each of these equations into first-order form and stacking them together results in the following matrix equation:

$$
\begin{bmatrix}
\dot{y}_1(t) \\
\ddot{y}_1(t) \\
\dot{y}_2(t) \\
\ddot{y}_2(t) \\
\vdots \\
\dot{y}_N(t) \\
\ddot{y}_N(t)
\end{bmatrix}
=
\underbrace{
\begin{bmatrix}
0 & 1 & 0 & 0 & \cdots & 0 & 0 \\
-\frac{2k}{m} & 0 & k & 0 & \cdots & 0 & 0 \\
0 & 0 & 0 & 1 & \cdots & 0 & 0 \\
k & 0 & -\frac{2k}{m} & 0 & \cdots & 0 & 0 \\
\vdots & & & & & & \vdots \\
0 & 0 & 0 & 0 & \cdots & 0 & 1 \\
0 & 0 & 0 & 0 & \cdots & -\frac{2k}{m} & 0
\end{bmatrix}
}_{\triangleq A}
\begin{bmatrix}
y_1(t) \\
\dot{y}_1(t) \\
y_2(t) \\
\dot{y}_2(t) \\
\vdots \\
y_N(t) \\
\dot{y}_N(t)
\end{bmatrix}.
$$

This equation looks just like Eq. (12.13) from the previous section. Nothing we did there presupposed that the system had a single degree of freedom. Thus we still expect our solution to consist of complex exponentials and the natural frequencies should be the eigenvalues of the $2N \times 2N$ matrix A. That is, the $2N$ complex exponentials are the roots of the $2N$th-order polynomial found from

$$|\lambda I - A| = 0.$$

The general homogeneous solution is then

$$\begin{bmatrix} y_1(t) \\ \dot{y}_1(t) \\ \vdots \\ y_N(t) \\ \dot{y}_N(t) \end{bmatrix} = c_1 \begin{bmatrix} a_1 \\ a_2 \\ \vdots \\ a_{2N-1} \\ a_{2N} \end{bmatrix} e^{\lambda_1 t} + c_2 \begin{bmatrix} a_1 \\ a_2 \\ \vdots \\ a_{2N-1} \\ a_{2N} \end{bmatrix} e^{\lambda_2 t} + \cdots + c_{2N} \begin{bmatrix} a_1 \\ a_2 \\ \vdots \\ a_{2N-1} \\ a_{2N} \end{bmatrix} e^{\lambda_{2N} t},$$

where the constants c_i, $i = 1, \ldots, 2N$, are found from initial conditions. Remember that this solution is for any linear system. That is, the trajectory is a sum of weighted complex exponentials, where the natural frequencies are the eigenvalues of A. When the eigenvalues are complex, they will always appear in complex pairs and the corresponding eigenvectors will also be complex conjugates, so that the physical response has no imaginary part. We call each of these pairs the *modes* of the response and the eigenvectors provide the *mode shapes* (Figure 12.7). That is, the eigenvector describes how each coordinate is summed to provide the shape associated with a modal frequency. An example is helpful.

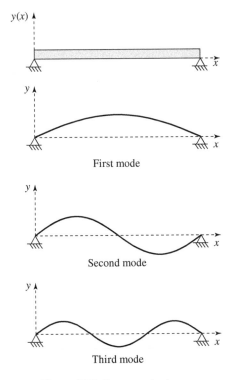

Figure 12.7 Beam mode shapes.

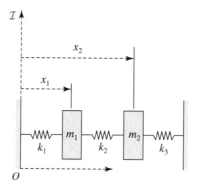

Figure 12.8 Two masses connected by springs between two fixed walls.

Example 12.2 Two Masses Coupled by Springs

Figure 12.8 shows two masses constrained to move in one dimension connected on both sides by springs with spring constants $k_1 = k_2 = k_3 = 1,000$ N/m. The two particles have masses $m_1 = m_2 = 4$ kg. The two coordinates for this two-degree-of-freedom problem are x_1 and x_2, the horizontal displacement of each mass from the origin of the inertial frame. The springs are unstretched when $x_1 = x_{1_0}$ and $x_2 = x_{2_0}$. The two equations of motion are thus

$$\ddot{x}_1 = -\frac{k_1}{m_1}(x_1 - x_{1_0}) - \frac{k_2}{m_1}\left[(x_1 - x_{1_0}) - (x_2 - x_{2_0})\right]$$

$$\ddot{x}_2 = -\frac{k_2}{m_2}\left[(x_2 - x_{2_0}) - (x_1 - x_{1_0})\right] - \frac{k_3}{m_2}(x_2 - x_{2_0}).$$

In first-order matrix form, we have

$$\begin{bmatrix} \dot{x}_1 \\ \ddot{x}_1 \\ \dot{x}_2 \\ \ddot{x}_2 \end{bmatrix} = \begin{bmatrix} 0 & 1 & 0 & 0 \\ -\frac{k_1+k_2}{m_1} & 0 & \frac{k_2}{m_1} & 0 \\ 0 & 0 & 0 & 1 \\ \frac{k_2}{m_2} & 0 & -\frac{k_2+k_3}{m_2} & 0 \end{bmatrix} \begin{bmatrix} x_1 \\ \dot{x}_1 \\ x_2 \\ \dot{x}_2 \end{bmatrix} + \begin{bmatrix} 0 \\ \frac{k_1+k_2}{m_1}x_{1_0} - \frac{k_2}{m_1}x_{2_0} \\ 0 \\ -\frac{k_2}{m_2}x_{1_0} + \frac{k_2+k_3}{m_2}x_{2_0} \end{bmatrix}.$$

The unstretched displacements set the position about which the masses oscillate (they enter the equations of motion like constant forces), so we can ignore them to study the homogeneous response. Plugging in the values for the masses and spring constants shows that the homogeneous solution must satisfy

$$\begin{bmatrix} \dot{x}_1 \\ \ddot{x}_1 \\ \dot{x}_2 \\ \ddot{x}_2 \end{bmatrix} = \begin{bmatrix} 0 & 1 & 0 & 0 \\ -500 & 0 & 250 & 0 \\ 0 & 0 & 0 & 1 \\ 250 & 0 & -500 & 0 \end{bmatrix} \begin{bmatrix} x_1 \\ \dot{x}_1 \\ x_2 \\ \dot{x}_2 \end{bmatrix}.$$

This system has two complex pairs of eigenvalues, $\lambda_{1,2} = \pm 27.4i$ and $\lambda_{3,4} = \pm 15.8i$. The eigenvectors corresponding to the first pair of eigenvalues are

$$\begin{bmatrix} -0.026i \\ 0.71 \\ 0.026i \\ -0.71 \end{bmatrix} \quad \text{and} \quad \begin{bmatrix} 0.026i \\ 0.71 \\ -0.026i \\ -0.71 \end{bmatrix}.$$

The eigenvectors associated with the second pair of eigenvalues are

$$\begin{bmatrix} -0.045i \\ 0.71 \\ -0.045i \\ 0.71 \end{bmatrix} \quad \text{and} \quad \begin{bmatrix} 0.045i \\ 0.71 \\ 0.045i \\ 0.71 \end{bmatrix}.$$

By examining the eigenvectors, we see that this system has two modes. The slow mode (small eigenvalue) consists of the masses moving together either to the right or to the left by the same amount at 15.8 rad/s. The fast (large eigenvalue) mode consists of the masses oscillating in opposite directions at 27.4 rad/s. The general motion, depending on the initial conditions, consists of a weighted sum of these two modes. The eigenvectors also indicate how to put the system into one of its modes. For instance, if we displace both masses together with no initial velocity, they will oscillate in the slow mode. If we displace them in opposite directions by the same amount, they will oscillate in the fast mode.

The discussion in this section gives you just an introduction to vibrations. This is a rich subject with much more to be learned, including the extension to continuous models of structures. It turns out that specific formulas can be found for the modal frequencies for different geometries and boundary conditions of both discrete and continuous models. Nevertheless, with only the material in this book, you are equipped to develop sophisticated models of complex systems and understand their resulting behavior. The most important thing to remember from this discussion is that *any dynamical system that can be described by a set of linear differential equations of motion behaves as a sum of simple harmonic oscillators.* The next section discusses how to find linear models of more complex systems.

12.2 Linearization and the Linearized Dynamics of an Airplane

The last section showed how every linear dynamical system can be solved as the sum of simple harmonic motions. Although that section focused on vibration, the fundamental result—that the modal frequencies are given by the eigenvalues of the equation-of-motion matrix and the modal shapes are given by its eigenvectors— applies to every linear system. This powerful result allows us to learn a great deal about the motion of a dynamical system if we can write its equations of motion in linear form. We have already considered the linear form of many examples in the book, such as the small-angle approximation of the equations of motion of the simple pendulum. This section discusses more generally under what situations it makes sense to linearize nonlinear equations of motion. We then illustrate this process by returning to the equations of motion for the airplane derived in Chapter 11.

12.2.1 Equilibrium

Chapter 1 provided a qualitative definition of the special solution to a differential equation called an *equilibrium point*. This section returns to that idea and treats it more carefully and with more mathematical rigor. Recall that an equilibrium point is simply a solution that does not change over time, like a simple pendulum hanging straight down. Here we study equilibrium solutions for first-order differential equations that are invariant to shifts in the initial time (meaning that the equations of motion are the same no matter when we start the system moving). Since we typically deal with second-order equations of motion, we first review how to transform a scalar second-order ODE into a system of two first-order ODEs. This system is called the *state-space form.*

We start with a second-order ODE in the variable x,

$$\ddot{x} = f(x, \dot{x}), \tag{12.14}$$

where $f(x, \dot{x})$ is a function of x, $\dot{x}$, and possibly some parameters as well. For example, $\ddot{x} = -(g/l) \sin x$ is the equation of motion of a simple pendulum of length l; and $\ddot{x} = -(k/m)(x - x_0) - (b/m)\dot{x}$ is the equation of motion of a mass-spring-damper system with mass m, spring constant k, rest length x_0, and damping constant b.

To convert Eq. (12.14) into state-space form (also called *first-order form* in Appendix C), let $y = [y_1 \ y_2]^T$, where $y_1 = x$ and $y_2 = \dot{x}$. Then Eq. (12.14) becomes

$$\dot{y} = \begin{bmatrix} \dot{y}_1 \\ \dot{y}_2 \end{bmatrix} = \begin{bmatrix} y_2 \\ f(y_1, y_2) \end{bmatrix} \triangleq g(y), \tag{12.15}$$

where $g_1(y) = y_2$ and $g_2(y) = f(y_1, y_2)$.

Definition 12.1 An **equilibrium point** y^* of the system $\dot{y} = g(y)$ satisfies the equation $g(y^*) = 0$.

As a direct consequence of Definition 12.1, we immediately observe that $\dot{y}^* = 0$, which implies that $y^*(t) = y^*(0)$; that is, an equilibrium solution is constant over time. For the example in Eq. (12.15), the equilibrium condition $g(y^*) = 0$ is equivalent to

$$g_1(y^*) = y_2^* = 0$$
$$g_2(y^*) = f(y_1^*, y_2^*) = 0.$$

This example is a special case in which all equilibrium points have $y_2^* = 0$, which is not always true for systems of the form $\dot{y} = g(y)$. Also note that a system of this form can have multiple equilibrium points, as illustrated in the following example.

Example 12.3 Equilibrium Points of the Simple Pendulum

For the simple pendulum, the equation of motion is $\ddot{x} = -(g/l)\sin x$, where x is the pendulum angle measured counterclockwise from down ($x = 0$). In state-space form with $y = [x \ \dot{x}]^T$, we have

$$\dot{y}_1 = y_2$$

$$\dot{y}_2 = -\frac{g}{l}\sin y_1. \qquad (12.16)$$

An equilibrium point y^* must satisfy $y_2^* = 0$ and $\sin y_1^* = 0$. Therefore the points $(0, 0)$, $(\pi, 0)$, and $(-\pi, 0)$ are all equilibrium solutions.

Example 12.4 Equilibrium Points of a Mass-Spring-Damper System

The equation of motion of the mass-spring-damper system is $\ddot{x} = -(k/m)(x - x_0) - (b/m)\dot{x}$. The state-space form using $y = [x \ \dot{x}]^T$ is

$$\dot{y}_1 = y_2 \qquad (12.17)$$

$$\dot{y}_2 = -\frac{k}{m}(y_1 - x_0) - \frac{b}{m}y_2. \qquad (12.18)$$

Thus there is only one equilibrium point at $(x_0, 0)$.

12.2.2 Linearization

The previous examples showed how some systems, like the simple pendulum, can have multiple equilibrium points, while others, like the mass-spring-damper, have only one such point. The mass-spring-damper system is an example of a linear system, which cannot have multiple isolated equilibrium points. The simple pendulum is a nonlinear system because its equations of motion contain a nonlinear term—$\sin x$. We can study the behavior of a nonlinear system near an equilibrium point using the concept of linearization. Just as the term implies, linearization means constructing a linear approximation to the nonlinear equations of motion in the vicinity of the equilibrium point. As long as solutions stay close to the equilibrium point, the behavior of the linear system is usually (but not always!) a good predictor of the behavior of the original nonlinear system. Thus we can use the results from Section 12.1 to describe small motions about the equilibrium point.

Consider again the state-space form in Eq. (12.15) with equilibrium point $y^* = [y_1^* \ y_2^*]^T$. Expanding the right-hand side of the state-space form in a Taylor series about y^* yields (see Appendix A)

$$\dot{y}_1 = y_2$$

$$\dot{y}_2 = \underbrace{f(y_1^*, y_2^*)}_{=0} + a_1(y_1 - y_1^*) + a_2(y_2 - y_2^*) + \ldots,$$

where we have dropped the quadratic and higher terms. The coefficients a_1 and a_2 are

$$a_1 = \left.\frac{\partial f}{\partial y_1}\right|_{y=y^*} \quad \text{and} \quad a_2 = \left.\frac{\partial f}{\partial y_2}\right|_{y=y^*}. \qquad (12.19)$$

Since we are interested in the system behavior near the equilibrium point y^*, we introduce a new variable, $z = y - y^*$. Because $\dot{y}^* = 0$, we have $\dot{z} = \dot{y}$. The linearized equations in terms of z are

$$\dot{z}_1 = z_2$$
$$\dot{z}_2 = a_1 z_1 + a_2 z_2.$$

These equations can be written in matrix form as

$$\dot{z} = Az, \quad A = \begin{bmatrix} 0 & 1 \\ a_1 & a_2 \end{bmatrix}, \tag{12.20}$$

where a_1 and a_2 are given by Eq. (12.19).

The matrix A in Eq. (12.20) is the *Jacobian matrix* of the system evaluated at y^*. Generally speaking, for an N-dimensional system, the state-space equations $\dot{y} = g(y)$ linearized about equilibrium point y^* using $z = y - y^*$ are

$$\boxed{\dot{z} = Az, \quad A = \left.\frac{\partial g}{\partial y}\right|_{y=y^*},} \tag{12.21}$$

where

$$\boxed{\frac{\partial g}{\partial y} = \begin{bmatrix} \dfrac{\partial g_1}{\partial y_1} & \cdots & \dfrac{\partial g_1}{\partial y_N} \\ \vdots & & \vdots \\ \dfrac{\partial g_N}{\partial y_1} & \cdots & \dfrac{\partial g_N}{\partial y_N} \end{bmatrix}.}$$

Eq. (12.21) is exactly the form studied in Section 12.1.4. That means we know how to solve for small motion about the equilibrium! (The motion is harmonic and is described by summing the sinusoidal responses at each of the eigenvalues of A and using the eigenvectors of A.)

Example 12.5 Linearization of the Simple Pendulum

Returning to the pendulum (Example 12.3), we now seek to linearize the equations of motion about the equilibrium points $(0, 0)$ and $(\pm\pi, 0)$. Referring to the state-space form in Eq. (12.16), we have

$$g_1(y) = y_2$$
$$g_2(y) = -\frac{g}{l}\sin y_1.$$

The Jacobian matrix is

$$\frac{\partial g}{\partial y} = \begin{bmatrix} 0 & 1 \\ -\frac{g}{l}\cos\theta & 0 \end{bmatrix}.$$

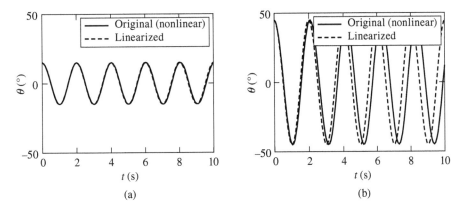

Figure 12.9 Solutions to the simple pendulum. (a) $\theta(0) = 15°$. (b) $\theta(0) = 45°$.

Next we evaluate the Jacobian matrix at each equilibrium point. At $(0, 0)$, A becomes

$$A_{(0,0)} = \left.\frac{\partial g}{\partial y}\right|_{(0,0)} = \begin{bmatrix} 0 & 1 \\ -\frac{g}{l} & 0 \end{bmatrix}, \tag{12.22}$$

and at $(\pm\pi, 0)$ it is

$$A_{(\pm\pi,0)} = \left.\frac{\partial g}{\partial y}\right|_{(\pm\pi,0)} = \begin{bmatrix} 0 & 1 \\ \frac{g}{l} & 0 \end{bmatrix}. \tag{12.23}$$

For each equilibrium point, the linearized equations of motion are $\dot{z} = Az$, where $z = y - y^*$ represents the angular displacement from the corresponding equilibrium point. A comparison between solutions of the original nonlinear system and the linearized system is shown in Figure 12.9. The solution to the linearized system approximates that to the original system well for $\theta(0) = 15°$, whereas the two solutions diverge for $\theta(0) = 45°$.

As a final note on linearization, we briefly illustrate the process of linearizing about a solution other than an equilibrium point. That is, what if we are looking for the behavior of small deviations about a steady-state trajectory that solves the equations of motion? Let $\bar{y}(t)$ be a nonequilibrium solution to $\dot{y} = g(y)$, which means $\dot{\bar{y}}(t) = g(\bar{y})$. Let $z = y - \bar{y}(t)$ be the small motion about the solution $\bar{y}(t)$. We have

$$\dot{z} = \dot{y} - \dot{\bar{y}}(t) = g(z + \bar{y}(t)) - g(\bar{y}) \triangleq h(t, z). \tag{12.24}$$

Thus the time-varying system $\dot{z} = h(t, z)$ has $z^* = 0$ as an equilibrium point. The linearized equations of motion about $z = 0$ are

$$\dot{z} = A(t)z, \quad A(t) = \left.\frac{\partial h}{\partial z}\right|_{z=0}.$$

In many cases, $A(t)$ is a constant matrix, and we can apply the same analysis as in Section 12.1 to solve for the response of small deviations about the trajectory.

In addition to the example below, this approach will be useful when we discuss linearization of the airplane equations of motion in Section 12.2.4.

Example 12.6 Linearizing the Gyropendulum

Recall the solution to the gyropendulum given in Example 11.2. We found a steady-state motion in which the pendulum is inverted and rotating, or precessing, at a constant rate about the vertical. Here we use the methodology discussed above to linearize the equations of motion about that steady-state solution.

The gyropendulum is a two-degree-of-freedom system described by the spherical coordinates θ and ϕ. In first-order form, its state is thus described by

$$y = \begin{bmatrix} \theta \\ \dot{\theta} \\ \phi \\ \dot{\phi} \end{bmatrix}.$$

Example 11.2 found a steady-state solution at constant offset ϕ_0 and a constant precession rate $\dot{\theta}_0$, leaving the nonequilibrium solution $\bar{y}(t)$:

$$\bar{y}(t) = \begin{bmatrix} \dot{\theta}_0 t \\ \dot{\theta}_0 \\ \phi_0 \\ 0 \end{bmatrix}.$$

We use the common notation for small deviations about this steady-state solution, $\delta\theta$, $\delta\dot{\theta}$, $\delta\phi$, and $\delta\dot{\phi}$, so that the deviation state $z = y - \bar{y}$ is

$$z = \begin{bmatrix} \delta\theta \\ \delta\dot{\theta} \\ \delta\phi \\ \delta\dot{\phi} \end{bmatrix}.$$

The nonlinear equations of motion $g(y)$ for the gyropendulum are given in Eqs. (11.5) and (11.6). Using Eq. (12.24), we write them as equations of motion for $z(t)$:

$$\delta\ddot{\theta} + 2(\delta\dot{\theta} + \dot{\theta}_0)\delta\dot{\phi} \cot(\delta\phi + \phi_0) + \frac{h\delta\dot{\phi}}{m_p l^2 \sin(\delta\phi + \phi_0)} = 0$$

$$\delta\ddot{\phi} - (\delta\dot{\theta} + \dot{\theta}_0)^2 \sin(\delta\phi + \phi_0) \cos(\delta\phi + \phi_0) - \frac{g}{l} \sin(\delta\phi + \phi_0)$$

$$+ \frac{h(\delta\dot{\theta} + \dot{\theta}_0) \sin(\delta\phi + \phi_0)}{m_p l^2} = 0,$$

where we used the steady-state condition $g(\bar{y}) = 0$. We now linearize by taking a Taylor series expansion about the equilibrium condition $z = 0$, keeping only first-order terms, to find the linear equations of motion:

$$\delta\ddot{\theta} + \left(2\dot{\theta}_0 \cot\phi_0 + \frac{h}{m_p l^2 \sin\phi_0}\right)\delta\dot{\phi} = 0$$

$$\delta\ddot{\phi} - \left(\dot{\theta}_0^2 \cos 2\phi_0 + \frac{g}{l}\cos\phi_0 - \frac{h\dot{\theta}_0 \cos\phi_0}{m_p l^2}\right)\delta\phi - \left(\dot{\theta}_0 \sin 2\phi_0 - \frac{h\sin\phi_0}{m_p l^2}\right)\delta\dot{\theta} = 0,$$

where we again used the fact that $g(\bar{y}) = 0$. Putting this system in first-order form for z results in the constant matrix

$$A = \begin{bmatrix} 0 & 1 & 0 & 0 \\ 0 & 0 & 0 & -\left(\begin{array}{c}2\dot{\theta}_0 \cot\phi_0 \\ +\frac{h}{m_p l^2 \sin\phi_0}\end{array}\right) \\ 0 & 0 & 0 & 1 \\ 0 & \left(\begin{array}{c}\dot{\theta}_0 \sin 2\phi_0 \\ -\frac{h\sin\phi_0}{m_p l^2}\end{array}\right) & \left(\begin{array}{c}\dot{\theta}_0^2 \cos 2\phi_0 \\ +\frac{g}{l}\cos\phi_0 - \frac{h\dot{\theta}_0 \cos\phi_0}{m_p l^2}\end{array}\right) & 0 \end{bmatrix}.$$

12.2.3 Stability

This section discusses a property of the equilibrium solutions to a dynamical system known as *stability*. In short, if an equilibrium point is stable, then every solution starting near the equilibrium point stays near the equilibrium point. If an equilibrium point is unstable, then it is possible for a solution starting nearby to move away. Characterization of the stability of the equilibrium points of a system helps in understanding the behavior of solutions to the system. The behavior of even nonlinear systems can often be predicted by characterizing the stability of its equilibrium points by linearization.

We start by considering a one-dimensional system $\dot{x} = ax$, $a \neq 0$, which has one equilibrium point at $x^* = 0$. The trajectory $x(t) = e^{at}x(0)$ is a solution to our one-dimensional system, since

$$\frac{d}{dt}\left(e^{at}x(0)\right) = ae^{at}x(0) = ax.$$

Next consider the limiting behavior of the solution $x(t)$ as a function of the sign of the parameter a. If $a < 0$, then the quantity e^{at} shrinks exponentially as time increases. Consequently, regardless of the initial condition $x(0)$, the solution $x(t)$ tends toward $x = 0$ as t increases. Thus $x^* = 0$ is (asymptotically) stable if $a < 0$.[1] However, if $a > 0$, then the quantity e^{at} grows exponentially with time and the solution $x(t)$ tends toward $\pm\infty$ as t increases. Thus $x^* = 0$ is unstable if $a > 0$. (Note that if $a = 0$, then every point on the real line is an equilibrium solution.)

Now consider a two-dimensional linear system:

$$\dot{x}_1 = ax_1$$

$$\dot{x}_2 = bx_2.$$

---

[1] For a nonlinear system, we often distinguish between a stable equilibrium point and an *asymptotically stable* equilibrium point: every solution starting near a stable equilibrium point stays nearby, whereas every solution near an asymptotically stable equilibrium point stays nearby and eventually converges to the equilibrium point.

As long as $a \neq 0$ and $b \neq 0$, this system has a single equilibrium point at the origin $(0, 0)$. In fact, it represents two uncoupled one-dimensional systems, since $\dot{x}_1$ does not depend on x_2 and vice versa. Then from the previous discussion, the solution is $x_1(t) = e^{at}x_1(0)$ and $x_2(t) = e^{bt}x_2(0)$. Furthermore, if $a < 0$ and $b < 0$, then the origin is stable; if $a > 0$ or $b > 0$ then the origin is unstable. (If either $a = 0$ or $b = 0$, but not both, then there is a continuous line of equilibrium points along the corresponding axis.)

What if we have a coupled two-dimensional system? Using $y = [x_1 \ x_2]^T$, the solution to our uncoupled system can also be written as

$$y(t) = e^{at}y_1(0) \begin{bmatrix} 1 \\ 0 \end{bmatrix} + e^{bt}y_2(0) \begin{bmatrix} 0 \\ 1 \end{bmatrix},$$

and the system dynamics are $\dot{y} = Ay$, where

$$A = \begin{bmatrix} a & 0 \\ 0 & b \end{bmatrix}.$$

The significance of writing the solution and the dynamics this way is that a and b are the eigenvalues of A and $[1 \ 0]^T$ and $[0 \ 1]^T$ are the corresponding eigenvectors. Recall from Section 12.1 that the solution to the system $\dot{y} = Ay$ can be written as

$$y(t) = e^{\lambda_1 t}c_1 v_1 + e^{\lambda_2 t}c_2 v_2,$$

where λ_1 and λ_2 are the eigenvalues of A with corresponding eigenvectors v_1 and v_2. The constants c_1 and c_2 can be determined from the initial conditions $y(0)$ and $\dot{y}(0)$:

$$y(0) = c_1 v_2 + c_2 v_2$$
$$\dot{y}(0) = \lambda_1 c_1 v_1 + \lambda_2 c_2 v_2.$$

So now we can characterize the stability of the origin as an equilibrium point of the linear system $\dot{y} = Ay$ by using the eigenvalues of A. If $\lambda_1 < 0$ and $\lambda_2 < 0$ then $e^{\lambda_1 t}$ and $e^{\lambda_2 t}$ shrink exponentially as t increases. Consequently, the solution $y(t)$ tends toward the origin, regardless of the initial conditions and eigenvectors; in this case the origin is stable. If $\lambda_1 > 0$ or $\lambda_2 > 0$, then $e^{\lambda_1 t}$ or $e^{\lambda_2 t}$ grows exponentially as t increases and the origin is unstable. (If either $\lambda_1 = 0$ or $\lambda_2 = 0$, then we have a continuous line of equilibrium points along the corresponding eigenvector.)

It is also possible for eigenvalues to be complex numbers, that is, to have an imaginary component. In this case, the eigenvalue inequalities are applied to the real part of each eigenvalue. To see why, let $\lambda = \alpha \pm i\beta$ be a pair of complex eigenvalues, where $i = \sqrt{-1}$ is the imaginary unit. (Complex eigenvalues of real matrices always come in pairs.) Then $e^{\lambda t} = e^{\alpha t}e^{\pm i\beta t}$. Thus if $\alpha < 0$ ($\alpha > 0$), then the term $e^{\alpha t}$ shrinks (grows) exponentially as time increases, whereas the magnitude of the term $e^{\pm i\beta t}$ is equal to one for all time.

The characterization of the stability of an equilibrium point using eigenvalue analysis applies to a nonlinear system. Recall that the linearization about an equilibrium of a nonlinear system yields a linear system describing motion near the equilibrium point. If every eigenvalue of the linearized system has a negative real part, then the equilibrium point is asymptotically stable. If any of the eigenvalues has a positive real part, then the equilibrium point is unstable. This analysis really shows the power of linearization as a tool for understanding the behavior of a nonlinear system. However, although the explanation is beyond our scope, it is possible to conclude asymptotic stability of an equilibrium point of a nonlinear system only when the linearized system has no eigenvalues with zero real part.

Example 12.7 Stability of Equilibrium Points of the Simple Pendulum

The simple pendulum has equilibrium points at $(0, 0)$ and $(\pm\pi, 0)$. The Jacobian matrices evaluated at each equilibrium point are given in Eqs. (12.22) and (12.23). The eigenvalues of $A_{(0,0)}$ are thus the solution of

$$\begin{vmatrix} -\lambda & 1 \\ -\frac{g}{l} & -\lambda \end{vmatrix} = \lambda^2 + \frac{g}{l} = 0.$$

The eigenvalues are $\lambda_{1,2} = \pm\sqrt{-g/l} = \pm i\sqrt{g/l}$. Both eigenvalues have zero real parts, so we cannot predict the behavior of the nonlinear system about the equilibrium point using linearization. Nevertheless, since $(0, 0)$ is the "down" equilibrium point, we know intuitively that it is stable—we just can't conclude this from the linearization. As shown earlier, the linear approximation is actually a good approximation to the motion near $(0, 0)$.

However, the eigenvalues of $A_{(\pm\pi,0)}$ are the solutions to

$$\begin{vmatrix} -\lambda & 1 \\ \frac{g}{l} & -\lambda \end{vmatrix} = \lambda^2 - \frac{g}{l} = 0,$$

which are $\lambda_1 = -\sqrt{g/l} < 0$ and $\lambda_2 = \sqrt{g/l} > 0$. Consequently, both "up" equilibrium points $(\pm\pi, 0)$ are unstable, which agrees with intuition. A solution starting near the "up" equilibrium of the linearized pendulum is shown in Figure 12.10a.

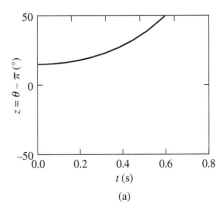
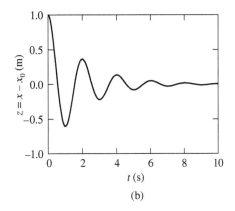

Figure 12.10 Stability of equilibrium points. (a) Starting near the "up" equilibrium of the simple pendulum. (b) Mass-spring-damper system.

Example 12.8 Stability of the Equilibrium Point
of the Mass-Spring-Damper System

Recall that the only equilibrium point of the mass-spring-damper system is the origin $(0, 0)$, and the system is already linear with the state-space equations given by Eqs. (12.17) and (12.18). The eigenvalues of $A = \frac{\partial g}{\partial y}$ are the solutions to

$$\begin{vmatrix} -\lambda & 1 \\ -\frac{k}{m} & \frac{-b}{m} - \lambda \end{vmatrix} = \lambda^2 + \frac{b}{m}\lambda + \frac{k}{m} = 0.$$

Using the quadratic formula, the eigenvalues are

$$\lambda_1 = \frac{-b + \sqrt{b^2 - 4mk}}{2m} \quad \text{and} \quad \lambda_1 = \frac{-b - \sqrt{b^2 - 4mk}}{2m}.$$

Although these eigenvalues can be real or complex, the real part is always negative for both (check this!), which implies that the origin is stable. In the original coordinates, then, the spring length will converge to its rest length x_0. A solution starting near the origin is shown in Figure 12.10b.

12.2.4 The Linearized Equations of Motion
of an Airplane—Straight and Level Flight

As an illustration of using linearization to solve for small motion about a steady-state trajectory, we study the motion of an airplane in straight and level flight. By this we mean that the velocity is nominally constant and parallel to the ground and the airplane holds a fixed orientation in absolute space. Chapters 10 and 11 discussed the kinematics and kinetics of an airplane in both translation and rotation. As a reminder, Figure 12.11a shows the geometry and relevant frames of reference. The equations of motion presented in Example 11.11 are nonlinear. In this section we linearize them about straight and level flight and show that the resulting linear equations of motion describe rather interesting and well-known trajectories.

Straight and level flight has the following conditions: (a) the angular-velocity components are all zero (i.e., $p_0 = q_0 = r_0 = 0$) since the aircraft is not rotating; (b) the aircraft is slightly pitched but otherwise level (i.e., $\theta_0 = $ constant and $\psi_0 = \phi_0 = 0$); (c) the velocity is level (perpendicular to gravity) and in the plane of symmetry of the plane; and (d) the airplane has a nonzero angle of attack (i.e., $v_0 = 0$ and u_0 and w_0 are constant). (The angle of attack is the angle between the velocity and the forward axis $\mathbf{b}_1$.) The equilibrium angle of attack is

$$\alpha_0 = \arctan\left(\frac{w_0}{u_0}\right).$$

The perturbed state variables $y = z + \overline{y}(t)$ are

$$[\, x \quad y \quad z_0 + \delta z \quad u_0 + \delta u \quad \delta v \quad w_0 + \delta w \quad \delta \psi \quad \theta_0 + \delta\theta \quad \delta\phi \quad \delta p \quad \delta q \quad \delta r \,]^T,$$

where the δ in front of each state variable represents the small perturbation from the nominal trajectory, what was called $z(t)$ in Section 12.2.2. Following the procedure outlined there, we substitute these perturbed states into the twelve equations of motion

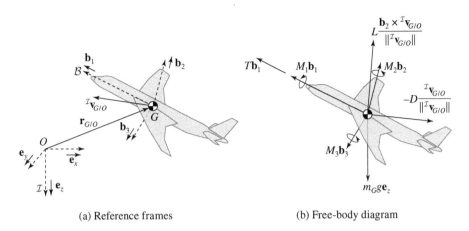

(a) Reference frames   (b) Free-body diagram

Figure 12.11 Airplane in flight.

in Example 11.11, perform a Taylor series expansion, and keep only first-order terms. The equations of motion for the airplane linearized about straight and level flight are

$$\delta \dot{p} = \frac{I_3 M_1 + I_{13} M_3}{I_1 I_3 - I_{13}^2} \tag{12.25}$$

$$\delta \dot{q} = \frac{M_2}{I_2} \tag{12.26}$$

$$\delta \dot{r} = \frac{I_{13} M_1 + I_1 M_3}{I_1 I_3 - I_{13}^2} \tag{12.27}$$

$$\delta \dot{\psi} = \frac{\delta r}{\cos \theta_0} \tag{12.28}$$

$$\delta \dot{\theta} = \delta q \tag{12.29}$$

$$\delta \dot{\phi} = \frac{\delta r}{\cos \theta_0} + \delta p \tag{12.30}$$

$$\delta \dot{u} = -\delta q w_0 + \frac{F_1}{m_G} \tag{12.31}$$

$$\delta \dot{v} = \delta p w_0 - \delta r u_0 + \frac{F_2}{m_G} \tag{12.32}$$

$$\delta \dot{w} = \delta q u_0 + \frac{F_3}{m_G} \tag{12.33}$$

$$\delta \dot{x} = \delta u \cos \theta_0 + \delta w \sin \theta_0 + \delta \theta (w_0 \cos \theta_0 - u_0 \sin \theta_0) \tag{12.34}$$

$$\delta \dot{y} = \delta \psi (u_0 \cos \theta_0 + w_0 \sin \theta_0) + \delta v - \delta \phi w_0 \tag{12.35}$$

$$\delta \dot{z} = -\delta u \sin \theta_0 + \delta w \cos \theta_0 - \delta \theta (u_0 \cos \theta_0 + w_0 \sin \theta_0), \tag{12.36}$$

where we used $^\mathcal{I} C^\mathcal{B}$ for a 3-2-1 Euler transformation (see Example 10.8).

We have not yet discussed the specific form of the forces and moments on an airplane or used the free-body diagram. Nevertheless, a close examination of these equations of motion shows that an interesting separation has occurred, resulting from the symmetry of the aircraft about the vertical plane. *The longitudinal and lateral motion of the aircraft can be considered separately.* Longitudinal refers to motion occurring only in the plane of symmetry of the aircraft (pitch, forward motion, and altitude); it is represented by the state variables x, z, u, w, q, and θ. Examination of the six equations for perturbations of these variables shows that they are completely independent of the other state variables and depend only on forces and moments in the longitudinal directions (F_1, F_3, and M_2). The remaining variables represent lateral motion of the plane (sideslip, roll, and yaw) represented by the variables y, v, p, r, ψ, and ϕ. In the remainder of this section we separately discuss the linearized trajectories for the longitudinal and lateral motions.

Longitudinal Equations of Motion

From Eqs. (12.25)–(12.36) the longitudinal equations of motion are

$$\delta \dot{q} = \frac{M_2}{I_2} \tag{12.37}$$

$$\delta \dot{\theta} = \delta q \tag{12.38}$$

$$\delta \dot{u} = -\delta q w_0 + \frac{F_1}{m_G} \tag{12.39}$$

$$\delta \dot{w} = \delta q u_0 + \frac{F_3}{m_G} \tag{12.40}$$

$$\delta \dot{x} = \delta u \cos \theta_0 + \delta w \sin \theta_0 + \delta \theta (w_0 \cos \theta_0 - u_0 \sin \theta_0) \tag{12.41}$$

$$\delta \dot{z} = -\delta u \sin \theta_0 + \delta w \cos \theta_0 - \delta \theta (u_0 \cos \theta_0 + w_0 \sin \theta_0), \tag{12.42}$$

with longitudinal forces and moments,

$$F_1 = T - m_G g (\sin \theta_0 + \delta \theta \cos \theta_0) - D \frac{u_0 + \delta u}{V} + L \frac{w_0 + \delta w}{V} + u_1$$

$$F_3 = m_G g (\cos \theta_0 - \delta \theta \sin \theta_0) - D \frac{w_0 + \delta w}{V} - L \frac{u_0 + \delta u}{V} + u_2$$

$$M_2 = u_3,$$

where T is the magnitude of the thrust force on the aircraft, L is the magnitude of the lift force (perpendicular to the flight path); D is the magnitude of the drag force (parallel to the flight path); V is the equilibrium speed, $V = \sqrt{u_0^2 + w_0^2}$, and u_1, u_2, and u_3 are control forces (throttle, elevator, and ailerons, respectively) (see Figure 12.11b).

Remember that the body axes are used to formulate the equations of motion because the aerodynamic forces and moments on the plane are functions of the body components of velocity and orientation. At the beginning of this section we discussed how to linearize a system of equations such as these by taking a Taylor series expansion of the nonlinear dependencies. Thus we write the lift, drag, and moment as a first-order Taylor series, assuming $q_0 = 0$:

$$L = L(u_0, w_0, \theta_0) + \frac{\partial L}{\partial u}\delta u + \frac{\partial L}{\partial w}\delta w + \frac{\partial L}{\partial q}\delta q + \frac{\partial L}{\partial \theta}\delta\theta$$

$$D = D(u_0, w_0, \theta_0) + \frac{\partial D}{\partial u}\delta u + \frac{\partial D}{\partial w}\delta w + \frac{\partial D}{\partial q}\delta q + \frac{\partial D}{\partial \theta}\delta\theta$$

$$M_2 = M_2(u_0, w_0, \theta_0) + \frac{\partial M_2}{\partial u}\delta u + \frac{\partial M_2}{\partial w}\delta w + \frac{\partial M_2}{\partial q}\delta q + \frac{\partial M_2}{\partial \theta}\delta\theta.$$

Note that the aerodynamic forces and moments don't usually depend on the absolute position (x and z) of the aircraft. The partial derivatives multiplying each perturbed variable are called *stability derivatives* and are typically written with a subscripted capital C. So, for example,

$$\frac{\partial L}{\partial u} = C_{L_u}.$$

It is beyond our scope to discuss the aerodynamic origins of the various stability derivatives and their values. Often the stability derivatives are not calculated but measured directly from wind-tunnel tests.

Because we are assuming a steady-state condition of straight and level flight (with a nonzero angle of attack), the nominal lift, drag, and moment balance the other forces from the free-body diagram. That is, the thrust must equal the component of gravity in the $\mathbf{b}_1$ direction plus the lift and drag in that direction. Likewise, the lift in the $\mathbf{e}_z$ direction must cancel gravity. Also, if the airplane is not nominally rotating, the total aerodynamic moment must be zero. These considerations allow us to re-write the forces and moments as linear functions of the small state variables:[2]

$$F_1 = \left(C_{L_\theta}\frac{w_0}{V} - C_{D_\theta}\frac{u_0}{V} - m_G\cos\theta_0\right)\delta\theta + \left(C_{L_u}w_0 - C_{D_u}u_0 - D_0\right)\frac{\delta u}{V}$$

$$+ \left(C_{L_w}w_0 - C_{D_w}u_0 + L_0\right)\frac{\delta w}{V} + \left(C_{L_q}\frac{w_0}{V} - C_{D_q}\frac{u_0}{V}\right)\delta q + u_1 \qquad (12.43)$$

$$F_3 = \left(C_{L_\theta}\frac{u_0}{V} - C_{D_\theta}\frac{w_0}{V} - m_G\sin\theta_0\right)\delta\theta + \left(C_{L_u}u_0 - C_{D_u}w_0 + L_0\right)\frac{\delta u}{V}$$

$$+ \left(C_{L_w}u_0 - C_{D_w}w_0 - D_0\right)\frac{\delta w}{V} + \left(C_{L_q}\frac{u_0}{V} - C_{D_q}\frac{w_0}{V}\right)\delta q + u_2 \qquad (12.44)$$

$$M_2 = C_{M_{2,\theta}}\delta\theta + C_{M_{2,u}}\delta u + C_{M_{2,w}}\delta w + C_{M_{2,q}}\delta q + u_3. \qquad (12.45)$$

Eqs. (12.43)–(12.45) are then substituted back into the equations of motion in Eqs. (12.37)–(12.42). Normally, since there is no force and moment dependence on

---

[2] Note that in many aircraft, there is also an F_1 and M_1 dependence on the upward acceleration $\delta\dot{w}$. We have dropped that here to simplify the presentation.

position (x and z), these states are dropped from the final linear equations of motion,

$$
\begin{bmatrix} \delta \dot{u} \\ \delta \dot{w} \\ \delta \dot{q} \\ \delta \dot{\theta} \end{bmatrix} = A_{\text{long}} \begin{bmatrix} \delta u \\ \delta w \\ \delta q \\ \delta \theta \end{bmatrix} + B_{\text{long}} \begin{bmatrix} u_1 \\ u_2 \\ u_3 \end{bmatrix},
$$

where A_{long} is a 4×4 matrix containing the coefficients from the equations of motion in Eqs. (12.37)–(12.42) and the stability derivatives from Eqs. (12.43)–(12.45). B_{long} is a 4×3 matrix that determines the linear response of the three longitudinal control inputs. The natural (uncontrolled) solution to this system of equations is the same as that in Section 12.1.4: a weighted sum of complex exponential responses, where the frequencies are given by the eigenvalues of A_{long} and the mode shapes are given by the eigenvectors. The natural motion of the airplane is stable as long as the real part of every eigenvalue is less than zero. Most airplanes, and all commercial jets and small general aviation aircraft, have stable natural responses. Some fighter jets, though, might be unstable to make them as responsive as possible. (These, of course, need autopilots to keep them from crashing.)

For stable aircraft, the longitudinal response consists of two oscillatory modes (two complex-conjugate pairs of roots). One is a rapid, quickly damped oscillation with a period of several seconds known as the *short-period mode*. It consists primarily of changes in vertical speed, pitch rate, and pitch angle. The second is a slowly damped long-period oscillation with period of roughly 30 s known as the *phugoid mode*. This mode consists primarily of changes in forward speed and pitch angle. In fact, it is possible to use the control forces (u_1, u_2, or u_3) to excite one or the other of these two modes. All commercial aircraft have autopilots that rapidly damp out the phugoid oscillation to increase passenger comfort.

Example 12.9 The Longitudinal Motion of a Typical Business Jet

As an example of the linearized longitudinal motion of an airplane, we consider the simple business jet described in Stengel (2004). The straight and level longitudinal motion of the jet is described by the linear equations of motion

$$
\begin{bmatrix} \delta \dot{u}/V \\ \delta \dot{w}/V \\ \delta \dot{q} \\ \delta \dot{\theta} \end{bmatrix} = \begin{bmatrix} -0.0121 & 0.096 & -0.0632 & -0.096 \\ -0.116 & -1.277 & 0.9882 & -0.0061 \\ 0.51 & -7.966 & -1.279 & 0 \\ 0 & 0 & 1 & 0 \end{bmatrix} \begin{bmatrix} \delta u/V \\ \delta w/V \\ \delta q \\ \delta \theta \end{bmatrix},
$$

where we have normalized the forward and vertical speeds by the equilibrium speed V, which in this case is 102 m/s. This normalization makes all state variables dimensionless.

This system has eigenvalues $\omega_{\text{sp}} = -1.277 \pm 2.81i$ and $\omega_{\text{ph}} = -0.0074 \pm 0.1256i$. Thus the short-period oscillation has a period of roughly 2 s and a damping constant of 0.4. The phugoid mode has a period of roughly 49 s and a damping constant of 0.054.

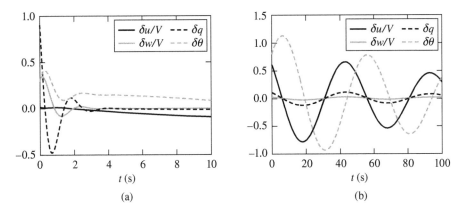

Figure 12.12 Modes of longitudinal motion of a typical business jet. Note the difference in time scales. (a) Short-period mode. (b) Phugoid mode.

The normalized eigenvector magnitudes for the short-period and phugoid modes, respectively, are

$$\begin{bmatrix} 0.015 \\ 0.317 \\ 0.902 \\ 0.2923 \end{bmatrix} \quad \text{and} \quad \begin{bmatrix} 0.603 \\ 0.023 \\ 0.099 \\ 0.791 \end{bmatrix}.$$

The short-period mode is dominated by sinusoidal motion in $\delta w/V$, δq, and $\delta\theta$. The phugoid mode is dominated by sinusoidal motion in $\delta u/V$ and $\delta\theta$. Figure 12.12 shows graphs of the four longitudinal states over time for initial conditions that excite each of the two modes. Note the large difference in time scales between the short-period and phugoid modes.

Lateral Equations of Motion

We can perform a similar analysis for the linearized lateral equations of motion in terms of the variables $\delta v, \delta p, \delta r, \delta\psi, \delta\phi$, and δy. We have, from Eqs. (12.25)–(12.36),

$$\delta\dot{v} = \delta p w_0 - \delta r u_0 + \frac{F_2}{m_G} \tag{12.46}$$

$$\delta\dot{p} = \frac{I_3 M_1 + I_{13} M_3}{I_1 I_3 - I_{13}^2} \tag{12.47}$$

$$\delta\dot{r} = \frac{I_{13} M_1 + I_1 M_3}{I_1 I_3 - I_{13}^2} \tag{12.48}$$

$$\delta\dot{\psi} = \frac{\delta r}{\cos\theta_0} \tag{12.49}$$

$$\delta\dot{\phi} = \frac{\delta r}{\cos\theta_0} + \delta p \tag{12.50}$$

$$\delta\dot{y} = \delta\psi(u_0\cos\theta_0 + w_0\sin\theta_0) + \delta v - \delta\phi w_0. \tag{12.51}$$

In this case, because of the symmetry of the aircraft, there are no nominal forces or moments in the perturbed lateral direction. In terms of stability derivatives, the perturbed lateral forces and moments on the aircraft are

$$F_2 = C_{F_{2,v}}\delta v + C_{F_{2,p}}\delta p + C_{F_{2,r}}\delta r + C_{F_{2,\psi}}\delta\psi + C_{F_{2,\phi}}\delta\phi$$

$$M_1 = C_{M_{1,v}}\delta v + C_{M_{1,p}}\delta p + C_{M_{1,r}}\delta r + C_{M_{1,\psi}}\delta\psi + C_{M_{1,\phi}}\delta\phi \tag{12.52}$$

$$M_3 = C_{M_{3,v}}\delta v + C_{M_{3,p}}\delta p + C_{M_{3,r}}\delta r + C_{M_{3,\psi}}\delta\psi + C_{M_{3,\phi}}\delta\phi.$$

Again, the stability derivatives are usually found from wind tunnel tests or sophisticated fluid dynamics modeling.

These expressions are substituted into the equations of motion (Eqs. (12.46)–(12.51)) to find the six-dimensional linear equations of motion for the lateral direction:

$$\begin{bmatrix} \delta\dot{v} \\ \delta\dot{p} \\ \delta\dot{r} \\ \delta\dot{\psi} \\ \delta\dot{\phi} \\ \delta\dot{y} \end{bmatrix} = A_{\mathrm{LD}} \begin{bmatrix} \delta v \\ \delta p \\ \delta r \\ \delta\psi \\ \delta\phi \\ \delta y \end{bmatrix} + B_{\mathrm{LD}} \begin{bmatrix} u_4 \\ u_5 \\ u_6 \end{bmatrix},$$

where A_{LD} is a 6×6 matrix containing the coefficients from the equations of motion in Eqs. (12.46)–(12.51) and the stability derivatives from Eq. (12.52). B_{LD} is a 6×3 matrix that determines the linear response of the three lateral control inputs.

As in the longitudinal case, the lateral motion consists of a number of different modes. In this case there are six modes consisting of four real eigenvalues (two of which are zero) and one pair of complex-conjugate eigenvalues. The slow, lightly damped oscillatory mode is known as the *Dutch roll* mode (see Figure 12.13). It is dominated by motion in δv, δp, δr, and $\delta\phi$. In it, the airplane slowly glides back and forth as it oscillates in roll and yaw. The four real roots correspond to the *spiral, crossrange, yaw,* and *roll* modes. The spiral mode is dominated by δv and $\delta\phi$ and is actually unstable. However, the time constant is on the order of minutes, making the mode easy to control by the pilot (or an autopilot). The roll mode is dominated by δv, δp, and $\delta\phi$ and represents a rapid damping of roll disturbances. The crossrange and yaw modes are dominated by δy and $\delta\psi$ and are both neutrally stable (i.e., they represent pure integrations). These, too, are easily controlled by the pilot.

Example 12.10 The Lateral Motion of a Typical Business Jet

As in Example 12.9, consider the lateral motion of the simple business jet described in Stengel (2004). Because in most planes the crossrange and yaw motion are neutral

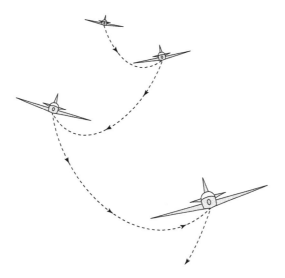

Figure 12.13 Dutch roll.

and decouple from the other modes, we usually drop them from the linearized model, leaving the fourth-order equations of motion:

$$
\begin{bmatrix} \delta\dot{v}/V \\ \delta\dot{p} \\ \delta\dot{r} \\ \delta\dot{\phi} \end{bmatrix} = \begin{bmatrix} -0.1567 & 0.0632 & -0.998 & 0.096 \\ -2.5194 & -1.1767 & 0.1823 & 0 \\ 1.7442 & -0.0112 & -0.0927 & 0 \\ 0 & 1 & 0.0634 & 0 \end{bmatrix} \begin{bmatrix} \delta v/V \\ \delta p \\ \delta r \\ \delta\phi \end{bmatrix},
$$

where, again, the sideways velocity has been normalized by the equilibrium velocity v_0.

This system has eigenvalues $\omega_{\mathrm{DR}} = -0.1158 \pm 1.39i$, $\omega_{\mathrm{spiral}} = -1.203$, and $\omega_{\mathrm{roll}} = -0.009$. Thus the Dutch roll oscillation has a period of roughly 4.5 s and a damping constant of 0.08. The spiral mode has an unstable time constant of roughly 112 s, and the roll mode has a time constant of 0.83 s.

The normalized eigenvector magnitudes for the Dutch roll, roll, and spiral modes, respectively, are

$$
\begin{bmatrix} 0.419 \\ 0.608 \\ 0.529 \\ 0.419 \end{bmatrix}, \begin{bmatrix} 0.008 \\ 0.769 \\ 0.005 \\ 0.639 \end{bmatrix}, \text{ and } \begin{bmatrix} 0.0056 \\ 0.0028 \\ 0.095 \\ 0.996 \end{bmatrix}.
$$

The Dutch roll mode contains all the states, δv, δp, δr, and $\delta\phi$. The roll mode is dominated by the roll angle and roll rate, δp and $\delta\phi$, and the spiral mode is almost entirely the roll angle $\delta\phi$. Figure 12.14 shows plots of the four lateral states over time in the three dominant modes for initial conditions of arbitrary magnitude.

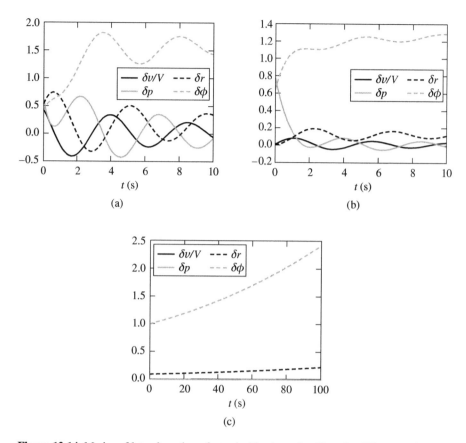

Figure 12.14 Modes of lateral motion of a typical business jet. Note the differences in time scales. (a) Dutch roll mode. (b) Roll mode. (c) Spiral mode.

12.3 Impacts of Finite-Sized Particles

This section returns to the discussion of collisions started in Chapter 6. In Section 6.2 we considered collisions between two infinitesimally small particles. Here we allow the colliding objects to have finite size, though they are still spherical in shape.

12.3.1 Determining the Collision Frame

This section explores more carefully the process of finding the collision frame and also considers the case of unequal-sized objects. The results in Section 6.2 show how to find the final velocities of the two particles in terms of their velocities just before the collision, assuming that we know the collision frame $\mathcal{C} = (O', \mathbf{e}_n, \mathbf{e}_t, \mathbf{e}_3)$. The origin O', which is the point of impact, is used only to determine the particle trajectories after the collision. However, we need to know the unit vectors $\mathbf{e}_t$ and $\mathbf{e}_n$, which determine the orientation of the collision frame. This is because, to use Eqs. (6.45) and (6.46), we must express the initial velocities as components in frame $\mathcal{C}$.

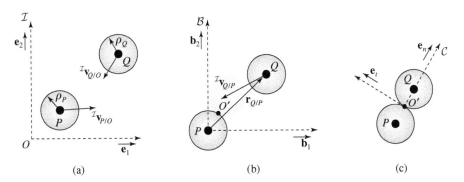

Figure 12.15 Constant-velocity collision frames. (a) Frame $\mathcal{I}$. (b) Frame $\mathcal{B}$. (c) Frame $\mathcal{C}$.

Recall that $\mathbf{e}_3 = \mathbf{e}_1 \times \mathbf{e}_2$ is known, and $\mathbf{e}_t = \mathbf{e}_3 \times \mathbf{e}_n$ can be calculated once we find $\mathbf{e}_n$. Therefore, all we need to determine frame $\mathcal{C}$ is the point of impact O' and the normal unit vector $\mathbf{e}_n$. In general, we need the particle absolute positions at impact, $\mathbf{r}_{P/O}(t_1)$ and $\mathbf{r}_{P/O}(t_2)$, to compute O', and we need the relative position $\mathbf{r}_{Q/P}(t_1)$ to compute $\mathbf{e}_n$. Thus we need to know the trajectories $\mathbf{r}_{P/O}(t)$ and $\mathbf{r}_{Q/O}(t)$ up to the impact time t_1. Let $\rho_P \geq 0$ and $\rho_Q \geq 0$ denote the radius of particles P and Q, respectively. The impact time t_1 is the first time t for which the condition $\|\mathbf{r}_{P/Q}(t)\| = \rho_P + \rho_Q$ is satisfied (under the assumption that the particles do indeed collide). One method to find O' and $\mathbf{e}_n$ is to numerically integrate the equations of motion of P and Q up to the impact time to find $\mathbf{r}_{P/O}(t_1)$ and $\mathbf{r}_{Q/O}(t_1)$. Sometimes we can find $\mathbf{e}_n$ and/or O' using analytical methods. Either way, once we have the positions at t_1, we can find the collision frame $\mathcal{C}$ and the final velocities using the methods of Chapter 6. We demonstrate this procedure in the following example.

Example 12.11 The Constant-Velocity Collision Frame

This example examines the collision between two spherical particles P and Q, as shown in Figure 12.15. (This model is excellent for describing the collisions between two billiard balls.) We assume here that there are no external forces acting on either particle before, during, or after the collision, which implies that the particle velocities are constant before and after the collision. We compute the collision frame $\mathcal{C}$ from the particle initial positions and initial velocities. Using the collision frame and the coefficient of restitution of the collision of particles P and Q, we can compute the final velocities after the collision.

Let $\rho_P \geq 0$ and $\rho_Q \geq 0$ denote the particle radii. Let $\mathcal{I} = (O, \mathbf{e}_1, \mathbf{e}_2, \mathbf{e}_3)$ denote an inertial frame and $\mathcal{B} = (P, \mathbf{b}_1, \mathbf{b}_2, \mathbf{b}_3)$ denote a body frame fixed to particle P. Since P moves at constant velocity and is not spinning, frame $\mathcal{B}$ is an inertial frame. We have

$$\mathbf{r}_{Q/O} = \mathbf{r}_{P/O} + \mathbf{r}_{Q/P},$$

which implies

$$\mathbf{r}_{Q/P} = \mathbf{r}_{Q/O} - \mathbf{r}_{P/O}$$

$${}^{\mathcal{I}}\mathbf{v}_{Q/P}(t) = {}^{\mathcal{I}}\mathbf{v}_{Q/O}(0) - {}^{\mathcal{I}}\mathbf{v}_{P/O}(0).$$

Let x and y denote Cartesian coordinates in frame $\mathcal{B}$, $(x, y)_{\mathcal{B}}$. We observe that

$${}^{\mathcal{I}}\mathbf{v}_{Q/P}(t) = \dot{x}_Q(0)\mathbf{b}_1 + \dot{y}_Q(0)\mathbf{b}_2$$

$$\mathbf{r}_{Q/P}(t) = (\dot{x}_Q(0)t + x_Q(0))\mathbf{b}_1 + (\dot{y}_Q(0)t + y_Q(0))\mathbf{b}_2.$$

The particles collide at the smallest time t_1 that satisfies $\|\mathbf{r}_{Q/P}(t_1)\| = \rho_A + \rho_B$. This observation leads us to the quadratic equation

$$(\dot{x}_Q(0)t_1 + x_Q(0))^2 + (\dot{y}_Q(0)t_1 + y_Q(0))^2 - (\rho_A + \rho_B)^2 = 0$$

$$\underbrace{(\dot{x}_Q(0)^2 + \dot{y}_Q(0)^2)}_{=A} t_1^2 + \underbrace{2(\dot{x}_Q(0)x_Q(0) + \dot{y}_Q(0)y_Q(0))}_{=B} t_1 +$$

$$\underbrace{x_Q(0)^2 + y_Q(0)^2 - (\rho_A + \rho_B)^2}_{=C} = 0,$$

which has the solution

$$t_1 = \frac{-A \pm \sqrt{B^2 - 4AC}}{2A}. \tag{12.53}$$

Note that if $B^2 - 4AC < 0$, then t_1 is imaginary and particles P and Q never collide.

Using the time of impact, we are now able to find frame $\mathcal{C}$. The normal vector $\mathbf{e}_n$ is the (normalized) position of particle Q relative to P at t_1, defined here as components in frame $\mathcal{B}$:

$$\mathbf{e}_n = \hat{\mathbf{r}}_{Q/P}(t_1) = \frac{(\dot{x}_Q(0)t_1 + x_Q(0))\mathbf{b}_1 + (\dot{y}_Q(0)t_1 + y_Q(0))\mathbf{b}_2}{\|(\dot{x}_Q(0)t_1 + x_Q(0))\mathbf{b}_1 + (\dot{y}_Q(0)t_1 + y_Q(0))\mathbf{b}_2\|},$$

where t_1 is given by Eq. (12.53). The point of impact O' is along the line between P and Q at time t_1. The position of O' in frame B is

$$\mathbf{r}_{O'/P} = \frac{\rho_P}{\rho_P + \rho_Q}\mathbf{r}_{Q/P}(t_1) = \frac{\rho_P}{\rho_P + \rho_Q}\left[(\dot{x}_Q(0)t_1 + x_Q(0))\mathbf{b}_1 + (\dot{y}_Q(0)t_1 + y_Q(0))\mathbf{b}_2\right],$$

where t_1 is given by Eq. (12.53). We find the tangent vector using $\mathbf{e}_t = \mathbf{e}_3 \times \mathbf{e}_n$.

If the equations of motion are not easily integrated, then we may have to resort to numerical integration to find the point of impact O'. In either case, knowledge of the two trajectories is required to find the relative position at the time of impact. There are some special cases, however, where we can define the impact frame knowing only the particle velocities (in particular, when the size of the bodies goes to zero). We explore these various categories of impacts next.

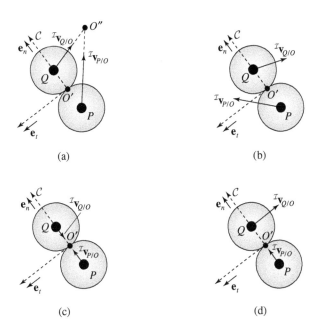

Figure 12.16 Center-hit and center-miss collisions. (a) Center-hit, oblique. (b) Center-miss, oblique. (c) Center-hit, head-on. (d) Center-miss, side-on.

12.3.2 Oblique, Head-On, and Side-On Collisions

Every collision between two spherical particles can be categorized as one of two types: a *center-hit* collision or a *center-miss* collision (see Figure 12.16). If a collision between two particles would have occurred even if the particles were arbitrarily small, then it is a center-hit collision. Otherwise, it is a center-miss collision. These concepts are formalized in the following definition.

Definition 12.2 Let P and Q be spherical particles with radii $\rho_P \geq 0$ and $\rho_Q \geq 0$, respectively. Assume the particles collide at time t_1, which means $\|\mathbf{r}_{P/Q}(t)\| > \rho_P + \rho_Q$ for $t < t_1$ and $\|\mathbf{r}_{P/Q}(t_1)\| = \rho_P + \rho_Q$. Let P' and Q' be arbitrarily small spherical particles initialized at the centers of P and Q at impact, that is, $\mathbf{r}_{P'/O}(t_1) = \mathbf{r}_{P/O}(t_1)$ and $\mathbf{r}_{Q'/O}(t_1) = \mathbf{r}_{Q/O}(t_1)$. Set $\mathbf{v}_{P'/O}(t) = \mathbf{v}_{P/O}(t_1)$ and $\mathbf{v}_{Q'/O}(t) = \mathbf{v}_{Q/O}(t_1)$ for $t > t_1$. If the virtual particles P' and Q' collide at some time $t'_1 > t_1$ then the collision between P and Q is a **center-hit**. Otherwise, it is a **center-miss** collision.

Every collision between two arbitrarily small particles is a center-hit collision. Some finite-sized particle collisions are center-hit collisions even if they would not have occurred for arbitrarily small particles because Definition 12.2 includes collisions that occur in the presence of external forces. In this case, propagating arbitrarily small particles along the original trajectories is not sufficient to determine whether a collision is a center-hit collision. (In the language of Definition 12.2, this insufficiency is because the arbitrarily small particles P' and Q' may not collide if their velocities are not held constant from t_1 to t'_1, even though they may collide if their velocities are held constant.)

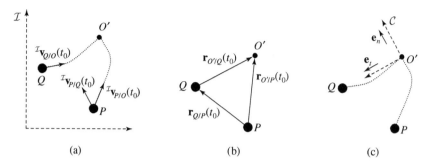

Figure 12.17 Determining the collision frame $\mathcal{C}$ for a center-hit collision between two arbitrarily small particles P and Q. The collision occurs at $t_1 > t_0$, where $\Delta t = t_1 - t_0$ is very small. (a) Particle trajectories. (b) Relative positions. (c) Collision frame.

There are two special cases of center-hit collisions and one special case of center-miss collisions for which we can identify the normal vector $\mathbf{e}_n$. (Collisions that are not one of the special cases are called *oblique*.) The first special case of center-hit collisions is the set of head-on collisions for which $\mathbf{e}_n = \pm\hat{\mathbf{v}}_{P/O}(t_1)$ (see Figure 12.16c). A special case of center-miss collisions is the set of side-on collisions for which $\mathbf{e}_n = \hat{\mathbf{v}}_{P/O}(t_1)$ or $\mathbf{e}_n = -\hat{\mathbf{v}}_{Q/O}(t_1)$ (see Figure 12.16d).

The second special case of center-hit collisions for which we can identify the normal vector $\mathbf{e}_n$ is the set of collisions of arbitrarily small particles, discussed in Section 6.2. In this case, the normal vector depends on the relative velocity at impact. To see this, consider Figure 12.17, which illustrates two arbitrarily small particles P and Q at time t_0 just before impact at O'. Assume time t_1 is the time of impact and $\Delta t = t_1 - t_0$ is small. We seek to find $\mathbf{e}_n$ and validate Eq. (6.25) in the presence of external forces on P and Q.

We observe from Figure 12.17b that

$$\mathbf{r}_{Q/P}(t_0) + \mathbf{r}_{O'/Q}(t_0) = \mathbf{r}_{O'/P}(t_0),$$

which means

$$\mathbf{r}_{Q/P}(t_0) = \mathbf{r}_{O'/P}(t_0) - \mathbf{r}_{O'/Q}(t_0), \qquad (12.54)$$

where

$$\mathbf{r}_{O'/i}(t_0) \triangleq \mathbf{r}_{i/O}(t_1) - \mathbf{r}_{i/O}(t_0), \quad i = P, Q.$$

Integrating Newton's second law, $\mathbf{F}_i = m_i {}^{\mathcal{I}}\mathbf{a}_{i/O}$ with respect to time from t_0 to t_1 yields

$$ {}^{\mathcal{I}}\mathbf{v}_{i/O}(t_1) = {}^{\mathcal{I}}\mathbf{v}_{i/O}(t_0) + \frac{1}{m_i}\int_{t_0}^{t_1} \mathbf{F}_i(\tau)d\tau $$

$$ = {}^{\mathcal{I}}\mathbf{v}_{i/O}(t_0) + \frac{1}{m_i}\underbrace{\overline{\mathbf{F}}_i(t_0, t_1)}_{O(\Delta t)}, $$

which implies ${}^{\mathcal{I}}\mathbf{v}_{i/O}$ is correct to within $O(\Delta t)$ during this interval.

Integrating the definition of the velocity, $^{\mathcal{I}}\mathbf{v}_{i/O} = \frac{^{\mathcal{I}}d}{dt}\mathbf{r}_{i/O}$, yields

$$\mathbf{r}_{i/O}(t_1) - \mathbf{r}_{i/O}(t_0) = \int_{t_0}^{t_1} {}^{\mathcal{I}}\mathbf{v}_{i/O}(\tau)d\tau$$

$$= {}^{\mathcal{I}}\mathbf{v}_{i/O}(t_0)\Delta t + \frac{1}{m_i}\underbrace{\int_{t_0}^{t_1}\overline{\mathbf{F}}_i(t_0, t_1)d\tau}_{O(\Delta t^2)}.$$

Returning to (12.54), we find

$$\mathbf{r}_{Q/P}(t_0) = {}^{\mathcal{I}}\mathbf{v}_{P/O}(t_0)\Delta t - {}^{\mathcal{I}}\mathbf{v}_{Q/O}(t_0)\Delta t + O(\Delta t^2)$$

$$= -{}^{\mathcal{I}}\mathbf{v}_{Q/P}(t_0)\Delta t + O(\Delta t^2),$$

which implies

$$\hat{\mathbf{r}}_{Q/P}(t_0) = -\frac{{}^{\mathcal{I}}\mathbf{v}_{Q/P}(t_0)}{\|{}^{\mathcal{I}}\mathbf{v}_{Q/P}(t_0) + O(\Delta t)\|} + O(\Delta t^2)$$

$$= -\frac{{}^{\mathcal{I}}\mathbf{v}_{Q/P}(t_1)}{\|{}^{\mathcal{I}}\mathbf{v}_{Q/P}(t_0)\|} + O(\Delta t).$$

Thus $\hat{\mathbf{r}}_{Q/P}(t_0)$ is parallel to $-{}^{\mathcal{I}}\mathbf{v}_{Q/P}(t_1)$ to within $O(\Delta t)$, and

$$\boxed{\mathbf{e}_n = \lim_{\Delta t \to 0} \hat{\mathbf{r}}_{Q/P}(t_1 - \Delta t) \approx {}^{\mathcal{I}}\hat{\mathbf{v}}_{P/Q}(t_1).} \qquad (12.55)$$

In the absence of external forces, the expression for $\mathbf{e}_n$ in Eq. (12.55) is exact.

Example 12.12 Bouncy Impact with a Pendulum

This example considers the same setting as Example 6.4; that is, we consider a moving particle Q colliding with a particle P that is attached to a pendulum of length l. The initial velocity of particle Q is known; particle P is initially at rest. Example 6.3 considered the case when P and Q stick together; here we assume that P and Q bounce apart, which implies the coefficient of restitution of the collision satisfies $e > 0$. We take to be true all of the assumptions of Section 6.2.1 other than Assumption 6.2. This assumption is violated by the presence of the force of gravity and the tension in the pendulum rod. The pendulum rod also constrains the motion of P. By carefully modeling the collision and the constraints, we find the velocities of P and Q after the collision.

Figure 12.18a shows the coordinates that describe the positions of P and Q. We use three distinct reference frames, as shown in Figure 12.18b: the inertial frame $\mathcal{I} = (O, \mathbf{e}_x, \mathbf{e}_y, \mathbf{e}_z)$, the body frame $\mathcal{B} = (O, \mathbf{e}_r, \mathbf{e}_\theta, \mathbf{e}_z)$, and the collision frame

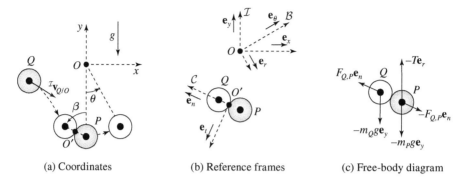

(a) Coordinates   (b) Reference frames   (c) Free-body diagram

Figure 12.18 Bouncy impact with a pendulum. (a) Before the collision, both particles are shaded gray. During the collision, particle Q is white and P is gray. After the collision, particle Q is not shown and P is white. (b) The orientation of the collision frame is determined by the angle of the line of impact between the particle centers. For equally sized particles, the origin of the collision frame is the midpoint of the line of impact. (c) Free-body diagram during the collision. The free-body diagrams for before and after the collision contain the tension and gravity forces but not the interaction force.

$\mathcal{C} = (O', \mathbf{e}_n, \mathbf{e}_t, \mathbf{e}_z)$. The orientation of the body frame is described by angle θ (the pendulum angle), and the orientation of the collision frame is described by angle β. Angle β is determined by the line of impact between the centers of the two particles; we assume that the line of impact is known and that $\beta \neq 0$.[3] The three reference frames are related by the transformations

$$
\begin{array}{c|cc}
 & \mathbf{e}_x & \mathbf{e}_y \\
\hline
\mathbf{e}_r & \sin\theta & -\cos\theta \\
\mathbf{e}_\theta & \cos\theta & \sin\theta \\
\hline
\mathbf{e}_n & -\sin\beta & \cos\beta \\
\mathbf{e}_t & -\cos\beta & -\sin\beta
\end{array}
\tag{12.56}
$$

and

$$
\begin{array}{c|cc}
 & \mathbf{e}_n & \mathbf{e}_t \\
\hline
\mathbf{e}_r & -\cos(\theta - \beta) & -\sin(\theta - \beta) \\
\mathbf{e}_\theta & \sin(\theta - \beta) & -\cos(\theta - \beta)
\end{array}
\tag{12.57}
$$

We can simplify the transformation in Eq. (12.57) for use during the collision. Let t_1 and t_2 denote the start and end times of the collision, where $\Delta t = t_2 - t_1$ is very small. By the fundamental theorem of calculus (see Appendix A), we have

$$
\theta(t_2) - \underbrace{\theta(t_1)}_{=0} = \int_{t_1}^{t_2} \dot{\theta}(\tau)d\tau = O(\Delta t).
$$

---

[3] $\beta = 0$ corresponds to Q hitting exactly the top of P, in which case P would not move.

Thus during the collision, $\theta = O(\Delta t)$; that is, θ stays very small. Using a Taylor series expansion (see Appendix A), we find $\sin \theta = \theta + O(\theta^3) = O(\Delta t)$ and $\cos \theta = 1 + O(\theta^2) = 1 + O(\Delta t^2)$. Consequently, during the collision, Eq. (12.57) can be approximated to within $O(\Delta t)$ by

	$\mathbf{e}_n$	$\mathbf{e}_t$
$\mathbf{e}_r$	$-\cos \beta$	$\sin \beta$
$\mathbf{e}_\theta$	$-\sin \beta$	$-\cos \beta$.

The kinematics of the two particles are

$$\mathbf{r}_{P/O} = l\mathbf{e}_r \tag{12.58}$$

$$^{\mathcal{I}}\mathbf{v}_{P/O} = l\dot\theta\mathbf{e}_\theta \tag{12.59}$$

$$^{\mathcal{I}}\mathbf{a}_{P/O} = l\ddot\theta\mathbf{e}_\theta - l\dot\theta^2\mathbf{e}_r \tag{12.60}$$

and

$$\mathbf{r}_{Q/O} = x_Q\mathbf{e}_x + y_Q\mathbf{e}_y$$

$$^{\mathcal{I}}\mathbf{v}_{Q/O} = \dot{x}_Q\mathbf{e}_x + \dot{y}_Q\mathbf{e}_y$$

$$^{\mathcal{I}}\mathbf{a}_{Q/O} = \ddot{x}_Q\mathbf{e}_x + \ddot{y}_Q\mathbf{e}_y.$$

The dynamics of the two particles before and after the collision are

$$-T\mathbf{e}_r - m_P g\mathbf{e}_y = m_P(l\ddot\theta\mathbf{e}_\theta - l\dot\theta^2\mathbf{e}_r)$$

$$-m_Q g\mathbf{e}_y = m_Q(\ddot{x}_Q\mathbf{e}_x + \ddot{y}_Q\mathbf{e}_y),$$

which yield the equations of motion:

$$\ddot\theta = -\frac{g}{l}\sin\theta \tag{12.61}$$

$$\ddot{x}_Q = 0 \tag{12.62}$$

$$\ddot{y}_Q = -g. \tag{12.63}$$

Recall that our goal is to determine the velocities of P and Q after the collision. Using the post-collision velocities (and positions) of P and Q as initial conditions, we can integrate the equations of motion (Eqs. (12.61)–(12.63)). Eq. (12.59) shows that the velocity of P is in the $\mathbf{e}_\theta$ direction; that is, particle P is subject to the motion constraint

$$^{\mathcal{I}}\mathbf{v}_{P/O} \cdot \mathbf{e}_r = 0. \tag{12.64}$$

By differentiating the constraint Eq. (12.64) and using Eq. (12.59), we find that

$$\frac{^{\mathcal{I}}d}{dt}\left(^{\mathcal{I}}\mathbf{v}_{P/O}\right) \cdot \mathbf{e}_r + \underbrace{^{\mathcal{I}}\mathbf{v}_{P/O} \cdot (\dot\theta\mathbf{e}_r)}_{=0} = \,^{\mathcal{I}}\mathbf{a}_{P/O} \cdot \mathbf{e}_r = 0. \tag{12.65}$$

Particle P still obeys the motion constraint Eq. (12.65) during the collision. Using Figure 12.18c, Eq. (12.65), and $\theta = O(\Delta t)$, the dynamics of P during the collision are

$$m_P{}^{\mathcal{I}}\mathbf{a}_{P/O} = -T\mathbf{e}_r - m_P g \mathbf{e}_y - F_{Q,P}\mathbf{e}_n$$

$$= \underbrace{(-T + m_P g + F_{Q,P}\cos\beta)}_{=0}\,\mathbf{e}_r + F_{Q,P}\sin\beta\mathbf{e}_\theta + O(\Delta t).\text{(12.66)}$$

Eq. (12.66) implies that the non-constraint force acting on P during the collision is

$$F_{Q,P}\sin\beta\mathbf{e}_\theta = -F_{Q,P}\sin\beta\cos\beta\mathbf{e}_t - F_{Q,P}\sin^2\beta\mathbf{e}_n + O(\Delta t). \quad (12.67)$$

Using Figure 12.18c, the force acting on Q during the collision is

$$F_{Q,P}\mathbf{e}_n - m_Q g\mathbf{e}_y = m_Q g\sin\beta\mathbf{e}_t + (F_{Q,P} - m_Q g\cos\beta)\mathbf{e}_n. \quad (12.68)$$

Take another look at Eq. (12.67). Quite remarkably, it states that, during the collision, there is an interaction force between P and Q in the tangential direction! Doesn't this violate our no-friction assumption? Actually, no. There is no such tangential interaction force on Q in Eq. (12.68). The tangential force on P during the collision comes from the tension and gravity forces (see Figure 12.18c). We used the motion constraint on P to express the tangential force in terms of the interaction force. Writing the total forces on P this way simplifies the following calculations.

We are now in a position to solve the collision problem using Newton's second law in its impulse form. Let u and v denote the tangential and normal speeds in the collision frame. Using Eqs. (12.67) and (12.68) and the transformation table Eq. (12.56) results in four scalar equations:

$$m_P u_P(t_2) = -\overline{F}_{Q,P}(t_1, t_2)\sin\beta\cos\beta + O(\Delta t) \qquad (12.69)$$

$$m_P v_P(t_2) = -\overline{F}_{Q,P}(t_1, t_2)\sin^2\beta + O(\Delta t) \qquad (12.70)$$

$$m_Q u_Q(t_2) = m_Q u_Q(t_1) + \underbrace{m_Q g\sin\beta\Delta t}_{O(\Delta t)} \qquad (12.71)$$

$$m_Q v_Q(t_2) = m_Q v_Q(t_1) - \underbrace{m_Q g\cos\beta\Delta t}_{O(\Delta t)} + \overline{F}_{P,Q}(t_1, t_2). \qquad (12.72)$$

Using Eq. (12.71), we observe that the tangential speed of Q is conserved to within $O(\Delta t)$; that is,

$$u_Q(t_2) = u_Q(t_1) + O(\Delta t). \qquad (12.73)$$

Furthermore, the tangential and normal speeds of P at time t_2 are related by

$$\frac{u_P(t_2)}{v_P(t_2)} = \cot\beta + O(\Delta t). \qquad (12.74)$$

In fact, Eq. (12.74) is simply an expression of the motion constraint Eq. (12.64) during the collision. To see this, we write Eq. (12.64) in the collision frame using Eq. (12.57):

$$(u_P \mathbf{e}_t + v_P \mathbf{e}_n) \cdot (-\sin(\theta - \beta)\mathbf{e}_t - \cos(\theta - \beta)\mathbf{e}_n)$$
$$= -u_P \sin(\theta - \beta) - v_P \cos(\theta - \beta) = 0,$$

which agrees with Eq. (12.74). Eq. (12.74) allows us to compute $u_P(t_2)$ from $v_P(t_2)$ using the orientation angle β.

Next we seek to find the final normal speeds $v_P(t_2)$ and $u_P(t_2)$ using Eqs. (12.70) and (12.72). To eliminate the interaction force, we use the coefficient of restitution e. Let t_c denote the time of maximum compression, which separates the deformation and restitution phases. By considering separately the deformation and restitution phases of the collision, Eqs. (12.70) and (12.72) become

$$m_P v_P(t_c) = -\overline{F}_{Q,P}(t_1, t_c) \sin^2 \beta + O(\Delta t)$$

$$m_P v_P(t_2) = m_P v_P(t_c) - \overline{F}_{Q,P}(t_c, t_2) \sin^2 \beta + O(\Delta t)$$

$$m_Q v_Q(t_c) = m_Q v_Q(t_1) + \overline{F}_{Q,P}(t_1, t_c) + O(\Delta t)$$

$$m_Q v_Q(t_2) = m_Q v_Q(t_c) + \overline{F}_{Q,P}(t_c, t_2) + O(\Delta t).$$

According to Definition 6.3, the coefficient of restitution is

$$e = \frac{\overline{F}_{Q,P}(t_c, t_2)}{\overline{F}_{Q,P}(t_1, t_c)} = \frac{v_P(t_c) - v_P(t_2) + O(\Delta t)}{-v_P(t_c) + O(\Delta t)} = \frac{v_Q(t_c) - v_Q(t_2) + O(\Delta t)}{v_Q(t_1) - v_Q(t_c) + O(\Delta t)}. \quad (12.75)$$

As in Section 6.2.1, we assume that, at the time of maximum compression, $v_Q(t_c) = v_P(t_c) = v_{PQ}(t_c)$. Consequently, after cross-multiplying in Eq. (12.75), we find

$$v_{PQ}(t_c) - v_P(t_2) = -e v_{PQ}(t_c) + O(\Delta t) \quad (12.76)$$

$$v_{PQ}(t_c) - v_Q(t_2) = e(v_Q(t_1) - v_{PQ}(t_c)) + O(\Delta t). \quad (12.77)$$

Subtracting Eq. (12.77) from Eq. (12.76) yields

$$v_Q(t_2) - v_P(t_2) = -e v_Q(t_1) + O(\Delta t). \quad (12.78)$$

Eq. (12.78) represents our first equation in the two unknowns $v_P(t_2)$ and $v_Q(t_2)$. If there were no external forces, then our second equation would come from conservation of total linear momentum. But total linear momentum is not conserved, since there are external forces (tension and gravity). To obtain a second equation in the two unknowns, we solve Eq. (12.70) for $\overline{F}_{P,Q}(t_1, t_2) = -m_P v_P(t_2) \csc^2 \beta + O(\Delta t)$ and substitute it into Eq. (12.72), which gives

$$m_Q v_Q(t_2) = m_Q v_Q(t_1) - m_P v_P(t_2) \csc^2 \beta + O(\Delta t). \quad (12.79)$$

Eq. (12.79) represents our second equation in the two unknowns.

We are now ready to solve for $v_P(t_2)$, $v_Q(t_2)$, and $u_P(t_2)$. Using Eqs. (12.78) and (12.79) yields

$$v_P(t_2) = \frac{(1+e)m_Q}{m_Q + m_P \csc^2 \beta} v_Q(t_1) + O(\Delta t) \tag{12.80}$$

$$v_Q(t_2) = \frac{m_Q \sin^2 \beta - e m_P}{m_Q \sin^2 \beta + m_P} v_Q(t_1) + O(\Delta t). \tag{12.81}$$

From Eq. (12.74), we have

$$u_P(t_2) = \frac{(1+e)m_Q \sin \beta \cos \beta}{m_Q \sin^2 \beta + m_P} v_Q(t_1) + O(\Delta t). \tag{12.82}$$

Using Eq. (12.73) and Eqs. (12.80)–(12.82) determines the post-collision velocities. Taking the limit $\Delta t \to 0$, we have

$$^{\mathcal{I}}\mathbf{v}_P(t_2) = \frac{(1+e)m_Q \sin \beta \cos \beta}{m_Q \sin^2 \beta + m_P} v_Q(t_1)\mathbf{e}_t + \frac{(1+e)m_Q}{m_Q + m_P \csc^2 \beta} v_Q(t_1)\mathbf{e}_n$$

$$\tag{12.83}$$

$$^{\mathcal{I}}\mathbf{v}_Q(t_2) = u_Q(t_1)\mathbf{e}_t + \frac{m_Q \sin^2 \beta - e m_P}{m_Q \sin^2 \beta + m_P} v_Q(t_1)\mathbf{e}_n.$$

These expressions complete the goal of this example. As a sanity check of these findings, evaluate Eq. (12.83) with a simple test case. For example, what if $\beta = \pi/2$, $e = 1$, and $m_P = m_Q$? These values correspond to a one-dimensional (horizontal) collision between two identical, perfectly elastic particles. In this case, the tension and gravity forces are tangential to the collision, and the normal (horizontal) component of total linear momentum should be conserved. Eq. (12.83) with $\beta = \pi/2$ and $e = 1$ yields

$$^{\mathcal{I}}\mathbf{v}_P(t_2) = v_Q(t_1)\mathbf{e}_n$$

$$^{\mathcal{I}}\mathbf{v}_Q(t_2) = u_Q(t_1)\mathbf{e}_t,$$

which implies

$$^{\mathcal{I}}\mathbf{v}_P(t_2) + {}^{\mathcal{I}}\mathbf{v}_Q(t_2) = u_Q(t_1)\mathbf{e}_t + v_Q(t_1)\mathbf{e}_n = {}^{\mathcal{I}}\mathbf{v}_Q(t_1),$$

as expected.

12.4 Key Ideas

- The equation of motion of the **unforced harmonic oscillator** is

$$\ddot{z} + 2\zeta\omega_0\dot{z} + \omega_0^2 z = 0,$$

where ω_0 is the natural frequency and ζ is the damping ratio. The solution with $\zeta = 0$ is **undamped**, $0 < \zeta < 1$ is **underdamped**, $\zeta = 1$ is **critically damped**, and $\zeta > 1$ is **overdamped**.

- The equation of motion of the **forced harmonic oscillator** subject to the (scalar) force F_P is

$$\ddot{z} + 2\zeta\omega_0\dot{z} + \omega_0^2 z = \frac{F_P(t)}{m_P}.$$

- **Unforced harmonic motion** can be described by the **linear system** $\dot{z} = Az$, where $z = [x \ \dot{x}]^T$ and

$$A = \begin{bmatrix} 0 & 1 \\ -\omega_0^2 & -2\zeta\omega_0 \end{bmatrix}.$$

- An **equilibrium point** y^* of the system $\dot{y} = g(y)$ satisfies the equation $g(y^*) = 0$.

- The behavior of solutions to the nonlinear equation $\dot{y} = g(y)$ can be approximated near the equilibrium point y^* by the **linearized equation** $\dot{z} = Az$, where $z = y - y^*$ is an $N \times 1$ matrix and

$$A = \begin{bmatrix} \frac{\partial g_1}{\partial y_1} & \cdots & \frac{\partial g_1}{\partial y_N} \\ \vdots & & \vdots \\ \frac{\partial g_N}{\partial y_1} & \cdots & \frac{\partial g_N}{\partial y_N} \end{bmatrix}\Bigg|_{y=y^*}$$

is the $N \times N$ **Jacobian** matrix evaluated at y^*.

- The **stability** of an equilibrium point y^* is characterized by the eigenvalues of the Jacobian matrix evaluated at y^*: if all eigenvalues have negative real parts, then the equilibrium point is **asymptotically stable** (solutions starting near y^* converge to y^*); if any eigenvalue has a positive real part, then the equilibrium point is **unstable** (solutions starting near y^* may diverge from y^*).

- Every collision between two spherical particles can be categorized as one of two types: if the collision would have occurred even if the particles were arbitrarily small, then it is a **center-hit** collision; otherwise, it is a **center-miss** collision.

12.5 Notes and Further Reading

Our goal in this chapter was to introduce a few examples of more advanced dynamics that show how the basic tools are applied. We also highlight some important conceptual ideas that you may use in more advanced courses and in practice. There are, of course, many other fascinating and interesting applications of the basic equations presented. Section 12.1 gave only the most introductory presentation of vibrations; the importance of this topic lies in the ubiquity of simple harmonic motion and the usefulness of linear analysis. There are many good texts on advanced vibration theory and continuum mechanics; a good comprehensive discussion is Meirovitch (2002). Our discussion of linearization and stability is also just an introduction to a central topic of dynamical and nonlinear systems. Two excellent advanced texts on the subject are Strogatz (2001) and Khalil (2002). There are also a number of texts available to explore airplane dynamics and control in more depth, including Anderson (1999), Nelson (1998), Pamadi (1998), and Stengel (2004). We used Stengel (2004) for our examples.

CHAPTER THIRTEEN

An Introduction to Analytical Mechanics

This final chapter is intended to act as a bridge to an advanced dynamics course. We introduce two alternative approaches to finding equations of motion: Lagrange's method and Kane's method. These approaches are among the techniques often referred to as *analytical mechanics*. In contrast to the vector-based approach taken throughout most of the book, these methods rely on a scalar description of a dynamical system. In other words, we work only with the scalar coordinates and formulate equations of motion directly in terms of them. The vector formulation has not disappeared completely; it just enters into the problem in more subtle ways. The elegance of this approach is that it simplifies the kinematics and eliminates the need to include constraint forces. Often the algebra required is also substantially reduced.

Analytical approaches to dynamics form the foundation of modern physics. In Lagrange's method, energy replaces acceleration as the starting point and the equations of motion are found directly from the kinetic and potential energies; it is particularly useful for conservative systems. In Kane's method, constraint forces are eliminated by "projecting" Newton's second law onto the constraint surfaces so that we solve only for the unconstrained degrees of freedom in the problem. Projection methods have the appeal that they are easily automated: there is software available that can find the equations of motion for almost any complex system.

Introducing you to these other approaches is not meant to replace the vector formulation of Newtonian mechanics. They are just additional tools in the arsenal of techniques for understanding dynamics. This chapter is only an introduction to the rich field of advanced mechanics, which we hope you will continue on to study. The art of engineering is not only being able to apply a particular problem-solving method but also being able to choose the best method for the problem at hand.

13.1 Generalized Coordinates

We begin by considering again the kinematics of a particle. It may help to review Chapters 1–3, where we discussed vectors, reference frames, and coordinates. The vector $\mathbf{r}_{P/O}$ locates particle P relative to origin O. A reference frame provides a point of view for referencing the position and motion of a particle. Chapters 1 and 3 showed that we can define a reference frame by three orthogonal unit vectors and the origin. Using the properties of vector algebra, we then introduced the concept of components of a vector:

$$\mathbf{r}_{P/O} = a_1\mathbf{a}_1 + a_2\mathbf{a}_2 + a_3\mathbf{a}_3.$$

In other words, we can write any vector as the sum of three component vectors in each of the three unit-vector directions of a reference frame. Given a reference frame, then, the scalar magnitudes of these components entirely determine the location of point P in that frame. This is, of course, our goal. We seek to describe the position of P as a function of time, which means finding three scalar functions of time: $a_1(t)$, $a_2(t)$, and $a_3(t)$. Likewise, from the definition of the vector derivative, we have an expression for the velocity of P in $\mathcal{A}$:

$$^{\mathcal{A}}\mathbf{v}_{P/O} = \dot{a}_1(t)\mathbf{a}_1 + \dot{a}_2(t)\mathbf{a}_2 + \dot{a}_3(t)\mathbf{a}_3.$$

We also defined a coordinate system in Chapter 1 as the set of scalars that locate a point relative to another point (the origin) of a reference frame. Thus the scalar magnitudes in the above equations constitute coordinates. In fact, they are just Cartesian coordinates, which we normally denote $(x, y, z)_{\mathcal{A}}$. This is what makes Cartesian coordinates so special in dynamics: they form the only coordinate system that has a one-to-one correspondence with the component magnitudes of the position. Nevertheless, Cartesian coordinates are not always the most convenient.

Much of the book has been devoted to understanding how to solve dynamics problems using non-Cartesian coordinate systems. In particular, we introduced polar and path coordinates for planar problems and cylindrical and spherical coordinates for three-dimensional problems. Many problems are more easily solved, or better understood, using one of these coordinate systems. This finding is particularly true in constrained problems, such as the simple pendulum, where a single scalar coordinate can be used to describe the only degree of freedom. The price we pay is that the vector-component magnitudes are complicated functions of the coordinates. This complexity makes finding the vector derivative in terms of the rate of change of the coordinates more involved, often needing a good amount of algebra. One way we simplified the finding of derivatives was by introducing new frames of reference and using the angular velocity.

Now we generalize this idea. The alternative coordinate systems introduced earlier (polar, path, cylindrical, and spherical) are just some of the infinite number of possible scalar coordinates one might use to locate particle P. For a single particle with three degrees of freedom, we can use any three independent scalar coordinates. In analytical mechanics we refer to such a coordinate as a *generalized coordinate* and label it $q_i(t)$, $i = 1, 2, 3$. In other words, we write the generic coordinates to locate particle P in $\mathcal{A}$ as $(q_1, q_2, q_3)_{\mathcal{A}}$, where the q_i's may be cylindrical coordinates, spherical coordinates, or any other set of three scalars that works nicely for the problem at hand. The position is then written in its most general form as

$$\mathbf{r}_{P/O}(t) = a_1(q_1, q_2, q_3, t)\mathbf{a}_1 + a_2(q_1, q_2, q_3, t)\mathbf{a}_2 + a_3(q_1, q_2, q_3, t)\mathbf{a}_3.$$

So, for instance, Eq. (10.4) is an example of writing the position $\mathbf{r}_{P/O}$ of a point in terms of spherical coordinates, where $q_1 = r$, $q_2 = \theta$, and $q_3 = \phi$. Note that the vector-component magnitudes can be nonlinear functions of the generalized coordinates.

We can also write a general expression for the velocity using the definition of the vector derivative and the chain rule:

$$\mathcal{A}\mathbf{v}_{P/O} = \left(\sum_{i=1}^{3} \frac{\partial a_1}{\partial q_i} \frac{dq_i}{dt} + \frac{\partial a_1}{\partial t} \right) \mathbf{a}_1 + \left(\sum_{i=1}^{3} \frac{\partial a_2}{\partial q_i} \frac{dq_i}{dt} + \frac{\partial a_2}{\partial t} \right) \mathbf{a}_2$$

$$+ \left(\sum_{i=1}^{3} \frac{\partial a_3}{\partial q_i} \frac{dq_i}{dt} + \frac{\partial a_3}{\partial t} \right) \mathbf{a}_3.$$

This approach is the same one used to find the velocity in polar coordinates in Eq. (3.18), in cylindrical coordinates in Eq. (10.2), and in spherical coordinates in Eq. (10.5). If there are multiple particles, then three generalized coordinates are required for each particle. For a rigid body, we need three translational generalized coordinates and three for rotation (usually the Euler angles).

In the remainder of this chapter we describe two alternative approaches to finding equations of motion for multiparticle systems and rigid bodies directly in terms of generalized coordinates.

Example 13.1 A Pendulum on a Spring

Tutorial 5.3 described how to find the equations of motion of a springy pendulum and compute its energy. The springy pendulum shown in Figure 13.1 is an example of a two-degree-of-freedom system. We can describe the position of the mass using Cartesian coordinates $(x, y)_{\mathcal{I}}$ in the inertial frame:

$$\mathbf{r}_{P/O} = x\mathbf{e}_x + y\mathbf{e}_y,$$

in which case the generalized coordinates would be $q_1 = x$, $q_2 = y$. Or we could describe the position of the pendulum using polar coordinates $(r, \theta)_{\mathcal{I}}$, which yields

$$\mathbf{r}_{P/O} = r \cos \theta \mathbf{e}_x + r \sin \theta \mathbf{e}_y = r\mathbf{e}_r,$$

as done in Tutorial 5.3, in which case the generalized coordinates would be $q_1 = r$, $q_2 = \theta$. Either choice is valid.

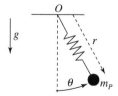

Figure 13.1 Pendulum on a spring.

13.2 Degrees of Freedom and Constraints

Before moving on to finding the equations of motion using Lagrange's method, we return to the concepts of degrees of freedom and constraints. It is for constrained systems that analytical methods have their biggest advantage. Recall the discussion in Chapter 2. The number of degrees of freedom of a particle is equal to the number of coordinates needed to describe its position. In the discussion of generalized coordinates above, we included three scalar coordinates for the three degrees of freedom of a particle. Remember, a single particle has three degrees of freedom unless there are constraints (and thus constraint forces), in which case the number of degrees of freedom decreases by the number of constraints. For example, the simple pendulum has only a single degree of freedom represented by the coordinate θ, the angle of the pendulum. There must, then, be two constraints: $z = 0$ (since the pendulum is confined to the plane) and $\sqrt{x^2 + y^2} = l$ is constant (since the pendulum length is fixed). The number of coordinates needed to describe the system is equal to the number of degrees of freedom and, for every degree-of-freedom reduction, there must be a constraint equation and corresponding constraint force. What the new methods will let us do is solve for equations of motion for the remaining coordinates without including the constraint forces.

13.2.1 Holonomic Constraints

How do we represent the concept of reduced degrees of freedom mathematically? For N particles, their configuration in $\mathcal{A}$ is given by the N positions $\mathbf{r}_{i/O}, i = 1, \ldots, N$. In the absence of constraints, these vectors are functions of $3N$ scalar coordinates. (For instance, each position is a function of the ith set of Cartesian coordinates.) Suppose, however, we have K constraints on the system. In that case, as discussed in Chapter 2, the number of degrees of freedom M reduces to

$$M = 3N - K,$$

where K is the number of constraints.

When the constraints can be described as restrictions on the possible positions of the N particles, they are known as *configuration constraints* or *holonomic constraints*. A holonomic constraint can always be written in the algebraic form

$$f(\mathbf{r}_{1/O}, \mathbf{r}_{2/O}, \ldots, \mathbf{r}_{N/O}, t) = 0.$$

For example, we might write the constraint in terms of the Cartesian coordinates $(x_i, y_i, z_i)_{\mathcal{A}}$ of each particle:

$$f(x_1, y_1, z_1, \ldots, x_N, y_N, z_N, t) = 0.$$

The pendulum constraints mentioned above are examples of holonomic constraints.

When there are K holonomic constraints, we must have K corresponding constraint equations of the form

$$\boxed{f_k(\mathbf{r}_{1/O}, \mathbf{r}_{2/O}, \ldots, \mathbf{r}_{N/O}, t) = 0, \quad k = 1, \ldots, K.}$$

When the K constraints are holonomic, we can always find a minimum set of $N_C = M = 3N - K$ generalized coordinates for the N particles, $(q_1, q_2, \ldots, q_{N_C})_\mathcal{A}$. This is because we can always invert the constraint equations to find the positions of the particles given the N_C generalized coordinate values. That is, we can use the constraint equations to rewrite the position of each particle in terms of only the N_C generalized coordinates,

$$
\begin{aligned}
\mathbf{r}_{j/O}(q_1, q_2, \ldots, q_{N_C}, t) = a_1(q_1, q_2, \ldots, q_{N_C}, t)\mathbf{a}_1 \\
+ a_2(q_1, q_2, \ldots, q_{N_C}, t)\mathbf{a}_2 \\
+ a_3(q_1, q_2, \ldots, q_{N_C}, t)\mathbf{a}_3, \quad (13.1)
\end{aligned}
$$

for $j = 1, \ldots, N$.

Note that we have explicitly included time t as an independent variable in the constraint equation. Holonomic constraints that depend explicitly on time are called *rheonomic* constraints (such as motion constrained to the surface of an inflating balloon). Holonomic constraints that are independent of time are called *scleronomic* constraints (such as motion constrained to the surface of a table). Because we are allowing for rheonomic constraints, writing each position in terms of the generalized coordinates introduced the possibility that the component magnitudes might now explicitly vary with time.

What about the velocity? We can use the chain rule to find the vector derivative of each position in terms of the rate of change of the generalized coordinates:

$$
{}^\mathcal{A}\mathbf{v}_{j/O} = \frac{{}^\mathcal{A}d}{dt}(\mathbf{r}_{j/O}) = \sum_{i=1}^{N_C} \frac{{}^\mathcal{A}\partial \mathbf{r}_{j/O}}{\partial q_i} \frac{dq_i}{dt} + \frac{{}^\mathcal{A}\partial \mathbf{r}_{j/O}}{\partial t}, \quad (13.2)
$$

where we explicitly include the partial derivative with respect to time. What are the partial derivatives of the position with respect to each generalized coordinate? These frame-dependent partial derivatives are defined exactly like the vector derivative (Definition 3.3): we simply take the partial derivative of the magnitudes of the vector components expressed in frame $\mathcal{A}$. Thus, for example, Eq. (13.2) can be written in component form using Cartesian coordinates to express $\mathbf{r}_{j/O}$ in $\mathcal{A}$, where $(x_j, y_j, z_j)_\mathcal{A}$ are the Cartesian coordinates of the jth particle in reference frame $\mathcal{A}$ (and thus also the component magnitudes of $\mathbf{r}_{j/O}$ in $\mathcal{A}$):

$$
\begin{aligned}
{}^\mathcal{A}\mathbf{v}_{j/O} = \left(\sum_{i=1}^{N_C} \frac{\partial x_j}{\partial q_i} \frac{dq_i}{dt} + \frac{\partial x_j}{\partial t} \right) \mathbf{a}_1 \\
+ \left(\sum_{i=1}^{N_C} \frac{\partial y_j}{\partial q_i} \frac{dq_i}{dt} + \frac{\partial y_j}{\partial t} \right) \mathbf{a}_2 \\
+ \left(\sum_{i=1}^{N_C} \frac{\partial z_j}{\partial q_i} \frac{dq_i}{dt} + \frac{\partial z_j}{\partial t} \right) \mathbf{a}_3.
\end{aligned}
$$

What if $\mathbf{r}_{j/O}$ is expressed in a rotating frame, as we often did throughout the book? In that case, we need to use the chain rule in Eq. (13.2) to include the time derivatives of the frame's unit vectors just as was done in Chapters 3 and 10. So, for example, we may write

$$^A\mathbf{v}_{j/O} = \left(\sum_{i=1}^{N_C} \frac{\partial x_j}{\partial q_i} \frac{dq_i}{dt} + \frac{\partial x_j}{\partial t} \right) \mathbf{b}_1 + x_j \frac{^A d}{dt} \mathbf{b}_1$$

$$+ \left(\sum_{i=1}^{N_C} \frac{\partial y_j}{\partial q_i} \frac{dq_i}{dt} + \frac{\partial y_j}{\partial t} \right) \mathbf{b}_2 + y_j \frac{^A d}{dt} \mathbf{b}_2$$

$$+ \left(\sum_{i=1}^{N_C} \frac{\partial z_j}{\partial q_i} \frac{dq_i}{dt} + \frac{\partial z_j}{\partial t} \right) \mathbf{b}_3 + z_j \frac{^A d}{dt} \mathbf{b}_2,$$

where now $(x_j, y_j, z_j)_B$ are the Cartesian coordinates of the jth particle in reference frame B. This looks just like the derivation of the transport equation and can indeed be written in the same shorthand:

$$^A\mathbf{v}_{j/O} = \sum_{i=1}^{N_C} \frac{^B\partial \mathbf{r}_{j/O}}{\partial q_i} \frac{dq_i}{dt} + \frac{^B\partial \mathbf{r}_{j/O}}{\partial t} + {}^A\boldsymbol{\omega}^B \times \mathbf{r}_{j/O}.$$

Example 13.2 A Simple Holonomic System

Figure 13.2 shows a simple holonomic system consisting of two particles, P and Q, free to move in the plane but connected by a massless rigid rod of length l. The position of each particle using Cartesian coordinates in $\mathcal{I}$ is

$$\mathbf{r}_{P/O} = x_P \mathbf{e}_x + y_P \mathbf{e}_y$$

$$\mathbf{r}_{Q/O} = x_Q \mathbf{e}_x + y_Q \mathbf{e}_y.$$

There is one scleronomic constraint given by the fixed length of the rod connecting the particles:

$$\|\mathbf{r}_{Q/O} - \mathbf{r}_{P/O}\| = \left[(x_Q - x_P)^2 + (y_Q - y_P)^2 \right]^{1/2} = l.$$

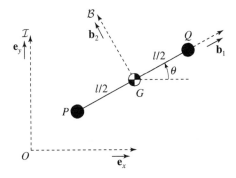

Figure 13.2 Simple holonomic system of two connected particles in the plane.

Of course, there are two additional constraints that restrict motion to the plane: $\mathbf{r}_{P/O} \cdot \mathbf{e}_z = \mathbf{r}_{Q/O} \cdot \mathbf{e}_z = 0$.

Since there are two particles ($N = 2$) and three constraints ($K = 3$), we have three degrees of freedom ($M = 3N - K = 3$). That means only three generalized coordinates are needed to describe the configuration of the system. There are many possible sets of coordinates we might use. One sensible set consists of the Cartesian coordinates of the center-of-mass position $(x_G, y_G)_{\mathcal{I}}$ with respect to O and the angle θ the rod makes relative to the $\mathbf{e}_x$ direction. That means $q_1 = x_G$, $q_2 = y_G$, and $q_3 = \theta$. The configuration for this choice of coordinates (the position of P and Q in Eq. (13.1)) is

$$\mathbf{r}_{P/O} = (q_1 - \frac{l}{2} \cos q_3)\mathbf{e}_x + (q_2 - \frac{l}{2} \sin q_3)\mathbf{e}_y \qquad (13.3)$$

$$\mathbf{r}_{Q/O} = (q_1 + \frac{l}{2} \cos q_3)\mathbf{e}_x + (q_2 + \frac{l}{2} \sin q_3)\mathbf{e}_y. \qquad (13.4)$$

The velocity of the two particles in terms of the rates of change of the generalized coordinates is found using the chain rule as in Eq. (13.2):

$$^{\mathcal{I}}\mathbf{v}_{P/O} = (\dot{q}_1 + \frac{l}{2}\dot{q}_3 \sin q_3)\mathbf{e}_x + (\dot{q}_2 - \frac{l}{2}\dot{q}_3 \cos q_3)\mathbf{e}_y$$

$$^{\mathcal{I}}\mathbf{v}_{Q/O} = (\dot{q}_1 - \frac{l}{2}\dot{q}_3 \sin q_3)\mathbf{e}_x + (\dot{q}_2 + \frac{l}{2}\dot{q}_3 \cos q_3)\mathbf{e}_y.$$

Alternatively, we might choose a different set of generalized coordinates to describe the configuration. For instance, we could choose the first two generalized coordinates to be the position of particle P, $q_1 = x_P$ and $q_2 = y_P$, with the third again being the angle of the rod. This choice yields the following for the position of each particle:

$$\mathbf{r}_{P/O} = q_1\mathbf{e}_x + q_2\mathbf{e}_y$$

$$\mathbf{r}_{Q/O} = (q_1 + l \cos q_3)\mathbf{e}_x + (q_2 + l \sin q_3)\mathbf{e}_y.$$

13.2.2 Nonholonomic Constraints

Perhaps not surprisingly, all other types of constraints are referred to as *nonholonomic*. These include inequality constraints (e.g., the particles may be confined to the inside of a sphere) and differential constraints (the velocities may be confined to certain directions, e.g., a skate blade cutting into a groove). Inequality constraints have the form

$$f(\mathbf{r}_{1/O}, \mathbf{r}_{2/O}, \ldots, \mathbf{r}_{N/O}, t) \le 0.$$

There is no specific formalism for handling inequality constraints. When satisfied, there is no reduction in degrees of freedom and the system is treated as if there were no constraint. Should the trajectory hit the boundary, or constraint surface, then care must be taken and constraint forces must be introduced at those times.

A common nonholonomic constraint is a nonintegrable differential constraint. In other words, the constraints are on the rate of change of the generalized coordinates rather than on the configuration of the particles. Such constraints have the general form

$$\sum_{i=1}^{N_C} a_{li} \frac{dq_i}{dt} + a_{lt} = 0, \quad l = 1, \dots, S, \tag{13.5}$$

where a_{li} $(q_i, \dots, q_{N_C}, t)$ and a_{lt} $(q_i, \dots, q_{N_C}, t)$ are functions of the generalized coordinates and S is the number of nonholonomic constraints.

Compare this expression to Eq. (13.2). We see that nonholonomic constraints of the form Eq. (13.5) are equivalent to constraints on combinations of the particle velocity components, where the coefficients are linear combinations of the partial derivatives of the position components with respect to the generalized coordinates. We often write the nonholonomic constraint in the following differential form:

$$\sum_{i=1}^{N_C} a_{li} dq_i + a_{lt} dt = 0 \quad l = 1, \dots, S. \tag{13.6}$$

The important observation about nonholonomic constraints is that they are non-integrable. They cannot be integrated to yield equations for the generalized coordinates themselves (as for holonomic constraints). So even though we have fewer degrees of freedom, we can't reduce the number of generalized coordinates. This is because there are no configuration constraints to invert that allow us to write the particle position in terms of the generalized coordinates, as in Eq. (13.1). Consequently, we must retain more generalized coordinates than there are degrees of freedom. In essence, we ignore the existence of the nonholonomic constraints when selecting generalized coordinates and then use special care when finding the equations of motion.

Mathematically, then, if we have K holonomic constraints and S nonholonomic constraints, the number of degrees of freedom in a system of N particles is

$$M = 3N - K - S,$$

but the number of generalized coordinates is still

$$N_C = 3N - K.$$

We have only used the holonomic constraints in choosing our generalized coordinates. We thus have S more coordinates than degrees of freedom. The expressions for the velocity in terms of the generalized coordinates are the same as in the discussion of holonomic constraints.

Example 13.3 The Rolling Wheel

Example 8.2 introduced the rolling wheel. The main idea is that the wheel rolls without slipping, implying that the contact point P has zero inertial velocity:

$$^{\mathcal{I}}\mathbf{v}_{P/O} = 0.$$

This is an example of an integrable motion constraint. We describe the position and orientation of the wheel by two coordinates, x and θ. Nevertheless, it is a

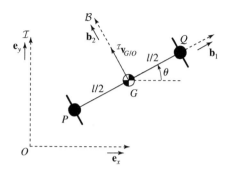

Figure 13.3 Simple nonholonomic system. The particles P and Q are constrained to move along the directions of the blades.

single-degree-of-freedom system because of the no-slip condition. The nonholonomic constraint equation is

$$\dot{x} = -r\dot{\theta},$$

where r is the radius of the wheel. This can be integrated with respect to time to obtain $x(t) - x(0) = -r(\theta(t) - \theta(0))$.

Example 13.4 Nonholonomic System

Figure 13.3 shows a simple nonholonomic system that is only a slight modification of that in Example 13.2. It consists of two particles, P and Q, connected by a massless rod of length l, lying flat on an icy surface. Each particle rests on a massless blade such that it can only move along the direction of the blade. This system has one holonomic constraint,

$$\|\mathbf{r}_{Q/O} - \mathbf{r}_{P/O}\| = l,$$

and one nonholonomic constraint,

$$^{\mathcal{I}}\mathbf{v}_{G/O} = \|^{\mathcal{I}}\mathbf{v}_{G/O}\|\mathbf{b}_2. \tag{13.7}$$

Eq. (13.7) states that the velocity of each particle or, equivalently, the velocity of the center of mass, is always directed along the blade direction.

In the absence of constraints, the two particles would have four degrees of freedom, represented by the two-dimensional position of each particle in $\mathcal{I}$. However, because there are two constraints ($K = 1$ holonomic and $S = 1$ nonholonomic), we know that the system has only two degrees of freedom (compared to the three in Example 13.2). Nevertheless, because one of the constraints is nonholonomic, the fewest coordinates we can use to describe the system is $N_C = 3$. As described in Example 13.2, there are many possible coordinates that could be used; we choose here the Cartesian coordinates $(x_G, y_G)_{\mathcal{I}}$ of the center of mass G in $\mathcal{I}$ and the orientation θ of $\mathcal{B}$ in $\mathcal{I}$. Thus our three generalized coordinates are again $q_1 = x_G$, $q_2 = y_G$, and $q_3 = \theta$. The configuration of the particles is given by Eqs. (13.3) and (13.4).

To write the nonholonomic constraint in terms of the generalized coordinates, we note that Eq. (13.7) is equivalent to

$$^{\mathcal{I}}\mathbf{v}_{G/O} \cdot \mathbf{b}_1 = 0.$$

Using the Cartesian expression for the velocity of the center of mass, ${}^{\mathcal{I}}\mathbf{v}_{G/O} = \dot{q}_1\mathbf{e}_x + \dot{q}_2\mathbf{e}_y$, we find

$$\dot{q}_1 \cos q_3 + \dot{q}_2 \sin q_3 = 0. \tag{13.8}$$

Comparing Eq. (13.8) to Eq. (13.5), we see that $a_{11} = \cos q_3$, $a_{12} = \sin q_3$, $a_{13} = 0$, and $a_{1t} = 0$.

13.3 Lagrange's Method

This section introduces our first analytical approach to finding equations of motion for a system of particles. The resulting formulation was introduced by Louis de Lagrange in the late eighteenth century in his seminal work *Mécanique Analytique*. Lagrange's goal was to unify the work of his predecessors (Newton, Euler, Bernoulli, and d'Alembert) into a coherent and consistent theory of mechanics. Even more so, he sought to reduce mechanics to "only the algebraic operations inherent to a regular and uniform process."[1] As shown here, Lagrange's formulation leads directly to scalar equations of motion in terms of the generalized coordinates using only scalar algebra while naturally incorporating the constraints. There are a number of modern approaches to deriving Lagrange's equations; we follow roughly the same procedure Lagrange did, building on d'Alembert's idea of *virtual work*.

13.3.1 D'Alembert's Principle and Virtual Work

As always, we begin our discussion of kinetics with Newton's second law, in this case for a system of particles. Let $\mathbf{F}_j$ be the total force on particle $j = 1, \ldots, N$. We have

$$\mathbf{F}_j - m_j \frac{{}^{\mathcal{I}}d}{dt}\left({}^{\mathcal{I}}\mathbf{v}_{j/O}\right) = 0. \tag{13.9}$$

Note that we have rewritten Newton's second law a bit differently; here it is presented as the difference between the net force and the rate of change of momentum on particle j. This version is known as d'Alembert's formulation.[2] Although this change, of course, has no effect on how we use Newton's second law to solve problems, it does provide insight into how we might proceed. By rewriting Newton's second law as in Eq. (13.9), d'Alembert recast it as a statics problem, where the mass times acceleration term is treated as an "inertia force." (We have avoided this language elsewhere in the book, but here it is useful.) D'Alembert's insight was that we can use the tools of statics, namely, the idea of virtual work, on dynamics problems. You may recall from a statics course that virtual work is a technique for finding the forces of static equilibrium. We apply it similarly here.

---

[1] Dugas (1988, p. 333).

[2] Jean Le Rond d'Alembert (1717–1783) was a French mathematician, physicist, and philosopher. In addition to his contributions to analysis, he is best known for his work in statics and mechanics, articulated in his book *Traite de dynamique* in 1743, where he formulated an alternative form of the laws of motion.

Consider a *virtual displacement* $^\mathcal{I}\delta\mathbf{r}_{j/O}$ of particle j in reference frame $\mathcal{I}$ that is consistent with the constraints on j. A virtual displacement is an infinitesimal, instantaneous change in the coordinates. Thus we can write the virtual displacement in terms of small virtual changes in the generalized coordinates:

$$^\mathcal{I}\delta\mathbf{r}_{j/O} = \sum_{i=1}^{N_C} \frac{^\mathcal{I}\partial\mathbf{r}_{j/O}}{\partial q_i}\delta q_i. \tag{13.10}$$

This procedure can be interpreted as "freezing" time so that no motion occurs. We infinitesimally displace the configuration. For now, it does not even have to be a change that solves the equation of motion. It only needs to be consistent with the configuration constraints. In other words, the configuration, after the virtual displacement, must still satisfy the constraint equations.

The next step is to take the dot product of the virtual displacement (Eq. (13.10)) with Eq. (13.9) to obtain

$$\left(\mathbf{F}_j - m_j\frac{^\mathcal{I}d}{dt}\left(^\mathcal{I}\mathbf{v}_{j/O}\right)\right)\cdot {}^\mathcal{I}\delta\mathbf{r}_{j/O} = 0. \tag{13.11}$$

Multiplying through the parentheses results in what is called the *virtual work*, or the dot product of the virtual displacement with the total force acting on j:

$$\delta W_j^{(\text{tot})} = \mathbf{F}_j \cdot {}^\mathcal{I}\delta\mathbf{r}_{j/O}.$$

We can make one more simplification by replacing the total force acting on particle j with the sum of the active forces and constraint forces, $\mathbf{F}_j = \mathbf{F}_j^{(\text{act})} + \mathbf{F}_j^{(\text{const})}$. Since the virtual displacement is consistent with the constraints, the constraint forces result in no virtual work (just as constraint forces do no work, as discussed in Chapter 5). The virtual work is thus due only to the active forces:

$$\delta W_j^{(\text{tot})} = \mathbf{F}_j^{(\text{act})} \cdot {}^\mathcal{I}\delta\mathbf{r}_{j/O}.$$

Eq. (13.11) becomes

$$\left(\mathbf{F}_j^{(\text{act})} - m_j\frac{^\mathcal{I}d}{dt}\left(^\mathcal{I}\mathbf{v}_{j/O}\right)\right)\cdot {}^\mathcal{I}\delta\mathbf{r}_{j/O} = 0.$$

Finally, summing over all particles yields an expression analogous to the virtual work principle in statics:

$$\sum_{j=1}^{N}\left(\mathbf{F}_j^{(\text{act})} - m_j\frac{^\mathcal{I}d}{dt}\left(^\mathcal{I}\mathbf{v}_{j/O}\right)\right)\cdot {}^\mathcal{I}\delta\mathbf{r}_{j/O} = 0. \tag{13.12}$$

Eq. (13.12) is known as *d'Alembert's principle*. The trajectories of the system of particles under the action of all active forces and constraints must satisfy Eq. (13.12). Note that by computing the virtual work, we have eliminated the constraint forces from the problem. We have projected the equations of motion into those directions where only the active forces do work. The next question is how to use d'Alembert's principle to find equations of motion for the generalized coordinates.

13.3.2 The Euler-Lagrange Equations

How can Eq. (13.12) be used to find the equations of motion? It is tempting to think that Eq. (13.12) is satisfied only if $\mathbf{F}_j^{(\text{act})} = m_j \frac{^{\mathcal{I}}d}{dt}(^{\mathcal{I}}\mathbf{v}_{j/O})$, which looks just like Newton's second law. This is not the case, however, since the $^{\mathcal{I}}\delta\mathbf{r}_{j/O}$ are not linearly independent variations; they must be consistent with the constraints. A bit more work is required.

There are two dot products in Eq. (13.12), and we work on each separately. We can rewrite the first dot product by substituting from Eq. (13.10) to obtain

$$\sum_{j=1}^{N}\mathbf{F}_j^{(\text{act})}\cdot{}^{\mathcal{I}}\delta\mathbf{r}_{j/O} = \sum_{i=1}^{N_C}\underbrace{\sum_{j=1}^{N}\mathbf{F}_j^{(\text{act})}\cdot\frac{^{\mathcal{I}}\partial\mathbf{r}_{j/O}}{\partial q_i}}_{\triangleq Q_i}\delta q_i,$$

where we have introduced the *generalized force*

$$Q_i \triangleq \sum_{j=1}^{N}\mathbf{F}_j^{(\text{act})}\cdot\frac{^{\mathcal{I}}\partial\mathbf{r}_{j/O}}{\partial q_i}. \tag{13.13}$$

A generalized force is the projection of the particle forces onto a single degree of freedom of the system.

The second dot product in Eq. (13.12) is

$$\sum_{j=1}^{N}m_j\frac{^{\mathcal{I}}d}{dt}\left(^{\mathcal{I}}\mathbf{v}_{j/O}\right)\cdot{}^{\mathcal{I}}\delta\mathbf{r}_{j/O} = \sum_{i=1}^{N_C}\sum_{j=1}^{N}m_j\frac{^{\mathcal{I}}d}{dt}\left(^{\mathcal{I}}\mathbf{v}_{j/O}\right)\cdot\frac{^{\mathcal{I}}\partial\mathbf{r}_{j/O}}{\partial q_i}\delta q_i. \tag{13.14}$$

The question is how to rewrite this equation in terms of the generalized coordinates only, as for the virtual work. To do so we use some foresight and note that

$$\frac{^{\mathcal{I}}d}{dt}\left(m_j{}^{\mathcal{I}}\mathbf{v}_{j/O}\cdot\frac{^{\mathcal{I}}\partial\mathbf{r}_{j/O}}{\partial q_i}\right) = m_j\frac{^{\mathcal{I}}d}{dt}\left(^{\mathcal{I}}\mathbf{v}_{j/O}\right)\cdot\frac{^{\mathcal{I}}\partial\mathbf{r}_{j/O}}{\partial q_i}$$

$$+ m_j{}^{\mathcal{I}}\mathbf{v}_{j/O}\cdot\frac{^{\mathcal{I}}d}{dt}\left(\frac{^{\mathcal{I}}\partial\mathbf{r}_{j/O}}{\partial q_i}\right). \tag{13.15}$$

The first term after the equal sign in Eq. (13.15) is what we are looking to replace in Eq. (13.14). The other two terms need a bit of adjusting. The second term on the right-hand side of Eq. (13.15) is modified using Eq. (13.2) to differentiate $^{\mathcal{I}}\partial\mathbf{r}_{j/O}/\partial q_i$

with respect to time, which has the appearance of exchanging the total and partial derivatives:

$$m_j{}^{\mathcal{I}}\mathbf{v}_{j/O} \cdot \frac{{}^{\mathcal{I}}d}{dt}\left(\frac{{}^{\mathcal{I}}\partial \mathbf{r}_{j/O}}{\partial q_i}\right) = m_j{}^{\mathcal{I}}\mathbf{v}_{j/O} \cdot \frac{{}^{\mathcal{I}}\partial}{\partial q_i}\left(\frac{{}^{\mathcal{I}}d\mathbf{r}_{j/O}}{dt}\right) = m_j{}^{\mathcal{I}}\mathbf{v}_{j/O} \cdot \frac{{}^{\mathcal{I}}\partial}{\partial q_i}\left({}^{\mathcal{I}}\mathbf{v}_{j/O}\right).$$

Summing over the particles and pulling the partial derivatives out of the sum yields

$$\sum_{j=1}^{N} m_j{}^{\mathcal{I}}\mathbf{v}_{j/O} \cdot \frac{{}^{\mathcal{I}}\partial}{\partial q_i}\left({}^{\mathcal{I}}\mathbf{v}_{j/O}\right) = \frac{{}^{\mathcal{I}}\partial}{\partial q_i}\underbrace{\left(\frac{1}{2}\sum_{j=1}^{N} m_j{}^{\mathcal{I}}\mathbf{v}_{j/O} \cdot {}^{\mathcal{I}}\mathbf{v}_{j/O}\right)}_{=T_O} = \frac{{}^{\mathcal{I}}\partial T_O}{\partial q_i}, \qquad (13.16)$$

where we have used the definition of the total kinetic energy for a system of particles from Eq. (7.21). Note that the kinetic energy can be written as a function of the generalized coordinates, their derivatives, and time:

$$T_O = T_O(q_1, q_2, \ldots, q_{N_C}, \dot{q}_1, \dot{q}_2, \ldots \dot{q}_{N_C}, t).$$

For the term on the left-hand side of Eq. (13.15), we use the fact that

$$\frac{{}^{\mathcal{I}}\partial}{\partial \dot{q}_i}\left({}^{\mathcal{I}}\mathbf{v}_{j/O}\right) = \frac{{}^{\mathcal{I}}\partial}{\partial \dot{q}_i}\left(\sum_{k=1}^{N_C} \frac{{}^{\mathcal{I}}\partial \mathbf{r}_{j/O}}{\partial q_k}\dot{q}_k + \frac{{}^{\mathcal{I}}\partial \mathbf{r}_{j/O}}{\partial t}\right) = \frac{{}^{\mathcal{I}}\partial \mathbf{r}_{j/O}}{\partial q_i},$$

where the expression for the velocity in terms of the generalized coordinates from Eq. (13.2) has been used. This expression allows us to make the substitution

$$\frac{{}^{\mathcal{I}}d}{dt}\left(m_j{}^{\mathcal{I}}\mathbf{v}_{j/O} \cdot \frac{{}^{\mathcal{I}}\partial \mathbf{r}_{j/O}}{\partial q_i}\right) = \frac{{}^{\mathcal{I}}d}{dt}\left(m_j{}^{\mathcal{I}}\mathbf{v}_{j/O} \cdot \frac{{}^{\mathcal{I}}\partial}{\partial \dot{q}_i}\left({}^{\mathcal{I}}\mathbf{v}_{j/O}\right)\right).$$

Again summing over the particles yields

$$\frac{{}^{\mathcal{I}}d}{dt}\left(\sum_{j=1}^{N} m_j{}^{\mathcal{I}}\mathbf{v}_{j/O} \cdot \frac{{}^{\mathcal{I}}\partial}{\partial \dot{q}_i}\left({}^{\mathcal{I}}\mathbf{v}_{j/O}\right)\right)$$

$$= \frac{{}^{\mathcal{I}}d}{dt}\frac{{}^{\mathcal{I}}\partial}{\partial \dot{q}_i}\left(\frac{1}{2}\sum_{j=1}^{N} m_j{}^{\mathcal{I}}\mathbf{v}_{j/O} \cdot {}^{\mathcal{I}}\mathbf{v}_{j/O}\right) = \frac{{}^{\mathcal{I}}d}{dt}\frac{{}^{\mathcal{I}}\partial T_O}{\partial \dot{q}_i}. \qquad (13.17)$$

Referring back to Eq. (13.14), the acceleration term in d'Alembert's principle can be rewritten by adjusting Eq. (13.15) and using Eq. (13.16) and Eq. (13.17):

$$\sum_{j=1}^{N} m_j \frac{^{\mathcal{I}}d}{dt}\left(^{\mathcal{I}}\mathbf{v}_{j/O}\right) \cdot \frac{^{\mathcal{I}}\partial \mathbf{r}_{j/O}}{\partial q_i} = \sum_{j=1}^{N} \frac{^{\mathcal{I}}d}{dt}\left(m_j \,^{\mathcal{I}}\mathbf{v}_{j/O} \cdot \frac{^{\mathcal{I}}\partial \mathbf{r}_{j/O}}{\partial q_i}\right)$$

$$- \sum_{j=1}^{N} m_j \,^{\mathcal{I}}\mathbf{v}_{j/O} \cdot \frac{^{\mathcal{I}}d}{dt}\left(\frac{^{\mathcal{I}}\partial \mathbf{r}_{j/O}}{\partial q_i}\right)$$

$$= \frac{^{\mathcal{I}}d}{dt} \frac{^{\mathcal{I}}\partial T_O}{\partial \dot{q}_i} - \frac{^{\mathcal{I}}\partial T_O}{\partial q_i}.$$

The end result is that we can now rewrite d'Alembert's principle in Eq. (13.12) solely in terms of the generalized coordinates and their rates:

$$\sum_{j=1}^{N} \left(\mathbf{F}_j^{(\text{act})} - m_j \frac{^{\mathcal{I}}d}{dt}(^{\mathcal{I}}\mathbf{v}_{j/O})\right) \cdot {}^{\mathcal{I}}\delta \mathbf{r}_{j/O}$$

$$= -\sum_{i=1}^{N_C} \left(\frac{^{\mathcal{I}}d}{dt} \frac{^{\mathcal{I}}\partial T_O}{\partial \dot{q}_i} - \frac{^{\mathcal{I}}\partial T_O}{\partial q_i} - Q_i\right) \delta q_i = 0. \qquad (13.18)$$

It is helpful and important to remember at this point that we have not introduced any new physics. This equation is still just Newton's second law summed over each particle and dotted with the virtual displacement. Through some rather lengthy manipulations, we have been able to rewrite it as a sum over generalized coordinates and in terms of the total kinetic energy. Eq. (13.18) is a rather remarkable and beautiful result.

At this point, we specialize to systems with only holonomic constraints. (Section 13.3.4 discusses the application to systems with nonholonomic constraints.) Eq. (13.18) is significant because the variations of the generalized coordinates are linearly independent, since we have the same number of coordinates as degrees of freedom and they have been varied to be consistent with the holonomic constraints. Since each δq_i is an independent variation, the only way for Eq. (13.18) to be true is if each term in the sum is zero:

$$\boxed{\frac{d}{dt} \frac{\partial T_O}{\partial \dot{q}_i} - \frac{\partial T_O}{\partial q_i} = Q_i, \quad i = 1, \ldots, N_C.} \qquad (13.19)$$

These are called the *Euler-Lagrange equations* or sometimes just *Lagrange's equations*. They are entirely equivalent to Newton's second law. The beauty of using this formulation is that it allows us to find equations of motion directly in terms of the generalized coordinates. This approach is particularly useful when there are many constraints in the problem because in that case there may be many fewer generalized coordinates than particles. This approach does not require introducing constraint forces and then algebraically eliminating them later.

The most common mistake made in using Lagrange's formulation of mechanics is to improperly use the kinematics to find the kinetic energy in terms of the generalized coordinates and their rates. The problem is still a vector problem and you must start with the velocity with respect to an inertial frame of reference. (Note we dropped the superscript $\mathcal{I}$ in Eq. (13.19).)

Example 13.5 A Spring-Damper System Using Lagrange's Equation

Tutorial 2.4 first introduced the simple mass-spring-damper system and derived the differential equation corresponding to simple harmonic motion (see Figure 2.7b and c). Here we re-solve this simple one-dimensional system using the Euler-Lagrange equations. As before, we choose to describe the position of the particle by the Cartesian x coordinate. Since this system has one degree of freedom, we need only a single generalized coordinate $q_1 = x$.

The kinetic energy for this system is $T_O = T_{P/O} = \frac{1}{2}m_P\dot{x}^2$. There are two forces acting on the mass: the spring force $-k(x - x_0)$ and the damping force $-b\dot{x}$. Since the position of the particle is $\mathbf{r}_{P/O} = x\mathbf{e}_1$, the generalized force from Eq. (13.13) is

$$Q_1 = -k(x - x_0)\frac{{}^{\mathcal{I}}\partial \mathbf{r}_{P/O}}{\partial q_1} - b\dot{x}\frac{{}^{\mathcal{I}}\partial \mathbf{r}_{P/O}}{\partial q_1} = -k(x - x_0) - b\dot{x}.$$

Plugging this expression into the Euler-Lagrange equations (Eq. (13.19)) gives the equation of motion

$$\frac{d}{dt}\frac{\partial}{\partial \dot{x}}\left(\frac{1}{2}m_P\dot{x}^2\right) + \frac{\partial}{\partial x}\left(\frac{1}{2}m_P\dot{x}^2\right) = -k(x - x_0) - b\dot{x}.$$

This expression simplifies to

$$\ddot{x} + \frac{b}{m_P}\dot{x} + \frac{k}{m_P}(x - x_0) = 0,$$

which is the equation for a damped harmonic oscillator from Tutorial 2.4 (Eq. (2.24)).

Example 13.6 A Spring on a Table with Friction

This example re-solves the problem of Example 3.12 using the Euler-Lagrange equations. Here a particle is connected to the origin by a spring but is free to move anywhere on a planar surface (see Figure 3.19). In addition to the spring force on the particle, we add a sliding (Coulomb) friction force with coefficient μ_c, so that the total force on P is given by

$$\mathbf{F}_P = -k(r - r_0)\mathbf{e}_r - \mu_c m_P g\,{}^{\mathcal{I}}\hat{\mathbf{v}}_{P/O}.$$

As in Example 3.12, it is most sensible to use polar coordinates for this problem, so that the particle is located in the inertial frame by the generalized coordinates $(q_1 = r, q_2 = \theta)_{\mathcal{I}}$. The position of P relative to O is

$$\mathbf{r}_{P/O} = r\cos\theta\mathbf{e}_1 + r\sin\theta\mathbf{e}_2 = r\mathbf{e}_r.$$

The velocity of P is

$${}^{\mathcal{I}}\mathbf{v}_{P/O} = \dot{r}\mathbf{e}_r + r\dot{\theta}\mathbf{e}_\theta,$$

which yields the kinetic energy

$$T_O = T_{P/O} = \frac{1}{2}m_P(\dot{r}^2 + r^2\dot{\theta}^2).$$

The generalized forces are calculated using Eq. (13.13). Since there are two degrees of freedom (and thus two generalized coordinates), we must solve for two generalized forces. The generalized force corresponding to $q_1 = r$ is

$$Q_1 = \left(-k(r - r_0)\mathbf{e}_r - \mu_c m_P g {}^{\mathcal{I}}\hat{\mathbf{v}}_{P/O}\right) \cdot \frac{{}^{\mathcal{I}}\partial}{\partial r}(r\cos\theta\mathbf{e}_1 + r\sin\theta\mathbf{e}_2)$$

$$= \left(-k(r - r_0)\mathbf{e}_r - \mu_c m_P g {}^{\mathcal{I}}\hat{\mathbf{v}}_{P/O}\right) \cdot \underbrace{(\cos\theta\mathbf{e}_1 + \sin\theta\mathbf{e}_2)}_{=\mathbf{e}_r}$$

$$= -k(r - r_0) - \frac{\mu_c m_P g \dot{r}}{\sqrt{\dot{r}^2 + r^2\dot{\theta}^2}}.$$

The second generalized force corresponding to $q_2 = \theta$ is

$$Q_2 = \left(-k(r - r_0)\mathbf{e}_r - \mu_c m_P g {}^{\mathcal{I}}\hat{\mathbf{v}}_{P/O}\right) \cdot \frac{{}^{\mathcal{I}}\partial}{\partial \theta}(r\cos\theta\mathbf{e}_1 + r\sin\theta\mathbf{e}_2)$$

$$= \left(-k(r - r_0)\mathbf{e}_r - \mu_c m_P g {}^{\mathcal{I}}\hat{\mathbf{v}}_{P/O}\right) \cdot \underbrace{(-r\sin\theta\mathbf{e}_1 + r\cos\theta\mathbf{e}_2)}_{=r\mathbf{e}_\theta}$$

$$= \frac{\mu_c m_P g r^2 \dot{\theta}}{\sqrt{\dot{r}^2 + r^2\dot{\theta}^2}}.$$

We now use these generalized forces in the two Euler-Lagrange equations to find the equations of motion. For $q_1 = r$ we have

$$Q_1 = \frac{d}{dt}\frac{\partial}{\partial \dot{r}}\left(\frac{1}{2}m_P(\dot{r}^2 + r^2\dot{\theta}^2)\right) - \frac{\partial}{\partial r}\left(\frac{1}{2}m_P(\dot{r}^2 + r^2\dot{\theta}^2)\right)$$

$$= \frac{d}{dt}\left(m_P\dot{r}\right) - m_P r\dot{\theta}^2,$$

which reduces to the first equation of motion:

$$\ddot{r} + \frac{k}{m_P}(r - r_0) + \frac{\mu_c g \dot{r}}{\sqrt{\dot{r}^2 + r^2\dot{\theta}^2}} - r\dot{\theta}^2 = 0.$$

Using the Euler-Lagrange equation, the second equation of motion for θ is similarly found to be

$$\frac{d}{dt}\frac{\partial}{\partial \dot{\theta}}\left(\frac{1}{2}m_P(\dot{r}^2 + r^2\dot{\theta}^2)\right) - \frac{\partial}{\partial \theta}\left(\frac{1}{2}m_P(\dot{r}^2 + r^2\dot{\theta}^2)\right) = Q_2$$

or

$$\frac{d}{dt}(m_P r^2\dot{\theta}) = Q_2,$$

which reduces to the second equation of motion:

$$\ddot{\theta} + 2\frac{\dot{r}}{r}\dot{\theta} - \frac{\mu_c g \dot{\theta}}{\sqrt{\dot{r}^2 + r^2\dot{\theta}^2}} = 0.$$

Setting $\mu_c = 0$ results in the same equations of motion as in Example 3.12 (Eqs. (3.46) and (3.47)).

13.3.3 Conservative Forces and the Lagrangian

The last step in developing Lagrange's formulation is to consider the situation when the active forces are conservative. (In the last two examples, the spring forces were conservative but the frictional or damping forces were not.) Remember that conservative forces can be found by using the gradient of a scalar potential energy. Let $U_{j/O}$ be the total potential energy of particle j. The total conservative force on j is

$$\mathbf{F}_j^{(c)} = -\nabla U_{j/O}(\mathbf{r}_{j/O}).$$

Substituting this expression into Eq. (13.13) for the generalized force yields

$$Q_i^{(c)} = -\sum_{j=1}^{N} \nabla U_{j/O}(\mathbf{r}_{j/O}) \cdot \frac{{}^{\mathcal{I}}\partial \mathbf{r}_{j/O}}{\partial q_i}. \tag{13.20}$$

Eq. (13.20) looks like a derivative of the potential taken in the direction of the generalized coordinate. It turns out to be equal to the partial derivative of the potential energy with respect to that generalized coordinate. To see why, write out the dot product in Eq. (13.20) by using the definition of the gradient in terms of Cartesian coordinates in $\mathcal{I}$ and the components of $\mathbf{r}_{j/O} = x(q_1, \ldots, q_{N_C}, t)\mathbf{e}_1 + y(q_1, \ldots, q_{N_C}, t)\mathbf{e}_2 + z(q_1, \ldots, q_{N_C}, t)\mathbf{e}_3$:

$$\nabla U_{j/O}(\mathbf{r}_{j/O}) \cdot \frac{{}^{\mathcal{I}}\partial \mathbf{r}_{j/O}}{\partial q_i} = \frac{\partial U_{j/O}}{\partial x}\frac{\partial x}{\partial q_i} + \frac{\partial U_{j/O}}{\partial y}\frac{\partial y}{\partial q_i} + \frac{\partial U_{j/O}}{\partial z}\frac{\partial z}{\partial q_i}.$$

This equation is just the chain rule for the derivative of $U_{j/O}$ with respect to q_i. That means we can rewrite the conservative generalized force in Eq. (13.20) as

$$Q_i^{(c)} = -\sum_{j=1}^{N} \frac{\partial U_{j/O}}{\partial q_i} = -\frac{\partial}{\partial q_i}\sum_{j=1}^{N} U_{j/O} = -\frac{\partial U_O}{\partial q_i},$$

where U_O is the total potential energy of the system of particles.

This expression can be used to update the Euler-Lagrange equations for the generalized coordinates in Eq. (13.19) by replacing the conservative generalized forces with the partial derivative of the potential energy:

$$\frac{d}{dt}\frac{\partial T_O}{\partial \dot{q}_i} - \frac{\partial T_O}{\partial q_i} + \frac{\partial U_O}{\partial q_i} = Q_i^{(nc)}.$$

We now make one last substitution. Since most often the potential is a function solely of the generalized coordinates and not of their rates, we introduce the *Lagrangian:*

$$L \triangleq T_O - U_O.$$

Note that the Lagrangian is the difference between the kinetic and potential energies. It is not the total energy. Using it simplifies the Euler-Lagrange equations for conservative forces (assuming that the potential energy is a function only of the generalized coordinates and time):

$$\frac{d}{dt} \frac{\partial L}{\partial \dot{q}_i} - \frac{\partial L}{\partial q_i} = Q_i^{(\text{nc})}, \quad i = 1, \ldots, N_C,$$

where $Q_i^{(\text{nc})}$ are the remaining nonconservative generalized forces given by

$$Q_i^{(\text{nc})} \triangleq \sum_{j=1}^{N} \mathbf{F}_j^{(\text{act,nc})} \cdot \frac{^{\mathcal{I}}\partial \mathbf{r}_{j/O}}{\partial q_i}.$$

These are the most general form of Lagrange's equations for a system with only holonomic constraints. They are most often used for completely conservative systems, where $Q_i^{(\text{nc})} = 0$. They are entirely equivalent to Newton's second law of motion. They provide an alternative approach to finding equations of motion, without explicitly including constraint forces, that starts with kinetic and potential energies rather than forces and accelerations. For some problems, using them can dramatically reduce the effort involved. It is also interesting to recall the work-energy formula in Eq. (5.16). We noted that, for a single-degree-of-freedom system, the total energy supplied all necessary information: we could differentiate and find the equations of motion, for instance. For a multiple-degree-of-freedom system, however, it was not enough. Simply differentiating did not give us enough information. Now, starting with the Lagrangian and using Lagrange's method, we have a tool that yields the equation of motion for each degree of freedom.

Example 13.7 The Simple Pendulum Using the Euler-Lagrange Equations

We return here to our canonical example and find the equations of motion for the simple pendulum using Lagrange's equations. This is a single-degree-of-freedom system, and the generalized coordinate is, of course, the angle θ of the pendulum from the vertical. Example 5.12 derived the total energy of the pendulum and showed how to integrate it to determine the trajectory or differentiate it to find the equation of motion. Here we use the Euler-Lagrange equations.

From Example 5.12, the kinetic and potential energies of the pendulum are

$$T_{P/O} = \frac{1}{2}m_P l^2 \dot{\theta}^2$$

$$U_{P/O} = -m_P g l \cos\theta.$$

The Lagrangian is thus

$$L = T_{P/O} - U_{P/O} = \frac{1}{2}m_P l^2 \dot{\theta}^2 + m_P g l \cos\theta.$$

We now find the equation of motion by substituting into the Euler-Lagrange equation for θ:

$$\frac{d}{dt}\frac{\partial L}{\partial \dot{\theta}} - \frac{\partial L}{\partial \theta} = \frac{d}{dt}(m_P l^2 \dot{\theta}) + m_P g l \sin\theta = 0,$$

where $Q_\theta^{(\mathrm{nc})} = 0$ for this problem. The result is, of course, the usual equation of motion:

$$\ddot{\theta} + \frac{g}{l}\sin\theta = 0.$$

Example 13.8 A Spinning Simple Pendulum

Although elegant, the Euler-Lagrange equations for the pendulum do not provide much advantage over finding the equation of motion by means of the angular momentum, as done in Chapter 7. Here we look at a slightly more complicated situation. Now the plane of the pendulum can rotate around the $\mathbf{e}_1$ axis at a fixed rate Ω. This device is not a spherical pendulum; it is still a single-degree-of-freedom system, where the pendulum coordinate is θ, as in Example 3.11. However, that plane now also rotates with respect to inertial space at a known rate Ω, as shown in Figure 13.4.

We begin by finding the inertial velocity and kinetic energy. We introduce a new frame $\mathcal{C} = (O, \mathbf{e}_x, \mathbf{e}_y, \mathbf{e}_z)$ that rotates about the $\mathbf{e}_1 = \mathbf{e}_x$ axis it shares with the inertial frame $\mathcal{I} = (O, \mathbf{e}_1, \mathbf{e}_2, \mathbf{e}_3)$. We also introduce the usual body frame, $\mathcal{B} = (O, \mathbf{e}_r, \mathbf{e}_\theta, \mathbf{e}_z)$ (see Figure 13.4b) that shares the $\mathbf{e}_z$ axis with the $\mathcal{C}$ frame and rotates about it by θ, the generalized coordinate associated with the single degree of freedom. The transformation table between $\mathcal{C}$ and $\mathcal{B}$ is

	$\mathbf{e}_r$	$\mathbf{e}_\theta$
$\mathbf{e}_x$	$\cos\theta$	$-\sin\theta$
$\mathbf{e}_y$	$\sin\theta$	$\cos\theta$.

The position of the pendulum in $\mathcal{B}$ is $\mathbf{r}_{P/O} = l\mathbf{e}_r$. To find the kinetic energy requires knowing the inertial velocity, which we find from the transport equation. Using the

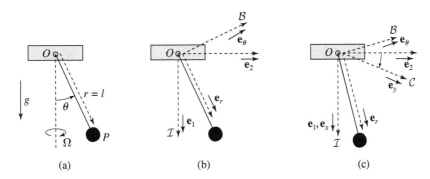

Figure 13.4 The rotating simple pendulum. (a) Coordinates. (b) Reference frames with $\mathcal{C}$ aligned with $\mathcal{I}$. (c) Reference frames with $\mathcal{C}$ and $\mathcal{B}$ rotating about $\mathbf{e}_1$.

addition property, the angular velocity of $\mathcal{B}$ in $\mathcal{I}$ is

$$^{\mathcal{I}}\boldsymbol{\omega}^{\mathcal{B}} = {^{\mathcal{I}}\boldsymbol{\omega}^{\mathcal{C}}} + {^{\mathcal{C}}\boldsymbol{\omega}^{\mathcal{B}}} = \dot{\theta}\mathbf{e}_z + \Omega\mathbf{e}_1. \tag{13.21}$$

Therefore

$$^{\mathcal{I}}\mathbf{v}_{P/O} = {^{\mathcal{I}}\boldsymbol{\omega}^{\mathcal{B}}} \times \mathbf{r}_{P/O} = (\dot{\theta}\mathbf{e}_z + \Omega\mathbf{e}_1) \times l\mathbf{e}_r$$

$$= l\dot{\theta}\mathbf{e}_\theta + l\Omega \sin\theta\mathbf{e}_z. \tag{13.22}$$

The kinetic energy is

$$T_{P/O} = \frac{1}{2}m_P{^{\mathcal{I}}\mathbf{v}_{P/O}} \cdot {^{\mathcal{I}}\mathbf{v}_{P/O}} = \frac{1}{2}m_P l^2(\dot{\theta}^2 + \Omega^2 \sin^2\theta).$$

The potential energy of the pendulum bob is the same as for the simple pendulum:

$$U_{P/O} = -m_P gl \cos\theta.$$

The Lagrangian is thus

$$L = T_{P/O} - U_{P/O} = \frac{1}{2}m_P l^2(\dot{\theta}^2 + \Omega^2 \sin^2\theta) + m_P gl \cos\theta.$$

Plugging into the Euler-Lagrange equation yields

$$\frac{d}{dt}\frac{\partial L}{\partial \dot{\theta}} - \frac{\partial L}{\partial \theta} = \frac{d}{dt}\left(m_P l^2\dot{\theta}\right) - (m_P l^2\Omega^2 \sin\theta \cos\theta - m_P gl \sin\theta) = 0,$$

which results in the following equation of motion:

$$\ddot{\theta} + \left(\frac{g}{l} - \Omega^2 \cos\theta\right)\sin\theta = 0. \tag{13.23}$$

Spinning the pendulum changes its natural frequency of oscillation. For small angles the pendulum behaves like a simple harmonic oscillator and the frequency lowers from $\sqrt{g/l}$ to $\sqrt{g/l - \Omega^2}$. For a spin rate $\Omega > \sqrt{g/l}$, the downward equilibrium point becomes unstable and a new equilibrium point appears at $\theta = \pm \arccos(g/l\Omega^2)$. That is, the pendulum revolves around the vertical axis at a fixed offset. (This behavior is also exhibited by the flyball governor (Tutorial 10.2).)

In this example, Lagrange's equations provided real benefit. Spinning the pendulum made it into a three-dimensional problem. To solve it using Newton's method would have required introducing additional constraint forces and eliminating them. Here, little additional work was needed above that for the simple pendulum. Of course, we have lost information—the constraint forces!

For our last example, we re-derive the equations of motion for the two-degree-of-freedom overhead crane with a rigid arm (Example 9.20). Note that the derivation of the Euler-Lagrange equations did not explicitly consider rigid bodies. Nevertheless, all our results apply to them. Remember, rigid bodies are just collections of many particles with many constraints. In general, the implied constraints of a rigid body collapse the number of degrees of freedom in a multiparticle system to six (three for translation of the center of mass plus three for rotation). As long as we pick appropriate generalized coordinates (e.g., the Euler angles) and properly write the kinetic and potential energies in terms of those generalized coordinates, the Euler-Lagrange equations can be used without modification. The whole point of the energy approach was to eliminate the need to consider constraints and constraint forces explicitly.

Example 13.9 The Overhead Crane Using Lagrange's Equations

As before, we begin with the kinematics. From Example 9.20 we have

$$\mathbf{r}_{Q/O} = x\mathbf{e}_x$$

$$^{\mathcal{I}}\mathbf{v}_{Q/O} = \dot{x}\mathbf{e}_x$$

for the block and, for P, the center of mass of the arm,

$$\mathbf{r}_{P/O} = x\mathbf{e}_x + l\mathbf{e}_r$$

$$^{\mathcal{I}}\mathbf{v}_{P/O} = \dot{x}\mathbf{e}_x + l\dot{\theta}\mathbf{e}_\theta.$$

The rigid arm rotates at angular velocity

$$^{\mathcal{I}}\boldsymbol{\omega}^{\mathcal{B}} = \dot{\theta}\mathbf{e}_z.$$

These expressions let us find the total kinetic energy using the separation property in Eq. (9.38):

$$T_O = T_{Q/O} + T_{P/O} + T_P = \frac{1}{2}m_Q\dot{x}^2 + \frac{1}{2}m_P\dot{x}^2 + \frac{1}{2}m_Pl^2\dot{\theta}^2 + m_Pl\dot{x}\dot{\theta}\mathbf{e}_x \cdot \mathbf{e}_\theta + \frac{1}{2}I_P\dot{\theta}^2$$

$$= \frac{1}{2}(m_Q + m_P)\dot{x}^2 + \frac{1}{2}(I_P + m_Pl^2)\dot{\theta}^2 + m_Pl\dot{x}\dot{\theta}\cos\theta,$$

where I_P is the moment of inertia of the arm about P.

The potential energy is due to raising the center of mass of the rod, so

$$U_O = U_{P/O} = -m_Pgl\cos\theta.$$

The Lagrangian is $L = T_O - U_O$. We now use the Euler-Lagrange equations separately on the generalized coordinates x and θ to find

$$\frac{\partial L}{\partial \dot{x}} = (m_Q + m_P)\dot{x} + m_Pl\dot{\theta}\cos\theta$$

$$\frac{\partial L}{\partial x} = 0$$

$$\frac{\partial L}{\partial \dot{\theta}} = (I_P + m_Pl^2)\dot{\theta} + m_Pl\dot{x}\cos\theta$$

$$\frac{\partial L}{\partial \theta} = -m_Pl\dot{x}\dot{\theta}\sin\theta - m_Pgl\sin\theta.$$

Substituting into the Euler-Lagrange equations yields the equations of motion,

$$(m_Q + m_P)\ddot{x} + m_Pl\ddot{\theta}\cos\theta - m_Pl\dot{\theta}^2\sin\theta = 0 \qquad (13.24)$$

$$(I_P + m_Pl^2)\ddot{\theta} + m_Pgl\sin\theta + m_Pl\ddot{x}\cos\theta = 0, \qquad (13.25)$$

which are identical to Eqs. (9.56) and (9.57).

Note that we could also introduce a horizontal control force $u\mathbf{e}_x$ on the block, as in Example 7.7. Since the force is in the direction of the generalized coordinate x, this force would add u to the right-hand side of Eq. (13.24).

13.3.4 Nonholonomic Constraints and Lagrange Multipliers

The essential step in our derivation of the Euler-Lagrange equations came right after Eq. (13.18). At that point we restricted the discussion to holonomic systems, where $N_C = M$, the number of degrees of freedom. Then the variations of the generalized coordinates were independent, thus making the only solution to Eq. (13.18) the one where all of the coefficients are zero, resulting in the Euler-Lagrange equations (Eq. (13.19)). What happens when there are nonholonomic constraints? We have already seen that, for such systems, there are more generalized coordinates than degrees of freedom ($N_C > M$). Thus the variations of the generalized coordinates in Eq. (13.18) need not be independent to be consistent with the constraints. We need to perform an additional step.

Returning to the differential form of the nonholonomic-constraint equation in Eq. (13.6), we see that, for virtual displacements of the generalized coordinates (where time is frozen), the constraint equation can be written as

$$\sum_{i=1}^{N_C} a_{li}\delta q_i = 0, \quad l = 1, \ldots, S. \tag{13.26}$$

We now introduce a *Lagrange multiplier* for each of the S nonholonomic constraints by multiplying Eq. (13.26) by λ_l and summing over all constraint equations:

$$\sum_{l=1}^{S}\sum_{i=1}^{N_C} \lambda_l a_{li}\delta q_i = 0. \tag{13.27}$$

Since Eq. (13.27) is equal to zero, we can add it to Eq. (13.18) without affecting the result. Factoring out the virtual change in the generalized coordinates gives the new form:

$$\sum_{i=1}^{N_C}\left(\frac{{}^\mathcal{I}d}{dt}\frac{{}^\mathcal{I}\partial T_O}{\partial \dot{q}_i} - \frac{{}^\mathcal{I}\partial T_O}{\partial q_i} - Q_i - \sum_{l=1}^{S}\lambda_l a_{li}\right)\delta q_i = 0. \tag{13.28}$$

Note that we have added an additional S unknowns to the problem, the λ_l. Thus even though the variations δq_i are not independent, we can choose the λ_l such that the only solution to Eq. (13.28) is the one in which each coefficient is zero. This choice results in the slightly modified form of the Euler-Lagrange equations:

$$\frac{d}{dt}\frac{\partial T_O}{\partial \dot{q}_i} - \frac{\partial T_O}{\partial q_i} = Q_i + \sum_{l=1}^{S}\lambda_l a_{li}, \quad i = 1, \ldots, N_C. \tag{13.29}$$

These equations can be used as before to find the equations of motion for the N_C generalized coordinates. How do we find the λ_l? There are only N_C equations, but we now have $N_C + S$ unknowns, $q_1, \ldots, q_{N_C}, \lambda_1, \ldots, \lambda_S$. The solution is to solve the equations of motion simultaneously with the constraint equations. Thus combining the Euler-Lagrange equations in Eq. (13.29) with the nonholonomic-constraint equation in Eq. (13.5) yields $N_C + S$ differential equations.

This approach should look familiar. Throughout the book we have developed equations of motion that contained unknown constraint forces and often had to use the constraint equations to eliminate them. The same is true here. Since there are more coordinates than degrees of freedom, we are implicitly introducing the constraint forces associated with the nonholonomic constraints and we need to use the constraint equations to eliminate them. In fact, the Lagrange multipliers λ_l are these constraint forces! Notice that they enter into Eq. (13.29) just as the generalized forces do. Thus we sometimes identify that term as the *generalized constraint force:*

$$R_i \triangleq \sum_{l=1}^{S}\lambda_l a_{li}.$$

The result is that using Lagrange's approach for nonholonomic systems does not have the same benefit of removing the constraint forces as with holonomic systems, but it does still provide a purely analytical approach to finding the equations of motion.

Example 13.10 Kinetics of a Simple Nonholonomic System

This example uses the Euler-Lagrange equations in Eq. (13.29) to find the equations of motion for the simple nonholonomic system described in Example 13.4 and shown in Figure 13.3. This system has two degrees of freedom but is described by the three coordinates $q_1 = x_G$, $q_2 = y_G$, and $q_3 = \theta$, where $(x_G, y_G)_\mathcal{I}$ are the Cartesian coordinates of the center of mass.

To find the total kinetic energy with respect to O, we use the separation property (Eq. (9.38)) and sum the kinetic energy of the center of mass with the kinetic energy of motion about the center of mass. The velocity of the center of mass is

$$^\mathcal{I}\mathbf{v}_{G/O} = \dot{q}_1\mathbf{e}_x + \dot{q}_2\mathbf{e}_y,$$

which makes the kinetic energy of the center of mass

$$T_{G/O} = \frac{1}{2}(m_P + m_Q)(\dot{q}_1^2 + \dot{q}_2^2).$$

For the kinetic energy about the center of mass, it is easiest to treat the system as a rigid body with angular velocity

$$^\mathcal{I}\boldsymbol{\omega}^\mathcal{B} = \dot{q}_3\mathbf{b}_3.$$

This expression makes the kinetic energy about the center of mass

$$T_G = \frac{1}{2}I_G\|^\mathcal{I}\boldsymbol{\omega}^\mathcal{B}\|^2 = \frac{1}{8}(m_P + m_Q)l^2\dot{q}_3^2,$$

where we have used the fact that the moment of inertia about G is $I_G = \frac{1}{4}(m_P + m_Q)l^2$.

Thus the total kinetic energy is

$$T_O = \frac{1}{2}(m_P + m_Q)(\dot{q}_1^2 + \dot{q}_2^2 + \frac{1}{4}l^2\dot{q}_3^2).$$

Plugging this expression into the Euler-Lagrange equations (Eq. (13.29)) with Lagrange multiplier λ corresponding to the single nonholonomic constraint yields three equations of motion (using $a_{11} = \cos q_3$, $a_{12} = \sin q_3$, and $a_{13} = a_{1t} = 0$ from Example 13.4):

$$(m_P + m_Q)\ddot{q}_1 = \lambda \cos q_3 \tag{13.30}$$

$$(m_P + m_Q)\ddot{q}_2 = \lambda \sin q_3 \tag{13.31}$$

$$(m_P + m_Q)\frac{l^2}{4}\ddot{q}_3 = 0. \tag{13.32}$$

Since λ is unknown, Eqs. (13.30) and (13.31) are solved simultaneously with the constraint equation (Eq. (13.8))

$$\dot{q}_1 \cos q_3 + \dot{q}_2 \sin q_3 = 0.$$

To eliminate λ, we first differentiate the constraint equation:

$$\ddot{q}_1 \cos q_3 + \ddot{q}_2 \sin q_3 - \dot{q}_1 \dot{q}_3 \sin q_3 + \dot{q}_2 \dot{q}_3 \cos q_3 = 0.$$

We then multiply Eq. (13.30) by $\cos q_3$ and Eq. (13.31) by $\sin q_3$ and add the result, giving us

$$\lambda = (m_P + m_Q)(\ddot{q}_1 \cos q_3 + \ddot{q}_2 \sin q_3) = (m_P + m_Q)(\dot{q}_1 \dot{q}_3 \sin q_3 - \dot{q}_2 \dot{q}_3 \cos q_3).$$

$$(13.33)$$

Eq. (13.33) can be substituted back into the equations of motion (Eqs. (13.30) and (13.31)) to get the final equations of motion for the system:

$$\ddot{q}_1 - (\dot{q}_1 \sin q_3 \cos q_3 - \dot{q}_2 \cos^2 q_3)\dot{q}_3 = 0$$

$$\ddot{q}_2 - (\dot{q}_1 \sin^2 q_3 - \dot{q}_2 \sin q_3 \cos q_3)\dot{q}_3 = 0$$

$$\ddot{q}_3 = 0.$$

Observe also that the Lagrange multiplier in Eq. (13.33) is the constraint force perpendicular to the skate blades.

13.3.5 Why Not Always Use Lagrange's Method?

At this point, many students wonder why anyone would bother finding equations of motion using Newton's method when Lagrange's technique seems so much simpler and more elegant. What could be more straightforward than finding the kinetic and potential energies and plugging them into the Euler-Lagrange equations? It is true that many problems are more easily solved this way. But not all. We close this introduction to Lagrange's method by emphasizing that this method is just one of the family of tools available for finding equations of motion. In particular, as we've seen, Lagrange's method is not as useful if there is a need to find the forces of constraint. It is also not nearly as convenient when there are nonconservative forces in a problem (e.g., Example 13.6).

Most significantly, to use the Euler-Lagrange equations, you must describe the system entirely in terms of the generalized coordinates and their rates $(q, \dot{q})$. Sometimes this restriction is undesirable. In many dynamical systems, we prefer to use the coordinates and some other speed-like variables to describe the system. For example, recall the development of the equations of motion of the airplane in Chapters 10 and 11. Although the position of the plane was described in the inertial frame by means of Cartesian coordinates, we chose to write the velocity of the plane as components in the body frame and use the scalar magnitudes there in the equations of motion. This was because the aerodynamic forces are best found in terms of the body-fixed speeds. This description is not possible using Lagrange's method. Likewise, we wrote the rigid-body equations in terms of the Euler angles and the angular velocity components rather than the Euler-angle rates. This strategy allowed us to use Euler's equations for the rotational equations of motion. In fact, many rigid-body problems are most easily solved using the angular-velocity components. Again, this description of rigid-body motion is not possible with Lagrange's equations: they produce only second-order differential equations in the Euler angles. In fact, most rigid-body problems are best solved by starting with angular momentum and angular velocity. Thus the approach you choose depends on the problem and how you wish to describe the system. Whatever your choice, the final equations of motion will be the same for the same coordinates.

13.4 Kane's Method

Thomas Kane introduced an alternative approach to finding equations of motion for systems of rigid bodies and particles. Part of a class of techniques known as *projection methods,* his approach has come to be called *Kane's method.* It is increasingly used for finding the equations of motion of a complex system. His technique is similar to Lagrange's method in that it formulates equations only for the generalized coordinates representing the degrees of freedom in the problem. The need to introduce constraint forces and later eliminate them using the constraint equations is gone. However, the method is not based on formulating the kinetic and potential energies but rather on projecting Newton's second law onto the space defined by the generalized coordinates. The result is a systematic method for finding equations of motion that is easily automated. In fact, there are computer software applications available that do just that.[3] It is also easily scaled and becomes very attractive for complex dynamic systems with many degrees of freedom, especially those combining rigid bodies and particles.

Another advantage of Kane's approach is its treatment of nonholonomic constraints. Kane's formalism is unchanged whether constraints are holonomic or nonholonomic. This is part of the appeal of his systematic presentation of the equations of motion. Nevertheless, for brevity, we only consider holonomic systems here. See Kane and Levinson (1985) for a description of how to incorporate nonholonomic constraints.

Finally, as Kane's method automatically produces the differential equations of motion in first-order form, it is easily combined with numerical-integration techniques.

13.4.1 Generalized Speeds

Recall that in the beginning of the book (and again in Chapter 10), we began by describing the configuration of a particle using Cartesian coordinates and determining its velocity by using their rates. The rate of change of each Cartesian coordinate was called the speed. However, we quickly discovered that other scalar coordinate systems are greatly beneficial for certain problems. This observation led to expressions for the velocity of a particle in terms of rates of change of these new coordinates, expressions that could get rather complicated. Nevertheless, we used these expressions for position and velocity to find the acceleration and then, from Newton's second law,

---

[3] Such as Autolev™, available from Online Dynamics, Inc.

the equations of motion. These equations of motion were second-order differential equations that could be integrated to find trajectories—solutions for each coordinate and its rate as functions of time. In dynamics, the state of the system is fully described by the scalar coordinates and their rates.

At the beginning of this chapter we extended this idea through the introduction of generalized coordinates. These are simply a symbolic representation of all possible coordinates that can be used to define the configuration of a system. The state of the system is fully defined by $q_i(t)$ and $\dot{q}_i(t)$. Lagrange's method is a technique for finding the second-order differential equations of motion for the generalized coordinates. These again can be integrated (analytically or numerically) to find the trajectories for $q_i(t)$ and $\dot{q}_i(t)$.

In Kane's approach we shift our thinking slightly. Rather than find M second-order differential equations of motion for the M generalized coordinates, we instead develop $2M$ first-order differential equations. By now you have tried to numerically integrate some of the equations of motion studied in the book, so this idea should be familiar. (You may want to review Appendix C.) When numerically integrating a set of second-order differential equations, we first convert them into twice as many first-order ones. In most cases, we define two new variables: the first is the coordinate of interest and the second is its rate. Thus the first of the two first-order equations is trivial; it just sets the time derivative of the first variable to the rate of change of the coordinate. The second equation is the equation of motion written in the new variables.

The first step in developing Kane's method is to generalize this idea. Rather than restrict ourselves to the generalized coordinates and their rates as the state of the system, we introduce the generalized speed u_r. For a system with M degrees of freedom and M generalized coordinates $q_1, \ldots, q_M$, we have the kinematic equations,

$$u_r \triangleq \sum_{s=1}^{M} Y_{rs}(q_1, \ldots, q_M, t)\dot{q}_s + Z_r(q_1, \ldots, q_M, t) \quad r = 1, \ldots, M, \quad (13.34)$$

where Y_{rs} and Z_r are functions of the generalized coordinates.

Eq. (13.34) is essentially a coordinate transformation. We are transforming the description of the system state from $(q_r, \dot{q}_r)$ to (q_r, u_r). As shown below, it is required that this transformation be invertible. That is, there must exist an inverse transformation:

$$\dot{q}_s = \sum_{r=1}^{M} W_{sr}(q_1, \ldots, q_M, t)u_r + X_s(q_1, \ldots, q_M, t), \quad s = 1, \ldots, M. \quad (13.35)$$

Kane's equations of motion produce first-order differential equations for the generalized speeds u_r. When combined with Eq. (13.35), there are $2M$ first-order differential equations to integrate for the state of the system described by $(q_r, u_r)_T$. Kane recognized that the objective of many engineering dynamics problems is to find equations of motion to be numerically integrated. By introducing the extra flexibility in the rate variable (by means of the generalized speeds), his method allows for a systematic derivation of compact equations of motion that may, for example, have a particularly efficient form.

You may have noticed that there is no specific rule for choosing the generalized speeds. They are, in fact, arbitrary. Just as we relied on our skill for choosing the best set of coordinates, so it is with the generalized speeds. Often there is a natural choice that makes finding the equations of motion most straightforward. For instance, for a simple system with no constraints, the generalized speeds are typically just the component magnitudes of the velocity of a particle. For example, if we are using Cartesian coordinates in the plane, the generalized speeds are $u_1 = \dot{x}$ and $u_2 = \dot{y}$. If, however, we choose polar coordinates, then we might choose $u_1 = \dot{r}$ and $u_2 = r\dot{\theta}$, the velocity component magnitudes in the polar frame.

We have already seen an example of the use of generalized speeds to find equations of motion. When studying three-dimensional rigid bodies in Chapter 10, Euler angles were used as generalized coordinates for describing the orientation of the rigid body. However, when developing the equations of motion of a rigid body in Chapter 11, we used the angular velocity rather than the Euler angles and their rates. This choice results in Euler's equations (Eq. (11.39)). The component magnitude of an angular velocity is an example of a generalized speed, with Euler's equations providing the corresponding equation of motion. In fact, we later use Kane's method to rederive Euler's equations. To find a complete trajectory of the rigid body, we integrate Euler's equations together with the three kinematic equations in Eq. (10.48) relating the angular velocities to the rates of change of the Euler angles (the generalized coordinates). Eq. (10.48) is simply a specific example of the general expression relating the rates of change of the generalized coordinates to the generalized speeds in Eq. (13.34). (Eq. (10.49) shows the inverse relationship corresponding to Eq. (13.35).)

Example 13.11 Generalized Speeds in Polar Coordinates

Section 3.3.3 introduced velocity and acceleration in various planar coordinate systems. In particular, Eq. (3.18) showed that the inertial velocity of a particle in polar coordinates and expressed as components in the inertial frame is

$$^{\mathcal{I}}\mathbf{v}_{P/O} = (\dot{r}\cos\theta - r\dot{\theta}\sin\theta)\mathbf{e}_1 + (\dot{r}\sin\theta + r\dot{\theta}\cos\theta)\mathbf{e}_2.$$

The generalized coordinates are the polar coordinates $(q_1 = r, q_2 = \theta)_{\mathcal{I}}$. We write the generalized speeds as

$$u_1 = \cos\theta\dot{r} - r\sin\theta\dot{\theta} \tag{13.36}$$

$$u_2 = \sin\theta\dot{r} + r\cos\theta\dot{\theta}, \tag{13.37}$$

so the inertial velocity in terms of the generalized speeds is $^{\mathcal{I}}\mathbf{v}_{P/O} = u_1\mathbf{e}_1 + u_2\mathbf{e}_2$. Eqs. (13.36) and (13.37) are the kinematic equations (Eq. (13.34)) with $Z_r = 0$ and

$$Y_{11} = \cos\theta$$

$$Y_{12} = -r\sin\theta$$

$$Y_{21} = \sin\theta$$

$$Y_{22} = r\cos\theta.$$

Alternatively, we may wish to write the inertial velocity in the polar frame in terms of generalized speeds: ${}^{\mathcal{I}}\mathbf{v}_{P/O} = u_1\mathbf{e}_r + u_2\mathbf{e}_\theta$. Now the generalized speeds are

$$u_1 = \dot{r} \tag{13.38}$$

$$u_2 = r\dot{\theta}. \tag{13.39}$$

In this case,

$$Y_{11} = 1$$
$$Y_{12} = 0$$
$$Y_{21} = 0$$
$$Y_{22} = r.$$

Example 13.12 The Generalized Speeds for a 3-1-3 Rotation

Example 10.7 described the orientation of a set of gimbals using a 3-1-3 set of Euler angles and derived the corresponding direction-cosine matrix. Here the generalized coordinates are the three angles $(q_1 = \psi, q_2 = \theta, q_3 = \phi)_{\mathcal{B}}^{\mathcal{I}}$ describing the orientation of body frame $\mathcal{B}$ in frame $\mathcal{I}$. Example 10.10 derived the kinematic equations for the angular velocity components in $\mathcal{B}$:

$$\omega_1 = \dot{\theta}\cos\phi + \dot{\psi}\sin\theta\sin\phi$$

$$\omega_2 = -\dot{\theta}\sin\phi + \dot{\psi}\sin\theta\cos\phi$$

$$\omega_3 = \dot{\phi} + \dot{\psi}\cos\theta.$$

Here the generalized speeds are the angular velocities ($u_1 = \omega_1$, $u_2 = \omega_2$, and $u_3 = \omega_3$), which gives

$$Y_{11} = \sin\theta\sin\phi$$
$$Y_{12} = \cos\phi$$
$$Y_{13} = 0$$
$$Y_{21} = \sin\theta\cos\phi$$
$$Y_{22} = -\sin\phi$$
$$Y_{23} = 0$$
$$Y_{31} = \cos\theta$$
$$Y_{32} = 0$$
$$Y_{33} = 1.$$

13.4.2 Partial Velocities and Partial Angular Velocities

In our discussions of kinematics in Chapters 3, 8, and 10, the objective was to find expressions for the velocity of a particle in terms of various coordinates and their rates. Sometimes these expressions were simple, but often component magnitudes were rather complex functions of the coordinates and their rates; this complexity often depended on which frame we chose to express them in. Nevertheless, it was always possible to find an expression for the velocity in terms of the coordinate system chosen. In fact, Section 13.2 developed general formulas for writing the position and velocity of a particle in terms of the generalized coordinates and their rates. Here our goal is to rewrite these formulas in terms of the generalized coordinates and the generalized speeds. This process is actually quite straightforward.

In principle, we can simply substitute for $\dot{q}_r$ from Eq. (13.35) in our previous expressions, thus replacing $\dot{q}_r$ with u_r. Let ${}^{\mathcal{I}}\mathbf{v}_r^P \triangleq \frac{{}^{\mathcal{I}}\partial}{\partial u_r}({}^{\mathcal{I}}\mathbf{v}_{P/O})$, which is called the rth *partial velocity of P in $\mathcal{I}$*. The partial velocities are functions of the generalized coordinates, $q_1, \ldots, q_M$, and the unit vectors of the frame in which we are expressing ${}^{\mathcal{I}}\mathbf{v}_{P/O}$. With this new notation, we can write the velocity of particle P as

$$
{}^{\mathcal{I}}\mathbf{v}_{P/O} \triangleq \sum_{r=1}^{M} {}^{\mathcal{I}}\mathbf{v}_r^P u_r + {}^{\mathcal{I}}\mathbf{v}_t^P. \tag{13.40}
$$

This is a bit different from the usual notation for vector quantities. Rather than distribute the unit vectors, Eq. (13.40) instead distributes the generalized speeds. The reason will become apparent later, when we develop the equations of motion.

Likewise, if there are rigid bodies in the system, the angular velocity of rigid body $\mathcal{B}$ with attached body frame B is

$$
{}^{\mathcal{I}}\boldsymbol{\omega}^B \triangleq \sum_{r=1}^{M} {}^{\mathcal{I}}\boldsymbol{\omega}_r^B u_r + {}^{\mathcal{I}}\boldsymbol{\omega}_t^B,
$$

where ${}^{\mathcal{I}}\boldsymbol{\omega}_r^B \triangleq \frac{{}^{\mathcal{I}}\partial}{\partial u_r}({}^{\mathcal{I}}\boldsymbol{\omega}^B)$ is the rth *partial angular velocity of B in $\mathcal{I}$*. It is worth noting again that there is always a unique expression for the velocity and angular velocity of the particles and rigid bodies in terms of the partial velocities, partial angular velocities, and chosen generalized speeds. Most of the time we don't define the generalized speeds first and then find the partial velocities as done here. Rather, we usually identify the generalized speeds based on the vector expressions for the velocities and angular velocities. This procedure automatically provides the kinematic equations in Eq. (13.34) and the resulting partial velocities by inspection.

Example 13.13 Partial Velocities in Polar Coordinates

Continuing Example 13.11, we can find the partial velocities associated with the generalized speeds in polar coordinates. Using the expression for the velocity as components in the inertial frame,

$$
{}^{\mathcal{I}}\mathbf{v}_{P/O} = (\dot{r}\cos\theta - r\dot{\theta}\sin\theta)\mathbf{e}_1 + (\dot{r}\sin\theta + r\dot{\theta}\cos\theta)\mathbf{e}_2,
$$

and the associated generalized speeds in Eqs. (13.36) and (13.37), the two partial velocities are, by inspection,

$$
{}^{\mathcal{I}}\mathbf{v}_1^P = \mathbf{e}_1
$$

$$
{}^{\mathcal{I}}\mathbf{v}_2^P = \mathbf{e}_2.
$$

If we instead write the velocity as components in the polar frame, $^{\mathcal{I}}\mathbf{v}_{P/O} = u_1 \mathbf{e}_r + u_2 \mathbf{e}_\theta$, then using the generalized speeds in Eqs. (13.38) and (13.39) gives

$$^{\mathcal{I}}\mathbf{v}_1^P = \mathbf{e}_r$$

$$^{\mathcal{I}}\mathbf{v}_2^P = \mathbf{e}_\theta.$$

Example 13.14 Generalized Speeds and Partial Velocities for a Spinning Pendulum

This example revisits the spinning pendulum of Example 13.8. This single-degree-of-freedom system is described by the generalized coordinate $q_1 = \theta$. The inertial velocity of the pendulum expressed as components in the rotating body frame is given by Eq. (13.22):

$$^{\mathcal{I}}\mathbf{v}_{P/O} = l\dot{\theta}\mathbf{e}_\theta + l\Omega \sin\theta \mathbf{e}_z.$$

A sensible choice for the generalized speed comes from the polar coordinate description, $u_1 = l\dot{\theta}$. Thus there are two partial velocities for this system. Using Eq. (13.40), we find

$$^{\mathcal{I}}\mathbf{v}_1^P = \mathbf{e}_\theta$$

$$^{\mathcal{I}}\mathbf{v}_t^P = l\Omega \sin\theta \mathbf{e}_z.$$

Example 13.15 Generalized Speeds and Partial Velocities for an Overhead Crane

This example determines the generalized speeds and partial velocities for the overhead crane of Example 13.9, a system combining a particle and a rigid body. This is another two-degree-of-freedom system with generalized coordinates $q_1 = x$ and $q_2 = \theta$. The velocity of the block is

$$^{\mathcal{I}}\mathbf{v}_{Q/O} = \dot{x}\mathbf{e}_x,$$

which means the generalized speed associated with q_1 is $u_1 = \dot{x}$ and the partial velocities for Q are

$$^{\mathcal{I}}\mathbf{v}_1^Q = \mathbf{e}_x \tag{13.41}$$

$$^{\mathcal{I}}\mathbf{v}_2^Q = 0 \tag{13.42}$$

$$^{\mathcal{I}}\mathbf{v}_t^Q = 0. \tag{13.43}$$

The second generalized speed comes from the angular velocity of the rigid arm:

$$^{\mathcal{I}}\boldsymbol{\omega}^B = \dot{\theta}\mathbf{e}_z.$$

Since the system is constrained to the plane, there is only a single degree of freedom in rotation of the rigid body. The generalized speed is thus $u_2 = \dot{\theta}$ and the partial angular velocities are

$$^{\mathcal{I}}\boldsymbol{\omega}_1^B = 0 \tag{13.44}$$

$$^{\mathcal{I}}\boldsymbol{\omega}_2^B = \mathbf{e}_z \tag{13.45}$$

$$^{\mathcal{I}}\boldsymbol{\omega}_t^B = 0. \tag{13.46}$$

Finally, we need to find the partial velocities associated with the translation of the rigid arm. We could simply use the partial velocities above for Q, where the arm is connected, or we could find the partial velocities associated with the center-of-mass motion of the arm. The latter comes from the velocity of the center of mass from Example 13.9:

$$^{\mathcal{I}}\mathbf{v}_{P/O} = \dot{x}\mathbf{e}_x + l\dot{\theta}\mathbf{e}_\theta.$$

The partial velocities of P are

$$^{\mathcal{I}}\mathbf{v}_1^P = \mathbf{e}_x \tag{13.47}$$

$$^{\mathcal{I}}\mathbf{v}_2^P = l\mathbf{e}_\theta \tag{13.48}$$

$$^{\mathcal{I}}\mathbf{v}_t^P = 0. \tag{13.49}$$

13.4.3 Generalized Active and Inertia Forces

The penultimate step before finding the equations of motion is to define the *generalized active* and *inertia forces*. These form the basis of Kane's equations of motion. Writing $\mathbf{F}_i$ as the total force on particle i in a collection of N particles (including all contact, constraint, and field forces), we define the generalized active force associated with the degree of freedom $r = 1, \ldots, M$ as

$$F_r \triangleq \sum_{i=1}^N {}^{\mathcal{I}}\mathbf{v}_r^i \cdot \mathbf{F}_i, \quad r = 1, \ldots, M, \tag{13.50}$$

where $^{\mathcal{I}}\mathbf{v}_r^i$ are the partial velocities of i associated with the generalized speeds $u_1, \ldots, u_r$.

The most important observation about the generalized active forces is that they are independent of the constraint forces. By taking the dot product with the partial velocities we have "projected out" the forces of constraint. In other words, we need only to consider the active forces in computing the generalized active forces in Eq. (13.50). A formal proof of this is somewhat involved, but we provide a sketch here.

We begin by noting that the forces of constraint on a particle must always be perpendicular to the velocity, which implies

$$^{\mathcal{I}}\mathbf{v}_{i/O} \cdot \mathbf{F}_i^{(\text{const})} = 0.$$

Dotting both sides of Eq. (13.40) with $\mathbf{F}_i^{(\text{const})}$ gives

$$\sum_{r=1}^{M} u_r \left({}^{\mathcal{I}}\mathbf{v}_r^i \cdot \mathbf{F}_i^{(\text{const})} \right) + {}^{\mathcal{I}}\mathbf{v}_t^i \cdot \mathbf{F}_i^{(\text{const})} = 0.$$

Now we make an argument similar to the one made when discussing virtual work in Lagrange's equations (Section 13.3.1). Because there are the same number of generalized speeds as degrees of freedom, each u_r can independently vary and remain consistent with the constraints. So, for instance, we can consider a stationary particle ($u_r = 0$), which implies the second dot product is zero, leaving

$$\sum_{r=1}^{M} u_r \left({}^{\mathcal{I}}\mathbf{v}_r^i \cdot \mathbf{F}_i^{(\text{const})} \right) = 0.$$

The only way for this equation to be true for all u_r is for the quantity in parentheses to be zero (remember, the u_r are linearly independent), which implies

$$ {}^{\mathcal{I}}\mathbf{v}_r^i \cdot \mathbf{F}_i^{(\text{const})} = 0.$$

This relation is what we set out to prove. The importance of this result will become clear in the next section.

The generalized active forces associated with a rigid body $\mathscr{B}$ and attached body frame B are similarly defined as

$$(F_r)_{\mathscr{B}} \triangleq {}^{\mathcal{I}}\boldsymbol{\omega}_r^B \cdot \mathbf{M}_Q + {}^{\mathcal{I}}\mathbf{v}_r^Q \cdot \mathbf{F}_Q, \quad r = 1, \ldots, M, \tag{13.51}$$

where ${}^{\mathcal{I}}\boldsymbol{\omega}_r^B$ is the rth partial angular velocity of B in $\mathcal{I}$, $\mathbf{M}_Q$ is the total moment acting on $\mathscr{B}$ relative to point Q (which may or may not be the center of mass), $\mathbf{F}_Q$ is the net force acting on $\mathscr{B}$ at the point Q, and ${}^{\mathcal{I}}\mathbf{v}_r^Q$ is the rth partial velocity of Q in $\mathcal{I}$. It can similarly be shown that the constraint forces on the rigid body have been projected out and the resulting generalized active force is independent of the constraints.

The generalized inertia forces are defined in a similar manner, taking the dot product of the partial velocities with the mass times acceleration (often called the *inertia force* on a particle):

$$F_r^* \triangleq \sum_{i=1}^{N} {}^{\mathcal{I}}\mathbf{v}_r^i \cdot (-m_i {}^{\mathcal{I}}\mathbf{a}_{i/O}), \quad r = 1, \ldots, M. \tag{13.52}$$

Unlike Lagrange's method, where the only kinematics required is to find the velocities, here we need to calculate the accelerations, as for Newton's method. To find the generalized inertia forces in terms of the rates of change of the generalized speeds, we typically find the acceleration by means of the derivative of the velocity (taking into account the rates of change of unit vectors) or we apply the general equation for the acceleration using rotating frames in Eq. (10.52). In either case we substitute for the rates of change of the generalized coordinates in terms of the generalized speeds from Eq. (13.35).

For a rigid body $\mathcal{B}$ the generalized inertia force is found by summing the *inertia torque* and the *inertia force*:

$$(F_r)_{\mathcal{B}}^* \triangleq {}^{\mathcal{I}}\boldsymbol{\omega}_r^{\mathcal{B}} \cdot \left(-\mathbb{I}_G \cdot {}^{\mathcal{I}}\boldsymbol{\alpha}^{\mathcal{B}} - {}^{\mathcal{I}}\boldsymbol{\omega}^{\mathcal{B}} \times (\mathbb{I}_G \cdot {}^{\mathcal{I}}\boldsymbol{\omega}^{\mathcal{B}}) \right)$$
$$+ {}^{\mathcal{I}}\mathbf{v}_r^G \cdot \left(-m_G {}^{\mathcal{I}}\mathbf{a}_{G/O} \right), \quad r = 1, \ldots, M, \qquad (13.53)$$

where Eq. (11.26) has been used to write the general form of the rate of change of angular momentum of $\mathcal{B}$.

As in Chapter 11, we can also consider motion about a point of the rigid body that is not the center of mass. Note that Eq. (13.51) computes the generalized active forces on the rigid body when the force acts at an arbitrary point Q and the moments are relative to Q. If that point is not the center of mass, we need to modify the generalized inertia forces to account for the coupling term in the angular momentum equation:

$$(F_r)_{\mathcal{B}}^* \triangleq {}^{\mathcal{I}}\boldsymbol{\omega}_r^{\mathcal{B}} \cdot \left(-\mathbb{I}_Q \cdot {}^{\mathcal{I}}\boldsymbol{\alpha}^{\mathcal{B}} - {}^{\mathcal{I}}\boldsymbol{\omega}^{\mathcal{B}} \times \left(\mathbb{I}_Q \cdot {}^{\mathcal{I}}\boldsymbol{\omega}^{\mathcal{B}} \right) + \mathbf{r}_{Q/G} \times m_G {}^{\mathcal{I}}\mathbf{a}_{Q/O} \right)$$
$$+ {}^{\mathcal{I}}\mathbf{v}_r^Q \cdot (-m_G {}^{\mathcal{I}}\mathbf{a}_{Q/O}), \quad r = 1, \ldots, M,$$

where the moment of inertia tensor about Q is found from the parallel axis theorem.

13.4.4 Kane's Equations

With the definition of the partial velocities, partial angular velocities, and the generalized active and inertia forces in hand, we can turn to Kane's simple statement of the M scalar equations of motion for the generalized speeds (corresponding to the M degrees of freedom):

$$\boxed{F_r + F_r^* = 0, \quad r = 1, \ldots, M,} \qquad (13.54)$$

where the F_r are the generalized active forces and the F_r^* are the generalized inertia forces.

Eq. (13.54) follows directly from Newton's second law by taking the dot product of d'Alembert's form (Eq. (13.9)) for each particle with each of the partial velocities and summing over all particles:

$$\sum_{i=1}^{N} \left(\mathbf{F}_i^{(\text{act})} - m_i {}^{\mathcal{I}}\mathbf{a}_{i/O} \right) \cdot {}^{\mathcal{I}}\mathbf{v}_r^i = 0. \qquad (13.55)$$

Distributing the dot product and using the definition of the generalized active forces from Eq. (13.50) and the generalized inertia forces from Eq. (13.52) yields Eq. (13.54). As before, because the constraint forces have been projected out by the dot product with the partial velocities, the equations of motion in Eq. (13.54) are independent of the constraint forces.

For systems with rigid bodies, the generalized active forces from Eq. (13.51) and the generalized inertia forces from Eq. (13.53) are simply included in Eq. (13.54). The derivation involves using the separation principle for angular momentum. Adding the angular-momentum form of Newton's second law in Eq. (11.26) to Newton's second law for the center of mass as in Eq. (13.55) yields

$$\left(\mathbf{M}_G^{(\mathrm{act})} - \mathbb{I}_G \cdot \frac{{}^{\mathcal{B}}d}{dt} \left({}^{\mathcal{I}}\boldsymbol{\omega}^{\mathcal{B}} \right) - {}^{\mathcal{I}}\boldsymbol{\omega}^{\mathcal{B}} \times \left(\mathbb{I}_G \cdot {}^{\mathcal{I}}\boldsymbol{\omega}^{\mathcal{B}} \right) \right) \cdot {}^{\mathcal{I}}\boldsymbol{\omega}_r^{\mathcal{B}}$$
$$+ \left(\mathbf{F}_G^{(\mathrm{act})} - m_G {}^{\mathcal{I}}\mathbf{a}_{G/O} \right) \cdot {}^{\mathcal{I}}\mathbf{v}_r^G = 0.$$

Again, this method works because the partial velocities are always orthogonal to the constraint forces. The proof for rigid-body motion relative to a point Q is similar and uses Eq. (11.50).

Recall that we showed earlier that the number of partial velocities (and partial angular velocities) corresponds to the number of degrees of freedom of the system (and thus the number of generalized coordinates), so that Eq. (13.54) automatically provides M differential equations for the M unknown generalized speeds. The kinematic equations in Eq. (13.35) provide the other M first-order equations to find the trajectories of the generalized coordinates.

Example 13.16 The Simple Pendulum Using Kane's Method

We return to the simple pendulum to illustrate using Kane's method for a single-degree-of-freedom system. We again use polar coordinates; the single generalized coordinate is $q_1 = \theta$. As in Example 3.11, the inertial kinematics expressed in the polar frame are

$$ {}^{\mathcal{I}}\mathbf{v}_{P/O} = l\dot{\theta}\mathbf{e}_\theta \tag{13.56}$$

$$ {}^{\mathcal{I}}\mathbf{a}_{P/O} = l\ddot{\theta}\mathbf{e}_\theta - l\dot{\theta}^2\mathbf{e}_r. \tag{13.57}$$

A logical choice for the generalized speed is $u_1 = \dot{q}_1 = \dot{\theta}$. From Eq. (13.56), the partial velocity is

$$ {}^{\mathcal{I}}\mathbf{v}_1^P = l\mathbf{e}_\theta. $$

From the free-body diagram in Figure 3.18c, the active force on the pendulum is $\mathbf{F}_P^{(\mathrm{act})} = -m_P g\mathbf{e}_1$. The generalized active force is thus

$$ F_1 = {}^{\mathcal{I}}\mathbf{v}_1^P \cdot \mathbf{F}_P^{(\mathrm{act})} = -m_P gl(\mathbf{e}_\theta \cdot \mathbf{e}_1) $$
$$ = -m_P gl \sin \theta, $$

where we have used the transformation table for the pendulum in Example 3.11 to find the dot product.

The generalized inertia force is found from the expression for the acceleration in Eq. (13.57):

$$ F_1^* = {}^{\mathcal{I}}\mathbf{v}_1^P \cdot (-m_P {}^{\mathcal{I}}\mathbf{a}_{P/O}) = -m_P l^2 (\dot{u}_1 \mathbf{e}_\theta \cdot \mathbf{e}_\theta + u_1^2 \mathbf{e}_\theta \cdot \mathbf{e}_r) $$
$$ = -m_P l^2 \dot{u}_1. $$

Substituting into Kane's equations, $F_1 + F_1^* = 0$, gives a set of first-order equations of motion for (q_1, u_1):

$$\dot{q}_1 = u_1$$

$$\dot{u}_1 = -\frac{g}{l} \sin q_1.$$

These are, of course, the usual equations of motion for the simple pendulum in first-order form. Kane's approach automatically produces a set of two first-order differential equations ready for numerical integration.

Example 13.17 A Spinning Simple Pendulum Using Kane's Method

Next we re-solve for the equation of motion from Example 13.8 using Kane's equations. Recall from Example 13.8 (Eq. (13.22)) that the velocity of the particle is

$$^{\mathcal{I}}\mathbf{v}_{P/O} = l\dot{\theta}\mathbf{e}_\theta + l\Omega \sin \theta \mathbf{e}_z. \tag{13.58}$$

The single generalized coordinate was $q_1 = \theta$. We showed in Example 13.14 that the generalized speed was $u_1 = l\dot{\theta}$ and that the partial velocities are

$$^{\mathcal{I}}\mathbf{v}_1^P = \mathbf{e}_\theta$$

$$^{\mathcal{I}}\mathbf{v}_t^P = l\Omega \sin \theta \mathbf{e}_z.$$

We find the acceleration of the particle in the usual way by taking the derivative of the velocity in Eq. (13.58) and using the angular velocity of the rotating frame:

$$^{\mathcal{I}}\mathbf{a}_{P/O} = \frac{^{\mathcal{I}}d}{dt}\left(^{\mathcal{I}}\mathbf{v}_{P/O}\right) = l\ddot{\theta}\mathbf{e}_\theta + l\dot{\theta}\,{}^{\mathcal{I}}\boldsymbol{\omega}^B \times \mathbf{e}_\theta + l\dot{\theta}\Omega \cos \theta \mathbf{e}_z + l\Omega \sin \theta\,{}^{\mathcal{I}}\boldsymbol{\omega}^B \times \mathbf{e}_z,$$

where the angular velocity is given in Eq. (13.21):

$$^{\mathcal{I}}\boldsymbol{\omega}^B = \dot{\theta}\mathbf{e}_z + \Omega\mathbf{e}_1.$$

Performing the cross products and simplifying leaves

$$^{\mathcal{I}}\mathbf{a}_{P/O} = l\ddot{\theta}\mathbf{e}_\theta - l\dot{\theta}^2\mathbf{e}_r + 2l\dot{\theta}\Omega \cos \theta \mathbf{e}_z - l\Omega^2 \sin \theta \mathbf{e}_y$$

$$= \dot{u}_1\mathbf{e}_\theta - \frac{u_1^2}{l}\mathbf{e}_r + 2u_1\Omega \cos \theta \mathbf{e}_z - l\Omega^2 \sin \theta \mathbf{e}_y, \tag{13.59}$$

where the inertial acceleration is written in terms of the generalized speed.

The only active force acting on the pendulum bob is gravity,

$$\mathbf{F}_P = m_P g \mathbf{e}_1,$$

which yields the generalized active force:

$$F_1 = m_P g \mathbf{e}_1 \cdot \mathbf{e}_\theta = -m_P g \sin \theta.$$

The generalized inertia force is found by taking the dot product of the (negative) acceleration in Eq. (13.59) with the partial velocity:

$$F_1^* = -m_P \left(\dot{u}_1 \mathbf{e}_\theta \cdot \mathbf{e}_\theta - \frac{u_1^2}{l} \mathbf{e}_r \cdot \mathbf{e}_\theta + 2u_1 \Omega \cos\theta \mathbf{e}_z \cdot \mathbf{e}_\theta - l\Omega^2 \sin\theta \mathbf{e}_y \cdot \mathbf{e}_\theta \right)$$

$$= -m_P(\dot{u}_1 - l\Omega^2 \sin\theta \cos\theta).$$

Finally, Kane's equation of motion gives us

$$F_1 + F_1^* = -m_P g \sin\theta - m_P \dot{u}_1 + m_P l\Omega^2 \sin\theta \cos\theta = 0.$$

Combining this expression with the kinematic equation for u_1 and simplifying yields

$$\dot{\theta} = \frac{u_1}{l}$$

$$\dot{u}_1 = -(g - l\Omega^2 \cos\theta)\sin\theta,$$

which is the equivalent of Eq. (13.23) in first-order form.

Example 13.18 The Overhead Crane Using Kane's Method

This example treats a two-degree-of-freedom problem and determines the equations of motion for the overhead crane using Kane's method. It combines particle and rigid body motion.

In Example 13.15 we selected generalized speeds and found the corresponding partial velocities for the overhead crane. The two generalized coordinates are the horizontal position of the block, $q_1 = x$, and the angle of the pendulum arm, $q_2 = \theta$. The generalized speeds are the rate of change of the two coordinates, $u_1 = \dot{x}$ and $u_2 = \dot{\theta}$. The partial velocities associated with translation of the block are thus given by Eqs. (13.41)–(13.43). The partial angular velocities of the arm are given by Eqs. (13.44)–(13.46). The partial velocities of P, the center of mass of the arm, are given by Eqs. (13.47)–(13.49).

The acceleration of the block is

$$^I\mathbf{a}_{Q/O} = \ddot{x}\mathbf{e}_x = \dot{u}_1\mathbf{e}_x.$$

The acceleration of P is given in Example 9.20 as

$$^I\mathbf{a}_{P/O} = \ddot{x}\mathbf{e}_x + l\ddot{\theta}\mathbf{e}_\theta - l\dot{\theta}^2\mathbf{e}_r$$

$$= \dot{u}_1\mathbf{e}_x + l\dot{u}_2\mathbf{e}_\theta - lu_2^2\mathbf{e}_r.$$

Because the motion is planar, the angular acceleration of the arm is

$$^I\boldsymbol{\alpha}^B = \ddot{\theta}\mathbf{e}_z = \dot{u}_2\mathbf{e}_z.$$

In this example we allow a horizontal control force on the block, $\mathbf{F}_Q = u\mathbf{e}_x$. Thus there are two active forces in the problem: the control force and the force of gravity on the pendulum, $\mathbf{F}_P = -m_P g\mathbf{e}_y$. There are no active moments about P, as the force of gravity acts at the center of mass of the pendulum arm. The generalized active forces are

$$F_1 = {}^{\mathcal{I}}\mathbf{v}_1^Q \cdot \mathbf{F}_Q + {}^{\mathcal{I}}\mathbf{v}_1^P \cdot \mathbf{F}_P = u$$

and

$$F_2 = {}^{\mathcal{I}}\mathbf{v}_2^Q \cdot \mathbf{F}_Q + {}^{\mathcal{I}}\mathbf{v}_2^P \cdot \mathbf{F}_P$$

$$= -m_P g l \mathbf{e}_\theta \cdot \mathbf{e}_y = -m_P g l \sin\theta,$$

where we used the transformation table in Eq. (9.53) to find the dot product.

The generalized inertia forces are found by taking the dot product of the partial velocities and the partial angular velocities with the accelerations of P and Q and the angular acceleration of the rod:

$$F_1^* = {}^{\mathcal{I}}\mathbf{v}_1^Q \cdot \left(-m_Q{}^{\mathcal{I}}\mathbf{a}_{Q/O}\right) + {}^{\mathcal{I}}\mathbf{v}_1^P \cdot \left(-m_P{}^{\mathcal{I}}\mathbf{a}_{P/O}\right) + {}^{\mathcal{I}}\boldsymbol{\omega}_1^B \cdot \left(-I_P{}^{\mathcal{I}}\boldsymbol{\alpha}^B\right)$$

$$= -m_Q\dot{u}_1 - m_P\dot{u}_1 - m_P l\dot{u}_2(\mathbf{e}_\theta \cdot \mathbf{e}_x) + m_P l u_2^2(\mathbf{e}_r \cdot \mathbf{e}_x)$$

$$= -(m_Q + m_P)\dot{u}_1 - m_P l \cos\theta\dot{u}_2 + m_P l \sin\theta u_2^2$$

and

$$F_2^* = {}^{\mathcal{I}}\mathbf{v}_2^Q \cdot \left(-m_Q{}^{\mathcal{I}}\mathbf{a}_{Q/O}\right) + {}^{\mathcal{I}}\mathbf{v}_2^P \cdot \left(-m_P{}^{\mathcal{I}}\mathbf{a}_{P/O}\right) + {}^{\mathcal{I}}\boldsymbol{\omega}_2^B \cdot \left(-I_P{}^{\mathcal{I}}\boldsymbol{\alpha}^B\right)$$

$$= -m_P l\dot{u}_1(\mathbf{e}_\theta \cdot \mathbf{e}_x) - m_P l^2\dot{u}_2 - I_P\dot{u}_2$$

$$= -m_P l \cos\theta\dot{u}_1 - m_P l^2\dot{u}_2 - I_P\dot{u}_2.$$

Kane's equations now give us two equations of motion:

$$F_1 + F_1^* = u - (m_Q + m_P)\dot{u}_1 - m_P l \cos\theta\dot{u}_2 + m_P l \sin\theta u_2^2 = 0$$

$$F_2 + F_2^* = -m_P g l \sin\theta - m_P l \cos\theta\dot{u}_1 - m_P l^2\dot{u}_2 - I_P\dot{u}_2 = 0.$$

Combining these expressions with the kinematic equations (the definitions of the generalized speeds) yields four first-order equations of motion:

$$\dot{x} = u_1$$

$$(m_Q + m_P)\dot{u}_1 + m_P l \cos\theta\dot{u}_2 - m_P l \sin\theta u_2^2 = u$$

$$\dot{\theta} = u_2$$

$$(I_P + m_P l^2)\dot{u}_2 + m_P l \cos\theta\dot{u}_1 + m_P g l \sin\theta = 0.$$

These are the equivalent of Eqs. (9.56) and (9.57) in Example 9.20, albeit in first-order form.

As with Lagrange's equations, Kane's method is not a replacement for other approaches to finding equations of motion but rather is another tool that has certain advantages for some dynamic systems. It is particularly useful for complex systems containing many connected particles and rigid bodies, where the objective is to find the equations of motion and integrate them. The method is also easily automated. However, it can often impede achieving a qualitative understanding of behavior (as done in Chapter 11) or making certain approximations, such as ignoring the inertia of a body and treating it as a point mass with angular momentum (as done for the gyropendulum in Chapter 11). The art of dynamics is knowing when to use which tool.

13.5 Key Ideas

- The **generalized coordinates** to locate particle P in frame $\mathcal{A} = (O, \mathbf{a}_1, \mathbf{a}_2, \mathbf{a}_3)$ are $(q_1, q_2, q_3)_{\mathcal{A}}$. The position of P with respect to O is

$$\mathbf{r}_{P/O} = a_1(q_1, q_2, q_3, t)\mathbf{a}_1 + a_2(q_1, q_2, q_3, t)\mathbf{a}_2 + a_3(q_1, q_2, q_3, t)\mathbf{a}_3.$$

- If a constraint on N particles can be expressed in the algebraic form

$$f(\mathbf{r}_{1/O}, \mathbf{r}_{2/O}, \ldots, \mathbf{r}_{N/O}, t) = 0,$$

then it is a **holonomic constraint**; otherwise, it is a **nonholonomic constraint**.

- The trajectories of the system of N particles under the action of all active (non-constraint) forces must satisfy **d'Alembert's principle:**

$$\sum_{j=1}^{N} \left(\mathbf{F}_j^{(act)} - m_j \frac{{}^{\mathcal{I}}d}{dt} \left({}^{\mathcal{I}}\mathbf{v}_{j/O} \right) \right) \cdot {}^{\mathcal{I}}\delta\mathbf{r}_{j/O} = 0,$$

where $\delta\mathbf{r}_{j/O}$ is the **virtual displacement** of particle j with respect to O.

- The **Euler-Lagrange equations** are used to find the equations of motion for each generalized coordinate from the total kinetic energy and the **generalized forces:**

$$\frac{d}{dt}\frac{\partial T_O}{\partial \dot{q}_i} - \frac{\partial T_O}{\partial q_i} = Q_i, \quad i = 1, \ldots, N_C,$$

where

$$Q_i \triangleq \sum_{j=1}^{N} \mathbf{F}_j^{(act)} \cdot \frac{{}^{\mathcal{I}}\partial\mathbf{r}_{j/O}}{\partial q_i}.$$

- **Lagrange's equations** for a system with N_C holonomic constraints are

$$\frac{d}{dt}\frac{\partial L}{\partial \dot{q}_i} - \frac{\partial L}{\partial q_i} = Q_i^{(nc)}, \quad i = 1, \ldots, N_C,$$

where $L \triangleq T_O - U_O$ is the **Lagrangian** and

$$Q_i^{(nc)} \triangleq \sum_{j=1}^{N} \mathbf{F}_j^{(act,\ nc)} \cdot \frac{{}^{\mathcal{I}}\partial\mathbf{r}_{j/O}}{\partial q_i}$$

is the **generalized nonconservative force.**

- The equations of motion for a system with nonholonomic constraints can be derived using the method of **Langrange multipliers**.

- **Kane's equations of motion** produce first-order differential equations for the **generalized speeds**

$$u_r \triangleq \sum_{s=1}^{M} Y_{rs}(q_1, \ldots, q_M, t)\dot{q}_s + Z_r(q_1, \ldots, q_M, t), \quad r = 1, \ldots, M,$$

where Y_{rs} and Z_r are functions of the generalized coordinates.

- **Kane's equations** for M degrees of freedom are

$$F_r + F_r^* = 0, \quad r = 1, \ldots, M,$$

where the F_r are the **generalized active forces** and the F_r^* are the **generalized inertia forces**.

13.6 Notes and Further Reading

Our objective in this chapter was not to be complete or to make you an expert in these new methods. Rather, it was to introduce you to two advanced techniques and make the connection to what you have learned in earlier chapters. Hopefully this exposition will be a stepping stone to the further study of dynamics. There is much we left out, including conservation theorems, Hamilton's principle, and canonical transformations. Many excellent advanced texts exist, including Goldstein (1980), Greenwood (1988), and Hand and Finch (1998), which cover these and other advanced topics in dynamical systems. For modern treatments using differential geometry, see Arnold (1989), Marsden and Ratiu (1999), and O'Reilly (2008). Our derivation of Lagrange's equations follows closely that of Goldstein (1980). For more on Kane's method—particularly how it is used for nonholonomic systems—the best source is Kane and Levinson (1985). It also contains excellent discussions of constraints and the various types of forces and how they come into play in Kane's equations. For an excellent discussion of how Kane's method is an example of a class of approaches known as projection methods, see Storch and Gates (1989).

Appendices

APPENDIX A

‒ ‒

A Brief Review of Calculus

Fundamental to the study of dynamics are differential and integral calculus. Dynamics is the study of bodies in motion. Calculus is the branch of mathematics that treats changing quantities. They are so deeply intertwined that Newton needed to invent calculus to complete his treatment of motion. In this appendix we review the fundamentals of scalar calculus and the key extensions to vectors (multivariable calculus). Clearly it is impossible to cover all of calculus in this short appendix; we assume the reader has had a previous course on scalar (and hopefully multivariable) calculus and is thus familiar with the basic theorems and definitions. Our goal here is only to review the most basic and important points relevant to an understanding of mechanics. For a more thorough treatment, we recommend the excellent books by Spivak (1980) and Williamson et al. (1972), which were our primary sources.

A.1 Continuous Functions

Before reviewing the definition of a derivative, it is useful to review the definition of a *function*. Recall from Chapters 1–3 that the vector description of motion is written in terms of some set of scalar coordinates. The components of the position in some frame are written as scalar functions of time (or a set of scalar functions) multiplied by the appropriate unit vector. The velocity is written in terms of the rates of change of those scalar coordinates with time. (We need three scalars for position and three scalars for velocity to completely describe the state of a point in our three-dimensional Euclidean universe.) It is therefore sufficient here to discuss the calculus of only a single scalar function (one of the coordinates); the generalization to vectors follows as in those chapters.

Consider, for instance, the Cartesian x-position of a particle (this could be the position in one dimension as in Chapter 2 or simply the $\mathbf{e}_x$ component of the three-dimensional position). The particle's position is a function of time, which we write $x(t)$. A function is a rule that assigns a real number to each member of a set of real

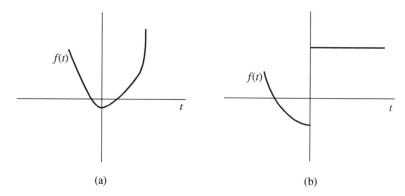

Figure A.1 (a) Continuous function $f(t)$. (b) Discontinuous function $f(t)$ with a discontinuity at $t = 0$.

numbers (the *domain* of the function). Thus at time t_1, the particle is at $x(t_1)$ and at time $t_2 = t_1 + \Delta t$, it is at $x(t_2)$.

It is a physical fact of our universe (or, if you like, a consequence of Newton's laws) that a particle cannot instantaneously change its position. The position $x(t)$ is a *continuous* function of time. This observation leads us to the mathematical definition of a continuous function.

Definition A.1 (Spivak 1980) A function $f(t)$ is **continuous** at $t = a$ if

$$\lim_{t \to a} f(t) = f(a).$$

Figure A.1a shows an example of a continuous function, and Figure A.1b shows a discontinuous function. Note that in Figure A.1b, the limit approaching from the right of $t = 0$ reaches a different value than the limit approaching from the left, thus violating the definition of continuity.

This definition is the mathematical statement of our intuitive understanding that a continuous function "contains no breaks, jumps, or wild oscillations" (Spivak 1980, p. 101). The position $x(t)$ of a particle is continuous for all t.[1] The velocity, however (and by extension the acceleration), need not be continuous. Chapter 4 discusses at length the situation where the velocity is discontinuous; this is our model of an impulsive force that acts over a very short time interval. (See, e.g., Figure 4.1.)

A.2 Differentiation

We all have an intuitive understanding of speed. For instance, if a particle travels between two points $x(t_1)$ and $x(t_2)$ in the time interval $\Delta t = t_2 - t_1 > 0$, then its average speed during the trip is

$$v = \frac{x(t_2) - x(t_1)}{t_2 - t_1}.$$

---

[1] The question of continuity on an interval $[a, b]$ is a bit more subtle and is not discussed here. However, note that if a function is continuous for all t, then it is continuous on any closed interval.

In other words, if it takes 30 minutes to travel 20 miles to the grocery store, then our average speed is 40 mph. If we stop at a number of red lights along the way, then we must also have been speeding some of the time for the speed to average 40 mph.

Assuming that the speed on the trip is a continuous function of time, then the mean value theorem of calculus states that at some point during the trip, our instantaneous speed must have been exactly 40 mph.

Theorem A.1 (Weisstein 2003) Let $f(t)$ be *differentiable* on the open interval (a, b) and *continuous* on the closed interval $[a, b]$. Then there is at least one point c in (a, b) such that

$$\dot{f}(c) = \frac{df}{dt}\bigg|_{t=c} = \frac{f(b) - f(a)}{b - a},$$

where we assume that $\dot{f}(t)$ is continuous at c.

We have not yet defined *differentiable* (see Definition A.2), but our example and Theorem A.1 should help physically understand the derivative. If we let the time interval shrink, then the average speed approaches the instantaneous speed exactly everywhere in the small interval. The mean value theorem also helps explain the meaning of velocity and how to measure it. In reality, instantaneous speed is an abstraction. Any measurement of speed involves measuring the change in position over some very small time interval and dividing; we simply hope that the interval is small enough that the speed is close to constant during the interval.

We are now ready to review the definition of the derivative.

Definition A.2 (Spivak 1980) The function $f(t)$ is **differentiable** at $t = a$ if

$$\lim_{h \to 0} \frac{f(a + h) - f(a)}{h}$$

exists and is finite. In this case the limit is denoted by $\frac{df}{dt}\big|_{t=a}$ or $\dot{f}(a)$ and is called the **derivative** of $f(t)$ at $t = a$.

Figure A.2 shows examples of continuous functions with continuous and discontinuous derivatives.

When $f(t) = x(t)$, we are speaking of position and the derivative $\dot{x}(t)$ is the speed $v(t)$. Likewise, we take the derivative of $v(t)$ to get the acceleration $a(t)$. Recall also that the derivative is the *slope of the tangent* to the function f. Thus the speed is always tangent to the trajectory at a given point, as illustrated in Figure A.2a.

It is true that all differentiable functions are continuous, but not all continuous functions are differentiable. Thus there are continuous trajectories $x(t)$ for which the speed $v(t)$ is not defined for all t. Let $f(t)$ in Figure A.2b represent the position $x(t)$. At time $t = a$, the trajectory reverses, corresponding to an instantaneous change in speed. The limit in Definition A.2 does not exist, since it has different values when $h > 0$ or $h < 0$. The speed is therefore not continuous at a; it is well defined everywhere else (e.g., on either side of a), as shown in Figure A.1b. Since the speed is not continuous there, the acceleration is undefined (in fact, it must be infinite). Although physically impossible, this abstraction is useful for solving many problems; we discuss it in Chapter 4.

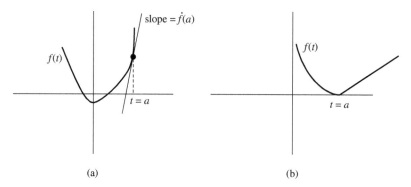

(a)   (b)

Figure A.2 (a) Continuous function $f(t)$ and its continuous derivative $\dot{f}(a)$ as the slope of the tangent at $t = a$. (b) A continuous function with discontinuous derivative at $t = a$. The derivative is well defined on either side of $t = a$ but is undefined at a.

A.3 Integration

Derivatives are used to formulate mechanics problems, but it is the integral that we ultimately use to solve them. Recall from Chapter 1 that dynamics is about defining the state of a particle at all times. By state, we mean the position and velocity of the particle. Newton's laws, however, provide tools for finding the acceleration of a particle (the derivative of the velocity). We need a method for going from the acceleration back to the velocity and from the velocity back to position. The integral calculus provides this method.[2]

A detailed mathematical treatment of the integral is quite complex and beyond the scope of this appendix. For our purposes, we recollect that, conceptually, the expression

$$\int_a^b f(t)dt$$

is read "the integral of f from a to b" and represents the "area" under the continuous function $f(t)$. This area is often represented as the limit of the sum of the area of many tall thin rectangles under the function $f(t)$ as the number of rectangles increases and their width decreases (see Figure A.3).

How does this definition of the integral relate to our problem in dynamics? The pertinent results are the two fundamental theorems of calculus.

Theorem A.2 If $f(t)$ is continuous on the open interval $[a, x)$ and

$$g(x) \triangleq \int_a^x f(t)dt,$$

then $\frac{dg}{dt} = f(t)$.

---

[2] Newton first developed differential calculus to formulate his laws. However, he was unable to solve certain problems (namely, proving that the spherical earth could be treated as a point mass). He therefore failed to publish his results for 20 years until he had developed integral calculus. In the meantime, Leibnitz published his version of calculus, and it is thus his notation we use today.

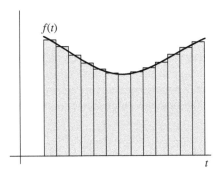

Figure A.3 The integral of $f(t)$ defined as the area under $f(t)$ and captured by the limit of the area of rectangles under the curve, called a *Riemann sum*.

Theorem A.3 If $f(t)$ is continuous on the closed interval $[a, b]$ and $f(t) \triangleq \frac{dg}{dt}$ for some function g, then

$$\int_a^b f(t)dt = g(b) - g(a).$$

Theorems A.2 and A.3 are remarkably powerful results. Because of Theorem A.3 we sometimes call the integral the *antiderivative* of a function. Theorem A.3 shows that given the acceleration $a(t)$ and the initial speed $v(t_0)$, we can find the speed at any later time by simply integrating the acceleration. (The discontinuous case is treated in Chapter 4.) Likewise, once the speed is determined, we can integrate it to find the position. The theorem also explains why numerical integration works. If we can't find a closed-form solution for a given integral, then numerically finding the area under $a(t)$ (what is called *quadrature*) will give us the antiderivative (i.e., the speed).

Of course, in most dynamics problems the acceleration is not given as a function of time but rather as a function of the position and speed. Such descriptions lead to the concept of the differential equation, the subject of Appendix C. Nevertheless, the concept of the integral as the antiderivative is extremely important.

A.4 Higher Derivatives and the Taylor Series

The previous section used the notation $\dot{f}(t)$ to denote the first derivative of $f(t)$; likewise $\ddot{f}(t)$ denotes the second derivative. What about the third derivative and the fourth? A compact notation for the nth derivative of $f(t)$ is $f^{(n)}(t)$. If all derivatives of $f(t)$ up to and including the nth derivative $f^{(n)}(t)$ exist and are continuous, then $f(t)$ is a *class C^n function*. A class C^0 function is merely continuous. In the limit $n \to \infty$, if all derivatives $f^{(n)}(t)$ exist and are continuous, then $f(t)$ is a *smooth* function.

One application of higher derivatives in dynamics is the approximation of a function.

Definition A.3 If $f(t)$ is smooth, then the **Taylor series** of $f(t)$ about the point $t = a$ exists and is

$$f(t) = f(a) + f^{(1)}(a)(t - a) + \frac{f^{(2)}(a)}{2!}(t - a)^2 + \cdots + \frac{f^{(n)}(a)}{n!}(t - a)^n + \cdots.$$

Suppose $\Delta t \triangleq t - a$ is very small. It then follows that $\Delta t \gg \Delta t^2 \gg \Delta t^n$ for any $n > 2$. Let $O(c)$ denote a quantity whose absolute value is less than or equal to a constant times a small positive quantity c (we say *order c* to refer to such a quantity). The second-order Taylor series approximation of $f(t)$ about the point $t = a$ is

$$f(t) = f(a) + f^{(1)}(a)(t - a) + \frac{f^{(2)}(a)}{2!}(t - a)^2 + O(\Delta t^3)$$

$$\approx f(a) + f^{(1)}(a)(t - a) + \frac{f^{(2)}(a)}{2!}(t - a)^2.$$

The first-order approximation, which is called the linearization of $f(t)$ at $t = a$,[3] is

$$f(t) = f(a) + f^{(1)}(a)(t - a) + O(\Delta t^2)$$

$$\approx f(a) + f^{(1)}(a)(t - a).$$

Note that when we drop the $(n + 1)$th-order terms, the $\approx$ symbol is used to remind us that the expression is only approximately true and holds only in the limit $\Delta t \to 0$.

Example A.1 The Binomial Expansion

One of the most common uses of the Taylor series is to expand the functions

$$f_1(x) = (1 \pm x)^n$$

$$f_2(x) = (1 \pm x)^{-n},$$

where x is small ($x^2 < 1$). This expansion about $x = 0$ is known as the *binomial expansion*. Expanding these functions is a straightforward application of the Taylor series. For instance, the first two derivatives of $f_1(x)$ about $x = 0$ are

$$f_1^{(1)}(x) = \pm n(1 \pm x)^{n-1}|_{x=0} = \pm n$$

$$f_1^{(2)}(x) = n(n - 1)(1 \pm x)^{n-2}|_{x=0} = n(n - 1).$$

The resulting Taylor series expansions of the two functions are

$$(1 \pm x)^n = 1 \pm nx + \frac{n(n - 1)x^2}{2!} \pm \frac{n(n - 1)(n - 2)x^3}{3!} + \cdots$$

$$(1 \pm x)^{-n} = 1 \mp nx + \frac{n(n + 1)x^2}{2!} \mp \frac{n(n + 1)(n + 2)x^3}{3!} + \cdots.$$

The binomial expansion is used when x is small enough that keeping terms only to first order is sufficient, in which case

$$(1 \pm x)^n \approx 1 \pm nx$$

$$(1 \pm x)^{-n} \approx 1 \mp nx.$$

---

[3] We discuss linearization using matrix notation in Section 12.2.

Example A.2 The Small Angle Approximation

Frequently in the book we use the Taylor series expansions of $\sin \theta$ and $\cos \theta$ about $\theta = 0$. Often, when θ is small, we replace the functions with the first term from their respective expansions, since when θ is measured in radians, the higher powers become exceedingly small. We call this the *small angle approximation*. In this example we find the Taylor series expansions of each function.

Letting $f_1(\theta) = \sin \theta$, we can find the first two derivatives evaluated at $\theta = 0$:

$$f_1^{(1)}(\theta) = \cos \theta|_{\theta=0} = 1$$

$$f_1^{(2)}(\theta) = -\sin \theta|_{\theta=0} = 0.$$

This pattern continues with odd derivatives equal to ± 1 and even derivatives equal to zero. Likewise for $\cos \theta$ we have

$$f_2^{(1)}(\theta) = -\sin \theta|_{\theta=0} = 0$$

$$f_2^{(2)}(\theta) = -\cos \theta|_{\theta=0} = -1.$$

Similarly, this pattern continues with even derivatives equal to ± 1 and odd derivatives equal to zero.

The resulting Taylor series expansions about $\theta = 0$ are thus

$$\sin \theta = \theta - \frac{\theta^3}{3!} + \frac{\theta^5}{5!} + \cdots$$

$$\cos \theta = 1 - \frac{\theta^2}{2!} + \frac{\theta^4}{4!} + \cdots.$$

Dropping higher powers of θ results in the small angle approximations:

$$\sin \theta \approx \theta$$

$$\cos \theta \approx 1.$$

A.5 Multivariable Functions and the Gradient

In many situations we employ functions of more than one independent variable, for example, a function $f(x, y, z)$ of the Cartesian coordinates $(x, y, z)_\mathcal{I}$. A common example is the potential energy (see Chapter 5), which is a function of the position coordinates. We may also consider motion on a surface $z(x, y)$, where z represents the height of the surface as a function of the Cartesian position $(x, y)_\mathcal{I}$. Unfortunately, the definition of the derivative in Definition A.2 does not directly apply to a function of more than one independent variable. We therefore introduce the *partial derivative*.

Definition A.4 (Williamson et al. 1972) The function $f(x)$ is a real-valued function of n variables, $x \triangleq (x_1, \ldots, x_i, \ldots, x_n)$. The **partial derivative of f with respect to x_i** is denoted $\partial f/\partial x_i$ and is, by definition

$$\frac{\partial f}{\partial x_i} \triangleq \lim_{h \to 0} \frac{1}{h} \left(f(x_1, \ldots, x_i + h, \ldots, x_n) - f(x_1, \ldots, x_i, \ldots, x_n) \right).$$

The partial derivative of f with respect to x_i is simply the ordinary derivative of f as a function of one variable while the remaining independent variables are held fixed; that is, f is considered to be a function of the ith variable only. All rules of differentiation and integration for single variables therefore apply.

For functions of one variable, the derivative evaluated at a point provides the slope of the tangent to the function at that point. The multivariable generalization is the *gradient*.

Definition A.5 (Williamson et al. 1972) If $f(x)$ is a differentiable, real-valued function of the coordinates of a point, $(x_1, x_2, \ldots, x_n)$, then the **gradient** ∇f of $f(x_1, x_2, \ldots, x_n)$ is a vector pointing in the direction of maximum increase of f and whose magnitude is the rate of increase.

The gradient expresses how a function changes in each of the orthogonal directions of a reference frame. If the function $f(x)$ is written as a function of the Cartesian coordinates $(x, y, z)_{\mathcal{I}}$ of a reference frame $\mathcal{I} = (O, \mathbf{e}_x, \mathbf{e}_y, \mathbf{e}_z)$, then the gradient of $f(x)$ is

$$\nabla f = \frac{\partial f}{\partial x}\mathbf{e}_x + \frac{\partial f}{\partial y}\mathbf{e}_y + \frac{\partial f}{\partial z}\mathbf{e}_z. \tag{A.1}$$

Note that the gradient of a function is a vector and thus has all the properties of a vector described in Appendix B. Also, like any vector, it can be written as components in the orthogonal unit directions of any frame; in Eq. (A.1) it is expressed as components in $\mathcal{I}$. Of course, we could always write the function in terms of the Cartesian coordinates of some other frame $\mathcal{A} = (O', \mathbf{e}_{x'}, \mathbf{e}_{y'}, \mathbf{e}_{z'})$ with coordinates $(x', y', z')_{\mathcal{A}}$. The vector would not change, but the components would, as we would now write them in the unit-vector directions in $\mathcal{A}$. For this new function $f'(x', y', z')$, the gradient $\nabla f'$ is

$$\nabla f' = \frac{\partial f'}{\partial x'}\mathbf{e}_{x'} + \frac{\partial f'}{\partial y'}\mathbf{e}_{y'} + \frac{\partial f'}{\partial z'}\mathbf{e}_{z'}.$$

The gradient vector itself has not changed, only how it is represented as components in a particular frame.

The gradient is used often in the text. For example, Chapter 5 uses the gradient operator to determine a vector conservative force from a scalar potential energy.

Example A.3 Gradient Operator

Let $U(x, y, z) = x^2 + 3y$ be a scalar function of the Cartesian coordinates $(x, y, z)_{\mathcal{I}}$. The gradient of $U(x, y, z)$ expressed as components in $\mathcal{I}$ is

$$\nabla U = 2x\mathbf{e}_x + 3\mathbf{e}_y.$$

For an intuitive perspective of the gradient, suppose $f(x, y)$ represents altitude on a topographical map (recall that topographical maps show curves of constant altitude) as a function of the Cartesian coordinates $(x, y)_{\mathcal{I}}$ in some frame $\mathcal{I}$. The gradient ∇f evaluated at a point (x_0, y_0) is the vector that is perpendicular to the constant-altitude curve that passes through (x_0, y_0). The gradient is oriented uphill and has magnitude proportional to the slope of the hill at (x_0, y_0). The gradient is zero at the top of a hill.

Recall that in dynamics we use a variety of scalar coordinates to represent the configuration of a point in a reference frame (some of them are discussed in Chapters 3 and 10). Thus the scalar function $f(x)$ can be written as a function of Cartesian coordinates $(x, y, z)_{\mathcal{I}}$, cylindrical coordinates $(r, \theta, z)_{\mathcal{I}}$, or any other set of scalar coordinates. It is thus convenient to develop an expression for the gradient vector in terms of partial derivatives with respect to such coordinates.

Eq. (A.1) provides the gradient in terms of Cartesian coordinates. We can convert the gradient to be with respect to cylindrical coordinates by using the chain rule of partial differentiation. Since the change of coordinates can be represented by r and θ as functions of x and y, we have

$$\frac{\partial f}{\partial x} = \frac{\partial f}{\partial r}\frac{\partial r}{\partial x} + \frac{\partial f}{\partial \theta}\frac{\partial \theta}{\partial x}$$

$$\frac{\partial f}{\partial y} = \frac{\partial f}{\partial r}\frac{\partial r}{\partial y} + \frac{\partial f}{\partial \theta}\frac{\partial \theta}{\partial y}.$$

The cylindrical coordinates $(r, \theta)_{\mathcal{I}}$ are provided by

$$x = r \cos \theta \tag{A.2}$$

$$y = r \sin \theta, \tag{A.3}$$

which implies

$$r^2 = x^2 + y^2 \tag{A.4}$$

$$\tan \theta = \frac{y}{x}. \tag{A.5}$$

Using Eq. (A.4), we compute the partial derivatives to be

$$\frac{\partial r}{\partial x} = \frac{x}{r} = \cos \theta$$

$$\frac{\partial r}{\partial y} = \frac{y}{r} = \sin \theta.$$

Using Eq. (A.5) and the identity $\frac{d}{d\theta} \tan \theta = \sec^2 \theta$ results in

$$\frac{\partial \theta}{\partial x} = -\frac{y}{x^2} \cos^2 \theta = -\frac{\sin \theta}{r}$$

$$\frac{\partial \theta}{\partial y} = \frac{1}{x} \cos^2 \theta = \frac{\cos \theta}{r}.$$

Therefore we have

$$\nabla f = \left(\frac{\partial f}{\partial r} \cos \theta - \frac{\partial f}{\partial \theta} \frac{\sin \theta}{r} \right) \mathbf{e}_x + \left(\frac{\partial f}{\partial r} \sin \theta + \frac{\partial f}{\partial \theta} \frac{\cos \theta}{r} \right) \mathbf{e}_y + \frac{\partial f}{\partial z} \mathbf{e}_z. \tag{A.6}$$

Eq. (A.6) can be simplified by expressing the gradient as components in a cylindrical frame $\mathcal{B} = (O, \mathbf{e}_r, \mathbf{e}_\theta, \mathbf{e}_z)$. The unit vectors of frames $\mathcal{B}$ and $\mathcal{I}$ are related by the transformation

	$\mathbf{e}_x$	$\mathbf{e}_y$
$\mathbf{e}_r$	$\cos \theta$	$\sin \theta$
$\mathbf{e}_\theta$	$-\sin \theta$	$\cos \theta$.

Using Eq. (A.6), we have

$$\nabla f = \left(\frac{\partial f}{\partial r} \cos \theta - \frac{\partial f}{\partial \theta} \frac{\sin \theta}{r} \right) (\cos \theta \mathbf{e}_r - \sin \theta \mathbf{e}_\theta)$$

$$+ \left(\frac{\partial f}{\partial r} \sin \theta + \frac{\partial f}{\partial \theta} \frac{\cos \theta}{r} \right) (\sin \theta \mathbf{e}_r + \cos \theta \mathbf{e}_\theta) + \frac{\partial f}{\partial z} \mathbf{e}_z,$$

which yields

$$\nabla f = \frac{\partial f}{\partial r} \mathbf{e}_r + \frac{1}{r} \frac{\partial f}{\partial \theta} \mathbf{e}_\theta + \frac{\partial f}{\partial z} \mathbf{e}_3.$$

The derivation of the gradient expressed as components in a spherical-reference frame follows the same procedure. Let $(r, \theta, \phi)_\mathcal{I}$ denote spherical coordinates, as shown in Figure 10.1b. The spherical coordinates are defined in terms of the Cartesian coordinates as

$$x = r \cos \theta \sin \phi \qquad (A.7)$$

$$y = r \sin \theta \sin \phi \qquad (A.8)$$

$$z = r \cos \phi, \qquad (A.9)$$

which implies

$$r^2 = x^2 + y^2 + z^2$$

$$\tan \theta = \frac{y}{x}$$

$$\cos \phi = \frac{z}{r}.$$

Let $\mathcal{C} = (O, \mathbf{e}_r, \mathbf{e}_\phi, \mathbf{e}_\theta)$ denote the corresponding spherical frame. The gradient ∇f in spherical coordinates expressed as components in frame $\mathcal{C}$ can be shown to be

$$\nabla f = \frac{\partial f}{\partial r} \mathbf{e}_r + \frac{1}{r} \frac{\partial f}{\partial \phi} \mathbf{e}_\phi + \frac{1}{r \sin \phi} \frac{\partial f}{\partial \theta} \mathbf{e}_\theta.$$

A.6 The Directional Derivative

The directional derivative is the change in a multivariable function along a particular direction. It is defined similarly to the scalar derivative.

Definition A.6 The **directional derivative** of real-valued function $f(x)$ with respect to unit vector $\mathbf{e}$, denoted $\frac{df}{d\mathbf{e}}$, is the real-valued function

$$\left. \frac{df}{d\mathbf{e}} \right|_x \triangleq \lim_{h \to 0} \frac{f(x + h\mathbf{e}) - f(x)}{h}.$$

It is possible to show that the directional derivative is equal to the dot product of the gradient of the function with the unit vector (Williamson et al. 1972):

$$\frac{df}{d\mathbf{e}} = \nabla f \cdot \mathbf{e}.$$

We can thus write the derivative of the function with respect to an arbitrary vector as a scaled version of the directional derivative,

$$\frac{df}{d\mathbf{v}} = \nabla f \cdot \mathbf{v},$$

since $\mathbf{v} = \|\mathbf{v}\|\hat{\mathbf{v}}$.

We often have occasion to examine the quantity

$$\nabla f \cdot {}^{\mathcal{I}}d\mathbf{r},$$

where ${}^{\mathcal{I}}d\mathbf{r}$ is the vector $[dx \;\; dy \;\; dz]_{\mathcal{I}}^{T}$. This expression is the directional derivative of f in the direction of ${}^{\mathcal{I}}d\mathbf{r}$ (the change in $\mathbf{r}$ in frame $\mathcal{I}$) times the infinitesimal displacement $\|{}^{\mathcal{I}}d\mathbf{r}\|$. From the definition of the directional derivative, it is thus equal to the differential change of f in that direction,

$$\nabla f \cdot {}^{\mathcal{I}}d\mathbf{r} \triangleq df = \frac{\partial f}{\partial x}dx + \frac{\partial f}{\partial y}dy + \frac{\partial f}{\partial z}dz,$$

where df is called the *total derivative* of f.

A.7 Differential Volumes and Multiple Integration

We have many occasions to perform multiple integrations over two- and three-dimensional domains. It might be to compute the area or the volume of a shape, find the center of mass of a collection of particles, or compute the moment of inertia of a rigid body. Using Cartesian coordinates, the multiple integral is given by a straightforward generalization of the single integral. In two dimensions over domain S, we have

$$A = \int\int_{S} f(x, y)dxdy,$$

and in three dimensions the integral becomes

$$V = \int\int\int_{S} f(x, y, z)dxdydz.$$

We say that $dA \triangleq dxdy$ is a *differential area* and $dV \triangleq dxdydz$ is a *differential volume*.

What about multiple integrals in different coordinate systems? That is, in many cases it will be easier to describe the boundaries of S in another coordinate system, say, polar or spherical. How do we change the differential area or volume element from Cartesian to the new coordinates? The general result is given by the following theorem.

Theorem A.4 Consider a general change of coordinates from Cartesian coordinates to a new set of three coordinates (q_1, q_2, q_3), such that $x(q_1, q_2, q_3)$, $y(q_1, q_2, q_3)$, and $z(q_1, q_2, q_3)$ are continuously differentiable functions. Then the multiple integral over a domain S in terms of the new coordinates is

$$\int \int \int_S f(x, y, z)dxdydz = \int \int \int_S f(q_1, q_2, q_3)|J|dq_1dq_2dq_3,$$

where J is the Jacobian matrix of the transformation:

$$J = \begin{bmatrix} \frac{\partial x}{\partial q_1} & \frac{\partial x}{\partial q_2} & \frac{\partial x}{\partial q_3} \\ \frac{\partial y}{\partial q_1} & \frac{\partial y}{\partial q_2} & \frac{\partial y}{\partial q_3} \\ \frac{\partial z}{\partial q_1} & \frac{\partial z}{\partial q_2} & \frac{\partial z}{\partial q_3} \end{bmatrix}.$$

The proof of this theorem is complicated and involves additional restrictions on the type of domain and the function f that we have not described. We do not include it here. (See, e.g., Williamson et al. 1972.) However, the theorem holds for the types of problems we encounter in dynamics. Thus the differential area element in any new coordinate system can be written as

$$dA = |J|dq_1dq_2,$$

and the differential volume element in three dimensions is

$$dV = |J|dq_1dq_2dq_3.$$

For polar coordinates, the transformation is given by differentiating Eqs. (A.2) and (A.3), which yields the Jacobian matrix,

$$J_{\text{polar}} = \begin{bmatrix} \cos\theta & -r\sin\theta \\ \sin\theta & r\cos\theta \end{bmatrix}.$$

This expression gives the differential area element

$$dA = rdrd\theta.$$

For spherical coordinates, given by Eqs. (A.7)–(A.9), the Jacobian matrix is

$$J_{\text{spherical}} = \begin{bmatrix} \cos\theta\sin\phi & -r\sin\theta\sin\phi & r\cos\theta\cos\phi \\ \sin\theta\sin\phi & r\cos\theta\sin\phi & r\sin\theta\cos\phi \\ \cos\phi & 0 & -r\sin\phi \end{bmatrix},$$

resulting in a differential volume element of

$$dV = r^2\sin\phi drd\theta d\phi.$$

APPENDIX B

Vector Algebra and Useful Identities

Chapter 1 defines a vector as a quantity that has both magnitude and direction in space (recall Qualitative Definition 1.1 and Figure 1.1). It also introduces the notation used throughout the book to identify position $\mathbf{r}_{P/O}$, which is the vector locating point P relative to point O. The concept of the vector has important consequences and forms the bridge between the physical meaning of dynamics and the mathematics. This appendix presents the mathematical definition of a vector, summarizes various important properties of vectors, and reviews some useful vector identities that are encountered throughout the book.

B.1 The Vector

We begin by introducing the mathematical definition of a vector.

Definition B.1 (Weisstein 1999) A **vector** is a mathematical object that is an element of a **vector space**, which implies that it obeys the rules of addition, subtraction, and scalar multiplication.

Don't be intimidated by this definition. It simply means that the result of the addition of vectors must also be a vector, and that of scalar multiplication of a vector is also still a vector (i.e., it is still in the vector space). Mathematically, we say that the vector space is *closed* under vector addition and scalar multiplication. In other words, if we form the sum

$$\mathbf{r}_{Q/O} = \mathbf{r}_{P/O} + \mathbf{r}_{Q/P}, \tag{B.1}$$

where $\mathbf{r}_{P/O}$ and $\mathbf{r}_{Q/P}$ are vectors, then the quantity $\mathbf{r}_{Q/O}$ is also a vector. The geometric meaning of this sum is shown in Figure B.1a. We call this equation a *vector triad*.

Scalar multiplication is determined by a scalar times a vector and is also a vector:

$$\mathbf{r}_{Q/O} = c\mathbf{r}_{P/O}, \tag{B.2}$$

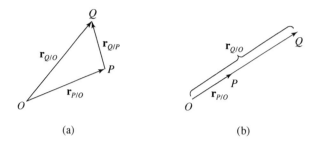

Figure B.1 (a) Addition of two vectors and (b) scalar multiplication of a vector to obtain another vector.

where c is a positive or negative scalar. This operation is shown geometrically in Figure B.1b. Since c is arbitrary, Eqs. (B.1) and (B.2) together show that vector subtraction also holds.

For convenience, we drop the subscripts on vectors for the remainder of the appendix (but keep the convention of using boldface).

Definition B.1 states that a vector is defined by its properties. These properties are very important to the development in the book; this appendix summarizes all of the properties used. It is this definition that ties the geometrical meaning of vectors discussed in Chapter 1 to its mathematical properties.

Consider any three vectors $\mathbf{a}$, $\mathbf{b}$, $\mathbf{c}$ and any two scalars r, s. Then $\mathbf{a}$, $\mathbf{b}$, $\mathbf{c}$ satisfy the following properties (Weisstein 1999):

a. Commutativity of vector addition:

$$\mathbf{a} + \mathbf{b} = \mathbf{b} + \mathbf{a};$$

b. Associativity of vector addition:

$$(\mathbf{a} + \mathbf{b}) + \mathbf{c} = \mathbf{a} + (\mathbf{b} + \mathbf{c});$$

c. Additive identity:

$$\mathbf{0} + \mathbf{a} = \mathbf{a} + \mathbf{0} = \mathbf{a};$$

d. Additive inverse:

$$\mathbf{a} + (-\mathbf{a}) = \mathbf{0};$$

e. Associativity of scalar multiplication:

$$r(s\mathbf{a}) = (rs)\mathbf{a};$$

f. Distributivity of scalar sums:

$$(r + s)\mathbf{a} = r\mathbf{a} + s\mathbf{a};$$

g. Distributivity of vector sums:

$$r(\mathbf{a} + \mathbf{b}) = r\mathbf{a} + r\mathbf{b};$$

h. Scalar identity:

$$1\mathbf{a} = \mathbf{a}.$$

B.2 Vector Magnitude

The scalar multiplication property in Eq. (B.2) allows us to introduce the *unit vector*. A unit vector is a vector of length one (and arbitrary direction). Thus any vector can be written as a scalar multiplication of a unit vector:

$$\mathbf{a} = a\mathbf{e}.$$

We thus define the magnitude[1] $\|\mathbf{a}\|$ of vector $\mathbf{a}$ as the scalar length of the vector. In this case we have $\|\mathbf{a}\| = |a|$, since $\|\mathbf{e}\| = 1$.

B.3 Vector Components

The scalar multiplication and summation properties also allow us to write any vector as the weighted sum of other vectors. In particular, we usually write a vector as the weighted sum of three orthogonal unit vectors $(\mathbf{e}_1, \mathbf{e}_2, \mathbf{e}_3)$ as follows:

$$\mathbf{a} = a_1\mathbf{e}_1 + a_2\mathbf{e}_2 + a_3\mathbf{e}_3. \tag{B.3}$$

This operation is shown geometrically in Figure B.2. The vectors $a_i\mathbf{e}_i$ are the *components* of $\mathbf{a}$. Chapter 3 discusses components in more detail, where the three orthogonal unit vectors define the basis of a reference frame.

Chapter 3 also introduces a matrix notation for writing the magnitudes of the components of a vector:

$$[\mathbf{a}]_{\mathcal{I}} \triangleq \begin{bmatrix} a_1 \\ a_2 \\ a_3 \end{bmatrix}_{\mathcal{I}},$$

where $\mathcal{I}$ is a reference frame defined by the three unit vectors $(\mathbf{e}_1, \mathbf{e}_2, \mathbf{e}_3)$.

---

[1] Also called the *norm.*

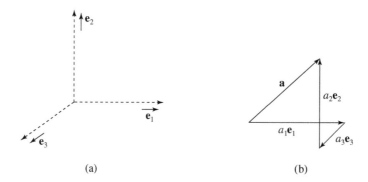

Figure B.2 (a) Right-handed triad of unit vectors. (b) Vector **a** as a sum of components.

B.4 Vector Multiplication

In contrast to the summation properties discussed in Section B.1, vector multiplication is not a necessary property of a vector space. Thus we can introduce definitions of multiplication without requiring that the result still be a vector. There are three operations that are important in dynamics, the *scalar dot product*, the *vector cross product*, and the *tensor product*.

B.4.1 Scalar Dot Product

Definition B.2 The **scalar dot product** (or **inner product**) of two vectors is the product of their magnitudes and the cosine of the angle θ between them:

$$\mathbf{a} \cdot \mathbf{b} \triangleq \|\mathbf{a}\| \|\mathbf{b}\| \cos \theta.$$

The geometry of the two vectors is shown in Figure B.3. Definition B.2 allows us to write the dot products of the three orthogonal unit vectors in Figure B.2:

$$\mathbf{e}_1 \cdot \mathbf{e}_1 = \mathbf{e}_2 \cdot \mathbf{e}_2 = \mathbf{e}_3 \cdot \mathbf{e}_3 = 1 \tag{B.4}$$

$$\mathbf{e}_1 \cdot \mathbf{e}_2 = \mathbf{e}_2 \cdot \mathbf{e}_3 = \mathbf{e}_3 \cdot \mathbf{e}_1 = 0. \tag{B.5}$$

These properties also mean that the magnitude of a vector satisfies

$$\mathbf{a} \cdot \mathbf{a} = \|\mathbf{a}\|^2 = a^2.$$

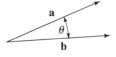

Figure B.3 Two vectors, **a** and **b**, with angle θ between them.

The scalar dot product obeys the following useful properties, where again $\mathbf{a}$, $\mathbf{b}$, $\mathbf{c}$ are vectors and s is a scalar:

a. Commutativity:

$$\mathbf{a} \cdot \mathbf{b} = \mathbf{b} \cdot \mathbf{a};$$

b. Distributivity:

$$\mathbf{a} \cdot (\mathbf{b} + \mathbf{c}) = \mathbf{a} \cdot \mathbf{b} + \mathbf{a} \cdot \mathbf{c};$$

c. Scalar multiplication:

$$s(\mathbf{a} \cdot \mathbf{b}) = (s\mathbf{a}) \cdot \mathbf{b} = \mathbf{a} \cdot (s\mathbf{b}).$$

We can use the component form of a vector in Eq. (B.3) and the properties of unit vectors in Eqs. (B.4) and (B.5) to find the scalar dot product of two vectors in terms of their component magnitudes in a particular frame. For reference frame $\mathcal{I} = (O, \mathbf{e}_1, \mathbf{e}_2, \mathbf{e}_3)$ and two vectors, $\mathbf{a}$ and $\mathbf{b}$, written as components in that frame,

$$\mathbf{a} = a_1\mathbf{e}_1 + a_2\mathbf{e}_2 + a_3\mathbf{e}_3 \tag{B.6}$$

$$\mathbf{b} = b_1\mathbf{e}_1 + b_2\mathbf{e}_2 + b_3\mathbf{e}_3, \tag{B.7}$$

the dot product of the two vectors is

$$\mathbf{a} \cdot \mathbf{b} = a_1b_1 + a_2b_2 + a_3b_3.$$

This equation can also be written compactly in matrix notation as

$$\mathbf{a} \cdot \mathbf{b} = [\mathbf{a}]_{\mathcal{I}}^{T}[\mathbf{b}]_{\mathcal{I}}.$$

The component form can be used to find that the square magnitude of a vector is given by the sum of the squares of the magnitudes of its components:

$$\mathbf{a} \cdot \mathbf{a} = \|\mathbf{a}\|^2 = a_1^2 + a_2^2 + a_3^2.$$

B.4.2 Vector Cross Product

Definition B.3 The **vector cross product** of $\mathbf{a}$ and $\mathbf{b}$ is the vector perpendicular to the plane containing $\mathbf{a}$ and $\mathbf{b}$ and equal to the product of their magnitudes and the sine of the angle θ between them:

$$\mathbf{a} \times \mathbf{b} \triangleq \|\mathbf{a}\|\|\mathbf{b}\| \sin\theta\mathbf{e},$$

where $\mathbf{e}$ is the unit vector perpendicular to the plane containing $\mathbf{a}$ and $\mathbf{b}$.

One consequence of this definition is that the cross product of a vector with itself is zero:

$$\mathbf{a} \times \mathbf{a} = 0.$$

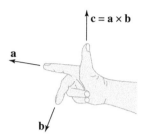

Figure B.4 Visualization of the cross product.

Definition B.3 allows us to write the various cross products of the three orthogonal unit vectors in Figure B.2a:

$$\mathbf{e}_1 \times \mathbf{e}_2 = \mathbf{e}_3, \quad \mathbf{e}_2 \times \mathbf{e}_3 = \mathbf{e}_1, \quad \mathbf{e}_3 \times \mathbf{e}_1 = \mathbf{e}_2 \tag{B.8}$$

$$\mathbf{e}_1 \times \mathbf{e}_1 = \mathbf{e}_2 \times \mathbf{e}_2 = \mathbf{e}_3 \times \mathbf{e}_3 = 0. \tag{B.9}$$

Note that it is these properties of the unit vectors that define them as a right-handed set. That is, use your right hand to curl your fingers from $\mathbf{e}_1$ to $\mathbf{e}_2$, for example. *Your thumb should now point in the direction of* $\mathbf{e}_3$. Figure B.4 shows an alternative approach to visualizing the right-hand rule.

The vector cross product obeys the following properties, where $\mathbf{a}$, $\mathbf{b}$, $\mathbf{c}$ are vectors and s is a scalar:

   a. Antisymmetry:

$$\mathbf{a} \times \mathbf{b} = -\mathbf{b} \times \mathbf{a};$$

   b. Distributivity:

$$\mathbf{a} \times (\mathbf{b} + \mathbf{c}) = \mathbf{a} \times \mathbf{b} + \mathbf{a} \times \mathbf{c};$$

   c. Scalar multiplication:

$$s(\mathbf{a} \times \mathbf{b}) = (s\mathbf{a}) \times \mathbf{b} = \mathbf{a} \times (s\mathbf{b});$$

   d. Scalar triple product:

$$\mathbf{a} \cdot (\mathbf{b} \times \mathbf{c}) = (\mathbf{a} \times \mathbf{b}) \cdot \mathbf{c} = \mathbf{b} \cdot (\mathbf{c} \times \mathbf{a});$$

   e. Vector triple product:

$$\mathbf{a} \times (\mathbf{b} \times \mathbf{c}) = \mathbf{b}(\mathbf{c} \cdot \mathbf{a}) - \mathbf{c}(\mathbf{a} \cdot \mathbf{b}).$$

Note that the vector cross product *does not* satisfy the commutative and associative properties.

The vector triple product can be derived by breaking the vector into components and using the other properties and Eqs. (B.8) and (B.9). Because the vector cross product is not associative, the placement of parentheses in the vector triple product matters. The result of permuting the vectors is the *Jacobi identity:*

$$\mathbf{a} \times (\mathbf{b} \times \mathbf{c}) + \mathbf{b} \times (\mathbf{c} \times \mathbf{a}) + \mathbf{c} \times (\mathbf{a} \times \mathbf{b}) = 0.$$

As with the scalar dot product, we can use the component form of a vector in Eq. (B.3), the distributive property, and the cross product properties of the unit vectors in Eqs. (B.8) and (B.9) to write the cross product vector as components in frame $\mathcal{I} = (O, \mathbf{e}_1, \mathbf{e}_2, \mathbf{e}_3)$:

$$\mathbf{a} \times \mathbf{b} = (a_2 b_3 - b_2 a_3)\mathbf{e}_1 + (a_3 b_1 - b_3 a_1)\mathbf{e}_2 + (a_1 b_2 - b_1 a_2)\mathbf{e}_3,$$

where the two vectors, $\mathbf{a}$ and $\mathbf{b}$, have been expressed as components in $\mathcal{I}$, as in Eqs. (B.6) and (B.7).

We can also develop a matrix notation for the cross product. Letting $\mathbf{c} = \mathbf{a} \times \mathbf{b}$, the magnitudes of the components of $\mathbf{c}$ in reference frame $\mathcal{I} = (O, \mathbf{e}_1, \mathbf{e}_2, \mathbf{e}_3)$ in matrix notation are

$$[\mathbf{c}]_{\mathcal{I}} = [\mathbf{a} \times]_{\mathcal{I}} [\mathbf{b}]_{\mathcal{I}},$$

where $[\mathbf{a} \times]_{\mathcal{I}}$ is the *cross product equivalent matrix:*

$$[\mathbf{a} \times]_{\mathcal{I}} \triangleq \begin{bmatrix} 0 & -a_3 & a_2 \\ a_3 & 0 & -a_1 \\ -a_2 & a_1 & 0 \end{bmatrix}_{\mathcal{I}}.$$

The cross product equivalent matrix is a *skew symmetric* matrix. That means it is the negative of its transpose:

$$[\mathbf{a} \times]_{\mathcal{I}} = -[\mathbf{a} \times]_{\mathcal{I}}^T.$$

Although these expressions for the cross product are true and can be very useful for coding the cross product in a computer language, it is almost always better to form the cross product directly by finding the cross product of each of the components separately and using Eqs. (B.8) and (B.9). It is also easier to remember its definition in terms of these equations.

B.4.3 Tensor Product

Definition B.4 The **tensor product** of the two vectors $\mathbf{a}$ and $\mathbf{b}$ is expressed as $\mathbf{a} \otimes \mathbf{b}$ and results in a second-rank **tensor:**

$$\mathbb{T} \triangleq \mathbf{a} \otimes \mathbf{b}.$$

A tensor is a new mathematical object. Unlike a vector, it does not have a clear geometric interpretation. However, mathematically it satisfies the same definition of a vector as in Definition B.1; that is, it is a member of a vector space because it obeys addition, subtraction, and scalar multiplication. In fact, it satisfies the same eight properties that vectors do. Let $\mathbb{O}$ denote the zero tensor and $\mathbb{U}$ denote the unity tensor, which satisfy

$$\mathbb{O} \cdot \mathbf{a} = 0$$
$$\mathbb{U} \cdot \mathbf{a} = \mathbf{a}.$$

The following properties hold for any tensors $\mathbb{T}$ and $\mathbb{S}$:

a. Commutativity of tensor addition:

$$\mathbb{T} + \mathbb{S} = \mathbb{S} + \mathbb{T};$$

b. Associativity of tensor addition:

$$(\mathbb{T} + \mathbb{S}) + \mathbb{R} = \mathbb{T} + (\mathbb{S} + \mathbb{R});$$

c. Additive identity:

$$\mathbb{O} + \mathbb{T} = \mathbb{T} + \mathbb{O} = \mathbb{T};$$

d. Additive inverse:

$$\mathbb{T} + (-\mathbb{T}) = \mathbb{O};$$

e. Associativity of scalar multiplication:

$$r(s\mathbb{T}) = (rs)\mathbb{T};$$

f. Distributivity of scalar sums:

$$(r + s)\mathbb{T} = r\mathbb{T} + s\mathbb{T};$$

g. Distributivity of tensor sums:

$$r(\mathbb{T} + \mathbb{S}) = r\mathbb{T} + r\mathbb{S};$$

h. Scalar identity:

$$1\mathbb{T} = \mathbb{T}.$$

As with the scalar dot product and the vector cross product, the tensor product obeys certain multiplicative properties:

a. Distributivity:

$$\mathbf{a} \otimes (\mathbf{b} + \mathbf{c}) = \mathbf{a} \otimes \mathbf{b} + \mathbf{a} \otimes \mathbf{c};$$

b. Scalar multiplication:

$$s(\mathbf{a} \otimes \mathbf{b}) = (s\mathbf{a}) \otimes \mathbf{b} = \mathbf{a} \otimes (s\mathbf{b});$$

c. Scalar tensor triple product (more commonly known as the left and right dot products, or vector dyadic dot products):

$$\mathbf{a} \cdot (\mathbf{b} \otimes \mathbf{c}) = (\mathbf{a} \cdot \mathbf{b})\mathbf{c}$$
$$(\mathbf{a} \otimes \mathbf{b}) \cdot \mathbf{c} = (\mathbf{b} \cdot \mathbf{c})\mathbf{a};$$

d. Vector tensor triple product:

$$\mathbf{a} \times (\mathbf{b} \otimes \mathbf{c}) = (\mathbf{a} \times \mathbf{b}) \otimes \mathbf{c}$$

$$(\mathbf{a} \otimes \mathbf{b}) \times \mathbf{c} = \mathbf{a} \otimes (\mathbf{b} \times \mathbf{c}).$$

Note that, for these properties, the order of operation matters. That is,

$$\mathbf{a} \otimes \mathbf{b} \neq \mathbf{b} \otimes \mathbf{a}$$

$$(\mathbf{a} \otimes \mathbf{b}) \cdot \mathbf{c} \neq \mathbf{c} \cdot (\mathbf{a} \otimes \mathbf{b})$$

$$(\mathbf{a} \otimes \mathbf{b}) \times \mathbf{c} \neq \mathbf{c} \times (\mathbf{a} \otimes \mathbf{b}).$$

These properties also imply that the following combinations of scalar and vector products hold:

$$\mathbf{a} \cdot (\mathbf{b} \otimes \mathbf{c}) \cdot \mathbf{d} = (\mathbf{a} \cdot \mathbf{b})(\mathbf{c} \cdot \mathbf{d})$$

$$\mathbf{a} \cdot (\mathbf{b} \otimes \mathbf{c}) \times \mathbf{d} = (\mathbf{a} \cdot \mathbf{b})(\mathbf{c} \times \mathbf{d})$$

$$\mathbf{a} \times (\mathbf{b} \otimes \mathbf{c}) \cdot \mathbf{d} = (\mathbf{a} \times \mathbf{b})(\mathbf{c} \cdot \mathbf{d})$$

$$\mathbf{a} \times (\mathbf{b} \otimes \mathbf{c}) \times \mathbf{d} = (\mathbf{a} \times \mathbf{b}) \otimes (\mathbf{c} \times \mathbf{d}).$$

Combining these properties (scalar multiplication, scalar triple product, and distributivity) shows that the scalar product of a tensor obeys the following product rules:

a. Distributivity of the scalar tensor triple product:

$$\mathbb{T} \cdot (\mathbf{a} + \mathbf{b}) = \mathbb{T} \cdot \mathbf{a} + \mathbb{T} \cdot \mathbf{b};$$

b. Associativity of the scalar product:

$$\mathbb{T} \cdot (s\mathbf{a}) = s\mathbb{T} \cdot \mathbf{a}.$$

The tensor is a useful notational tool because of these multiplicative properties. It allows us to compactly write certain operations that arise in rigid-body dynamics.

The multiplicative properties of the tensor product can be used to find a component representation of a tensor. Expressing the two vectors $\mathbf{a}$ and $\mathbf{b}$ as components in $\mathcal{I} = (O, \mathbf{e}_1, \mathbf{e}_2, \mathbf{e}_3)$ allows us to write their tensor product as

$$\mathbb{T} = \mathbf{a} \otimes \mathbf{b} = (a_1\mathbf{e}_1 + a_2\mathbf{e}_2 + a_3\mathbf{e}_3) \otimes (b_1\mathbf{e}_1 + b_2\mathbf{e}_2 + b_3\mathbf{e}_3).$$

Distributing the tensor product then expresses the tensor in terms of the scalar magnitudes of its components in $\mathcal{I}$:

$$\mathbb{T} = \sum_{i=1}^{3} \sum_{j=1}^{3} T_{ij}\mathbf{e}_i \otimes \mathbf{e}_j$$

where

$$T_{ij} = a_i b_j.$$

The scalar T_{ij} are the nine component magnitudes of the tensor (in contrast to the three component magnitudes of a vector).

The quantity $\mathbf{e}_i \otimes \mathbf{e}_j$ is the basis of the tensor in $\mathcal{I}$, just as the unit vector $\mathbf{e}_i$ is the basis of a vector expressed as components in $\mathcal{I}$. Thus, just as the magnitudes of the components of a vector in a particular frame are given by the dot product of the vector with the basis unit vector,

$$a_i = \mathbf{a} \cdot \mathbf{e}_i,$$

the scalar magnitudes of the components of the tensor in $\mathcal{I}$ are given by the scalar tensor product,

$$T_{ij} = \mathbf{e}_i \cdot \mathbb{T} \cdot \mathbf{e}_j = a_i b_j.$$

For example, the components of the unit tensor are

$$U_{ij} = \begin{cases} 1 & i = j \\ 0 & i \neq j. \end{cases}$$

These expressions suggest a matrix representation of a tensor, as we had for the component magnitudes of vectors. As with vectors, the matrix representation is frame dependent because the component magnitudes of a tensor change depending on the frame specified. The matrix representation of the vector component magnitudes is a column vector, but the matrix representation of the tensor components is a 3×3 matrix:

$$[\mathbb{T}]_{\mathcal{I}} = \begin{bmatrix} T_{11} & T_{12} & T_{13} \\ T_{21} & T_{22} & T_{23} \\ T_{31} & T_{32} & T_{33} \end{bmatrix}_{\mathcal{I}}.$$

The tensor product is thus given by matrix multiplication:

$$[\mathbb{T}]_{\mathcal{I}} = [\mathbf{a} \otimes \mathbf{b}]_{\mathcal{I}} = [\mathbf{a}]_{\mathcal{I}}[\mathbf{b}]_{\mathcal{I}}^T.$$

We can also write the scalar and vector triple products as the following matrix operations:

$$\begin{aligned} [\mathbf{a} \cdot \mathbb{T}]_{\mathcal{I}} &= \left([\mathbf{a}]_{\mathcal{I}}^T[\mathbb{T}]_{\mathcal{I}}\right)^T = [\mathbb{T}]_{\mathcal{I}}^T[\mathbf{a}]_{\mathcal{I}} \\ [\mathbb{T} \cdot \mathbf{a}]_{\mathcal{I}} &= [\mathbb{T}]_{\mathcal{I}}[\mathbf{a}]_{\mathcal{I}} \\ [\mathbf{a} \times \mathbb{T}]_{\mathcal{I}} &= [\mathbf{a} \times]_{\mathcal{I}}[\mathbb{T}]_{\mathcal{I}} \\ [\mathbb{T} \times \mathbf{a}]_{\mathcal{I}} &= [\mathbb{T}]_{\mathcal{I}}[\mathbf{a} \times]_{\mathcal{I}}. \end{aligned}$$

APPENDIX C

Differential Equations

The typical end product of almost all dynamics problems is a set of differential equations called the *equations of motion*. To find the actual trajectories of a system over time, we must solve these equations, usually for the position and velocity. Doing so analytically, in closed form, is often a formidable, if not impossible, task, though certain equations do admit very important solutions. Fortunately, with the advent of the digital computer, it has become common and straightforward to find solutions numerically. This appendix provides a rudimentary introduction to differential equations and their solution, focusing specifically on numerical techniques for solving them. The mathematical theory of differential equations is broad and fascinating; we clearly have the space only to touch on it here. One recommended text on differential equations is Boyce and DiPrima (1977). There are also many good books on the numerical solution of ordinary differential equations, including Gear (1971), Lapidus and Seinfeld (1971), Lambert (1973, 1993), and Press et al. (1986).

We feel it is helpful to introduce students to some of the background material and basic theory on finding the solutions to differential equations. We do so as briefly as possible in this appendix. However, the reader interested only in learning how to use the available tools for finding numerical solutions can skip to Section C.5 with no loss of continuity.

C.1 What Is a Differential Equation?

A differential equation is simply an algebraic equation in terms of a function $y(x)$ and its derivatives $y^{(1)}(x)$, $y^{(2)}(x)$, ... For example, the equation

$$\left(\frac{d^2y}{dx^2}\right)^3 - bx\sqrt{\frac{dy}{dx}} + cy^2 = 0$$

is a differential equation in $y(x)$. In this case, y is the dependent variable and x is the independent variable.

A differential equation involving a function of a single independent variable is known as an ordinary differential equation or ODE. In contrast, a differential equation of a function of many independent variables is called a partial differential equation or PDE. We are not concerned with solving PDEs in this book. In fact, almost all problems that interest us involve functions of time t, which is a single scalar independent variable. We are typically interested in functions of time that describe the state of a dynamical system, that is, the coordinates and their rates. Differential equations arise in this setting in the form of the equations of motion. For example, the equation of motion for the simple pendulum is

$$\ddot{\theta} = -\frac{g}{l}\sin\theta. \tag{C.1}$$

Eq. (C.1) is a typical, albeit rather simple, differential equation. Our objective is to find a solution trajectory $\theta(t)$ that, when inserted in Eq. (C.1), will satisfy the differential equation. If t_0 is the initial time, then $\theta(t_0) = \theta_0$ is the initial value of θ; we call Eq. (C.1) an *initial value problem*. Under relatively mild conditions, an initial value problem has a unique solution. All problems we examine have unique solutions.[1]

Unfortunately, the vast majority of differential equations, such as the one in Eq. (C.1), do not have a closed, analytic solution (i.e., a solution in terms of known elementary functions of t). We call a differential equation *nonlinear* when, for example, the function or its derivatives appear in the equation raised to a power or as an argument to a transcendental[2] function. Except for very special cases, nonlinear ODEs rarely have a known solution. For such ODEs we usually resort to numerical integration, which is the subject of Section C.4.

One special class of differential equations that do have known solutions are linear ODEs, in which the function and its derivatives appear linearly (that is, by themselves or multiplied by a constant).[3] For example, for small θ, the equation of motion in Eq. (C.1) can be approximated by the linear ODE

$$\ddot{\theta} = -\omega_0^2\theta, \tag{C.2}$$

where $\omega_0 = \sqrt{g/l}$. Eq. (C.2) is the equation for a simple harmonic oscillator (see Tutorial 2.3), which is a very common linear ODE. Chapter 12 discusses the solution to this equation and its physical significance in vibration theory.

The next section briefly discusses a few differential equations with known solutions, focusing in particular on scalar linear differential equations. For a more thorough and detailed discussion, the reader is directed to Boyce and DiPrima (1977) or a similar text. Section C.4 introduces numerical techniques for solving an arbitrary nonlinear set of differential equations. A set of differential equations contains $k > 1$ coupled differential equations in k scalar functions and one independent variable (e.g., time t).

---

[1] See Boyce and DiPrima (1977), for example, for a detailed discussion and proof of existence and uniqueness of solutions.

[2] Exponential and trigonometric functions are transcendental.

[3] The strict mathematical definition of linearity is that, if $y(x)$ is a solution and a is a constant, then $ay(x)$ is also a solution. Likewise, if the functions $y_1(x)$ and $y_2(x)$ both are solutions, then their sum, $y_1(x) + y_2(x)$ is also a solution.

C.2 Some Common ODEs and Their Solutions

The order of a differential equation is the order of the highest derivative present in the equation. Thus differential equations involving only first derivatives are called *first-order* differential equations. A general form for a first-order nonlinear differential equation is

$$\frac{dy}{dt} = f(y, t). \tag{C.3}$$

There is no general solution to Eq. (C.3). There are two cases, however, for which we can often find solutions: when the equation is *separable* and when it is linear. We discuss each of these cases in the following two subsections.

C.2.1 Separability

Consider the special case where $f(y, t)$ separates into a product (or quotient) of functions of a single variable, that is, $f(y, t) = h(t)/g(y)$. In this case, we can separate the terms that depend on t from the terms that depend on y and integrate. Multiplying both sides of Eq. (C.3) by $g(y)dt$ and integrating yields

$$\int_{y(t_0)}^{y(t_1)} g(y)dy = \int_{t_0}^{t_1} h(t)dt. \tag{C.4}$$

Direct integration such as this is sometimes referred to as *quadrature*. If this integral exists, then we can find an analytical solution to the ODE in terms of the initial conditions. Even if the integral does not exist in closed form, we can often solve half of the equation and use a simple numerical quadrature routine, such as *Simpson's rule*, to find a trajectory. For instance, if $g(y) = 1$, Eq. (C.4) becomes $y(t_1) = y(t_0) + \int_{t_0}^{t_1} h(t)dt$, which can be integrated numerically.

We use separability in the solution to Tutorial 2.2. Here is another example.

Example C.1 A Simple, Separable ODE

Suppose we wish to solve the initial value problem:

$$\frac{dy}{dx} = \frac{y \cos x}{1 + 2y^2},$$

where y is the dependent variable and x is the independent variable.[4] Let $y(0) = 1$ be the initial condition. By separating terms involving y from terms involving x, we obtain the integral equation:

$$\int \frac{1 + 2y^2}{y} dy = \int \cos x \, dx.$$

---

[4] This example is adapted from Boyce and DiPrima (1977).

These integrals can be solved analytically to find the general solution:[5]

$$\ln |y| + y^2 = \sin x + c.$$

The integration constant c is found by substituting the initial condition to obtain the final result:

$$\ln |y| + y^2 = \sin x + 1. \tag{C.5}$$

This example is interesting because we can't solve Eq. (C.5) for y in terms of x. Eq. (C.5) is an example of an *implicit* representation of y with x as the independent variable.

Since Newton's second law involves acceleration, which is a second-order derivative, the equations of motion in this book are usually second-order differential equations. However, in many cases, for a particular form of the equations of motion, we can employ a simple change of variables to turn them into a set of first-order differential equations. Suppose, for instance, we have a second-order differential equation with the following general form:

$$\ddot{y} = f(y)g(\dot{y}). \tag{C.6}$$

The right-hand side of Eq. (C.6) is separable in y and $\dot{y}$; it is also independent of time. Often, the right-hand side of Eq. (C.6) represents a "force" and $g(\dot{y}) = 1$.

To solve Eq. (C.6), we multiply both sides by $\dot{y}/g(\dot{y})$. Then we employ the change of variables,

$$v \triangleq \dot{y} = \frac{dy}{dt},$$

which implies $\ddot{y} = dv/dt$. In terms of the variables v and y, Eq. (C.6) transforms to a first-order differential equation:

$$\frac{v\dot{v}}{g(v)} = f(y)\dot{y}. \tag{C.7}$$

Multiplying both sides of Eq. (C.7) by dt and integrating from $t = t_0$ to $t = t_1$, we obtain

$$\int_{v(t_0)}^{v(t_1)} \frac{v\,dv}{g(v)} = \int_{y(t_0)}^{y(t_1)} f(y)dy. \tag{C.8}$$

Eq. (C.8) is identical in form to Eq. (C.4). We thus converted the original second-order differential equation in Eq. (C.6) into a separable first-order one. The differential equation in first-order form can be integrated by quadrature.

The cost of converting a second-order differential equation into a separable first-order differential equation is the elimination of an independent variable (e.g., time).

---

[5] The term *general solution* means that the equation has been integrated to within an arbitrary integration constant. The value of that integration constant is found by applying the initial condition.

Thus the solution will not be a function of time but rather a function of one of the state variables (and sometimes only an implicit function), as in Tutorial 4.2. To find the complete trajectory, another approach is usually needed. Nevertheless, this approach to integrating the equations of motion once (called finding the *first integral of the motion*) is a common one used throughout the book.

C.2.2 Linearity

The general form for a linear first-order ODE involving only one scalar function of a single independent variable is

$$\frac{dy}{dt} = -ay. \tag{C.9}$$

Eq. (C.9) has a particularly simple solution. In fact, linear ODEs of any order always have an analytical solution.

One way to find a solution of Eq. (C.9) is to guess a general form and substitute it into the ODE. Guessing a solution of the form $y(t) = ce^{-at}$ and substituting it into Eq. (C.9) yields

$$-ace^{-at} = -ace^{-at},$$

which means that the solution works. The arbitrary constant c is found by evaluating $y(t)$ at $t = 0$, from which we obtain $c = y(0)$. Thus the complete solution of the initial value problem in Eq. (C.9) is

$$y(t) = y(0)e^{-at}.$$

The minus sign in the exponent is important. If the constant a is positive, then the solution asymptotically converges to zero. If $a < 0$, the exponential solution blows up; we call such a case *unstable*.

As mentioned above, every linear ODE is solvable. (This is the only class of ODEs for which we can make such a general statement.) It turns out that the solution to a linear ODE is a sum of weighted, possibly complex, exponentials. Let us revisit the second-order linear ODE in Eq. (C.2):

$$\frac{d^2y}{dt^2} + \omega_0^2 y = 0. \tag{C.10}$$

Assuming a solution of the form $y(t) = ce^{-at}$, we find

$$a^2 ce^{-at} + \omega_0^2 ce^{-at} = (a^2 + \omega_0^2)y(t) = 0.$$

If $y \neq 0$ is indeed the solution, then the following condition on a must be satisfied:

$$a^2 + \omega_0^2 = 0. \tag{C.11}$$

We call Eq. (C.11) the *characteristic equation* of the differential equation in Eq. (C.10). Eq. (C.11) has two solutions: $a = -i\omega_0$ and $a = +i\omega_0$. Since the original differential equation is linear, a sum of solutions is also a solution. So the general solution to Eq. (C.2) is

$$y(t) = c_1 e^{-i\omega_0 t} + c_2 e^{i\omega_0 t}.$$

Because the original differential equation is second order, we need to specify two initial conditions, $y(0)$ and $\dot{y}(0)$. Determining the constants c_1 and c_2 and the final solution requires using algebra and Euler's equation $e^{i\omega_0 t} = \cos \omega_0 t + i \sin \omega_0 t$. We obtain

$$y(t) = y(0) \cos \omega_0 t + \frac{\dot{y}(0)}{\omega_0} \sin \omega_0 t. \tag{C.12}$$

Note that the complex terms cancel out, as they should, since this solution represents a physical system. The solution in Eq. (C.12) is called *harmonic* because it involves sinusoidal functions of time.

We can make Eq. (C.10) slightly more complicated by adding a term proportional to $\dot{y}(t)$. Such a term adds damping and causes the magnitude of the harmonic solution to asymptotically converge to zero. We could also add known forcing functions to the right-hand side. Tutorials 2.3 and 2.4 and Section 12.1 discuss the solution of these differential equations.

C.3 First-Order Form

Many dynamic systems we examine involve a set of coupled second-order differential equations. For example, if we specify the position of a particle in the inertial frame with Cartesian coordinates x, y, and z, the particle motion under the action of forces is governed by three second-order differential equations:

$$\ddot{x} = f(x, \dot{x}, y, \dot{y}, z, \dot{z}, t)$$
$$\ddot{y} = g(x, \dot{x}, y, \dot{y}, z, \dot{z}, t)$$
$$\ddot{z} = h(x, \dot{x}, y, \dot{y}, z, \dot{z}, t).$$

Such a system can be particularly difficult to solve, even if the functions are linear. Whether solving it exactly or numerically, it may be convenient to perform a change of variables to simplify the problem. We introduce the matrix $Y \triangleq [y_1, y_2, y_3, y_4, y_5, y_6]^T$, where

$$\begin{bmatrix} y_1 \\ y_2 \\ y_3 \\ y_4 \\ y_5 \\ y_6 \end{bmatrix} \triangleq \begin{bmatrix} x \\ \dot{x} \\ y \\ \dot{y} \\ z \\ \dot{z} \end{bmatrix}.$$

The equations of motion can now be written in a convenient matrix form

$$\dot{Y} = \begin{bmatrix} \dot{y}_1 \\ \dot{y}_2 \\ \dot{y}_3 \\ \dot{y}_4 \\ \dot{y}_5 \\ \dot{y}_6 \end{bmatrix} = \begin{bmatrix} y_2 \\ f(y_1, y_2, y_3, y_4, y_5, y_6, t) \\ y_4 \\ g(y_1, y_2, y_3, y_4, y_5, y_6, t) \\ y_6 \\ h(y_1, y_2, y_3, y_4, y_5, y_6, t) \end{bmatrix} \triangleq F(Y, t). \qquad (C.13)$$

Every set of coupled differential equations of second or higher order can be rewritten as a first-order matrix differential equation (i.e., as a larger set of coupled first-order differential equations in new variables). This representation is particularly convenient for linear ODEs, since it is known how to solve such a linear matrix differential equation (as discussed in Section 12.1). But even for nonlinear systems, this rewrite is convenient, as the numerical algorithms discussed in Section C.4 apply to a set of first-order equations.

Another reason the first-order form in Eq. (C.13) is convenient is that it lends itself naturally to the process of linearization. Since we know how to solve a set of linear differential equations (i.e., a matrix linear ODE), we often try to approximate a nonlinear ODE by a linear one. As long as the states stay close to some equilibrium or steady-state solution, the approximation is usually quite good and thus can give great insight into the motion. Section 12.2 discusses this process in more detail.

C.4 Numerical Integration of an Initial Value Problem

As emphasized in the previous section, the vast majority of differential equations encountered in solving dynamics problems have no analytical solution. We must find the solution trajectory numerically. That is, we use a computer to find approximate values for the solution $y(t)$ at a set of discrete time points $t_1, \ldots, t_N$. This section introduces basic techniques for numerically integrating such differential equations as Eq. (C.1) or Eq. (C.13) given the initial conditions.

This is a rich subject with a long history, and many fine books have been written about it. Our only objective here is to give you an introduction to the material and to provide several common algorithms for numerically integrating ODEs. It is our view that you will be a much more informed user of commercial software, and better able to both understand the results and find mistakes, if you are familiar with the basic structure of the algorithms being employed. For a deeper treatment, we refer you to Gear (1971), Lapidus and Seinfeld (1971), Lambert (1973, 1993), Boyce and DiPrima (1977), and Press et al. (1986). Alternatively, you can always skip to Section C.5, which explains how to use the built-in differential equation solvers in MATLAB.

C.4.1 The Problem

What do we mean by numerical integration? Consider again the first-order scalar differential equation,

$$\dot{y} = f(y, t). \qquad (C.14)$$

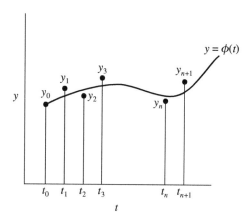

Figure C.1 Schematic of the computed sequence of solution values approximating the exact solution $y = \phi(t)$ of an ordinary differential equation. (After Boyce and DiPrima 1977.)

Figure C.1 is a sketch of the problem at hand. The solid line represents the (unknown) exact solution trajectory of the ODE. The goal is to find the sequence of points $(y_1, y_2, \ldots, y_n, y_{n+1})$ at the corresponding times $(t_1, t_2, \ldots, t_n, t_{n+1})$ that are as close as possible to the exact solution trajectory, $y = \phi(t)$, knowing only the original differential equation and the initial conditions (y_0, t_0). For all algorithms discussed here, the time points are equally spaced with an interval of $h = t_{n+1} - t_n$.

There are four criteria that are used to measure the effectiveness of a given algorithm: *local truncation error*, *global truncation error*, *roundoff error*, and *stability*.

The local truncation error is the difference between the true and approximated solution after one step. That is, if the value at y_n is known to be exact, the local truncation error is the difference between y_{n+1} and $\phi(t_{n+1})$. The global truncation error is the difference between the approximate and exact solution at any later time given only a common initial condition (i.e., $y_0 = y(t_0)$). It is almost always true that the global error is larger than the local error. We compare numerical algorithms by the proportionality of their corresponding local truncation errors with powers of the step size h. A technique is said to be of order h^2, for example, if the local truncation error is proportional to h^3. Often the global truncation error is proportional to one lower power of h than the local truncation error (it is also much harder to find expressions for the global error). It is possible to create algorithms with almost arbitrary accuracy, but at the expense of many function evaluations. Selection of an algorithm becomes a trade-off between accuracy and computer resources.

Roundoff error refers to the limited ability of a digital computer to represent a number. That is, the solution of the differential equation at any time can only be represented by a finite number of digits. Normally this error is much less than the truncation error. However, it can be very important. For example, a dynamic system that conserves energy can be difficult to simulate, as the round-off error will often act as an energy source or damping term.

Stability of a numerical algorithm refers to its ability to stay bounded; that is, to stay within some finite (and hopefully small) region of the exact solution. We normally require that algorithms be stable for a reasonable range of step sizes. (Almost all algorithms become unstable for a large enough step size!)

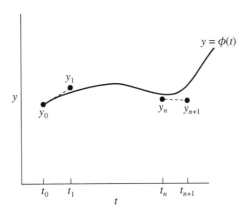

Figure C.2 The Euler integrator approximates the solution at t_{n+1} by projecting forward from t_n using the local slope of the solution at t_n.

The remainder of this section summarizes several common algorithms and highlights their respective accuracy. All these techniques (except the one in Section C.4.4) are designed for first-order differential equations. This restriction is not significant, however, since we showed above that any high-order differential equation can be rewritten as a collection of coupled first-order ODEs. All techniques described are easily generalized to multidimensional sets of differential equations. Nevertheless, we focus on the scalar case in Eq. (C.14), as it minimizes the algebra and lends itself to graphical depictions.

C.4.2 Euler and Improved Euler Integration

The simplest possible integration algorithm is the *Euler method* (also called the *tangent line method*), illustrated in Figure C.2. In this method we approximate the solution of the equation at time t_{n+1} by moving along the tangent $\frac{d\phi}{dt}$ to the solution at t_n. Thus, using the point-slope equation for a line, we have

$$y_{n+1} = y_n + \dot{\phi}(t_n)(t_{n+1} - t_n).$$ (C.15)

Since the slope $\dot{\phi}$ at t_n is given by $f(y_n, t_n)$, the Euler integrator is

$$y_{n+1} = y_n + hf(y_n, t_n),$$ (C.16)

where $h \triangleq t_{n+1} - t_n$.

For the Euler method, the local truncation error is the error in propagating from t_n to t_{n+1} using only the slope as an approximation to the actual solution. The global truncation error is the error incurred by assuming that y_n is the solution value at t_n when we used it in the evaluation of $f(y, t)$.

The Euler method is an example of a first-order integration technique. This is easily seen by examining an alternative, and more algebraic, derivation. Consider a Taylor series expansion of the solution trajectory $\phi(t)$ about its value at t_n:

$$\phi(t_n + h) = \phi(t_n) + \dot{\phi}(t_n)h + \ddot{\phi}(t_n)\frac{h^2}{2!} + O(h^3).$$

This expression can be rewritten using Eq. (C.14) to obtain

$$\phi(t_{n+1}) = \phi(t_n) + f(\phi(t_n), t_n)h + \ddot{\phi}(t_n)\frac{h^2}{2!} + O(h^3). \qquad (C.17)$$

The first two terms are our Euler integrator. The third is the error term. It is evident that the error in this method is proportional to h^2.

The Euler method is also the simplest example of the *Taylor-series method*. It should be evident from Eq. (C.17) that we could find higher-order methods by simply using more terms in the Taylor series. This could be done by replacing the higher derivatives of $\phi(t)$ with multiple partial derivatives of $f(y, t)$. Although in principle this technique works, it is highly impractical, as it can require a great amount of work to find these higher derivatives and can also result in many complex function evaluations (resulting in a slow algorithm). Instead, the objective is to find higher-order methods utilizing only evaluations of the function in Eq. (C.14).

One modification that falls into this category is to replace the slope at t_n in Eq. (C.15) by the average slope at t_n and t_{n+1},

$$y_{n+1} = y_n + \frac{1}{2}\left(f(y_n, t_n) + f(y_{n+1}, t_{n+1})\right)h. \qquad (C.18)$$

Eq. (C.18) is an example of an *implicit method* because the desired quantity y_{n+1} appears on both sides of the equation. There are many families of implicit algorithms; a discussion of implicit integration is beyond the scope of this appendix. However, we can make this algorithm explicit by simply replacing y_{n+1} on the right-hand side by its value from the Euler integration in Eq. (C.16):

$$y_{n+1} = y_n + \frac{1}{2}\left(f(y_n, t_n) + f(y_n + hf(y_n, t_n), t_n + h)\right)h.$$

This algorithm is known as the *modified* or *improved* Euler (it is also sometimes called the *Heun formula*). It can be shown that it is a second-order algorithm (i.e., the error is proportional to h^3). Note that we achieved an increase in order, but now we need to evaluate the function twice at each step rather than only once as in the Euler method.

The modified Euler is a member of a class of methods known as *Runge-Kutta* algorithms. In fact, it is a second-order Runge-Kutta integrator. (Euler is a first-order Runge-Kutta integrator.) We discuss this important family of algorithms next.

C.4.3 The Runge-Kutta Method

In the modified Euler method we obtained one order of improvement in accuracy by evaluating the function at both of the endpoints of the interval. The Runge-Kutta algorithm generalizes this approach to achieving improvements in accuracy by means of multiple evaluations of $f(y, t)$ at points in the interval (t_n, y_n) and (t_{n+1}, y_{n+1}). The interested reader is directed to Gear (1971) and Lapidus and Seinfeld (1971) for excellent discussions of many different Runge-Kutta algorithms and their relative trade-offs.

The most common Runge-Kutta algorithm is the symmetric fourth-order algorithm *RK4*:

$$y_{n+1} = y_n + \frac{h}{6}(k_1 + 2k_2 + 2k_3 + k_4),$$

where

$$k_1 = f(y_n, t_n)$$

$$k_2 = f(y_n + \frac{h}{2}k_1, t_n + \frac{1}{2}h)$$

$$k_3 = f(y_n + \frac{h}{2}k_2, t_n + \frac{1}{2}h)$$

$$k_4 = f(y_n + hk_3, t_n + h).$$

This algorithm is simple to code and is very effective. It represents a balanced compromise between speed and accuracy. Almost all differential equations you will encounter will be adequately simulated with this routine.

C.4.4 The Leapfrog Algorithm

Here we consider a special class of algorithms for the second-order differential equation

$$\ddot{y} = f(y, t). \tag{C.19}$$

Note the special form of this equation: the function $f(y, t)$ does not depend on the first derivative of y. In fact, Eq. (C.19) is extremely common in dynamics. For example, any system with only conservative forces acting on it has this form.

A simple algorithm for integrating second-order ODEs of this type was introduced by Feynman et al. (1963). It has since come to be called the *leapfrog algorithm*. To see how it works, we rewrite Eq. (C.19) as two first-order equations in the "position" y and "velocity" $\dot{y}$:

$$\dot{y} = v \tag{C.20}$$

$$\dot{v} = f(y, t). \tag{C.21}$$

We now recognize, from our experience with the Euler and modified Euler algorithms, that the best approximation to the slope in Eq. (C.20) is given by its value at the midpoint rather than at either end. We can thus write an Euler-like update for the position,

$$y_{n+1} = y_n + hv_{n+1/2},$$

leaving alone for the moment the problem of how to find the velocity at the midpoint. Likewise, we use the value y_{n+1} to update the velocities using another Euler-like equation:

$$v_{n+3/2} = v_{n+1/2} + hf(y_{n+1}, t_{n+1}).$$

It is easy to see why this is called the leapfrog algorithm; the estimates for y and v leapfrog over each other at each step.

One of the remarkable features of the leapfrog algorithm is that it is a second-order formula, despite there being only a single function evaluation (see Hairer et al. 2002). It works quite well for many problems, primarily because of its important global properties.

The leapfrog algorithm is the simplest member of a family of integrators called *symplectic algorithms*. It is far beyond the scope of this book to discuss these formulas in any detail; the interested reader is directed to Sanz-Serna and Calvo (1994), Hut et al. (1995), and Hairer et al. (2002) for excellent discussions. However, we do want to point out two important properties that make a symplectic algorithm useful for dynamical systems. First, they are *time reversible*. That is, if after integrating forward in time the same algorithm is used to numerically integrate backward from the final state (using $-h$ time steps), then the values return to their initial conditions. This is a property of all Newtonian dynamical systems and is thus desirable in an integrator. (It is not a property of the general Runge-Kutta integrator.)

Second, the leapfrog algorithm (and all symplectic integrators) conserves certain properties of the trajectory. In particular, for bounded or periodic motion, the numerical solution is guaranteed to also stay bounded. Of most significance are its properties for a conservative system, that is, a system whose solution has constant total energy. The leapfrog integrator, while not guaranteeing a constant energy, does guarantee that the total energy associated with the numerical solution stays bounded and close to the actual total energy. (This is not true for the other Runge-Kutta integrators.) It is this property that makes symplectic integrators particularly attractive to researchers performing long-time simulations of the solar system, for example.

There are two remaining open questions with the leapfrog algorithm. The first is how to initialize the velocity at $v_{1/2}$ when starting with initial conditions at t_0. Initialization is normally done with a simple half-step Euler update:

$$v_{1/2} = v_0 + \frac{h}{2} f(y_0, t_0).$$

The second question is how to synchronize the velocity and position values. The output of the leapfrog algorithm is a set of states y and v separated in time by $h/2$. It is generally useful to have the position and velocity values at the same time points (to compute energy, for instance). This is usually done by splitting the velocity update into two equivalent steps of width $h/2$. The resulting algorithm is called the *velocity Verlet:*

$$v_{n+1/2} = v_n + \frac{h}{2} f(y_n, t_n) \tag{C.22}$$

$$y_{n+1} = y_n + h v_{n+1/2} \tag{C.23}$$

$$v_{n+1} = v_{n+1/2} + \frac{h}{2} f(y_{n+1}, t_{n+1}). \tag{C.24}$$

Note that this algorithm is the same as the original leapfrog algorithm and thus is still a second-order method. It also does not require any more function evaluations, since the function value from the previous step can be stored and used in the next one.

The velocity Verlet algorithm can be rewritten without velocities by combining Eqs. (C.22)–(C.24) over two time steps. After a bit of algebra, we obtain the position-only algorithm:

$$y_{n+2} - 2y_{n+1} + y_n = h^2 f(y_{n+1}, t_{n+1}).$$

C.5 Using MATLAB to Solve ODEs

For our purposes, most ODEs can be integrated using the numerical integration
algorithms in MATLAB. If you are not familiar with basic MATLAB usage, you can
refer to one of the many demonstrations that are included with MATLAB. To choose
a demonstration to view, simply type demo or doc demo in the MATLAB Command
Window. For a description of numerical integration algorithms included in MATLAB,
type doc ode45 in the MATLAB Command Window.

   The most common algorithm, ODE45, is based on an explicit Runge-Kutta method.
This section describes how to use ODE45 by considering the equation of motion of the
simple pendulum in Eq. (C.1). The steps are as follows:

   a. The first step actually does not involve a computer. We need to write the ODE
      in first-order form, as described in Section C.3. Using the notation of that
      section, we write the ODE in the first-order matrix form: $\dot{Y} = F(Y, t)$.

      For example, consider the simple pendulum equation of motion in
      Eq. (C.1). Let $Y = [y_1, y_2]^T \triangleq [\theta, \dot{\theta}]^T$, which means

      $$\dot{Y} = F(Y, t) = \left[\begin{array}{c} y_2 \\ -\frac{g}{L} \sin y_1 \end{array} \right].$$

   b. The second step is to create a MATLAB function MYODEFUN that defines the
      first-order ODE. The inputs to MYODEFUN are, in the following order: time t,
      the matrix Y at time t, and any parameters used in $F(Y, t)$. The output of
      MYODEFUN is $\dot{Y}$.

      For the simple pendulum, we create a new file myodefun.m that contains:

      ```
      function ydot = myodefun(t,y,g,L)
      ydot(1,1) = y(2);
      ydot(2,1) = -g/L*sin(y(1));
      ```

   c. The third step is to call ODE45. To view the inputs and outputs of ODE45, type
      help ode45 in the Command Window. The form of ODE45 that we will use is
      [T,Y] = ODE45(ODEFUN,TSPAN,Y0,OPTIONS,P1,P2,...).[6] The inputs are
      as follows: ODEFUN is a *handle* to the function that contains the first-order
      ODE; TSPAN is a matrix of at least two entries whose first entry corresponds
      to the start time of the integration and whose last entry corresponds to the
      end time; Y0 is a matrix of the initial conditions; OPTIONS is an argument
      either created by the ODESET function or set to the empty brackets []; and
      P1,P2,... are optional parameters used by ODEFUN.

      To integrate the pendulum equation of motion from $t = 0$ s to $t = 10$ s
      with $L = 0.1$ m, $\theta(0) = \frac{\pi}{2}$ rad, $\dot{\theta}(0) = 0$ rad/s, and $g = 9.81$ m/s^2 we type the
      following in the Command Window:

      ```
      >> [t,y]= ode45(@myodefun,[0 10],[pi/2 0],[],9.81,0.1);
      ```

---

[6] This form may not be explicitly listed in the ODE45 documentation, but, as of MATLAB R2010a, it still
exists!

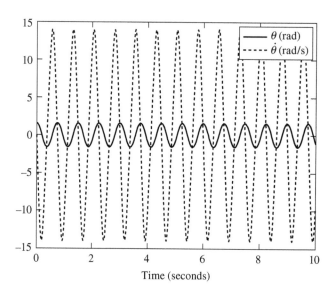

Figure C.3 Output of MATLAB function ODE45 for the simple pendulum equation of motion in Eq. (C.1) with $L = 0.1$ m, $\theta(0) = \frac{\pi}{2}$ rad, and $\dot{\theta}(0) = 0$ rad/s.

d. The fourth step is to plot the output (see Figure C.3). The first output variable of ODE45 is the time t, which has one column and N rows. The first and last entry of t correspond to the first and last entry of TSPAN. The second output of ODE45 is a matrix y whose *rows* are $Y(t_n)$, where $n = 1, \ldots, N$ (N is determined by MATLAB, unless TSPAN is a $1 \times N$ matrix).

For the simple pendulum, the nth row of y is $[\theta(t_n) \ \dot{\theta}(t_n)]$, where t_n is the nth entry in t. To plot the output, we type[7]

```
>> plot(t,y)
```

in the Command Window.

An alternative to plotting the solution versus time is to animate it. To animate a solution, we first need to construct a plotting command that illustrates the configuration of the model (e.g., the pendulum) at a single instant in time. Iterating over the entire time span produced by ODE45 yields the animation. For example, to plot the pendulum at t(ii), where ii is in the range $1, \ldots, N$, we could use the command

```
>> h = plot([0 L*sin(y(ii,1))],[0 -L*cos(y(ii,1))],'o-');
```

(The plot function returns a *graphics handle* h, which we will use when animating; the last argument 'o-' indicates the plot style.) Note, you may need to first use the AXIS and HOLD commands to keep the plot steady.

Figure C.4 depicts MATLAB plots of the pendulum at two different times. The following code animates the solution to the pendulum or, for that matter, any dynamical system integrated with ode45:

---

[7] For more detailed plotting options, type help plot.

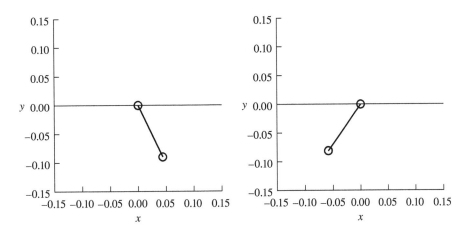

Figure C.4 Snapshots of a MATLAB animation of a swinging pendulum.

```
axis(<axis size>*[-1 1 -1 1]); hold on
h = [];
for ii=1:length(t),
    delete(h)
    h = <insert command here to plot at t(ii)>;
    drawnow
end
```

APPENDIX D

Moments of Inertia of Selected Bodies

All bodies have mass m, G is the center of mass, and all bodies have uniform density.

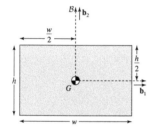

Flat rectangular plate

$$[\mathbb{I}_G]_\mathcal{B} = \frac{m}{12} \begin{bmatrix} h^2 & 0 & 0 \\ 0 & w^2 & 0 \\ 0 & 0 & (h^2 + w^2) \end{bmatrix}_\mathcal{B}$$

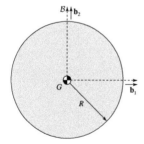

Flat circular plate

$$[\mathbb{I}_G]_\mathcal{B} = \frac{mR^2}{4} \begin{bmatrix} 1 & 0 & 0 \\ 0 & 1 & 0 \\ 0 & 0 & 2 \end{bmatrix}_\mathcal{B}$$

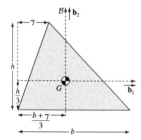

Flat triangular plate

$[\mathbb{I}_G]_\mathcal{B}$

$$= \frac{m}{36} \begin{bmatrix} 2h^2 & h(b - 2\gamma) & 0 \\ h(b - 2\gamma) & 2(b^2 - b\gamma + \gamma^2) & 0 \\ 0 & 0 & 2(b^2 + h^2 - b\gamma + \gamma^2) \end{bmatrix}_\mathcal{B}$$

Thin rod

$$[\mathbb{I}_G]_{\mathcal{B}} = \frac{ml^2}{12} \begin{bmatrix} 1 & 0 & 0 \\ 0 & 0 & 0 \\ 0 & 0 & 1 \end{bmatrix}_{\mathcal{B}}$$

Note: Applies only when rod radius $\ll l$. Otherwise use the equations for cylinders.

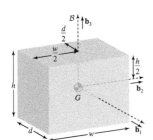

Solid rectangular parallelepiped

$$[\mathbb{I}_G]_{\mathcal{B}} = \frac{m}{12} \begin{bmatrix} (h^2 + w^2) & 0 & 0 \\ 0 & (h^2 + d^2) & 0 \\ 0 & 0 & (d^2 + w^2) \end{bmatrix}_{\mathcal{B}}$$

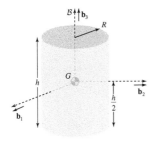

Solid circular cylinder

$$[\mathbb{I}_G]_{\mathcal{B}} = \frac{m}{12} \begin{bmatrix} (3R^2 + h^2) & 0 & 0 \\ 0 & (3R^2 + h^2) & 0 \\ 0 & 0 & 6R^2 \end{bmatrix}_{\mathcal{B}}$$

Thin circular cylindrical shell (no end caps)

$$[\mathbb{I}_G]_{\mathcal{B}} = \frac{m}{12} \begin{bmatrix} (6R^2 + h^2) & 0 & 0 \\ 0 & (6R^2 + h^2) & 0 \\ 0 & 0 & 12R^2 \end{bmatrix}_{\mathcal{B}}$$

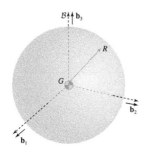

Solid sphere

$$[\mathbb{I}_G]_{\mathcal{B}} = \frac{2m}{5} \begin{bmatrix} R^2 & 0 & 0 \\ 0 & R^2 & 0 \\ 0 & 0 & R^2 \end{bmatrix}_{\mathcal{B}}$$

Thin spherical shell

$$[\mathbb{I}_G]_{\mathcal{B}} = \frac{2m}{3} \begin{bmatrix} R^2 & 0 & 0 \\ 0 & R^2 & 0 \\ 0 & 0 & R^2 \end{bmatrix}_{\mathcal{B}}$$

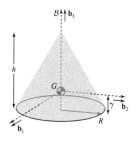

Solid cone

$$\gamma = \frac{h}{4}, \quad [\mathbb{I}_G]_\mathcal{B} = \frac{3m}{80} \begin{bmatrix} (4R^2 + h^2) & 0 & 0 \\ 0 & (4R^2 + h^2) & 0 \\ 0 & 0 & 8R^2 \end{bmatrix}_\mathcal{B}$$

Thin conical shell (no cap)

$$\gamma = \frac{h}{3}, \quad [\mathbb{I}_G]_\mathcal{B} = \frac{m}{36} \begin{bmatrix} (9R^2 + 2h^2) & 0 & 0 \\ 0 & (9R^2 + 2h^2) & 0 \\ 0 & 0 & 18R^2 \end{bmatrix}_\mathcal{B}$$

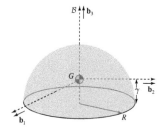

Solid hemisphere

$$\gamma = \frac{3R}{8}, \quad [\mathbb{I}_G]_\mathcal{B} = \frac{mR^2}{320} \begin{bmatrix} 83 & 0 & 0 \\ 0 & 83 & 0 \\ 0 & 0 & 128 \end{bmatrix}_\mathcal{B}$$

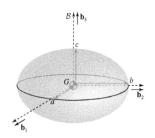

Solid ellipsoid

$$[\mathbb{I}_G]_\mathcal{B} = \frac{m}{5} \begin{bmatrix} (b^2 + c^2) & 0 & 0 \\ 0 & (a^2 + c^2) & 0 \\ 0 & 0 & (a^2 + b^2) \end{bmatrix}_\mathcal{B}$$

BIBLIOGRAPHY

Anderson, J. D., Jr. 1999. *Aircraft Performance and Design*. WCB/McGraw-Hill.

Antman, S. S. 1998. The Simple Pendulum Is Not So Simple. *SIAM Review* 40(4): 927–930.

Arnold, V. I. 1989. *Mathematical Methods of Classical Mechanics*. Springer.

Bate, R. R., D. D. Mueller, and J. E. White. 1971. *Fundamentals of Astrodynamics*. Dover Publications.

Bedford, A., and W. Fowler. 2002. *Engineering Mechanics, Dynamics,* third edition. Prentice-Hall.

Beer, F. P., R. R. Johnston, Jr., and W. E. Clausen. 2007. *Vector Mechanics for Engineers, Dynamics,* eighth edition. McGraw-Hill.

Boyce, W. E., and R. C. DiPrima. 1977. *Elementary Differential Equations and Boundary Value Problems,* third edition. John Wiley & Sons.

Bryson, A. E., Jr. 1994. *Control of Spacecraft and Aircraft*. Princeton University Press.

Chobotov, V. A. 1991. *Orbital Mechanics*. American Institute of Aeronautics and Astronautics.

Diacu, F., and P. Holmes. 1996. *Celestial Encounters*. Princeton University Press.

Dickinson, H. W. 1936. *James Watt: Craftsman & Engineer*. Cambridge University Press.

Dugas, R. 1988. *A History of Mechanics*. Dover Publications.

Fetter, A. L., and J. D. Walecka. 1980. *Theoretical Mechanics of Particles and Continua*. McGraw-Hill.

Feynman, R. P., R. B. Leighton, and M. Sands. 1963. *The Feynman Lectures on Physics*, vol. 1. Addison-Wesley.

Gear, C. W. 1971. *Numerical Initial Value Problems in Ordinary Differential Equations*. Prentice-Hall.

Ginsberg, J. 2008. *Engineering Dynamics*. Cambridge University Press.

Gleick, James. 2003. *Isaac Newton*. Pantheon Books.

Goldstein, H. 1980. *Classical Mechanics,* second edition. Addison-Wesley.

Gómez, G., J. Llibre, R. Martínez, and C. Simó. 2001. *Dynamics and Mission Design near Libration Points*, vols. I–IV. World Scientific.

Greenwood, D. T. 1988. *Principles of Dynamics*, second edition. Prentice-Hall.

Guckenheimer, J., and P. Holmes. 2002. *Nonlinear Oscillations, Dynamical Systems, and Bifurcations of Vector Fields*. Applied Mathematical Sciences 42. Springer.

Hairer, E., C. Lubich, and G. Wanner. 2002. *Geometric Numerical Integration*. Springer.

Hand, L. N., and J. D. Finch. 1998. *Analytical Mechanics*. Cambridge University Press.

Hibbeler, R. C. 2003. *Engineering Mechanics, Dynamics*, tenth edition. Prentice-Hall.

Hut, P., J. Makino, and S. McMillan. 1995. Building a Better Leapfrog. *Astrophysical Journal* 443(April 20): L93–L96.

Kane, T. R. 1978. *Dynamics*. Holt, Reinhart and Winston.

Kane, T. R., and D. A. Levinson. 1985. *Dynamics: Theory and Applications*. McGraw-Hill.

Kane, T. R., P. W. Likins, and D. A. Levinson. 1983. *Spacecraft Dynamics*. McGraw-Hill.

Kaplan, M. H. 1976. *Modern Spacecraft Dynamics and Control*. John Wiley & Sons.

Kaplan, W. 1973. *Advanced Calculus*, second edition. Addison-Wesley.

Khalil, H. K. 2002. *Nonlinear Systems*, third edition. Prentice-Hall.

Koon, W. S., M. W. Lo, J. E. Marsden, and S. D. Ross. 2007. *Dynamical Systems, the Three-Body Problem, and Space Mission Design*. Springer.

Lambert, J. D. 1973. *Computational Methods in Ordinary Differential Equations*. John Wiley & Sons.

————. 1993. *Numerical Methods for Ordinary Differential Systems*. John Wiley & Sons.

Lapidus, L., and J. H. Seinfeld. 1971. *Numerical Solution of Ordinary Differential Equations*. Academic Press.

Marsden, J. E., and T. S. Ratiu. 1999. *Introduction to Mechanics and Symmetry*, second edition. Springer.

Meirovitch, L. 2002. *Fundamentals of Vibrations*. McGraw-Hill.

Meriam, J. L., and L. G. Kraige. 2001. *Engineering Mechanics, Dynamics*, fifth edition. John Wiley & Sons.

Moon, F. C. 1998. *Applied Dynamics*. John Wiley & Sons.

National Academy of Sciences. 1999. *Science and Creationism: A View from the National Academy of Sciences*, second edition. National Academy Press.

Nelson, R. C. 1998. *Flight Stability and Automatic Control*. WCB/McGraw-Hill.

Newton, I. 1934. *Mathematical Principles of Natural Philosophy and His System of the World*. University of California Press. Translated into English by Andrew Motte in 1729 and revised by Florian Cajori.

O'Reilly, O. M. 2008. *Intermediate Dynamics for Engineers*. Cambridge University Press.

Pamadi, B. N. 1998. *Performance, Stability, Dynamics, and Control of Airplanes*. AIAA.

Peterson, I. 1993. *Newton's Clock: Chaos in the Solar System*. W. H. Freeman and Company.

Press, W. H., B. P. Flannery, S. A. Teukolsky, and W. T. Vetterling. 1986. *Numerical Recipes: The Art of Scientific Computing*. Cambridge University Press.

Pytel, A., and J. Kiusalaas. 1999. *Engineering Mechanics: Dynamics*, second edition. Brooks/Cole Publishing.

Rao, A. V. 2006. *Dynamics of Particles and Rigid Bodies: A Systematic Approach*. Cambridge University Press.

Routledge, R. 1900. *Discoveries and Inventions of the Nineteenth Century*, thirteenth edition. George Routledge and Sons.

Sanz-Serna, J. M., and M. P. Calvo. 1994. *Numerical Hamiltonian Problems*. Chapman and Hall.

Smits, A. 1999. *Fluid Mechanics*. John Wiley & Sons.

Sobel, D. 1999. *Galileo's Daughter*. Walker and Company.

Spivak, M. 1980. *Calculus*, second edition. Publish or Perish.

Stengel, R. F. 2004. *Flight Dynamics*. Princeton University Press.

Storch, J., and S. Gates. 1989. Motivating Kane's Method for Obtaining Equations of Motion for Dynamic Systems. *AIAA Journal of Guidance, Control, and Dynamics* 12(4): 593–595.

Strogatz, S. 2001. *Nonlinear Dynamics and Chaos*. Westview Press.

Szebehely, V. 1967. *Theory of Orbits*. Academic Press.

Tenenbaum, R. A. 2004. *Fundamentals of Applied Dynamics*. Springer.

Thornton, S. T., and J. B. Marion. 2004. *Classical Dynamics of Particles and Systems*, fifth edition. Thomson Brooks/Cole.

Tongue, B. H., and S. D. Sheppard. 2005. *Dynamics: Analysis and Design of Systems in Motion*. John Wiley & Sons.

Truesdell, C. 1968. *Essays in the History of Mechanics*. Springer.

Vallado, D. A. 2001. *Fundamentals of Astrodynamics and Applications*, second edition. McGraw-Hill.

Varieschi, B., and K. Kamiya. 2003. Toy Models for the Falling Chimney. February 17. arXiv:physics/0210033v2.

Weisstein, E. W. 1999. Vector Space. http://mathworld.wolfram.com/VectorSpace.html.

―――. 2003. Mean Value Theorem. http://mathworld.wolfram.com/Mean-ValueTheorem .html.

Wertz, J. R. (ed). 1978. *Spacecraft Attitude Determination and Control*. D. Reidel.

Wie, B. 1998. *Space Vehicle Dynamics and Control*. American Institute of Aeronautics and Astronautics.

Wilczek, F. 2004. Whence the Force of $F = ma$? I: Culture Shock. *Physics Today* 57(10): 11–12.

Williamson, R. E., R. H. Crowell, and H. F. Trotter. 1972. *Calculus of Vector Functions*. Prentice-Hall.

INDEX

www.ingramcontent.com/pod-product-compliance
Ingram Content Group UK Ltd.
Pitfield, Milton Keynes, MK11 3LW, UK
UKHW011007270125
454197UK00001B/3